Neil R Bowlly

ANNUAL REVIEW OF
PLANT PHYSIOLOGY
AND PLANT
MOLECULAR BIOLOGY

EDITORIAL COMMITTEE (1990)

ANNUAL REVIEW OF PLANT PHYSIOLOGY AND PLANT MOLECULAR BIOLOGY

VOLUME 41, 1990

WINSLOW R. BRIGGS, *Editor*

Carnegie Institution of Washington, Stanford, California

RUSSELL L. JONES, *Associate Editor*

University of California, Berkeley

VIRGINIA WALBOT, *Associate Editor*

Stanford University

ANNUAL REVIEWS INC.   4139 EL CAMINO WAY   PO BOX 10139   PALO ALTO, CALIFORNIA 94303-0897 USA

Ⓡ ANNUAL REVIEWS INC.
Palo Alto, California, USA

International Standard Serial Number: 1040-2519
International Standard Book Number: 0-8243-0641-4
Library of Congress Catalog Card Number: A-51-1660

∞ The paper used in this publication meets the minimum requirements of American National Standard for Information Sciences—Permanence of Paper for Printed
Library Materials, ANSI Z39.48-1984.

Annual Reviews Inc. and the Editors of its publications assume no responsibility
for the statements expressed by the contributors to this *Review*.

Typesetting by Kachina Typesetting Inc., Tempe, Arizona; John Olson, President
Typesetting Coordinator, Janis Hoffman

PRINTED AND BOUND IN THE UNITED STATES OF AMERICA

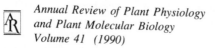

*Annual Review of Plant Physiology
and Plant Molecular Biology
Volume 41 (1990)*

CONTENTS

RELATED ARTICLES OF INTEREST TO READERS

From the *Annual Review of Biochemistry,* Volume 59 (1990)

Recent Topics in Pyridoxal 5'-Phosphate Enzyme Studies, *H. Hayashi, H. Wada, T. Yoshimura, N. Esaki, and K. Soda*

Phytochelatins, *W. E. Rauser*

Structure and Function of Cytochrome *c* Oxidase, *R. A. Capaldi*

Defense-Related Proteins in Higher Plants, *D. J. Bowles*

The Mitochondrial Protein Import Apparatus, *N. Pfanner and W. Neupert*

DNA Recognition by Proteins with the Helix-Turn-Helix Motif, *S. C. Harrison and A. K. Aggarwal*

cAMP-Dependent Protein Kinase: Framework for A Diverse Family of Regulatory Enzymes, *S. S. Taylor, J. A. Beuchler, and W. Yonemoto*

RNA Polymerase B(II) and General Transcription, *M. Sawadogo and A. Sentenac*

Intermediates in the Folding Reactions of Small Proteins, *P. S. Kim and R. L. Baldwin*

The Classification and Origins of Protein Folding Patterns, *C. Chothia and A. Finkelstein*

The Bacterial Phosphoenolpyruvate-Glucose Phosphotransferase System, *N. D. Meadow, D. K. Fox, and S. Roseman*

From the *Annual Review of Cell Biology,* Volume 5 (1989)

The Chloroplast Chromosomes in Land Plants, *M. Sugiura*

Simple and Complex Cell Cycles, *F. Cross, J. Roberts, and H. Weintraub*

Initiation of Eukaryotic DNA Replication in Vitro, *B. Stillman*

Control of Protein Exit from the Endoplasmic Reticulum, *H. R. B. Pelham*

Endoplasmic Reticulum in Protein, *S. M. Hurtley and A. Helenius*

Origin and Evolution of Mitochondrial DNA, *M. W. Gray*

From the *Annual Review of Genetics,* Volume 23 (1990)

The Molecular Genetics of 21-Hydroxylase Deficiency, *W. Miller and Y. Morel*

Habituation: Heritable Variation in the Requirement of Cultured Plant Cells for Hormones, *F. Meins*

Maize Transposable Elements, *A. Gierl, H. Saedler, and P. A. Peterson*

The Molecular Genetics of Self-Incompatibility Systems in *Brassica, J. B. Nasrallah and M. E. Nasrallah*

Genetic Analysis of Protein Stability and Function, *R. T. Sauer and A. A. Pakula*

From the *Annual Review of Microbiology,* Volume 43 (1989)

Evolution of a Biosynthetic Pathway: The Tryptophan Paradigm, *I. P. Crawford*

From the *Annual Review of Phytopathology,* Volume 28 (1990)

Quantifying Pesticide Behavior in Soil, *J. Wagenet and J. L. Hutson*

Jack Dainty

Annu. Rev. Plant Physiol. Plant Mol. Biol. 1990. 41:1–20

PREFATORY CHAPTER

Jack Dainty

Department of Botany, University of Toronto, Toronto, Ontario, Canada M5S 3B2

This is the story of how I became and what I have done as a plant biophysicist, and what I consider to be the function and importance, if any, of such people in plant physiology.

I was born and grew up in a depressed area of South Yorkshire, England. We lived on the edge of D. H. Lawrence country, surrounded by miserable, poverty-stricken, coal-mining towns, near the steel-making city of Sheffield. The waters and soil were heavily polluted, and the unpleasant smell of industrial gaseous waste hung in the air. It was not then, nor is it today, a place anybody would choose to live. The pollution has greatly decreased because many of the coal mines and all the steel works have been shut down, with resulting heavy unemployment.

The small mining towns were practically contiguous and equally depressing. "Deprived" is the word that best describes them, a characterization I first became aware of only after I had gone away to university. Nearly everybody was poor; many—like us—were very poor. The "aristocracy" of the towns consisted of the small shopkeepers and school teachers. The vast majority of us voted Labor (socialist)—indeed the area produced the largest Labor majorities in the country. I grew up, and have remained, a socialist.

My father having died young (when I was three), my brother, my mother and I lived with my grandparents in poverty. We often received "welfare"—an even more distressing experience then than now. Inevitably, the atmosphere of home was devoid of culture, literature, and learning. Nobody in the immediate family had been to school beyond the age of 14, and my many cousins, uncles, and aunts had the same background. By some quirk, however, I was precocious, learning to read by the time I was three. I thus early discovered the world of books—still the world in which I like to live.

At that time (the 1920s and 1930s), a child such as I entered the educational system through the so-called elementary school; at the age of 10 or 11 one

1

took an examination for entry to the high school—in my town of Mexborough it was called the secondary school. Fewer than one child in ten passed this scholarship examination; those who did not pass were condemned to stay in the elementary school until age 14. I passed, but to keep going in the secondary school I needed the charity of free school meals, clothes (hated) given by the local Rotary Clubs, and the encouragement of my mother.

At the end of the first four years, during which we studied a broad range of subjects—French, Latin, English literature, mathematics, physics (no chemistry or biology), geography, history, art, music, workshop—there was a national examination. If one was reasonably successful in this, one could pass on to a specialized program for another two years. This was a crucial point in the educational system; it involved a narrow choice of specialized subjects: not merely whether to do arts (i.e. humanities) or science but what kind of arts or sciences. I chose science because I believed I would rank high up in the class, and I chose to do mathematics, physics, and chemistry rather than botany, zoology, and chemistry (the only two options offered by the school) both because I thought I was quite good at mathematics and because boys did mathematics while girls did Biology! Such are the determining factors in a situation such as mine was, with good advice unavailable.

I was not uninterested in biology. In fact, from an early age I had been a dedicated naturalist, spending much of my time going for long, lonely walks in the neighboring countryside (which despite the pollution was not unattractive) and on the moors and hills of Derbyshire and Yorkshire. I read deeply in the English naturalist literature: Belt, Wallace, Bates, Waterton, Gilbert White, Darwin, and others. It did not occur to me to translate this great interest into the formal study of biology, and there was nobody to suggest such a strategy. However, my much later move from physics to biology was based on this early interest in natural history.

Thus I was set, at the age of 15, on the mathematics-physics-chemistry track in the early specialization characteristic of English higher education. After two years of such specialization there was another national examination, and in this I did well in all three subjects. Throughout my education, in fact, I was an accomplished examinee. (This trait was of course useful, but it is not, I believe, closely related to ability in scientific research, which needs different qualities.) After this success it was suggested to me that I might consider the great heights (for Mexborough) of the University of Cambridge and that I should stay at the school for another year in order to take the scholarship examination for university entry. I spent this final year doing mathematics only and was successful in getting a scholarship.

I went to Cambridge to study mathematics with more than adequate financial support from scholarships. It seemed—it *was*—an extraordinary place. The faculty were outstanding. Most of the students were male, the vast majority from well-to-do families. They had gone to private schools where the standards of education had often been excellent. Both their families and these prior schools had influence in the Cambridge colleges, and there were long traditions behind the entry of these students into the University. Many seemed arrogant and rather stupid; some were brilliant; almost none would fail: the system did not allow it. The percentage of students who had arrived there by my route was small. I felt thoroughly out of place, with my Yorkshire accent and lack of social know-how—a feeling that certainly did nothing to weaken my socialist sympathies.

I settled down, made one or two friends, and had some success as a football (soccer) player; indeed, as I had also done at the secondary school, I spent rather a lot of time playing football. I was "reading" mathematics, which meant doing only mathematics. At the secondary school in the backwoods of South Yorkshire, I had thought myself rather good at mathematics, but I discovered in Cambridge that many were much better than I. Both because I did not like this and because I felt that mathematics only was too narrow, I switched to physics. By this time the Second World War was starting, but I was allowed, indeed instructed by the military authorities, to finish my physics undergraduate degree at Cambridge. By the end of this course and the final examinations it was June 1940 and Hitler's armies were overrunning France, the Netherlands, Luxembourg, Belgium, Denmark, and Norway. Almost all the rest of the graduating physics class had been recruited to work on radar and other urgent applications of technology to war. The recruiters must have thought little of my technical competence for I was not asked to take part in this important work. However, after a short time I was asked to return to Cambridge to join one of the two small teams (the other being at Liverpool University) being then set up to work on nuclear fission—i.e. to carry out studies leading to the eventual construction of an atomic bomb.

We were a mixed lot in the Cambridge team: two French physicists who had escaped from France with a large stock of heavy water, a Swiss who was working in Cambridge, two or three German and Austrian refugees, and a handful of Britishers who had not been swept up by the radar net. We soon realized that the making of an atomic bomb was more a technological and industrial problem than an intellectual one—that it involved either the separation of large amounts of ^{235}U or the construction of large reactors for the production of fissionable plutonium. Under wartime conditions, only the United States could set aside sufficient industrial power to make such a bomb, but we carried on with our fundamental studies of nuclear fission in Cambridge. The team was greatly reduced in number when some were sent to help the Americans in Los Alamos and others to Montreal to join with Canadians in initiating the Canadian Atomic Energy effort, the latter being solely concerned with peaceful applications of the release of nuclear energy.

I remained with the Cambridge group, indeed became head of the cyclotron team, and spent the war carrying out rather academic nuclear research on fission. Because of the shortage of faculty, I also did a great deal of teaching during those years. I gave an advanced course of lectures on atomic and nuclear physics, instructed in the practical classes, and did a lot of individual tutorial teaching, according to the Cambridge system. This research and particularly this teaching strengthened my understanding of physics, which was of great importance later when I switched to biological problems. I stayed in Cambridge until the fall of 1946. I retained my interest in natural history, chiefly birds at this time, and improved my understanding of some aspects of biology, chiefly that of evolution, by study of the works of Darwin, Huxley, Mayr, Dobzhansky, Haldane, and others. I also read the little book *What is Life* by Schrödinger. My love of literature developed greatly during these Cambridge wartime years, a love that has strengthened as the years pass.

In 1946 I made a half-hearted attempt to move from nuclear physics to biology by joining the Division of Biology and Medicine of the Canadian Atomic Energy Laboratories at Chalk River, Ontario; but the problem I worked on was really pure nuclear physics. I set out to develop a method of measuring the dose (energy per gram) delivered to biological tissue by the fast neutrons in, for instance, a nuclear reactor where the radiation would be a mixture of fast neutrons and gamma rays. The solution, differential ionization chambers, one with walls of polystyrene [$(CH)_n$, approximately] filled with acetylene and the other with walls of polyethylene filled with ethylene, was relatively simple, but I spent about two years on it. Again I carried on with my natural history (largely birds) and my reading of the classics of evolution. Deep River, the newly built town on the banks of the Ottawa River, was then a stimulating place for an amateur naturalist.

In early 1949 I returned to the United Kingdom to join the faculty of the Department of Natural Philosophy (physics) at the University of Edinburgh. I had finished, or so I thought, with any dabbling in biological problems, and was going back to "proper" (nuclear) physics. And so I did for two or three years. I lectured on relativity, quantum theory, and the theory of errors; taught in practical classes; and built up a laboratory around a small accelerator. I supervised graduate students in nuclear physics.

Around 1952 the department asked me if I would take on the onerous task of teaching elementary physics to a joint class of 300 medical, dental, and veterinary students. (In the United Kingdom such students are recruited directly from high school. In Scotland at that time all had to spend their first year on physics, chemistry, and biology). Initially I refused. In order to handle such a body of students, who refused to take physics seriously no matter how relevant one made it to physiology and medicine, one needed the qualities of an army sergeant rather than those of a university lecturer. But the head of the department, the nuclear physicist Norman Feather, knew of my interest in biology and got the university to promise that I could develop a department of biophysics if I would undertake this teaching task. I finally agreed, with considerable misgivings arising not only from the nature of the teaching task but also because I was comfortable in (nuclear) physics. I knew the literature, many of the techniques, and many of the leading people involved. After all, I had effectively been a nuclear physicist for 12 years or so.

Thus I moved into biology almost by accident. I'd had a long interest in certain aspects of the field, but not in those where I felt I could use my knowledge of physics. I had been interested not in how living organisms work but in their diversity and evolution. Therefore I had first to find out what I could do. Many people in Edinburgh were prepared to help. Since I was teaching medical students, some clinicians wanted my help, as a kind of high-grade technician, with the more physical aspects of their problems. C. H. Waddington, the geneticist and embryologist, tried to interest me in the physical problems of morphogenesis, and I carried out some experiments on early *Xenopus* embryos. The zoologists also had me doing experiments on heat production as a function of the cell cycle in developing sea urchin eggs. But the work that really started me moving in the direction I was ultimately to take was a collaborative study in 1953, with K. Krnjevic of the physiology department, of the exchange of sodium across the membranes of nerve bundles of the cat. It was this that led me to the work of Hodgkin, Huxley, Katz, Cole and others on membrane transport, particularly that on the squid axon. I did not exactly enjoy working with the nerves of cats, but membrane transport seemed to me a physiological process that one had to think of in terms of physics, at least in its overall aspects. In choosing a membrane system to investigate, I certainly did not want to compete with Hodgkin et al. It soon struck me that almost nobody was looking at membrane transport in plant cells from a biophysical point of view. And plants were not cats!

I was fortunate that at about this time a very good physics student, Enid MacRobbie, wanted to do biophysical research with me. The early work on plant cells was a good collaboration between the two of us. Our ideas were guided—perhaps too much, as will become clear later—by the work of animal physiologists on the squid axon. We had no botanical contacts, and we decided that it would be best to work with a marine alga—i.e. with cells immersed in a well-defined medium of fairly high concentration: sea water. Mostly in the 1930s and predating the squid axon work, excellent work on algal cell membranes had been done by Osterhout and Blinks on *Valonia*, *Halicystis*, and *Nitella*. It seemed to us that this work had been forgotten by plant physiologists.

We did not at first take up *Nitella* and the Characeae in general because they are mostly freshwater species and we had this idea that we needed a concentrated bathing medium, sea water. *Valonia* and *Halicystis,* being tropical, were hardly likely to occur on the icy coasts of Scotland, so we chose to work with a red alga, *Rhodymenia palmata,* and to a lesser extent with a green alga, *Ulva lactuca.* In these early experiments we measured concentrations of Na, K, and Cl in the cells and, using radioactive isotopes, the fluxes out of and into the algal discs we were using. What we did not realize (and did not even look at sections to find out—such a poor biologist was I) was that we were dealing with complex tissues and a very inhomogeneous cell population. We could not easily express our fluxes on a meaningful moles per square meter (square centimeter in those days) per second basis and could not move forward in any biophysical sense. Our minds were thus focused on finding a homogeneous cell population or, better, a large single cell of simple geometry. And since we had no *Valonia* in Scotland, we looked at the Characeae. We were still obsessed by the notion of a concentrated, well-defined external medium for our cell. We came across some work being carried out in Finland by Collander & Wartiovaara on *Nitellopsis,* an ecorticate Characean species living in the Baltic at a salt concentration of about 25 mM. This was pretty dilute sea water, but it was well defined and concentrated enough for our purposes. *Nitellopsis obtusa* therefore became our first real "guinea pig," and ion transport in this organism was the object of Enid MacRobbie's PhD thesis, the first one produced at the Biophysics Department at the University of Edinburgh.

The *Nitellopsis* studies were basically kinetic, aimed at measuring the influx and efflux of Na^+, K^+, and Cl^- across the plasmalemma and tonoplast. We sought to interpret the measurements in the light of the concentrations of these ions in the protoplasm and vacuole, taking into account some early measurements of the electric potential between vacuole and external solution by our colleagues E. J. Williams and R. Johnston. Although hindsight reveals that the compartmental analysis of the efflux curves (which seemed to split nicely into two exponentials and a fast diffusion curve) was somewhat flawed, the general conclusion was valid: There had to exist a sodium extrusion pump at the plasmalemma and an inwardly directed chloride pump, which we located at the tonoplast. (We missed the point that there must also be one at the plasmalemma.) We also deduced that most of the potential difference between vacuole and tonoplast was likely to be across the plasmalemma. This agreed with earlier measurements by Alan Walker in Australia of a low, positive potential across the tonoplast and a high, negative one across the plasmalemma. We located the fast-diffusion so-called free-space fraction exclusively in the cell wall and thus came down against the then prevalent idea, put forward by Briggs, that the protoplasm was part of the apparent free space of the cell. This latter idea then, I think, began to die. We did not consider the hydrogen ion concentrations and any possible proton pump; this came a decade or more later.

While we were developing our biophysical approaches to plant membrane transport during the 1950s in Edinburgh, somewhat similar ideas were being followed in Australia, particularly by Alex Hope. Alex had been a member of a remarkable group in the Physics Department of the University of Tasmania at Hobart. This group was the brainchild of Lester MacAulay, the head of the department, who had an interest in how plants function and encouraged some graduate students to share it. Bruce Scott, Alex Hope, Alan Walker, Geoff Findlay, and Ian Newman are all from this school and have helped to make plant biophysical physiology eminent in Australia. In 1958 Alex was in the C.S.I.R.O. Plant Physiology Unit at the University of Sydney, which was headed by the Australian plant physiologist R. N. (Bob) Robertson, who himself had a major interest in membrane transport but was more biochemically oriented than Alex and me. Robertson was the ideal person to head this unit and to encourage a biophysical approach.

I did the obvious thing and applied for four or five months of research leave in 1958 to work with Alex in Sydney. This turned out to be an extremely fruitful collaboration. We came up with what I believe to be one or two important concepts from the experiments we carried out on the water permeability of *Chara australis* (now *corallina*) and on the ion exchange properties of the *Chara* cell wall.

The water permeability of a cell can be measured, indeed specified, in two ways: The net flow of water across a membrane separating two solutions of different osmotic pressure gives, for a semipermeable membrane, the hydraulic conductance, L_p; the exchange flux of labeled water under conditions of equilibrium gives the so-called diffusional permeability coefficient, P_d. If these two measures are expressed in the same units and compared, then if there is no equality, or specifically if $L_p > \bar{V}_w P_d/RT$, where $\bar{V}_w$ is the partial molar volume of water, this indicates that water crosses the membrane via aqueous pores; the relative magnitudes give some idea of the diameter of these pores. We were inspired to make water permeability measurements in these two ways on *Chara australis* by similar work on animal cells, in particular that of A. K. Solomon and his associates on red blood cells. Measuring L_p was no problem using the method of transcellular osmosis invented by Kamiya & Tazawa a year or two previously. Measuring P_d likewise appeared to be straightforward, using heavy water as the tracer and following the exchange of D_2O for H_2O by a simple weighing technique. However, when we repeated the P_d measurements on a cylinder of agar of exactly the same dimensions, the exchange kinetics were precisely the same; the agar cylinder had of course no membrane(s), and thus we could only conclude that the water exchange was rate controled by diffusion in the unstirred layers of water external (and to some extent internal) to the cell membrane! We thus rather dramatically "discovered" the importance of the unstirred layer in membrane transport, although of course it had long been well known to chemical engineers.

We were able to point out that P_d is always underestimated to a greater degree than L_p, because of the unstirred-layer effect, and thus to throw much doubt on some of the estimates of membrane pore size made by animal physiologists. We also used the unstirred layer to "explain" the difference between the L_p values when water is flowing into a cell and those when it is flowing out. It was impossible to correct for the unstirred layer in the P_d experiment with *Chara australis* because the effect was so dominant. No estimate of pore size could be made. In fact only one good experiment has been carried out on L_p and P_d for a plant cell: In my laboratory some years later John Gutknecht found in *Valonia* that after correcting P_d for the unstirred layer effect, $L_p = P_d \bar{V}_w/RT$. The osmotic and diffusional permeabilities when expressed in the same units were the same. Thus there are *no* aqueous pores in the plasma membrane of *Valonia*; water molecules must go through the membrane one by one.

I consider the little experiment on *Chara* one of the most important with which I have been involved, for it brought to the notice of physiologists in general, even if it took a long time to be recognized, the potentially great importance of the unstirred layer in transport processes.

We met the unstirred layer again in our studies in Sydney on ion exchange with the plant cell wall. I think these studies were initiated to resolve definitely the question of whether any part of the apparent free space (AFS) was in the protoplasm or whether it was all in the cell wall. The experiments certainly resolved this question: The AFS is all in the wall. In addition it was shown that the exchanges of Na^+ in the wall for Na^+ in the bathing solution and of Ca^{2+} in the wall for Ca^{2+} in the bathing solution were rate controled by diffusion in the unstirred layers of solution adjacent to the wall. Indeed, calculations of the thickness of the unstirred layer from the two separate experiments (Na^+ exchange and Ca^{2+} exchange) gave similar values of about 100 μm, a reasonable value under the conditions of stirring and cell wall size in the experiments. Naturally we were delighted with this reinforcement of the importance of the unstirred layer in a quite different example.

This cell wall work was also important in several other respects. For instance, we showed that the negative fixed charge of about 1.3 meq/g dry weight probably arose from the uronic acid groups on the pectins with a pK of about 2.2. Mobile anions such as Cl^- or I^- were not excluded from the wall water as much as expected by such a high fixed-charge density, if the latter were uniform. Thus a reasonable simple-minded view would be that this heterogeneous wall could be divided into a water free space (WFS), not under the influence of the negative fixed charges, and a Donnan free space (DFS) in which the fixed charges were uniformly distributed and the Donnan theory applied. We made this view more plausible (to us) by imagining the fixed charges giving rise to electrical double layers in the wall water. Such a picture "explained" rather naturally the idea of WFS and DFS, although it has been recently superceded by the rather *more* natural picture of linear chains of charges.

This experience of working in Australia with Alex Hope was of the greatest benefit to me, not only owing to the ideas we had and the experiments we did. It was the first time I had worked in a biological milieu, and I believe I derived much from my close contacts with Bob Robertson, Joe Wiskich, and others who were more biochemically inclined than I and were trained as biologists, not physicists.

Back in Edinburgh later in 1958 I started something that turned out to be useful. We developed a series of graduate courses in biophysics designed to attract physics graduates. We hoped to orient such students as well as possible toward biology without preventing them from thinking as physicists. For a year's set of courses plus a small three- or four-month research project they were awarded a Diploma in Biophysics. We had five or six students each year, and the best of them stayed on to do a PhD. I think the program was rather successful, attracting such students as Peter Anderson, Roger Spanswick, Regis Kelly, Julian Collins, Randall House, Peter Kohn, David Aikman, and John Sinclair. One part of the course not so well appreciated by the students was a trip to the top of a Scottish mountain, Ben Lawers, to see the alpine flora there.

Early in the 1960s I was asked to write a review on "Ion transport and electrical potentials in plant cells" for the *Annual Review of Plant Physiology*. For me this was an important means of introducing to plant physiologists a biophysical approach to plant membrane transport, and it became a piece of pedagogy rather than strictly a review. Rereading it now I am struck by the excessive influence of the work of such animal physiologists as Hodgkin and Katz; indeed, I think I subconsciously looked upon single cells of the Characeae as green squid axons. Thus although I discussed the possible electrogenicity of ion pumps, I preferred to consider them neutral, as they at that time appeared to be in animal cells, despite evidence from Noe Higinbotham's group of the electrogenic nature of some ion pump(s) in higher plant cells. And again, although I discussed in some detail the great discrepancy between the resistance of the *Chara* or *Nitella* plasmalemma as calculated from the ion fluxes of Na^+, K^+ and Cl^- and as measured directly, I tried to interpret this in terms of some kind of ion interaction. It did not occur to me at that time to consider the hydrogen ion, presumably because the animal physiologists had not thought about its possible importance in membrane transport. Nevertheless I think this "review" did play some part in familiarizing plant physiologists with a more biophysical understanding of transport.

Immediately after this paper on ion transport, I wrote another "review" article on "Water relations of plant cells" for the first volume of *Advances in Botanical Research* in 1963. This was perhaps an even more didactic piece of work than the ion transport "review" and was, rather deliberately, quite heavily theoretical. At the outset I strongly supported the suggestions made a year or two previously by Taylor & Slatyer that the ancient concepts of "suction pressure" (European) or "diffusion pressure deficit" (American) should be replaced by a more thermodynamic approach to the driving forces of plant water relations; these old concepts should be replaced by that of water potential, which is a rough expression for the free energy per unit volume of water in a system. I regret now that I did not go further than Taylor & Slatyer (who, I believe, wished to keep some continuity with the past) and recommend that we use not water potential but the chemical potential of water—i.e. the free energy per *mole* of water, in water relations considerations.

I used this article also to introduce to plant physiologists the theory of irreversible thermodynamics as it applies in water transport, and in particular to derive the two equations for water flow and for solute flow in a real, nonsemipermeable membrane. These equations bring out the importance of the three basic parameters of water transport—the hydraulic conductance, L_p; the solute permeability, ω; and the reflection coefficient, σ. The latter parameter, the reflection coefficient, is the "new" solute-water interaction coefficient brought out by the irreversible thermodynamic formulation. I discussed it, following some work Benz Ginzburg and I had done together, in terms of a frictional model of membrane transport in which a specific frictional solute-water interaction is invoked. Perhaps this is no longer important. Certainly little is heard now of another application of irreversible thermodynamic theory I discussed: electro-osmosis, where the solute (ion)-water interaction is partly electrical. My interest in this, at that time, arose from the visit to Edinburgh of David Fensom and also from the suggestion by Bennet-Clark and others that active transport of water could take place through some kind of electro-osmotic "pump." I think we were able to show by both experiment and theory that electro-osmosis was unlikely to play any great physiological role.

I took advantage of this water-relations "review" to discuss two of my hobbyhorses: the mechanism of osmosis and the importance of unstirred layers.

I have already discussed my realization with Alex Hope in Australia that membrane transport processes could be seriously affected by diffusion in the unstirred layers of solution adjacent to the membrane. I took the opportunity in this article to set out the theory of unstirred layers and their effects on measurements of solute and water permeability, of reflection coefficients, and of their unequal effects on measurement of L_p depending on whether the water was entering or leaving a cell. (In the ion transport review I had also discussed unstirred-layer effects on membrane potentials, particularly in nonequilibrium situations.)

The mechanism of osmosis has long been an interest of mine, stimulated from time to time by discussions with and articles by Alex Mauro, Peter Ray, Patrick Meares, Alan Walker, Jack Ferrier, and others. From a thermodynamic point of view the mechanism presents no problem, as Alex Mauro has demonstrated. In the pores of a porous but semipermeable membrane separating two solutions at the same hydrostatic pressure but of different concentration there must be a hydrostatic pressure gradient that is the actual driving force on the water within the membrane. The hydrostatic pressure difference within the pores must be equal to RT multiplied by the concentration difference in the solutions. This can be shown from the necessity that the chemical potential of the water, which depends on both pressure and solute concentration, must be continuous (i.e. exhibit no sharp jumps) from one solution to the other. A molecular picture of osmosis, however, is more difficult to envisage and has proved more contentious. In the water relations article I put forward an "explanation," developed from an idea of Peter Ray, based on the unbalanced molecular jumping of water across the mouth of the pore, leading to a sharp drop in hydrostatic pressure across this mouth. (I return to the mechanism of osmosis below.)

We had academic visitors to the small Biophysics Department in Edinburgh. I have already mentioned David Fensom, who made us think about electro-osmosis and helped us demonstrate that it seems unlikely to be of major physiological importance. (We missed an important unstirred-layer effect associated with electro-osmosis, later demonstrated by Peter Barry and Alex Hope: The inequality between the transport numbers of ions in the membrane and in the bathing solutions leads to the creation of a local osmotic pressure difference, and hence osmotic flows, across the membrane.) Other important visitors were Karel Janáček from Czechoslovakia, Jan Stolarek from Poland, Bud Etherton, who brought with him insights obtained in the laboratory of Noe Higinbotham at Washington State University in Pullman, and Benz Ginzburg from Jerusalem.

Through Benz Ginzburg I came to know rather well Aharon Katchalsky, whose life was cut short by terrorists in the airport at Tel Aviv. It was Aharon, of course, together with his colleague Ora Kedem, who developed the basic membrane transport equations from Stavermann's formulation of the theory of irreversible thermodynamics. Benz and I embarked on some experiments and theorizing on transport across the membranes of single cells of *Chara australis* and *Nitella translucens,* inspired by the approach of Kedem & Katchalsky. Our exercise in basic theory was to calculate the membrane transport parameters L_p and σ for a lipid/pore model of a cell membrane in terms of frictional interactions between water and solute in an aqueous pore, between water and pore wall, between solute and pore wall, between water and the lipid part of the membrane, and between solute and the lipid part of the membrane. We were successful in obtaining suitable equations for L_p and σ, compatible with other equations derived by Katchalsky & Durbin, and we were able to make use of them in our experimental work. This experimental work in part consisted of showing that the hydraulic conductance, L_p, and the permeability of the plasmalemma of *Nitella* to urea both decreased markedly as the ambient osmotic pressure, due to sucrose, increased. We suggested that a kind of dehydration of the membrane was responsible for this permeability decrease. The other experiments were concerned with the somewhat difficult task of measuring permeabilities and reflection coefficients of *Nitella translucens* for rapidly penetrating solutes such as methanol, ethanol, and isopropanol. In these cases the unstirred layer was of great significance and we had to handle a situation in which the concentration profile of the solutes in the unstirred layer varied with time. It was an interesting exercise; we were also able to consider the reflection coefficients in the light of our frictional model equation for σ.

Benz Ginzburg left Edinburgh in 1962, I think, to spend some time in Thimann's laboratory at Harvard. Ron Poole came to Edinburgh from Harvard, but unfortunately we did not overlap much because in 1963 I was on my way to taking up a position at the new University of East Anglia, Norwich.

Before leaving the topic of Edinburgh, so important to me, there are one or two more things to say. Peter Mitchell was at Edinburgh, in the Zoology Department, during the time I was there. He was then developing his ideas on chemiosmosis and spent much time discussing various aspects of them with me. He would come to me to try out the physical and physicochemical aspects of his hypothesis that a proton gradient across the inner mitochondrial membrane drives a vectorial chemical reaction and results in the synthesis of ATP. I really missed a great chance here of realizing very early the significance of these momentous developments; then I might have enlarged my research to become more biochemical. But I was still too much an isolated physicist, working with physicists, and far too ignorant of biochemistry, for which I had no feeling. Thus I did not sufficiently realize the great importance of what Peter Mitchell was doing!

An event of considerable importance took place in the summer of 1960: the first (biophysical) membrane transport meeting held in Prague. There were about 50 invited delegates, all except me working on animal or microbial systems. I found it a great stimulus to meet the great names in the field. Peter Mitchell there gave the first (I think) exposition of his ideas. It was a notable event in the history of membrane biophysics. For me it underlined the influence of the animal biologists—not necessarily too good a thing.

The University of East Anglia (UEA) was a new University. It began operation in 1963, in wooden huts. It was to be a strong center for biology, an emphasis to be stressed by having a School of Biological Sciences—a name then new, I believe, at least in the United Kingdom. Thomas Bennet-Clark, a distinguished British plant physiologist, was made first head (called "Dean") of the School and given the task of building it up. There were to be four "chairs"—botany, zoology, biochemistry and biophysics—and four corresponding sections of the school. It was to be a loose arrangement with room for section-spanning disciplines such as genetics and developmental biology. B-C, as we called Bennet-Clark, asked me if I would take the chair of biophysics. Despite the fact that I had been able to build up a nice small department in Edinburgh, I accepted the chance to start again at UEA. I was excited by the concept of the School, by the promise that biology would be the strong science, and particularly by the thought that I was going to be part of a biology environment, involved in the teaching of undergraduate biologists. Another attraction was the pleasant city of Norwich and the surrounding county of Norfolk—particularly beloved by naturalists, among whom I still counted myself.

I took with me to UEA some of the people, including graduate students, from Edinburgh and rapidly built up a small group that included Alan Walker for four years or so before he decided to go back to Sydney. We had excellent financial support from what is now the Science and Engineering Research Council and from the Nuffield Foundation. We taught physics and mathematics to biology undergraduates through biology (i.e. using biological examples) and offered a third year (final year) option in biophysics. At the graduate level we carried on with the Edinburgh tradition and gave a series of courses in biophysics. These plus a small research project led to the degree of MSc in Biophysics.

The years at UEA in Norwich were good ones. There were excellent MSc and PhD students such as Julian Collins, John Sinclair, David Aikman, George Duncan, Mel Tyree, Ed Tarr, and Jim Barber. We had a small research group on root physiology led by Randall House and Peter Anderson. And we had many academic visitors and postdoctoral fellows—e.g. Alex Hope, Geoff and Nele Findlay, John Gutknecht, David Clarkson, John Cram, Jack Hanson, Bob Lannoye, and Tom Crossett.

This period 1963–1969 still saw relatively little research—at least from a biophysical point of view—on plant cell membranes. Perhaps the dominant transport idea at that time among plant physiologists was that of Emanuel Epstein—i.e. that of comparing the kinetics of uptake of a solute (by a plant root), via a carrier, with enzyme kinetics. This was a concept that we biophysicists—wrongly—did not take seriously, for it seemed to ignore any possible passive movement of solutes, the effects of membrane potential and cell inhomogeneities. However, we should have taken this idea—as Clifford Slayman, Dale Sanders, Peter Hansen, and others later did—and elaborated it into a proper kinetic model of the carrier-mediated transport of a solute, for I am now convinced that such a model will greatly enhance our understanding of this aspect of membrane transport. This was a typical error of judgment on my part; I was still too much under the influence of the animal physiologists, who were working with systems in which passive ion transport was more important than it is in plant systems.

The kinds of studies we made in Norwich involved in general this biophysical approach. Julian Collins tackled the problem of the ion-exchange capacity of plant cell walls using *Sphagnum* as material. John Sinclair studied the ionic relations of a moss, *Hookeria lucens,* chosen for its large cells and relatively large leaves, one cell thick. David Aikman studied the ionic relations of *Valonia ventricosa,* and Jim Barber those of *Chlorella.* (Jim Barber, now of photosynthesis fame, was the first to measure the membrane potential of *Chlorella* using a microelectrode.) Alan Walker carried on his electrophysiological studies of the Characeae. Some nice experiments on water and ion uptake by maize roots were carried out by Randall House and Peter Anderson. Our postdoctoral visitor, John Gutknecht, did the important experiment, to which I referred earlier, of measuring the two water permeabilities, L_p and P_d, of *Valonia;* after correcting for unstirred layers he showed that the two permeabilities were equivalent and thus that there was no evidence that water-filled pores in the membrane(s) of *Valonia* were an important route for the movement of water across the membrane(s). Another visitor, Bob Lannoye, together with Ed Tarr did some good work on the resting and active electrophysiological properties of *Chara corallina.*

I remember with considerable chagrin that at this time ($\sim$1964) Alan Walker and I tried to see if we could pick up any net movement of protons in or out of a *Chara* cell under conditions that I can no longer remember. The reason for such an experiment was our constant worry about the great discrepancy between the electrical conductance of the plasmalemma of *Chara* as calculated from the ion fluxes (of Na^+, K^+ and Cl^-) and as directly measured. The latter was 10–15 times that of the former! Was there a missing moving ion? If so, it could only be H^+. But for reasons we cannot now understand we could detect no change in the pH of the medium, and decided that H^+ did not play any significant role in the ionic relations of *Chara.* Thus the idea of the H^+ extrusion pump, developed from the work of Kitasato in 1968, and the electrogenic nature of some pump, was missed by us.

Towards the end of the 1960s I was persuaded to move from the University of East Anglia to the University of California, Los Angeles—at first to the Laboratory of Nuclear Medicine and Radiation Biology and then to the Department of Botany. As it turned out, I did not stay long at UCLA, but moved after two years to become chairman of the Department of Botany at the University of Toronto. My short time at UCLA was certainly pleasant, thanks to George Laties, Park Nobel, Jacob Biale, and others.

From the point of view of my own scientific development it was a period in which I asked myself whether I could make any contributions to plant ecology. In part this query arose because when I arrived Hal Mooney had just moved from UCLA to Stanford, leaving behind one or two graduate students and the course in plant ecology. I volunteered both to look after the graduate students and to give the course. This was truly "the blind leading the blind," but I think we all enjoyed it—even though I put a biophysical slant on everything. Since my early interest in biology had been as a naturalist, I wondered whether I could fruitfully tap that early enthusiasm and combine it with my training as a physicist and my experience as a plant cell membrane biophysicist. I tried, but it didn't work. I didn't know what to do. I couldn't think of the right questions to ask. Ecology was too complex, quite the wrong milieu for a physicist who had to work with simple things for his approach to be fruitful. (Park Nobel, on the other hand, has made important biophysical contributions to plant ecology.) In any case, after moving to Toronto in the summer of 1971, I went back to being a biophysical plant physiologist.

Being chairman of what I believe is the largest botany department in North America leaves little time for one's own research. But I was fortunate in that Mel Tyree joined the faculty in Toronto at the same time as I did. He initiated some fundamental research on the water relations of plants as investigated by the so-called pressure bomb. (This device had been reinvented by Scholander & Hammel after Dixon, who had invented it around 1920 but made it of the wrong material—glass.) Our chosen plant was hemlock, *Tsuga canadensis*. I cooperated to a limited extent in this work, which was quite extensive and exposed, or so we thought, the advantages and the limitations of the technique. At least with the material we used, the technique seemed reasonable for measuring such "static" parameters as water potential as a function of volume of water expressed from a shoot and the more "indirect" parameters such as turgor pressure, osmotic pressure, and volumes of water both in the cells and in the apoplast. But our attempts to obtain "kinetic" parameters such as hydraulic conductance, L_p, were unsatisfactory. Perhaps the best thing about this research was that it initiated Tyree's interest in the physiology, and particularly the water relations, of trees, which he still pursues with great success.

Another interest of Tyree's at this time was phloem translocation. He theorized that the flow might be driven by hydrostatic pressure gradients—i.e.

he helped develop a mathematical model of flow according to the Munch hypothesis. In this work he became associated with a physicist from Ohio State University, Jack Ferrier. Jack Ferrier joined us in Toronto, initially as a postdoctoral fellow. (For the last few years he has been with a group in the University of Toronto Dental School, where he works on the electrophysiology of cells important in bone.) During his time in the Department of Botany, Jack Ferrier made a number of important biophysical contributions. At first he worked on developing this computer model of phloem transport. Later he became interested in plant water relations and we tried to develop a so-called external force method for measuring hydraulic conductivities and elastic coefficients for higher plant cells. (The idea involved squeezing water out of cells.) The theoretical aspects of this method proved horrendous, and we eventually dropped it.

We had a momentous visitor, first as a postdoctoral fellow and then as research associate, in Bill Lucas, who was a very good experimenter indeed. He worked at this time on the spatial distribution of the apparent H^+ pumps, OH^- carriers, and CO_2 or HCO_3^- uptake regions of the *Chara corallina* plasmalemma. Bill Lucas and Jack Ferrier (and I to some extent) became associated through this work, Jack Ferrier in particular developing realistic models for the flow of ions externally between the acid and alkaline bands.

Claudine Morvan came to us from Rouen on a year's postdoctoral fellowship. She had worked in Rouen on the plant cell wall, but in Toronto, working with Jack Ferrier and Bill Lucas, she helped us initiate studies on electrical noise, both voltage and current fluctuations, in the plasma membrane of *Chara corallina*. At the time we were not able to say exactly what was the origin of the noise: within (passive) channels, the fluctuations in the numbers of open channels or active transport fluctuations? This study of electrical noise in *Chara* was later taken up by a graduate student, Stephen Ross, who managed to convince himself that the low-frequency noise was indeed attributable to active transport, probably the H^+ pump. By injecting white noise current and measuring the consequent voltage noise Stephen Ross also measured the impedance of the *Chara* plasma membrane, obtaining the apparently strange result that at very low frequencies (<0.5 Hz) the membrane capacitance becomes large and negative. We (Stephen Ross, Jack Ferrier, and I) recognized that this phenomenon is related to the diffusion-driven movement of protons in the unstirred layer. This is perhaps an amusing rather than an important effect, but it illustrates the necessity of a sound physical understanding in some aspects of membrane electrophysiology.

Jack Ferrier has recently had a very nice idea about the mechanism of osmosis and we have worked on it a little together. If a porous semipermeable membrane separates, say, pure water from a not-too-concentrated solution, then only occasionally, and for a brief time, would there be a solute molecule opposite the pore mouth on the solution side. Only at such an instant would the water in such a pore "realize" it was facing a solution—and not pure water; and only at such instants would there be a force, between solute molecule and water molecule, pulling water through the pore from the pure-water side. (The solute molecule could not move: It would be stopped by the membrane.) The theory, made quantitative by means of some simple physical ideas, *works*—i.e. gives the right answer that the equivalent pressure driving the water through the membrane is RT multiplied by the solute concentration. Even though the concept is still in rough form, I feel we have an adequate theory of osmotic flow at last. Note that although a pressure difference of $RT\Delta C$ will produce exactly the same amount of flow as a concentration difference of ΔC, the pressure difference would give a smooth flow while the flow produced by concentration difference would comprise spurts—i.e. would be noisy. It seems unlikely that these different flow types could be demonstrated experimentally. Such is a "hobbyhorse" of mine—of no real importance, but intellectually challenging and satisfying.

During the past 7 or 8 years I have returned to another old interest of mine: the ion exchange properties of the plant cell wall. Marie Kleinová, who worked with me for two years, started the work on the isolated cell walls of *Chara* and of some higher plants. She also carried out studies on the water sorption properties of such walls, an important aspect of their physical chemistry. The ion exchange work was taken over by my graduate student, Conrad Richter, who used cell wall material from the moss *Sphagnum*. He was able to interpret the ion exchange properties in terms of the Manning model of linear systems of fixed negative charges comprising the pectin fraction of the wall.

But now the plant biophysics group in Toronto, as such, is practically finished. Mel Tyree left 3 or 4 years ago. Jack Ferrier, although still in Toronto, devotes his time to his bone cell studies. I have retired, to play only a minor role in the research of the department. However plant membrane studies are more vigorous than ever, although now permeated with a more biochemical approach, led by my successor, Eduardo Blumwald. It is appropriate that there should now be an integrated approach—biophysical, biochemical, cell physiological, molecular biological—to plant membrane transport; and this is what is happening in Toronto.

Of course, my activities in Toronto were not confined to encouraging biophysical membrane research. Teaching is the other pillar of a university, and I wanted to inculcate in at least a few students some feeling for the biophysical approach to biology—to plant physiology in particular. After all, living organisms obey the laws of physics (and of chemistry, which is also physics).

As an administrator I did not repeat the system I had developed in Edinburgh and Norwich of offering a MSc in Biophysics that involved a lot of course work plus a little research. Instead I attempted to inject my ideas, my approach, into the undergraduate teaching program. For example, I offered a course to first-year biologists on the "Design of Organisms" in which we discussed questions of size and shape—how these were determined by the simple physics of diffusion, circulation of fluids, and/or mechanical constraints. The textbook for this course was *Gulliver's Travels!* After their shocked discovery that the course treated not only physics but also English Literature, most students quite enjoyed it; but whether it had any subsequent influence on their biology, I very much doubt. I also handled the more biophysical aspects of ionic and water relations in the basic plant physiology course, taken in the second year in Toronto. Students clearly found this difficult, despite the fact that they had all had physics, chemistry, and mathematics during their first year. They seem to keep their individual courses in little boxes which, once closed upon completion of the term are not opened and looked into ever again. The North American System involving frequent assessment of progress in individual courses, examinations confined to the content of each course in isolation, and a transcript of courses taken with grades given may encourage this tendency to keep one's knowledge in separate compartments. The other course I gave, for advanced undergraduate and graduate students, was entitled "Cellular Transport Processes"—really a course on the biophysics of membrane transport in plant cells. I shared with Mel Tyree advanced and graduate courses on "Physical Environment of Plants" and "Plant Biophysics." The former treated the interactions among heat, light, humidity, etc, and the plant; the latter contained more membrane biophysics, some soil physics, and a lot on translocation.

Did these courses have much impact? I confess that I feel they had little. In some ways most biology students have preselected themselves as feeling hostile to mathematics and the physical sciences; they can almost be described as refugees from the physical sciences. Therefore the impact of one or two biophysical courses on their general outlook is negligible. All their other courses tend to be given by people who share their distaste for, and incompetence in, physics and physical chemistry. This is a somewhat depressing conclusion from so many years of attempting to explain, to put over, a physical viewpoint that I fervently believe is essential. My greatest success in teaching, and perhaps the only thing that has justified my academic career, was in bringing people from a background in physics or physical chemistry into biology through the Diploma in Biophysics at Edinburgh and the MSc in Biophysics at East Anglia. Does this mean that the best hope for plant biophysics, if it is deemed of some importance, is to attract students already trained—or perhaps training—in physics and/or physical chemistry? Maybe this will prove to be the only way; perhaps the demand for plant biophysicists is not great enough to encourage what is really a ridiculous idea—revolutionizing the training of biologists.

My years in Canada have coincided with a great change in the study of plant membrane transport. From being a field in which only a handful of people worked, particulary from a biophysical point of view, it has developed into a sophisticated and rapidly changing subject. No longer is it a poor relation of the study of animal and microbial cell membrane transport. One important coordinating feature of this development has been the series of International Workshops on Plant Membrane Transport we have been holding every three years or so for the past 20 years. These started in 1968 at Schloss Reinhardsbrunn in East Germany but perhaps really got going at the Workshop organized by Peter Anderson in Liverpool in 1971. The third workshop was held in Jülich, West Germany, in 1974, organized by Ernst Steudle and Ulrich Zimmermann, and the fourth in 1976, in Rouen and Paris, the organizer being Michel Thellier. The fifth took place in Toronto in 1979; then came Prague in 1983, Sydney in 1986, and this year, 1989, Venice. These workshops have been attended by essentially all workers in the field. The proceedings of each of them have been published and given a clear record of the rapid increase in our understanding of transport processes across plant cell membranes. The International Workshops have acted both as a periodic focus of interest and as an important stimulus to work in this field. I think it fair to say that from the outset the guiding approach was biophysical.

In addition to these Workshops I have enjoyed being an Associate Editor of the journal *Plant Physiology* for two periods of two or three years, handling the more biophysical papers—especially but not exclusively those on membrane transport. As has been said many times the ASPP and the journal owe a tremendous debt to Martin Gibbs, who has now been Chief Editor for more than 25 years. When Marty asked me if I would succeed Noe Higinbotham as an Associate Editor, I didn't hesitate to accept the important job. I thus succeeded the person who did more to move forward the field of plant membrane transport, particularly in North America, than anyone else I can think of. I found being an editor of scientific papers a difficult but rewarding task. Accepting a paper rarely involved a straightforward decision, for the assessments of referees often disagreed. Sometimes a paper had to be declined, never a pleasant decision to take, but I never had an angry "customer"—at least authors never vented any anger at me. I look upon this associate editorship as one of the most useful things I have done, always made pleasant by the cordial relations with Marty Gibbs and his secretary.

Now 70, retired, and able to spend more time reading and studying my beloved literature (not scientific), how can I summarize what I think I have done for plant physiology? Whatever I have done, however it may be judged, I see it purely as an accident or series of accidents, as should be clear from what I have so far written: the accidents of my year of birth, my study of physics, my change from physics to biology (because the University of Edinburgh wanted me to teach physics to medical students!); and then the taking up of the study of membrane transport in plants for reasons that certainly did not include a burning desire to do so. At that time the field was nearly empty. Whatever one did—particularly from a biophysical point of view—was new and exciting. One could not help but make some progress, succeed in a kind of way. I myself have made only one or two small contributions, nothing that justifies my writing about them as I have done here. My main contribution has been, as I see it, the introduction to the field of plant biophysics of a number of outstanding people, among them Enid MacRobbie, Roger Spanswick, and Jim Barber. They and their students have made *major* contributions. Somehow or other the biophysical approach to plant physiology has become respectable, and for this I have shared with others some responsibility.

Annu. Rev. Plant Physiol. Plant Mol. Biol. 1989. 40:21–53

SALINITY TOLERANCE OF EUKARYOTIC MARINE ALGAE

G. O. Kirst

Department of Marine Botany, University of Bremen, D-2800 Bremen 33, Federal Republic of Germany

KEY WORDS: osmoregulation, turgor pressure regulation, compatible solutes, ionic relations, osmolytes

CONTENTS

INTRODUCTION

General Scope

Growth and distribution of marine algae are primarily controlled by light, temperature, nutrients, water movement, and salinity. Salinity is typically a local, rather than a global, parameter and is highly variable in coastal regions, especially in the intertidal zones, estuaries, and rock pools. Our understanding of the physiology of salinity tolerance in marine algae has made substantial progress in the last two decades. This is reflected in a wide variety of

21

1040-2519/89/0601-0021$02.00

review articles with emphasis on various aspects of the ecology and physiology of osmotic adjustment and turgor pressure regulation. General features are covered in the articles by Gessner & Schramm (46), Hellebust (69), Cram (30), Wyn Jones & Gorham (175) and Munns et al (109). Ionic relations during osmotic stress and membrane transport were reviewed by Gutknecht et al (67), Raven et al (125), Glass (59), Tazawa et al (157), and with additional emphasis on the physical aspects by Zimmermann & Steudle (181), and Zimmermann (179). The effect of salinity stress on metabolism and its regulation was investigated almost exclusively on microalgae and is summarized by Brown (21), and Kauss (80, 81). In this context the halotolerant genus *Dunaliella* deserves special mention as one of the most thoroughly investigated organisms (18, 22, 58, 166; including aspects of industrial application: 8). Finally the phylogenetic and evolutionary implications of salinity tolerance have been presented by Yancey et al (176), Russell (141), and Raven (124).

This review reports mainly on salinity tolerance of marine macroalgae and phytoplankton species. Although the chrysophyte *Poterioochromonas malhamensis* is a freshwater flagellate (80) and *Dunaliella* spp. inhabit extreme saline environments (58) rather than typical marine habitats, relevant results from studies of these organisms are also included here, because most of our knowledge on metabolic responses to salt stress has been obtained with these species.

Changes of salinity affect organisms in three ways: (*a*) osmotic stress with direct impact on the cellular water potential; (*b*) ion (salt) stress caused by the inevitable uptake or loss of ions, which simultaneously is a part of the acclimation, of course; and (*c*) change of the cellular ionic ratios due to the selective ion permeability of the membrane.

Both of the ion effects (*b* and *c*) may be amplified depending on different experimental approaches used to subject the algae to osmotic stresses: Media salinity is changed either by controlling the NaCl concentration, thus keeping the other ion contents constant, or by concentrating or diluting seawater and so maintaining all ion ratios constant. The latter procedure simulates natural conditions.

The ranges of salinity encountered in marine habitats differ greatly with respect to space and time: In the open oceans salinity varies between 33 and $37^{\circ}/_{\circ\circ}$ (about 970–1060 mosmol/kg; 2.5–2.73 MPa), gradually decreasing from the tropics towards the polar seas, and thus for a given region may be considered constant. In near-shore waters and estuaries the seawater is diluted by river water. The degree of dilution largely depends on the rate at which fresh water is flushed out of an estuary (118)—e.g. the water of the Amazon can be recognized at the surface for some 300 km into the open ocean (142).

Saline fluctuations are further complicated by tidal actions, as a result of which the salinity may range from 0 °/oo to full seawater strength.

The osmotic stresses acting on algal populations exposed under low tide to either desiccation or rainwater may be less than anticipated: The salinity of water samples taken from between the layers of *Fucus* or the filaments of *Cladophora* exhibited only small changes (159; G. O. Kirst, unpublished observations).

In semi-enclosed seas with only small connections to the open oceans the seasons have a marked influence on the salinity. Examples include the Baltic Sea (4–24 °/oo; 46, 141) and such lagoons as the Coorong, South Australia (up to three times that of seawater; 17).

Definition of Turgor Pressure Regulation and Osmotic Adaptation

The physiological processes behind salinity tolerance are frequently termed "osmoregulation" or "turgor regulation." Kesseler (83) discriminated clearly between (a) the turgor pressure regulation, typically observed in marine algae, that adjusts cellular osmotic potential to keep the cell turgor constant, and (b) osmoregulation in organisms (mostly animals) that aspire to maintain the internal osmotic potential constant. Osmoregulation is an inappropriate and misleading term for the adjustment of cellular osmotic potential following changes in external salinity. A precise definition was given by Cram (30), and Reed (128) suggested the terms osmotic adjustment, osmoacclimation, turgor pressure, or volume (wall-less cells) regulation to describe biochemical and physiological modifications in cellular structure and function resulting from salt stress, while the term osmoadaptation was proposed for genetic modifications evolving under osmotic stress (e.g. the selection of ecotypes; 141). These terms are not mutually exclusive; a plant can both be osmoadapted and possess osmoacclimation properties.

GENERAL RESPONSES OF THE ALGAE AND RANGE OF TOLERANCE

Since the early days of research on salinity stress (46, 141), growth or rates of survival, photosynthesis, and respiration have commonly been used to describe the range of tolerance.

Growth Rates

Growth reflects the balance between photosynthesis and respiration. The growth rates of microalgal cultures are frequently estimated from cell counts (division rate per day), cell density (transmission), or on a chlorophyll basis.

In macroalgae, growth may be expressed as increase in fresh or dry weight. Survival rates have also been measured by microscopically counting surviving and dead cells after applying specific dyes (131).

In phytoplankton species when growth was reduced it was at the lower salinities ($< 5\ °/_{oo}$) (105). Lower and upper limits of growth depend largely on the adaptation of the species: Typical estuarine algae tolerate low salinities better than the oceanic species; coastal phytoplankters take an intermediate place (20). This observation is also valid for macroalgae—e.g. salinity tolerances of upper-shore algae are much broader than those from lower levels, with sublittoral species having the narrowest tolerance limits (141). Salinities commonly encountered in areas of abundance are most favorable for growth (10, 11, 46, 127, 141). Growth at low salinities may be governed by the availability of certain ions—e.g. in *Enteromorpha intestinalis* a high internal K^+ concentration is required to maintain metabolic activity (140). The presence of Ca^{2+} plays an important role, increasing the tolerance limits of algae and higher plants as well (46, 72, 82, 156, 157). Comparing parts of thalli differing in age reveals that young distal cells in algae with apical growth (*Cladophora, Ceramium, Phycodrys,* and *Plumaria*) are more sensitive to low salinity (141). It may be speculated that this is a secondary effect of Ca^{2+} availability, since fast-growing cells depend on Ca^{2+}—e.g. for cell wall formation.

However, if not only growth but also survival is considered, most algae exhibit a remarkable physiological potential—e.g. *Porphyra umbilicalis* grew optimally in a range of $7°/_{oo}$–$52°/_{oo}$ and survived without cell division in salinity 6-times that of seawater (170). Similar observations have been recorded in studies of elongation growth in *Valonia macrophysa* (67). Near the limits of salinity tolerance growth may be sacrificed in order to maintain osmotic adjustment, which guarantees survival at least for short periods.

Several processes have been suggested that may act singly or in combination to reduce growth under extreme salinity stresses. These include (*a*) adverse effects of high ion concentrations subsequently reducing water potential and affecting metabolism, and (*b*) factors related to osmotic adjustment, with organic osmolytes causing a drain on metabolites needed for cell growth (109). This is reflected in obvious changes in size and morphology of plants under long-term salt stress (46, 141).

If additional factors that greatly affect the growth rate are included in the investigation of the growth-salinity relationship (e.g. light, temperature, and nutrients), the analysis becomes very complex. However, data sets such as these are needed to predict maximum growth and possible bloom conditions for phytoplankton populations. Only a few species have been investigated consequently in this respect: *Nitzschia americana* (106–108), *Thalassiosira rotula* (151, 152), *Dunaliella* spp. (58, and literature quoted there), and—as examples for macroalgae—*Cladophora* spp. (159) and *Polysiphonia* (127, 129).

Thus with some reservations the following generalization may be allowed: Growth primarily depends on light and temperature; the more these parameters approach species-specific optimal demands, the broader is the salinity range tolerated.

Photosynthesis and Respiration

Under conditions of extreme hypo- or hyperosmotic stresses, photosynthesis and respiration are inhibited in all algae investigated. Whether a stress is extreme depends on the long-term adjustment of the organism. The responses following exposure to moderate and high changes in salinity are inconsistent. Frequently a transient stimulation of respiration and a stimulatory or inhibitory effect on photosynthesis have been observed. The time required for a more or less complete recovery is variable and lasted from minutes for most microalgae up to several hours for thalloid species (46, 58). A possible explanation is that microalgae with their greater cytoplasmic/vacuolar ratios are faster in their metabolic performance to regulate the internal osmotic potential. In the unicellular flagellate *Tetraselmis (Platymonas)* subjected to hyperosmotic shocks photosynthesis was reduced for 10–120 min (70, 86, 89) while respiration was not affected (92). In contrast, the recovery period of photosynthesis in the giant-celled charophyte *Lamprothamnium papulosum* lasted 8 hr (94).

In microalgae, extensive hypoosmotic shocks have been reported to have little effect on photosynthesis in species with strong cell walls (e.g. *Stichococcus bacillaris)*, while most wall-less species appear to be more sensitive (71).

Salt tolerance studies that use impaired photosynthesis of macroalgae as an indication of cell damage during or after osmotic stress find an interesting tendency with respect to the three major taxonomic groups [see Table 1, summarized from a literature survey by D. N. Thomas and J. S. Collins (unpublished) and 46]: Rhodophyceae and Chlorophyceae are more sensitive to hypoosmotic treatment than are the Phaeophyceae. This corresponds— referring to the red algae—to the observations on growth behavior noted above (10).

The effects of osmotic stresses on the primary processes of photosynthesis have been studied by chlorophyll fluorescence kinetics mainly on *Dunaliella tertiolecta* (48–50); *Prasiola, Porphyra* spp., *Enteromorpha, Ulva,* and *Petalonia* species from the upper intertidal range (173); and *Porphyra perforata* (43, 144, 155). These investigations were aimed primarily at measuring the effects of desiccation on photosynthesis. The two stresses (increasing salinity and desiccation) are comparable since they result in a reduction of the cellular water potential. During desiccation, however, cellular ionic concentrations

Table 1 Photosynthetic responses of marine macroalgae subjected to osmotic shocks.

Class	Inhibition (%)	No effect (%)	Stimulation (%)
Chlorophycea (12 species)			
a. hyposaline	80	6	14
b. hypersaline	78	7	14
Phaeophyceae (16 species)			
a. hyposaline	45	23	32
b. hypersaline	50	25	25
Rhodophyceae (34 species)			
a. hyposaline	88	9	3
b. hypersaline	66	14	20

increase and the ion ratios remain constant. In contrast, during salinity stress algal cells may increase ionic concentrations but also undergo changes in ion ratios owing to selective uptake. This has to be taken into account when comparing the results obtained with species under salt or desiccation stress. The results with *Dunaliella* suggest that the initial charge separations at the reaction centers of photosystems I and II are inhibited more by ionic stress than by osmotic stress. It is assumed that osmotic stress increases the permeability of the thylakoids to ions—e.g. Na^+ and Cl^-, which subsequently inhibit photosystem I and II (47, 50). In macroalgae the primary photosynthetic mechanism is affected at the electron transport stage between PS I and PS II. The sensitive site in *Porphyra, Ulva,* and *Enteromorpha* species is most likely between plastoquinone and P 700 (173). Further details are available for *Porphyra perforata* (144): There are at least three sites in the photosynthetic apparatus that are inhibited by high salinity. First, the photoactivation of electron flow on the reducing side of PS I is impaired. The second sensitive site is the electron flow on the water side of PS II. Finally, the transfer of light energy between the pigment complexes may be impaired. Satoh et al (144) stress that an inhibition of electron flow at all three sites is essential to avoid photodamage, which will occur if only one site is blocked: Accumulation of free reductants or oxidants quickly destroys the reaction centers.

Effect on Fine Structure

Reduction in size and striking changes in morphology are well known from marine macroalgae growing in brackish water habitats (46, 141, and literature cited there; 127, 178). Variable osmotic pressures directly affect the cell wall morphogenesis of diatoms (113, 148, 153). Investigations at the submicroscopic level showing the effect of osmotic stress on the cellular fine structure are comparatively rare, however.

Shortly after a hypoosmotic shock the whole cell of *Dunaliella salina,* including the cytoplasmic structures, was swollen (160). Hyperosmotic conditions caused a shrinkage of the cells, with the membranes and the thylakoids appressed. *Dunaliella* acclimated to high salinity exhibited compression of the thylakoids over large portions of the membrane system. Using X-ray microanalysis in *Dunaliella parva,* Hajibagheri et al (68) showed that the concentrations of Na^+ and Cl^- were much higher in the vacuoles than in the cytoplasm. After exposure of the alga to an increased salinity the ion concentrations increased in both compartments immediately. The marine flagellate *Tetraselmis (Platymonas) subcordiformis* formed a large vacuolar system by fusion of Golgi vesicles following a hypoosmotic shock (93). The chloroplast was swollen and the thylakoid stacking transiently disturbed. After 5–10 min the chloroplast regained its original size and the stacked organization of the thylakoids, but the total cell volume remained enlarged because of the persisting vacuoles. The changes in the structures coincided with a transient inhibition of photosynthesis. Respiration was not affected, a lack of response consistent with the mitochondrial structure remaining intact throughout the treatment. It is assumed that the cytoplasmic compartment, rather than the whole cell, is under volume control.

Under normal conditions the red alga *Porphyra umbilicalis* is devoid of vacuoles. Subjected to hypo- or hyperosmotic stress the cells developed vacuoles (98, 171). As in *Tetraselmis* the protoplasts recover their original size. The vacuoles formed during the hypersaline treatment may serve as compartments to sequester ions, mainly Na^+ and Cl^-. Freeze-fracture studies of the tonoplast in *Porphyra* revealed an increase in size and density of intramembranous particles probably correlated with the transport function of the membrane (172). In *Porphyra purpurea,* however, vacuolation is primarily a response to hypoosmotic conditions (133, 134).

Compartmentation by vacuole development is the most striking change in fine structure after osmotic stress of unicells as well as multicellular algae.

PROCESSES OF OSMOTIC ACCLIMATION

The response of algae to moderate changes of salinity is a well-regulated biphasic process. The first phase is characterized by rapid changes in turgor pressure (walled cells) or volume changes (wall-less cells) caused by massive water fluxes in or out of the organism following the osmotic gradient. The second phase represents the osmotic adjustment: Cellular concentrations of the osmotically active solutes (osmolytes) change until a new steady state is achieved. Both phases are part of a feedback loop that comprises the osmotic acclimation (30, 67). The change of salinity is recognized by either a turgor pressure or volume sensing detector. This in turn triggers an effector to alter the turgor pressure or the volume by means of reactions. The adjustment finally has to be regulated via a feedback mechanism between detector and effector.

Sensing of Turgor Pressure: The Detector Mechanisms

Identification of the turgor pressure sensing mechanism is one of the most challenging problems that faces researchers investigating the physiological mechanisms of salinity tolerance. It has been speculated that subcellular structures such as the cytoskeleton may be involved in the sensing of changes in cell turgor pressure (145), but at present little experimental evidence supports this idea. Three models have been suggested to describe the transformation of an external stimulus caused by the change of salinities into regulation of ion transport by the cells. The disturbance of the membrane structure plays a role in all three models, which involve (a) anisotropic changes in the membrane due to tension or stretch, (b) the electro-mechanical compression of the membrane, and (c) stretch-activated ion channels.

THE ANISIOTROPIC MEMBRANE CHANGE MODEL   Gutknecht et al (67) observed in experiments using *Valonia, Codium,* and *Halicystis* that absolute hydrostatic pressure has no effect on turgor regulation. Therefore, it is the pressure gradient that causes the anisotropic changes in the membrane. The sensor must be located in the plasmalemma since pressure gradients are unlikely to exist within the cytoplasm and between organelles and the cytoplasm (14).

The turgor transduction process works as follows: The plasmalemma is pressed against the cell wall by turgor pressure and forced into the gaps between the cellulose microfibrils, thereby causing an asymmetrical curvature of the membrane lipid bilayer. Subsequently there is a shift in the distribution of lipids, and especially of proteins within the membrane. This shift affects active ion transport, changes permeability, or releases a messenger of some kind to activate enzyme systems involved in osmotic adaptation, such as the membrane-derived proteinase in *Poterioochromonas* (81). There is little experimental evidence supporting this model. However, in freeze fractures of the plasmalemma of *Pelvetia* cells, imprints of the cell wall microfibrils are clearly visible in turgid cells and disappear in plasmolyzed cells (116).

THE ELECTRO-MECHANICAL MODEL   Measurements of the effect of turgor pressure on dielectric breakdown in plant membranes led to the formulation of the electro-mechanical model (29, 179, 182). In this model the membrane is considered as an electrical capacitor filled with an elastic, dielectric material.

The elastic properties of this material counterbalance the mechanical compressive forces deriving from the turgor pressure and the electrical compressive forces arising from the membrane potential. The first step in sensing changes of turgor pressure is the change of membrane thickness as a consequence of the alteration in mechanical compression. In turn the transport through the membrane is affected. This can derive either from changes in the intrinsic electric field or from changes of the active pumps or channels.

For the giant celled alga *Valonia utricularis* it was calculated that the membrane thickness will be reduced by 0.2–0.4 nm when the turgor pressure is changed by 0.1 MPa (183). The effect of changes in turgor pressure on transport properties of the plasmalemma and the tonoplast have been verified experimentally with *Valonia macrophysa* (67) and *V. utricularis* (179). These studies have shown that active uptake of potassium is regulated by pump activity. Increasing turgor pressure under hypoosmotic conditions stimulated, while decreasing turgor pressure under hyperosmotic treatment inhibited, pump activity.

Using a different experimental approach Zimmermann and coworkers (26, 27, 165, 183) found further evidence for the compressibility of the plasmalemma and the tonoplast in *V. utricularis*. Results from charge-pulse experiments that estimated the capacity of the membrane strongly support the assumption that there are negative mobile charges in the membranes of *Valonia*—i.e. plasmalemma and tonoplast. The translocation rate of the mobile charges increases with increasing turgor pressure, a finding that agrees with the predictions for a compressible membrane. The response of the translocation rate was observed in the same range of turgor pressures over which potassium transport is affected. From experiments with inhibitors of protein synthesis, alkaline earth ions, and anaesthetics it was concluded that the negative mobile charges are connected to a protein component. This may be an integrated part of a carrier system, of the channels or pumps, involved in potassium transport. In summary, according to this hypothesis changes of salinity result in a change of turgor pressure that in turn alters membrane thickness. This physical parameter controls the translocation rate of mobile charges in the membrane, which are the link to the regulatory biochemical and biophysical processes of the cell.

In the electro-mechanical model absolute pressure as well as a pressure gradient (turgor pressure) is effective. The model therefore is valid for the plasmalemma and the tonoplast, since membrane compression does not require a pressure gradient (179). However, according to the experiments by Gutknecht et al (67) described above, absolute pressure is not an essential element in the turgor-sensing process. Furthermore, the primary assumption of this model—i.e. a homogeneous, electrically neutral membrane composition as plasmalemma—is difficult to imagine.

THE STRETCH-ACTIVATED ION CHANNELS MODEL   By means of patch-clamp measurements the characteristics of stretch-activated ion channels were investigated first in tissue-cultured chick skeletal muscle (63). In animal cells it is suggested that those stretch-operated channels are involved in mechanoreception and osmoreception—e.g. in volume regulation (161). The frequency of channel opening increases with increasing force applied to the membrane. Similar patch-clamp studies on plant cells are beset with many more difficulties because of the rigid cell wall. Using protoplast preparations of tobacco cell suspension cultures, Falke et al (42) presented evidence for a stretch-activated, anion-selective channel in the plasmalemma. Stretch-activated ion channels have also been reported in yeast (64) and may also play a role in stomatal movement (120).

The basic features of the mechanoreceptor-operated ion channels may well serve as a model for transduction of mechanical signals in plant cells (38). It is straightforward to imagine this system as part of a turgor-sensing mechanism: Channel proteins distributed in the membrane are connected with inelastic filaments ("spectrin-like linkers"; 63) parallel to the inner surface of the membrane. A change in the tension of the membrane would cause the filaments to tug at their sites of attachment to the channel proteins, resulting in opening or closure of the conduit. The network of filaments connected via the channel proteins works as an amplifier, gathering force from a large area of membrane.

Applying this model to what is known in osmotic adjustment of algae the following two possible cascades of responses were suggested by M. A. Bisson (private communication): (a) distortion of plasmalemma → opening of Ca^{2+} channels → influx of Ca^{2+} → rise of cytosolic Ca^{2+} → effect on transport functions (membrane ATPases; other transport proteins) and metabolism (activity of enzymes); (b) distortion of plasmalemma → opening of Cl^- channels → change of cellular Cl^- and/or alteration of electrical properties with subsequent effects on transport.

The opening of Ca^{2+} channels was demonstrated in *Lamprothamnium succinctum* (111, 157). Turgor pressure regulation failed when *Lamprothamnium* was subjected to hypotonic stresses in media containing less than 0.01 mM Ca^{2+}. If the external Ca^{2+} concentration exceeded 1 mM the elevated turgor pressure was restored owing to a transient increase of cytoplasmic Ca^{2+} that stimulated membrane permeability to K^+ and Cl^- (112). Thus either an abrupt increase in turgor pressure affects the Ca^{2+} channel directly—i.e. the channel itself is the pressure transducer—or the membrane depolarization that accompanies this process activates the Ca^{2+} channel (157).

As in the case of the electro-mechanical model, the model of stretch-activated ion channels is appropriate for plant cells with cell walls as well as for naked cells such as *Dunaliella*. The valve properties of the channel proteins may also explain the rapid loss of ions and organic osmolytes observed with many algae subjected to hypoosmotic shocks. However, it remains unclear how the selectivity of the perception is achieved—e.g. how a distortion deriving from a change in turgor pressure will be distinguished from other external mechanical stimuli such as bending.

None of these models explains sufficiently how the biochemical responses of the cells might be triggered. Ca^{2+} may act as a link between the membrane properties and the biochemical pathways (see 82).

Phytohormones called "turgorines" have recently been isolated and chemically identified. These cause leaf movements based on turgor pressure mechanisms in higher plants such as *Mimosa pudica* and *Acacia karro* (146, 147). These compounds (e.g. 4-O-[3,5-Dihydroxybenzoic acid]-β-D-glucoside-6'-sulfate) may alter the permeability of the membranes, resulting in a rapid turgor pressure decrease in the cells. Algae have not yet been tested for the presence of similar compounds.

Changing the Cellular Concentrations of Osmolytes: The Effector Processes and Their Regulation

WATER FLUXES   The water fluxes observed as the first phase are rapid processes with half times in microalgae in the range of 5–10 sec (119) and in macroalgae lasting from minutes to hours (179, 182). As a result of the water influx under hypoosmotic treatment or water efflux during hyperosmosis the osmotic stress is mitigated at least transiently. These processes are not under immediate metabolic control. As passive "osmometer" behavior they depend on physico-chemical properties of the cell-wall–membrane complex such as the hydraulic conductivity (water permeability) and the elasticity. [For information on the pressure probe technique for measuring the water-relations parameter see 179, and 182.] Plastic (irreversible) cell wall expansion by turgor pressure, which plays a role during growth, will be neglected because it is less important during the much faster process of osmotic adjustment (67). The hydraulic conductivity in *Valonia utricularis* and in some freshwater and brackish-water Characeae was little affected by small turgor pressure changes but increased steeply with decreasing turgor (156, 182). The elastic property of the cell wall described by the volumetric elastic modulus (ϵ) (179) is of more importance for the osmotic adjustment: ϵ values rise with the rigidity of the cell wall. For many marine and freshwater algal cells they are on the order of 5–60 MPa. *Halicystis parvula* is an example exhibiting a very low ϵ value of 0.06–0.2 MPa (67, 180). Wall-less cells may be considered to be at the lowest end of the scale with values for ϵ close to 0 MPa (179). Low ϵ values provide the cells with a high resistance to short-term fluctuations in salinity.

High values ensure an immediate response to osmotic stress since even small changes in water flux result in turgor pressure differences large enough to trigger the detector. The elastic modulus increases with increasing turgor pressure and cell volume. This feature supports the tolerance to fluctuating salinities: Under hypoosmotic conditions the increase of ϵ prevents further cell expansion and water influx; decreasing ϵ values during hyperosmotic stresses retard plasmolytic effects.

In contrast to the first phase the second phase is a slow process lasting 40–120 min in microalgae and up to two to three days in thalloid or giant algal cells. As a result the internal osmotic potential becomes adjusted by changing the concentrations of ions and organic osmolytes so as to restore the turgor pressure or the volume. These processes are under direct metabolic control.

IONIC RELATIONS   The main ions involved are K^+, Na^+, Cl^- and to a lesser extent sulfate. Concentrations of other ions, especially Mg^{2+} and Ca^{2+}, are not influenced; but Ca^{2+} may function as a kind of messenger (82, 157). The ionic composition of algal cells and in particular of the vacuolar sap varies widely (66, 125). In marine macroalgae representing the three main classes (Chlorophyceae, Phaeophyceae, and Rhodophyceae) the change of Cl^- concentration usually parallels the fluctuation of the salinity. With respect to the K^+ and Na^+ relation, three characteristic types could be distinguished [(90, 91); H. Kesseler (84) distinguished two types]: 1. species with high Na^+ content, increasing together with Cl^- under hyperosmotic conditions while K^+ is almost unaffected; 2. those with K^+ as the main cation and with low Na^+ concentrations that steeply increase to exceed K^+ when under extreme salinity stresses; and 3. algae containing both cations in about equal amounts. There is no obvious correlation between these cation types and taxonomic classes nor a connection with thallus morphology. For example, the Siphonales, which exhibit a huge vacuolar system (e.g. *Caulerpa, Halimeda,* and *Bryopsis*) are typically Na^+ algae; but *Valonia, Dictyosphaeria,* and the rhodophyte *Griffithsia monilis,* giant-celled algae also consisting of 95°/$_{oo}$ vacuole, are representatives of the K^+ type. *Codium* shows both types in the same genus (12, 13).

In most thalloid and all giant-celled algae the vacuole dominates the osmotic relations during salinity stress because of its large portion of the cell volume. Ionic relations in the cytoplasm are not so well investigated, and most of our knowledge is based either on measurements with suitable objects such as *Acetabularia mediterranea* (143) and Characean internodal cells where cytoplasm can be centrifuged into one end of the cell, or on more indirect methods—e.g. flux analysis (121, 124, 157). From these results it is concluded that the cytoplasm is usually (but not always) high in K^+ and low in Na^+ and Cl^-. This resembles the typical ionic composition observed in marine microalgae with a high cytoplasm/vacuole ratio (87, 88). However, the absolute ion content in microalgae is difficult to determine and varies with the methods used for washing the algal pellets free of adhering medium, estimating the cell volume, and correcting for the free space between the packed cells (109). Excellent examples of conflicting data in this respect are the ionic values published for *Dunaliella* species: The Na^+ content ranges from 7 mM–800 mM in cells cultivated in 1.5 M NaCl solutions (58 and literature cited there). Despite this difference, all measurements agree in finding less Na^+ and Cl^- and more K^+ than in the medium. Using a direct method, energy-dispersive X-ray microanalysis in freeze-sectioned *Dunaliella tertiolecta*, Wegmann (166) determined the highest local Na^+ concentration to be about 65 mM in cells subjected to an external concentration of 3 M NaCl. After removal of extracellular ions by ion-exchange minicolumns the ion content of *Dunaliella salina* was 30–50 mM for Cl^- and 20–40 mM for Na^+ (76).

An apparent anion deficit is frequently observed: Usually Cl^- and other anions don't balance the combined internal K^+ and Na^+ levels. This imbalance is pronounced in cytoplasm-rich microalgae (35, 87, 88) but also known from macroalgae (90, 122, 131, 177). A satisfactory explanation is still lacking and, hence, the balance for the excess positive charge is attributed to fixed negative charges or inorganic polyanions such as polyphosphate.

Ion concentrations in algae are regulated by ion-selective carriers driven by the membrane potential. In addition, facilitated diffusion via ion-selective channels plays a role during rapid changes and recovery of ionic composition (see above; e.g. 6, 158). The energy requirements for ion transport in algae are in the range of about 10–30% of the energy gained from respiration (123, 137).

In most freshwater algae and higher plants the membrane potential is generated by an active H^+ pump (67, 124, 164). In marine algae, membrane potentials and the active ion pumps that may be involved in generating the potential difference are more complex. A typical proton pump–driven potential exhibiting negative values in the range of -130 to -240 mV has been described up to now only for the euryhaline charophyte *Lamprothamnium papulosum* (15). The membrane potential of this species may also be found in a second state characterized by a K^+ diffusion potential. The passive K^+ diffusion potential in the range of -40 to -100 mV is common within the marine species (66, 122, 124). The membrane potential of *Ulva lactuca* was -39 mV (light) and -25 mV (dark), and calculations based on measurements of permeabilities of K^+, Na^+, and Cl^- indicated a diffusion potential (33, 137). In some algae—e.g. *Valonia* spp. (65, 67)—the potential is maintained by active inward K^+ pumps; unlike the diffusion potential it is usually positive: $+6$ to $+15$ mV. Negative potentials were observed in *Acetabularia* spp. ranging from -170 to -190 mV, caused primarily by an electrogenic Cl^- pump (61, 167). It has to be emphasized that the membrane potentials are vacuolar potentials. They represent the sum of both membranes—e.g. the potential of *Valonia* cited above is a composite of a plasmalemma potential of about -70 mV and a tonoplast potential of about $+80$ mV. In addition, the mechanisms involved in active Na^+ efflux (such as Na^+-ATPase or H^+/Na^+ antiport) and passive diffusion pathways controlled by K^+ and other ions contribute to the potential.

During osmotic adjustment the changes in ion composition are achieved by regulating the activity of one or several of these transport components. This may be recognized by shifts in the membrane potentials or in the membrane resistance accordingly. Comparing resting potentials of algae adapted to various salinities, no significant difference could be observed under steady state: In *Codium decorticatum* the vacuolar potential of -76 mV, which is predominantly a K^+ diffusion potential, was constant over a salinity range from 23°/₀₀ to 37°/₀₀ (12). In this species the decrease of the internal osmotic pressure under hyperosmotic conditions was achieved by reduction of the active Cl^- influx. This was accompanied by a depolarization of about 7 mV (13). Subjecting the alga to hyperosmotic treatment resulted in an increase of Cl^- influx accompanied by a slight hyperpolarization. Similar results were obtained with the brackish-water charophyte *Lamprothamnium succinctum*, the potential of which also depolarized after hypotonic treatment and regained its original level after about 6 hr (110). The membrane potentials of *Polysiphonia lanosa* from populations grown in an estuarine and a marine site exhibited the same response: Hypersaline treatment produced a depolarization, while hypoosmotic conditions hyperpolarized (129). Another example of how turgor pressure is controlled by the activity of the potassium pump in *Valonia* spp. has already been reported above.

Usually the ions involved in osmotic adjustment are far out of their electrochemical equilibrium; their internal concentrations must therefore be controlled by pump and leak mechanisms. Hence, one may speculate about how the three cation types described above could be explained by assuming that Cl^- transport is directly affected by turgor pressure and that the cation which accumulates or is excluded as counter-ion with Cl^- is determined by other factors: Na^+ is actively transported out of the cell (including "out" into the vacuole), while K^+ is near equilibrium (14, 90). Conspicuous shifts in the cation ratios as described for the K^+ type (changing into a N^+-dominated cell) are observed frequently when tolerance limits of osmotic adjustment are reached in hyperosmotic treatment of marine and estuarine algae such as *Dictyosphaerea versluysii*, *Ecklonia radiata*, *Dictyota dichotoma* (90), *Enteromorpha prolifera*, *E. intestinalis* (177, 178), and *Pilayella littoralis* (131); of marine Characeae (15); and of microalgae such as *Tetraselmis*

(Platymonas) subcordiformis (88) and *Dunaliella* spp. (41, 56, 57). In *Polysiphonia lanosa* a decrease of K^+ was reported under hypoosmotic stress (127). The loss of K^+ is often correlated with a general decrease in viability, including cessation of growth and decrease in photosynthesis (127, 131).

With respect to membrane potentials in microalgae, measurements are not yet reliable (51); ion transport is explained mainly by drawing conclusions by analogy to what is known from macroalgae.

ORGANIC OSMOLYTES—COMPATIBLE SOLUTES   In addition to ions, certain low-molecular-weight organic solutes are accumulated or degraded in algae in response to changes of salinity (8, 69, 80, 81, 166). In many cases these compounds are identical with the main photosynthetic product; hence, there are preferences in taxonomic classes (8, 100, 124): polyols such as glycerol (*Dunaliella:* 58), mannitol (most Phaeophyceae: 90, 135; Prasinophyceae: 70, 86), sorbitol, the amino acid proline (most diatoms: 102, 150; but also found in Chlorophyceae), sucrose (Chlorophyceae and Charophyceae: 16, 90, 95, 130), floridoside and digeneaside (Rhodophyceae: 90, 134), and isofloridoside (*Poterioochromonas malhamensis:* 80). Several solutes derive from the quaternary type ammonium compounds such as glycinebetaine [some algae: *Tetraselmis (Platymonas)* (35), *Chaetomorpha capillaris, Cladophora rupestris* (130); more important in higher plants: 175] and tertiary sulphonium compounds—e.g. β-(dimethylsulphonio)-propionate (DMSP: present mainly in phytoplankton; abundant in Chlorophyceae, in some Rhodophyceae; in Phaeophyceae in traces only: 33, 34, 36, 37, 79, 126, 168). Lists summarizing the osmolytes and their abundance in algae are presented in reviews recently published (8, 69, 71, 124).

Molal solutions of these osmolytes have approximately the same osmotic potential of about -2.5 to -2.8 MPa. However, there is a substantial difference in energy costs and in the amounts of carbon and nitrogen required to achieve these potentials. Nevertheless there is an advantage that may be explained by the double function of the organic osmolytes: These compounds act not only as osmolytes but also as "compatible solutes." This term was introduced to describe the protective function of some organic solutes that becomes obvious using in vitro enzyme assays (18, 21, 22): In the presence of ion concentrations exceeding 100 mM most enzyme activities are severely inhibited, while compatible solutes at isosmotic concentrations are not or are much less inhibitory (32, 136). Even at high concentrations they cause minimal inhibition of metabolism and membrane-dependent processes, exerting rather a stabilizing effect on macromolecules (176) and plant ribosomes (19). Among osmolytes tested for their protective capacity, proline and glycerol were the most effective. However, not all organic osmolytes are suitable as compatible solutes: Sucrose, for example, is fairly poor in this respect. The physico-chemical properties that bring about the stabilization of macromolecules at very low water potentials are not yet fully understood. Schobert (149) suggested two possibilities: Polyols, especially glycerol, may act as "water-like" substances that mimic the water structure and maintain an artificial water sphere around the macromolecule. Proline, on the other hand, protects as an amphiphilic molecule the hydrophobic parts of proteins, which suffer first when water potential is lowered. By forming associations with the hydrophobic portions of macromolecules proline converts them into hydrophilic parts.

Another benefit of this system was pointed out by Yancey et al (176), who compared modes of coping with high salinity: Halophilic prokaryotes such as *Halobacterium* evolved adapted proteins characterized by massive amino acid substitutions that restrict these organisms to saline environments. In contrast, algal species such as *Dunaliella* are more flexible because they are able to live under the same conditions and under low salinities as well merely by adjusting the concentration of compatible solutes, without a genetically fixed modification of their enzymes.

The key role of the compatible solutes is especially obvious in the most extreme halotolerant species. Accumulation of proline and glycerol allows growth at salinities close to saturation of NaCl (71). The osmolytes are highly soluble, but even then the limits under severe osmotic stresses are usually reached and a combination of several organic solutes may be needed. An increasing number of micro- and macroalgal species have been found to contain more than one organic osmolyte: Typical combinations are sucrose/proline (*Chlorella emersonii*: 154; *Enteromorpha intestinalis*: 39, 40; *Blidingia minima*: 78) or sorbitol/proline (*Stichococcus bacillaris* and *Klebsormidium marinum*: 23, 24). Combinations with exotic polyols are often reported in brown algae; mannitol/altritol was found in *Himanthalia elongata* (174); and mannitol/volemitol was found in the top-shore alga *Pelvetia canaliculata* (135). The tertiary sulphonium compound dimethylsulfoniopropionate (DMSP) is frequently found together with other organic osmolytes in a wide variety of marine algae (36, 37, 78, 79, 126). In most cases this compound is not involved in short-term osmotic adjustment but changes with long-term stress and especially under high salinities—e.g. in *Tetraselmis (Platymonas) subcordiformis* (35), *Enteromorpha intestinalis* (40, 126), and *E. prolifera* (177). In contrast to related *Enteromorpha* spp., *Blidingia minima* accumulated DMSP under gentle hyperosmotic conditions (78). DMSP has drawn interest in the last years as precursor of the most important biogenic volatile sulfur compound, dimethylsulfide (DMS), in the oceanic surface waters (1, 2, 162). It is speculated that the ability to accumulate DMSP was evolved during the last ice age, when the salinity of the seas was higher (28). In the short time since then the genetic potential has not been lost and is therefore still present in many phytoplankton organisms in the salinity-stable open oceans, where, according to the Gaia hypothesis (28), it is now used "in controlling the climate." This may explain the observation that up to now DMSP has been detected in none of the typical freshwater micro- and macroalgae.

The term "compatible solute" must be used with care, taking into account that it is based on findings with in vitro systems. Furthermore, a closer look at the various enzymes tested reveals that the inhibition by electrolytes at moderate concentrations is less than anticipated—at least the remaining activity is often sufficient for maintenance metabolism. Half-maximal inhibition of pyruvate kinase in extracts from *Porphyra umbilicalis* was at 300–400 mM added NaCl (170). In *Dunaliella parva* the in vitro salt resistance of various enzymes ranged from 100 mM up to the molar range (55). In this species salt inhibition was primarily due to Cl^- rather than to K^+ or Na^+. Similar results were published for various enzymes from other species of *Dunaliella* (58). Such metabolic "overcapacity" is required to bridge the "lag phase" that usually occurs after osmotic shock until the content of osmolytes starts to be adjusted.

Although many of the organic osmolytes are typically photosynthetic products, accumulation proceeds in the light and in the dark (7, 58, 60, 69, 86, 89). In the dark the organic solutes are remobilized from storage products, although accumulation is usually not as fast as in the light. In *Tetraselmis (Platymonas) subcordiformis* treated with hyperosmotic shocks, mannitol accumulation commenced after a lag phase and at a reduced rate in the dark, as it did in algae under light but inhibited by DCMU (89). *Dunaliella parva* synthesized glycerol at the expense of starch in the dark, but with severe osmotic stress the volume regulation was incomplete (54).

ENZYME ACTIVITIES AND METABOLIC PATHWAYS   The wall-less unicellular alga *Poterioochromonas malhamensis* accumulates isofloridoside [IF, α-galactosyl-(1,1)-glycerol] as an osmolyte during shrinkage caused by increasing external osmotic potential (80–82 and references cited there). The cellular concentration of IF is regulated at the site of production via activation of enzymes involved in synthesis as well as at the site of degradation—i.e. the transfer of IF into the osmotically inactive reserve polymer chrysolaminarin $(1,3-\beta-D-glucan)$.

The synthesis of IF starts within one minute after the onset of the hyperosmotic stress. This and results of studies using inhibitors of protein synthesis suggested that the regulation commences with activation of preexisting enzyme systems instead of de novo synthesis of enzymes. The cascade that leads to the activation of the key enzyme of IF metabolism, the isofloridoside-phosphate-synthase (IFP-synthase), may begin with a change in the membrane lipid composition resulting from the shrinkage process (81). It is speculated that this triggers a signal transfer that involves the formation of a calcium/calmodulin complex to activate a membrane-bound serine proteinase.

The proteinase then becomes soluble and subsequently activates IFP-synthase by proteolysis, probably at a specific peptide sequence, from smaller inactive subunits that aggregate to larger enzyme complexes. Thus, the synthesis of IF increases and the cell volume will be adjusted. During the initial phase the activation of IFP-synthase is roughly proportional to the degree of cell shrinkage and may be stopped by re-swelling the algal cells.

In *Tetraselmis (Platymonas) subcordiformis* the regulation of the mannitol pool during osmotic acclimation depends on enzyme properties rather than on activation (136): The key enzyme of the mannitol synthesis is the mannitol-1-phosphate dehydrogenase, which exhibits increased activity rates in the presence of high NaCl concentrations. A reverse reaction with mannitol-1-phosphate as substrate could not be detected. In contrast, the degradative pathway starts with mannitol dehydrogenase, an enzyme sensitive to NaCl. This suggests that the mannitol pool is directly regulated via alternative pathways, with different activities dependent on the osmotic pressure. The glycerol cycle that tunes the glycerol content in *Dunaliella* demonstrates similar features of regulating metabolic pathways: enzymes with irreversible actions combined with those of different substrate affinities and pH optima for the backward and forward reactions (58).

Metabolic pathways and their regulation under salinity stress have been investigated up to now primarily in unicellular algal species. Our knowledge of the biosynthesis of most of the organic components of marine macroalgae is scarce, as is our understanding of the regulation involved in formation and degradation of the compounds. Even less is known at the level of molecular biology: It is unlikely that in the near future the complex responses to osmotic stress may be traced to algal genes analogous to the *osm* genes in *Escherichia coli* that govern the production of proline and betaine to protect the cell against dehydration (101).

In summary the following mechanisms may participate in the regulation of the osmolyte pool size during osmotic adjustment:

(a) Hyperosmotic stress:
  1. de novo synthesis of enzymes
  2. activation of enzymes involved in synthesis of the osmolyte
  3. reduced degradation or metabolism
  4. remobilization of reserve products in light and dark
  5. redistribution within cellular compartments

(b) Hypoosmotic stress:
  1. transfer into polymeric reserve products
  2. inhibition of enzymes involved in synthesis
  3. stimulation of degradative pathways
  4. release of organic osmolytes into the medium: emergency reaction observed in micro- and macroalgae (71, 88, 132)

COMPARTMENTATION   It is obvious considering the contribution of ionic and organic osmolytes to the internal osmotic potential on a whole-cell basis, that with increasing vacuole size the proportion of the organics decreases. In microalgae having a high cytoplasm:vacuole ratio the fraction of organic compounds is 25–60% of the osmotic potential (70, 71, 88). In giant algal cells such as *Griffithsia monilis* (95–98% vacuole!) the overall concentration of organic solutes is negligible compared to ionic concentrations, although there is a clear correlation with salinity stress (15, 90). These observations, together with the arguments on energy costs and the inhibitory effects on metabolism of high cytoplasmic ion contents already mentioned above, have led to the hypothesis that in vacuoles the osmotic potential is primarily accounted for by ions. In the cytoplasm, in addition to a basis of mainly K^+, the compatible solutes regulate the osmotic value (15, 32, 90, 109, 127, 135). Admittedly the evidence is indirect, but estimates of absolute concentrations of osmolytes in the cytoplasm are scarce: In *Lamprothamnium succinctum* the change of cytoplasmic free amino acids and sucrose contents during turgor regulation were estimated by perfusion of the vacuole and thus separation of the cytoplasm (112). The cytoplasmic sucrose content increased linearly with increasing salinity, accounting for 40% of the enhancement of the osmotic potential.

Glycerol is known to permeate easily through membranes. Since it is sequestered in high concentrations within the cytoplasm of *Dunaliella* species the plasmalemma has to maintain large gradients of glycerol and NaCl in opposite directions. This may be achieved either by a low permeability of the membrane for glycerol or by an increased production to compensate for unavoidable leakage in the case of a conventional permeability. In *Dunaliella parva* the permeability coefficients of the plasmalemma were found to be very low for a variety of different compounds and in particular for glycerol (52; cf 123:346–47). The unusually low permeability may be the consequence of a uniquely high content of sterol derivatives of the plasma membrane.

Further compartmentation of osmolytes into organelles within the cytoplasm is discussed as a possibility for organisms containing more than one compatible solute (154; see above) or with respect to intracellular distribution of enzymes regulating synthesis or degradation of osmolytes (44, 53).

The cell wall as a compartment plays an important but relatively unexplored role in the ionic and osmotic relations of marine algae. All marine algae contain sulfated wall polysaccharides, with the larger portion of the highly acidic polymers located in the outer regions of the wall and in the outer cellular layers of the thallus (97, 103, 117). Sulfated polysaccharides may be involved passively in cellular ion transport by selective cation binding. The selective adsorption and enrichment of K^+ over the divalent cations Mg^{2+} and Ca^{2+} was demonstrated in the cell wall of *Valonia utricularis* (85). The cation-exchange capacity of cell walls of *Enteromorpha intestinalis* was 2.5 mmol g^{-1} dry weight (138) and those for the intertidal brown algae *Ascophyllum nodosum* and *Pelvetia canaliculata* were about 3.8 meq g^{-1} dry weight compared to about 1.0 meq g^{-1} dry weight of primary cell walls from land plants (96). As a consequence of the negative fixed charges in the cell wall one would assume the wall to be a barrier to anions at least in dilute solutions. However, the rates of Cl^- fluxes measured in *Enteromorpha* indicate that the exchange across the cell wall is facilitated by pores (139). The polyanion composition of the cell wall leads to a discrimination between monovalent and divalent cations as was shown for *Fucus virsoides* (103): K^+ and Na^+ were found in the outermost wall of the epidermal layer, while higher portions of Mg^{2+} and Ca^{2+} were in the inner part and inner wall layers of the surface cells. These gradients are caused by the different cation selectivities of alginate and sulfated fucans. Those selective properties of weak (carboxylic) and strong (sulfuric) acid moieties may also exert a substantial influence on active ion uptake driven by proton gradients and secondary proton pumps (see above).

COORDINATION OF CHANGES IN OSMOLYTE CONCENTRATIONS   During osmotic adjustment, ionic and organic solute concentrations do not change independently but are coordinated. Time courses reveal that changes in ionic contents generally precede those in organic osmolytes. Later as steady state is approached, the metabolically adverse ion concentrations are reduced again in favor of the then accumulated compatible solutes. After the application of hyperosmotic shocks, microalgae such as *Tetraselmis (Platymonas) subcordiformis* (88), *Dunaliella parva* (57), and *D. tertiolecta* (41) responded with a rapid increase of Na^+ concentrations—partially balanced by Cl^-—during the first 20–60 min, followed by a progressive decrease, while mannitol and glycerol reached their final concentrations after 60–90 min, respectively. The major osmolyte K^+ was hardly affected during these events. The transient increase of ions bridged the concentration gap until the organic solutes were synthesized or remobilized. It is not yet clear if the change in ionic content was brought about by a transitory facilitated membrane permeability (88) or driven by a distinct Na^+/K^+ exchange pump combined with a Na^+ pump (41). *Ulva lactuca* is an example of a macroalga exhibiting a similar increase in intracellular Na^+ with subsequent efflux and accumulation of DMSP under hyperosmotic conditions, but on a longer time scale (hours) (34).

Under constant salinity, a change in the composition of inorganic and organic osmolytes was observed in some algae, depending on developmental stages and seasonal variations. The pronounced shifts in mannitol content of *Laminaria digitata* during the seasons were balanced by isotonic substitutions with ionic osmotica—e.g. K^+ and NO_3^-, so as to maintain a constant cell turgor pressure (31). In the marine charophyte *Lamprothamnium papulosum* high sucrose levels accompanied by an enhanced turgor pressure were observed in the vacuolar sap during the period of sexual reproduction (169). In another charophyte, *Chara vulgaris*, adapted to slightly brackish water, turgor pressure was kept constant with onset of the reproductive phase, and the increasing sucrose concentration was compensated by changing the ionic contents accordingly (95). If sucrose was present its concentration was increased under hyperosmotic conditions together with the ionic concentrations (15).

PARTIAL (INCOMPLETE) TURGOR PRESSURE REGULATION   Algae growing in the intertidal zone or in estuaries may be expected to exhibit a precise regulation of turgor pressure or cell volume in response to salinity changes since this is the most obvious environmental factor acting on plants in those habitats. Yet, the best examples of marine algae investigated and found to tolerate incomplete turgor pressure regulation were from these sites.

In specimens of *Polysiphonia lanosa* from a marine and an estuarine habitat, changes of the principal osmolytes, DMSP, K^+, and Cl^-, were insufficient to restore the original level of turgor pressure after osmotic distortion (127). In marine plants turgor pressure was higher than in estuarine species, but both sets of plants responded with the same tendency: With decreasing salinity the turgor pressure increased. The efficiency of turgor pressure regulation was 60–70% in estuarine *Polysiphonia* compared to about 50% in marine species. Partial recovery of turgor pressure was also observed in the red alga *Bostrychia scorpioides* growing in the supralittoral zone (77), in brown algae [e.g. *Dictyota dichotoma* (90) and *Pilayella littoralis* (131)], and in green algae such as *Ulva lactuca* (34) and *Enteromorpha* spp. (177, 178). In the last case, three separate populations were investigated: marine, rock pool, and estuarine. In dilute seawater, *Enteromorpha* did not alter the internal solute content significantly but rather tolerated an increase in turgor pressure.

In most cases strict turgor pressure regulation fails in algae subjected to hyposaline media. A satisfactory explanation for the partial turgor pressure regulation is at present not available. In all estuarine examples reported above, the adjustment of the internal osmotic potential is achieved by changes in cell volume in addition to the control of internal osmolytes. The cell walls are thinner compared to those of the marine counterparts, and the low elastic modulus gives the cells the capacity to swell or shrink. As mentioned above, this is a fast response and meets the demands needed to survive in rapid fluctuations of salinity under tidal regimes. In estuarine algae, changes in ionic concentrations, which are also fast effector mechanisms, contribute more to the regulation than do organic solutes. It is argued that this reduces the energy costs of generating turgor pressure (90, 131).

Considering the time aspect, it must be recognized that in macroalgae (not in microalgae!) the time lag to achieve a new steady state by adjustment of osmolyte concentrations is much too long to be of ecological relevance in a tidal rhythm. This is especially true for all organic osmolytes. In experiments simulating the fluctuating salinity regimes in a tidal time scale, the ionic contents of *Ulva lactuca* followed fairly closely the increasing and decreasing salinities; but turgor pressure also changed, indicating an incomplete regulation (33). The alterations in organic solute concentrations (DMSP and sucrose) were small. Under shock conditions, however, there was a substantial and rapid change. This short-term effect on DMSP levels was not observed in other *Ulva lactuca* samples (39).

The ability to tolerate variations in turgor pressure seems to be an important property, enabling the organism to endure the rapid fluctuations of salinity occurring during tidal cycles. The buffer capacities of the tissues and the compartments such as the cell walls and the vacuole delay and mitigate the immediate stress. An important factor controlling the rate of desiccation and contributing to water conservation in intertidal algae is the shrinkage of tissue during water loss. This shrinkage reduces both the conductance for water and the area of the evaporating surface (75). The creation of favorable microhabitats by overlapping and clump formation of thalli under natural conditions may also play an important role (see introduction; reviews: 46, 141). The dogma of strict turgor pressure regulation has to be modified to allow for more flexible and less energy-consuming mechanisms of short-term passive toleration.

A different aspect of "incomplete" turgor pressure regulation should be mentioned as a curiosity: The turgor pressure of *Acetabularia mediterranea* increased though the external salinity was constant. Exceeding a threshold in the range of 0.2–0.28 MPa, the turgor pressure started to oscillate between 0.22 and 0.26 MPa in intervals from 16–26 min (167). This type of turgor pressure regulation is accomplished by Cl^- efflux bursts, including a counterion via a reversible change in membrane permeability.

MARINE ALGAE FROM AN UNUSUAL HABITAT: SEA ICE ALGAE

In polar regions, microalgae trapped or growing in and at the bottom or surface of the sea ice are important as primary producers in the food web, as well as an inoculum for phytoplankton blooms after the spring ice melt. Diatoms are the most conspicuous component of the ice algal assemblages; other algal groups such as dinoflagellates, chrysophytes, prasinophytes, chlorophytes, euglenophytes, and cyanophytes are also abundant (73). A standard terminology for ice algal assemblages has been proposed recently (74). Three major assemblages are recognized, according to their distribution in the ice; these can be further divided into subassemblages. The salinity ranges that the three assemblages encounter are quite different and may vary considerably during the seasons. The surface assemblages are to be found in the seawater-infiltrated snow-ice interface and the pools on the ice surface. It is obvious that the algae in these habitats have to tolerate salinities from seawater strength down to about $3^{\circ}/_{\circ\circ}$ during the ice melt, while extreme hyperosmotic conditions may occur in winter with low temperatures. The latter condition is characteristic also for the interior assemblages consisting of bottom ice algal layers frozen into the ice and trapped in cracks, brine pockets, and channels. The salinity of the brine and the temperature of the surrounding ice are in thermodynamic equilibrium. With decreasing temperature the salinity increases in accordance with the equations given by Assur (3). In winter the ice temperature may be as low as -5 °C resulting in a salinity of $100^{\circ}/_{\circ\circ}$ and above (104). With further ice formation under low temperature the decreasing volume of the brine channels limits growth and survival of larger microalgae, especially those possessing spines (e.g. *Chaetoceros* species). In spite of these adverse conditions the brine channels are usually densely populated with algae. Because of the enrichment during freezing out of water, there is no depletion of nutrients. The third algal group comprises the bottom assemblages: algae growing between ice crystals and platelets or as mats and strands attached to the bottom surface of the ice. The salinity is close to that of seawater, fluctuating from about $29^{\circ}/_{\circ\circ}$ to $34^{\circ}/_{\circ\circ}$.

Until now, investigations on sea ice algae have been focused primarily on growth and metabolism with respect to low temperature and light intensity as limiting conditions (45, 115). Only a few investigations include effects of salinity stress on the growth and chemical composition of ice algae. Under laboratory conditions algal samples from the McMurdo Sound photosynthesized over a salinity range up to $36^{\circ}/_{\circ\circ}$ but exhibited a narrow range of $14–18^{\circ}/_{\circ\circ}$ for growth (25). Growth rates of diatom species collected in the Arctic were the same, ranging from 10 to $50^{\circ}/_{\circ\circ}$ at 5 °C with an average cell division of 0.6–0.8 per day (62). In recent investigations hypoosmotic conditions have proved to be more harmful than high salinities: Growth rates and metabolic activity were drastically reduced in salinities below $20^{\circ}/_{\circ\circ}$ (4, 99, 163). Bates & Cota used photosynthesis vs irradiance experiments and fluorescence induction to study the short-term effects on the photosynthetic response of Arctic algae (5). The optimum was near $30^{\circ}/_{\circ\circ}$, close to that of the ambient sea water. However, the dilution of the ice samples by melting during preparation for physiological experiments had an adverse effect on the algae. From the low variable fluorescence it was concluded that the transfer of light energy to the PS II reaction centers and the noncyclic flow of electrons were affected by the low salinity.

The responses of diatom clones (*Nitzschia cylindrus* and an unidentified pennate species) from ice algal samples to the characteristic changes of the major environmental factors (decreasing light and temperature, and increasing salinity to simulate summer-winter transitions) were investigated (114). During incubation for 30 days, the diatoms reduced their metabolism as indicated by a decline of growth and photosynthesis, decrease of cellular ATP content, and utilization of storage products.

The limits of stress tolerance have also been tested with cultures of *Amphiprora kufferathii*, *Nitzschia* and *Thalassiosira antarctica* isolated from ice cores in the Weddell Sea (4): Cell division was observed up to a salinity of $90°/_{oo}$ at a temperature of -5.5 °C. The diatoms survived 20 days at $150°/_{oo}$ and -7.5 °C. In experiments that mimicked natural conditions by subjecting the algae to increasing salinities together with the correspondingly decreasing equilibrium temperatures, the decrease of the division rates did not differ from those in control conditions at constant (-1 °C) temperature. This indicated that salinity was more effective than temperature.

Osmotic acclimation in sea ice algae proceeds with the same reactions as known for other microalgae: *Chaetoceros neogracile* isolated from a Weddell Sea sample was adapted to 34, 49, and $68°/_{oo}$ at 4 °C (G. O. Kirst, J. Nothnagel, M. Wanzek. unpublished results). The content of K^+, Na^+, and Cl^- increased with salinity; Na^+ and K^+ yielded similar concentrations in the range of 200–400 mM, balanced partially by Cl^-. Proline increased fourfold, clearly acting as an organic osmolyte. It should be noted that the antifreeze properties of compounds such as proline and glycerol may also play a role advantageous to ice algae. Osmotica and compounds protective against freezing are chemically identical.

CONCLUSIONS AND FURTHER RESEARCH

The mechanisms and features involved in salinity tolerance at the cellular and organismic level as reported in this review are summarized in Table 2, where (at the risk of oversimplification) an attempt has also been made to indicate the possible relevance of these to natural conditions.

The following aspects have emerged in recent years and need further research:

- The dogma of strict turgor pressure regulation has to be modified. Incomplete (partial) regulation is tolerated by a variety of marine algae, and may be the standard adaptation of species subjected to frequent short-term fluctuations of salinities in intertidal zones.
- Numerous species accumulate more than one organic osmolyte. The characteristic of a "compatible solute" is not necessarily linked with all of those compounds detected recently.

Table 2 Summary of the characteristics and the ecological significance of the mechanisms and features involved in osmotic acclimation.

Processes and properties	Characteristics	Possible relevance under natural conditions
1. Water flux	very fast; passive; inherent result of any change in salinity	may be a sufficient response to balance small osmotic gradients; shock experiences
2. Ion transport	fast; low energy cost; selective uptake or release; vacuole and partially cytoplasm	major short-term response in intertidal zones and estuaries
3. Organic osmolytes 　a. synthesis or degradation	slow; high energy costs; mainly in cytoplasm	long-term adjustment to extreme osmotic stresses; seasonal changes in lagoons; desiccation in supralittoral habitats (salt marshes and rock pools); ice algae; too slow for tidal zones
b. accumulation of several osmolytes	slow; high energy costs; possibly in several cellular compartments	as above (3a)
c. buffer capacity of high contents of organics	compounds accumulated due to developmental stages	hyperosmotic shocks
4. Morphological and anatomical features: 　a. buffer effects: large vacuoles; massive thalli 　b. cell walls: ion exchange capacity 　c. elastic properties of cell walls	passive; energy costs during construction	supralittoral, tidal zones and estuaries
5. Endurance of salinity fluctuations	passive; partial (incomplete) turgor pressure regulation	intertidal zones; time limited
6. Life cycles	gametophyte and sporophyte differ in resistance; dormant states (zygotes; spores)	survival of extreme conditions; ice algae
7. Retention of seawater between tufts and thallus layers	passive; depends on population density	intertidal zones

- Under in vitro conditions, enzyme systems tolerate fairly high ionic concentrations. The presence of "compatible solutes" seems to be not a "conditio sine qua non." Our knowledge of in vivo conditions and actual enzyme activities, especially in marine macroalgae, is scarce.
- The study of cell wall characteristics in the context of salinity tolerance is just starting and may be a promising research area because of detailed knowledge on chemical composition already available.
- Progress both in NMR techniques and in X-ray microprobe analysis is likely to provide an increasing amount of information on compartmentation of osmolytes at the cellular level. Together with more investigations on fine structure, such progress will close a gap that is still amazingly large.
- The detection of *osm* genes in *Escherichia* will certainly stimulate work on algae, even though algal genotypes and their functions are likely to be more complex.
- Investigations of salinity tolerance as a tool to describe the selection of ecotypes and even the evolution from subspecies to species are in progress. By studying complete life cycles, we may obtain more insight into the ecological advantages, distribution, and seasonality of gametophyte and sporophyte formation.
- In spite of the increasing power of data analysis, investigations on complex interrelations (e.g. among the effects of temperature, salinity, and irradiance) are disappointingly rare.

ACKNOWLEDGMENTS

I am grateful to M. A. Bisson, J. C. Collins, J. A. Raven, G. Russell, G. Thiel, and D. N. Thomas for critically reading the manuscript and for valuable comments. Financial support by the Deutsche Forschungsgemeinschaft over many years has enabled the research in my laboratory.

Literature Cited

1. Ackman, R. G., Tocher, C. S., McLachlan, J. 1966. Occurrence of dimethyl-β-propiothetin in marine phytoplankton. *J. Fish. Res. Board Can.* 23:357–64
2. Andreae, M. D., Raemdonck, H. 1983. Dimethyl sulfide in the surface ocean and the marine atmosphere: a global view. *Science* 221:744–47
3. Assur, A. 1958. Composition of sea ice and its tensile strength. In *Arctic Sea Ice. Natl. Res. Coun. Publ. 598*, pp. 105–38. Washington, DC: Natl. Acad. Sci.
4. Bartsch, A., 1989. *Die Eisalgenflora des Weddellmeeres (Antarktis). Artenbestand und Biomasse, sowie Ökophysiologie ausgewählter Arten.* PhD thesis. Univ. Bremen (In German)
5. Bates, S. S., Cota, G. F. 1986. Fluorescence induction and photosynthetic responses of Arctic ice algae to sample treatment and salinity. *J. Phycol.* 22: 421–29
6. Beilby, M. J. 1985. Potassium channels and different states of *Chara* plasmalemma. *J. Exp. Bot.* 6:228–39

7. Belmans, D., van Laere, A. 1987. Glycerol cycle enzymes and intermediates during adaptation of *Dunaliella tertiolecta* cells to hyperosmotic stress. *Plant, Cell Environ.* 10:184–90
8. Ben-Amotz, A., Avron, M. 1983. Accumulation of metabolites by halotolerant algae and its industrial potential. *Annu. Rev. Microbiol.* 37:95–119
9. Deleted in press
10. Bird, C. J., McLachlan, J. 1986. The effect of salinity on distribution of species of *Gracilaria* (Rhodophyta, Gigartinales): an experimental assessment. *Bot. Mar.* 29:231–38
11. Bird, N. L., Chen, L. C.-M., McLachlan, J. 1979. Effects of temperature, light and salinity on growth in culture of *Chondrus crispus, Furcellaria lumbricalis, Gracilaria rikvahiae* (Gigartinales, Rhodophyta), and *Fucus serratus* (Fucales, Phaeophyta). *Bot. Mar.* 22: 521–27
12. Bisson, M. A., Gutknecht, J. 1975. Osmotic regulation in the marine alga, *Codium decorticatum.* I. Regulation of

turgor pressure by control of ionic composition. *J. Membr. Biol.* 24:183–200

13. Bisson, M. A., Gutknecht, J. 1977. Osmotic regulation in the marine alga, *Codium decorticatum*. II. Active chloride influx exerts negative feedback control on the turgor pressure. *J. Membr. Biol.* 37:85–98

14. Bisson, M. A., Gutknecht, J. 1980. Osmotic regulation in algae. In *Plant Membrane Transport: Current Conceptual Issues*, ed. R. M. Spanswick, W. J. Lucas, J. Dainty, pp. 131–42. Amsterdam/New York/Oxford: Elsevier/North-Holland Biomedical Press

15. Bisson, M. A., Kirst, G. O. 1979. Osmotic adaptation in the marine alga *Griffithsia monilis* (Rhodophyceae): the role of ions and organic compounds. *Aust. J. Plant Physiol.* 6:523–38

16. Bisson, M. A., Kirst, G. O. 1980. *Lamprothamnium*, a euryhaline charophyte. I. Osmotic relations and membrane potential at steady state. *J. Exp. Bot.* 31:1223–35

17. Bisson, M. A., Kirst, G. O. 1983. Osmotic adaptations of charophyte algae in the Coorong, South Australia and other Australian lakes. *Hydrobiologia* 105:45–51

18. Borowitzka, L. J. 1981. Solute accumulation and regulation of cell water activity. In *Physiology and Biochemistry of Drought Resistance in Plants*, ed. L. G. Paleg, D. Aspinall, pp. 97–130. Melbourne: Academic

19. Brady, C. J., Gibson, T. S., Barlow, E. W. R., Spiers, J., Wyn Jones, R. G. 1984. Salt tolerance in plants. I. Ions, compatible organic solutes and the stability of plant ribosomes. *Plant, Cell Environ* 7:571–78

20. Brand, L. E. 1984. The salinity tolerance of forty-six marine phytoplankton isolates. *Estuarine, Coastal Shelf Sci.* 18:543–56

21. Brown, A. D. 1976. Microbial water stress. *Bacteriol. Rev.* 40:803–46

22. Brown, A. D., Borowitzka, L. J. 1979. Halotolerance of *Dunaliella*. In *Biochemistry and Physiology of Protozoa*, ed. M. Levandowsky, S. H. Hutner, 1:139–90. New York: Academic

23. Brown, L. M., Hellebust, J. A. 1978. Sorbitol and proline as intracellular osmotic solutes in the green alga *Stichococcus bacillaris*. *Can. J. Bot.* 56:676–79

24. Brown, L. M., Hellebust, J. A. 1980. The contribution of organic solutes to osmotic balance in some green and eustigmatophyte algae. *J. Phycol.* 16:265–70

25. Bunt, J. S. 1964. Primary productivity under sea ice in the Antarctic waters. II. Influence of light and other factors on photosynthetic activities of Antarctic marine microalgae. *Antarct. Res. Ser.* 1:27–31

26. Büchner, K.-H., Rosenheck, K., Zimmermann, U. 1985. Characterization of the mobile charges in the membrane of *Valonia utricularis*. *J. Membr. Biol.* 88:131–37

27. Büchner, K.-H., Walter, L., Zimmermann, U. 1987. Influence of anaesthetics on the movement of the mobile charges in the algal cell membrane of *Valonia utricularis*. *Biochim. Biophys. Acta* 903:241–47

28. Charlson, R. J., Lovelock, J. E., Andreae, M. O., Warren, S. G. 1987. Oceanic phytoplankton, atmospheric sulphur, cloud albedo and climate. *Nature* 326:655–61

29. Coster, H. G. L., Steudle, E., Zimmermann, U. 1976. Turgor pressure sensing in plant cell membranes. *Plant Physiol.* 58:636–43

30. Cram, W. J. 1976. Negative feedback regulation of transport in cells. The maintenance of turgor, volume and nutrient supply. In *Encyclopedia of Plant Physiology (N S), Vol. 2A. Transport in Plants II*, ed. U. Lüttge, M. G. Pitman, pp. 283–316. Berlin/Heidelberg/New York: Springer

31. Davison, I. R., Reed, R. H. 1985. Osmotic adjustment in *Laminaria digitata* (Phaeophyta) with particular reference to seasonal changes in internal solute concentrations. *J. Phycol.* 21:41–50

32. Davison, I. R., Reed, R. H. 1985. The physiological significance of mannitol accumulation in brown algae: the role of mannitol as a compatible cytoplasmic solute. *Phycologia* 24:449–57

33. Dickson, D. M. J., Wyn Jones, R. G., Davenport, J. 1980. Steady state osmotic adaptation in *Ulva lactuca*. *Planta* 150:158–65

34. Dickson, D. M. J., Wyn Jones, R. G., Davenport, J. 1982. Osmotic adaptation in *Ulva lactuca* under fluctuating salinity regimes. *Planta* 155:409–15

35. Dickson, D. M. J., Kirst, G. O. 1986. The role of dimethylsulphoniopropionate, glycine betaine and homarine in the osmoacclimation of *Platymonas subcordiformis*. *Planta* 167:536–43

36. Dickson, D. M., J., Kirst, G. O. 1987. Osmotic adjustment in marine eukaryotic algae: the role of inorganic ions, quaternary ammonium, tertiary sulfonium and carbohydrate solutes: I. diatoms and a rhodophyte. *New Phytol.* 106:645–55

37. Dickson, D. M. J., Kirst, G. O. 1987. Osmotic adjustment in marine eukaryotic algae: the role of inorganic ions, quaternary ammonium, tertiary sulfonium and carbohydrate solutes: II. prasinophytes and haptophytes. *New Phytol.* 106:657–66

38. Edwards, K. L., Pickard, B. G. 1987. Detection and transduction of physical stimuli in plants. In *The Cell Surface in Signal Transduction, NATO ASI Series,* ed. E. Wagner, H. Greppin, B. Millet, 12:41–66. Heidelberg: Springer

39. Edwards, D. M., Reed, R. H., Chudek, J. A., Foster, R., Stewart, W. D. P. 1987. Organic solute accumulation in osmotically-stressed *Enteromorpha intestinalis. Mar. Biol* 95:583–92

40. Edwards, D. M., Reed, R. H., Stewart, W. D. P. 1988. Osmoacclimation in *Enteromorpha intestinalis:* long-term effects of osmotic stress on organic solute accumulation. *Mar. Biol.* 98:467–76

41. Ehrenfeldt, J., Cousin, J.-L. 1984. Ionic regulation of the unicellular green alga *Dunaliella tertiolecta:* response to hypertonic shock. *J. Membr. Biol.* 77: 45–55

42. Falke, L. C., Edwards, K. L., Pickard, B. G., Misler, S. 1988. A stretch-activated anion channel in tobacco protoplasts. *FEBS Lett.* 237:141–44

43. Fork, D. C., Öquist, G. 1981. The effects of desiccation on excitation energy transfer at physiological temperatures between the two photosystems of the red alga *Porphyra perforata. Z. Pflanzenphysiol.* 104:385–93

44. Frank, G., Wegmann, K. 1974. Physiology and biochemistry of glycerol biosynthesis in *Dunaliella. Biol. Zbl.* 93:707–23

45. Garrison, D. L., Sullivan, C. W., Ackley, S. F. 1986. Sea ice microbial communities in Antarctica. *BioScience* 36: 243–50

46. Gessner, F., Schramm, W. 1971. Salinity: Plants. In *Marine Ecology, Vol. 1 (2) Environmental Factors,* ed. O. Kinne, pp. 705–1083. London: Wiley Interscience

47. Gilmour, D. J., Hipkins, M. F., Boney, A. D. 1982. The effect of salt stress on the primary processes of photosynthesis in *Dunaliella tertiolecta. Plant Sci. Lett.* 26:325–30

48. Gilmour, D. J., Hipkins, M. F., Boney, A. D. 1984. The effect of osmotic and ionic stress on the primary processes of photosynthesis in *Dunaliella tertiolecta. J. Exp. Bot.* 35:18–27

49. Gilmour, D. J., Hipkins, M. F., Boney, A. D. 1984. The effect of decreasing salinity on the primary processes of photosynthesis in *Dunaliella tertiolecta. J. Exp. Bot.* 35:28–35

50. Gilmour, D. J., Hipkins, M. F., Webber, A. N., Baker, N. R., Boney, A. D. 1985. The effect of ionic stress on photosynthesis in *Dunaliella tertiolecta.* Chlorophyll fluorescence kinetics and spectral characteristics. *Planta* 163:250–56

51. Gimmler, H., Greenway, H. 1983. Tetraphenylphosphonium (TTP$^+$) is not suitable for the assessment of electrical potentials in *Chlorella emersonii. Plant, Cell Environ.* 6:739–44

52. Gimmler, H., Hartung, W. 1988. Low permeability of the plasma membrane of *Dunaliella parva* for solutes. *J. Plant Physiol.* 133:165–72

53. Gimmler, H., Lotter, G. 1982. The intracellular distribution of enzymes of the glycerol cycle in the unicellular alga *Dunaliella parva. Z. Naturforsch.* 37c: 1107–14

54. Gimmler, H., Möller, E. M. 1981. Salinity-dependent regulation of starch and glycerol metabolism in *Dunaliella parva. Plant, Cell Environ.* 4:367–75

55. Gimmler, H., Kaaden, R., Kirchner, U., Weyand, A. 1984. The chloride sensitivity of *Dunaliella parva* enzymes. *Z. Pflanzenphysiol.* 114:131–50

56. Ginzburg, M. 1981. Measurements of ion concentrations and fluxes in *Dunaliella parva. J. Exp. Bot.* 32:321–32

57. Ginzburg, M. 1981. Measurements of ion concentrations in *Dunaliella parva* subjected to hypertonic shock. *J. Exp. Bot.* 32:333–40

58. Ginzburg, M. 1987. *Dunaliella:* a green alga adapted to salt. *Adv. Bot. Res.* 14:93–183

59. Glass, A. D. M. 1983. Regulation of ion transport. *Annu. Rev. Plant. Physiol.* 34:311–26

60. Goyal, A., Brown, A. D., Lilley, R. McC. 1988. The response of green halotolerant alga *Dunaliella* to osmotic stress: effects on pyridine nucleotide contents. *Biochim. Biophys. Acta* 936: 20–28

61. Gradmann, D., Mummert, H. 1980. Plant action potentials. See Ref. 14, pp. 333–44

62. Grant, W. S., Horner, R. A. 1976.

Growth responses to salinity variation in four Arctic ice diatoms. *J. Phycol.* 12:180–85

63. Guharay, F., Sachs, F. 1984. Stretch-activated single ion channel currents in tissue-cultured embryonic chick skeletal muscle. *J. Physiol.* 352:685–701

64. Gustin, M. C., Zhou, X.-L., Martinac, B., Kung, C. 1988. A mechanosensitive ion channel in the yeast plasma membrane. *Science* 242:762–65

65. Gutknecht, J. 1966. Sodium, potassium, and chloride transport and membrane potentials in *Valonia ventricosa*. *Biol. Bull.* 130:331–44

66. Gutknecht, J., Dainty, J. 1968. Ionic relations of marine algae. *Oceanogr. Mar. Biol. Annu. Rev.* 6:163–200

67. Gutknecht, J., Hastings, D. F., Bisson, M. A. 1978. Ion transport and turgor pressure regulation in giant algal cells. In *Membrane Transport in Biology III: Transport Across Biological Membranes*, ed. G. Giebisch, D. C. Tosteson, H. H. Ussing, pp. 125–74. Berlin/Heidelberg/New York: Springer

68. Hajibagheri, M. A., Gilmour, D. J., Collins, J. C., Flowers, T. J. 1986. X-ray microanalysis and ultrastructural studies of cell compartments of *Dunaliella parva*. *J. Exp. Bot.* 37:1725–32

69. Hellebust, J. A. 1976. Osmoregulation. *Annu. Rev. Plant Physiol.* 27:485–505

70. Hellebust, J. A. 1976. Effect of salinity on photosynthesis and mannitol synthesis in the green flagellate *Platymonas suecica*. *Can. J. Bot.* 54:1735–41

71. Hellebust, J. A. 1985. Mechanisms of response to salinity in halotolerant microalgae. *Plant and Soil* 89:69–81

72. Hoffmann, R., Bisson, M. A. 1986. *Chara buckellii*, a euryhaline charophyte from an unusual saline environment. I. Osmotic relations at steady state. *Can. J. Bot.* 64:1599–605

73. Horner, R. A. 1985. Taxonomy of sea ice microalgae. In *Sea Ice Biota,* ed. R. A. Horner, pp. 147–58. Boca Raton, Fla: CRC Press

74. Horner, R. A., Syvertsen, E. E., Thomas, D. P., Lange, C. 1988. Proposed terminology and reporting units for sea ice algal assemblages. *Polar Biol.* 8: 249–53

75. Jones, H. G., Norton, T. A. 1979. Internal factors controlling the rate of evaporation from fronds of some intertidal algae. *New Phytol.* 83:771–81

76. Karni, L., Avron, M. 1988. Ion content of the halotolerant alga *Dunaliella salina*. *Plant Cell Physiol.* 29:1311–14

77. Karsten, U., Kirst, G. O. 1989. Incomplete turgor pressure regulation in the "terrestrial" red alga, *Bostrychia scorpioides*. *Plant Sci.* 61:29–36

78. Karsten, U., Kirst, G. O. 1989. Intracellular solutes, photosynthesis and respiration of the green alga *Blidingia minima* in response to salinity stress. *Bot. Acta* 102:123–28

79. Karsten, U., Wiencke, C., Kirst, G. O. 1989. Dimethylsulphoniopropionate content of macroalgae from Antarctica and southern Chile. *Bot. Mar.* 32. In press

80. Kauss, H. 1977. Biochemistry of osmotic regulation. In *International Review of Biochemistry, Plant Biochemistry II,* ed. D. H. Northcote, 13:119–40. Baltimore: University Park Press

81. Kauss, H. 1986. A membrane-derived proteinase capable of activating a galactosyl-transferase involved in volume regulation of *Poterioochromonas*. In *Plant Proteolytic Enzymes II,* ed. M. J. Dalling, pp. 91–102. Boca Raton, Fla: CRC Press

82. Kauss, H. 1987. Some aspects of calcium-dependent regulation in plant metabolism. *Annu. Rev. Plant Physiol.* 38:47–72

83. Kesseler, H. 1959. Mikrokryoskopische Untersuchungen zur Turgorregulation von *Chaetomorpha linum*. *Kieler Meeresforsch.* 15:51–73

84. Kesseler, H. 1965. Turgor, osmotisches Potential und ionale Zusammensetzung des Zellsaftes einiger Meeresalgen verschiedener Verbreitungsgebiete. *Bot. Gothob.* 3:103–11

85. Kesseler, H. 1980. On the selective adsorption of cations in the cell wall of the green alga *Valonia utricularis*. *Helgoländer Meeresunters.* 34:151–58

86. Kirst, G. O. 1975. Correlation between content of mannitol and osmotic stress in the brackish-water alga *Platymonas subcordiformis*. *Z. Pflanzenphysiol.* 76: 316–25

87. Kirst, G. O. 1977. Ion composition of unicellular marine and freshwater algae with special reference to *Platymonas subcordiformis* cultivated in media with different osmotic strength. *Oecologia* 28:177–89

88. Kirst, G. O. 1977. Coordination of ionic relations and mannitol concentrations in the euryhaline unicellular alga, *Platymonas subcordiformis* after osmotic shocks. *Planta* 135:69–75

89. Kirst, G. O. 1980. Mannitol accumulation in *Platymonas subcordiformis* after osmotic stresses and the effect of inhibitors. *Z. Pflanzenphysiol.* 98:35–42

90. Kirst, G. O., Bisson, M. A. 1979. Regulation of turgor pressure in marine algae: ions and low-molecular-weight organic compounds. *Aust. J. Plant Physiol.* 6:539–56

91. Kirst, G. O., Bisson, M. A. 1980. Osmotic adaptation in marine algae. See Ref. 14, pp. 485–86

92. Kirst, G. O., Keller, H. J. 1976. Der Einfluss unterschiedlicher NaCl-Konzentrationen auf die Atmung der einzelligen Alge *Platymonas subcordiformis*. *Bot. Mar.* 19:241–44

93. Kirst, G. O., Kramer, D. 1981. Cytological evidence for cytoplasmic volume control in *Platymonas subcordiformis* after osmotic stress. *Plant, Cell Environ.* 4:455–62

94. Kirst, G. O., Wichmann, F. 1988. Adaptation of the euryhaline charophyte *Lamprothamnium papulosum* to brackish and freshwater: photosynthesis and respiration. *J. Plant Physiol.* 131:413–22

95. Kirst, G. O., Janssen, M. I. B., Winter, U. 1988. Ecophysiological investigations of *Chara vulgaris* grown in a brackish water lake: ionic changes and accumulation of sucrose in the vacuolar sap during sexual reproduction. *Plant, Cell Environ.* 11:55–61

96. Kloareg, B. 1984. Isolation and analysis of cell walls of the brown marine algae *Pelvetia canaliculata* and *Ascophyllum nodosum*. *Physiol. Veg.* 22:47–56

97. Kloareg, B., Quatrano, R. S. 1988. Structure of cell walls of marine algae and ecophysiological functions of the matrix polysaccharides. *Oceanogr. Mar. Biol. Annu. Rev.* 26:259–315

98. Knoth, A., Wiencke, C. 1984. Dynamic changes of protoplasmic volume and of fine structure during osmotic adaptation in the intertidal alga *Porphyra umbilicalis*. *Plant, Cell Environ.* 7:113–19

99. Kottmeier, S. T., Sullivan, C. W. 1988. Sea ice microbial communities. 9. Effects of temperature and salinity on rates of metabolism and growth of autotrophs and heterotrophs. *Polar Biol.* 8:293–304

100. Kremer, B. P., Kirst, G. O. 1982. Biosynthesis of photosynthates and taxonomy of algae. *Z. Naturforsch.* 73C:761–71

101. Le Redulier, D., Strom, A. R., Dandekar, A. M., Smith, L. T., Valentine, R. C. 1984. Molecular biology of osmoregulation. *Science* 224:1064–68

102. Liu, M. S., Hellebust, J. A. 1976. Effects of salinity changes on growth and metabolism of the marine centric diatom *Cyclotella cryptica*. *Can. J. Bot.* 54:930–37

103. Mariani, P., Tolomio, C., Braghetta, P. 1985. An ultrastructural approach to the adaptive role of the cell wall in the intertidal alga *Fucus versoides*. *Protoplasma* 128:208–17

104. Maykut, G. A. 1985. The ice environment. See Ref. 75, pp. 21–82

105. McLachlan, J. 1961. The effect of salinity on growth and chlorophyll content in representative classes of unicellular marine algae. *Can. J. Microbiol.* 7:399–406

106. Miller, R. L., Kamykowsky, D. L. 1986. Effects of temperature, salinity, irradiance and diurnal periodicity on growth and photosynthesis in the diatom *Nitzschia americana:* light-limited growth. *J. Plankton Res.* 8:215–28

107. Miller, R. L., Kamykowsky, D. L. 1986. Short-term photosynthetic response in the diatom *Nitzschia americana* to a simulated salinity environment. *J. Plankton Res.* 8:305–15

108. Miller, R. L., Kamykowsky, D. L. 1986. Effects of temperature, salinity, irradiance and diurnal periodicity on growth and photosynthesis in the diatom *Nitzschia americana:* light-saturated growth. *J. Phycol.* 22:339–48

109. Munns, R., Greenway, H., Kirst, G. O. 1983. Halotolerant eukaryotes. In *Encyclopedia of Plant Physiology (NS), Vol. 12C. Physiological Plant Ecology III*, ed. O. L. Lange, P. S. Nobel, C. B. Osmond, H. Ziegler, pp. 59–135. Berlin / Heidelberg / New York: Springer

110. Okazaki, Y., Shimmen, T., Tazawa, M. 1984. Turgor regulation of the brackish charophyte, *Lamprothamnium succinctum*. II. Changes in K^+, Na^+, and Cl^- concentrations, membrane potential and membrane resistance during turgor regulation. *Plant & Cell Physiol.* 25:573–81

111. Okazaki, Y., Tazawa, M. 1986. Involvement of calcium ion in turgor regulation upon hypotonic treatment in *Lamprothamnium succinctum*. *Plant, Cell Environ.* 9:185–90

112. Okazaki, Y., Yoshimoto, Y., Hiramoto, Y., Tazawa, M. 1987. Turgor regulation and cytoplasmic free Ca^{2+} in the alga *Lamprothamnium*. *Protoplasma* 140:67–71

113. Paasche, E., Johansson, S., Evensen, D. L. 1975. An effect of osmotic pressure on the valve morphology of the diatom *Skeletonema subsalsum*. *Phycologia* 14:205–11

114. Palmisano, A. C., Sullivan, C. W.

1982. Physiology of sea ice diatoms. I. Response of three polar diatoms to a simulated summer-winter transition. *J. Phycol.* 18:489–98

115. Palmisano, A. C., Sullivan, C. W. 1985. Growth, metabolism, and dark survival in sea ice microalgae. See Ref. 75, pp. 131–46

116. Peng, H. B., Jaffe, L. F. 1976. Cell-wall formation in *Pelvetia* embryos. A freeze-fracture study. *Planta* 133:57–71

117. Percival, E. 1979. The polysaccharides of green, red and brown seaweeds: their basic structure, biosynthesis and function. *Br. Phycol. J.* 14:103–17

118. Postma, H. 1975. Hydrography and hydrochemistry of brackish waters. *Hydrobiol. Bull.* 8:40–45

119. Rabinowitch, S., Grover, N. B., Ginzburg, B. Z. 1975. Cation effects on volume and water permeability in the halophilic alga *Dunaliella parva*. *J. Membr. Biol.* 22:211–30

120. Raschke, K., Hedrich, R., Beckmann, U., Schroeder, J. I. 1988. Exploring biophysical and biochemical components of the osmotic motor that drives stomatal movement. *Bot. Acta* 101:283–94

121. Raven, J. A. 1975. Algal cells. In *Ion Transport in Plants Cells and Tissues*, ed. D. A. Baker, J. A. Hall, pp. 125–60. Amsterdam: North-Holland

122. Raven, J. A. 1976. Transport in algal cells. See Ref. 30, pp. 129–88

123. Raven, J. A. 1985. *Energetics and Transport in Aquatic Plants*. New York: Alan R. Liss, Inc.

124. Raven, J. A. 1987. Biochemistry, biophysics and physiology of chlorophyll *b*–containing algae: implications for taxonomy and phylogeny. *Prog. Phycol. Res.* 5:1–122

125. Raven, J. A., Smith, F. A., Smith, S. E. 1980. Ions and osmoregulation. In *Genetic Engineering of Osmoregulation, Basic Life Sciences*, ed. D. W. Rains, R. C. Valentine, A. Hollaender, 14:101–18. New York/London: Plenum

126. Reed, R. H. 1983. Measurement and osmotic significance of dimethylsulphoniopropionate in marine macroalgae. *Mar. Biol. Lett.* 4:173–81

127. Reed, R. H. 1983. The osmotic response of *Polysiphonia lanosa* from marine and estuarine sites: evidence for incomplete recovery of turgor. *J. Exp. Mar. Biol. Ecol.* 68:169–93

128. Reed, R. H. 1984. Use and abuse of osmo-terminology. *Plant, Cell Environ.* 7:165–70

129. Reed, H. R. 1984. The effects of extreme hyposaline stress upon *Polysiphonia lanosa* from marine and estuarine sites. *J. Exp. Mar. Biol. Ecol.* 76:131–44

130. Reed, R. H. 1989. Osmotic adjustment and organic solute accumulation in *Chaetomorpha capillaris*. *Br. Phycol. J.* 24:21–37

131. Reed, R. H., Barron, J. A. 1983. Physiological adaptation to salinity change in *Pilayella littoralis* from marine and estuarine sites. *Bot. Mar.* 26:409–16

132. Reed, R. H., Wright, P. J. 1986. Release of mannitol from *Pilayelle littoralis* (Phaeophyta: Ectocarpales) in response to hypoosmotic stress. *Mar. Ecol. Prog. Ser.* 29:205–8

133. Reed, R. H., Collins, J. C., Russell, G. 1980. The effects of salinity upon cellular volume of the marine red alga *Porphyra purpurea*. *J. Exp. Bot.* 31:1521–37

134. Reed, R. H., Collins, J. C., Russell, G. 1980. The effects of salinity upon galactosyl-glycerol content and concentration of the marine red alga *Prophyra purpurea*. *J. Exp. Bot.* 31:1539–54

135. Reed, R. H., Davison, L. R., Chudek, J. A., Foster, R. 1985. The osmotic role of mannitol in the Phaeophyta: an appraisal. *Phycologia* 24:35–47

136. Richter, D. F. E., Kirst, G. O. 1987. Mannitol dehydrogenase and mannitol-1-phosphate dehydrogenase in *Platymonas subcordiformis:* some characteristics and their role in osmotic adaptation. *Planta* 170:528–34

137. Ritchie, R. J. 1988. The ionic relations of *Ulva lactuca*. *J. Plant Physiol.* 133:183–92

138. Ritchie, R. J., Larkum, A. W. D. 1982. Cation exchange properties of the cell walls of *Enteromorpha intestinalis* (Ulvales, Chlorophyta). *J. Exp. Bot.* 33:125–39

139. Ritchie, R. J., Larkum, A. W. D. 1982. Ion exchange fluxes of the cell walls of *Enteromorpha intestinalis* (Ulvales, Chlorophyta). *J. Exp. Bot.* 33:140–53

140. Ritchie, R. J., Larkum, A. W. D. 1985. Potassium transport in *Enteromorpha intestinalis*. II. Effects of medium composition and metabolic inhibitors. *J. Exp. Bot.* 36:394–412

141. Russell, G. 1987. Salinity and seaweed vegetation. In *The Physiological Ecology of Amphibious and Intertidal Plants*, ed. R. M. M. Crawford, pp. 35–52. Oxford: Blackwell

142. Ryther, J. H., Menzel, D. W., Corwin, N. 1967. Influence of the Amazon river outflow on the ecology of the western tropical Atlantic. I. Hydrography and

nutrient chemistry. *J. Mar. Res.* 25:69–83

143. Saddler, H. D. W. 1970. The ionic relations of *Acetabularia mediterranea*. *J. Exp. Bot.* 21:345–59

144. Satoh, K., Smith, C. M., Fork, D. C. 1983. Effects of salinity on primary processes of photosynthesis in the red alga *Porphyra perforata*. *Plant Physiol.* 73:643–47

145. Saxton, M. J., Breitenbach, R. W., Lyons, J. M. 1980. Membrane dynamics: effects of environmental stress. See Ref. 127, pp. 203–33

146. Schildknecht, H., Bender, W. 1983. Chemonastisch wirksame Leaf Movement Factors aus *Mimosa pudica*. *Chem. Z.* 107:111–14

147. Schildknecht, H., Schumacher, K. 1981. Ein hochwirksamer Leaf Movement Factor aus *Acacia karroo*. *Chem. Z.* 105:287–90

148. Schmid, A. M. 1979. Influence of environmental factors on the development of the valve in diatoms. *Protoplasma* 99:99–115

149. Schobert, B. 1977. Is there an osmotic regulatory mechanism in algae and in higher plants? *J. Theor. Biol.* 68:17–26

150. Schobert, B. 1980. Proline catabolism, relaxation of osmotic strain and membrane permeability in the diatom *Phaeodactylum tricornutum*. *Physiol. Plant.* 50:37–42

151. Schöne, H. K. 1972. Experimentelle Untersuchungen zur Ökologie der marinen Kieselalge *Thalassiosira rotula*. I. Temperatur und Licht. *Mar. Biol.* 13:284–91

152. Schöne, H. K. 1974. Experimentelle Untersuchungen zur Ökologie der marinen Kieselalge *Thalassiosira rotula*. II. Der Einfluss des Salzgehaltes. *Mar. Biol.* 27:287–98

153. Schultz, M. E. 1971. Salinity-related polymorphism in the brackish-water diatom *Cyclotella cryptica*. *Can. J. Bot.* 49:1285–89

154. Setter, T. L., Greenway, H. 1983. Changes in the proportion of endogenous osmotic solutes accumulated by *Chlorella emersonii* in the light and dark. *Plant, Cell Environ.* 6:227–34

155. Smith, C. M., Satoh, K., Fork, D. C. 1986. The effects of osmotic tissue dehydration and air drying in morphology and energy transfer in two species of *Porphyra*. *Plant Physiol.* 80:843–47

156. Tazawa, M., Shimmen, T. 1987. Cell motility and ionic relations in Characean cells as revealed by internal perfusion

and cell models. *Int. Rev. Cytol.* 109:259–312

157. Tazawa, M., Shimmen, T., Mimura, T. 1987. Membrane control in the Characeae. *Annu. Rev. Plant Physiol.* 38:95–117

158. Tester, M. 1988. Pharmacology of K^+ channels in the plasmalemma of the green alga *Chara corallina*. *J. Membr. Biol.* 103:159–69

159. Thomas, D. N., Collins, J. C., Russell, G. 1988. Interactive effects of temperature and salinity upon net photosynthesis of *Chladophora glomerata* and *C. rupestris*. *Bot. Mar.* 31:73–77

160. Trezzi, F., Galli, M. G., Bellini, E. 1965. L'osmo-resistanza di *Dunaliella salina* ricerchi ultra-strutturali. *G. Bot. Ital.* 72:255–63

161. Ubl, J., Murer, H., Kolb, H.-A. 1988. Ion channels activated by osmotic and mechanical stress in membranes of *Opossum* kidney cells. *J. Membr. Biol.* 104:223–32

162. Vairavamurthy, A., Andreae, M. O., Iverson, R. L. 1985. Biosynthesis of dimethylsulfide and dimethylpropiothetin by *Hymenomonas carterae* in relation to sulfur source and salinity variations. *Limnol. Oceanogr.* 30:59–70

163. Vargo, G. A., Fanning, K., Heil, C., Bell, L. 1986. Growth rates and salinity response of an antarctic ice microflora community. *Polar Biol.* 5:241–47

164. Walker, N. A. 1980. The transport systems of charophyte and chlorophyte giant algae and their integration into modes of behaviour. See Ref. 14, pp. 287–300

165. Walter, L., Büchner, K.-H., Zimmermann, U. 1988. Effect of alkaline earth ions on the movement of mobile charges in *Valonia utricularis*. *Biochim. Biophys. Acta* 939:1–7

166. Wegmann, K. 1986. Osmoregulation in eukaryotic algae. *FEMS Microbiol. Rev.* 39:37–43

167. Wendler, S., Zimmermann, U., Bentrup, F.-W. 1983. Relationship between cell turgor pressure, electrical membrane potential, and chloride efflux in *Acetabularia mediterranea*. *J. Membr. Biol.* 72:75–84

168. White, R. H. 1982. Analysis of dimethylsulfonium compounds in marine algae. *J. Mar. Res.* 40:529–36

169. Wichmann, F., Kirst, G. O. 1989. Adaptation of the euryhaline Charophyte *Lamprothamnium papulosum* to brackish and freshwater: turgor pressure and vacuolar solute concentrations during

steady-state culture and after hypo-osmotic treatment. *J. Exp. Bot.* 40:135–41

170. Wiencke, C. 1984. The response of pyruvate kinase from the intertidal red alga *Porphyra umbilicalis* to sodium and potassium ions. *J. Plant Physiol.* 116:447–53

171. Wiencke, C., Läuchli, A. 1980. Growth, cell volume, and fine structure of *Porphyra umbelicalis* in relation to osmotic tolerance. *Planta* 150:303–11

172. Wiencke, C., Läuchli, A. 1983. Tonoplast fine structure and osmotic regulation in *Porphyra umbilicalis*. *Planta* 159:342–46

173. Wiltens, J., Schreiber, U., Vidaver, W. 1978. Chlorophyll fluorescence induction: an indicator of photosynthetic activity in marine algae undergoing desiccation. *Can. J. Bot.* 56:2787–94

174. Wright, P. J., Reed, R. H. 1985. The effects of osmotic stress on intracellular hexitols in the marine brown alga *Himanthalia elongata*. *J. Exp. Mar. Biol. Ecol.* 93:183–90

175. Wyn Jones, R. G., Gorham, J. 1983. Osmoregulation. See Ref. 111, pp. 35–58

176. Yancey, P. H., Clark, M. E., Hand, S.

C., Bowlus, R. D., Somero, C. N. 1982. Living with water stress: evolution of osmolyte systems. *Science* 217:1214–22

177. Young, A. J., Collins, J. C., Russell, G. 1987. Solute regulation in the euryhaline marine alga *Enteromorpha prolifera*. *J. Exp. Bot.* 38:1298–1308

178. Young, A. J., Collins, J. C., Russell, G. 1987. Ecotypic variation in the osmotic response of *Enteromorpha intestinalis*. *J. Exp. Bot.* 38:1309–24

179. Zimmermann, U. 1978. Physics of turgor- and osmoregulation. *Annu. Rev. Plant Physiol.* 29:121–48

180. Zimmermann, U., Hüsken, D. 1980. Elastic properties of the cell wall of *Halicystis parvula*. See Ref. 14, pp. 469–70

181. Zimmerman, U., Steudle, E. 1978. Physical aspects of water relations of plant cells. *Adv. Bot. Res.* 6:45–117

182. Zimmermann, U., Steudle, E. 1980. Fundamental water relations parameter. See Ref. 14, pp. 113–27

183. Zimmermann, U., Büchner, K.-H., Benz, R. 1982. Transport properties of mobile charges in algal membranes: Influence of pH and turgor pressure. *J. Membr. Biol.* 67:183–97

Annu. Rev. Plant Physiol. Plant Mol. Biol. 1990. 41:55–75

SOME CURRENT ASPECTS OF STOMATAL PHYSIOLOGY

T. A. Mansfield, A. M. Hetherington, and C. J. Atkinson

Division of Biological Sciences, Institute of Environmental & Biological Sciences, Lancaster University, Lancaster LA1 4YQ, United Kingdom

KEY WORDS: guard cells, calcium ions, carbon dioxide, leaf conductance, ionic relations of cells

CONTENTS

INTRODUCTION

The mechanisms behind stomatal movements have been studied for well over a century, and the literature on the subject now contributes to many central areas of plant science—e.g. photobiology, ionic relations of cells, and hormonal mechanisms. A recent book (113) dealt with many of these subjects in detail, and we do not attempt here to cover the same ground again. Instead we emphasize two topics that have assumed importance only after new discoveries in the last five years—i.e. the regulation of guard cell turgor by calcium, and the heterogeneity of stomatal aperture over the leaf surface. We also review the responses of stomata to CO_2 because this is an area where progress

55

has been been slow and where much more research will be needed if we are to assess the impact of increasing global CO_2 concentrations on land plants.

THE ROLE OF CALCIUM IONS IN GUARD CELL PHYSIOLOGY

It has been recognized for some 30 years that calcium ions can control the aperture of the stomatal pore (48). The addition of calcium salts to the medium in which isolated epidermis is incubated can stimulate stomatal closure and inhibit opening in *Commelina communis* (23, 39, 50, 60, 108) and tobacco (102). Although initial studies (36, 108) revealed no response to calcium ions in epidermal strips of *Vicia faba,* inhibition of opening was subsequently demonstrated (37, 77). Calcium ions also inhibit K^+-induced swelling of guard cell protoplasts (38, 96).

In seeking a mechanistic explanation for the observed effects of calcium ions on stomata it is useful to distinguish between the activity of calcium in the apoplast and cytosol.

Apoplastic Calcium Ions

Variations in the supply of rhizospheric calcium occur as part of standard agricultural practice—for example, when the soil is limed to reduce acidity, or indirectly through the addition of basic slag or superphosphate-based inorganic fertilizers. There is also considerable natural variation in the calcium concentrations in soil (15, 95).

It is believed that the transport of calcium ions in the xylem is by mass flow, and that the rate of delivery to the leaves is related to the intensity of transpiration (13, 47). Solutes within the transpiration stream have been shown to accumulate within or near the stomatal complex, and it has been assumed that this distribution reflects the terminal point for the movement of liquid water that has passed from the xylem into the apoplast of the epidermis (63, 100). The role of calcium nutrition in influencing stomatal behavior in the whole plant has received little attention. However, measurements of calcium ions in xylem sap (41, 54) suggest that considerable fluctuations occur at and above the concentrations of calcium (50–250 μM) that have been shown to influence stomata on isolated epidermis (see references above). Furthermore, a detailed examination of epidermis from *Commelina communis* has shown that the bulk calcium concentration of this tissue changes with external supply, and also that calcium accumulates to a greater extent in the epidermis than in the remainder of the leaf (3). These findings have potentially important implications for the regulation of gas exchange. A calcium-induced reduction in stomatal conductance might enhance the ratio of carbon fixed relative to water lost by transpiration. However, such an enhancement of the efficiency of water use may not always be beneficial, because there are costs in terms of biomass production when stomatal conductance is reduced enough to cause decline in photosynthesis.

Studies using guard cell protoplasts (38, 96) suggest that calcium-induced alterations in the elasticity of the guard cell wall do not make an appreciable contribution to the response. How might apoplastic calcium influence stomatal opening and closure? In order to provide a possible answer to this question it is necessary to consider some additional experimental data. Atkinson et al (3) grew *Commelina communis* on different concentrations of calcium nitrate and then used invasive (epidermal strip) and non-invasive (diffusion porometry) techniques to estimate stomatal aperture. Calcium nutrition had little effect on abaxial stomatal aperture when determinations were made on the intact plants using diffusion porometry, and it was concluded that rhizospherically derived calcium had not adversely influenced the gas exchange of the whole plant or its biomass production. Nevertheless, the measurements of stomatal apertures on epidermal strips, after pre-incubation in calcium-free media, suggested that plants grown in concentrations of calcium nitrate greater than 1 mM might have experienced a 25% reduction in the *maximum* stomatal aperture they could achieve. Thus it is possible that some aspects of stomatal functioning are influenced by the supply of calcium to the whole plant. More detailed studies are needed.

To summarize, the effects of external calcium on stomata in intact leaves may be different from those observed in isolated epidermal strips, and the basis of any differences between whole leaves and epidermal strips awaits thorough investigation by plant cell biophysicists.

Cytosolic Calcium Ions

THE SECOND MESSENGER HYPOTHESIS   This hypothesis (4, 12, 45, 86) provides a framework for examining the regulation of stomatal aperture by cytosolic free calcium ions. Abscisic acid (ABA) (21–23), darkness (91), and cytokinins (49) have all been nominated as stimuli that might employ calcium ions as second messengers. Implicit in these suggestions is the belief that cytosolic free calcium concentrations ($[Ca^{2+}]_{cyt}$) in the guard cells increase in response to such signals, and that the increases trigger the intracellular machinery responsible for either stomatal closure (ABA and darkness) or opening (cytokinins). As these hypotheses have not been rigorously tested, the experimental evidence upon which they are based must be carefully examined.

The $[Ca^{2+}]_{cyt}$ in the unstimulated guard cell has been estimated to be approximately 100 nM (C. Brownlee, A. M. Hetherington, unpublished observations). This assignment of guard cell cytosolic calcium to the nM range is important in order to fulfil one of the prerequisites of a signal transduction system based on calcium. A stimulus-induced increase in $[Ca^{2+}]_{cyt}$ may be achieved through activation of plasma membrane calcium channels and/or the release of calcium from internal stores stimulated by a second messenger. There is some evidence to support both these possibilities; however, it should be stressed that the important question of whether guard cells employ stimulus-induced hydrolysis of phosphoinositides leading to the deployment of intracellular second messengers (51, 67) as a mechanism for signal transduction has not yet been addressed.

We have shown that guard cells reduce their aperture in response both to a calcium channel agonist, the dihydropyridine BAY K 8644 (M. McAinsh and A. M. Hetherington, unpublished observations), and to the divalent cation ionophore A23187 (23). Additionally, it is known that the chelator EGTA and various inorganic and organic calcium channel blockers will partially inhibit stomatal responses mediated by ABA and darkness (22, 23, 91) and that these effects are reversible (M. McAinsh, A. Webb, and A. M. Hetherington, unpublished observations). It is important to emphasize that the calcium channel antagonists only appear to inhibit the responses partially, which may support the suggestion that intracellular stores are involved. Alternatively, it may imply that the responses to ABA and darkness result from the operation of a multicomponent pathway in which calcium features but is not necessarily the key regulator. A third possibility would be that each of these signals achieves its response by more than one mechanism and that the above experimental manipulations result in the blocking of the calcium-dependent pathway. Finally, it is possible that these compounds are unable to effect total blockage of calcium influx systems located in the plasma membrane.

There are several mechanisms by which an increase in $[Ca^{2+}]_{cyt}$ could trigger alterations in stomatal aperture. There is evidence from other systems that the plasma membrane H^+ ATPase may be regulated by phosphorylation, and the protein kinase responsible is in turn stimulated by μM concentrations of free calcium (92). Calcium-dependent protein kinases are widely distributed in higher plants (11, 46, 81), and we have detected such activity in guard cell protoplasts (S. Holland and A. M. Hetherington, unpublished observations). Although few in vivo substrates have as yet been identified in plant cells, the regulatory potential of this mechanism is enormous. In animal cells the phosphorylation-dephosphorylation couplet provides a mechanism known to control a wide range of cellular activities (16, 30, 56).

An interesting demonstration is that in both CAM and C_4 plants phosphorylation contributes to the regulation of phosphoenolpyruvate carboxylase. Specifically, phosphorylation results in both an increase in the activity of the enzyme and a reduction in its feedback inhibition by malate (53, 74). A recent report also describes regulation by both calcium and calmodulin in vitro (29). As malate levels are known to increase during stomatal opening, these results may support the involvement of increased $[Ca^{2+}]_{cyt}$ in the control of stomatal opening; but the above extrapolations should be treated with caution until the physiology and biochemistry of this enzyme have been studied in stomatal guard cells in more detail (76). Although, it is known that in C_4 plants increased phosphorylation of the enzyme occurs in the light and that in stomatal guard cells light is an opening signal, it is also known that in *Nitellopsis* light decreases $[Ca^{2+}]_{cyt}$ (68). Finally, there is some evidence that the chloroplastic form of fructose-1,6-bisphosphatase (which is involved in the reconversion of malate to starch) is also activated by calcium ions (57). The same mechanisms may not apply to stomata, but it is clear that the interaction of calcium ions with guard cell carbon metabolism requires some further investigation.

The calcium-binding protein calmodulin may also participate in reactions mediated by calcium (80). Immunofluorescence studies indicate that this protein is present in stomatal guard cells (103), and there is evidence that it may participate in the regulation of stomatal aperture. Donovan et al (24) and Nejidat (73) found that treatment of epidermal strips with calmodulin inhibitors resulted in increased stomatal apertures, while De Silva et al (21) found that compounds with a similar pharmacology inhibited ABA-induced stomatal closure. A possible interpretation of these results is that calmodulin is required to effect closure. However, it must always be remembered that although these compounds are useful pharmacological tools for investigating the role of calmodulin in well-defined in vitro systems their use in vivo may produce artefactual responses (1, 87). Among the enzymes known to be regulated by calcium and calmodulin in plants are calcium ATPases. It can be predicted that an important function of these enzymes is to return the cytosolic calcium concentration to its resting level after agonist-induced increases (33). However, these enzymes have not been studied in guard cells.

Important evidence that supports a primary function for the calcium ion in the regulation of stomatal aperture comes from the work of Schroeder and colleagues, who have demonstrated that the inwardly conducting K^+ channel is inhibited by μM concentrations of cytosolic calcium (89). This is important because it provides a link between stimulus-induced increases in $[Ca^{2+}]_{cyt}$ and the inhibition of stomatal opening. Stomatal closure, on the other hand, may be initiated by the calcium-mediated release of anions and cations from the vacuole. These processes have been observed in sugarbeet vacuoles (44). If similar events take place in guard cells, then the concurrent opening of calcium-activated chloride channels (89) may depolarize the plasma membrane sufficiently to activate the outwardly conducting K^+ channel (88). This would provide a mechanism whereby a stimulus-induced increase in $[Ca^{2+}]_{cyt}$

could initiate stomatal closure (90). It is important to emphasize that although these are attractive mechanisms conceptually, further biophysical investigation is required before they can be fully substantiated.

To summarize, a stimulus-induced increase in guard cell $[Ca^{2+}]_{cyt}$ may trigger a loss of turgor in the guard cells of open stomata or inhibit the increase of turgor required to open closed stomata by the mechanisms discussed above.

Testing the hypothesis   In order to test this hypothesis rigorously, it is necessary to monitor simultaneously the cytosolic concentrations of potassium ions, protons, and free calcium ions. If calcium has a regulatory role to play in guard cell physiology then its concentration should rise after a closing stimulus, and this rise should precede changes in guard cell pH and potassium concentration. It would also be desirable to monitor alterations to guard cell morphology during the course of the experiment.

There have been several attempts to test aspects of the hypothesis. Clarkson et al (14) measured $[Ca^2]_{cyt}$ in root hair cells and found no alteration in response to ABA. However, it is not known whether these cells are competent to respond to ABA using mechanisms comparable with those in guard cells. Smith & Willmer (96) concluded that the effects of ABA on guard cell protoplasts were independent of extracellular calcium, but they were careful to note that ABA-induced release of calcium from intracellular stores may still contribute to the response. MacRobbie (61, 62) monitored ^{45}Ca fluxes in guard cells previously exposed to ABA. In some experiments fluxes increased in response to ABA while in others there were either no changes in the fluxes or they decreased. However, this technique would not resolve a stimulus-induced release of calcium from internal reservoirs and, as MacRobbie pointed out, would be unlikely to resolve short-lived transient alterations to $[Ca^{2+}]_{cyt}$. Transient increases in $[Ca^{2+}]_{cyt}$ are known to be of great importance in the coupling of extracellular stimuli to intracellular responses in animals (5, 6, 109).

The techniques used to study these events in single animal cells rely on the use of molecules that fluoresce on binding the ion of interest. With these indicators it is possible to quantify spatial and temporal aspects of stimulus-induced alterations to individual ion concentrations in single cells (105). These techniques have also been applied to the eggs of the marine alga *Fucus serratus* (9, 10) and to root hairs (14). Currently we are using the calcium indicators INDO 1 and FURA 2, and the proton indicator BCECF, to investigate the molecular basis of stimulus-response coupling in stomatal guard cells. Using FURA 2 we have demonstrated recently that ABA evokes a transient increase in $[Ca^{2+}]_{cyt}$ from 100 nM to 1 μM in stomatal guard cells (59).

RESPONSES OF STOMATA TO CO_2

The responses of plants and plant communities to rising atmospheric concentrations of CO_2 is a topic being extensively discussed at the present time. It is generally believed that two of the major consequences will be (a) increased rates of CO_2 assimilation and (b) partial stomatal closure and decreased rates of transpiration per unit area of leaf. Conjunction of these two effects leads to the conclusion that there will be an increased efficiency of water use. There are, however, deficiencies in our knowledge of the responses of stomata to CO_2 that must be remedied before firm conclusions can be drawn. Our mechanistic understanding of the responses of photosynthesis to CO_2 far outstrips that concerning the action of CO_2 on stomatal guard cells.

Morison (69, 70) has pointed out that although stomatal reactions to CO_2 have been described for over 50 different species, there are sometimes bewildering variations in response; it is, however, possible to conclude that "normally" stomata close to varying degrees in enhanced atmospheric CO_2 concentrations, and "exceptionally" they show insensitivity or open in response to CO_2. Morison succeeded in providing a useful general picture of the closing reaction that is probably valid for many different species. He drew together data from the literature covering nine C_4 and sixteen C_3 species to produce the relationship in Figure 1 showing the fall in stomatal conductance caused by a doubling of the atmospheric CO_2 concentration.

Functional Significance of the CO_2 Responses of Stomata

Discussions of the way the CO_2 responses of stomata contribute to the regulation of leaf conductance have often focused on the tendency for the intercellular $[CO_2]$ (c_i) to be nearly constant in the light—around 100 and 230 $\mu mol\ mol^{-1}$ for C_4 and C_3 plants, respectively (69). Morison argued (70), however, that the fact that c_i is generally conservative does not permit the conclusion that it is precisely regulated, and he cited evidence from several sources that it can deviate from the expected value.

It appears appropriate to refer to c_i as the effective CO_2 concentration controlling stomata. A study by Mott (72) has recently confirmed earlier deductions by Heath (42) that stomata respond to c_i and not to the concentration at the surface of the leaf. It also appeared that they were insensitive to the concentration within the stomatal pore—i.e. the surfaces of the guard cells facing the substomatal cavity appear to be the sites of perception.

With respect to the provision of CO_2 for photosynthesis, Farquhar & Sharkey (35) deduced that the stomata of C_4 plants should ideally not respond to intercellular $[CO_2](c_i)$ when it is below the saturation level for photosynthesis; but they should become very sensitive to c_i above this level, because then as they close transpiration but not assimilation will be reduced. Similar

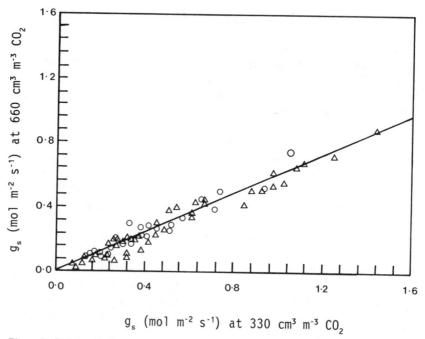

Figure 1 Relationship between stomatal conductance at 330 μmol mol^{-3} CO_2 (abscissa) and 660 μmol mol^{-3} CO_2 (ordinate) for different plant species. $\bigcirc$ = C_4 species, $\triangle$ = C_3 species. Based on data extracted from the literature. Taken from Morison (69), used by permission of Blackwell Scientific Publications Limited.

reasoning can be applied to C_3 plants, but in their case there is no level of c_i at which partial stomatal closure will curtail transpiration without some obstruction of assimilation.

Davies & Mansfield (20) pointed out that this situation in C_3 plants means that there are functional advantages if the CO_2 sensitivity of stomata is variable according to the water relations of the plant. When there is a plentiful supply of water it is an advantage to have a small response to c_i, because this reduces the tendency of the stomata to restrict the supply of CO_2 for photosynthesis. However, if water supplies are limited the plant's priority changes from maximizing assimilation to restricting transpiration while maintaining as much assimilation as possible. When a leaf is photosynthesizing in light a major factor determining c_i is wind speed, which increases total leaf conductance by reducing the boundary layer resistance. When the amount of solar energy is low, wind speed can increase transpiration greatly if stomatal aperture remains constant (40). A closing response to c_i could therefore provide a valuable mechanism for water conservation under such conditions, and it is under low irradiances that stomata are particularly sensitive to CO_2 (71).

If we are correct in deducing that changes in the magnitude of the CO_2 response of stomata have functional significance as part of a strategy for water conservation, then the mechanism behind the changes must be considered.

Variations in the CO_2 Response

Figure 1 shows remarkable consistency in behavior between different species even though the data were obtained by several different authors. It is important, however, to note that physiologists making determinations of this type usually take care to avoid major sources of variation, such as wide differences in plant water potentials, or in ages of the plants or leaves being studied. There is a considerable amount of evidence available to suggest that such factors can profoundly alter the type of relationship displayed in Figure 1.

Raschke and his group (e.g. 27, 82) have shown that if plants are grown under conditions in which they experience negligible stress from water deficits, their stomata can display zero sensitivity to CO_2. If such plants are provided with an external supply of ABA, or if production of endogenous ABA is stimulated by rapid transpiration or chilling (83, 85), then the stomata become sensitive to CO_2.

More recent studies have suggested that auxins and/or cytokinins are probably involved in modulating the stomatal response to CO_2, in conjunction with ABA. Using isolated epidermis of *Commelina communis,* Snaith & Mansfield (97) showed that high concentrations of IAA in the incubation medium reduced or eliminated the inhibitory effect of CO_2 on the stomata, but this effect of IAA was counteracted by supplying ABA. A complicated interplay among ABA, IAA, and CO_2 was revealed by these experiments.

Research in the last 10 years has also shown that cytokinins can have a profound influence on the way stomata behave. Like IAA (20), cytokinins can act as promotors of opening.

Jewer & Incoll (52) reported that ten different cytokinins, some natural and some synthetic, promoted stomatal opening by up to 50% in the grass *Anthephora pubescens,* and many other species have since been shown to respond in essentially the same way (see 7, 8). As is the case with IAA, however, cytokinins applied to leaves on their own sometimes have little effect on stomatal aperture; but in the presence of inhibitory agents such as ABA and CO_2, they can restore full stomatal opening. Application of zeatin to maize leaf pieces could completely overcome the closing effect of CO_2 and partly reverse that caused by ABA. When closure was induced by CO_2 and ABA in combination, however, there was no detectable response to zeatin.

Although it still remains to be proved that endogenous hormones produce changes in stomatal sensitivity to CO_2, the most realistic hypothesis we can offer is that hormones cause environmentally induced differences in stomatal responses such as those originally identified by Raschke and his colleagues (82).

Action of CO_2 on Guard Cells

If we are to explain such "flexibility" in the CO_2 responses of stomata we need an understanding of the cellular events controlled by CO_2. Raschke (83) suggested that CO_2 is required by the guard cells to produce malate, which is used as a counter ion to K^+ (malate may accompany Cl^- in this role). The dilemma here is that CO_2 generally causes stomatal closure but opening is accompanied by increased amounts of malate in guard cells. Raschke attempted to resolve this problem by suggesting that at low $[CO_2]$ the malate formed in the cytoplasm can be transferred as quickly as it is produced into the vacuole where, along with K^+, it is osmotically active. At high $[CO_2]$, on the other hand, the amounts of malate and H^+ in the guard cells will increase to a point at which there is a slowdown in malate production, perhaps using regulatory mechanisms involving PEP carboxylase and malate dehydrogenase. Raschke cited experimental evidence from careful studies with sub-ambient CO_2 concentrations in which stomata did not open maximally in CO_2-free air but as CO_2 increased from zero to 100 μmol mol^{-1} the familiar closing response to CO_2 occurred. Thus a critical amount of exogenous CO_2 appeared to be required to allow maximal stomatal opening, and this, Raschke suggested, was related to malate production.

Even today too little is known about the carbon metabolism of guard cells and its regulation to evaluate this ingenious hypothesis critically. In a recent summary of knowledge about regulatory aspects of PEP metabolism in stomata, Outlaw (76) cautioned against making assumptions that PEP carboxylases in guard cells have the same properties as those in other locations. Nevertheless, there is some support for Raschke's suggestion that malate formation plays a role in the CO_2 responses of stomata, perhaps as only part of the basic regulatory mechanism.

Travis & Mansfield (104) found that the fungal toxin fusicoccin, already known to stimulate stomatal opening, could reverse the response of stomata to CO_2. In the presence of 10 μM fusicoccin, increasing the CO_2 concentration from zero to 350 μmol mol^{-1} increased both stomatal aperture and the epidermal content of malate. It was suggested that CO_2 exerts two different effects on guard cells, one of which is normally obscured by the other. Stomatal closure in response to increased $[CO_2]$ usually predominates; but when fusicoccin stimulates abnormally high turgor in the guard cells there is greater demand for malate, and the role of CO_2 as a substrate for malate formation is revealed.

Later work in other laboratories has shown that further factors can also be manipulated to reverse the response of stomata to CO_2. Wardle & Short (107) found that 10^{-4} M kinetin both stimulated stomatal opening in *Vicia faba* and produced a positive response to CO_2, there being greater opening in ambient than in CO_2-free air. Spence et al (99) showed that the nature of the response of stomata to CO_2 in *Vicia faba* depended on whether the guard cells were functionally connected to the other leaf tissues, a finding consistent with a role for endogenous hormones transported to the stomatal complex from elsewhere. Eamus & Wilson (28) discovered that the stomata of *Phaseolus vulgaris* closed in response to ambient CO_2 at 22°C but opened at 5°C, there being a 2–3 times greater drop in diffusion resistance in ambient than in CO_2-free air.

Reversal of a physiological response in this manner is best explained in terms of two (or more) opposing processes that respond in a different manner to the controlling agents (i.e. temperature/kinetin/fusicoccin).

In view of the experimental data of Travis & Mansfield (104) and the importance that can now be attached to the role of malate in the control of guard cell turgor (2), it seems reasonable to propose that dark CO_2 fixation to generate malate is one of these processes; but since stomata normally close in response to CO_2 this process is usually obscured by another (or others) that dominates the situation. Locating other activities regulated by CO_2 to *suppress* stomatal opening is more difficult.

Based on observations by Melis & Zeiger (66), Zeiger (113) put forward a hypothesis that CO_2 modulates photophosphorylation in guard cells. The functional significance of the chloroplasts in guard cells has been a matter of much debate (e.g. 75), but recent evidence suggests that photosynthetic capacity, including carbon reduction, is present (94). This gives increased credibility to Zeiger's hypothesis but does not represent a direct test of it.

In discussing the role of modulation of photophosphorylation in the CO_2 responses of stomata, Zeiger (112) pointed out that this mechanism could not operate in darkness. There have been several reports of stomatal opening in response to low CO_2 concentrations in darkness (cf 64), and in those species whose stomata open at night (e.g. members of the *Crassulaceae*) this may be a fundamental mechanism behind the diurnal pattern of movements. Zeiger suggested that different mechanisms could operate in light and darkness, but once again we lack substantive evidence that this is the case; indeed, little attention has been paid to establishing the characteristics of stomatal opening in the absence of light signals. Morison & Jarvis (71) (see their Figure 7) replotted some data from a study many years ago by Heath & Russell (43) showing the responses of stomata to CO_2 in darkness and at three different light intensities. The picture that emerged was of light's modifying a basic response that could be identified in the dark, a situation not in disagreement with Zeiger's proposals. Some recent work (92a) leads to the suggestion that CO_2 may regulate ATP levels in guard cells via its effect on oxidative phosphorylation, which may be the additional mechanism that operates in darkness.

There have been alternative suggestions concerning the mechanism behind the effect of CO_2 on stomata, but they also lack sufficient experimental support. Edwards & Bowling (31) suggested that the site of action of CO_2 is most likely to be the plasma membrane of the guard cell. They used micro-electrodes to measure the membrane potentials while a 2.5-min pulse of CO_2 was applied to open and closed stomata. During the treatment of open or closed stomata there was a depolarization of 5.5 mV; immediately after its cessation there was a polarization of about 20 mV, followed by a return to the original potential difference. They suggested that CO_2 inhibited the proton efflux pump in the plasma membrane of guard cells, and drew attention to work of Spanswick & Miller (98) showing that CO_2 inhibits the proton pump and Cl^- influx in *Nitella*. The experiments by Edwards & Bowling were only performed using pulses of 100% CO_2. Even though these were applied for short periods, it is difficult to accept without further evidence that the depolarization bore any relation to physiological events that are driven by changes in CO_2 concentration over the range 0–0.1%.

The following suggestions are made in the hope of stimulating more interest in a neglected area of stomatal physiology. We propose that guard cells respond to CO_2 in two ways that are in direct opposition:

1. There is a mechanism by which guard cell turgor can increase as $[CO_2]$ increases. This is believed to occur because high $[CO_2]$ favors the formation of malate, which functions as a counter ion for K^+ (104).

2. There is an opposing mechanism (or mechanisms) causing guard cell turgor to fall as $[CO_2]$ increases. This may involve modulation of photophosphorylation and/or oxidative phosphorylation by CO_2, or other effects still to be defined, such as direct action on the plasma membrane.

In normal circumstances where CO_2 causes stomata to close, the second of these mechanisms is assumed to be dominant. However, the first can assume dominance in some exceptional conditions as defined above (p. 65). However, the significance of the first mechanism may lie not in such circumstances but in the part it plays in producing variable CO_2 responses of stomata according to the physiological state of the plant. As we have noted earlier, the CO_2 responses are indeed highly variable, and they do seem to be related to the previous history of the plant as well as to the prevailing conditions.

Acclimation of Leaf Conductance to Different CO_2 Concentrations

Enrichment of the atmosphere with CO_2 to improve biomass production has been practiced in glasshouses for many years. It appears that the sensitivity of stomata to CO_2 does not acclimate appreciably to a moderately elevated CO_2 concentration. For example, Jones & Mansfield (55) found that after lettuce had been grown for 4 weeks in 1000 μmol mol^{-1} CO_2, the stomatal closing response to CO_2 over the range 0–2400 μmol mol^{-1} was the same as in plants grown in ambient CO_2 (330 μmol mol^{-1}). In carnation, Enoch & Hurd (32) observed a 28% decrease in transpiration as the CO_2 concentration increased from 300 to 1000 μmol mol^{-1}. At very high CO_2 concentrations stomatal malfunctioning may cause further physiological injuries. Cucumbers grown in 5000 μmol mol^{-1} CO_2 during both day and night showed permanent stomatal opening, and the loss of control of transpiration is thought to be the cause of damage to this crop when excessive CO_2 is provided (79).

Determinations of the long-term effects of CO_2 enrichment on stomata have usually depended on porometers that measure leaf conductance but do not provide any guide to changes in structure of the leaf epidermis. Woodward (110) and Woodward & Bazzaz (111) have shown that stomatal density (number of stomata per unit area of epidermis) is very sensitive to changes in CO_2 concentration, decreasing in the range 225–700 μmol mol^{-1}. The effect is nonlinear, the percentage change being greater below 340 μmol mol^{-1} than above. Experiments on *Nardus stricta* originating from different altitudes revealed genetic differences between populations, those from a higher altitude showing a greater decrease in stomatal density as the CO_2 concentration was increased.

The existence of such phenotypic plasticity, which enables a plant to develop a stomatal density that leads to greater water use efficiency as CO_2 concentration increases, is an important new discovery, the ecological significance of which has yet to be fully explored. There is no information on the underlying mechanism controlling stomatal density during plant development, although several other factors such as light intensity and water deficits can bring about changes comparable to those caused by CO_2.

HETEROGENEITY OF STOMATAL CONDUCTANCE

Studies of the effects of applying ABA to leaves initially suggested that it did not have any inhibitory action on photosynthesis per se, though its suppression of stomatal opening could exert an indirect effect by restricting the diffusion of CO_2 (18, 27, 58, 65, 93). The conclusions were based not only on analyses of gas exchange but also on data acquired using an array of physiological and biochemical techniques. It was therefore surprising when further analyses of the relationship between the rate of photosynthesis and the calculated value for the intercellular CO_2 concentration (c_i), for leaves supplied with sufficient exogenous ABA to cause stomatal closure, suggested that c_i remained constant or even increased; c_i would have been expected to fall as stomata closed and restricted the gaseous diffusion of CO_2 into the leaf.

As this appeared not to be the case, it was implied that a short-term exposure to ABA might have a direct inhibitory effect on photosynthesis, this being independent of ABA-induced stomatal closure (17, 84, 106).

It now appears that the apparent contradiction arose because of errors resulting from the conventional method of estimating c_i from gas exchange data. The rate of transpiration is calculated as an average for the entire leaf surface, and this provides the basis for estimating the conductance of the leaf to CO_2 (g_c). One then calculates c_i using the relationship

$$C_i = C_a - \frac{1.6A}{g_c},$$

where c_a is the CO_2 concentration in the air outside the leaf and A is net assimilation rate.

This method of estimating c_i becomes invalid if stomata on different parts of a leaf behave very differently. Let us suppose that the stomata over half the area of a leaf are tightly closed, while those over the other half are wide open. In this situation, measurements made on the whole leaf produce values for g_c and A that are averages for the two halves. Figure 2 presents data showing the relationship between net assimilation rate and stomatal conductance, and we have calculated c_i in two ways, as shown in the legend. Use of the average value for g_c for the whole leaf leads to a spurious value of c_i of 249 μmol mol^{-1}. This is much higher than the "real" value (114 μmol mol^{-1}) obtained as indicated. If c_i is calculated for the two halves separately we obtain (a) 46 μmol mol^{-1} for the portion with closed stomata (at the CO_2 compensation point) and (b) 249 μmol mol^{-1} for the portion with open stomata; it will be seen that calculation b is the same as the spurious value of 249.

The simple explanation for this is that if 50% of the leaf surface has zero A and zero g_c, then for the whole leaf the values of both A and g_c fall by 50%. Hence c_i calculated from the above equation is the same for the whole leaf as for the half with open stomata.

The overestimate of c_i not only gives a misleading measure of intercellular CO_2 concentration but also creates the impression that mesophyll photosynthesis has been impaired more than can be explained by stomatal closure.

The possibility that there might be wide differences in g_c over different parts of a leaf, leading to overestimates of this kind, came to light in several recent studies (25, 34, 101). Downton et al (25) found that if detached leaves of *Vitis vinifera* and *Helianthus annuus* were supplied with ABA via their petioles, stomatal closure occurred in distinct patches over the leaf surfaces. They used measurements of chlorophyll fluorescence quenching to make independent estimates of c_i, and these showed that the response to ABA was

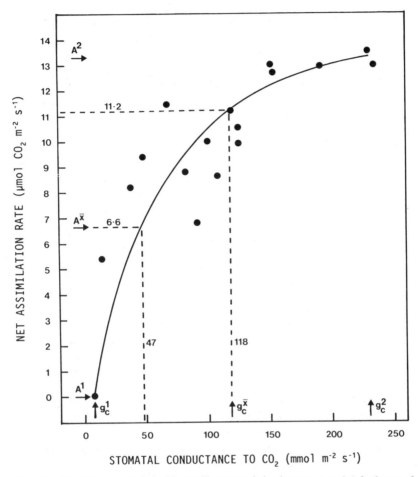

Figure 2   Calculations made from this curvilinear correlation between g_c and A for leaves of *Commelina communis* demonstrate how errors can arise in the calculation of c_i if the stomata over half the area of a leaf are tightly closed and the remainder are wide open.

If c_i is calculated using an average ($g_c^{\bar{x}}$) of the two extreme values of g_c (g_c^1 and g_c^2) and an average ($A^{\bar{x}}$) of the two extreme values of A (A^1 and A^2) we obtain 249 μmol mol^{-1} [$c_i = 340 - (6.6 \times 10^3 \times 1.6)/118$]. By comparison the actual value for c_i for a leaf with an average value of A of 6.6 μmol m^{-2} sec^{-1} should be 114 μmol mol^{-1} [$c_i = 340 - (6.6 \times 10^3 \times 1.6)/47$].

sufficiently heterogeneous to cause overestimates when gas exchange methods were used to determine c_i. A further study by the same authors (26) showed that when increased amounts of endogenous ABA were induced by withholding water from *Vitis vinifera*, *Nerium oleander*, and *Eucalyptus ficifolia*, non-uniform stomatal closure occurred in all three species.

There is thus a satisfactory explanation for the apparent nonstomatal inhibition of photosynthesis by ABA. This research does, however, leave an intriguing question unanswered—i.e. why do stomata respond in such a non-uniform manner? In many leaves the network of vascular bundles causes the effective isolation of sections of intercellular spaces, and there is restricted lateral diffusion of CO_2 between the sections. Leaves with such "heterobaric" anatomy are believed to be prone to non-uniform stomatal closure, while those with well-connected intercellular spaces ("homobaric") providing high gaseous mobility, are not (101).

The uneven distribution of a substance moving in the xylem could result from local differences in transpiration, or from differences in resistance of the xylem conduits so that water flow is favored to some areas of the leaf. This may be more likely to occur in a heterobaric than a homobaric leaf. The uneven distribution of material is not, however, the only possibility that needs to be investigated. It is conceivable that changes in tissue sensitivity to ABA are also involved. There is a highly sensitive regulatory relationship between the stomatal guard cells and their physical and chemical environments. It is known that the sensitivity of stomata to leaf water potential and photon flux are different on adaxial and abaxial leaf surfaces (19), perhaps as a result of differences in auxin concentrations or tissue sensitivity to auxin (78). Our own unpublished work has shown that potassium concentrations may vary greatly in different parts of a leaf, and the availability of K^+ appears to have a considerable influence on the magnitude of the stomatal response to ABA (97).

Heterogeneity of stomatal behavior is a topic worthy of more detailed investigation in the future. If the phenomenon is found to occur generally under experimental and natural conditions there will be important questions about the validity of various physiological and biochemical determinations on heterobaric leaves. "Patchiness" of the kind displayed in recent studies (25, 26, 101) would lead not only to variations in c_i, but also to major differences in carbon metabolism resulting from changes in CO_2 fixation and photorespiration. A high degree of patchiness is likely to represent an inefficiency in leaf functioning that is worthy of consideration in relation to crop productivity.

Acknowledgments

We are grateful to the Agricultural and Food Research Council, the Gatsby Charitable Foundation, and Shell Research Ltd. for financial support. We thank Prof. E. A. C. MacRobbie (Botany Department, University of Cambridge) and Dr. C. Brownlee (Marine Biological Association of the U.K., Plymouth) for critical comments on part of the review.

Literature Cited

1. Asono, M., Stull, J. T. 1985. Effects of calmodulin antagonists on smooth muscle contraction and myosin phosphorylation. In *Calmodulin Antagonists*, ed. H. Hidaka, D. J. Hartshorne, pp. 225–60. Orlando: Academic
2. Assman, S. M., Zeiger, E. 1987. Guard cell bioenergetics. See Ref. 113, pp. 163–93
3. Atkinson, C. J., Mansfield, T. A., Kean, A. M., Davies, W. J. 1989. Control of stomatal aperture by calcium in isolated epidermal tissue and whole leaves of *Commelina communis* L. *New Phytol.* 111:9–17
4. Berridge, M. J. 1988. Inositol lipids and calcium signalling. *Proc. R. Soc. London Ser. B* 234:1–9
5. Berridge, M. J., Cobbold, P. H., Cuthbertson, K. S. R. 1988. Spatial and temporal aspects of cell signalling. *Philos. Trans. R. Soc. London Ser. B* 320:325–43
6. Berridge, M. J., Galione, A. 1988. Cytosolic calcium oscillators. *FASEB J.* 2:3074–82
7. Blackman, P. G., Davies, W. J. 1984. Modification of the CO_2 responses of maize stomata by abscisic acid and by naturally occurring and synthetic cytokinins. *J. Exp. Bot.* 35:174–79
8. Blackman, P. G., Davies, W. J. 1985. Root-to-shoot communication of the effects of soil drying in maize. *J. Exp. Bot.* 36:39–48
9. Brownlee, C., Hetherington, A. M., Woods, J. W. 1989. Cytoplasmic calcium and calcium transport in *Fucus serratus* rhizoids. In *8th International Workshop on Plant Membrane Transport*. New York: Elsevier. In press
10. Brownlee, C., Pulsford, A. L. 1988. Visualisation of the calcium gradient in *Fucus serratus* rhizoids: correlation with cell ultrastructure and polarity. *J. Cell Sci.* 91:249–56
11. Budde, J. A., Randall, D. D. 1989. Protein kinases in higher plants. In *Second Messengers in Plant Growth and Development,* ed. W. F. Boss, D. S. More. New York: Liss. In press
12. Carafoli, E. 1987. Intracellular calcium homeostasis. *Annu. Rev. Biochem.* 56: 395–433
13. Clarkson, D. T. 1984. Calcium transport between tissues and its distribution in the plant. *Plant Cell Environ.* 7:449–56
14. Clarkson, D. T., Brownlee, C., Ayling, S. M. 1988. Cytoplasmic calcium measurements in intact higher plant cells: results from fluorescence ratio imaging of FIRA-2. *J. Cell Sci.* 91:71–80
15. Clymo, R. S. 1962. An experimental approach to part of the calcicole problem. *J. Ecol.* 50:707–31
16. Cohen, P. 1988. Protein phosphorylation and hormone action. *Proc. R. Soc. London Ser. B* 234:115–44
17. Cornic, G., Miginiac, E. 1983. Nonstomatal inhibition of net CO_2 uptake by ($\pm$) abscisic acid in *Pharbitis nil. Plant Physiol.* 73:529–33
18. Cummings, W. R., Kende, H., Raschke, K. 1971. Specificity and reversibility of the rapid stomatal response to abscisic acid. *Planta* 99:347–51
19. Davies, W. J. 1977. Stomatal responses to water stress and light in plants grown in controlled environments and in the field. *Crop Sci.* 17:735–40
20. Davies, W. J., Mansfield, T. A. 1987. Auxins and stomata. See Ref. 113, pp. 293–309
21. De Silva, D. L. R., Cox, R. C., Hetherington, A. M., Mansfield, T. A. 1985. Suggested involvement of calcium and calmodulin in the responses of stomata to abscisic acid. *New Phytol.* 101:555–63
22. De Silva, D. L. R., Cox, R. C., Hetherington, A. M., Mansfield, T. A. 1986. The role of abscisic acid and calcium in determining the behaviour of adaxial and abaxial stomata. *New Phytol.* 104:41–51
23. De Silva, D. L. R., Hetherington, A. M., Mansfield, T. A. 1985. Synergism between calcium ions and abscisic acid in preventing stomatal opening. *New Phytol.* 100:473–82
24. Donovan, N., Martin, S., Donkin, M. E. 1985. Calmodulin binding drugs trifluoperazine and compound 48/80 modify stomatal responses of *Commelina communis* L. *J. Plant Physiol.* 118:177–87
25. Downton, W. J. S., Loveys, B. R., Grant, W. J. R. 1988a. Stomatal closure fully accounts for the inhibition of photosynthesis by abscisic acid. *New Phytol.* 108:263–66
26. Downton, W. J. S., Loveys, B. R., Grant, W. J. R. 1988b. Non-uniform stomatal closure induced by water stress causes putative non-stomatal inhibition of photosynthesis. *New Phytol.* 110: 503–9
27. Dubbe, D. R., Farquhar, G. D., Raschke, K. 1978. Effect of abscisic acid on the gain of the feedback loop involving

carbon dioxide and stomata. *Plant Physiol.* 62:406–17

28. Eamus, D., Wilson, J. M. 1984. A model for the interaction of low temperature ABA, IAA and CO_2 in the control of stomatal behaviour. *J. Exp. Bot.* 35:91–98

29. Echevarria, C., Vidal, J., Lemarechal, P., Brulfert, J., Ranjeva, R., Gadal, P. 1988. The phosphorylation of sorghum leaf phosphoenolpyruvate carboxylase is a calcium-calmodulin dependent process. *Biochem. Biophys. Res. Commun.* 155:835–40

30. Edelman, A. M., Blumenthal, D. K., Krebs, E. G. 1987. Protein serine/threonine kinases. *Annu. Rev. Biochem.* 56:567–613

31. Edwards, A., Bowling, D. J. F. 1985. Evidence for a CO_2 inhibited proton extrusion pump in the stomatal cells of *Tradescantia virginiana. J. Exp. Bot.* 36:91–98

32. Enoch, H. Z., Hurd, R. G. 1979. The effect of elevated CO_2 concentrations in the atmosphere on plant transpiration and water use efficiency. A study with potted carnation plants. *Int. J. Biometeorol.* 23:343–51

33. Evans, D. E. 1988. Regulation of cytoplasmic free calcium by plant cell membranes. *Cell Biol. Int. Rep.* 12:383–95

34. Farquhar, G. D., Hubrick, K. T., Terashima, I., Condon, A. G., Richards, R. A. 1987. Genetic variation in the relationship between photosynthetic CO_2 assimilation rate and stomatal conductance to water loss. In *Progress in Photosynthesis Research,* ed. J. Biggins, 4:209–12. Amsterdam: Martinus Nishoff

35. Farquhar, G. D., Sharkey, T. D. 1982. Stomatal conductance and photosynthesis. *Annu. Rev. Plant Physiol.* 33:317–45

36. Fischer, R. A. 1968. Stomatal opening in isolated epidermal strips of *Vicia faba* 1. Response to light and CO_2 free air. *Plant Physiol.* 43:1953–58

37. Fischer, R. A. 1972. Aspects of potassium accumulation by stomata of *Vicia faba. Aust. J. Biol. Sci.* 25:1107–23

38. Fitzsimons, P. J., Weyers, J. D. B. 1986. Volume changes of *Commelina communis* guard cell protoplasts in response to K^+, light and CO_2. *Physiol. Plant.* 66:463–68

39. Fujino, M. 1967. Role of adenosinetriphosphate and adenosinetriphosphatase in stomatal movement. *Sci. Bull. Fac. Edu. Nagasaki Univ.* 18:1–47

40. Grace, J. 1977. *Plant Response to Wind.* London: Academic. 204 pp.

41. Hanson, J. B. 1984. The functions of calcium in plant nutrition. In *Advances in Plant Nutrition,* ed. P. B. Tinker, A. Lauchli, 1:149–208. New York: Praeger

42. Heath, O. V. S. 1949. Studies in stomatal behaviour. V. The role of carbon dioxide in the light response of stomata. *J. Exp. Bot.* 1:29–62

43. Heath, O. V. S., Russell, J. 1954. Studies in stomatal behaviour. VI. An investigation of the light responses of wheat stomata with the attempted elimination of control by the mesophyll. Part II. Interactions with external CO_2 and general discussion. *J. Exp. Bot.* 5:269–92

44. Hedrich, R., Neher, E. 1987. Cytoplasmic calcium regulates voltage dependent ion channels in plant vacuoles. *Nature* 329:833–36

45. Hepler, P. K., Wayne, R. O. 1985. Calcium and plant development. *Annu. Rev. Plant Physiol.* 36:397–439

46. Hetherington, A. M., Battey, N. H., Millner, P. A. 1989. Protein kinase. In *Methods in Plant Biochemistry,* ed. P. J. Lea, Vol. 7A. London: Academic. In press

47. Ho, L. C. 1989. Environmental effects on the diurnal accumulation of ^{45}Ca by young fruit and leaves of tomato plants. *Ann. Bot.* 63:281–88

48. Iljin, W. S. 1957. Drought resistance in plants and physiological processes. *Annu. Rev. Plant Physiol.* 8:257–74

49. Incoll, L. D., Jewer, P. C. 1987. Cytokinins and stomata. See Ref. 113, pp. 281–92

50. Inoue, H., Katoh, Y. 1987. Calcium inhibits ion-stimulated stomatal opening in epidermal strips of *Commelina communis* L. *J. Exp. Bot.* 38:142–49

51. Irvine, R. F., Moor, R. M., Pollock, W. K., Smith, P. M., Wreggett, K. A. 1988. Inositol phosphates: Proliferation, metabolism and function. *Philos. Trans. R. Soc. London Ser. B* 320: 281–98

52. Jewer, P. C., Incoll, L. D. 1980. Promotion of stomatal opening in the grass *Anthephora pubescens* Nees. by a range of natural and synthetic cytokinins. *Planta* 150:218–21

53. Jiao, J. A., Chollett, R. 1989. Regulatory seryl phosphorylation of C_4 phosphoenolpyruvate carboxylase by a soluble protein kinase from maize leaves. *Arch. Biochem. Biophys.* 269:526–35

54. Jones, H. G., Higgs, K. H., Samuelson, T. J. 1983. Calcium uptake by developing apple fruits. I. Seasonal changes in calcium content of fruits. *J. Hortic. Sci.* 58:173–82

55. Jones, R. J., Mansfield, T. A. 1970. Increases in the diffusion resistances of leaves in a carbon-dioxide enriched atmosphere. *J. Exp. Bot.* 21:951–58

56. Krebs, E. G. 1985. The phosphorylation of proteins: a major mechanism for biological control. *Biochem. Soc. Trans.* 13:813–20

57. Kreimer, G., Melkonian, M., Holtum, J. A. M., Latzko, E. 1988. Stromal free calcium concentrations and light mediated activation of chloroplast fructose-1-6 bisphosphate. *Plant Physiol.* 86:423–28

58. Kriedemann, P. E., Loveys, B. R., Downton, W. J. S. 1975. Internal control of stomatal physiology and photosynthesis. II. Photosynthetic response to phaseic acid. *Aust. J. Plant Physiol.* 2:553–67

59. McAinsh, M. R., Brownlee, C., Hetherington, A. M. 1990. Abscisic acid induced elevation of cytoplasmic calcium precedes stomatal closure in *Commelina communis.* Submitted

60. MacRobbie, E. A. C. 1986. Calcium effects in stomatal guard cells. In *Molecular and Cellular Aspects of Calcium in Plant Development,* ed. A. J. Trewavas, pp. 383–84. New York: Plenum

61. MacRobbie, E. A. C. 1989a. Calcium influx at the plasma membrane of isolated guard cells of *Commelina communis.* Effects of abscisic acid. *Planta* 178:231–41

62. MacRobbie, E. A. C. 1989b. Effects of ABA on ion fluxes in guard cells. See Ref. 9

63. Maier-Maercker, U. 1983. The role of peristomatal transpiration in the mechanism of stomatal movement. *Plant Cell Environ.* 6:369–80

64. Mansfield, T. A., Travis, A. J., Jarvis, R. G. 1981. Responses to light and carbon dioxide. In *Stomatal Physiology,* ed. P. G. Jarvis, T. A. Mansfield. *Soc. Exp. Biol.* (Seminar Ser.) 8:119–35. Cambridge: Cambridge Univ. Press. 295 pp.

65. Mawson, B. T., Colman, B., Cummins, W. R. 1981. Abscisic acid and photosynthesis in isolated leaf mesophyll cells. *Plant Physiol.* 67:233–36

66. Melis, A., Zeiger, E. 1982. Chlorophyll *a* fluorescence transients in mesophyll and guard cells. Modulation of guard cell photophosphorylation by CO_2. *Plant Physiol.* 69:642–47

67. Michell, R. H., Kirk, C. J., MacCullum, S. H., Hunt, P. A. 1988. Inositol lipids: receptor-stimulated hydrolysis and cellular lipid pools. *Philos. Trans. R. Soc. London Ser. B* 320:239–46

68. Miller, A. J., Sanders, D. 1987. Depletion of cytosolic free calcium induced by photosynthesis. *Nature* 326:397–400

69. Morison, J. I. L. 1985. Sensitivity of stomata and water use efficiency to high CO_2. *Plant Cell Environ.* 8:467–74

70. Morison, J. I. L. 1987. Intercellular CO_2 concentration and stomatal response to CO_2. See Ref. 113, pp. 229–51

71. Morison, J. I. L., Jarvis, P. G. 1983. Direct and indirect effects of light on stomata. II. In *Commelina communis* L. *Plant Cell Environ.* 6:103–9

72. Mott, K. A. 1988. Do stomata respond to CO_2 concentrations other than intercellular? *Plant Physiol.* 86:200–3

73. Nejidat, A. 1987. Effect of ophiobolin on stomatal movement: role of calmodulin. *Plant Cell Physiol.* 28:455–60

74. Nimmo, G. A., McNaughton, G. A. L., Fewson, C. A., Wilkins, M. B., Nimmo, H. G. 1987. Changes in the kinetic properties and phosphorylation state of phosphoenolpyruvate carboxylase in *Zea mays* leaves in response to light and dark. *FEBS Lett.* 213:18–22

75. Outlaw, W. H. Jr. 1982. Carbon metabolism in guard cells. *Recent Adv. Phytochem.* 16:185–222

76. Outlaw, W. H. Jr. 1987. An introduction to carbon metabolism in guard cells. See Ref. 113, pp. 115–23

77. Pallaghy, C. K. 1970. The effect of calcium on the ion specificity of stomatal opening in epidermal strips of *Vicia faba.* *Z. Pflanzenphysiol.* 62:58–62

78. Pemadasa, M. A. 1982. Differential abaxial and adaxial stomatal responses to indole-3-acetic acid in *Commelina communis* L. *New Phytol.* 90:209–19

79. Pfeufer, B., Krug, H. 1984. Effects of high CO_2 concentrations on vegetables. *Acta Hortic.* 162:37–44

80. Piazza, G. J. 1988. Calmodulin in plants. In *Calcium Binding Proteins,* ed. M. P. Thompson, 1:127–43. Boca Raton: CRC Press

81. Ranjeva, R., Boudet, A. M. 1987. Phosphorylation of proteins in plants: regulatory effects and potential involvement in stimulus-response coupling. *Annu. Rev. Plant Physiol.* 38:73–93

82. Raschke, K. 1975. Stomatal action. *Annu. Rev. Plant Physiol.* 26:309–40
83. Raschke, K. 1977. The stomatal turgor mechanism and its responses to CO_2 and abscisic acid: observations and a hypothesis. In *Regulation of Cell Membrane Activities in Plants,* ed. E. Marre, O. Ciferri, pp. 173–83. Amsterdam: Elsevier/North Holland Biomedical Press
84. Raschke, K., Hedrich, R. 1985. Simultaneous and independent effects of abscisic acid on stomata and the photosynthetic apparatus in whole leaves. *Planta* 163:105–18
85. Raschke, K., Pierce, M., Popiela, C. C. 1976. Abscisic acid content and stomatal sensitivity to CO_2 in leaves of *Xanthium strumarium* L. after pretreatments in warm and cold growth chambers. *Plant Physiol.* 57:115–21
86. Rasmussen, H., Barrett, P. 1984. Calcium messenger system: an integrated view. *Physiol. Rev.* 64:938–84
87. Roufogalis, B. D., Minocherhomjee, A. E. V. F., Al Jobore, A. 1983. Pharmocological antagonism of calmodulin. *Can. J. Biochem. Physiol.* 61:927–33
88. Schroeder, J. I. 1988. K^+ transport properties of K^+ channels in the plasma membrane of *Vicia faba* guard cells. *J. Gen. Physiol.* 92:667–83
89. Schroeder, J. I., Hagiwara, S. 1989. Cytosolic calcium regulates ion channels in the plasma membrane of *Vicia faba* guard cells. *Nature* 338:427–30
90. Schroeder, J. I., Hedrich, R. 1989. Involvement of ion channels and active transport in osmoregulation and signalling of higher plant cells. *Trends Biochem. Sci.* 14:187–92
91. Schwartz, A. 1985. Role of Ca^{2+} and EGTA on stomatal movements in *Commelina communis* L. *Plant Physiol.* 79:1003–5
92. Serrano, R. 1989. Structure and function of plasma membrane ATPase. *Annu. Rev. Plant Physiol.* 40:61–94
92a. Shaish, A., Roth-Bejerano, N., Itai, C. 1989. The response of stomata to CO_2 relates to its effect on respiration and ATP level. *Physiol. Plant.* 76:106–11
93. Sharkey, T. D., Raschke, K. 1980. Effects of phaseic acid and dihydrophaseic acid on stomata and the photosynthetic apparatus. *Plant Physiol.* 65:291–97
94. Shimazaki, K., Zeiger, E. 1987. Red light-dependent CO_2 uptake and oxygen evolution in guard cell protoplasts of

Vicia faba L.: Evidence for photosynthetic CO_2 fixation. *Plant Physiol.* 84:7–9
95. Smith, C. J. 1980. *Ecology of the English Chalk.* London: Academic. 573 pp.
96. Smith, G. N., Willmer, C. M. 1988. Effects of calcium and abscisic acid on volume changes of guard cell protoplasts of *Commelina communis. J. Exp. Bot.* 30:1529–39
97. Snaith, P. J., Mansfield, T. A. 1982. Control of the CO_2 responses of stomata by indol-3-ylacetic acid and abscisic acid. *J. Exp. Bot.* 33:360–65
98. Spanswick, R. M., Miller, A. G. 1977. The effect of CO_2 on the Cl^- influx and electrogenic pump in *Nitella translucens.* In *Transmembrane Ionic Exchanges in Plants,* ed. M. Thellier, A. Monnier, M. Demarty, J. Dainty, pp. 239–45. Paris/Rouen: C. N. R. S. 607 pp.
99. Spence, R. D., Sharpe, P. J. H., Powell, R. D. 1984. The role of epidermal cells in the carbon dioxide response of stomata of *Vicia faba* L. *New Phytol.* 97:145–54
100. Tanton, T. W., Crowdy, S. H. 1972. Water pathways in higher plants. III. The transpiration stream within leaves. *J. Expo. Bot.* 23:619–25
101. Terashima, I., Wong, S. C., Osmond, C. B., Farquhar, G. D. 1989. Characterisation of non-uniform photosynthesis induced by abscisic acid in leaves having different mesophyll anatomies. *Plant Cell Physiol.* 29:385–94
102. Thomas, D. A. 1970. The regulation of stomatal aperture in tobacco leaf epidermal strips. *Aust. J. Biol. Sci.* 23:961–79
103. Thomspon, M. P., Brower, D. P. 1984. Immunofluorescent localization of calmodulin and tubulin in guard cells of higher plants. *Fed. Proc. Fed. Am. Soc. Exp. Biol.* 43:3477 (Abstr.)
104. Travis, A. J., Mansfield, T. A. 1979. Reversal of the CO_2-responses of stomata by fusicoccin. *New Phytol.* 83:607–14
105. Tsien, R. Y. 1989. Fluorescent probes of cell signaling. *Annu. Rev. Neurosci.* 12:227–54
106. Ward, D. A., Bunce, J. A. 1987. Abscisic acid simultaneously decreases carboxylation efficiency and quantum yield in attached soybean leaves. *J. Exp. Bot.* 38:1182–92
107. Wardle, K., Short, K. C. 1981. Responses of stomata in epidermal strips of

Vicia faba to carbon dioxide and growth responses when incubated on potassium chloride and potassium iminodiacetate. *J. Exp. Bot.* 32:303–9

108. Willmer, C. M., Mansfield, T. A. 1969. A critical examination of the use of detached epidermis in studies of stomatal physiology. *New Phytol.* 68:363–75

109. Woods, N. M., Cuthbertson, K. S. R., Cobbold, P. H. 1986. Repetitive transient rises in cytosollic free calcium in hormone—stimulated hepatocytes. *Nature* 319:600–2

110. Woodward, F. I. 1987. Stomatal numbers are sensitive to increases in CO_2 from pre-industrial levels. *Nature* 327:617–18

111. Woodward, F. I., Bazzaz, F. A. 1988. The responses of stomatal density to CO_2 partial pressure. *J. Exp. Bot.* 39:1771–81

112. Zeiger, E. 1983. The biology of stomatal guard cells. *Annu. Rev. Plant Physiol.* 34:441–75

113. Zeiger, E., Farquhar, G. D., Cowan, I. R. 1987. *Stomatal Function.* Stanford: Stanford Univ. Press. 503 pp.

Annu. Rev. Plant Physiol. Plant Mol. Biol. 1990. 41:77–107

KINETIC MODELING OF PLANT AND FUNGAL MEMBRANE TRANSPORT SYSTEMS

Dale Sanders

Biology Department, University of York, York YO1 5DD, England

KEY WORDS: proton pump, gradient-coupled transport, ion channel, rate constant, current-voltage relationship

CONTENTS

Transport systems are essential elements of biological membranes, facilitating the passage of ions and other solutes across otherwise impermeable barriers. The plasma membrane and tonoplast (which bounds the vacuole) are the two major transport membranes of plant and fungal cells, and the general energetic principles that govern the organization of transport at these membranes are now well established. Primary transport systems, which derive their energy

77

1040-2519/90/0601-0077$02.00

from hydrolysis of phosphoanhydride bonds (78, 104), pump H^+ electrogenically from the cytosol, thereby creating an H^+ electrochemical gradient ($\Delta\bar{\mu}_{H^+}$) directed into the cytosol. Secondary, gradient-coupled transport systems are specific for a wide variety of inorganic ions and organic solutes (87) and are energized by coupling solute transport to passive reflux of H^+: The proton and solute fluxes can, depending on the transport system, be in the same or opposing directions (symport and antiport, respectively). Finally, ionic channels catalyze dissipative transport (48).

Significant advances in our understanding of the mechanisms by which all three classes of system execute transport are being achieved by the application of reaction kinetic models. Reaction kinetic modeling seeks to take the description of transport kinetic data one stage beyond the simple fitting of a phenomenological relationship, such as a rectangular hyperbola. The underlying principle is that important information is embedded in phenomenological relationships and should be extracted. Reaction kinetic modeling has enabled:

- *determination of the kinetic effects of a ligand at its transport site*, which can potentially be distinguished from allosteric action. Such studies, in turn, can demonstrate how transport rate is controlled and coordinated with respect to cellular demand (88);
- *in vivo electrical assays of transport activity* (91)—an especially useful facility in the case of the plasma membrane H^+ pump for which an alternative radioisotopic assay is not possible;
- *derivation of rate constants* for transitions between different states of a transport system. Since the ratio of the forward/backward rate constants for any given state transition itself defines the free energy change associated with that transition, the major sites of endergonic and exergonic partial reactions associated with transport can be identified (106);
- *description of the on/off transitions of ionic channels* by stochastic models on the basis of analysis of the frequency distribution of open and closed times of single channels (11). Such models make it possible to describe the response of channel currents to voltage or to agonists and antagonists in terms of quantitative effects on transitions between open and closed states.

GENERAL APPROACHES TO MODELING TRANSPORT SYSTEMS

Carriers and Channels

Transport systems in biological membranes have conventionally been classified as either carrier or channel systems. Transport binding site(s) on carriers are envisaged (by definition) to undergo alternate exposure to each side of the membrane as a result of conformational changes in the transport protein. Carriers turn over at a relatively low rate (10^2–10^4 s^{-1}: see 49), and fluxes through them saturate with respect to ligand concentration and (if a current is carried) with respect to transmembrane electrical potential ($\Delta\Psi$). Additional properties, which can be described by reaction kinetic models, include transinhibition and isotope exchange (or "exchange diffusion"). Typical carriers include all metabolically coupled and H^+-gradient-driven transport systems.

Channels, on the other hand, have turnover numbers in the region of 10^7-10^8 s^{-1} (69). These high turnover numbers enable resolution with the patch clamp technique of electrical currents flowing through single channel molecules, and the resultant observations reveal that individual channels undergo discrete transitions between open and closed states: Channels are gated. A complete model for a temporally or spatially averaged channel-mediated flux across a membrane therefore comprises two components. First, the current through a single open channel can be described with respect to such factors as ionic selectivity, responses to $\Delta\Psi$, and concentration. Second, the factors governing the open-state probability can be characterized in kinetic terms.

Practical Aspects

Kinetic data on transport systems are normally obtained either as radioisotopic flxues or, for electrically active systems, as currents. In the latter case the measured flux will be a net one, whereas radioisotopes enable an estimate of the unidirectional flux. Ideally, a complete kinetic description will embody data from both sources, though in practice (and often for good technical reasons) most transport systems have been modeled utilizing only one class of data.

There are essentially two conceptually different approaches to modeling the kinetics of membrane transport. The first and simplest takes a specific model and explores its behavior in terms of general algebraic characteristics. Properties such as the overall form of the flux-vs-concentration relationship and the dependence of the relationship on $\Delta\Psi$ can be investigated using this approach. The results can then be compared with biological data to establish whether the model is physiologically reasonable. In this context, negative conclusions can be more significant than positive ones: Failure of a model to behave in qualitative accord with observations allows a firm (albeit negative) conclusion concerning transport system behavior, whereas the capability of a model to replicate data cannot, alone, demonstrate that the correct model has been found.

A second approach, which is really the inverse of the first, comprises least squares fitting of experimental data by a reaction kinetic model. This statistical method is, of course, required if the quantitative aspects of transport are to be explored in any depth (e.g. in ascribing finite values to rate constants). However, there is an element of risk associated with adoption of a statistical approach in isolation, since the investigator has to guard against choice of an incorrect model, even though it may be statistically reasonable. From this point of view, it is clearly essential to appreciate the general algebraic properties of models, and, in the event that more than one provides satisfactory fits to the data, to proceed with the analysis by using that which provides the simplest mechanistic or biological description.

This review focuses on the kinetic properties of transport systems as they relate to transmembrane reactions per se. Extensive kinetic analysis of the intracellular control factors that act upon transport has also been undertaken, and the interested reader is referred to the recent work of Hansen and collaborators (29, 44, 110, 116–118) for coverage of this area.

KINETIC MODELS OF CARRIERS FROM THEIR ELECTRICAL PROPERTIES

Experimental Approach

Plant and fungal cells provide ideal material for studying the electrokinetics of plasma membrane carriers because the high background specific resistance of the membrane (in comparison with that, for example, of animal cells; see 93) results in readily measurable electrical manifestation of transport system activity. The central experimental goal of such studies has normally been to measure the steady-state current voltage (I–V) relationship of the carrier. This relationship describes the response of carrier-mediated current (net flux) as a function of $\Delta\Psi$. In many ways, an I–V relationship can be viewed as analogous to the familiar Michaelis-Menten function in enzyme kinetics, since voltage can be loosely regarded as a participant in the transport reaction. The two relationships therefore exhibit a similar tendency to saturation, though (as discussed below) their precise forms differ.

To date, reaction kinetic models describing the electrokinetics of carriers have been developed only for those systems residing at the plasma membrane and have been based on data obtained with classical electrophysiological techniques involving intracellular impalement by a microelectrode. An electrical measurement of the carrier I–V relationship can, in principle, be obtained via two measurements of the I–V relationship of the entire membrane. The first measurement is made in control conditions in the absence of substrate, while the second is obtained as soon as possible after substrate is applied. In vivo measurements on the primary H^+ pump, for which rapid introduction of the ion is not possible, require that the first I–V relationship is measured with the pump working fully, and the second in conditions in which the pump is subject to metabolic restriction. In either case, the difference between the two measurements is often taken to indicate the I–V relationship of the transport system under investigation. This difference I–V method has technical drawbacks in the context of application to the H^+ pump, since it is obviously neither convenient nor desirable to subject cells to metabolic restriction for each measurement of activity: Time resolution is limited by the rate which the inhibitor can be applied and removed. Furthermore, inhibitors should be shown to result in a change only in the activity of the H^+ pump, and not in that of other transport systems.

There are also general theoretical flaws in the difference I–V method if applied uncritically either to primary pumps or to secondary transporters. Residual transport activity in the presence of a reversible inhibitor not only results in underestimation of true activity (37) but can also distort the shape of the derived difference I–V relationship (23, 38). The errors associated with identification of the difference I–V relationship as that for the transport system become particularly acute in the vicinity of the reversal potential (13).

An alternative approach relies on a "group-fitting" strategy. (In practice this approach has been applied almost exclusively to primary pumps, though in principle secondary systems can also be accommodated.) For group-fitting, two or more membrane I–V relationships are measured during an ex-perimental treatment designed to modulate pump activity. The ensemble of I–V relationships is then fitted with a model that takes into account the fact that the pump operates in parallel with other electrically active (leak) path-ways in the membrane (16, 38). This then allows dissection of the membrane I–V relationship into separate pump and leak components that are additive with respect to their currents at any particular value of $\Delta\Psi$. Either of the two components can be held constant (but are statistically optimized) for the series of I–V curves, so far as the data permit. As a variation of the group-fitting approach, the difference I–V relationships can be formally analyzed in the context of unique diagnostic features (e.g. multiple conductance maxima) not present in the bona fide I–V relationship of the transport system itself (14, 16).

The Pseudo-Two-State Model

Algebraic analysis of the I–V relationships of carrier type transport systems that possess just one charge-carrying transmembrane pathway has resulted in an important conclusion (42): Regardless of the actual number of carrier states (Figure 1A), any given I–V relationship can be described in terms of a two-state model (Figure 1B). This model comprises one pair of voltage-sensitive reaction constants and one pair of voltage-insensitive reaction con-stants. The corollary of this conclusion is that there is not enough information in a single I–V relationship to merit construction of a reaction kinetic model that embodies more than two carrier states, even though such states might exist.

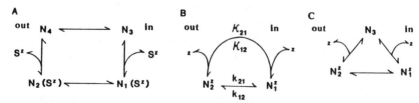

Figure 1 A: Four-state reaction kinetic model for carrier-mediated transport. An ion, S, of valence z reacts with the carrier (N). The four states of the carrier—binding sites unloaded or loaded with S and either facing the interior or the exterior of the membrane—are identified with the subscripts 1 through 4. Binding of S is envisaged as discrete from the transport reaction. Rate constants connecting the carrier states $N_i \rightleftharpoons N_j$ can be designated k_{ij} (forward reaction) and k_{ji} (backward reaction): see text. Note that in this and suceeding figures (Figures 2, 4), the unloaded carrier is arbitrarily depicted as being uncharged, though application of such models to biological transport systems normally takes account of the possibility that transmembrane charge transloca-tion occurs on the unloaded carrier. B: Reduced, pseudo-two-state model, a reduction of a "real" multi-state model. The voltage-insensitive rate constants, which cannot be individually dis-tinguished in any one I–V relationship, are lumped together into the reaction constants labeled κ, with the voltage-sensitive reactions designated by the k reaction constants. C: Expansion of the pseudo-two-state model to a three-state model in which ligand binding to the inside or outside surfaces of the carrier is made explicit.

The current through any transport system in which k_{12} and k_{21} represent the sole charge-carrying reactions can be described as

$$i = zF(N_1 k_{12} - N_2 k_{21}),$$
1.

where N_1 and N_2 are the respective carrier-state densities (units: $mol \cdot m^{-2}$), k_{12} and k_{21} represent the magnitudes of the charge-carrying reaction constants (units: s^{-1}), z is the number of charges (with appropriate sign) carried per turn of the carrier cycle, and F is the Faraday constant. For the steady-state condition ($dN_i/dt = 0$), the current through the so-called pseudo-two-state model can be expressed independently of the carrier states simply as

$$i = zFN \frac{\kappa_{21} k_{12} - \kappa_{12} k_{21}}{\kappa_{21} + \kappa_{12} + k_{12} + k_{21}},$$
2.

where N is the total density of carrier and the reaction constants κ and k are defined as in Figure 1B. Assuming a symmetric Eyring barrier (62), voltage sensitivity can be introduced into the equation as

$$k_{12} = k_{12}^o \exp(zF\Delta\Psi/2RT),$$
3a.

and

$$k_{21} = k_{21}^o \exp(-zF\Delta\Psi/2RT),$$
3b.

in which R and T have their usual meanings. The symmetric Eyring barrier implies a simple potential energy profile across the membrane with a single high-energy barrier located in the center of the membrane dielectric. It should be emphasized that, although this barrier structure is probably the simplest way of incorporating voltage sensitivity into rate equations, the validity of the underlying assumption has yet to be established in the case of plant and fungal transport systems. Alternative models incorporating several local potential energy maxima (61) or a single asymmetric barrier (30) should therefore be explored.

Figure 2 illustrates the I–V characteristics described by Equation 2 for two extreme cases in which the voltage-sensitive (k) reaction constants are either much larger or are much smaller than the voltage-insensitive (κ) constants. In the former case, where the voltage-insensitive limb (Figure 1B) will become rate limiting more rapidly as $\Delta\Psi$ moves from the point of zero current flow, the curve approximates a tanh function, with a limiting slope that is only exceeded if $|z| > 1$. The two cases represented in Figure 2 can be referred to, respectively, as k- and κ-type I–V relationships (36). The saturation currents are simple functions of the κ reaction constants, κ_{21} defining the positive and κ_{12} the negative saturation current. The point at which each curve crosses the voltage axis indicates the reversal potential (E_r) for the transport system—that is, the value of $\Delta\Psi$ at which the direction of transport changes from influx of positive charge (at potentials more negative than E_r) to efflux of positive charge (at potentials positive of E_r). The E_r of primary transport systems lies more negative than the resting $\Delta\Psi$, whereas for gradient-coupled transport, E_r is more positive than $\Delta\Psi$.

It is important to note that E_r is a thermodynamically determined parameter. In the case of the primary pumps, for example, E_r is given as the $\Delta\Psi$ at which the driving force from hydrolysis of phosphoanhydride bonds is balanced by the potential energy stored in $\Delta\bar\mu_{H^+}$. (Full relationships for primary and secondary transport systems at the plasma membrane and tonoplast are given in references 37, 78, 87.) Since E_r can also be calculated from the ratio of the product of the counterclockwise:clockwise reaction constants (see 42), the thermodynamic parameter provides an internal control on the accuracy of reaction constants derived from kinetic modeling. In other words, kinetic models must conform to the dictates of thermodynamics.

Reference to Figure 2 shows that, regardless of the relative values of the k and κ reaction constants, the current through the transport system saturates as a function of either negative or positive applied voltage. Thus, although the carrier catalyzes an electrogenic reaction, transport can, depending on the voltage range over which it is measured, be virtually insensitive to $\Delta\Psi$. The magnitude of the current through the carrier in both regions of voltage

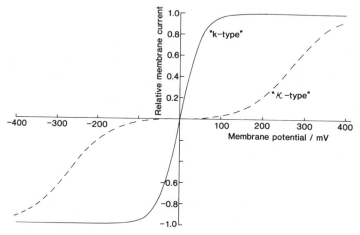

Figure 2 Two extreme forms of I–V relationship for carrier-type transport systems. The relationships were calculated by substituting the following parameter values into Equation 2: $FN = 1$; $z = +1$; $\kappa_{12} = \kappa_{21} = 1$; $k^o_{12} = k^o_{21} = 10^3$ ("k-type" curve) or 10^{-2} ("κ-type" curve). The reversal potentials ($= 0$ mV) and the saturation currents ($= -1$ and $+1$) are identical.

saturation is directly related to the respective voltage-insensitive (κ) reaction constants, which become rate limiting.

Derivation of absolute values of reaction constants is not possible using steady-state reaction kinetic analysis because the density of pump sites (N in Equation 2) is not known a priori. For most purposes, a quantitative assessment of the relative values of the reaction constants is sufficient, and the pump density can be set to some arbitrary value (e.g. 1 pmol·cm^{-2}: see 38).

Effects of Ligands: Higher-State Models

The investigator is normally interested in characterizing the response of transport to a particular ligand, which involves measuring least two I–V relationships for the system. It is here, however, that the pseudo-two-state model is of limited utility. Although this embodiment of multi-state models in a two-state model is legitimate—indeed required if only a single I–V relationship is analyzed—it must be recognized that the carrier will exist in several states. The consequence of model reduction is that the reactions that connect hidden states in the pseudo-two-state model are subsumed not only in the "lumped" voltage-insensitive rate constants (κ_{21}, κ_{12}) but also in the rate constants that describe charge translocation. For example, it can be shown (36, 39, 86) that if a "real" three-state model is reduced to two states, the resultant pseudo-two-state reaction constant k_{12} is a function not only of the three-state reaction constant k_{12} but also of k_{31}, k_{32}, and k_{13} (reaction constants defined as in Figure 1B and 1C). Correspondingly, real changes in the rate of voltage-insensitive reactions (here in a three-state model) can manifest themselves as rate changes in more than one of the reaction constants of a pseudo-two-state model, including, paradoxically, the voltage-sensitive (k) reaction constants. However, working with the assumption that a given ligand reacts at a discrete step in the reaction cycle, and therefore affects the rate of just one of the partial reactions of the carrier, it is normally possible to fit higher state (3- and 4-state) models to I–V data, particularly if a series of I–V relationships is available for a progressive change in ligand concentration. For this purpose, other, invariant, rate constants are statistically optimized, but are held in common for the series of I–V relationships.

The relevant equations for currents carried by higher state models can be derived from expansions of Equation 1 by expressing the carrier state densities $N_1, N_2 \ldots N_n$ solely in terms of N and the component rate constants. This substitution can be achieved simply, even for topologically complex models, with the diagrammatic method of King & Altman (52).

Experimental Applications

Among plants and fungi, the most detailed electrical studies of carrier kinetics have been performed on the plasma membrane H^+-ATPase of the mycelial fungus *Neurospora*. With respect to its protein chemistry and molecular biology, the fungal H^+-ATPase closely resembles that of higher plants (47). The major conclusions of the analysis, which has been performed with the group-fitting strategy (described above) to account for the independent behavior of leak pathways, are as follows:

1. The H^+/ATP stoichiometric ratio (z) is unity, whether estimated from the derived value of E_r or from the quality of the statistically optimized fits of the I–V relationships during ATP depletion by cyanide (38) or during modulation of cytosolic pH (91). This value of z is in accord with direct estimates of pump stoichiometry from quantitative in vitro analysis of H^+ pumping (76).

2. The effects of CN^- can satisfactorily be accounted for in terms of a single reaction constant (κ_{21}) with a pseudo-two-state model (38). None of the other three reaction constants in Figure 1B is capable of describing the response of the pump I–V relationship to cyanide. The experimental finding accords nicely with naive prediction, since it is this lumped reaction constant that can be taken to subsume ATP binding.

3. The values of the two-state reaction constants are estimated (38) in control conditions as: $k_{12}^o = 2.8 \cdot 10^6 \text{ s}^{-1}$; $k_{21}^o = 0.18 \text{ s}^{-1}$; $\kappa_{21} = 350 \text{ s}^{-1}$; $\kappa_{12} = 2.4 \cdot 10^4 \text{ s}^{-1}$. These determinations allow the major site of energy release during forward operation of the pump to be identified as occurring at (or close

to) the voltage-sensitive reaction ($\Delta G = -RT\ln(k_{12}^o/k_{21}^o) = -41$ kJ mol^{-1}). The reaction chemistry of the H$^+$-ATPase is very similar to that of other so-called "P-class" ion-motive ATPases, including the formation of an aspartylphosphoryl (E-P) intermediate (75). Hydrolysis of the E-P intermediate is thought to be the major site of energy transduction among this class of ATPase, and it is therefore interesting to note that kinetic studies on other members of the P-class ATPases have also resulted in the suggestion that hydrolysis of the E-P intermediate is kinetically closely associated with the transport event (103).

4. Based on goodness of fit for carrier-modeled I–V curves during modulation of pump activity by CN$^-$ and pH, the carrier appears to translocate positive charge outwards rather than negative charge inwards (38, 91). Thus, charge translocation appears to coincide with H$^+$ transport rather than with reorientation of empty binding sites. The result has parallels with others on the animal plasma membrane Na$^+$,K$^+$-ATPase, which is also a member of the P class of ATPases. In that case, translocation of 3 Na$^+$ ions outwards— which occurs in a part of the reaction cycle analogous to H$^+$ translocation—is proposed to be accompanied by movement of a single positive charge, whereas reorientation of binding sites loaded with 2 K$^+$ ions is electroneutral (2).

5. The cyanide-induced fall in κ_{21} is by a factor of 2.7, and this has a proportional effect on the pump current at any measured voltage, since, in the $\Delta\Psi$ range from -200 mV to 0 mV, pump current is approximately proportional to κ_{21}. (This point can be appreciated by inspection of Equation 2, since, as long as the membrane is more depolarized than about -200 mV, the numerator is dominated by the term $k_{12}\kappa_{21}$ and the denominator by k_{12} for the numerical values listed above.) Despite this large kinetic effect, the cyanide-induced change in pump E_r amounts only to some 8%. This finding then highlights the possibility that kinetic features of transport systems are likely to be controlled independently from the overall driving forces, such that sensitivity to some ligands (here ATP level) occurs in a homeostatically relevant manner.

6. A relatively small (0.5 unit) decrease in cytosolic pH (pH$_c$) results in two-fold stimulation of pump current at most membrane potentials (91). Again, these marked kinetic effects are accompanied by the predicted minor changes in thermodynamic properties. The effects of pH$_c$ appear on two separate reaction constants (κ_{21} and k$_{12}$) in the two-state model.

7. The effects of external pH (pH$_o$) are in marked contrast to those of changed pH$_c$ and depletion of ATP (106). Major (2 unit) decreases in pH$_o$ result in the anticipated change in pump E_r [around 120 mV (30%) positive as the external pH declines from 5.8 to 3.8]. However, the sensitivity of the pump current to pH$_o$ over the range -160 to 0 mV is small, being less than

two-fold for the 100-fold change in external $[H^+]$. As with analysis of the pump response to pH_c, the effects of pH_o on the pump I-V curve can only be replicated if two (here, κ_{12} and k_{21}) of the four reaction constants in the two-state model are allowed to vary.

8. The effects of both pH_o and pH_c on the pump current can be localized at single rate constants by expanding the model (106). Data sets from the two classes of experiment can be jointly fitted with a four-state model (see Figure 1A) in which the effects of pH_o are manifested only on the rate constant k_{42} and the effects of pH_c only on k_{31}. Evaluation of the ratio of the off:on rate constants for H^+ (i.e. k_{24}/k_{42} and k_{13}/k_{31} for external and internal H^+ binding, respectively, with both on rate constants extrapolated to 1 M H^+) enables the effective pK of the external binding site to be determined as 2.9 and that for the internal binding site as 5.4. These values have a significant impact on our understanding of how the homeostatic demands on the proton pump can be aided quite simply by the pump's reaction kinetics. With pKs significantly lower than the pHs normally encountered on each side of the membrane, the proton pump is able to operate (in a clockwise direction according to the four-state scheme in Figure 1A) with near-maximal theoretical sensitivity to internal pH, but with relative insensitivity to external pH, even at pH 4.

Reaction kinetic analyses of the proton pumps of plant cells have also been performed, albeit in less detail. The I–V relationship of the H^+ pumps of *Chara* and of *Vicia* guard cells has been derived (5, 14–16) from analysis of membrane I–V relationships obtained in the presence of cyanide or of diethylstilbestrol (a presumed pump inhibitor, though subsequently shown to be not entirely specific: 6, 109). Either two-state pump models were fitted to the difference I–V relationships or a group-fitting strategy was adopted. Despite the differences in method used to derive the pump I–V relationship, the conclusions regarding the thermodynamic and kinetic behavior of the pump in each of the plant studies were broadly similar to those derived for *Neurospora*: A model in which $z = 1$ was the only one to generate sufficiently low pump conductances to fit the data; the major site of energy dissipation during the pump cycle is located in the transmembrane charge translocating reactions ($\Delta G = -22$ kJ mol^{-1}): the plant pumps also exhibit pump I–V relationships intermediate between the purely k- or κ-type models of Figure 2. Figure 3 presents a typical example showing a diagnostic conductance ($\equiv$ slope) maximum some way from E_r (5).

Several subtle differences between the *Neurospora* and *Chara* pumps are apparent, however. The kinetic effects of ATP depletion are more complex in *Chara*, being distributed among at least two reaction constants in the psuedo-two-state model (16). The discrepancy appears to arise because the κ reaction constants (Figure 1B) tend to rate-limit carrier cycling to a greater extent in

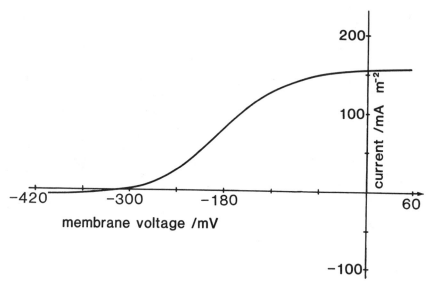

Figure 3   Current-voltage relationship for the plasma membrane H^+ pump of *Chara*. The curve is constructed using Equation 2 from the two-state reaction constants derived by Blatt et al (16) at an external pH = 5.5. Reaction constants are as follows: $N = 2.10^{-8}$ mol·m^{-2}; $k_{21}^0 = 0.6$ s^{-1}; $k_{12}^0 = 4180$ s^{-1}; $\kappa_{21} = 81$ s^{-1}; $\kappa_{12} = 2.6$ s^{-1}.

Neurospora than in *Chara*. Furthermore, a three-state model was insufficient to isolate the effects of pH$_o$ at a single rate constant, though the results from four-state modeling suggested satisfactory fits if H^+ is taken to bind externally at k_{34} (Figure 2C), with k_{42} representing a slow back-reaction separating binding from charge translocation (16). Fusicoccin, an elicitor of stomatal opening, generates a slow increase in k_{12} or, jointly, k_{12} and k_{21} of the H^+ pump of *Vicia* guard cells (15).

Further kinetic modeling of the *Chara* H^+ pump has been undertaken by Kishimoto and colleagues (50, 53, 54, 111). Analysis was performed using either a pseudo-two-state model or a five-state model to derive the pump reaction constants from I–V curves. In contrast to conclusions from work on *Chara* (5, 16) and *Neurospora* (91), the modeling predicted $z = 2$. Although some studies on the *Neurospora* H^+ pump have suggested that occasionally a stoichiometry of $2H^+$ per ATP might occur, particularly in conditions of metabolic restriction (119), it seems more likely that the fundamental discrepancy over the value of z arises from the methods used to dissect the pump I–V relationship from that of the membrane as a whole. Thus Kishimoto and colleagues used dicyclohexylcarbodiimide (DCCD) to inhibit the H^+ pump on the basis that DCCD would deplete cellular ATP levels via its action on mitochondrial and chloroplastic ATP production. However, DCCD also directly inhibits the plasma membrane H^+-ATPase of plant cells (73). Furthermore, the difference I–V relationships (minus, then plus DCCD) were measured as much as 100 min apart. It is possible that during this slow period of stabilization of the membrane electrical parameters, inhibitor-elicited controls acting from within the cell served to adjust the rate of other electrically active transport processes, thereby invalidating the subtraction procedure (see 14). Nevertheless, it is perhaps noteworthy that, in accord with previous findings, the major site of energy dissipation was associated with transmembrane charge transit, and that lowering pH_o results in a predicted increase in κ_{12} and a decrease in k_{21}, with an additional (smaller) contribution from an increase in k_{12} (111).

The primary electrogenic pumps of some marine algae *(Acetabularia, Halicystis)* are specific for Cl^- rather than H^+. The anion is pumped into the cell, and thus, like the H^+ pumps, the Cl^- pumps pass an outward current and tend to hyperpolarize the membrane. The I–V relationship for the pump in *Acetabularia* can be derived by the difference-curve method, using low temperature to inhibit the pump (35). Reaction kinetic analysis of the pump I–V relationship has resulted in a model that contrasts in several respects with those derived for the primary H^+ pumps (36, 38). First, a stoichiometry of two Cl^- per reaction cycle is indicated by the relatively low reversal potential (more positive than -200 mV), by the slope of the I–V relationship (too steep for a value of $z = 1$), and by the quality of the kinetic fits. Second, the shape of the I–V relationship describes a classic k-type function, with $k_{12}^o =$ $9.6 \cdot 10^{-2}$ s^{-1}, $k_{21}^o = 6.8 \cdot 10^5$ s^{-1}, $\kappa_{12} = 420$ s^{-1}, and $\kappa_{21} = 320$ s^{-1}. The near equality of the κ reaction constants results in a fairly symmetric I–V curve. However, inspection of these reaction constants reveals that, as with the H^+ pumps, energy transduction occurs primarily in the gross reaction constants embedding the charge translocation reactions. Third, physiological inhibition of the Cl^- pump by darkness results in a change not in individual reaction constants but in the total apparent number of operational pumps. The same effect results from partial replacement of extracellular Cl^- with other anions such as I^-: it appears, then, that I^- is able to bind to the carrier but is not transported (43).

Mummert et al (71) confirmed the quanitative conclusions of the electrophysiological analysis by showing that pump-mediated Cl^- efflux displays a $\Delta\Psi$-dependence in excellent agreement with that predicted by the two-state reaction constants derived from the I–V relationship.

The effects of extracellular Cl^- were investigated by model expansion to three states (Figure 1C). The best fits were obtained, using a model in which $z = -2$, by allowing k_{32} as the rate constant which, when varied, was able to describe the dependence of the I-V relationships on external $[Cl^-]$. Furthermore, this reaction constant displayed approximately second-order dependence on the Cl^- concentration, as would be anticipated with a stoichiometry of $2Cl^-$ per reaction cycle (40). It was also possible from the analysis to eliminate from consideration a (positively) charged carrier, with Cl^- transport occurring electroneutrally, since with $z = +2$, the changes in external Cl^- concentration would manifest themselves through an effect on the *inner* reaction constants (k_{31} and k_{13}).

Further insights into the reaction kinetics of the Cl^--ATPase have been gained from non-steady-state studies in which a voltage-sensitive component of membrane capacitance (C_p) has been identified as emanating from redistribution of charged states of the electrogenic pump (46, 114). Measurement of C_p, its associated series conductance, and the steady-state slope conductance of the pump enables absolute values to be ascribed to the four pseudo-two-state reaction constants, and hence to pump density (N in Equation 2). The value of N lies between 27 and 60 $nmol \cdot m^{-2}$ ($=$ between $1.6 \cdot 10^4$ and $3.6 \cdot 10^4$ $pumps \cdot \mu m^{-2}$). The rather high estimate of N appears reasonable in the case of *Acetabularia,* for which the existence of a pump contribution to membrane capacitance is unusual. Similarly high transport-system density has been postulated on the basis of charge-pulse relaxation studiers on another marine alga, *Valonia* (8, 121).

Among gradient-coupled transport systems in plants and fungi, the only system to have been rigorously analyzed with respect to its electrical properties is the high-affinity $K^+ - H^+$ symporter of *Neurospora,* which translocates K^+ and H^+ into the cell with a $1:1$ stoichiometric ratio (80). Current through the system obeys Michaelis-Menten kinetics with respect to changes in the external concentrations of the transport substrates K^+ and H^+ (18). Furthermore, whereas increases in either external $[H^+]$ or $[K^+]$ lead to a decrease in the Michaelis constant (K_m) for the other transport substrate (measured under voltage-saturating conditions), only in the case of variable $[K^+]$ were effects on the maximum current (J_{max}) observed. Additional observations (18) of note were that, with the kinetics for one substrate measured at saturating concentrations of the other, membrane hyperpolarization resulted in increases in J_{max} for both transport substrates, accompanied by an increase in K_m for K^+ and a decrease in K_m for H^+.

Three classes of model were considered for charge translocation: a neutral unloaded carrier with translocation of two positive charges occurring on the loaded form of the carrier; a carrier bearing a single negative charge, and translocating inwards a positive charge when loaded; and a carrier bearing two negative charges and translocating no net charge when transporting K^+ and H^+. Since internal ion concentrations were not manipulated in these experiments, the "full" six-state models describing ordered binding of H^+ and K^+ on each side of the membrane can be reduced to four- or five-state models. Within each of the three classes there therefore exist two subclasses of model that describe the order of external binding of K^+ and H^+. On the basis of their kinetic observations and algebraic analysis of model behavior. Blatt et al (18) eliminated four of the six models, since their theoretical behavior could not be reconciled with observation. The two remaining plausible models for topological behavior of the K^+-H^+ symporter were either (a) that in which two positive charges are carried on the loaded form of the carrier with external K^+ binding before H^+ or (b) that in which full charge translocation is on the unloaded form of the carrier and external K^+ binds after H^+. However, even these models were inconsistent with some observations. The latter model, for example, could not satisfactorily explain the observation that increasing external $[K^+]$ decreases the K_m and increases the J_{max} of the system for H^+, since for the K_m effect the lumped ligand dissociation reactions must proceed faster than the external K^+ binding reaction, whereas the J_{max} effect demands the opposite size-ordering of the rate constants. One explanation for the discrepancy may simply be that the current kinetics with respect to external $[H^+]$ are independent of H^+ binding at a transport site, and that to model the effect of $[H^+]$ as though it acts solely on a voltage-insensitive surface reaction is an over simplification. Statistical fitting of the I–V relationships to multi-state models would enable an independent check on the identity of the reaction constant influenced by $[H^+]$.

KINETIC MODELS OF CARRIERS FROM UNIDIRECTIONAL FLUXES

Experimental Approach

It is essential for the purposes of accurate reaction kinetic analysis of unidirectional flux data that the experimental design is one that allows a true initial rate determination. Thus, progressive changes in internal concentration of the ion should not occur during the measurement uptake period, and the effective internal specific activity of radioisotope should be kept at a value negligible with respect to that in the bathing medium. For most higher plant cells, in which concentration changes can occur rapidly in the comparatively small cytosolic compartment, only a measurement lasting less than about 10 min is likely to satisfy the initial rate criterion (24), unless linearity of total uptake with time can be independently demonstrated for longer periods.

Analytical Methods

The smallest number of model states with which it is practicable to describe a unidirectional flux is four, since decrease in specific activity of cis-supplied isotope by unlabeled ions in the trans compartment has to be taken into account. The isotopic flux of S from left to right through a four-state model (see Figure 1A) can be written as

$$*J_S = *N_1 \cdot k_{13}, \qquad\qquad 4.$$

in which $*N_1$ is the concentration of radioisotopically labeled carrier state immediately preceding S release on the *trans* side and k_{13} is the value of the rate constant that describes release of S. Since the specific activity of N_1 is not known a priori, it is more convenient to express Equation 4 in terms of the carrier state that immediately precedes isotope binding—N_4 in the case of the model in Figure 1A. For the steady state

$$\frac{d*N_1}{dt} = 0 = *N_2 k_{21} - *N_1(k_{12} + k_{13}), \qquad\qquad 5a.$$

and

$$\frac{d*N_2}{dt} = 0 = *N_1 k_{12} + N_4 k_{42} *S - *N_2(k_{24} + k_{21}), \qquad\qquad 5b.$$

with the asterisks * denoting radioisotopically labeled forms and k_{42}^0 the rate constant for binding of external $*S$ at a concentration of 1 M. Substituting Equations 5 into Equation 4 results in

$$*J_S = \frac{N_4 *S k_{42}^0 k_{21} k_{13}}{k_{13}(k_{24} + k_{21}) + k_{12}k_{24}}. \qquad\qquad 6.$$

N_4 can then be expressed in terms of N and the eight rate constants of the carrier cycle using the King & Altman (52) method. This expansion results in a relationship between $*J_S$ and $*S$ that invariably has the form of a rectangular hyperbola. The kinetic parameters can therefore be expressed in the conventional Michaelis-Menten formalism as $*J_{max}$ and K_m. *Trans*-stimulation (exchange diffusion) will occur when the return of the unloaded carrier rate-limits progress through the carrier reaction cycle, whereas *trans*-inhibition results if the transmembrane reactions of the loaded carrier are rate limiting.

Gradient-Driven Transport: Flexibility of Ordered Binding Models

In the case of a solute flux coupled to that of H^+, it is useful to expand the four-state model to one comprising at least six states (Figure 4A). The algebraic properties of this model (and that of its three congeners in which binding of H^+ and of S at the two membrane surfaces exhibit alternate orders) have been thoroughly explored (90). As with the four-state model, Michaelis-Menten kinetics will always be observed with respect to variation of *S on the

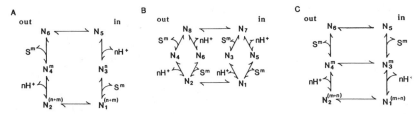

Figure 4   Reaction kinetic models for H^+-coupled transport. A: "First-on-first-off" ordered-binding six-state model for a substrate (S, valence m) and protons (H^+). The charge:reaction cycle stoichiometry ($= z$, Figure 1) is given as ($n + m$). The transport system is envisaged to operate in the counterclockwise direction, bringing S into the cell. B: Generalized random-binding variant of ordered-binding model, in which ligand binding to the carrier is not sterically constrained. C: Ordered-binding model (first-on-last-off) incorporating a slip pathway for (efflux of) S via the carrier state transition $N_3 \rightleftharpoons N_4$.

cis side. Since the vast majority of H^+-coupled transport systems also transport charge (89), the effect of $\Delta\Psi$ on transport must also be taken into account. $\Delta\Psi$ effects can be introduced via Equations 3.

The most noteworthy feature of the simple six-state models concerns the impressive array of kinetic responses that can result from changes of ligand concentration or of $\Delta\Psi$. Thus, the response of each of the four ordered-binding models to an increase in *cis* $[H^+]$ or $\Delta\Psi$ can include a selective decrease in K_m for S, a selective increase in $*J_{max}$, a joint rise in both $*J_{max}$ and K_m, or opposing effects on $*J_{max}$ and K_m. Which of these responses actually occurs in any given instance depends on specific size-ordering of the 12 individual rate constants. The models are equally versatile in describing a range of kinetic effects that result from variation of $[H^+]$ or $[S]$ on the *trans* side. This algebraic versatility presents problems for the experimentalist who is interested in deriving the correct reaction scheme from kinetic data.

A priori assumptions designed to limit flexibility, such as that in which ligand binding is in equilibrium with the carrier as a result of rate-limiting transmembrane reactions, are clearly unwarranted at this stage. Such assumptions can result in incorrect conclusions regarding carrier properties (e.g., regarding binding order, or voltage sensitivity of ligand binding: see 90). Nevertheless, some workers are still using these assumptions (105).

Random Binding and "Dual Isotherms"

The reaction scheme of H^+-coupled transport in Figure 4A is rather restrictive in portraying the binding of H^+ and S as being obligatorily ordered. A more general model for coupled transport, in which binding of H^+ and S is not constrained to any particular order but instead is taken to be functionally random, is shown in Figure 4B. The greater topological complexity of this model in comparison with ordered-binding models results, of course, in even greater kinetic versatility, and it might therefore be asked what justifies extension of the analysis from that of ordered-binding models. In fact, random-binding models display one property of potentially great relevance for studies of solute transport in plants: The dependence of $*J_S$ on external $[S]$ ($[S]_o$) can describe not one but two distinct Michaelian phases (85). It is therefore possible that the molecular basis for the commonly observed biphasic kinetics ("dual isotherms") for transport of many solutes with respect to solute concentration (21) resides in a single transport system—a point made some years ago in the specific context of transport of sugars and amino acids (79). A careful kinetic analysis of H^+ and SO_4^{2-} symport by the red-blood-cell Cl^- transporter has already established the feasibility of random-order mechanisms in transport (70).

Several other hypotheses have also been advanced to explain the wide occurrence of dual isotherm kinetics for transport of solutes at the plasma membrane of plant cells. However, those invoking control of transport at more than one membrane (57) or unstirred layers (26) can be discounted, at least as general explanations, by the observations that dual isotherms are also observed in isolated, evacuolate cells (51; see also 67, 68, 115). Gerson & Poole (31) have considered ionic effects on $\Delta\Psi$ as a possible determinant of dual isotherms. Although this explanation might be relevant in some cases (18), it also appears unlikely to be generally applicable: E_r of most H^+-coupled transport systems is significantly positive of values of $\Delta\Psi$ normally encountered in plant cells, so that transport becomes effectively voltage insensitive (27, 45, 94).

An alternative reaction kinetic explanation for dual isotherm kinetics (56) envisages a separate "slip" pathway for the binary carrier-ion complex, as well as the pathway for the ternary form that includes H^+ (Figure 4C). Indeed, the slip model has also found favor in the context of regulation of transport, since the existence of a slip pathway would prevent the gradient-coupled pathway from producing undesirably high levels of solute within the cell (25). However, this regulatory characteristic can also be attained with models in which coupling between the H^+ and ion flows is tight, since *trans*-inhibition can effectively halt net flow through the transport system even in the presence of a substantial driving force (90, 120).

The random binding and slip hypotheses as explanations for dual isotherms can in principle be experimentally discriminated. First, the slip hypothesis predicts that net translocation of the ion against its electrochemical gradient should not occur as a result of operation of the low-affinity/high-velocity phase. There is, however, an isolated report that Cl^- accumulation by mung bean root tips occurs well into the range of the low-affinity phase (32). Second, transport current can be compared with net flux of S (J_S). The random-binding model requires a fixed ratio between J_S and carrier current for all values of $[S]_o$, whereas the slip model predicts that the ratio will decline at higher $[S]_o$. To date, the relevant experiments have not been performed. However, both random binding and slip models can be distinguished kinetically from the trivial alternative in which parallel operation of two carriers generates the two isotherms. The single-carrier models predict that, as the external $[H^+]$ is raised, the two isotherms should merge into just one high-affinity isotherm (85). In the sole case investigated in detail, that of sugar transport in *Chlorella* (56), a single saturable function does result at low external pH (85).

Experimental Applications

As discussed above, the behavior of even the simplest reaction kinetic model for gradient-coupled transport is so versatile that, at first sight, solution of the inverse problem—that of using experimental data to derive a unique model in terms of transport substrate binding order and site of charge translocation—might appear to be impossible. However, two circumstances, if applied jointly, enable realization of this goal. The first experimental condition is access to both sides of the membrane. The second condition is one of experimental simplification of conditions. Briefly, the rate equations can be simplified considerably, with resultant restriction of their kinetic flexibility, by application of two or more of the following conditions: *trans* $[S]$ set to zero; *cis* $[H^+]$ saturating; $\Delta\Psi$ saturating (90).

Investigation of the form of the carrier (loaded or unloaded with substrate) responsible for translocation of charge provides an example of the value of this simplified approach. If net positive charge is translocated on the loaded form of the carrier, than $*J_{max}$ will fall towards zero as the *trans* $[H^+]$ is raised, providing *trans* $[S]$ is zero and $\Delta\Psi$ is saturating. In contrast, for transport systems carrying macroscopic current in the same direction but via translocation of negative charge on the unloaded form of the carrier, $*J_{max}$ plateaus at a stable value. These considerations apply regardless of whether binding of H^+ and S is ordered or random (85).

Naturally, most plant material does not readily afford manipulation of solute composition on each side of the membrane, although inaccessibility to the cytosolic phase does not necessarily preclude a reaction kinetic approach (18). Indeed, general statements concerning the site of charge translocation are possible from simple studies during the approach to a steady state (net flux zero).

The free energy stored in plasma membrane $\Delta\bar{\mu}_{H^+}$ and available for gradient-coupled transport is considerable (87, 94). H^+-coupled transport systems must therefore function far from equilibrium in order to prevent accumulation of unphysiologically high levels of solutes via symport (94). Kinetic control of transport appears to be exerted by cytosolic solute concentrations, and Sauer et al (96) have drawn attention to two different system-specific modes of control during the approach to a steady state. In Mode I, typified by amino acid transport systems as well as that for Cl^- (see below), accumulation of solute results in straightforward *trans*-inhibition of unidirectional influx. Mode II, common among sugar transport systems, exhibits progressively increasing exchange diffusion as solutes are accumulated, with the result that although net influx declines to a value close to zero, unidirectional influx is actually stimulated. A simple reaction kinetic interpretation that distinguishes the two modes of control is that charge is carried on the loaded form of the carrier in Mode I but on the unloaded form in Mode II (34, 79, 94). Thus, in the former case, the influence of negative $\Delta\Psi$ forces k_{21} to be much larger than k_{12} (Figure 4A), thereby effectively preventing efflux. In the latter case, negative charge is carried outwards between carrier states 5 and 6 (Figure 4A) with a resultant high ratio for k_{56}/k_{65}. Solute efflux is therefore possible and will increase as cytosolic solute concentration rises, though the only kinetic pathway for recycling of binding sites to the cytosolic side is after they are loaded with extracellular solute.

Since the net result is identical regardless of the mode by which kinetic control is achieved, it is apposite to ask why two mechanisms exist. One possibility is simply that an ancestral sugar transport system, but not its amino acid counterpart, possessed a negative charge at the H^+ binding site. The argument is supported by recent studies suggesting a common evolutionary origin for some bacterial, animal (3), and plant (97) sugar transport systems.

Several techniques are available that, by allowing experimental control on both sides of a membrane, will enable us to undertake full reaction kinetic analysis of gradient-coupled transport systems. Patch clamp, carried out in whole-cell recording mode, will readily allow exchange of the pipette contents with those of the cytosolic phase (102), though to date the method has not been applied for the study of gradient-coupled transport systems. Membrane vesicles also permit experimental control of solution composition on both sides of the membrane, and transport kinetic data are beginning to emerge from work on tonoplast preparations (12, 19, 20, 77, 101). The most detailed studies on plasma membrane have, however, been undertaken on intracellularly perfused, tonoplast-free preparations of Characean algae.

Cl^- influx in *Chara* is H^+ coupled (stoichiometric ratio $2 H^+ : 1 Cl^-$; see 7, 82, 84) and obeys Michaelis-Menten kinetics (88). The flux is inhibited by elevated cytoplasmic $[Cl^-]$ (81) and $[H^+]$, and is stimulated by raised external $[H^+]$ (82). Kinetically, cytoplasmic Cl^- and H^+ are noncompetitive inhibitors of influx, measured in conditions in which $\Delta\Psi$ and external $[H^+]$ can both be taken as saturating in their effects on transport.

There exist eight models that may reasonably describe the reaction kinetics of H^+-coupled Cl^- transport: two classes describing the form of the carrier on which charge is carried, with each class comprising the four possible permutations of transport substrate binding order on either side of the membrane. The properties of these models have been analyzed (88, 90); only two models are competent to describe the observation that cytoplasmic Cl^- is a noncompetitive inhibitor of transport. In both models, Cl^- binds to the carrier before H^+ externally, and charge translocation is on the loaded form of the carrier; the models are distinguished by the order in which ligand dissociation occurs internally. The same two models are also capable of replicating the observation that cytoplasmic H^+ behaves similarly as a noncompetitive inhibitor. However, only the one in which Cl^- leaves the carrier first at the inner surface is competent to describe the observation that internal H^+ is a noncompetitive inhibitor even in the absence of internal Cl^-.

Three additional and independent observations are predicted by the unique first-on-first-off model (88). First, the K_i for internal Cl^- (actual value 0.5–2.0 mM, depending on cytoplasmic $[H^+]$) is greater than the K_m for transport of external Cl^- (about 40 μM), despite the fact that the dissociation constants for internal and external binding sites are given identical values in the model, thus representing vectorial rearrangement of a single site. Second, the K_i for cytoplasmic Cl^- is predicted to increase with decreasing cytoplasmic $[H^+]$, and this was observed. Third, the model is able to replicate the impressive sensitivity of Cl^- influx to changes in cytoplasmic pH: 10- to 20-fold stimulation of flux for a mere 0.75-unit increase in pH (88, 95).

The simple first-on-first-off reaction scheme succeeds not only in accounting quantitatively for the observed kinetics, but also demonstrates how homeostatic control of transport can be achieved at the elementary level of reaction kinetics. The quantitative parameters in the reaction scheme, as well as possibly the scheme itself, have evidently been subject to selection that enables the cell to control the levels of cytosolic Cl^- and H^+ in a quasi-steady manner. Thus, for example, the second-order sensitivity of the flux to cytosolic pH does not follow axiomatically from the simple fact that two H^+ are translocated. Rather, the value of the pK_a (at 7.85, a little above the normal cytosolic pH) is critical. This sensitivity to cytosolic pH is likely to be of importance in sustaining transport at high external pH where supply of H^+ as substrate for the transport system is obviously limited, since the slight elevation of cytosolic pH in these conditions releases the *trans*-inhibitory effects normally exerted by cytosolic $[H^+]$ (92).

Some other H^+-coupled transport systems in plants and fungi also display a marked sensitivity to cytosolic pH (4, 55), and it appears likely that the homeostatic significance of these observations can be explained on a similar reaction kinetic basis (4, 90).

It is apparent, then, that simple *trans*-inhibition by transport system substrates should be thoroughly investigated before invoking more complex control at allosteric sites. Glass (33) has pointed out that, since *trans*-inhibition is presumed to occur via binding to the transport site, the model taken in isolation will not explain commonly reported phenomena in which solute uptake through a transport system with narrow solute specificity appears to be under the control of intracellular levels of a far broader range of ions. However, an obvious point is frequently ignored in many such studies on higher plants: The intracellular concentrations measured are primarily representative of those prevailing in the vacuole. Thus, as discussed previously (83), it is conceivable that simple *trans*-inhibition models for control of transport at the plasma membrane could apply so long as more general control of transport (acting via cytosolic ion concentrations) resides at the tonoplast. Until cytosolic ion concentrations are measured directly in these nutritional experiments, the point will remain unresolved.

KINETIC MODELS OF IONIC CHANNELS

Unitary Currents

Ionic channels in plant membranes all exhibit a degree of ionic selectivity (48, 113). This implies the presence of ionic binding sites, and with it the existence of kinetic saturation phenomena (49). Indeed, some years ago, Läuger (60) pointed out that, although different in detail, the kinetic characteristics of both carriers and channels can be treated with an identical formalism. Accordingly, a unifying rate-theory approach has been advocated (59–61). Noting the tendency of unitary currents through plant channels to saturate as a function of $\Delta\Psi$ (66, 100), Gradmann et al (39) have developed a general framework for the analysis of unitary currents by classical carrier-type models of the kind illustrated in Figure 1.

A change in $[K^+]$ on just one side of the membrane results in an effect on both the negative and positive saturation currents of a plasma membrane K^+ channel in *Vicia faba* guard cells (39). This dual response of the saturation currents is compatible only with a model in which K^+ binds to a neutral carrier. Quantitatively, the I–V data are described well by a three-state model in which the effect of $[K^+]$ is taken to be solely on its binding reaction, although the effect of $[K^+]$ on this reaction is somewhat less than first order. Unitary current studies and reaction kinetic modeling of a K^+ channel in the plasma membrane of *Acetabularia* have revealed that the relative magnitudes of the three-state reaction constants are remarkably similar to those derived for the guard cell K^+ channel, with ligand binding and dissociation being the fastest reactions in both systems (10). However, the absolute magnitudes of the reaction constants are larger by factors of between two and five in *Acetabularia*.

In some conditions, particularly at high external $[K^+]$, the electrical properties of plant plasma membranes are dominated by a K^+ conductance (K state: see 6). In these circumstances, it is sometimes possible to apply a modeling methodology to the study of membrane I–V relationships with little or no correction for other electrically active pathways, though, of course, unitary currents are not measured. Fisahn et al (28) have demonstrated that the electrical behavior of the plasma membrane of *Nitella* in the K state is also described excellently by a three-state model in which external K^+ binding is explicit: The results are in accord with studies on *Vicia* and *Acetabularia* in suggesting that charge translocation occurs simultaneously with ion transport and that K^+ binding occurs rapidly in comparison with translocation reactions. At concentrations of K^+ between 1 mM and 100 mM, the rate constant subsuming K^+ binding (k_{32} in Figure 1B) displays a linear dependence on $[K^+]$, thus suggesting that a single ion is translocated for each reaction cycle. However, it should be noted that studies on *Chara* (108) have failed to find voltage saturation of plasma membrane K^+ currents that are diagnostic of the reaction kinetic models used by Fisahn et al (28).

Further studies on the guard cell channel, in which K^+ was substituted by Na^+ on each side of the membrane, have enabled derivation of a six-state model in which Na^+ and K^+ modes of the channel (each comprising binding and transport reactions) are connected via reorientation of the unloaded transport site (39). Channel selectivity for K^+ over Na^+ appears to be achieved through larger rate constants in the K^+ mode than in the Na^+ mode, even though the apparent dissociation constant for K^+ is larger than for Na^+. At zero voltage, all transmembrane reactions are poised close to equilibrium, and the major sites of energy transduction at physiological concentrations of ions occur in the binding reactions, which are poised in favor of the uncomplexed carrier on each side of the membrane. A similar study in a K^+ channel in the tonoplast of *Chara* (9) has revealed that Na^+ is not measurably translocated at all, though it is able to inhibit the K^+ current by virtue of binding. An analogous six-state model was derived for the *Chara* channel, but with the added complication that the Na^+ binding reactions are markedly sensitive to voltage. This might be taken to imply that the Na^+ binding site is positioned within the membrane dielectric, in agreement with the conclusions from studies showing voltage-dependent blockage of K^+ currents in *Chara* by Cs^+ (112).

For both the guard cell and *Chara* K^+ channels functioning solely in the presence of K^+, the slowest reaction constants derived are those for "reorientation" of the empty binding site. A resultant prediction is that the channel should be competent to catalyze exchange diffusion, at least in the region of 0 mV where the asymmetry in the charge translocation reaction constants is minimal. Radiometric studies to test this proposal have not been performed for either of the channels, though a detailed study of K^+ transport at the plasma membrane of *Chara* failed to find any evidence for flux interaction (107, 108).

An alternative (and more conventional) approach for modeling the kinetics of the *Chara* tonoplast K^+ channel has been elegantly developed by Laver et al (64). The observations that underlie the analysis are qualitatively identical to those used for reaction kinetic modeling by Bertl (9): The unitary current saturates as a function of both voltage and K^+ concentration. The model developed by Laver et al envisages a pore through which ions flow essentially electrodiffusively, but which nevertheless contains an effective binding site for K^+. At high K^+ concentration, current flow through the pore is limited by saturation of pore occupancy by K^+, just as K^+ binding sites are envisaged to do in the reaction kinetic model. However, saturation of current flow at extreme voltages is explained in the pore model by the additional constraint that diffusion to the mouth of the pore becomes limiting, whereas the reaction kinetic model invokes slow "reorientation" of binding sites as an explanation for the same observation. Sucrose (which reduces the diffusion coefficient of K^+) inhibits unitary current flow, and this observation has been taken to support the diffusion-limited pore model (64).

With respect to explaining the concentration dependence of currents, the pore and reaction kinetic models are really variants on the same theme, since "pore occupancy" can be defined by a voltage-insensitive equilibrium constant (58). However, despite finding a common phenomenological K_m for K^+ (in the region 30–70 mM in the two studies), other differences between the models that relate to binding site equilibria yield markedly divergent estimates for the dissociation constant between the channel and K^+: around 50 mM for the diffusion-limited pore, and 8–16 M [sic] for the reaction kinetic model.

Gating

Quantitative studies on channel transitions between open and closed states are in their infancy in plant biology.

K^+ currents across the plasma membrane of plant cells generally show outward rectification—that is, in the range of membrane potentials more positive than the equilibrium potential for K^+, the efflux of K^+ rises as a steeper function of voltage than does K^+ influx at more negative potentials. How is this type of I–V relationship generated from essentially sigmoid unitary I–V relationships? Bertl & Gradmann (10) analyzed a simple model for the outwardly rectifying K^+ channel in *Acetabularia* in which the channel exists in an open state (O), a closed, activated state (C_1) and a closed, inactivated state (C_2), with interconversions among the states occurring in the linear sequence $C_2 \rightleftharpoons C_1 \rightleftharpoons O$. The transition between states C_1 and C_2 is taken to be voltage sensitive, with positive voltage favoring the formation of C_1, and hence channel opening. Unidirectional rate constants can be derived for the state transition reactions, either from non-steady-state measurements of macroscopic currents or from probability density functions of steady-state single channel data (11). The model provides a good quantitative description of the steady-state I–V relationship for K^+ at the plasma membrane of *Acetabularia*. It is predicted that, even at the most positive voltages, the channel exists for only around 50% of the time in the O State, the remaining time being spent in the C_1 State. The existence of closed states of the K^+ channel in the plasma membrane of *Nitella* has also been inferred from the effects of temperature on the channel I–V relationship (41), although in this particular case there is no evidence for voltage gating of the channel.

The gating properties of the outward rectifier of stomatal guard cells have been modeled from macroscopic (whole cell) currents by Schroeder (99), principally with respect to the kinetics of voltage-sensitive activation and inactivation. Activation is a sigmoidal function of time, and can be well fitted over a range of $\Delta\Psi$ with a Hodgkin-Huxley-type model involving two gating particles. A linear three-state model for gating (as for the *Acetabularia* K^+ channel) was proposed, although the voltage sensitivity was embedded in rate constants describing the $C_1 \rightleftharpoons 0$ transition as well as the $C_2 \rightleftharpoons C_1$ reaction.

Schroeder (99) has noted that the three-state gating model is a minimal one. It does not, for example, account for the observation (17, 98) that gating of the outward rectifier is dependent on external K^+ concentration, as well as voltage. Furthermore, although the three-state model describes kinetic features of macroscopic K^+ currents well, more complex kinetics are apparent at the level of single channels (99). Analysis of mean open lifetimes of the tonoplast K^+ channel from *Chara* has revealed just such a complex pattern (65). The channel exhibits the greatest propensity for opening between -100 and -150 mV. Although the mean channel open time can be fitted with a single exponential function (indicating the presence of just one open state), as many as four exponentials are required to fit the closed-time distributions in the hyperpolarizing (-250 mV) direction, and three exponentials for the depolarization-induced closure (at $+50$ mV). Since each exponential can be taken to represent the existence of a closed state, the kinetics indicates the presence of four closed states on the hyperpolarizing side and three closed states on the depolarizing side of the open state, respectively. As a minimum model, the eight states were taken to be arranged in series, and values for the 14 state transition rate constants were identified on the basis of open and closed state frequency distributions. Voltage sensitivity was ascribed to the two rate constants leading from the open-state as well as to one closed-state transition on either side of the open state. Overall, the channel could be characterized as closing at hyperpolarizing potentials as a result of a decrease in the mean open time, and as closing at more positive (physiological) potentials as a result of the increased lifetime of closed states.

CONCLUDING REMARKS

An ultimate goal of research in membrane transport is a physical description of the molecular mechanism and control of transport. Knowledge of the structure of native and engineered transport proteins will constitute an essential component of this description. A full understanding will only result, however, when structure can be related to function, as described by kinetic models. In this regard, significant insights into gating mechanisms of ionic channels in animal cells are now emerging from the combined approaches of in vitro mutagenesis and kinetic analysis. Channel inactivation can be mediated by discrete, identifiable domains of the channel protein (1).

A kinetic framework is now in place for functional description of all major classes of plant transport system: primary pumps, gradient-coupled systems, and channels. In many cases refinement of kinetic models is necessary. For example, all reaction kinetic models for carrier-type kinetics to date have introduced voltage dependence via the assumption of a symmetric Eyring barrier (Equations 3). This assumption must be utilized more critically and (ideally) tested, particularly before more detailed kinetic analysis can proceed. It is now thought likely that ion transport in the light-driven H^+ pump bacteriorhodopsin proceeds through a series of rather small energy maxima and minima (binding sites) provided by the protein (22), rather than across just one barrier. It is possible that non-steady-state kinetic approaches (62, 72) will clarify the problem of transmembrane energy barriers in plant transport systems.

Many of the crucial pilot studies described here have been carried out on "model" systems: those that can be intracellularly perfused (giant-celled algae), or that do not exhibit extensive vacuolation or intercellular connections (fungi, microalgae, guard cells). We now have techniques for extending these detailed studies to higher plants, either through the use of suspension cultures (for which membrane electrical currents can be quantified on a surface-area basis: see 74), or via the application of patch clamp to protoplasts (which facilitates measurement of specific membrane currents in conditions of more defined cytosolic composition: see 48). Despite some uncertainties over details, modeling performed thus far suggests that most plant and fungal transport systems can at least be described quantitatively in terms of relatively simple kinetic models. The challenge now will be to relate the rate constants of such models to the molecular dynamics of transport systems.

ACKNOWLEDGMENTS

I am very grateful to Mike Blatt, Julia Davies, Phil Rea, and Mark Tester for constructive and critical comments on an earlier version of the manuscript, and to Dietrich Gradmann, Peter Hansen, and Clifford Slayman for many patient and productive hours spent discussing reaction kinetics during the past decade. Work in my laboratory in this area has been supported by the Agricultural and Food Research Council.

Literature Cited

1. Aldrich, R. W., Zagotta, W. N., Hoshi, T. 1989. Biophysical and molecular aspects of voltage-dependent potassium channel gating. In *Smith, Kline and French Symposium on Ion Transport*, ed. G. H. Poste, T. J. Rink. Orlando: Academic. In press
2. Bahinski, A., Nakao, M., Gadsby, D. C. 1988. Potassium translocation by the Na^+/K^+ pump is voltage insensitive. *Proc. Natl. Acad. Sci. USA* 85:3412–16
3. Baldwin, S. A., Henderson, P. J. F. 1989. Homologies between sugar transporters from eukaryotes and prokaryotes. *Annu. Rev. Physiol.* 51:459–71
4. Ballarin-Denti, A., den Hollander, J. A., Sanders, D., Slayman, C. W., Slayman, C. L. 1984. Kinetics and pH-dependence of glycine-proton symport in *Saccharomyces cerevisiae. Biochim. Biophys. Acta* 778:1–16
5. Beilby, M. J. 1984. Current-voltage characteristics of the proton pump at *Chara* plasmalemma: I. pH dependence. *J. Membr. Biol.* 81:113–25
6. Beilby, M. J. 1986. Factors controlling the K^+ conductance in *Chara. J. Membr. Biol.* 93:187–93
7. Beilby, M. J., Walker, N. A. 1981. Chloride transport in *Chara* I. Kinetics and current-voltage curves for a probable proton symport. *J. Exp. Bot.* 32:43–54
8. Benz, R., Zimmermann, U. 1983. Evidence for the presence of mobile charges in the cell membrane of *Valonia utricularis. Biophys. J.* 43:13–26
9. Bertl, A. 1989. Current-voltage relationships of a sodium-sensitive potassium channel in the tonoplast of *Chara corallina. J. Membr. Biol.* 109:9–19
10. Bertl, A., Gradmann, D. 1987. Current-voltage relationships of potassium channels in the plasmalemma of *Acetabularia. J. Membr. Biol.* 99:41–49
11. Bertl, A., Klieber, H.-G., Gradmann, D. 1988. Slow kinetics of a potassium channel in *Acetabularia. J. Membr. Biol.* 102:141–52
12. Blackford, S., Rea, P. A., Sanders, D. 1990. Role of H^+/Ca^{2+} antiport at higher plant tonoplast in vacuolar calcium accumulation. *J. Biol. Chem.* Submitted
13. Blatt, M. R. 1986. Interpretation of steady-state current-voltage curves: consequences and implications of current subtraction in transport studies. *J. Membr. Biol.* 92:91–110
14. Blatt, M. R. 1987. Electrical characteristics of stomatal guard cells: the contribution of ATP-dependent, "electrogenic" transport revealed by current-voltage and difference-current-voltage analysis. *J. Membr. Biol.* 98:257–74
15. Blatt, M. R. 1988. Mechanisms of fusicoccin action: a dominant role for secondary transport in a higher-plant cell. *Planta* 174:187–200
16. Blatt, M. R., Beilby, M. J., Tester, M. 1990. Voltage-dependence of the *Chara* proton pump revealed by current-voltage measurement during rapid metabolic blockade with cyanide. *J. Membr. Biol.* In press
17. Blatt, M. R., Clint, G. M. 1989. Mechanisms of fusicoccin action: kinetic modification and inactivation of K^+ channels in guard cells. *Planta* 178:509–23
18. Blatt, M. R., Rodriguez-Navarro, A., Slayman, C. L. 1987. Proton-potassium symport in *Neurospora:* kinetic control by pH and membrane potential. *J. Membr. Biol.* 98:169–89
19. Blumwald, E., Poole, R. J. 1985. Na^+/H^+ antiport in isolated tonoplast vesicles from storage tissue of *Beta vulgaris. Plant Physiol.* 78:163–67
20. Blumwald, E., Poole, R. J. 1986. Kinetics of Ca^{2+}/H^+ antiport in isolated tonoplast vesicles from storage tissue of *Beta vulgaris* L. *Plant Physiol.* 80:727–31
21. Borstlap, A. C. 1983. The use of model-fitting in the interpretation of "dual" uptake isotherms. *Plant Cell Environ.* 6:407–16
22. Braiman, M. S., Mogi, T., Marti, T., Stern, L. J., Khorana, H. G., Rothschild, K. J. 1988. Vibrational spectroscopy of bacteriorhodopsin mutants: light-driven proton transport involves protonation changes of aspartic acid res-

idues 85, 96, and 212. *Biochemistry* 27:8516–20

23. Chapman, J. B., Johnson, E. A., Kootsey, J. M. 1983. Electrical and biochemical properties of an enzyme model of the sodium pump. *J. Membr. Biol.* 80:405–24

24. Cram, W. J. 1973. Chloride fluxes in cells of the isolated root cortex of *Zea mays*. *Aust. J. Biol. Sci.* 26:757–79

25. Eddy, A. A. 1980. Slip and leak models of gradient-coupled transport. *Trans. Biochem. Soc. Lond.* 8:271–73

26. Ehwald, R., Meshcheryakov, A. B., Kholodova, V. P. 1979. Hexose uptake by storage parenchyma of potato and sugar beet at different concentrations and different thicknesses of tissue slices. *Plant Sci. Lett.* 16:181–88

27. Felle, H. 1983. Driving forces and current-voltage characteristics of amino acid transport in *Riccia fluitans*. *Biochim. Biophys. Acta* 730:342–50

28. Fisahn, J., Hansen, U.-P., Gradmann, D. 1986. Determination of charge, stoichiometry and reaction constants from I–V curve studies on a K^+ transporter in *Nitella*. *J. Membr. Biol.* 94:245–52

29. Fisahn, J., Mikschl, E., Hansen, U.-P. 1986. Separate oscillations of a K^+-channel and of a current-source in *Nitella*. *J. Exp. Bot.* 37:34–47

30. Gadsby, D. C., Nakao, M. 1989. Steady-state current-voltage relationship of the Na/K pump in guinea pig ventricular myocytes. *J. Gen. Physiol.* 95:511–37

31. Gerson, D. F., Poole, R. J. 1971. Anion absorption by plants. A unary interpretation of "dual mechanisms". *Plant Physiol.* 48:509–11

32. Gerson, D. F., Poole, R. J. 1972. Chloride accumulation by mung bean root tips. A low affinity active transport system at the plasmalemma. *Plant Physiol.* 50:603–7

33. Glass, A. D. M. 1983. Regulation of ion transport. *Annu. Rev. Plant Physiol.* 34:311–26

34. Gogarten, J. P., Bentrup, F.-W. 1989. The electrogenic proton/hexose carrier in the plasmalemma of *Chenopodium rubrum* suspension cells: effects of Δc, ΔpH and $\Delta \Psi$ on hexose exchange diffusion. *Biochim. Biophys. Acta* 978:43–50

35. Gradmann, D. 1975. Analog circuit of the *Acetabularia* membrane. *J. Membr. Biol.* 25:183–208

36. Gradmann, D. 1984. Electrogenic Cl^- pump in the marine alga *Acetabularia*. In *Chloride Transport Coupling in Biological Membranes and Epithelia*, ed. G. A. Gerencser, pp. 13–61. Amsterdam: Elsevier. 451 pp.

37. Gradmann, D., Hansen, U.-P., Long, W. S., Slayman, C. L., Warncke, J. 1978. Current-voltage relationships for the plasma membrane and its principal electrogenic pump in *Neurospora crassa*: I. Steady-state conditions. *J. Membr. Biol.* 29:333–67

38. Gradmann, D., Hansen, U.-P., Slayman, C. L. 1982. Reaction-kinetic analysis of current-voltage relationships for electrogenic pumps in *Neurospora* and *Acetabularia*. *Curr. Top. Membr. Transp.* 16:257–76

39. Gradmann, D., Klieber, H.-G., Hansen, U.-P. 1987. Reaction kinetic parameters for ion transport from steady-state current-voltage curves. *Biophys. J.* 51:569–85

40. Gradmann, D., Tittor, J., Goldfarb, V. 1982. Electrogenic Cl^- pump in *Acetabularia*. *Philos. Trans. R. Soc. Lond. Ser. B* 299:447–57

41. Hansen, U.-P., Fisahn, J. 1987. I/V-curve studies of the control of a K^+ transporter in *Nitella* by temperature. *J. Membr. Biol.* 98:1–13

42. Hansen, U.-P., Gradmann, D., Sanders, D., Slayman, C. L. 1981. Interpretation of current-voltage relationships for "active" ion transport systems: I. Steady-state reaction-kinetic analysis of Class I mechanisms. *J. Membr. Biol.* 63:165–90

43. Hansen, U.-P., Gradmann, D., Tittor, J., Sanders, D., Slayman, C. L. 1982. Kinetic analysis of active transport: reduction models. In *Plasmalemma and Tonoplast: Their Functions in the Plant Cell*, ed. D. Marmé, E. Marré, R. Hertel, pp. 77–84. Amsterdam: Elsevier. 446 pp.

44. Hansen, U.-P., Kolbowski, J., Dau, H. 1987. Relationship between photosynthesis and plasmalemma transport. *J. Exp. Bot.* 38:1965–81

45. Hansen, U.-P., Slayman, C. L. 1978. Current-voltage relationships for a clearly electrogenic cotransport system. In *Membrane Transport Processes*, ed. J. F. Hoffman, 1:141–54. New York: Raven

46. Hansen, U.-P., Tittor, J., Gradmann, D. 1983. Interpretation of current-voltage relationships for "active" ion transport systems: II. Nonsteady-state reaction kinetic analysis of Class-I mechanisms with one slow time constant. *J. Membr. Biol.* 75:141–69

47. Harper, J. F., Surowy, T. K., Sussman,

M. R. 1989. Molecular cloning and sequence of cDNA encoding the plasma membrane proton pump (H^+-ATPase) of *Arabidopsis thaliana. Proc. Natl. Acad. Sci. USA* 86:1234–38

48. Hedrich, R., Schroeder, J. I. 1989. The physiology of ion channels and electrogenic pumps in higher plants. *Annu. Rev. Plant Physiol. Plant Mol. Biol.* 40:539–69

49. Hille, B. 1984. *Ionic Channels of Excitable Membranes.* Sunderland, Mass: Sinauer Associates. 426 pp.

50. Kami-Ike, N., Ohkawa, T., Kishimoto, U., Takeuchi, Y. 1986. A kinetic analysis of the electrogenic pump of *Chara corallina:* IV. Temperature dependence of pump activity. *J. Membr. Biol.* 94:163–71

51. Kannan, S., 1971. Plasmalemma: the seat of dual mechanisms of ion absorption in *Chlorella pyrenoidosa. Science* 173:927–29

52. King, E. L., Altman, C. 1956. A schematic method of deriving the rate laws for enzyme-catalysed reactions. *J. Phys. Chem.* 60:1375–78

53. Kishimoto, U., Kami-Ike, N., Takeuchi, Y., Ohkawa, T. 1984. A kinetic analysis of the electrogenic pump of *Chara corallina:* I. Inhibition of the pump by DCCD. *J. Membr. Biol.* 80:175–83

54. Kishimoto, U., Takeuchi, Y., Ohkawa, T., Kami-Ike, N. 1985. A kinetic analysis of the electrogenic pump of *Chara corallina:* III. Pump activity during the action potential. *J. Membr. Biol.* 86:27–36

55. Komor, E., Schwab, W. G. W., Tanner, W. 1979. The effect of intracellular pH on the rate of hexose uptake in *Chlorella. Biochim. Biophys. Acta* 555:524–30

56. Komor, E., Tanner, W. 1975. Simulation of a high- and low-affinity sugar uptake system in *Chlorella* by a pH-dependent change in the K_m of the uptake system. *Planta* 123:195–98

57. Laties, G. G. 1969. Dual mechanisms of salt uptake in relation to compartmentation and long-distance transport. *Annu. Rev. Plant Physiol.* 20:89–116

58. Läuger, P. 1973. Ion transport through pores: a rate-theory analysis. *Biochim. Biophys. Acta* 311:423–41

59. Läuger, P. 1979. A channel mechanism for electrogenic pumps. *Biochim. Biophys. Acta* 552:143–61

60. Läuger, P. 1980. Kinetic properties of ion carriers and channels. *J. Membr. Biol.* 57:163–78

61. Läuger, P. 1984. Thermodynamic and kinetic properties of electrogenic pumps. *Biochim. Biophys. Acta* 779:307–41

62. Läuger, P., Apell, H.-J. 1988. Transient behaviour of the Na^+/K^+-pump: microscopic analysis of nonstationary ion-translocation. *Biochim. Biophys. Acta* 944:451–64

63. Läuger, P., Stark, G., 1970. Kinetics of carrier-mediated ion transport across lipid bilayer membranes. *Biochim. Biophys. Acta* 211:458–66

64. Laver, D. R., Fairley, K. A., Walker, N. A. 1989. Ion permeation in a K^+ channel in *Chara australis:* direct evidence for diffusion limitation of ion flow in a maxi-K channel. *J. Membr. Biol.* 108:153–64

65. Laver, D. R., Walker, N. A. 1987. Steady-state voltage-dependent gating and conduction kinetics of single K^+ channels in the membrane of cytoplasmic drops of *Chara australis. J. Membr. Biol.* 100:31–42

66. Lühring, H. 1986. Recording single K^+ channels in the membrane of cytoplasmic drop of *Chara australis. Protoplasma* 133:19:28

67. McDaniel, C. N., Lyons, R. A., Blackman, M. S. 1981. Amino acid transport in suspension-cultured plant cells: IV. Biphasic saturable uptake kinetics of L-leucine in isolates from six *Nicotiana tabacum* plants. *Plant Sci. Lett.* 23:17–23

68. Mettler, I. J., Leonard, R. T. 1979. Ion transport in isolated protoplasts from tobacco suspension cells: II. Selectivity and kinetics. *Plant Physiol.* 63:191–94

69. Moczydlowski, E. 1986. Single-channel enzymology. In *Ion Channel Reconstitution,* ed. C. Miller, pp. 75–113. New York: Plenum. 577 pp.

70. Milanick, M. A., Gunn, R. B. 1982. Proton-sulfate co-transport: mechanism of H^+ and sulfate addition to the chloride transporter of human red blood cells. *J. Gen. Physiol.* 79:87–113

71. Mummert, H., Hansen U.-P., Gradmann, D. 1981. Current-voltage curve of electrogenic Cl^- pump predicts voltage-dependent Cl^- efflux in *Acetabularia. J. Membr. Biol.* 62:139–48

72. Nakao, M., Gadsby, D. C. 1986. Voltage dependence of Na translocation by the Na/K pump. *Nature* 323:628–30

73. Oleski, N. A., Bennett, A. B. 1987. H^+-ATPase activity from storage tissue of *Beta vulgaris:* IV. *N,N'*-dicyclohexylcarbodiimide binding and inhibition of the plasma membrane H^+-ATPase. *Plant Physiol.* 83:569–72

74. Parsons, A., Sanders, D. 1989. Electri-

cal properties of soybean plasma membrane measured in heterotrophic suspension cells. *Planta* 177:499–510

75. Pedersen, P. L., Carafoli, E. 1987. Ion motive ATPases. Part I. Ubiquity, properties and significance to cell function. *Trends Biochem. Sci.* 12:146–50

76. Perlin, D. S., San Francisco, M. J. D., Slayman, C. W., Rosen, B. P. 1986. H$^+$/ATP stoichiometry of proton pumps from *Neurospora crassa* and *Escherichia coli*. *Arch. Biochem. Biophys.* 248:53–61

77. Pope, A. J., Leigh, R. A. 1988. The use of a chloride-sensitive fluorescent probe to measure chloride transport in isolated tonoplast vesicles. *Planta* 176:451–60

78. Rea, P. A., Sanders, D. 1987. Tonoplast energization: two H$^+$ pumps, one membrane. *Physiol. Plant.* 71:131–41

79. Reinhold, L., Kaplan, A. 1984. Membrane transport of sugars and amino acids. *Annu. Rev. Plant Physiol.* 35:45–83

80. Rodriguez-Navarro, A., Blatt, M. R., Slayman, C. L. 1986. A potassium-proton symport in *Neurospora crassa*. *J. Gen. Physiol.* 87:649–74

81. Sanders, D. 1980. Control of Cl$^-$ influx in *Chara* by cytoplasmic Cl$^-$ concentration. *J. Membr. Biol.* 52:51–60

82. Sanders, D. 1980. The mechanism of Cl$^-$ transport at the plasma membrane of *Chara corallina*: I. Cotransport with H$^+$. *J. Membr. Biol.* 53:129–42

83. Sanders, D. 1981. Physiological control of chloride transport in *Chara corallina* II. The role of chloride as a vacuolar osmoticum. *Plant Physiol.* 68:401–6

84. Sanders, D. 1984. Gradient-coupled chloride transport in plant cells. In *Chloride Transport Coupling in Biological Membranes and Epithelia*, ed. G. A. Gerencser, pp. 63–120. Amsterdam: Elsevier. 451 pp.

85. Sanders, D. 1986. Generalized kinetic analysis of ion-driven cotransport systems: II. Random ligand binding as a simple explanation for non-Michaelian kinetics. *J. Membr. Biol.* 90:67–87

86. Sanders, D. 1988. Steady-state kinetic analysis of chemiosmotic proton circuits in microorganisms. In *Physiological Models in Microbiology*, ed. M. J. Bazin, J. I. Prosser, 1:49–74. Boca Raton: CRC Press. 139 pp.

87. Sanders, D., Davies, J. D., Rea, P. A. 1990. Gradient-coupled transport in plants. *Physiol. Plant.* In press

88. Sanders, D., Hansen, U.-P. 1981. Mechanism of Cl$^-$ transport at the plasma membrane of *Chara corallina*. II:

Transinhibition and the determination of H$^+$/Cl$^-$ binding order from a reaction kinetic model. *J. Membr. Biol.* 58:139–53

89. Sanders, D., Hansen, U.-P., Gradmann, D. 1990. Electrical properties of plant membranes. *Biochim. Biophys. Acta.* In press

90. Sanders, D., Hansen, U.-P., Gradmann, D., Slayman, C. L. 1984. Generalized kinetic analysis of ion-driven cotransport systems: a unified interpretation of selective ionic effects on Michaelis parameters. *J. Membr. Biol.* 77:123–52

91. Sanders, D., Hansen, U.-P., Slayman, C. L. 1981. Role of the plasma membrane proton pump in pH regulation in non-animal cells. *Proc. Natl. Acad. Sci. USA* 78:5903–7

92. Sanders, D., Hopgood, M., Jennings, I. R. 1989. Kinetic response of H$^+$-coupled transport to extracellular pH: critical role of cytosolic pH as a regulator. *J. Membr. Biol.* 108:253–61

93. Sanders, D., Slayman, C. L. 1989. Transport at the plasma membrane of plant cells: a review. In *Membrane Transport in Plants. Proceedings of the 8th International Workshop on Membrane Transport in Plants*. Amsterdam: Elsevier. In press

94. Sanders, D., Slayman, C. L., Pall, M. L. 1983. Stoichiometry of H$^+$/amino acid cotransport in *Neurospora crassa* revealed by current-voltage analysis. *Biochim. Biophys. Acta* 735:67–76

95. Sanders, D., Smith, F. A., Walker, N. A. 1985. Proton/chloride cotransport in *Chara*: mechanism of enhanced influx after rapid acidification. *Planta* 163:411–18

96. Sauer, N., Komor, E., Tanner, W. 1983. Regulation and characterization of two inducible transport systems in *Chlorella vulgaris*. *Planta* 159:404–10

97. Sauer, N., Wolf, K., Schnelbögl, G., Tanner, W. 1989. Cloning of the inducible H$^+$/glucose cotransporter from *Chlorella kessleri*. In *Membrane Transport in Plants. Proceedings of the 8th International Workshop on Membrane Transport in Plants*. Amsterdam: Elsevier. In press

98. Schroeder, J. I. 1988. Potassium transport properties of the plasma membrane of *Vicia faba* guard cells. *J. Gen. Physiol.* 92:667–83

99. Schroeder, J. I. 1989. Quantitative analysis of outward rectifying K$^+$ channel currents in guard cell protoplasts from *Vicia faba*. *J. Membr. Biol.* 107:229–35

100. Schroeder, J. I., Hedrich, R., Fernan-

dez, J. M. 1984. Potassium-selective single channels in guard cell protoplasts of *Vicia faba*. *Nature* 312:361–62

101. Schumaker, K. S., Sze, H. 1985. A Ca^{2+}/H^+ antiport system driven by the proton electrochemical gradient of a tonoplast H^+-ATPase from oat roots. *Plant Physiol.* 79:1111–17

102. Serrano, E., Zeiger, E., Hagiwara, S. 1988. Red light stimulates an electrogenic proton pump in *Vicia* guard cell protoplasts. *Proc. Natl. Acad. Sci. USA* 85:436–40

103. Serrano, R. 1988. Structure and function of proton translocating ATPase in plasma membranes of plants and fungi. *Biochim. Biophys. Acta* 947:1–28

104. Serrano, R. 1989. Structure and function of plasma membrane ATPase. *Annu. Rev. Plant Physiol. Plant Mol. Biol.* 40:61–94

105. Severin, J., Langel, P., Höfer, M., 1989. Analysis of the H^+/sugar symport in yeast under conditions of depolarized plasma membrane. *J. Bioenerg. Biomembr.* 21:321–34

106. Slayman, C. L., Sanders, D. 1985. Steady-state kinetic analysis of an electroenzyme. In *The Molecular Basis of Movement Through Membranes*, ed. P. J. Quinn, C. A. Pasternak, pp. 11–29. London: Biochem. Soc. 267 pp.

107. Smith, J. R. 1987. Potassium transport across the membranes of *Chara* II. ^{42}K fluxes and the electrical current as a function of membrane voltage. *J. Exp. Bot.* 38:752–77

108. Smith, J. R., Smith, F. A., Walker, N. A. 1987. Potassium transport across the membranes of *Chara*. I. The relationship between radioactive tracer influx and electrical conductance. *J. Exp. Bot.* 38:731–51

109. Smith, J. R., Walker, N. A., Smith, F. A. 1987. Potassium transport across the membranes of *Chara*. III. Effects of pH, inhibitors and illumination. *J. Exp. Bot.* 38:778–87

110. Stein, S., Hansen, U.-P. 1988. Involvement of photosynthesis in the action of temperature on plasmalemma transport in *Nitella*. *J. Membr. Biol.* 103:149–58

111. Takeuchi, Y., Kishimoto, U., Ohkawa, T., Kami-Ike, N. 1985. A kinetic analysis of the electrogenic pump of *Chara*

corallina. II. Dependence of the pump activity on external pH. *J. Membr. Biol.* 86:17–26

112. Tester, M. 1988. Potassium channels in the plasmalemma of *Chara corallina* are multi-ion pores: voltage-dependent blockade by Cs^+ and anomalous permeabilities. *J. Membr. Biol.* 105:87–94

113. Tester, M. 1990. Plant ion channels: whole cell and single channel studies. *New Phytol.* In press

114. Tittor, J., Hansen, U.-P., Gradmann, D. 1983. Impedance of the electrogenic Cl^- pump in *Acetabularia*: electrical frequency entrainments, voltage-sensitivity, and reaction kinetic interpretations. *J. Membr. Biol.* 75:129–39

115. van Bel, A. J. E., Borstlap, A. C., van Pinxteren-Bazuine, A., Ammerlaan, A. 1982. Analysis of valine uptake by *Commelina* mesophyll cells in a biphasic active and a diffusional component. *Planta* 155:335–41

116. Vanselow, K. H., Dau, H., Hansen, U.-P. 1988. Indication of transthylakoid proton-fluxes in *Aegopodium podagraria L.* by light-induced changes of plasmalemma potential, chlorophyll fluorescence and light-scattering. *Planta* 176:351–61

117. Vanselow, K. H., Hansen, U.-P. 1989. Rapid light effect on the K^+ channel in the plasmalemma of *Nitella*. *J. Membr. Biol.* 110:175–87

118. Vanselow, K. H., Kolbowski, J., Hansen, U.-P. 1989. Further evidence for the relationship between light-induced changes of plasmalemma transport and of transthylakoid proton uptake. *J. Exp. Bot.* 40:239–45

119. Warncke, J., Slayman, C. L. 1980. Metabolic modulation of stoichiometry in a proton pump. *Biochim. Biophys. Acta* 591:224–33

120. Wright, J. K. 1989. Product inhibition during ion:solute cotransport is an alternative to leaks as a cause of ion accumulations. *J. Membr. Biol.* 109:1–8

121. Zimmermann, U., Büchner, K.-H., Benz, R. 1982. Transport properties of mobile charges in algal membranes: influence of pH and turgor pressure. *J. Membr. Biol.* 67:183–97

Annu. Rev. Plant Physiol. Plant Mol. Biol. 1990. 41:109–25

GENETICS AND MOLECULAR BIOLOGY OF ALTERNATIVE NITROGEN FIXATION SYSTEMS

Paul E. Bishop and Rolf D. Joerger

Department of Microbiology and United States Department of Agriculture, Agricultural Research Service, North Carolina State University, Raleigh, North Carolina 27695-7615

KEY WORDS: *vnf* genes, *anf* genes, nitrogenase, molybdenum, vanadium

CONTENTS

INTRODUCTION

The aerobic soil bacterium *Azotobacter vinelandii* is able to reduce atmospheric nitrogen to ammonia. The process of N_2 fixation has been studied for many years in this diazotroph but it was not realized until recently that this organism harbors three genetically distinct nitrogenase complexes (17, 38, 52). One of these enzyme complexes is the well-characterized, conventional

109

molybdenum-containing nitrogenase (nitrogenase 1). Nitrogenase 1 is only expressed when bacteria are grown in medium containing molybdenum (Mo). The enzyme has two components: dinitrogenase reductase 1 (also called component II or Fe protein) and dinitrogenase 1 (also designated component I or MoFe protein). Dinitrogenase reductase 1, which serves as an electron donor to dinitrogenase 1, is a dimer of two identical subunits with a M_r of approximately 60,000. A single [4Fe-4S] cluster is bridged between the two subunits (27). Dinitrogenase 1 is a tetramer with a M_r of about 220,000; it is made up of two pairs of nonidentical subunits (α and β). Dinitrogenase 1 contains two types of metal centers involved in the redox reactions of the N_2 reduction process: P centers that might be organized as four unusual 4Fe-4S clusters (21, 46), and two identical FeMo cofactors (FeMoco) that are almost certainly the sites for N_2 binding and reduction (67).

Nitrogenase 2 is a vanadium-containing enzyme complex synthesized when bacteria are grown in N-free medium lacking Mo but containing vanadium (V) (28, 29, 59). This enzyme complex consists of two components, dinitrogenase reductase 2, a dimer of two identical subunits, and dinitrogenase 2, now thought to be a hexamer (M_r of about 240,000) of two dissimilar pairs of large subunits (α and β) and a pair of small subunits (δ) (61). Dinitrogenase reductase 2 has a M_r of about 62,000 and contains four Fe atoms and four acid-labile sulfide groups per dimer (28). Dinitrogenase 2 contains two V atoms, 23 Fe atoms, and 20 acid-labile sulfide groups per molecule (23). A cofactor (FeVaco) analogous to FeMoco has been extracted from dinitrogenase 2 using N-methylformamide (1, 68).

Nitrogenase 3 does not appear to contain either Mo or V; it is made under Mo- and V-deficient conditions (17). This enzyme is composed of two components, dinitrogenase reductase 3 and dinitrogenase 3. Dinitrogenase reductase 3 is a dimer (M_r of approximately 65,000) of two identical subunits, while dinitrogenase 3 is a tetramer (M_r of about 216,000) composed of two dissimilar pairs of subunits (α and β). Dinitrogenase 3 may actually be a hexamer, however, because the structural gene operon for nitrogenase 3 contains an ORF (open reading frame) that may encode a protein similar to the δ subunit of dinitrogenase 2 (38). Dinitrogenase reductase 3 contains four Fe atoms and four acid-labile sulfide groups per dimer. Dinitrogenase 3 contains approximately 24 Fe atoms and 18 acid-labile sulfide groups per molecule; however, it lacks significant amounts of Mo or V. It is also interesting that dinitrogenase 3 can be isolated in at least two active configurations; $\alpha_2\beta_2$ and $\alpha_1\beta_2$.

Nitrogenase 2 and nitrogenase 3 were unknown prior to 1980. Until that time it was generally thought that Mo was absolutely required for N_2 fixation, even though scattered reports indicated that some diazotrophs could grow slowly in N-free medium lacking Mo. These low rates of N_2 fixation were usually attributed to the incorporation of trace amounts of contaminating Mo into nitrogenase (for a more detailed account of these observations, see 36). The early work of Bortels (11) on *Azotobacter* species established the enhancement of N_2 fixation upon the addition of small amounts of Mo to growth media. Bortels (12) also showed that low concentrations of V stimulated the growth of *Azotobacter* species under diazotrophic conditions. For many years it seemed possible that V could substitute for Mo in dinitrogenase 1, because many of the chemical properties of these two metals are similar (16, 45). By the mid 1970s it appeared that stimulation by V might be explained by other hypotheses such as the incorporation of V into dinitrogenase 1 with consequent stabilization of the enzyme and a more effective utilization of the small amount of Mo found in Mo-starved cells (3).

In the early 1980s we presented evidence indicating that *A. vinelandii* contained at least two nitrogenase systems; the conventional Mo-containing nitrogenase system and an alternative nitrogenase system expressed in the absence of Mo. The evidence was built around the core observation that Nif⁻ (unable to fix N_2) mutant strains underwent phenotypic reversal (i.e. Nif⁻ to Nif⁺) under conditions of Mo deprivation. These reports (8, 9, 50, 54) were received with skepticism because they challenged the long-held beliefs that Mo was absolutely required for nitrogen fixation and that nitrogenases were essentially the same regardless of their source. The latter notion was further bolstered by the results of Southern blot experiments by Ruvkun & Ausubel (62) which indicated that structural genes encoding nitrogenases from diverse diazotrophic organisms were highly conserved at the nucleotide sequence level.

Because the Nif⁻ strains of *A. vinelandii* first used to demonstrate phenotypic reversal contained point mutations, it was considered possible that phenotypic reversal resulted from increased leakiness of the mutant phenotypes under conditions of Mo starvation and not from derepression of an alternative N_2-fixation system. This hypothesis was ruled out when mutant strains carrying deletions in the structural genes for nitrogenase 1 were unequivocally shown to undergo phenotypic reversal under Mo-deficient conditions (7, 10, 58). Strains with deletions in the structural genes (*nifHDK*) have also facilitated the isolation of nitrogenase 2 from *A. chroococcum* (59) and nitrogenases 2 and 3 from *A. vinelandii* (17, 28, 29).

In this review we focus on the rapidly emerging genetics and molecular biology of the alternative nitrogen fixation systems. For other reviews, see references 22, 24, 36, 42, 51.

BRIEF OVERVIEW OF THE GENETICS OF MOLYBDENUM-CONTAINING NITROGENASES

The genetics of Mo-containing nitrogenases are best understood in the free-living diazotroph, *Klebsiella pneumoniae*. In this organism the 20 genes known to be involved in N_2 fixation (*nif* genes) are clustered in a 24.2-kbp region of the genome (2). The function of many of the products of these genes is poorly understood. Eighteen of these 20 *nif* genes are known to have counterparts in *A. vinelandii* (33). The sequential arrangement of the *A. vinelandii nif* genes follows that of *K. pneumoniae;* however, many ORFs are interspersed among the *nif* genes of *A. vinelandii*. It should also be noted that linkage of the *nifA-nifB* region to the major *nif* gene cluster has not yet been established in *A. vinelandii*.

The sequences of structural genes encoding the nitrogenase subunits are known for many diazotrophs. In *A. vinelandii* as in *K. pneumoniae,* these genes are organized in a single operon, *nifHDKTY* (Figure 1, System 1) (33). The subunits for dinitrogenase reductase 1 are encoded by *nifH;* the α-subunit of dinitrogenase 1 is encoded by *nifD;* and the β-subunit of dinitrogenase 1 is encoded by *nifK*. The functions of the products of *nifT* and *nifY* are unknown and are apparently dispensable for diazotrophic growth by *A. vinelandii* in the presence of Mo (33).

Genes known to be involved in the synthesis of FeMoco (the cofactor that can be extracted from dinitrogenase 1) are: *nifE, nifN, nifV, nifB,* and *nifQ* (65). Dinitrogenase reductase 1 (encoded by *nifH*) is also required for the synthesis and insertion of active cofactor (25, 57). Just how all of these gene products are involved in synthesis of FeMoco (and insertion into apo-dinitrogenase 1) is not well understood, but this problem is under intensive investigation (65). It has been speculated that NifE and NifN proteins may form a complex that serves as a scaffold for the synthesis of FeMoco (14, 18). *nifV* apparently encodes a homocitrate synthase that catalyzes the synthesis of homocitrate which in turn is incorporated into FeMoco during its formation (30, 65). The *nifQ* gene product plays a role in the transformation of Mo for insertion into FeMoco (32). The function of the NifB protein is not known, but it has been speculated that it is involved in the early steps of FeMoco synthesis (2). The *nifM* gene product is required for maturation of dinitrogenase reductase (31). In *K. pneumoniae, nifLA* forms an operon whose gene products regulate the expression of the other *nif* transcriptional units in response to nitrogen and oxygen status (47). *A. vinelandii* also possesses a *nifA* gene, but the presence of a gene resembling *nifL* has not yet been demonstrated (4). The *nifF* gene encodes a flavodoxin that is the immediate electron donor to dinitrogenase reductase in *K. pneumoniae* (19, 49, 66). In *A. vinelandii* NifF$^-$ mutants are phenotypically Nif$^+$ with respect to N_2 fixation; thus the role of the *nifF* gene product in *A. vinelandii* remains uncertain (5, 44). Functions for *nifY, nifX, nifU, nifS, nifW, nifT,* and *nifZ* remain undetermined, as are the roles played by the 14 ORFs interspersed within and among *nif* operons in the sequenced regions of the *A. vinelandii* genome (33).

System 1

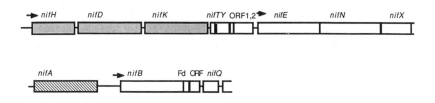

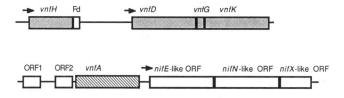

System 2

System 3

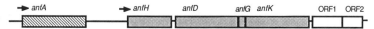

Figure 1 Organization of some genes involved in the three nitrogenase systems of *A. vinelandii*. Systems 1, 2, and 3 refer to genes required for nitrogenases 1, 2, and 3, respectively. Arrows indicate promoters and direction of transcription.

IDENTIFICATION AND ORGANIZATION OF GENES REQUIRED FOR MOLYBDENUM-INDEPENDENT NITROGENASES

nif *Genes Shared by the Three Nitrogenase Systems in* Azotobacter vinelandii

Two *nif* genes are known to be shared by the three nitrogenase systems. One of these is *nifB*, because NifB⁻ mutants (strains CA30 and UW45) are unable to grow under Mo-sufficient, Mo-deficient plus V, and Mo-deficient diazotrophic conditions (35, 39). In the wild type a *nifB*-hybridizing transcript

(4 kb in size) is also observed under all three diazotrophic conditions (35). It is interesting that these transcripts also include *nifQ*, which is not required for growth in Mo-deficient plus V or Mo-deficient N-free medium. Although the role that NifQ might play during growth under the latter conditions is unknown, we have previously speculated that NifQ could make the expression of nitrogenases 2 and 3 less sensitive to repression by trace amounts of Mo. This speculation is founded on the observation that *nifQ* mutations (and mutations polar to *nifQ*) cause strains to exhibit lag periods longer than those seen with NifQ$^+$ strains during diazotrophic growth (35). Because NifB is known to be required for FeMoco synthesis, it is probably safe to assume that this gene product plays an analogous role in the formation of FeVaco [the V-containing cofactor of dinitrogenase 2 (68)]. In the case of dinitrogenase 3, NifB must be involved in some function other than one that relates to Mo or V. At this time it is not certain that native dinitrogenase 3 has a cofactor; however, if it does, it would be surprising if the cofactor contained a metal other than iron. The other *nif* gene shared by the three nitrogenase systems is *nifM*. NifM is required for maturation of dinitrogenase reductase 1, and a NifM$^-$ mutant (strain MV21) was unable to grow diazotrophically in either the presence or absence of Mo or V (41). This is not surprising in view of the high degree of similarity between dinitrogenase reductases 1, 2, and 3 at the amino acid sequence level (Table 1).

In future studies other genes will undoubtedly be found that are shared

Table 1 Sequence comparisons between *nif*, *vnf*, and *anf* genes of *A. vinelandii*[a]

Gene comparisons	Percent nucleotide sequence identity	Percent amino acid sequence identity of predicted gene products
nifH × *vnfH*	88.5	91.0
nifH × *anfH*	69.3	62.8
vnfH × *anfH*	70.1	63.5
nifD × *vnfD*	52.6	33.0
nifD × *anfD*	49.6	32.7
vnfD × *anfD*	65.8	54.4
vnfG × *anfG*	55.9	39.8
nifK × *vnfK*	51.5	31.1
nifK × *anfK*	50.7	32.1
vnfK × *anfK*	69.8	57.4
nifD × *nifK*	—	19.8
vnfD × *vnfK*	—	24.5
anfD × *anfK*	—	21.5

[a] The values for *vnfH*, *vnfD*, *vnfG*, and *vnfK* are derived from unpublished sequence data (R. Joerger, T. Loveless, R. Pau, L. Mitchenall, and P. Bishop). The sources of sequence data for *nif* and *anf* genes are references 33 and 38, respectively.

among the three nitrogenase systems or between any two systems. This has already been observed with some regulatory genes—e.g. *nfrX* (63) and *ntrD* (C. Kennedy, personal communication) are required for growth under Mo-sufficient or Mo-deficient conditions. Now that *A. vinelandii* strains carrying mutations in most of the *nif* genes are available (33), testing of these strains for diazotrophic growth under Mo-deficient conditions in the presence and absence of V should indicate whether or not other *nif* genes are shared with the alternative Mo-independent systems.

Genes Encoding Nitrogenase 2

The structural genes encoding dinitrogenase 2 and dinitrogenase reductase 2 have been cloned, sequenced, and mutagenized for both *A. chroococcum* (60, 61) and *A. vinelandii* (55; R. Joerger, T. Loveless, R. Pau, L. Mitchenall, and P. Bishop, unpublished results). Robson et al (61) have designated these genes as *vnf* (*v*anadium *n*itrogen *f*ixation). In contrast to the single operon (*nifHDK*) encoding the subunits for nitrogenase 1, the genes encoding the nitrogenase 2 proteins are split between two operons (Figure 1). *vnfH* encodes the dinitrogenase reductase 2 subunits and is part of a two-gene operon. The ORF 3' to *vnfH* encodes a ferredoxin-like protein that has not been ascribed a function. *vnfH* is preceded by a potential promoter sequence that would be predicted to interact with core RNA polymerase containing the sigma 54 factor (*ntrA, rpoN,* or *glnF* gene product). The *vnfDGK* operon, located 1.0 kbp *(A. vinelandii)* or 2.5 kbp *(A. chroococcum)* downstream from the *vnfH-Fd* operon, encodes the subunits for dinitrogenase 2. *vnfD* encodes the α-subunit and *vnfK* encodes the β-subunit. In *A. vinelandii* the 1.0-kbp region between the *vnfH-Fd* and *vnfDGK* operons does not appear to contain any identifiable ORFs (R. Joerger, T. Loveless, R. Pau, L. Mitchenall, and P. Bishop, unpublished results).

The third subunit, δ, (M_r of 13,274) is encoded by *vnfG* (61). This gene does not have a counterpart in the nitrogenase 1 system (System 1 in Figure 1); but it does exhibit some sequence similarity to *anfG*, an ORF located between *anfD* and *anfK* (Figure 1). Whether or not the δ subunit is required for full activity of dinitrogenase 2 is presently unknown. The *vnfDGK* genes appear to be cotranscribed, and transcription is NH_4^+ repressible. The largest and most abundant transcript is 3.4 kb long; the two transcripts present in lesser amounts are 1.9 and 1.7 kb long (61).

The removal of a 1.4-kbp *Bgl*II fragment, which spans all of *vnfG* and the 3' and 5' ends of *vnfD* and *vnfK*, from the genomes of both *A. chroococcum* (61) and *A. vinelandii* (52) results in deletion strains that lack dinitrogenase 2. When this deletion was transferred to *A. chroococcum* strain MCD1155 (carrying a deletion of the structural genes for nitrogenase 1), the resulting double deletion strain was unable to grow under any N_2-fixing condition. This result indicates that *A. chroococcum* does not contain a third nitrogenase (61).

A similar double deletion strain of *A. vinelandii* (strain RP206) grew in N-free media lacking Mo (52). This finding provided genetic evidence for the expression of a third nitrogenase (nitrogenase 3) in *A. vinelandii* that lacks Mo and V; it also supported previously published results describing the isolation and partial characterization of nitrogenase 3 (17).

Genes Encoding Nitrogenase 3

The structural genes encoding nitrogenase 3 in *A. vinelandii* have been cloned, sequenced, and mutagenized (38). These genes have been designated *anf* (*alternative nitrogen fixation*) and they are organized in a single operon, *anfHDGK*,ORF1,ORF2 (Figure 1). The subunits of dinitrogenase reductase 3 are encoded by *anfH* while the α and β subunits of dinitrogenase 3 are encoded by *anfD* and *anfK*, respectively. *anfG* probably encodes a third subunit (δ) for dinitrogenase 3 (R. Pau, personal communication). The *anfHDGK* operon is preceded by a potential promoter sequence that would interact with RNA polymerase containing sigma 54. The predicted protein products of the two ORFs that are located 3' to *anfK* (Figure 1) do not show overall similarity to any *nif* gene products. However, the predicted ORF1 product contains some sequence identity to the NH_2-terminal part of dinitrogenase reductase, and another region exhibits identity to presumed heme-binding domains of P-450 cytochromes. The predicted product of ORF2 does not show significant similarity to other amino acid sequences in the Bionet data base (38). Deletions and insertions placed in several regions of the *anfHDGK* operon resulted in Anf⁻ mutants that were unable to grow in N-free, Mo-deficient medium. Growth in media containing Mo or V was normal (38). The absence of nitrogenase 3 proteins in these Anf⁻ mutants was also confirmed by two-dimensional gel electrophoresis (38). Transcription from the *anfHDGK*,ORF1,ORF2 operon results in NH_4^+-repressible transcripts that are 6.6, 4.2, and 2.6 kb long (R. Premakumar, M. Jacobson, and P. Bishop unpublished results).

Sequence Comparisons Between Structural Genes

In Table 1, sequence comparisons are shown for the structural genes and their presumed protein products. Overall sequence identity is greater for *nifH*, *vnfH*, and *anfH* and their products (dinitrogenase reductases) than for the genes encoding the three dinitrogenases. Five cysteine residues are conserved across the three dinitrogenase reductase proteins, and two of these conserved residues appear to serve as ligands for the Fe-S center which is bound symmetrically between the dinitrogenase reductase 1 subunits (38). A motif characteristic of nucleotide binding domains (Gly-X-Gly-XX-Gly) is also present in the three gene products (38). The percent identity between *nifH* and *vnfH* is high (88.5%) and suggests that these two genes may have diverged relatively recently (in evolutionary time) from an ancestral gene or that functional constraints are similar for the two dinitrogenase reductases. In contrast, *anfH* seems to be more distantly related to *nifH* and *vnfH* (about 70% identity). This correlates well with the inability of dinitrogenase reductase 3 to yield high nitrogenase activity in a complementation assay with dinitrogenase 1. In contrast, dinitrogenase reductase 2 produces high activity in the same assay (17).

The *vnfDK* and *anfDK* genes are more similar to each other than either set of genes is to *nifDK*, with *vnfK* and *anfK* sharing slightly more identity (69.8%) than *vnfD* and *anfD* (65.8%). The products of *nifDK*, *vnfDK*, and *anfDK* contain cysteine and histidine residues that are conserved in nearly all dinitrogenase proteins thus far examined. These highly conserved Cys and His residues are thought to be coordinating ligands for Fe-S centers. Some of these Cys residues are essential for dinitrogenase activity, as demonstrated by site-directed mutagenesis experiments (13).

As previously mentioned, *vnfG* and *anfG* show identity with respect to both nucleotide sequence (55.9%) and amino acid sequences of the presumed products (39.8%). This may indicate that these gene products function as subunits of their respective dinitrogenases in a similar fashion.

Finally, the percentage identity between amino acid sequences of the α and β subunits of each dinitrogenase (Table 1) suggests that the genes encoding these two subunits may have evolved from a common ancestral gene, as previously suggested for Mo-containing dinitrogenases (70). Based on these identity comparisons it has been speculated that *nifD* and *nifK* may have diverged somewhat earlier during evolution than the genes encoding the subunits of dinitrogenases 2 and 3 (38, 61).

An observation that may relate to the occurrence of alternative nitrogenases in other organisms is the finding that predicted products of *nifH*-like genes from two very different diazotrophs exhibit a high degree of identity with the *anfH* product (Table 2). One of these is NifH3 (81.7% identity) from the obligate anaerobe *Clostridium pasteurianum* (Cp), and the other is NifH1 (72.4% identity) from the thermophilic archaebacterium *Methanococcus thermolithotrophicus* (Mt). It remains to be seen, however, whether or not these identities signify functional similarity to the nitrogenase 3 system. In the case of Cp *nifH3*, no evidence for transcription could be found under Mo-sufficient diazotrophic conditions (72). Thus it would be interesting to look for expression of Cp *nifH3* transcripts under Mo-deficient conditions. Although the predicted product of Mt *nifH1* shows a fairly high degree of identity (72.4%) with the *anfH* product, it is clear that the Mt *nifD* product shows much less identity (38.3%); therefore it can be concluded that the degree of identity between amino acid sequences of dinitrogenase reductase proteins does not necessarily correlate with the percentage identity observed for the α and β

Table 2 Sequence comparisons of predicted products of *nif*, *vnf*, and *anf* genes from *A. vinelandii* and of *nif* genes from *Methanococcus thermolithotrophicus* and *Clostridium pasteurianum*[a]

Gene product comparison	Percent amino acid identity
Av NifH × Cp NifH3	63.2
Av NifH × Mt NifH1	63.4
Av VnfH × Cp NifH3	62.3
Av VnfH × Mt NifH1	63.0
Av AnfH × Cp NifH3	81.7
Av AnfH × Mt NifH1	72.4
Cp NifH3 × Mt NifH1	70.3
Av NifD × Mt NifD	40.4
Av VnfD × Mt NifD	41.5
Av AnfD × Mt NifD	38.3

[a] Source of sequence: *A. vinelandii*, Av (see footnote for Table 1); *Clostridium pasteurianum*, Cp (72); and *Methanococcus thermolithotrophicus*, Mt (69).

subunits of different dinitrogenases (also see Table 1). This higher degree of variability is also observed with the conventional Mo-containing dinitrogenases (21).

Nonstructural Genes

Recently two *nifA*-like ORFs were identified in DNA cloned from *A. vinelandii*. One of these ORFs was recognized in the DNA sequence flanking the Tn*5* insertion carried by a Vnf⁻ mutant (strain CA46) that is unable to express nitrogenase 2 when derepressed in N-free medium containing V (37, 39). Because this mutant synthesizes both nitrogenases 1 and 3 under Mo-deficient N-free conditions in the presence or absence of V, the *nifA*-like ORF was designated *vnfA* (37). The other *nifA*-like ORF was located approximately 700 bp upstream from *anfH* (Figure 1). A mutant (strain CA66) carrying a deletion plus insertion in this *nifA*-like ORF synthesized only nitrogenase 2 proteins after derepression in Mo-deficient media with or without V. Thus this ORF was designated *anfA* (37). The highest degree of similarity between the predicted products of *nifA*, *vnfA*, and *anfA* is in the C-terminal half of the proteins where the potential RNA polymerase-sigma 54 interaction sites and ATP-binding domains are located (20). A potential DNA-binding domain (20, 26) that includes a helix-turn-helix motif is present in both predicted products of *vnfA* and *anfA* (37).

nifENX-like ORFs are located immediately downstream from *vnfA* (73). Preliminary results with a strain containing Tn*5-lacZ* inserted in the 3'-terminal end of the *nifN*-like ORF indicate that the *nifN*- and/or *nifX*-like ORF is required for diazotrophic growth in Mo-deficient media with or without V. In the presence of Mo this strain shows wild-type growth (E. Wolfinger and P. Bishop, unpublished results). These results imply that these *nifENX*-like genes are required for functional nitrogenases 2 and 3 but not for nitrogenase 1.

REGULATION OF EXPRESSION OF THE ALTERNATIVE N_2 FIXATION SYSTEMS

Knowledge of how nitrogenases 2 and 3 are regulated at the level of gene expression is rudimentary. Transcription of genes involved in systems 2 and 3 appears to initiate at potential promoter sites that conform to the RNA polymerase-sigma 54 recognition sequence [C$\underline{TGG}$-N_8-T$\underline{TGCA}$, where the underlined nucleotides are invariant (6)]. Such sites are found in the 5' noncoding region of all *vnf* and *anf* operons examined to date except for *vnfA* where one potential promoter is situated within the 3' end of ORF2 (Figure 1) and another is located in the region between ORF1 and ORF2. As might be expected from these observations, *ntrA* is required for all three nitrogenase systems (71). Expression of nitrogenase 1 in *A. vinelandii* is activated by NifA (4). This gene product is thought to recognize upstream activating sequences (UAS) [TGT-N_{10}-ACA (15)]. A characteristic feature of alternative system operons is the apparent lack of this UAS. This is not particularly surprising because NifA is not required for expression of nitrogenase 2 or 3 (37). VnfA and AnfA may function as activators for these operons, and the binding sites for these proteins may be quite different from the binding site for NifA.

The factors that regulate expression and activity of these activator proteins from *A. vinelandii* are not yet known. It is assumed that both ammonia and metals are involved in the regulation process. Early on, it was observed that proteins attributed to alternative nitrogenases were absent in *A. vinelandii* cells grown in the presence of NH_4^+ or Mo (8, 9). Nitrogenase 2 (previously called N_2ase B_1) was present in cells grown in Mo-deficient medium containing V, whereas cells grown in Mo-deficient medium in the absence of V expressed nitrogenase 3 (formerly designated N_2ase B_2). Dinitrogenase reductase 2 was also expressed in cells grown in Mo-deficient medium, even in the absence of V (9, 53). In another study, Jacobson et al (34) found that *nifH*-hybridizing transcripts from the *nifHDK* operon were undetectable in *A. vinelandii* cells derepressed under Mo-deprived conditions. Rather, under these conditions, a different set of *nifH*-hybridizing transcripts was observed that probably originated from the *vnfH-Fd* operon. Although the details of how the expression of nitrogenases 2 and 3 is modulated by NH_4^+ and metals remain obscure, progress on these regulatory aspects can be expected to be rapid now that transcriptional *lacZ* fusions, which are integrated into the *A. vinelandii* genome, have been constructed for *vnfH, vnfD*, and *anfH* (C. Kennedy, personal communication, and our unpublished results). Experiments with an *anfH-lac* fusion strain indicate that transcription initiated from the *anfH* promoter is repressed by NH_4^+, Mo, and V. Furthermore, experiments with this *lac* fusion strain indicate that dinitrogenase reductase 2 is required for in vivo transcription of the *anfHDGK* operon (R. Joerger, E. Wolfinger, and P. Bishop, unpublished results). This may be one reason why dinitrogenase reductase 2 is always present under Mo-deficient conditions (as noted previously) even though dinitrogenase 2 is absent and dinitrogenase reductase 2 is unable biochemically to complement dinitrogenase 3 (17).

A new gene, *ntrD*, is required for growth on nitrate and for diazotrophic growth under Mo-sufficient and Mo-deficient conditions. NtrD$^-$ mutants are blocked in transcription from the *anfH* promoter, but not from the *vnfH* and *vnfD* promoters (C. Kennedy, personal communication). Another regulatory gene, *nfrX*, is necessary for growth under both Mo-sufficient and Mo-deficient conditions, but not in the presence of V (63). Thus at least four gene products (AnfA, dinitrogenase reductase 2, NtrD, and NfrX) may be involved in the regulation of nitrogenase 3.

CONCLUDING REMARKS

More than 50 genes are probably involved in some aspect of nitrogen fixation by *A. vinelandii*. The nucleotide sequence has been determined for 48 of these genes and ORFs. Mutations in many of the ORFs and some of the *nif* genes (which have counterparts in *K. pneumoniae*) fail to give identifiable phenotypes using standard laboratory culture procedures (33). This has led to the conclusion that these gene products are dispensable for diazotrophy under these conditions. It is possible, however, that these gene products serve an important function when *A. vinelandii* is fixing nitrogen in a more natural environment such as soil. A related question is the role that alternative nitrogenases play under natural conditions. One intriguing possibility is that low temperature may be a factor, as suggested by a recent report (48) showing that low temperature favors N_2 reduction by nitrogenase 2. This suggestion is supported by the observation that β-galactosidase expression was not affected by Mo in *vnfH-lac* and *vnfD-lac* fusion strains at 10°C, whereas at 30°C expression was considerably diminished by the presence of Mo (C. Kennedy, personal communication). Thus nitrogenase 2 could be ideally suited for low-temperature diazotrophy. A more obvious role for Mo-independent nitrogenases might be N_2 fixation in low-pH soils with high iron oxide contents where Mo is known to be biologically limiting. Such soils are common to the southeastern part of the United States and to many tropical regions of the world. Other Mo-deficient environments might be locations where Mo is removed by other organisms or where the natural abundance is low.

Little information is currently available on the presence of alternative nitrogenase systems in diazotrophs other than the azotobacters. With the availability of nucleotide sequence data for the structural genes encoding the alternative nitrogenases, it should now be possible to construct specific hybridization probes that can be utilized to screen diazotrophs for alternative system genes. Recently the cyanobacterium *Anabaena variabilis* has been reported to possess an alternative V-containing nitrogenase (43); however, no supporting genetic evidence is available except for previous reports of multiple *nifH*-like sequences in *Anabaena* (56) and *Calothrix* (40). The archaebacterium *Methanosarcina barkeri* has been shown to require either Mo or V for diazotrophic growth, and thus this diazotroph may synthesize a V-containing nitrogenase that is distinct from the Mo-containing enzyme (64). It should be mentioned that although it is convenient to classify nitrogenases according to their metal content, it is entirely possible that some Mo-containing nitrogenases may share more similarity with either nitrogenase 2 or 3 than with the conventional Mo-containing nitrogenase 1. Souillard & Sibold (69) have raised this possibility by suggesting that methanogens may have an unusual Mo-containing nitrogenase that has biochemical properties similar to those of nitrogenase 3.

The genetics and molecular biology of alternative N_2 fixation systems are still in their infancy, but we can expect to see many interesting developments in the next few years—particularly if these findings are extended to diazotrophs not closely related to the azotobacters.

ACKNOWLEDGMENTS

We thank Richard Pau and Christina Kennedy for providing unpublished information.

Literature Cited

1. Arber, J. M., Dobson, B. R., Eady, R. R., Stevens, P., Hasnain, S. S., et al. 1987. Vanadium K-edge X-ray absorption spectrum of the VFe protein of the vanadium nitrogenase of *Azotobacter chroococcum*. *Nature* 325:372–74
2. Arnold, W., Rump, A., Klipp, W., Priefer, U. B., Pühler, A. 1988. Nucleotide sequence of a 24,206-base-pair DNA fragment carrying the entire nitrogen fixation gene cluster of *Klebsiella pneumoniae*. *J. Mol. Biol.* 203:715–38
3. Benemann, J. R., McKenna, C. E., Lie, R. F., Traylor, T. G., Kamen, M. D. 1972. The vanadium effect in nitrogen fixation by *Azotobacter*. *Biochim. Biophys. Acta* 264:25–38
4. Bennett, L. T., Cannon, F., Dean, D. R. 1988. Nucleotide sequence and mutagenesis of the *nifA* gene from *Azotobacter vinelandii*. *Mol. Microbiol.* 2:315–21
5. Bennett, L. T., Jacobson, M. R., Dean, D. R. 1988. Isolation, sequencing and mutagenesis of the *nifF* gene encoding flavodoxin from *Azotobacter vinelandii*. *J. Biol. Chem.* 263:1364–69

6. Beynon, J., Cannon, M., Buchanan-Wollaston, V., Cannon, F. 1983. The *nif* promoters of *Klebsiella pneumoniae* have a characteristic primary structure. *Cell* 34:665–71

7. Bishop, P. E., Hawkins, M. E., Eady, R. R. 1986. Nitrogen fixation in Mo-deficient continuous culture by a strain of *Azotobacter vinelandii* carrying a deletion of the structural genes for nitrogenase *(nifHDK)*. *Biochem. J.* 238:437–42

8. Bishop, P. E., Jarlenski, D. M. L., Hetherington, D. R. 1980. Evidence for an alternative nitrogen fixation system in *Azotobacter vinelandii. Proc. Natl. Acad. Sci. USA* 77:7342–46

9. Bishop, P. E., Jarlenski, D. M. L., Hetherington, D. R. 1982. Expression of an alternative nitrogen fixation system in *Azotobacter vinelandii. J. Bacteriol.* 150:1244–51

10. Bishop, P. E., Premakumar, R., Dean, D. R., Jacobson, M. R., Chisnell, J. R., et al. 1986. Nitrogen fixation by *Azotobacter vinelandii* strains having deletions in structural genes for nitrogenase. *Science* 232:92–94

11. Bortels, H. 1930. Molybdaen als Katalysator bei der biologischen Stickstoffbindung. *Arch. Mikrobiol.* 1:333–42

12. Bortels, H. 1936. Weitere Untersuchungen ueber die Bedeutung von Molybdaen, Vanadium, Wolfram und andere Erdaschenstoffe fuer stickstoffbindende und andere Mikroorganismen. *Zentralbl. Bakteriol. Parasitenkd. Infektionskr. Abt.2*, 95:193–218

13. Brigle, K. E., Setterquist, R. A., Dean, D. R., Cantwell, J. S., Weiss, M. C., Newton, W. E. 1987. Site-directed mutagenesis of the nitrogenase MoFe protein of *Azotobacter vinelandii. Proc. Natl. Acad. Sci. USA* 84:7066–69

14. Brigle, K. E., Weiss, M. C., Newton, W. E., Dean, D. R. 1987. Products of the iron-molybdenum cofactor-specific biosynthetic genes, *nifE* and *nifN*, are structurally homologous to the products of the nitrogenase molybdenum-iron protein genes, *nifD* and *nifK. J. Bacteriol.* 169:1547–53

15. Buck, M., Miller, S., Drummond, M., Dixon, R. 1986. Upstream activator sequences are present in the promoters of nitrogen fixation genes. *Nature* 320:374–78

16. Burns, R. C., Fuchsman, W. H., Hardy, R. W. F. 1971. Nitrogenase from vanadium-grown *Azotobacter:* isolation,

characteristics, and mechanistic implications. *Biochem. Biophys. Res. Commun.* 42:353–58

17. Chisnell, J. R., Premakumar, R., Bishop, P. E. 1988. Purification of a second alternative nitrogenase from a *nifHDK* deletion strain of *Azotobacter vinelandii. J. Bacteriol.* 170:27–33

18. Dean, D. R., Brigle, K. E. 1985. *Azotobacter vinelandii nifD*- and *nifE*-encoded polypeptides share structural homology. *Proc. Natl. Acad. Sci. USA* 82:5720–23

19. Deistung, J., Cannon, F. C., Cannon, M. C., Hill, S., Thornley, R. N. F. 1985. Electron transfer to nitrogenase in *Klebsiella pneumoniae. Biochem. J.* 231:743–53

20. Drummond, M., Whitty, P., Wootton, J. 1986. Sequence and domain relationships of *ntrC* and *nifA* from *Klebsiella pneumoniae:* homologies to other regulatory proteins. *EMBO J.* 5:441–47

21. Eady, R. R. 1986. Enzymology in free-living diazotrophs. In *Nitrogen Fixation*, ed. W. J. Broughton, A. Puehler, 4:1–49. Oxford: Clarendon. 321 pp.

22. Eady, R. R. 1988. The vanadium-containing nitrogenase of *Azotobacter. Biofactors* 1:111–16

23. Eady, R. R., Robson, R. L., Richardson, T. H., Miller, R. W., Hawkins, M. 1987. The vanadium nitrogenase of *Azotobacter chroococcum.* Purification and properties of the VFe protein. *Biochem. J.* 244:197–207

24. Eady, R. R., Robson, R. L., Smith, B. E. 1988. Alternative and conventional nitrogenases. In *The Nitrogen and Sulfur Cycles*, ed. J. A. Cole, S. J. Ferguson, pp. 363–82. Cambridge / New York / New Rochelle/Melbourne/Sydney: Cambridge Univ. Press. 490 pp.

25. Filler, W. A., Kemp, R. M., Ng, J. C., Hawkes, T. R., Dixon, R. A., Smith, B. E. 1986. The *nifH* gene product is required for the synthesis or stability of the iron-molybdenum cofactor of nitrogenase from *Klebsiella pneumoniae. Eur. J. Biochem.* 160:371–77

26. Fischer, H.-M., Bruderer, T., Hennecke, H. 1988. Essential and nonessential domains in the *Bradyrhizobium japonicum* NifA protein: identification of indispensable cysteine residues involved in redox reactivity and/or metal binding. *Nucleic Acids Res.* 16:2207–24

27. Gillum, W. O., Mortenson, L. E., Chen, J. S., Holm, R. H. 1977. Quantitative extrusion of the Fe_4S_4 cores of

active sites of ferredoxins and the hydrogenase of *Clostridium pasteurianum*. *J. Am. Chem. Soc.* 99:584–95

28. Hales, B. J., Langosh, D. J., Case, E. E. 1986. Isolation and characterization of a second nitrogenase Fe-protein from *Azotobacter vinelandii*. *J. Biol. Chem.* 261:15301–6

29. Hales, B. J., Morningstar, E. E., Dzeda, M. F., Mauterer, L. A. 1986. Isolation of a new vanadium-containing nitrogenase from *Azotobacter vinelandii*. *Biochemistry* 25:7251–55

30. Hoover, T. R., Robertson, A. D., Cerny, R. C., Hayes, R. N., Imperial, J., et al. 1987. Identification of the V factor needed for synthesis of the iron-molybdenum cofactor of nitrogenase as homocitrate. *Nature* 329:855–957

31. Howard, K. S., McLean, P. A., Hansen, F. B., Lemley, P. V., Koblan, K. S., Orme-Johnson, W. H. 1986. *Klebsiella pneumoniae nifM* gene product is required for stabilization and activation of nitrogenase iron protein in *Escherichia coli*. *J. Biol. Chem.* 261:772–78

32. Imperial, J., Ugalde, R. A., Shah, V. K., Brill, W. J. 1984. Role of the *nifQ* gene product in the incorporation of molybdenum into nitrogenase in *Klebsiella pneumoniae*. *J. Bacteriol.* 158:187–94

33. Jacobson, M. R., Brigle, K. E., Bennett, L. T., Setterquist, R. A., Wilson, M. S., Cash, V. L., et al. 1989. Physical and genetic map of the major *nif* gene cluster from *Azotobacter vinelandii*. *J. Bacteriol.* 171:1017–27

34. Jacobson, M. R., Premakumar, R., Bishop, P. E. 1986. Transcriptional regulation of nitrogen fixation by molybdenum in *Azotobacter vinelandii*. *J. Bacteriol.* 167:480–86

35. Joerger, R. D., Bishop, P. E. 1988. Nucleotide sequence and genetic analysis of the *nifB-nifQ* region from *Azotobacter vinelandii*. *J. Bacteriol.* 170:1475–87

36. Joerger, R. D., Bishop, P. E. 1988. Bacterial alternative nitrogen fixation systems. *Crit. Rev. Microbiol.* 16:1–14

37. Joerger, R. D., Jacobson, M. R., Bishop, P. E. 1989. Two *nifA*-like genes required for expression of alternative nitrogenases of alternative nitrogenases by *Azotobacter vinelandii*. *J. Bacteriol.* 171:3258–67

38. Joerger, R. D., Jacobson, M. R., Premakumar, R., Wolfinger, E. D.,

Bishop, P. E. 1989. Nucleotide sequence and mutational analysis of the structural genes *(anfHDGK)* for the second alternative nitrogenase from *Azotobacter vinelandii*. *J. Bacteriol.* 171:1075–86

39. Joerger, R. D., Premakumar, R., Bishop, P. E. 1986. Tn*5*-induced mutants of *Azotobacter vinelandii* affected in nitrogen fixation under Mo-deficient and Mo-sufficient conditions. *J. Bacteriol.* 168:673–82

40. Kallas, T., Rebiere, M.-C., Rippka, R., de Marsac, N. T. 1983. The structural *nif* genes of the cyanobacteria *Gloeothece* sp. and *Calothrix* sp. share homology with those of *Anabaena* sp., but the *Gloeothece* genes have a different arrangement. *J. Bacteriol.* 155:427–31

41. Kennedy, C., Gamal, R., Humphrey, R., Ramos, J., Brigle, K., Dean, D. 1986. The *nifH*, *nifM*, and *nifN* genes of *Azotobacter vinelandii*: characterization by Tn*5* mutagenesis and isolation from pLAFR1 gene banks. *Mol. Gen. Genet.* 205:318–25

42. Kennedy, C., Toukdarian, A. 1987. Genetics of azotobacters: applications to nitrogen fixation and related aspects of metabolism. *Annu. Rev. Microbiol.* 41:227–58

43. Kentemich, T., Danneberg, G., Hundeshagen, B., Bothe, H. 1988. Evidence for the occurrence of the alternative, vanadium-containing nitrogenase in the cyanobacterium *Anabaena variabilis*. *FEMS Microbiol. Lett.* 51:19–24

44. Martin, A. E., Burgess, B. K., Iisma, S. E., Smartt, C. T., Jacobson, M. R., Dean, D. R. 1989. Construction and characterization of an *Azotobacter vinelandii* strain with mutations in genes encoding flavodoxin and ferredoxin I. *J. Bacteriol.* 171:3162–67

45. McKenna, C. E., Benemann, J. R., Traylor, T. G. 1970. A vanadium-containing nitrogenase preparation: implications for the role of molybdenum in nitrogen fixation. *Biochem. Biophys. Res. Commun.* 41:1501–8

46. McLean, P. A., Papaefthymiou, V., Münck, E., Orme-Johnson, W. H. 1988. Use of isotopic hybrids of the MoFe protein to study the mechanism of nitrogenase catalysis. In *Nitrogen Fixation: Hundred Years After*, ed. H. Bothe, F. J. de Bruijn, W. E. Newton, pp. 101–6. Stuttgart: Gustav Fischer Verlag. 878 pp.

47. Merrick, M. J. 1988. Organization and regulation of nitrogen fixation genes in *Klebsiella* and *Azotobacter*. See Ref. 46, pp. 293–302

48. Miller, R. W., Eady, R. R. 1988. Molybdenum and vanadium nitrogenases of *Azotobacter chroococcum:* low temperature favours N_2 reduction by vanadium nitrogenase. *Biochem. J.* 256:429–32

49. Nieva-Gomez, D., Roberts, G. P., Klevickis, S., Brill, W. J. 1980. Electron transport to nitrogenase in *Klebsiella pneumoniae*. *Proc. Natl. Acad. Sci. USA* 77:2555–58

50. Page, W. J., Collinson, S. K. 1982. Molybdenum enhancement of nitrogen fixation in a Mo-starved *Azotobacter vinelandii* Nif⁻ mutant. *Can. J. Microbiol.* 28:1173–80

51. Pau, R. N. 1989. Nitrogenases without molybdenum. *Trends Biochem. Sci.* 14:183–86

52. Pau, R. N., Mitchenall, L. A., Robson, R. L. 1989. Genetic evidence for an *Azotobacter vinelandii* nitrogenase lacking molybdenum and vanadium. *J. Bacteriol.* 171:124–29

53. Premakumar, R., Chisnell, J. R., Bishop, P. E. 1989. A comparison of the three dinitrogenase reductases expressed by *Azotobacter vinelandii*. *Can. J. Microbiol.* 35:344–48

54. Premakumar, R., Lemos, E. M., Bishop, P. E. 1984. Evidence for two dinitrogenase reductases under regulatory control by molybdenum in *Azotobacter vinelandii*. *Biochim. Biophys. Acta* 797:64–70

55. Raina, R., Reddy, M. A., Ghosal, D., Das, H. K. 1988. Characterization of the gene for the Fe-protein of the vanadium dependent alternative nitrogenase of *Azotobacter vinelandii* and construction of a Tn*5* mutant. *Mol. Gen. Genet.* 214:121–27

56. Rice, D., Mazur, B. J., Haselkorn, R. 1982. Isolation and physical mapping of nitrogen fixation genes from the cyanobacterium *Anabaena* 7120. *J. Biol. Chem.* 257:13157–63

57. Robinson, A. C., Dean, D. R., Burgess, B. K. 1987. Iron-molybdenum cofactor biosynthesis in *Azotobacter vinelandii* requires the iron protein of nitrogenase. *J. Biol. Chem.* 262:14327–32

58. Robson, R. L. 1986. Nitrogen fixation in strains of *Azotobacter chroococcum* bearing deletions of a cluster of genes coding for nitrogenase. *Arch. Microbiol.* 146:74–79

59. Robson, R. L., Eady, R. R., Richardson, T. H., Miller, R. W., Hawkins, M., Postgate, J. R. 1986. The alternative nitrogen fixation system of *Azotobacter chroococcum* is a vanadium enzyme. *Nature* 322:388–90

60. Robson, R. L., Woodley, P. R., Jones, R. 1986. Second gene *(nifH*)* coding for a nitrogenase iron-protein in *Azotobacter chroococcum* is adjacent to a gene coding for a ferredoxin-like protein. *EMBO J.* 5:1159–63

61. Robson, R. L., Woodley, P. R., Pau, R. N., Eady, R. R. 1989. Structural genes for the vanadium nitrogenase from *Azotobacter chroococcum*. *EMBO J.* 8:1217–24

62. Ruvkun, G. B., Ausubel, F. M. 1980. Interspecies homology of nitrogenase genes. *Proc. Natl. Acad. Sci. USA* 77:191–95

63. Santero, E., Toukdarian, A., Humphrey, R., Kennedy, C. 1988. Identification and characterization of two nitrogen fixation regulatory genes, *nifA* and *nfrX*, in *Azotobacter vinelandii* and *Azotobacter chroococcum*. *Mol. Microbiol.* 2:303–14

64. Scherer, P. 1989. Vanadium and molybdenum requirement for the fixation of molecular nitrogen by two *Methanosarcina* strains. *Arch. Microbiol.* 151:44–48

65. Shah, V. K., Hoover, T. R., Imperial, J., Paustian, T. D., Roberts, G. P., Ludden, P. W. 1988. Role of *nif* gene products and homocitrate in the biosynthesis of iron-molybdenum cofactor. See Ref. 46, pp. 115–20

66. Shah, V. K., Stacey, G., Brill, W. J. 1983. Electron transport to nitrogenase. *J. Biol. Chem.* 258:12064–68

67. Smith, B. E., Bishop, P. E., Dixon, R. A., Eady, R. R., Filler, W. A., et al. 1985. The iron-molybdenum cofactor of nitrogenase. In *Nitrogen Fixation Research Progress*, ed. H. J. Evans, P. J. Bottomley, W. E. Newton, pp. 597–603. Dordrecht/Boston/Lancaster: Martinus Nijhoff. 731 pp.

68. Smith, B. E., Eady, R. R., Lowe, D. J., Gormal, C. 1988. The vanadium-iron protein of vanadium nitrogenase from *Azotobacter chroococcum* contains an iron-vanadium cofactor. *Biochem. J.* 250:299–302

69. Souillard, N., Sibold, L. 1989. Primary structure, functional organization and expression of nitrogenase structural genes of the thermophilic archaebacterium *Methanococcus thermolithotrophicus*. *Mol. Microbiol.* 3:441–552

70. Thöny, B., Kaluza, K., Hennecke, H. 1985. Structural and functional homology between the α and β subunits of the nitrogenase MoFe protein as revealed by sequencing the *Rhizobium japonicum nifK* gene. *Mol. Gen. Genet.* 198:441–48

71. Toukdarian, A., Kennedy, C. 1986. Regulation of nitrogen metabolism in *Azotobacter vinelandii:* isolation of *ntr* and *glnA* genes and construction of *ntr* mutants. *EMBO J.* 5:399–407

72. Wang, S.-Z., Chen, J.-S., Johnson, J. L. 1988. The presence of five *nifH*-like sequences in *Clostridium pasteurianum:* sequence divergence and transcriptional properties. *Nucleic Acids Res.* 16:439–54

73. Wolfinger, E. D., Pau, R. N., Bishop, P. E. 1989. *Annu. Meet. Am. Soc. Microbiol.* 185. (Abstr. H-97)

Annu. Rev. Plant Physiol. Plant Mol. Biol. 1990. 41:127–51

THE PHYSIOLOGY AND BIOCHEMISTRY OF PARASITIC ANGIOSPERMS

George R. Stewart

Striga Research Group, Department of Biology, University College, London, WC1E 6BT, England

Malcolm C. Press

Department of Environmental Biology, The University, Manchester, M13 9PL, England

KEY WORDS: ecophysiology of parasites, haustorium structure and development, carbon/ nitrogen relations of parasites, *Striga*

CONTENTS

127

1040-2519/90/0601-0127$02.00

INTRODUCTION

Witchweed, broomrape, little fire, stealer of bread, and devil's thread—these are among the common names given to a remarkable group of flowering plants, the parasitic angiosperms. Such names recognize the damage caused to man's cultivated plants by some parasitic plants. Over 3000 species of flowering plants utilize a parasitic mode of nutrition, yet basic knowledge about their physiology and biochemistry is limited. About some aspects of these organisms we know nothing.

Parasitic angiosperms are generally separated into holo- and hemiparasites, although only at the extremes of this range is the distinction readily made. Holoparasitic species are always obligate parasites, devoid of chlorophyll and having little independent capacity to assimilate carbon and inorganic nitrogen. Hemiparasites may be facultative or obligate; they are chlorophyllous and are traditionally thought to rely on their host only for water and minerals. Parasitic flowering plants are further subdivided on the basis of their site of attachment to the host. There are stem parasites, such as the holoparasitic dodders and the hemiparasitic mistletoes. There are root parasites, such as the holoparasitic broomrapes and hemiparasitic witchweeds. Parasitic angiosperms encompass small herbaceous species such as *Thesium humile* and large trees such as *Santalum album,* both in the Santalaceae. The distinguishing feature of all parasitic plants is the haustorium, a novel organ that functions in attachment, penetration, and solute transfer. Parasitic plants vary greatly with respect to host range. At one extreme are species such as *Conopholis americana* (Orobanchaceae), which parasitises only *Quercus borealis* (71); at the other are species like *Olax phyllanthi* (Olaceae), which can parasitize a wide range of species—herbaceous and woody, annual and perennial, as well as other species of root hemiparasites (70).

Some parasites are a serious agricultural problem, particularly in countries of the Third World where they attack and devastate cereal and legume crops of subsistence farmers. Stunting of shoot growth, severe wilting, chlorosis, and yield reductions of up to 100% have been reported in sorghum infected with *Striga hermonthica* (4, 19). *Striga asiatica* was introduced into the Carolinas as a contaminant of maize seed in the 1950s and is still a threat to maize, sugar cane, and sorghum crops, which have an annual value in excess of $23 billion (US Agricultural Statistics 1982 in Ref. 86). This weed could cost farmers an estimated $1 billion in annual control costs plus yield losses of about 10% (86).

PARASITE GROWTH AND DEVELOPMENT

Seed Germination

In contrast to almost all other plants, parasitic angiosperms depend completely on the host for signals that control their initial stages of development. Genera in the Scrophulariaceae, Orobanchaceae, Balanophoraceae, Rafflesiaceae, Hydnoraceae, and Lennoaceae all require a chemical signal from the host root prior to germination. Following this signal, most parasitic genera will only develop a functional haustorium in the presence of a second chemical signal derived from the host.

The germination stimulants have proved elusive to investigators for at least two reasons. First, they are active at low concentrations and second, the structures that have been elucidated to date are extremely labile molecules—characteristics that are probably critical to their successful operation as germination cues. Both blocking or intercepting these signals and providing signals in the absence of a host plant provide means to control parasitic weeds, and the elucidation and synthesis of germination stimulants have attracted the attention of a number of research groups.

The first naturally occurring germination stimulant was identified from cotton (15, 16), a plant that stimulates seed germination but will not support the developing parasite. The absolute structure of this molecule, a sesquiterpene given the trivial name strigol, was not established until 1985 (10). Strigol is active at concentrations as low as 10^{-15} mol m^{-3} in the soil solution. The molecule consists of four rings: a six-carbon ring (ring A), a five-carbon ring (B), and a four-carbon lactone (C) are coupled to another lactone (ring D) via a =C-O- connecting unit. Following the identification of strigol in 1972, structure-bioactivity studies led to the synthesis of a number of analogues and precursors (so called GR compounds) (see e.g. 44, 45). Such investigations shed little light on the mechanism by which stimulants operate but indicated the importance of the D ring for bioactivity.

More recently Zwannenburg and coworkers (60, 118, 119) have revived this approach, and strigol analogues have been synthesized based on (a) the ABCD-ring framework, but with different substituted groups; (b) systematic clipping of the strigol skeleton; (c) the concept of bio-isosterism; and (d) the molecular shape of strigol. These studies suggest that the linkage between the two lactones is critical for activity and that the stimulant is activated at the receptor site by a nucleophilic group (eg. HS- or H$_2$N-) that causes the ester function to cleave, with ring D acting as a leaving group with the oxygen atom from the connecting unit. Thus the D ring is not a critical part of the molecule, and good activity has been obtained by replacing the D ring in GR24 with other leaving groups.

A different stimulant has been identified from the roots of sorghum, a host for *Striga*. Biologically active hydrophobic droplets on the root hairs of sorghum contain a number of *p*-benzoquinones present in both oxidized and reduced (dihydroquinone) form (13, 65). The major component has been named sorgoleone-358(2-hydroxy-5-methoxy-3-[(8'Z,11'Z)-8',11',14'pentadecatriene]-p-benzoquinone), and the biological activity of the droplets is directly related to the concentration of the dihydroquinone of sorgoleone-358. This compound is very unstable and is rapidly converted to its oxidized form. The latter has been characterized, rather than the germination stimulant (13). Sorgoleones are active at higher concentrations than strigol (10^{-10} mol m^{-3}). They are not the only germination stimulant present in sorghum root exudate, since biological activity can also be detected in hydrophilic fractions of sorghum root exudate (115). Their extremely low solubility in water together with their absence from other Gramineae hosts (although the simpler alkylresorcinols have been reported from the bran fraction of other cereals) suggest that their primary role in vivo may be as allelochemicals (see 65) rather than germination stimulants.

A third compound has been tentatively identified from the root exudate of cowpea, which is a host for both *Striga* and *Alectra* (40, 106). The molecule is cyclic, consisting of three parts: a xanthine ring, an unsaturated C_{12}-carboxylic acid, and a dipeptide of glycine and aspartic acid.

The germination stimulants from cotton, sorghum, and cowpea all appear very different in structure, and together with these molecules a number of other compounds have been reported to stimulate germination *in vitro* at concentrations orders of magnitude greater than those of the natural stimulants. These include kinetin, zeatin, abscisic acid, scopoletin, inositol, methionine, sodium hypochlorite, and ethylene (115). However, there is no evidence that any of these compounds operates *in vivo*. Zwannenburg's hypothesis suggests a mechanism whereby a large diversity of compounds might effect germination via a common reaction. It is possible that compounds with molecular shapes not dissimilar to strigol's might operate in cereals as well.

Haustorium Initiation

A different complement of chemicals is responsible for initiating haustorial development in parasitic plants. The first haustorium inducers isolated were phenylpropanoids, named xenognosin A and B (7,2'-hydroxy-4'-methoxyisoflavone), extracted from a nonhost plant, *Astragalus gummifer* (55). A number of analogues were synthesized, which showed that *m*-methoxyphenol functionality was a strict requirement for haustorium-inducing activity in *Agalinis purpurea* (95) and other root parasites. Smaller phenolic inducing molecules such as some phenolic and benzoic acid derivatives have considerably less activity.

Striga asiatica will only develop a haustorium in response to a single chemical in sorghum root extracts. The active molecule is a quinone, 2,6-dimethoxy-*p*-benzoquinone (2,6-DMBQ); as with the xenognosins, a methoxy functionality appears to be critical (12). However, 2,6-DMBQ cannot be detected in sorghum root exudate; its presence can only be detected after vigorous shaking or sonication of root tissue, which suggests that the molecule is tightly bound to, or inside the host root. 2,6-DMBQ is a product of lignin degradation by fungi that possess extracellular laccases, and laccase activity has been demonstrated at the root tip of both *Striga* and *Agalinis* (12). This observation suggests that parasitic plants may be able to degrade phenylpropanoid components at the surface of host roots. If these molecules are degraded to the appropriately substituted quinone then the parasite would respond by initiating a haustorium. Thus xenognosin and other propanoids may show activity as a result of their conversion to active quinones.

A pentacyclic triterpene haustorium inducer isolated from the roots of *Lespedeza sericea,* which is parasitized by *Agalinis,* has been identified as soyasapogenol B (3β, 22β, 24-trihydroxy-olean-12-ene) (96). This molecule has a biosynthetic origin very different from that of the phenylpropanoids and is at least an order of magnitude less active than the xenognosins. Haustorium inducers are active over the concentration range 10^{-8}–10^{-10} mol m^{-3}. Full activity of soyasapogenol B may depend on the presence of other components of root exudate (96).

Both the germination stimulants and the haustorium initiators provide a means of host selection and location in root parasites, and it has been proposed that, like herbivorous insects, parasitic plants utilize host defense compounds as recognition cues (2). These chemicals are often toxic to the parasite at concentrations above those normally encountered, and reports of the inhibitory effects of germination stimulants at superoptimal concentrations are abundant in the literature. Many of the chemicals are either allelopathic [e.g. sorgoleones (65)] or biosynthetically related to phytoalexins [e.g. xenognosin B is a direct precursor of medicarpin (96)].

Haustorial Function

The haustorium, unique to parasitic angiosperms, has three functions: attachment, penetration, and water and solute acquisition (104). Although functionally similar in different species it is highly variable, morphologically and anatomically. Following attachment and penetration the haustorium functions primarily to transfer nutrients and water from host to parasite (104). The existence of apoplastic continuity between host and parasite (although not necessarily xylem-to-xylem contact—see below) has led some to conclude that the haustorium of hemiparasites has a passive role in solute transfer (80). Ultrastructural studies of the haustoria of several species indicate the presence of parenchyma cells with a high density of cell organelles such as mitochondria, ribosomes, dictyosomes, and well-developed endoplasmic reticulum (53, 57, 104, 105). These cells are in contact with xylem elements either within the haustorium, as in *Striga* (57, 93, 105), or at the haustorial interface, as in *Olax* (53) and the mistletoes (see 26). Histochemical studies have shown the presence of high enzyme activities in the haustorial cells of *Striga hermonthica* (5). These features of haustorial ultrastructure imply an active metabolic role for the haustorium.

The carbohydrates, amino acids, and organic acids present in the xylem sap of *Striga hermonthica* are different from those in that of its host *Sorghum bicolor* (56, 73). The carbohydrate concentrations in the parasite xylem sap are five times those in the host, and the major component is mannitol, which is absent from the host xylem sap. In sorghum the major nitrogenous solute of the xylem is asparagine, while in *Striga* it is citrulline. There are also differences in organic acid composition. The main components of sorghum sap are malate and citrate. The latter is absent from *Striga;* but shikimic acid, which is absent from sorghum sap, is present in the sap of the parasite. These results suggest that the nitrogen and carbon compounds entering the haustorium from the root xylem are actively metabolized within the haustorial cells prior to entering the parasite shoot. Differences in metabolite composition of host and parasite xylem saps have also been reported for other species of root hemiparasites (33) and mistletoes (84).

We interpret these results, together with those from ultrastructural studies, as indicating that haustorial cells have specialized biochemical functions related to the regulation of solute transfer and that the haustorium plays an active metabolic role in the nutrition of parasitic plants.

Regulation of Parasite Growth

Parasitic plants tap into the xylem and in some cases also the phloem of their hosts, and therefore have access to the growth regulators transported within the host. Stem parasites such as the dodders and mistletoes lack root systems and could be dependent on their hosts for the synthesis of certain plant growth regulators. The endophytic strands of the Rafflesiaceae comprise the entire vegetative body, ramifying through the host and emerging only to flower. Clearly in species such as this, growth and reproductive responses may be mediated by host plant growth regulators. At present we have little information on the capacity of these highly specialized parasites to synthesize plant growth regulators. The evidence from in vitro culture experiments is conflicting. Okonkwo (67) found that the morphology and development of *Striga hermonthica* were normal when the parasite was grown with a source of carbon and mineral salts, there being no requirement for exogenous plant growth regulators. In contrast normal seedling development of *Striga asiatica* required the addition of cytokinins and auxin (117). Exogenously supplied cytokinin stimulates seedling leaf expansion of the mistletoe *Amyema miquelii* but not that of *A. pendulum* (37). At present there is no clear evidence that obligate parasites rely on their host for plant growth regulators, but more definitive studies are required.

An extraordinary phenomenon is seen in the leaves of some Australian mistletoes, which in external morphology strongly resemble those of their hosts. Barlow & Wiens (6) suggest this is an example of cryptic mimicry, evolved because it minimizes vertebrate herbivory. The underlying causes of this resemblance between host and parasite are unknown, although Atsatt (3) has suggested that host cytokinins may play a role. His hypothesis is that the composition of plant growth regulators in the host xylem regulates cell division and expansion of the mistletoe leaf, producing a close resemblance to the leaf of its host. Analysis of the cytokinins present in the xylem sap of mistletoes does indicate a closer relationship between the types and amounts of cytokinins in the sap of mimic species and their hosts than between that of nonmimics and their hosts (37). However, recent *in vitro* culture experiments of mimic and nonmimic species of *Amyema* indicate that the leaf shapes of seedlings were similar to those of plants in nature (37). *In vitro* cultured seedlings of *A. cambagei* had the narrow needle-like leaves that strikingly resembled those of its host, *Casuarina cunninghamiana*. Application of exogenous cytokinins, while affecting leaf expansion of some species, did not bring about the morphological responses that would be expected if Atsatt's hypothesis were correct. It seems probable that leaf morphology in mimic mistletoes is genetically determined. Culture on different host species might be a useful approach to understanding this intriguing phenomenon.

WATER RELATIONS

An outstanding physiological characteristic of most parasites is their very high rates of transpiration, which often exceed that of the host by an order of magnitude. This maintains a gradient in leaf water potential towards the parasite and thus facilitates the flux of resources to the parasite. This attribute is common to both stem (27, 103) and root parasites (74) from subarctic, temperate, and tropical regions. Rates of transpiration in parasitic plants are at the higher end of the range observed in angiosperms. Early studies suggested that mistletoes exerted little stomatal control over water loss (38, 39, 47); however, their stomata have been shown to respond directly to vapor pressure deficit (41). They close in response to decreases in leaf or shoot water content

(29), although lower leaf water potentials are required to induce stomatal closure in the parasite than in the host (17, 29, 110). It is significant that nighttime transpiration values for *Amyema linophyllum* (17) and *Pthirusa maritima* (31) were much greater than those of their hosts, *Casuarina obesa* and *Coccoloba ivifera,* respectively. This difference indicates that the parasite can extract water from its host throughout the night as well as during the day. A similar phenomenon is seen with many root hemiparasites in the Scrophulariaceae which maintain high rates of transpiration at night (74, 91). Withholding water from *Striga hermonthica* and *S. asiatica* has a surprisingly small effect on stomatal conductance, and complete stomatal closure does not occur until the relative water content of the leaves is reduced to about 70% (79). In line with these observations are reports that the stomata of *Striga* are less responsive to application of abscisic acid than those of *Antirrhinum majus,* a nonparasitic member of the Scrophulariaceae (91).

The high rates of day and night transpiration and the dampened responses of the stomata to irradiance and water deficit may be general characteristics of leafy parasites. The mechanism underlying this apparent uncoupling of stomata from the environment is not known. Recent studies using epidermal strips prepared from *Striga hermonthica* leaves indicate that their high potassium content may modify their response to changes in irradiance, CO_2, and abscisic acid (94). In general the leaves of mistletoes also have a high potassium content, and this could be responsible for the dampened responses of their stomata.

High transpiration rates are a major component contributing to the lower water potentials of parasites and generating the hydrostatic gradient that facilitates the transfer of solutes from host to parasite. Comparisons of water potential between host and parasite for a wide range of species and environmental conditions have found, for the most part, those of the latter to be more negative (27, 87). Recent studies of the mistletoe *Pthirusa maritima,* which parasitizes species of mangroves, show that it too maintains leaf water potentials lower than those of its hosts under various light and humidity conditions, on a daily and seasonal basis (31). Some mistletoes are obligate or facultative parasites of other mistletoes, and increasingly negative water potentials can be traced from the primary host through the first mistletoe to the epiparasite, accompanied by increasing rates of transpiration (30). Although the shoot water potential of *Rhinanthus serotinus* is lower before attachment, indicating a high resistance to water uptake from the roots (48), and becomes less negative after attachment, it is still more negative than that of the host plant, *Hordeum vulgare.*

The lower water potentials seen in parasites are, not surprisingly, accompanied by higher osmolarities; typically there are high tissue concentrations of inorganic ions, particularly potassium (54, 70, 72). The sum of water-soluble cations in mistletoe leaves was found to range from 300 to as high as 900 mol m^{-3}, the concentrations being greater in older leaves (72). The mangrove mistletoe *P. maritima* shows an accumulation of sodium (up to 10% of dry weight) as well as potassium ions, the latter being some 4–10 times the level found in its hosts (31).

High concentrations of soluble carbohydrates [e.g. glucose and fructose in mistletoes (72); mannitol in *Orobanche* (76, 108), *Lathraea* (76), *Striga* (66), and possibly other species (92)] have been implicated in the generation of an osmotic gradient between host and parasite. Popp (72) has shown that the low-molecular-weight carbohydrate fraction in mistletoes is, in part, derived from their hosts; components such as pinitol are only present in the parasite when they are present in the host.

These low-molecular-weight carbohydrates may also act as compatible cytosolic solutes, playing a role analogous to that described for these and similar compounds in another group of plants with high ion concentrations, namely halophytes (116).

Davidson et al (17) have drawn attention to the coordinated control of water relations that exists in the association between the mistletoe *Amyema linophyllum* and its host *Casuarina obesa*. Their results indicate that the parasite displays a marked sensitivity to the water status of its host, which implies that it responds to signals received, possibly from the host root. Integration and coordination of water and solute partitioning may be of particular importance in perennial associations in order to maintain host competence and parasite growth. Although the stomata of mistletoes are less sensitive to changes in leaf water potential than those of their hosts, they generally exhibit responses that parallel those of their host. Similarly, the water relations of perennial species in the Scrophulariaceae appear more in synchrony with their hosts than those of annual species (74). This type of coordinate behavior has also been observed in mistletoes parasitizing plants exhibiting crassulacean acid metabolism (H. Ziegler, personal communication). In these associations the mistletoes show diurnal stomatal closure and nocturnal stomatal opening mirroring that of the host plant.

Although there is little doubt that water potential differences between host and parasite are largely maintained by differential rates of transpiration, other factors such as the resistance across the host-parasite interface play a role. The haustorial junction represents the largest resistance component to transpiration-driven water flow into the parasite. In experiments with *Loranthus europaeus* parasitizing *Quercus robur,* Glatzel (30) estimated haustorial resistances to be 2.6–4.2 times those generally found for plant stems. Transpiration rates of *Amyema linophyllum* increase some 300 times from the night's minimum to the day's maximum, while water potential differences between host and parasite alter by only a factor of two (17). It is suggested that haustorial resistance to water flow increases disproportionately as flow rate decreases. Similarly it was found that the hydraulic conductivity of *Pthirusa maritima* was a function of the hydraulic properties of the host species. Resistance to water flow was lower when it parasitized a species with an efficient water transport system than when it parasitized a species with a relatively low water transport efficiency (31). The mechanism for this physiological acclimation is unknown, but Davidson and coworkers (17) suggest that under conditions of water deficiency a decrease in turgor of parenchyma cells at the haustorial interface might widen the apoplastic pathways between haustorial cells, leading to an increase in hydraulic conductivity at the interface.

The high rates of transpiration exhibited by many parasites have implications with respect to the energy budgets of their leaves. Evaporative cooling of leaves has been shown to make an appreciable contribution to their energy budget, particularly where daytime temperatures are high (34). The leaf temperatures of *Striga hermonthica* are appreciably below those of the air, and at an ambient temperature of 40°C they can be as much as 7°C lower (75). An interesting consequence of this marked transpirational cooling is that application of an antitranspirant, which mechanically impedes water loss, causes leaf temperature to rise; if ambient air temperature is >37°C the leaves then blacken and die. This suggests the plant is adapted or acclimated to temperatures several degrees below those of its environment. Consistent with this suggestion are observations that physiological processes such as photosynthesis have temperature optima below those of the growth temperature (75). Field trials indicate antitranspirants may have potential in the control of *Striga*. Application of an antitranspirant under field conditions in the Sudan led to increases in sorghum straw and grain yield and a reduction in parasite growth (75).

MINERAL NUTRITION

The nutrition of parasitic plants has been studied using several approaches. In vitro culture experiments have attempted to define the minimal nutrient requirements of the parasite; from these, host-derived nutritional factors have been inferred. Axenic cultures of obligate and facultative hemiparasites show some growth to the reproductive stage is possible in medium containing sugars and the essential inorganic mineral elements (67). Unfortunately there are few direct comparisons of growth quantity and quality between plants growing in the parasitic state and those cultured in vitro.

Nitrogen Metabolism

Nitrogen is a limiting factor in many ecosystems. Carnivory, parasitism, and other symbiotic associations involving plants can be viewed as strategies for enhancing nitrogen acquisition (54). Nitrogen may be present in the transpiration stream in both organic and inorganic forms, the proportion being dependent on the source of nitrogen assimilated and (for nitrate assimilation) the relative contribution of root and shoot to nitrate reduction (81). Use has been made of the substrate inducibility of nitrate reductase to investigate the availability and utilization of nitrate by parasitic plants. McNally & Stewart (59) demonstrated considerable variation in the capacity of parasitic plants to assimilate nitrate, but they found little evidence of any correlation between nitrate reductase activity and leaf nitrogen content. They found several mistletoe species with a very limited capacity for nitrate assimilation, and these were presumably largely dependent on reduced nitrogen in the host xylem fluid. The nitrate reductase activities of the loranthaceous mistletoe *Tapinanthus bangwensis* were found to be similar to those of the host species it was growing on (98). While parasite leaf nitrogen concentrations mirrored those of host leaves, they were not related to nitrate reductase activity. Govier et al (33) found lower concentration ratios of nitrate: organic nitrogen compounds in the bleeding sap of *Odontites* than in the sap of its hosts. This again indicates that nitrogen sources other than nitrate were available from some host species.

Studies with *Striga hermonthica* suggest that the in vitro nitrogen requirements are unusual insofar as growth was considerably stimulated when glutamine was added to medium containing high concentrations of inorganic sources such as nitrate or ammonium ions (67). Similar results have been obtained in in vitro culture experiments with mistletoe seedlings, growth being markedly better with reduced sources of nitrogen than with nitrate (37). Translocation of amino acids and amides from host to parasite has been demonstrated for *Cuscuta* (25), *Lathraea* (82), *Striga* (58), *Orobanche* (1), and *Cytinus* (101). Both growth and enzymological studies suggest that parasitic plants exhibit a preference for reduced sources of nitrogen and have a limited capacity to assimilate inorganic nitrogen sources such as nitrate.

It has been suggested that nitrogen acquisition is a key factor regulating transport processes between host and parasite. Estimates of the seasonal nitrogen requirement and the rate of solute supply via the transpiration stream show that high transpiration rates are necessary to meet the seasonal nitrogen requirements of the mistletoes *Viscum laxum* and *Loranthus europaeus* (88, 89). The water use efficiency (WUE) of mistletoes appears to be closer to that of their hosts when a more concentrated source of nitrogen is available from the hosts' xylem sap (23). This has been interpreted as indicating that stomatal conductance of the parasite is regulated by the nitrogen concentrations in the host xylem sap. Givnish (28) has criticized this interpretation, making the point that the reduction in the WUE differences found between host and parasite is largely accounted for by a decrease in the WUE of the host rather than an increase in that of the parasite. He suggests that this decrease in host

WUE is the result of increased stomatal conductance brought about by a higher nitrogen supply.

Ehleringer and coworkers (22) present data showing that mimic mistletoes have higher leaf nitrogen contents than nonmimic species. They suggest that this is consistent with the Barlow-Wiens hypothesis that mimicry is a device to reduce herbivore predation. However, the results presented by Glatzel (30) for *Dendrophthoe falcata* growing on different host species indicate that its nitrogen content is directly related to that of its host species. None of the mimic and nonmimic species analyzed by Ehleringer and coworkers were found on the same host species. The relationship between nitrogen content and mimicry needs to be reexamined.

The activities of the ammonia-assimilating enzymes glutamine synthetase and glutamate synthase are relatively low in parasitic plants (58, 99). In holoparasitic species only the cytosolic isoform of glutamine synthetase was detected using ion exchange chromatography and immunoprecipitation (58, 59), although immunocytochemical localization studies indicated the presence of glutamine synthetase activity in the amyloplasts of *Lathraea clandestina* (100). Most hemiparasitic species exhibit surprisingly low levels of the chloroplastic isoform (58, 59). Although this has been interpreted as indicating metabolic reductionism associated with a preference among parasitic species for reduced nitrogen (76), it seems more likely that it relates to the photorespiratory capacity of these plants. There is now much evidence that the major function of chloroplastic glutamine synthetase is the reassimilation of ammonia produced in photorespiration (46). Mutants have been isolated that, although unable to grow under photorespiratory conditions because they lack chloroplastic glutamine synthetase, assimilate inorganic nitrogen and grow normally when photorespiration is suppressed (7). The low rates of photosynthesis reported for many hemiparasites (see below) suggest photorespiration will also be low and that this would reduce the requirement for chloroplastic glutamine synthetase.

Holoparasites that have access to a supply of organic nitrogen compounds from the host phloem as well as xylem nitrogen have been reported to lack the enzyme serine/threonine dehydratase (64), making them heterotrophic for the amino acid isoleucine. However, there is little evidence that the hemiparasites *Striga hermonthica* and *S. asiatica* exhibit a reduced capacity for amino acid biosynthesis (76). The biosynthetic potential of parasitic plants would prove an interesting area for future research.

Cation Accumulation and Transfer

As discussed above, the concentration of some ions in the leaves of mistletoes is often markedly higher than in the leaves of their hosts (54). Differences between host and parasite in rates of growth and senescence, degree of herbivory, and capacity to store elements in bark tissue may introduce concentration differences in the mineral elements of their leaves. The concentrations of some elements, notably potassium, are particularly high in mistletoe leaves, and may be as much as 20 times those in their hosts. If uptake were passive, via the transpiration stream, the relative proportions of ions in host and parasite leaves should be the same. The concentration of ions would be determined by the relative rates of transpiration in parasite and host. Differences in element ratios between host and parasite have prompted the suggestion that there is selective uptake of ions into parasite tissues via the haustorium (3, 54). This idea has been criticized (30) on the grounds that no account is taken of the export of an ion such as potassium that occurs from host leaves via the phloem (see 51). The high potassium concentrations in mistletoe leaves and those of other xylem-tapping parasites may simply reflect the ion's lack of mobility in these species because of the absence of host-parasite phloem connections. If this reasoning is correct then a phloem-immobile ion such as calcium should show little enrichment in parasite tissues. The analytical data for a wide range of xylem parasites shows that the calcium levels are comparable with those of their hosts (29, 31, 70, 72).

In a detailed study of *Dendrophthoe falcata* parasitizing 48 different host species a good correlation between host and parasite concentrations of nitrogen, potassium, magnesium, and chloride was found; the leaf potassium concentrations of the parasite were, on almost all hosts, enriched (29). These results are consistent with a passive rather than a selective flux of ions from host to parasites. However, results obtained with the mistletoe *Pthirusa maritima* parasitic on the mangrove *Conocarpus erectus,* which has high sodium chloride concentrations in its xylem stream (nearly three times that in the parasite), indicate that there may be some selectivity of ion transfer (31). This mistletoe does not possess salt glands and has a transpiration rate considerably higher than its host's, yet the amount of sodium in its leaves is only 20% greater than that in host leaves. Goldstein et al (31) speculate that the haustorium might act as an ion exchange column where cations could be sequestered.

There is also evidence for selective transfer of resources from host to root hemiparasites. In *Odontites verna* growing on *Hordeum* and *Stellaria,* the potassium/calcium quotient in the parasite xylem fluid was higher than in either of the host xylem streams (33). In recent work on the *Striga*-sorghum association (56), the xylem fluids of host and parasite were found to have markedly different ion, amino acid, organic acid, and carbohydrate profiles.

The Apoplastic Continuum

Anomalies in mineral element composition of host and parasite tissues and xylem saps put into question the traditional view that there is direct xylem-to-xylem continuity between host and parasite (27, 50). Several lines of evidence from anatomical and physiological studies indicate that such direct connections may not be as widespread as once thought (54). Recent studies of the dwarf mistletoe *Korthalsella lindsayi* indicate that there is an apoplastic continuum between it and its host and that this comprises the walls of the haustorial parenchyma rather than direct tracheary element contact (14). It was concluded from a study of the ultrastructure and physiology of the root hemiparasite *Olax phyllanthi* that both vascular and nonvascular routes were involved in the apoplastic transfer of xylem sap from host to parasite (53). However, the movement and distribution of apoplastic tracers such as lanthanum nitrate indicated a major pathway for water flow through host xylem pits and into the haustorium via the terminal and lateral walls of the contact parenchyma. Although direct apoplastic continuity via xylem elements has been shown in the association between *Striga hermonthica* and its host *Sorghum bicolor,* it was suggested that this was not the normal pathway of solute transfer but that solutes are unloaded into haustorial parenchyma cells prior to being transferred to the parasite shoot (56).

CARBON ASSIMILATION

Photosynthetic Capacity

Obligate hemiparasites are generally assumed to rely on their hosts only for water and mineral elements, the presence of chlorophyll implying their ability to assimilate carbon dioxide. In many of these parasites, rates of photosynthesis are in fact rather low (0.5–5.0 μmol m^{-2} s^{-1}) and are towards the bottom of the range found in C3 plants (18, 74, 91). Moreover, they are often coupled with high rates of respiration, the net result being little net carbon gain, certainly too little to support growth (79).

Low rates of photosynthesis in species of *Striga* are in part related to the relatively undifferentiated leaf mesophyll and the low number of plastids per mesophyll cell (102). Chlorophyll concentrations and the activity of 1,5-ribulose bisphosphate carboxylase-oxygenase (Rubisco) were also found to be low (76). A further cause of low photosynthetic activity may be the low photosystem II activity exhibited by isolated chloroplasts (85). Polypeptide analysis of the thylakoids indicated differences in the organization of the photosystem II antennae in *Striga hermonthica* compared with nonparasitic plants.

It is interesting that Rubisco activity can be detected in the tissues of the holoparasite *Lathraea clandestina* even though it is entirely devoid of chlorophyll (9). The metabolic significance of this is unclear since in another holoparasite, *Orobanche ramosa,* no net carbon dioxide assimilation is detectable (18). The activities of the alternative carboxylating enzyme, phosphoenolpyruvate (PEP) carboxylase, in hemiparasitic species of the Loranthaceae, Viscaceae, Santalaceae, and Scrophulariaceae and in holoparasites of the Orobanchaceae (82, 83) are comparatively low and are within the range reported for C3 species. However, very low levels of PEP carboxylase were found in *Cytinus hypocistus*, a member of the Rafflesiaceae, a family of parasites exhibiting extreme morphological reductionism (83).

Carbon Transfer from Host to Parasite

Several studies have shown import of carbon from the host, although not the quantitative extent of this transfer. Hull & Leonard (42, 43) found that the dwarf mistletoe *Arceuthobium* had a low rate of autotrophic carbon dioxide fixation coupled with a large transfer of ^{14}C-labeled photosynthate from host to parasite. For the green leafy mistletoe *Phoradendron* there was a low, almost negligible transfer of ^{14}C-labeled assimilate but a relatively high photosynthetic capacity. [14]C transfer has been detected from *Populus nigra* to the green leafy mistletoe *Viscum album* (90).

The form in which carbon is transferred is not well established. After exposing the leaves of barley to [14]CO_2 for two hours, 88% of the label was present as sugars in the host, and in the bleeding sap of the parasite *Odontites* 72% of the label recovered was in the form of amino acids (33). After a further 24 hr a greater proportion of the label was recovered as sugars in the parasite following the metabolism of amino acids.

Press et al (77) employed differences in the distribution of naturally occurring ^{12}C and ^{13}C between a C4 host and a C3 parasite to determine the amount of carbon transferred during the life cycle of the sorghum-*Striga* association. In a C3 parasite growing on a C4 host, the $\delta^{13}C$ value of the parasite will be less negative by an amount proportional to the quantity of carbon transferred from the host. This technique indicated that 28–35% of the carbon in the parasite leaves was derived from sorghum photosynthate. In a similar study of the *Pennisetum typhoides-Striga hermonthica* association it was found that in the root, leaf, and stem of the parasite 87, 70, and 49%, respectively, of the carbon was host derived (36). Another approach to quantify carbon transfer has been to construct a carbon balance model from measurements of photosynthesis, respiration, and dry mass of all plant parts of the sorghum- and millet-*Striga hermonthica* associations (35, 36). From the carbon balance model the amount of host-derived carbon that would be required to account for the observed *Striga* growth was determined. These results indicated that around 38% and 85% of the parasite carbon was obtained from sorghum and millet, respectively—estimates in reasonable agreement with those determined using $\delta^{13}C$ analysis.

In the predominantly phloem-feeding holoparasite *Cuscuta*, enhanced unloading rates from host phloem are essential for the transfer of carbohydrate.

Unloading occurs directly into the parasite apoplast and is under metabolic control, since it occurs neither at 0 °C nor in the presence of metabolic inhibitors (112, 114). Assimilate transport from host phloem to *Cuscuta* haustorial cells is similar to the transport from maternal tissues to the embryo in developing legume seeds (114). When developing fruits and *Cuscuta* are competing for assimilates, the latter benefits at the expense of the former (111). The mechanism controlling unloading into the apoplasm is unknown. One possibility is that this occurs via carrier-mediated transmembrane fluxes. Alternatively, there may be some kind of mass flow through pores or a mechanism akin to the granulocrine secretion seen in secretory cells of nectaries (113).

Obligate Heterotrophic Carbon Supply

Parasites with higher rates of carbon assimilation may also obtain a substantial passive flux of carbon from the host in the form of organic nitrogen compounds. Raven (80) calculated that for a stem or root parasite on a host that assimilates ammonium or nitrate ions in its roots, about 20% of the carbon in the parasite could be derived from the host as organic nitrogen (assuming 3 C for every 1 N transported from the host and an overall C/N quotient of 15 in the parasite). Xylem connections may be another source of carbon for predominantly phloem-feeding parasites. The contribution from this source may be limited by the relatively low transpiration rates of such plants (80). Glandular secretion or guttation may substitute for transpiration in some of them. *Orobanche* species have a high density of glands on their scales (102), and glands have been implicated in both guttation and the excretion of ions from the hemiparasite *Odontites verna* (32).

Raven (80) suggests that the low rates of carbon transfer reported for mistletoes and other xylem-feeding species may be a function of the separation in time and space between ^{14}C incorporation in host leaves and its eventual movement back up the xylem in the form of organic nitrogen compounds. Certainly the low nitrate reductase activities reported for many parasitic plants suggests that they obtain much of their nitrogen in an organic form (59). If this is correct then even species that have an apparently adequate photosynthetic capacity will import substantial amounts of host carbon. At present we lack quantitative measurements of growth and carbon requirements in most species and are unable to determine the general significance of heterotrophic carbon gain.

Much of the physiological work with mistletoes has concentrated on the relationship between transpiration and nitrogen acquisition (89). However, a plausible case can be made for a transpiration-driven heterotrophic carbon gain. Rather than interpreting the high transpirational fluxes as a mechanism for maximizing nitrogen gain, we suggest they might maximize carbon gain and thereby reduce the demand for nitrogen. The import of substantial carbon from the host has implications with respect to the nitrogen requirements of these species since the photosynthetic apparatus accounts for a major portion of plant nitrogen. Access to a heterotrophic carbon source would seem to reduce the demand for plant nitrogen, and in nitrogen limited soils this could be important. It would be of much interest to examine the investment of nitrogen in photosynthetic capital in a facultative parasite grown with and without a host.

In the ecophysiological literature great emphasis is placed on the interrelationships between CO_2 assimilation and leaf conductance, and for a great range of species CO_2 assimilation increases with increasing leaf conductance over a wide range of light intensities and mineral nutrient status. The theory developed from such studies is the optimization hypothesis of Cowan & Farquhar (see references in 74). Stomatal responses to humidity and temperature maximize daily water use efficiency. The water and carbon economy of many parasitic plants appear to depart markedly from the tenets of the optimization hypothesis, with the combination of high transpiration rates and, in many species, low rates of carbon assimilation giving rise to extremely low water use efficiencies.

PHYSIOLOGY OF INFECTED HOSTS

The response of host plants to infection varies from very spectacular growth abnormalities to an almost complete absence of visible symptoms. Probably alterations in the balance of growth regulators result in the development of witches' brooms as seen in conifers infected with the dwarf mistletoe *Arceuthobium* and hardwood trees infected with *Viscum* or *Loranthus* (52). The effects on host growth can be devastating; grain yields of sorghum, maize, and millet can be completely eliminated by infection with species of *Striga* (20). In contrast, some leafy mistletoes may live for decades with their host trees, inducing little apparent damage (88).

Competition for water, inorganic ions, and metabolites is the simplest explanation for losses in host production. The type and extent of the competition are determined by the autotrophic capacity of the parasite, which regulates the demand for resources. We suggest also that the structure and metabolic activity of the haustorium control the flux between host and parasite and therefore further limit host-parasite competition for resources.

Competition for Water

The high transpiration rate of xylem-feeding parasites affects the availability of water to their hosts, and whether or not this leads to water stress in the host depends on the availability of water in the environment. Under conditions of low water availability stomatal closure was induced in *Acacia grasbyi* while the mistletoe *Amyema nestor* continued to transpire at a high rate (39). Consistent with this, damage to mistletoe-infected trees has been attributed to localized parasite-induced water deficits (50).

Unrestricted water use by the parasite may, however, be disadvantageous if the host is severely damaged. In mistletoes it was found that stomatal closure occurred in parallel with that of the host under arid conditions, although somewhat later, thereby conserving water in the association as a whole (103). As discussed above, similar behavior is seen in some perennial root parasitic associations (74). Water losses via the parasite are expected to be low in associations involving species that are principally phloem feeders. Initially *Orobanche*-induced yield reductions are not due to competition for water; however, this factor does become important later, when the capacity for host-root water uptake is reduced as a consequence of carbohydrate loss to the parasite (109). In general it would appear that competition for water is a secondary cause of yield reduction and only becomes important when water availability is restricted.

Nutrient Competition

Knutson (50) suggests that parasite-induced nitrogen deficiency may be responsible for many of the symptoms of parasitic infection. As discussed above, nitrogen acquisition may play a central role in the regulation of assimilate transport in mistletoes. However, there is little evidence for parasite-induced nitrogen deficiencies in host tissues. No differences in leaf nitrogen concentration were found between *Striga*-infected and -uninfected plants of sorghum (97). Lower concentrations of major cations have been observed in the branches of hosts infected with mistletoe, but it has not been established that this causes any ion-deficiency diseases (30). More conclusively, tobacco infected with *Orobanche ramosa* exhibits a 30% growth reduction attributable to potassium deficiency brought about by the high potassium demand of the parasite (24).

Among the mistletoes, the greater pathogenic effect of the *Arceuthobium* species has been attributed to their reliance on host-derived carbohydrates (42, 43). Export of carbon from host to parasite was responsible for 20% of the growth reductions observed in sorghum and for about 16% of the yield loss in millet infected with *S. hermonthica* (35, 36). In addition to direct effects on growth, the loss of carbon compounds may bring about secondary damage to plant processes, as discussed above for *Vicia faba* infected with *Orobanche crenata*. It appears that while competition for inorganic solutes plays a small role in determining host productivity, competition for organic solutes may be an important determinant of host performance. In hemiparasitic associations this accounts for a small but nevertheless significant reduction in host biomass; it is clearly more important in holoparasitic associations. In the absence of detailed growth models it is not possible to determine whether carbohydrate drain is the only major growth limitation on the host.

Metabolic and Physiological Dysfunction

Many of the changes observed in host morphology following infection are consistent with the hypothesis that parasite infection results in an imbalance of growth-regulatory compounds. Reduced internode expansion, lack of floral initiation, and the occurrence of witches' brooms could all reflect changes in host growth-regulator balance. Drennan & El Hiweris (21) have shown considerable changes in concentration of growth promoters and inhibitors in the sap and tissues of sorghum infected with *Striga*. However, split-pot experiments indicated that *Striga* brought about a similar reduction in host growth even when infection was confined to only 30% of the root system (78). This result suggests that a largely healthy root system is unable to compensate for the effects of limited infection and argues for a more direct effect of the parasite on host growth processes.

There is much speculation in the literature that parasitic plants directly perturb the metabolic processes of their hosts, and the idea that they produce toxins has been proposed by a number of authors. It has been suggested that a toxin is responsible for the growth effects of *Striga* on cereals (61, 68), though direct experimental evidence for this is lacking.

In pot experiments, disruption of host photosynthesis appears to account for 80–85% of the growth reduction in sorghum (35) and millet infected with *S. hermonthica* (36). Studies in this laboratory (M. C. Press, J. D. Graves, P. Weigel, and F. A. Mansfield, unpublished) have shown that the water and ionic relations of infected and uninfected sorghum are similar. There were no differences in chlorophyll concentration or activities of the carboxylating enzymes Rubisco and phosphoenolpyruvate carboxylase. However, measurements of quantum yield, carbon dioxide fixation at high intercellular concentrations of carbon dioxide, and host $\delta^{13}C$ values suggest that photosynthetic dysfunction results from impairments to electron transport and/or metabolite shuttling between mesophyll and bundle sheath chloroplasts.

Similar alterations in host photosynthetic characteristics have been reported for fungal parasites (see, e.g., 11) and have in some instances been attributed to the action of toxins. Tentoxin (49), the tetrapeptide produced by *Alternaria*, induces symptoms similar to those seen in *Striga*-infected cereals.

For *Striga* infections, the alterations in host architecture and the reduction in yield reflect physiological and metabolic perturbation induced by the parasite. The mechanisms that bring these changes about are unknown.

CONCLUDING REMARKS

Parasitic plants are sometimes regarded as something of a byway, a small group of curious but not really important flowering plants that receive better coverage in the comic books than in the scientific literature. Such a perspective not only neglects the problems they cause in some agricultural systems but more importantly overlooks their considerable experimental potential in exploring basic physiological and biochemical phenomena. A major area of uncertainty at present is the nature of the complex mechanisms underlying the interactions between host and parasite. Much progress has been made in the last few years regarding the chemistry of events leading from seed germination, through host recognition, to the establishment of the parasitic association. Although we can describe and quantify host responses to parasitic infection we know nothing of the signals and receptors that initiate the sometimes dramatic changes in host development and physiology. Questions of how host and parasite are modified in the association are particularly relevant since this may entail specific changes in gene expression and protein synthesis. An understanding of the mechanisms regulating the stomatal apparatus in parasitic plants and in particular the role played by host-derived signals will contribute generally to our understanding of stomatal behavior.

This fundamental information is needed if we are to develop effective control strategies against parasitic plants that threaten agricultural production in many developing countries.

ACKNOWLEDGMENTS

We thank our colleagues in the UCL *Striga* Research Group for their contributions to the Group's work and for useful discussions. The Group's studies described here were funded by grants from the Overseas Development Natural Research Institute, the Leverhulme Trust, and the Science and Engineering Research Council.

Literature Cited

1. Aber, M., Fer, A., Salle, G. 1983. Etude du transfert des substances organiques de l'hôte (*Vicia faba* L.) vers le parasite (*Orobanche crenata* Forsk) *Z. Planzenphysiol.* 112:297–308
2. Atsatt, P. R. 1977. The insect herbivore as a predictive model in parasitic seed plant biology. *Am. Nat.* 111:579–86
3. Atsatt, P. R. 1983. Host-parasite interactions in higher plants. In *Physiological Plant Ecology III: Responses to the Chemical and Biological Environment. Encyclopedia of Plant Physiology, N.S.,* ed. O. L. Lange, P. S. Nobel, C. B. Osmond, H. Ziegler, 12C: 519–35. Berlin: Springer-Verlag
4. Ayensu, E. S., Doggett, H., Keynes, R. D., Marton-Lefevre, J., Musselman, L. J., Parker, C., Pickering, A., eds. 1984. Striga *Biology and Control.* Paris: ICSU/IDRC
5. Ba, A. T., Kahlem, G. 1979. Mise en évidence d'activatés enzymatiques au niveau de l'haustorium d'une phanérogame parasite; *Striga hermonthica* (Scrophulariaceae). *Can. J. Bot.* 57:2564–67
6. Barlow, B. A., Wiens, D. 1977. Host-

parasite resemblances in Australian mistletoes: the case for cryptic mimicry. *Evolution* 31:69–84

7. Blackwell, R. D., Murray, A. J. S., Lea, P. J., Joy, K. W. 1988. *J. Exp. Bot.* 39:845–58

8. ter Borg, S. J., ed. 1986. *Biology and Control of* Orobanche Wageningen: LH/VPO. 206 pp.

9. Bricaud, C. H., Thalouarn, P., Renaudin, S. 1986. Ribulose 1,5-biphosphate carboxylase activity in the holoparasite *Lathraea clandestina* L. *J. Plant Physiol.* 125:367–70

10. Brooks, D. W., Bevinakatti, H. S., Powell, D. R. 1985. The absolute structure of (+)-strigol. *J. Org. Chem.* 50:3779–81

11. Buchanan, B. B., Hutchinson, S. W., Magyarosy, A. C., Montalbini, P. 1981. Photosynthesis in healthy and diseased plants. In *Effects of Disease on the Physiology of the Growing Plant*, ed. P. G. Ayres, pp. 13–28. Cambridge: Cambridge Univ. Press

12. Chang, M., Lynn, D. G. 1986. Haustoria and the chemistry of host recognition in parasitic angiosperms. *J. Chem. Ecol.* 12:561–79

13. Chang, M., Netzly, D. H., Butler, L. G., Lynn, D. G. 1986. Chemical regulation of distance: characterization of the first natural host germination stimulant for *Striga asiatica*. *J. Am. Chem. Soc.* 108:7858–60

14. Coetzee, J., Fineran, B. A. 1987. The apoplastic continuum, nutrient absorption and plasma tubules in the dwarf mistletoe *Korthalsella lindsayi* (Viscaceae). *Protoplasma* 136:145–53

15. Cook, C. E., Whichard, L. P., Turner, B., Wall, M. E., Egley, G. H. 1966. Germination of witchweed (*Striga lutea* Lour): isolation and properties of a potent stimulant. *Science* 154:1189–90

16. Cook, C. E., Whichard, L. P., Wall, M. E., Egley, G. H., Coggan, P., et al. 1972. Germination stimulants II. The structure of strigol—a potent seed germination stimulant for witchweed (*Striga lutea* Lour.). *J. Am. Chem. Soc.* 94:6198–99

17. Davidson, N. J., True, K. C., Pate, J. S. 1989. Water relations of the parasite: host relationship between the mistletoe *Amyema linophyllum* (Fenzl) Tieghem and *Casuarina obesa* Miq. *Oecologia*. In press

18. de la Harpe, A. L., Visser, J. H., Grobbelaar, N. 1981. Photosynthetic characteristics of some South African parasitic flowering plants. *Z. Pflanzenphysiol.* 103:265–75

19. Doggett, H. 1965. *Striga hermonthica* on sorghum in East Africa. *J. Agric. Sci.* 65:83–194

20. Doggett, H. 1982. Factors reducing sorghum yields, *Striga* and birds. In *Sorghum in the Eighties*, ed. ICRISAT, pp. 313–20. Patancheru: ICRISAT

21. Drennan, D. S. H., El Hiweris, S. O. 1979. Changes in growth regulating substances in *Sorghum vulgare* infected with *Striga hermonthica*. In *2nd International Symposium on Parasitic Weeds*, ed. L. J. Musselman, A. D. Worsham, R. E. Eplee, pp. 144–55. Raleigh: North Carolina State Univ.

22. Ehleringer, J. R., Schulze, E.-D., Ziegler, H., Lange, O. L., Farquhar, G. D., Cowan, I. R. 1985. Xylem-tapping mistletoes: water or nutrient parasites? *Science* 227:1479–81

23. Ehleringer, J. R., Ullmann, I., Lange, O. L., Farquahar, G. D., Cowan, I. R., et al. 1986. Mistletoes; a hypothesis concerning morphological and chemical avoidance of herbivory. *Oecologia* 70:234–37

24. Ernst, W. H. O. 1986. Mineral nutrition of *Nicotiana tabacum* cv. Bursana during infection by *Orobanche ramosa*. See Ref. 8, pp. 80–85

25. Fer, A. 1979. *Contribution a la physiologie de la nutrition des phanérogames parasites; etude du genre* Cuscuta L. *(Convolvulaceae)*. PhD thesis. Univ. Grenoble

26. Fineran, B. A. 1985. Graniferous tracheary elements in haustoria of root parasitic angiosperms. *Bot. Rev.* 51:389–441

27. Fisher, J. T. 1983. Water relations of mistletoes and their hosts. In *The Biology of Mistletoes*, ed. D. M. Calder, P. Bernhardt, pp. 161–83. New York: Academic

28. Givnish, T. J. 1986. Optimal stomatal conductance, allocation of energy between leaves and roots, and the marginal cost of transpiration. In *On the Economy of Plant Form and Function*, ed. T. J. Givnish, pp. 171–213. Cambridge: Cambridge Univ. Press

29. Glatzel, G. 1983. Mineral nutrition and water relations of hemiparasitic mistletoes: a question of partitioning. Experiments with *Loranthus europaeus* on *Quercus petraea* and *Quercus robur*. *Oecologia* 56:193–201

30. Glatzel, G. 1987. Hautorial resistance, foliar development and mineral nutrition in the hemiparasitic mistletoe *Loranthus*

europaeus Jacq. (Loranthaceae). See Ref. 107, pp. 253–62

31. Goldstein, G., Rada, F., Sterngerg, L., Burguera, J. L., Burguera, M., et al. 1989. Gas exchange and water balance of a mistletoe species and its mangrove hosts. *Oecologia*. In press

32. Govier, R. N., Brown, J. G., Pate, J. S. 1968. Hemiparasitic nutrition in angiosperms II. Root haustoria and leaf glands of *Odontites verna* (Bell.) Dum. and their relevance to abstraction of solutes from the host. *New Phytol.* 67:963–72

33. Govier, R. N., Nelson, M. D., Pate, J. S. 1967. Hemiparasitic nutrition in angiosperms I. The transfer of organic compounds from host to *Odontites verna* (Bell.) Dum. (Scrophulariaceae). *New Phytol.* 66:285–97

34. Grace, J. 1983. *Plant-Atmosphere Relations*. London: Chapman & Hall

35. Graves, J. D., Press, M. C., Stewart, G. R. 1989. A carbon balance model of the sorghum-*Striga hermonthica* host-parasite association. *Plant Cell Environ.* 12:101–7

36. Graves, J. D., Wylde, A., Press, M. C., Stewart, G. R. 1990. Growth and carbon allocation in the *Pennisetum typhoides* infected with the parasitic angiosperm *Striga hermonthica*. *Plant Cell Environ.* 13: In press

37. Hall, P. J., Badeboch-Jones, J., Parker, C. W., Letham, D. S., Barlow, B. A. 1987. Identification and quantification of cytokinins in the xylem sap of mistletoes and their hosts in relation to leaf mimicry. *Aust. J. Plant Physiol.* 14:429–38

38. Hartel, O. 1956. Der Wasserhaushalt der Parasiten. In *Handbuch der Planzenphysiologie III*, ed. O. Stocker, pp. 951–60. New York: Springer

39. Hellmuth, E. O. 1971. Ecophysiological studies on plants in arid and semi-arid regions in Western Australia IV. Comparison of the field physiology of the host, *Acacia grasbyi*, and its hemiparasite, *Amyema nestor*, under optimal and stress conditions. *J. Ecol.* 59:5–17

40. Herb, R., Visser, J. H., Schildknecht, H. 1987. Recovery, isolation and preliminary structural investigation of germination stimulants produced by *Vigna unguiculata* Walp. cv Saunders upright. See Ref. 107, pp. 351–66

41. Hollinger, D. Y. 1983. Photosynthesis and water relations of the mistletoe *Phoradendron villosum*, and its host the Californian valley oak, *Quercus lobata*. *Oecologia* 60:396–400

42. Hull, R. J., Leonard, O. A. 1964. Physiological aspects of parasitism in mistletoes (*Arceuthobium* and *Phoradendron*) I. The carbohydrate nutrition of mistletoes. *Plant Physiol.* 39:996–1007

43. Hull, R. J., Leonard, O. A. 1964. Physiological aspects of parasitism in mistletoes (*Arceuthobium* and *Phoradendron*) II. The photosynthetic capacity of mistletoe. *Plant Physiol.* 39:1008–17

44. Johnson, A. W., Gowda, G., Hassanali, A., Knox, J., Monaco, S., et al. 1981. The preparation of synthetic analogues of strigol. *J. Chem. Soc. Perkin Trans.* I, pp. 1734–43

45. Johnson, A. W., Rosebury, G., Parker, C. 1976. A novel approach to *Striga* and *Orobanche* control using synthetic germination stimulants. *Weed Res.* 16:223–27

46. Joy, K. W. 1988. Ammonia, glutamine, and asparagine: a carbon-nitrogen interface. *Can. J. Bot.* 66:2103–9

47. Kammerling, Z. 1910. Verdunstungsversuche mit tropischen Loranthacean. *Ber. Dtsch. Bot. Ges.* 32:17–24

48. Klaren, C. H., van de Dijk, S. J. 1976. Water relations of the hemiparasite *Rhinanthus serotinus* before and after attachment. *Physiol. Plant.* 38:121–25

49. Klotz, M. G. 1988. The action of tentoxin on membrane processes in plants. *Physiol. Plant.* 74:575–82

50. Knutson, D. M. 1979. How parasitic seed plants induce disease in other plants. In *Plant Disease: An Advanced Treatise*, ed. J. G. Horsfall, E. B. Cowling, 4:293–312. New York: Academic

51. Kramer, P., Kozlowski, T. T. 1979. *Physiology of Woody Plants*. New York: Academic

52. Kuijt, J. 1969. *The Biology of Parasitic Flowering Plants*. Berkeley: Univ. Calif. Press

53. Kuo, J., Pate, J. S., Davidson, N. J. 1989. Ultrastructure of the haustorial interface and apoplastic continuum between host and the root hemiparasite *Olax phyllanthi* (Labill.) R. Br. (Olacaceae). *Protoplasma* 150:27–39

54. Lamont, B. 1983. Mineral nutrition of mistletoes. In *The Biology of Mistletoes*, ed. D. M. Calder, P. Bernhardt, pp. 185–204. New York: Academic

55. Lynn, D. G., Steffens, J. C., Kamut, V. S., Graden, D. W., Shabanowitz, J., Riopel, J. L. 1981. Isolation and characterization of the first host recognition substances for parasitic angiosperms. *J. Am. Chem. Soc.* 103:1868–70

56. Mallaburn, P. S., Press, M. C., Stewart, G. R. 1990. Haustorial structure and function in *Striga hermonthica*. *J. Exp. Bot.* In press

57. Mallaburn, P. S., Stewart, G. R. 1987. Haustorial function in *Striga:* comparative anatomy of *S. asiatica* (L.) Kuntz and *S. hermonthica* (Del.) Benth. (Scrophulariaceae). See Ref. 107, pp. 523–36

58. McNally, S. F., Orebamjo, T. O., Hirel, B., Stewart, G. R. 1983. Glutamine synthetase isoenzymes of *Striga hermonthica* and other angiosperm root parasites. *J. Exp. Bot.* 34:610–19

59. McNally, S. F., Stewart, G. R. 1987. Inorganic nitrogen assimilation by parasitic angiosperms. See Ref. 107, pp. 539–46

60. Mhehe, G. L. 1987. A novel chemical approach to the control of witchweed (*Striga asiatica* (L.) Kuntze) and other *Striga* spp. (Scrophulariaceae). See Ref. 107, pp. 563–74

61. Musselman, L. J. 1980. The biology of *Striga, Orobanche,* and other root-parasitic weeds. *Annu. Rev. Phytopathol.* 18:463–89

62. Musselman, L. J., ed. 1987. *Parasitic Weeds in Agriculture I. Striga.* Florida: CRC Press

63. Musselman, L. J., Wegmann, K., eds. 1990. *Recent Advances in* Orobanche *Research.* In press

64. Nandakumar, S., Kachru, D. N., Krishnan, P. S. 1976. Threonin-serine dehydratase activity in angiospermous parasites. *New Phytol.* 18:613–18

65. Netzly, D. H., Riopel, J. L., Ejeta, G., Butler, L. G. 1988. Germination stimulants of witchweed *(Striga asiatica)* from hydrophobic root exudate of sorghum *(Sorghum bicolor). Weed Sci.* 36:441–46

66. Nour, J. J., Todd, P., Yaghmaie, P., Panchal, G., Stewart, G. R. 1984. The role of mannitol in *Striga hermonthica.* See Ref. 68, pp. 81–89

67. Okonkwo, S. N. C. 1987. Developmental studies on witchweeds. See Ref. 62, pp. 63–74

68. Parker, C. 1984. The physiology of *Striga* spp.: present state of knowledge and priorities for future research. See Ref. 68, pp. 179–93

69. Parker, C., Musselman, L. J., Polhill, R. M., Wilson, A. K., eds. 1984. *Proceedings of the Third International Symposium on Parasitic Weeds.* Aleppo: ICARDA/IPSPRG

70. Pate, J. S., Kuo, J., Davidson, N. J. 1989. Morphology and anatomy of the haustorium of the root hemiparasite *Olax phyllanthi* (Labill.) R. Br. (Olaceae), with special reference to the haustorial interface. *Ann. Bot.* In press

71. Percival, W. C. 1931. The parasitism of *Conophilis americana* on *Quercus borealis. Am. J. Bot.* 18:817–37

72. Popp, M. 1987. Osmotica in *Amyema miquelii* (Lehm. ex Mig.) Tieghem. and *Amyema pendulum* (Sieber ex Sprengel) Tieghem, (Loranthaceae) on different hosts. See Ref. 107, pp. 621–30

73. Press, M. C. 1989. Autotrophy and heterotrophy in root hemiparasites. *Trends Ecol. Evol.* 4:258–63

74. Press, M. C., Graves, J. D., Stewart, G. R. 1988. Transpiration and carbon acquisition in root hemiparasites. *J. Exp. Bot.* 39:1009–14

75. Press, M. C., Nour, J. J., Bebawi, F. F., Stewart, G. R. 1989. Antitranspirant-induced heat stress in the parasitic plant *Striga hermonthica*—a novel method of control. *J. Exp. Bot.* 40:585–91

76. Press, M. C., Shah, N., Stewart, G. R. 1986. The parasitic habit: trends in metabolic reductionism. See Ref. 8, pp. 96–107

77. Press, M. C., Shah, N., Tuohy, J. M., Stewart, G. R. 1987. Carbon isotope ratios demonstrate carbon flux from C_4 host to C_3 parasite. *Plant Physiol.* 85:1143–45

78. Press, M. C., Stewart, G. R. 1987. Growth and photosynthesis in *Sorghum bicolor* infected with *Striga hermonthica. Ann. Bot.* 60:657–62

79. Press, M. C., Tuohy, J. M., Stewart, G. R. 1987. Gas exchange characteristics of the sorghum-*Striga* host-parasite association. *Plant Physiol.* 84:814–19

80. Raven, J. A. 1983. Phytophages of xylem and phloem: a comparison of animal and plant sap-feeders. *Adv. Ecol. Res.* 13:135–234

81. Raven, J. A., Smith, F. A. 1976. Nitrogen assimilation and transport in vascular land plants in relation to intracellular pH regulation. *New Phytol.* 76:415–31

82. Renaudin, S., Larher, F. 1981. The transfer of organic substances from host (*Alnus glutinosa* Gaertn.) to the holoparasite (*Lathraea clandestina* L.). *Z. Planzenphysiol.* 104:71–80

83. Renaudin, S., Vidal, J., Larher, F. 1982. Characterization of phosphoenolpyruvate carboxylase in a range of parasitic phanerogames. *Z. Pflanzenphysiol.* 106:229–37

84. Richter, A., Popp, M. 1987. Patterns of

organic acids and solutes in *Viscum album* L. on 12 different hosts. See Ref. 107, pp. 709–14

85. Salle, G., Dembele, B., Raynal-Roques, A., Hallais, M. F. 1987. Biological aspects of *Striga* species, pest of food crops (Scrophulariaceae). See Ref. 107, pp. 719–31

86. Sand, P. F. 1987. The American witchweed quarantine and eradication program. See Ref. 107, pp. 207–23

87. Scholander, P. F., Hammel, H. T., Bradstreet, E. D., Hemmingson, E. A. 1965. Sap pressure in vascular plants. Negative hydrostatic pressure can be measured in plants. *Science* 148:339–46

88. Schulze, E.-D., Ehleringer, J. R. 1984. The effect of nitrogen supply on growth and water-use efficiency of xylem-tapping mistletoes. *Planta* 162:268–75

89. Schulze, E.-D., Turner, N. C., Glatzel, G. 1984. Carbon, water and nutrient relations of two mistletoes and their hosts: a hypothesis. *Plant Cell Environ.* 7:293–99

90. Seledzhanu, N., Galan-Fabian, D. 1962. Nutrition of common mistletoe. *Fiziol. Rastenii* 8:436–42

91. Shah, N., Smirnoff, N., Stewart, G. R. 1987. Photosynthesis and stomatal characteristics of *Striga hermonthica* in relation to its parasitic habit. *Physiol. Plant.* 69:699–703

92. Smith, D. C., Muscatine, L., Lewis, D. H. 1969. Carbohydrate movement from autotrophs to heterotrophs in parasitic and mutualistic symbiosis. *Biol. Rev.* 44:17–90

93. Smith, P. L., Stewart, G. R. 1987. *Striga gesnerioides* (Willd.) Vatke : haustorial ontogeny and a role for ergastic substances in the maintainance of physiological integrity. See Ref. 107, pp. 763–74

94. Smith, S., 1990. Stomatal behaviour of *Striga hermonthica*. See Ref. 63. In press

95. Steffens, J. C., Lynn, D. G., Kamat, V., Riopel, J. L. 1982. Molecular specificity of haustorial induction in *Agalinis purpurea* (L.) Raf. (Scrophulaariaceae). *Ann. Bot.* 50:1–7

96. Steffens, J. C., Lynn, D. G., Riopel, J. L. 1986. An haustorial inducer for the root parasite *Agalinis purpurea*. *Phytochemistry* 25:2291–98

97. Stewart, G. R., Nour, J. J., MacQueen, M., Shah, N. 1984. Aspects of the biochemistry of *Striga*. See Ref. 1, pp. 161–78

98. Stewart, G. R., Orebamjo, T. O. 1980. Nitrogen status and nitrate reductase

activity of the parasitic angiosperm *Tapinanthus bangwensis* (Engl. & K. Krause) Danser growing on different hosts. *Ann. Bot.* 45:587–89

99. Thalourn, P., Philouze, V., Renaudin, S. 1988. Nitrogen metabolism key enzymes in a Scrophulariaceae holoparasite *Lathraea clandestina* L. *J. Plant Physiol.* 132:63–66

100. Thalourn, P., Rey, L., Hirel, B., Renaudin, S., Fer, A. 1987. Activity and immunocytochemical localisation of glutamine synthetase in *Lathraea clandestina* L. *Protoplasma* 141:95–100

101. Thalouarn, P., Rey, L., Renaudin, S. 1986. Carbon nutrition in a Rafflesiaceae holoparasite *Cytinus hypocitis* L. fixed on or experimentally isolated from the host *Cistus monspeiliensis* L. *J. Plant Physiol.* 123:271–81

102. Tuohy, J. M., Smith, E. A., Stewart, G. R. 1986. The parasitic habit: trends in morphological and ultrastructural reductionism. See Ref. 8, pp. 86–95

103. Ullmann, I., Lange, O. L., Ziegler, H., Ehleringer, J. R., Schulze, E.-D., Cowan, I. R. 1985. Diurnal courses of leaf conductance and transpiration of mistletoes and their hosts in central Australia. *Oecologia* 67:577–87

104. Visser, J. H., Dorr, I. 1987. The haustorium. See Ref. 62, pp. 91–106

105. Visser, J. H., Dorr, I., Kollmann, R. 1984. The hyaline body of the root parasite *Alectra orobanchoides* Benth. (Scrophulariaceae)—its anatomy, ultrastructure and histochemistry. *Protoplasma* 121:146–56

106. Visser, J. H., Herb, R., Schildknecht, H. 1987. Recovery and preliminary chromatographic investigation of germination stimulants produced by *Vigna unguiculata* Walp. cv. Saunders upright. *J. Plant Physiol.* 129:375–81

107. Weber, H. C., Forstreuter, W., eds. 1987. *Parasitic Flowering Plants. Proceedings of the 4th International Symposium on Parasitic Flowering Plants.* Marburg: Phillips Universität

108. Wegmann, K. 1986. Biochemistry of osmoregulation and possible biochemical reasons of resistance against *Orobanche*. See Ref. 8, pp. 107–17

109. Whitney, P. J. 1972. The carbohydrate and water balance of beans *(Vicia faba)* attacked by broomrape *(Orobanche crenata)*. *Ann. Appl. Biol.* 70:59–66

110. Whittington, J., Sinclair, R. 1988. Water relations of the mistletoe *Amyema miquelii* and its host *Eucalyptus fasiculosa*. *Aust. J. Bot.* 36:239–56

111. Wolswinkel, P. 1974. Complete inhibi-

tion of setting and growth of fruits of *Vicia faba* L., resulting from the draining of the phloem by *Cuscuta* species. *Acta Bot. Neerl.* 23:48–60

112. Wolswinkel, P. 1978. Phloem unloading in stem parts by *Cuscuta:* the release of ¹⁴C and K⁺ to the free space at 0 °C and 25 °C. *Physiol. Plant.* 42:167–72

113. Wolswinkel, P. 1985. Phloem unloading and turgor-sensitive transport: factors involved in sink control of assimilate partitioning. *Physiol. Plant.* 65:331–39

114. Wolswinkel, P., Ammerlaan, A., Peters, H. F. C. 1984. Phloem unloading of amino acids at the site of *Cuscuta europeae*. *Plant Physiol.* 75:13–20

115. Worsham, A. D. 1987. Germination of witchweed seeds. See Ref. 62, pp. 45–61

116. Wyn Jones, R. G. 1984. Phytochemical aspects of osmotic adaptation. *Rec. Adv. Phytochem.* 18:55–78

117. Yoshikawa, F., Worsham, A. D., Moreland, D. E., Eplee, R. E. 1978. Biochemical requirements for seed germination and shoot development of witchweed *(Striga asiatica)*. *Weed Sci.* 26:119–23

118. Zwanenburg, B. 1990. Design and synthesis of germination stimulants for *Striga* and *Orobanche*. See Ref. 63. In press

119. Zwannenburg, B., Mhehe, G. L., 't Lam, G. K., Dommerholt, F. J., Kishimba, M. A. 1986. The search for new germination stimulants of *Striga* spp. See Ref. 107, pp. 35–41

Annu. Rev. Plant Physiol. Plant Mol. Biol. 1990. 41:153–85

FRUCTOSE-2,6-BISPHOSPHATE AS A REGULATORY MOLECULE IN PLANTS

Mark Stitt

Lehrstuhl für Pflanzenphysiologie, Universität Bayreuth, 8580 Bayreuth, West Germany 0921-55-2624

CONTENTS

Fru2,6bisP is a signal metabolite found universally in eukaryotes (91, 143). It usually acts to regulate glycolysis and gluconeogenesis via its action as an activator of PFK and an inhibitor of the Fru1,6Pase. Accordingly, it has been proposed that Fru2,6bisP can be viewed as a "glycolytic signal" (56, 143). Fru2,6bisP is synthesized and degraded by two specific enzymes, called Fru6P,2-kinase and Fru2,6Pase (91, 143), and its concentration can be altered by changing the activity of these enzymes. This signal system can

153

1040-2519/90/0601-0000$02.00

transmit intra- and extracellular information. The main intracellular information in animals and fungal systems relates to the balance between 6-carbon and 3-carbon phosphorylated intermediates. This information is transmitted because selected representatives of these pools modulate Fru6P,2-kinase and Fru2,6Pase. Extracellular information is transmitted in liver and yeast, where these proteins are regulated by cAMP-dependent protein phosphorylation.

In plants, many elements of this system are preserved, but there have also been some important changes. The most clearly defined role for Fru2,6bisP is in the regulation of photosynthetic sucrose synthesis (125, 131). The role of Fru2,6bisP during glycolytic metabolism is less well defined; it clearly functions differently in plants than in other eukaryotes. These differences involve unique features in plant primary metabolism. Instead of activating PFK, Fru2,6bisP activates an enzyme termed PFP, which catalyses a reversible phosphorylation of Fru6P using PPi as a phosphoryl donor (24, 103, 145). Although PFP was originally thought to be a glycolytic enzyme, I argue here that it may have a more complex role. The study of Fru2,6bisP and PFP has been complicated by two further unusual aspects of plant metabolism, namely the extensive compartmentation of metabolism between the cytosol and the plastid (1, 27, 28), and the unusual role of PPi as an energy donor in the cytosol (3, 14, 97, 140, 149).

I first discuss the in vitro properties of the enzymes responsible for the synthesis and degradation of Fru2,6bisP, and then relate their properties to in vivo measurements of Fru2,6bisP and metabolites to illustrate what information is being transmitted via changes of Fru2,6bisP. I next discuss how Fru2,6bisP acts on its target enzymes in vitro. In the last section I discuss how Fru2,6bisP regulates sucrose synthesis in vivo and, more controversially, the possible significance of PFP. The reader is referred to earlier reviews of Fru2,6bisP in plants (22, 53, 54, 114), to reviews of its role during photosynthetic sucrose metabolism (125, 131), and to reviews of its role in nonphotosynthetic metabolism (3, 13, 14, 140, 141).

THE ENZYMES THAT SYNTHESIZE AND DEGRADE FRU2,6BISP

The initial stages of research on Fru2,6bisP (51) were dominated by studies on liver, where Fru6P,2-kinase and Fru2,6Pase (the enzyme activities that synthesize and degrade this signal metabolite) are located on a special bifunctional protein (91, 143). However, it has subsequently become clear that there is considerable variation, depending on the tissue, and that this bifunctional enzyme is not always present (143). Research in plants has mirrored these developments.

Presence and Location of Fru2,6bisP-Metabolizing Enzymes

The earliest studies in plants indicated that they also contained a bifunctional enzyme (23, 120). Subsequently, Larondelle et al (72) showed that a high-affinity Fru2,6Pase ($K_m \sim 30$ nM) copurified with spinach leaf Fru6P,2-kinase during a 9300-fold purification. As with the liver bifunctional enzyme, incubation with [2^{32}P]-Fru2,6bisP leads to the formation of a ^{32}P-labeled phosphoprotein. The spinach leaf enzyme is probably a tetramer with a subunit molecular mass of 90–100 kD (72, 81), whereas the enzyme from liver is a dimer with a subunit molecular mass of 59 kD (91). The ratio of Fru2,6Pase:Fru6P,2-kinase activity in the spinach enzyme (1:10) is rather low (72).

Plants also contain a monofunctional Fru2,6Pase (72, 73, 79–81) similar to the enzyme found in yeast (41). This monofunctional enzyme can be separated from Fru6P,2-kinase by anion-exchange (79, 81) or blue Sepharose chromatography (73). It has been purified to homogeneity and shown to be a dimer (subunit molecular mass 33 kD) that lacks Fru6P,2-kinase activity (80). The monofunctional Fru2,6bisP is highly specific for the osyl-link at the carbon-2 position of Fru2,6bisP, is specific for Fru2,6bisP compared to other substrates (73, 80), and has a neutral pH optimum (80) but a rather high K_m of about 30μM (73, 80). Fru6P,2-kinase and the monofunctional Fru2,6Pase are located in the cytosol in spinach leaves (80), as are Fru2,6bisP itself (112, 124) and the target enzymes Fru1,6Pase and PFP (24, 134).

Spinach leaves contain other low-affinity Fru2,6bisP-hydrolyzing enzymes, which are not specific for the carbon-2 position (73). A similar activity has also been purified from yeast (93). These enzymes hydrolyze other substrates and probably represent nonspecific phosphatases (41, 73). The presence of these enzymes means that it is difficult to interpret measurements of Fru2,6bisP disappearance in crude extracts, especially when they are carried out at high Fru2,6bisP concentrations and without checking that Fru6P is the only product.

Regulation by Metabolites

I now discuss how these enzymes are affected by metabolites in vitro. The Fru2,6bisP concentration in vivo depends on the relative activity of Fru6P,2-kinase and Fru2,6Pase. Metabolites that modulate these enzymes therefore represent potential inputs into this signaling system. Most of the following information has been obtained using the enzymes from spinach leaves.

FRU6P,2-KINASE   Spinach leaf Fru6P,2-kinase is activated by Pi, which increases the V_{max} and the affinity for Fru6P (21, 72, 120). Indeed Fru6P,2-kinase from all eukaryotes is dependent on Pi (143). Fru6P,2-kinase is also activated by Fru6P. The enzyme shows a sigmoidal saturation curve for Fru6P, especially when Pi is low, or when inhibitory 3-carbon phosphoesters are present (21, 72, 120). This property is also shared with the liver and yeast enzymes.

Fru6P,2-kinase is inhibited by several 3-carbon phosphorylated intermediates including 3PGA, 2PGA, PEP, dihydroxyacetone P, and the 2-carbon intermediate glycollate-2-P (21, 72, 120). These compounds act antagonistically to Pi, except for dihydroxy-acetone-P, which is a poor inhibitor unless high (10 mM) Pi concentrations are present (120, 121). The K_i values for 3PGA, 2 PGA, and PEP are similar (72). However, owing to the equilibrium constants of the reactions catalyzed by phosphoglycerate mutase and enolase, these metabolites are normally present in a 10:1:3 ratio, and 3PGA is therefore the most important inhibitor in vivo (84). Spinach leaf Fru6P,2-kinase is also inhibited by PPi, which acts competitively to ATP and induces sigmoidal ATP saturation kinetics (72, 81).

Preliminary investigations in maize leaves (110), germinating castor bean cotyledons (66), and a range of storage tissues (121) suggest that these tissues contain a Fru6P,2-kinase resembling that from spinach leaves, but a detailed purification and study of Fru6P,2-kinase from different tissues have not yet been carried out.

BIFUNCTIONAL FRU2,6PASE   Studies carried out with the highly purified spinach leaf bifunctional enzyme (72, 73) have confirmed earlier studies with a less highly purified preparation (23, 120) and shown that Fru6P and Pi both inhibit Fru2,6bisP hydrolysis. Fru6P is a powerful noncompetitive inhibitor. At 0.2 mM, Fru6P produces a 50% inhibition. In contrast, Pi is a relatively weak competitive inhibitor (72, 120). For example, when the mixed preparation was assayed in conditions (10μM Fru2,6bisP in the presence of MgCl$_2$) that would only register the bifunctional enzyme, 20 mM Pi was needed to produce a 50% inhibition (120). Pi also decreases the sensitivity to inhibition by Fru6P (72, 120). An identical pattern of product inhibition is found for the liver bifunctional enzyme (91, 143).

MONOFUNCTIONAL FRU2,6PASE   Fru6P (80) and Pi (73, 80) are mixed-type inhibitors of the monofunctional Fru2,6Pase. Fru6P is a rather weak inhibitor. At 10 uM Fru2,6bisP, 1 and 4 mM Fru6P only reduce activity by 10% and 30%, respectively (80). In contrast, Pi is a powerful inhibitor of the monofunctional enzyme. At 10 μM Fru2,6bisP, activity was inhibited 50% by 0.5mM Pi (80). Comparison of the data published for the bifunctional (72) and monofunctional (73) Fru2,6Pase suggests that Pi may be a 1000-fold better inhibitor (relative to the K_m for Fru2,6bisP) of the monofunctional enzyme than the bifunctional enzyme. The monofunctional enzyme also differs from the bifunctional enzyme in being inhibited by Fru1,6bisP (73) and by divalent cations including Mg^{2+}, Ca^{2+}, and Mn^{2+}, which act competitively to the substrate (73, 79–81). Activity is also inhibited by AMP (80) but only at unphysiologically high concentrations.

RELATIVE CONTRIBUTION OF THE BIFUNCTIONAL AND MONOFUNCTIONAL FRU2,6PASE   The monofunctional enzyme has an extremely high activity. Its V_{max} is 1000- and 300-fold higher than the V_{max} of the bifunctional Fru2,6Pase in spinach leaves (73, 80) and artichoke tubers (73), respectively. However, three factors are likely to decrease the activity of the monofunctional Fru2,6Pase in vivo. First, the physiological Fru2,6bisP concentration usually lies between 1 and 15 uM (66, 83, 84, 124, 129, 132), well below the K_m of the monofunctional Fru2,6Pase. Second, the monofunctional enzyme will be strongly inhibited in vivo by Pi, which has been estimated to be about 10 mM in leaves (77) and 1–10 mM in nonphotosynthetic root tips (99) and sycamore cell suspension cultures (100). Third, the monofunctional enzyme will be inhibited in vivo by divalent cations, especially Mg^{2+}. On the other hand, the bifunctional Fru2,6Pase is likely to be always partially inhibited by the cytosolic concentration of Fru6P [0.1–1 mM (see 45, 125)].

Larondelle et al (73) have compared the V_{max} activities of the bifunctional and the monofunctional Fru2,6Pase with the rate of Fru2,6bisP disappearance after darkening spinach leaves (124) or adding mannose to artichoke tubers (73). They concluded that the bifunctional enzyme could just about cope with the rate of Fru2,6bisP disappearance. However, since this enzyme will be partially inhibited by Fru6P (see above) I suggest that the monofunctional enzyme is likely to be needed too. The differing properties of the two Fru2,6Pases suggest, as a working hypothesis, that the bifunctional enzyme catalyzes Fru2,6bisP breakdown in response to falling Fru6P, while the monofunctional enzyme catalyzes Fru2,6bisP removal in response to falling Pi.

POSSIBLE CONTRIBUTION OF COVALENT MODIFICATION   There is still no evidence for phosphorylation of Fru6P,2-kinase in plants. Although an ATP-dependent phosphorylation of Fru6P,2-kinase (8) and activation of a novel partially purified Fru6P,2-kinase (146) have been reported, these observations could not be confirmed (54); and other groups have reported negative results after incubating spinach leaf Fru6P,2-kinase with protein kinases (21, 72). Some confusion may have been introduced by the susceptibility of Fru6P,2-kinase to proteolytic degradation. It is known that partial proteolysis of liver Fru6P,2-kinase leads to a selective loss of Fru6P,2-kinase activity without affecting Fru2,6Pase activity (91, 143). Several groups working on spinach leaf Fru6P,2-kinase have described different "forms" for Fru6P,2-kinase, which differ in the relative amount of Fru2,6Pase activity (72, 133, 146). These can be suppressed if protease inhibitors are included (72, 79, 80).

REGULATION OF THE FRU2,6BISP CONCENTRATION IN VIVO: WHAT INFORMATION IS THE SIGNAL METABOLITE TRANSMITTING?

These in vitro properties of Fru6P,2-kinase and Fru2,6Pase suggest that Fru2,6bisP could be acting as a signal for several metabolic parameters including (a) the level of 3-carbon and (b) 6-carbon phosphorylated intermediates, (c) the metabolic pool of phosphate and (d) the balance between PPi and ATP. In the following section I consider evidence that these potential inputs actually operate to modify the Fru2,6bisP concentration in vivo. I also discuss whether, despite the absence of in vitro evidence, there may be additional "higher" mechanisms regulating the Fru2,6bisP concentration.

Authentication of In Vivo Metabolite Measurements

Before considering the in vivo relation between Fru2,6bisP and other metabolites I comment on the precautions needed in making such measurements. It is essential that metabolism is rapidly quenched and that the metabolite is adequately extracted and assayed. Unfortunately, many published studies have not taken these precautions.

Metabolite pools can change rapidly, with half times as short as 0.1 sec for ADP or PPi or about 1 sec for PGA (131). Even the relatively sluggish Fru2,6bisP pool can change dramatically within 30 sec (126). It is therefore essential to quench metabolism in liquid N_2. If the tissue is carrying out photosynthesis, constant illumination should be maintained. For bulky storage tissues, freeze-clamping is essential. Great care is needed in interpreting measurements in sections cut out of bulky tissues, because slicing induces rapid changes of fluxes and metabolite levels (84).

Methods for extracting Fru2,6bisP have been reviewed (117). The reliability of the extraction and assay should be checked by carrying out recovery experiments. A small amount of metabolite, similar to the amount in the tissue, is added to the tissue in the killing mixture and its recovery is monitored through the extraction process (2, 6, 60, 108, 128, 134). It is essential to check each tissue studied, including different developmental stages if these are being studied. Substantial and selective losses of Fru2,6bisP can occur in many plant tissues, leading to very erroneous conclusions (compare 6 and 83).

Fru3,6bisP is measured in a bioassay, which exploits its ability to activate PFP (24, 145). It is essential to include an internal standard curve in each Fru2,6bisP bioassay (117, 128), because other compounds in the extracts can affect PFP activity or its sensitivity to Fru2,6bisP. It may also be helpful to treat the extract with activated charcoal or albumin to remove phenols (88) or with micro anion-exchange columns to remove high concentrations of organic acids (36).

Fru2,6bisP is a Signal for the Ratio of 3-Carbon Phosphorylated Intermediates : Pi

A SIGNAL FOR THE AVAILABILITY OF FIXED CARBON DURING PHOTOSYNTHESIS   When photosynthesis increases in response to rising light intensity or CO_2 concentration, the immediate products of CO_2 fixation—3PGA and triose-phosphates—increase (5, 37, 87, 127, 129) and Pi probably decreases (107). These changes would be expected to inhibit Fru6P,2-kinase and activate Fru2,6bisP hydrolysis (see above). In agreement, Fru2,6bisP decreases 3–5 fold when the light intensity or CO_2 concentration is increased (87, 108, 127, 129). Experiments with inhibitors have provided further evidence that Fru2,6bisP acts as a signal for the availability of fixed carbon. Addition of uncouplers and tentoxin to inhibit the ATP synthase (95), or methyl viologen to interfere with the supply of NADPH (87), all lead to a 3–5-fold increase of Fru2,6bisP. In every case, the increase of Fru2,6bisP could be explained in terms of a decrease in the pools of 3-carbon intermediates, in particular 3PGA. As discussed below, this decrease of Fru2,6bisP relieves the inhibition of the cytosolic Fru1,6Pase and contributes to the activation of sucrose synthesis.

It was previously suggested that 3PGA and triose-phosphates both contribute to this decrease of Fru2,6bisP (120, 129), without specifying which plays the major role. Both metabolites usually increase in conditions where photosynthesis is increasing, but their effectiveness as inhibitors of Fru6P,2-kinase depends on the cytosolic Pi concentration (see above), which is not known with certainty. Two lines of evidence now show that the rising 3PGA : Pi ratio is the crucial signal. First, after supplying methyl viologen the normal response of Fru2,6bisP to increasing rates of photosynthesis is abolished (87). Methyl viologen leads to a large increase of triose phosphates, but PGA no longer increases, suggesting this is the signal that normally drives Fru2,6bisP down. Second, the latest ^{31}P-NMR estimate of the cytoplasmic Pi concentration in leaves suggests it may be as low as 10 mM, even in the dark (77). Earlier estimates were considerably higher (29, 39, 123), probably owing to technical problems in quantifying this signal in leaves (77). The lower Pi concentration favors a dominant role for 3PGA in regulating Fru6P,2-kinase activity.

This conclusion simplifies our understanding of the recycling of Pi and the regulation of photosynthate partitioning. When the rate of photosynthesis increases, sucrose synthesis must be activated to recycle the Pi that has been incorporated into the various phosphorylated intermediates in the Calvin cycle (131). It is well established from studies with isolated chloroplasts that a decreased supply of Pi leads to a decreased ATP:ADP ratio and thus inhibits the reduction of 3PGA (49, 92). Considerable evidence has accumulated that a similar chain of events occurs in leaves, from experiments using Pi-sequestering agents like mannose (47) or glycerol (75), or using mutants with a lowered rate of sucrose synthesis (88), or studying transients during which the rate of photosynthesis is temporarily inhibited by a low rate of sucrose synthesis (74, 88, 106, 126). Thus, a rising 3PGA:Pi ratio signals that photosynthesis is occurring and end-product synthesis is needed to recycle Pi. This rising ratio then acts to stimulate sucrose synthesis by driving the Fru2,6bisP concentration downwards and activating the cytosolic Fru1,6Pase. It also acts to stimulate starch synthesis by activating ADP glucose pyrophosphorylase (92).

A SIGNAL FOR RESPIRATORY ACTIVITY   Changes of the 3PGA:Pi ratio also regulate Fru2,6bisP during respiratory metabolism. This is seen most clearly when short-term metabolic changes are studied—for example, after inducing wound respiration by slicing artichoke (84), carrot, or potato tubers (119); during short-term anaerobiosis (84); or after adding uncouplers, glucose, or transported glucose analogs (47b,c). In these experiments, the increased rate of O_2 uptake or dark respiration was due to activation of pyruvate kinase or PEP carboxylase, and there was a rapid decrease of 3PGA. This was followed, 2–5 minutes later, by an increase of Fru2,6bisP. In other words, Fru2,6bisP concentration rises in response to an increase in the rate of respiration or biosynthesis, which is signalled via a decrease in the pools of the 3-carbon intermediates.

Rising Fru2,6bisP Signals when Hexose-Phosphates are Accumulating

FEEDBACK CONTROL OF SUCROSE SYNTHESIS   The properties of Fru6P,2-kinase and the bifunctional Fru2,6Pase suggest that an increase of Fru6P should lead to an increase of Fru2,6bisP. This has been tested in two experimental systems. (a) When sucrose accumulates in spinach leaves there is a deactivation of SPS (138), probably due to covalent protein modification (147, 148). The resulting increase of the cytosolic Fru6P pool (45) is accompanied by an increase of Fru2,6bisP (124, 125). (b) Mutants of Clarkia xantiana with reduced activity of the cytosolic phosphoglucose isomerase have been used to increase specifically the Fru6P pool (64) and show that this is accompanied by an increase of Fru2,6bisP (88). Thus, rising Fru2,6bisP signals that hexose-phosphates are being produced too rapidly. The increased Fru2,6bisP then acts to restrict the cytosolic Fru1,6Pase and redirect photosynthate towards starch synthesis in the chloroplasts.

In both experimental systems the increase of Fru2,6bisP was 2–3 fold larger than the increase of Fru6P (88, 115, 125). This provides a measure of the amplification available in this regulator cycle, in which Fru6P is acting simultaneously to stimulate the synthesis and inhibit the degradation of Fru2,6bisP. For comparison, a theoretical analysis suggested that an amplification of four could easily be attained (115).

Several factors will be acting to dampen the in vivo response of Fru2,6bisP to rising Fru6P. First, when Fru6P and other phosphorylated intermediates increase, we must expect that the cytosolic Pi will decrease (at least in the short term; see below), and the monofunctional bisphosphatase will be activated (79, 80). Second, inhibition of the cytosolic Fru1,6Pase will lead to an increase of 3PGA (88), which will reinhibit Fru6P,2kinase. Third, inhibition of the Fru1,6Pase will decrease the extent to which Fru6P accumulates, thus acting to dampen the initial peturbation. This example serves to illustrate how the Fru2,6bisP concentration at any one time reflects a balance between different inputs.

FEEDFORWARD CONTROL OF RESPIRATION?   There is less evidence that rising Fru6P plays a decisive role in regulating the Fru2,6bisP concentration in respiratory tissues. Although Fru2,6bisP increases after adding glucose to cell suspension cultures, this can be explained by the changes of 3PGA (47b,c). Starvation of roots led to a decrease of Fru2,6bisP (119), but long-term experiments are rather unsuitable for assessing the significance of fine control by metabolites. There was no increase of Fru2,6bisP in other respiring tissues where hexose phosphates accumulate, such as cold sweetening of potatoes, (86, 119). More studies are needed of the effect of a rising supply of carbohydrate on the concentration of Fru2,6bisP and other metabolites, and on the rate of respiration.

Further Possible Mechanisms to Regulate the Fru2,6bisP Concentration

FRU2,6BISP INCREASES IN RESPONSE TO AN INCREASE OF THE TOTAL METABOLIC PHOSPHATE POOL   In the short term, the total phosphate pool in the cytosol will be a conserved moiety. An increase in the general level of phosphorylated metabolites is therefore accompanied by a decrease of the concentration of free Pi. However, the total metabolic Pi pool could change in the mid or the long term, allowing both free Pi and the major phosphorylated intermediates to increase. The hexose-phosphates are the major pool of phosphorylated intermediates in most tissues (34, 45, 48, 98, 99, 123, 131). Since Fru6P and Pi both activate Fru6P,2-kinase and inhibit Fru2,6bisPase (see above), an increase in the total metabolic phosphate pool could produce a dramatic increase of Fru2,6bisP.

In agreement, there is a large increase of Fru2,6bisP after supplying Pi to leaves (135), cell suspension cultures (W.-D. Hatzfeld and M. Stitt, unpublished), and carrot and potato storage tissue (M. Stitt, unpublished). An increase of Fru2,6bisP in water-stressed leaves (96) and cell suspension cultures (J. E. Dancer, M. David, and M. Stitt, unpublished) could also be explained by the rising concentration of phosphate in the metabolic compartments when the cell volume shrinks. Alternatively, when Pi is sequestered as mannose-6-P after adding mannose to leaves (124, 135) or artichoke tubers (73), or as 2-deoxyglucose-P after adding 2-deoxyglucose to cell suspension cultures (W.-D. Hatzfeld and M. Stitt, unpublished) there is a dramatic decrease of Fru2,6bisP.

At present, little is known about the factors that control the size of the phosphate pool in the cytosol. However ^{31}P-NMR studies leave little doubt that this pool is regulated by storing surplus Pi in the vacuole and then slowly withdrawing it, as required (15, 29, 40, 98, 100). More research is needed to elucidate how carbohydrate and phosphate status interact and how this affects respiration, biosynthesis, and growth.

THE RELATION BETWEEN PPi AND ATP   Rising PPi could inhibit Fru6P,2-kinase and lead to a decrease of Fru2,6bisP. However, the PPi concentration is remarkably constant (2, 3, 25, 149), and no experimental systems have yet been found in which this potential input can be tested. Its potential significance is indicated by the observation that PPi inhibits in the concentration range of 0–0.5 mM (72, 79), which compares well with the PPi concentration in the cytosol of spinach leaves (149).

IN VIVO EVIDENCE FOR REGULATION BY PROTEIN MODIFICATION OR TURNOVER   Although there is no evidence from in vitro studies that Fru6P,2-kinase is regulated by protein phosphorylation, there are indications that some "higher" level of control is operating in vivo. First, the activity of Fru6P,2-kinase that can be extracted from spinach or soybean leaves increases gradually during the photoperiod and decreases again in the dark (60, 133). Second, ethylene leads to an increase of Fru6P,2-kinase activity in carrot storage roots (122). Third, adding K$^+$ to guard cells to induce swelling leads to an increase of Fru2,6bisP (48) and an increase in the extractable activity of Fru6P,2-kinase (M. Stitt, unpublished). However, the underlying mechanisms are unclear.

THE TARGETS: IN VITRO EFFECT ON THE ENZYMES

In most eukaryotes, Fru2,6bisP activates PFK and inhibits the Fru1,6Pase (91, 143). Since these enzymes catalyze irreversible reactions, it is relatively easy to predict that rising Fru2,6bisP will favor glycolysis and restrict gluconeogenesis. In plants, the picture is more complicated. The cytosolic Fru1,6Pase is inhibited by Fru2,6bisP, so Fru2,6bisP should act to inhibit gluconeogenesis, at least in tissues like leaves where the cytosolic Fru1,6Pase predominates. However, the other target enzyme, PFP, catalyzes a near-equilibrium reaction. A feeling that this activation of PFP has replaced the more usual activation of PFK has encouraged many investigators to search for evidence that Fru2,6bisP somehow selectively favors the "glycolytic" reaction of PFP or, as we will see in the next section, to search for evidence that rising Fru2,6bisP or PFP activity is correlated with increased respiration. However, I argue that there is little evidence supporting this admittedly attractive idea. The role of Fru2,6bisP and PFP in plant metabolism is probably more complex.

The Cytosolic Fru1,6Pase

The spinach leaf cytosolic Fru1,6Pase is a tetramer with a subunit molecular mass of 37 kD (156). In the absence of Fru2,6bisP it has a very high affinity for Fru1,6bisP ($K_m \sim 4~\mu M$) and is weakly inhibited by AMP. AMP acts noncompetitively to Fru1,6bisP and competitively to Mg^{2+} (52, 130, 156).

Low concentrations (nM to μM) of Fru2,6bisP have a dramatic effect on these properties. The affinity for Fru1,6bisP is decreased and the substrate saturation curve becomes sigmoidal, with a Hill number between 2.5 and 3 (52). The sensitivity to inhibition by AMP is increased, especially at low Mg^{2+} or neutral pH (52, 130). The inhibition by Fru2,6bisP is therefore strongest at low Fru1,6bisP, and is synergistic with AMP. The sensitivity to product inhibition by Pi is increased, while the product inhibition by Fru6P is abolished (130). This resembles the effect of Fru2,6bisP on the binding constants for the two products (44). Rb^+ and K^+ increase the sensitivity to Fru2,6bisP, while Li^+ protects against Fru2,6bisP (135).

The cytosolic Fru1,6Pase from castor bean endosperm has similar properties (65). Indeed, the plant cytosolic Fru1,6Pase has a striking resemblance to Fru1,6Pase from liver, kidney, and yeast (143). However, the properties are very different from those of the chloroplast Fru1,6Pase, which has a much lower Mg^{2+} and Fru1,6bisP affinity, is not inhibited by AMP, is only weakly inhibited by Fru2,6bisP (24, 134), but is regulated via thioredoxin (19).

PFP

POTENTIAL ROLES   PFP catalyzes a near-equilibrium reaction [$K_{eq} = 3.3$ (34, 149)] and could, in principle, catalyze a net flux of carbon in the direction of glycolysis or gluconeogenesis. PFP could also operate in a cycle with either PFK or Fru1,6Pase to catalyze the net production or removal of PPi, respectively. These potential reactions are summarized in Figure 1.

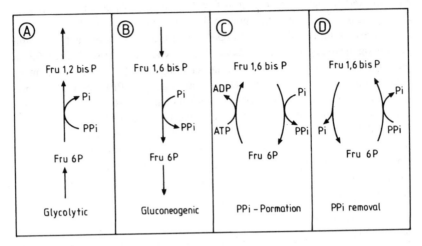

Figure 1   Summary of the potential roles of PFP.

MOLECULAR PROPERTIES   PFP is a tetramer with a molecular mass of about 250 kD (69, 150, 155) containing two different kinds of subunit, termed α and β. Both subunits are found in a wide variety of plants (150). There is some evidence (155) that the β subunit (molecular mass 60 kD) may be responsible for catalysis, and the α subunit, whose molecular mass apparently varies between 60 and 67 kD depending on the source (69, 150, 155), may be involved in regulation.

REACTION CHARACTERISTICS   Several studies have shown that the substrates bind in random order, and that the reaction mechanism is rapid equilibrium random (12, 17, 61, 62, 116). This resembles PFP from *Entamoeba* (90) and *Proprionibacteria* (11). One report of a ping-pong mechanism for pineapple PFP (13) might have arisen because these kinetic studies were, apparently, carried out in the absence of Fru2,6bisP, and interpretation may have been complicated by traces of Fru2,6bisP in commercial preparations of Fru6P.

Since PFP catalyzes a readily reversible reaction, the product inhibition pattern is of considerable interest (116). Significantly, PPi is a powerful inhibitor of the reverse (Fru6P-forming) reaction of PFP from mung bean seedlings (12) and potato tubers (116), inhibiting competitively to Fru1,6bisP. The K_i (PPi) is 10–15 μM, which is identical to the K_m (PPi) for the forward (Fru1,6bisP-forming) reaction. Pi is an effective product-inhibitor of the forward reaction. Pi has been reported to inhibit PFP from cucumber and bean seeds (17, 18) or potato tubers (116) noncompetitively to Fru6P and PPi. A mixed inhibition was found for castor bean endosperm PFP (61, 62) and a competitive inhibition for mung bean seedling PFP (12). Interpretation of the Pi inhibition pattern may sometimes have been complicated by indirect effects of Pi on the affinity for Fru6P or PPi, which arise because Pi decreases the affinity for Fru2,6bisP (see below).

FRU2,6BISP: EFFECT ON ENZYME KINETICS   PFP has been partially purified from a wide range of plant tissues including mung bean seedlings (12, 103), potato tubers (116, 145), spinach leaves (24), pea seeds (152, 153), castor bean endosperm (61, 62), wheat seedlings (155), black gram seeds (4), pineapple (13), cucumber seeds (18), bean seeds (16, 17), soybean cells (82), and carrot roots (150). PFP was always activated by Fru2,6bisP. The extent of the activation varied. However, these differences should not be over-interpreted, because sensitivity to Fru2,6bisP depends on the conditions (see below), which varied among the studies. Apparent differences have also arisen because commercial preparations of Fru6P often contain enough Fru2,6bisP to allow a partial activation of PFP and thus increase the "basal" rate artefactually. Nevertheless, there is general agreement on the major qualitative effects of Fru2,6bisP, which I now discuss.

Fru2,6bisP activates the forward (Fru1,6bisP-forming) reaction by increasing the V_{max} and lowering the K_m (Fru6P), often by a factor of 10 or more (12, 16, 62, 82, 103, 116, 145). The effect on the K_m (PPi) is smaller and more varied, with reports of a decrease (12, 82), no effect (116, 145, 155), or a small increase (13, 62), Fru2,6bisP also relieves the inhibitory effect of high (5 mM) PPi (24, 103, 145).

Fru2,6bisP activates the reverse (Fru6P-forming) reaction by decreasing the K_m (Fru1,6bisP) more than 10 fold (12, 17, 62, 116, 145). The affinity for Pi may be slightly increased (17) or decreased (12, 62, 116, 145). There is no change of the apparent V_{max}, but this is an artefact. Activation actually leads to a considerable increase of the V_{max} of the reverse reaction (see below).

The affinity for Fru2,6bisP depends markedly on the conditions. It is increased by Fru6P and Fru1,6bisP (24, 62, 63, 116, 145), K_a^{app} (Fru2,6bisP) decreasing 15–20 fold as Fru6P or Fru1,6bisP is increased (116). It is decreased by Pi (16, 62, 63, 116, 145) and by many phosphorylated intermediates (62), organic anions, and inorganic anions (62, 116, 145).

Do the kinetic properties of PFP provide any clues to its function in vivo? Many investigators (53, 62, 63, 82, 116, 137) have concluded that the properties are consistent with a glycolytic or a gluconeogenic role. However, other investigators (7, 109, 144, 145, 152, 153) have suggested that the in vitro properties provide evidence that PFP is involved in glycolysis. I now consider three lines of argument, namely (a) observations indicating that Fru2,6bisP has an asymmetric effect on the forward and reverse reactions, (b) reports of a metabolite-mediated interconversion of PFP and PFK, and (c) reports of various PPi- or Fru2,6bisP-dependent conversions between different forms of PFP.

SYMMETRICAL ACTIVATION OF BOTH REACTIONS   The following two observations have sometimes been taken as evidence that the forward (Fru6P-forming) reaction is more sensitive to activation by Fru2,6bisP than the reverse (Fru1,6bisP forming) reaction: (a) Fru2,6bisP increases the V_{max} of the forward reaction but does not seem to affect the V_{max} of the reverse direction (62, 145). (b) Higher concentrations of Fru2,6bisP are apparently needed to activate catalysis in the reverse direction (62, 145). However, reexamination of these observations reveals there is actually a very symmetrical activation of both reaction directions.

First, several investigators (12, 116, 145) have now concluded that activation is actually accompanied by a considerable increase of the reverse reaction V_{max}. This increase is normally masked, because Fru1,6bisP is itself a weak activator of PFP (12, 102). This means that the low "basal" V_{max} of the reverse reaction cannot be measured, because high Fru1,6bisP will eventually activate PFP even when Fru2,6bisP is absent. Strong evidence for this argument is provided by the observation that the apparent K_m (Fru1,6bisP) in the absence of Fru2,6bisP is similar to the K_a (Fru1,6bisP) (12).

Second, phosphonate analogs of Fru2,6bisP have been used to show that PFP only has one type of Fru2,6bisP binding site, and that the K_a for activation of the forward and the reverse reactions is identical for a given activator (137). The apparent differences in the sensitivity of the two reactions arises because Pi, which (see above) decreases the affinity for Fru2,6bisP, has to be included in the assay for the reverse reaction but not in the assay for the forward reaction.

INTERCONVERSION OF PFP AND PFK?   It was originally proposed that PFP and PFK are a single protein, and that Fru2,6bisP and other metabolites lead to a "metabolite-mediated" catalyst conversion between these two activities (7). However the appearance of "PFK" activity in a PFP preparation was subsequently attributed to the PPi that contaminates commercial ATP preparations (68). "PFP" activity appeared in a PFK preparation after adding UDPGlc because the commercial preparation of glycerol-3-P dehydrogenase used as a coupling enzyme in the PFP assay is contaminated with UDP-glucose dehydrogenase. This enzyme will convert UDPGlc and PPi into UTP, which acts as a substrate for PFK (67, 151). It has now been unambiguously established that PFP and PFK are different proteins in potato tuber (69, 70) and carrot roots (150, 151). The subunits are of different molecular mass and show different protease digestion patterns (69, 70); there is no antigenic crossreaction (70, 150).

CONVERSIONS BETWEEN DIFFERENT FORMS OF PFP?   Several workers have reported that PFP can exist in vitro in forms with different molecular masses (7, 69, 152, 153, 155). Yan & Tao (155) showed that wheat seedling PFP can be isolated as a large-molecular-mass form that contains two α (67-kD) and two β (60-kD) subunits, or as a small-molecular-mass form that contains only the two β subunits. The high-molecular-mass form had a 4-fold higher affinity for Fru2,6bisP, while the lower basal activity of the small-molecular-mass form meant it showed a larger relative activation by Fru2,6bisP. Potato PFK also has the $\alpha_2\beta_2$ structure (69), and pretreatment leads to dissociation into a small-molecular-mass form (129 kD) of undefined subunit composition. Both groups of workers were cautious about ascribing a physiological significance to these changes.

Black and coworkers (13, 152, 153) have proposed that changes of the aggregation state provide the mechanism by which rising Fru2,6bisP activates glycolysis. They observed that Fru2,6bisP converts pea seedling PFP from a small (6.3 S)-molecular-mass form, which is strongly dependent on Fru2,6bisP and has a low ratio of forward:reverse activity, into a large (12.7 S)-molecular-mass form, which is largely Fru2,6bisP independent and has a high ratio of forward:reverse activity. Unfortunately, no information is available on the subunit composition of these two forms, so it is impossible to compare their results directly with the studies in wheat seedlings or potato. Indeed, the reported reversion of the 6.3 S into a 12.7 S form after adding Fru2,6bisP (153) is difficult to understand if the α subunits are absent in the small-molecular-mass form as in wheat seedlings. There are also differences in the detailed kinetic properties of PFP from pea seedlings and wheat seedlings. Some of the results in pea seedlings might be explained by the assay conditions (152, 153). For example, the reverse reaction was assayed with 10 mM Fru1,6bisP and this (see above) will substitute for Fru2,6bisP as an activator. The selective increase of the forward reaction activity when Fru2,6bisP is added is therefore trivial.

Although it remains possible that Fru2,6bisP induces changes in the aggregation state of PFP that favor the forward reaction, the evidence is inconclusive, even in vitro. It is also extremely difficult to extrapolate from changes of protein aggregation in a dilute extract or enzyme preparation, to the way in which a protein will behave in vivo. Anyway, even if Fru2,6bisP does sometimes lead to changes in the molecular mass of PFP, these changes cannot be essential for activation, because three independent investigations (12, 82, 150) have shown that activation can occur without any change of the molecular mass.

PROPERTIES OF PFP IN DIFFERENT TISSUES   It is also of interest to ask whether there are tissue- or species-specific differences in the properties of

PFP. These might reflect a varying role for PFP in differnt tissues. Only data obtained with saturating Fru2,6bisP can be compared, because it is impossible meaningfully to compare values obtained in the absence of Fru2,6bisP (see above). Almost all preparations of PFP have a K_m (Fru6P) of 100–300 μM, and a K_m (PPi) of 5–20 μM for the forward reaction, and a K_m (Fru1,6bisP) of 10–70μM and a K_m (Pi) of 200–600 μM for the reverse reaction (4, 12, 13, 16–18, 24, 61, 62, 82, 103, 116, 145, 150, 152, 153).

However, two potentially important differences in the response to Fru2,6bisP have recently been reported. First, PFP from algae is not activated by Fru2,6bisP (26). It will be important to establish at what stage in plant evolution PFP acquired this property, because this could provide important clues about the functional significance of this regulatory mechanism. Second, it has recently been reported that Fru2,6bisP decreases the Fru6P-affinity of carrot root PFP (150). This could be an adaptation to favor the reverse reaction and provide PPi for sucrose mobilization via sucrose synthase (see below for further discussion). More studies are needed to show if this change can be found in other sink tissues.

PFP: THE BOTTOM LINE   The pertinent effects of Fru2,6bisP on PFP can be summarized as follows: (a) Fru2,6bisP activates catalysis in both directions in a strictly symmetrical manner. (b) The affinity for Fru2,6P will be increased when the hexose-phosphate or triose-phosphates (Fru1,6bisP) are high. (c) The affinity for Fru2,6bisP will be decreased when Pi increases. (d) Low concentrations of PPi will strongly inhibit the reverse (PPi-producing) reaction. These properties are consistent with PFP operating as a glycolytic or as a gluconeogenic enzyme and would also allow PFP to contribute to the control of the cytosolic PPi concentration. Significantly, they indicate that PFP will be active in conditions when the cytosolic metabolite pools are high, especially since Fru2,6bisP often rises in these conditions (see above). These properties are completely different from those of PFK. PFK is inhibited by PEP and activated by Pi (27), as expected for an enzyme catalyzing a glycolytic flux, and responding to the need for respiratory substrate.

THE TARGETS: IN VIVO IMPACT OF FRU2,6BISP ON FLUXES

There are several steps in assessing the in vivo significance of a regulatory mechanism. Obviously, the pathway must first be defined. It can then be asked whether pathway flux is affected by, in this case, Fru2,6bisP. Then, as more becomes known about the pathway, it will probably become necessary to evaluate how Fru2,6bisP interacts with other regulatory mechanisms and, if possible, to quantitate its impact on flux. Research on the role of Fru2,6bisP in photosynthetic metabolism has now reached the stage where its interaction with other mechanisms is being clarified and quantified. In contrast, the pathways involved in nonphotosynthetic metabolism are still controversial, and we have not yet unambiguously defined which fluxes actually increase when Fru2,6bisP activates PFP.

Role of the Cytosolic Fru1,6Pase in the Control of Sucrose Synthesis

PHOTOSYNTHETIC SUCROSE SYNTHESIS   We have already seen (a) that Fru2,6bisP decreases when 3-carbon intermediates become available for sucrose synthesis, (b) that Fru2,6bisP increases when hexose-phosphates accumulate, and (c) that Fru2,6bisP is a potent inhibitor of the cytosolic Fru1,6Pase. I now discuss the significance of these changes for the control of sucrose synthesis in vivo.

The following aspects of the photosynthetic pathway must be kept in mind during this discussion (Figure 2). The chloroplast, in effect, converts Pi and CO_2 into triose-phosphates, which are then exported to the cytosol in strict counterexchange for Pi. This counterexchange is catalyzed by the phosphate translocator (38, 50). The triose-phosphates are then converted to sucrose in

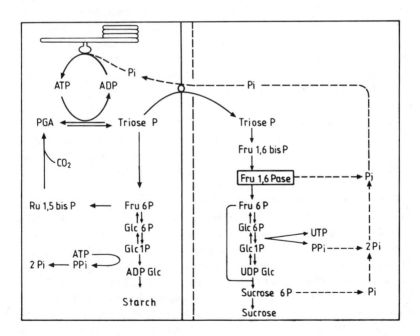

Figure 2   Fru1,6Pase and the photosynthetic pathway

the cytosol, releasing Pi that returns to the chloroplast in exchange for further triose-phosphates. There are two overriding reasons why sucrose synthesis has to be regulated. First, the rate of sucrose synthesis must be coordinated with the rate of CO_2 fixation to ensure that enough Pi is recycled to support the current rate of photosynthesis, but without depleting the triose-phosphate pool to a point where the regeneration of Ru1,6bisP (the CO_2-acceptor) is inhibited. Second, sucrose synthesis is regulated to allow partitioning to be altered.

REGULATION OF FLUXES   The cytosolic Fru1,6Pase catalyzes the first irreversible reaction leading to sucrose (45, 125, 131) and is, therefore, strategically placed to regulate the removal of triose-phosphates from the Calvin cycle. The response of the Fru1,6Pase to a rising supply of triose-phosphates has been described in a semi-empirical model (52, 113, 125, 131). This model used the measured concentrations of Fru2,6bisP and triose-phosphates to predict how Fru1,6Pase activity responds to a rising rate of photosynthesis, and agreed closely with the actual response of sucrose synthesis in vivo. The Fru1,6Pase is totally inhibited until a "threshold" concentration of triose-phosphate is exceeded. Activity then rises sharply, in response to further small increases of the triose-phosphate concentration. This sensitive response to a rising concentration of 3-carbon intermediates is generated because (a) the inhibitor (Fru2,6bisP) concentration is driven downwards and (b) the substrate (Fru1,6bisP) concentration rises. The amplification is increased even further because (c) the cytosolic Fru1,6Pase has a sigmoidal substrate-saturation curve, and falling Fru2,6bisP shifts the $S_{0.5}$ downwards (52). This response has two important consequences for the operation of the Calvin cycle. First, it ensures that the Calvin cycle intermediates are not removed for sucrose synthesis until a threshold concentration of triose-phosphates is exceeded, presumably corresponding to the levels needed for the Calvin cycle turnover. Second, it ensures that triose-phosphates are rapidly converted to sucrose once this threshold is exceeded. This allows Pi to be recycled to support further photosynthesis.

There is also strong evidence that Fru2,6bisP contributes to the feedback control of sucrose synthesis in vivo. In both spinach (124, 132) and soybean (50), there is a gradual increase of Fru2,6bisP during the photoperiod or after adding sucrose (39, 132). This is accompanied by a partial inhibition of sucrose synthesis and an increase of starch synthesis. In spinach leaves, sucrose accumulation leads to an inactivation of SPS (138) and thence to an increase of Fru6P in the cytosol (45). This explains why Fru2,6bisP increases. These experiments also demonstrated that Fru1,6bisP and triose-phosphates increase in the cytosol (45), providing direct evidence that the cytosolic Fru1,6Pase has indeed been inhibited.

REGULATION OF POOL SIZES   Not only must the *rate* of sucrose synthesis be regulated, but the steady-state concentrations of intermediates must also be maintained in a range that allows efficient operation of the other reactions in photosynthesis. This depends on the "threshold" response being correctly "tuned." The importance of Fru2,6bisP in controlling pool sizes is emphasized by three examples where the "threshold" for removal of triose-phosphate is changed as an adaptation to new circumstances.

First, feedback inhibition of sucrose synthesis (see above) actually acts by changing the "threshold" triose-phosphate concentration. When the cytosolic Fru6P increases, there is an additional "input" that shifts the Fru2,6bisP concentration upward. Consequently, higher triose phosphate and 3PGA levels are maintained (45, 88; H. E. Neuhaus, W. P. Quick and M. Stitt, unpublished), and starch synthesis is activated. Second, in maize leaves, sucrose is synthesized from triose-phosphates in the mesophyll cells (43). The cytosolic Fru1,6Pase in the mesophyll cells has a 10-fold higher K_m (Fru1,6bisP) and is less sensitive to Fru2,6bisP (128). This allows triose-phosphate concentrations of over 10 mM to be maintained in the mesophyll (128). Two thirds of the triose-phosphate must return to the Calvin cycle in the bundle sheath and, as this occurs by diffusion, it depends on these high concentrations (113). Third, there are "threshold" changes in response to temperature. At low temperatures, high concentrations of metabolites are maintained in the Calvin cycle (30, 74, 127). These enhanced metabolite pools partly counteract the effect of temperature on the turnover rate of the enzymes, and "buffer" the Calvin cycle against changes of temperature. They can be explained by a temperature-dependent shift of the properties of the Fru1,6Pase, which becomes more sensitive to Fru2,6bisP and AMP when the temperature is decreased (127).

INTERACTION WITH OTHER ENZYMES TO CONTROL FLUX   The cytosolic Fru1,6Pase is not the only enzyme involved in regulating sucrose synthesis. For example, SPS is regulated by a hierarchy of mechanisms including metabolites, covalent protein modification, and possibly protein turnover (138, 147, 148). Turnover of PPi could also control the rate of sucrose synthesis (94). This means we need a way of assessing the contributions of Fru2,6bisP and the cytosolic Fru1,6Pase to the control of flux through the pathway. A more detailed review of how a combination of genetics and biochemistry can be used to approach this problem is available (58, 118).

The contribution of Fru2,6bisP to control of flux has been quantified in experiments using *Clarkia xantiana* mutants with a progressive reduction of PGI activity in the cytosol (64, 88). This genetic change allowed a specific increase of Fru6P in the cytosol and (see above) a 2–3-fold larger increase of Fru2,6bisP. Significantly, the impact of this increase of Fru2,6bisP on flux varied with the conditions. At moderate rates of sucrose synthesis, a 50% increase of Fru2,6bisP led to a 30% inhibition of sucrose synthesis and a 30% stimulation of starch synthesis. In contrast, a doubling of the Fru2,6bisP led to less than a 10% change in the fluxes in saturating light and CO_2. Clearly, the effectiveness of Fru2,6bisP depends on the conditions. The weakened impact in high light and CO_2 resulted because the high rate of photosynthesis allowed triose-phosphate (and Fru1,6bisP) to increase and to counteract the increase of Fru2,6bisP (88, 118).

Decreased Fru1,6Pase activity affects other enzymes via changes in the concentration of shared metabolites, or other ligands. The response of an enzyme to its effectors can be defined as the elasticity coefficient (58). Briefly, the elasticity coefficient

$$\epsilon = \frac{\dfrac{dv}{v}}{\dfrac{ds}{s}}$$

where dv/v is the change of enzyme activity that results when the concentration of one of its effectors is changed (ds/s) and the others are kept constant. The in vivo elasticity coefficients of the cytosolic Fru1,6Pase for triose-phosphates and Fru2,6bisP were estimated to be $+1.5$ and -0.4, respectively (88). These are the values that would be expected if the Fru1,6Pase operates in vivo with a substrate concentration just over the $S_{0.5}$ (115). The values emphasize (a) that Fru2,6bisP is an effective inhibitor of the cytosolic Fru1,6Pase in vivo and (b) that this inhibition can be overcome by a 2–4-fold smaller increase of the triose-phosphates. This provides a quantitative basis for the flexible response of the Fru1,6Pase. This flexibility will be further increased by the balance of "inputs" that regulate the Fru2,6P, concentration itself (see above).

The estimated elasticity coefficients of the Fru1,6Pase have been incorporated into a model (115, 118) to illustrate how control is shared between the Fru1,6Pase and the reactions grouped around SPS. The flux control coefficient, C, quantifies the ability of a given enzyme to alter the flux through the entire pathway, and is defined (58) as

$$C = \frac{\dfrac{dJ}{J}}{\dfrac{dE}{E}}$$

where dJ/J is the change in pathway flux that results from dE/E, a change in the amount of the enzyme. The model suggested that control is shared equally between Fru1,6Pase and SPS at moderate flux rates but is redistributed towards SPS at high flux rates (115). Supporting data were obtained in studies of *Clarkia xantiana* PGI mutants (88). However, a definitive resolution of this problem will not be possible until isogenic mutants or transgenic plants with a changed complement of SPS or Fru1,6Pase are available.

On the Role of PFP in Plant Metabolism

It is not yet possible to assign a definitive role to PFP. The major problems in approaching this question are (*a*) uncertainties concerning the subcellular compartmentation of plant metabolism; (*b*) the unusual role of PPi as an energy donor in the cytosol of plants; and (*c*) the reversibility of the reaction catalyzed by PFP, which means it may catalyze a variety of different net fluxes. I first outline the background to these three problems and then consider the available evidence that PFP is involved in glycolysis, in gluconeogenesis, or in regulating the cytosolic PPi concentration.

METABOLIC COMPARTMENTATION   Plants metabolize sucrose in the cytosol and starch in the chloroplast, and each compartment contains a glycolytic sequence (1, 27, 28). Two opposing routes have been proposed for starch accumulation. One proposes that triose-phosphates are imported into the plastid via the phosphate translocator and are then converted to starch via a plastid Fru1,6Pase (33, 78, 85). The other proposes that hexose units can be imported directly into the plastid (35, 59, 142). Until these routes are distinguished it will not be possible to assess the significance of PFP during starch accumulation.

Four kinds of experiment have been carried out to distinguish between these routes. One approach is to investigate whether nonphotosynthetic plastids contain Fru1,6Pase. Low activities have been reported in plastids from sycamore suspension cells (78), cauliflower buds (57), maize endosperm (33), and potato tuber (85) but not in guard cells (48) or wheat endosperm plastids (35). A second approach is to establish the best (labeled) precursor for starch synthesis by isolated intact amyloplasts. Glc1P was the best precursor for starch in wheat endosperm amyloplasts (142). Fru1,6bisP was incorporated into an unidentified insoluble product but not into starch (142). Isolated amyloplasts from maize endosperm (33) and potato tubers (85) also convert Fru1,6bisP into an insoluble product. This was assumed to be starch, but no evidence was provided. Third, specifically labeled glucose supplied to whole tissues can subsequently be reisolated from starch to determine the randomization between the C-1 and C-6 position. Glucose is incorporated into starch without much randomization in wheat endosperm (59), *Chenopodium* cell cultures, potato tubers, and maize endosperm (47a). These results rule out an exclusive role via triose-phosphates in these tissues, because triose-phosphate isomerase would lead to extensive randomization between the upper and lower halves of the glucose moiety. Fourth, and most pertinent, the transport properties of plastids can be studied. It has recently been demonstrated that the phosphate translocator in pea root plastids will transport Glc6P (50). In conclusion, evidence is accumulating to suggest that the major route of starch accumulation involves export of hexose moieties, probably Glc6P. However, it is still too early to rule out the possibility of tissue-specific differences.

The contribution of different routes during starch mobilization is also unclear (136). In chloroplasts, starch can be degraded (a) via phosphorolysis and glycolysis, and exported as triose-phosphate and 3PGA via the phosphate translocator; or (b) via hydrolysis, and exported as hexose via the glucose carrier (105, 136). Their relative contribution in vivo is unknown, and starch mobilization has not yet been studied in nonphotosynthesis plastids.

PPi AS AN ENERGY DONOR IN THE CYTOSOL OF PLANT CELLS   PPi is usually present at low concentrations in cells, because it is hydrolyzed by pyrophosphatase to drive the synthesis of nucleotide-linked precursors during polymer biosynthesis (76). However, in plants, almost all the alkaline pyrophosphatase is restricted to the plastid (46, 149), and the cytosol contains about 250 μM PPi (149). The PPi level varies independently of ATP or UTP (26a, 47c), underlining its importance as a potential energy donor. Further, at least three cytosolic enzymes could utilize PPi as an energy source—namely, UDPGlc pyrophosphorylase during sucrose mobilization (3, 14, 55), a PPi-dependent H$^+$ pump on the tonoplast (97), and PFP itself.

EVIDENCE THAT PFP CATALYZES A NEAR-EQUILIBRIUM REACTION   The reaction catalyzed by PFP is readily reversible ($K_{eq} = 3.3$), and two lines of evidence show that the reaction is indeed close to equilibrium in vivo. First, the measured levels of PPi (149), Fru6P, Fru1,6bisP (45), and Pi (29, 40, 77) in the cytosol of spinach leaves suggest the mass action ratio is close to the theoretical keq (131, 149). A similar conclusion has been reached from the overall metabolite levels in wheat endosperm (34) and Chenopodium rubrum cell suspension cultures (47b). Second, when specifically labeled glucose is converted to sucrose in wheat endosperm (59), Chenopodium cells, potato tubers, or maize endosperm (47a), there is considerable randomization between the C-1 and C-6 position. This provides evidence for a rapid substrate cycle between triose-phosphates and hexose-phosphates in the cytosol. In Chenopodium cells, Fru1,6Pase activity was too low to account for the cycling, which showed the characteristics expected from a rapid reversible reaction catalyzed by PFP (47a,b).

This finding poses a practical and a theoretical problem. Practically, it makes it impossible to decide the direction of the net flux, because metabolite measurements are not accurate enough to allow this decision to be made on the basis of the estimated mass action ratio when reactions are close to their equilibrium. The theoretical problem arises because it is unusual for a "regulated" reaction to be close to equilibrium (89, 101). Indeed, it is widely thought that enzymes close to equilibrium are in "excess" and cannot have a dramatic effect on flux through the pathway. Instead, near-equilibrium reactions respond sensitively to alterations in the concentrations of their reactants, caused by changes in their use or production elsewhere in the pathway.

In this case, PFP might play a flexible role in metabolism, depending on the cell type and the conditions. I now discuss the evidence that PFP could be involved in (a) glycolysis, (b) gluconeogenesis, or (c) catalyzing substrate cycles with PFK or the Fru1,6Pase to regulate the cytosolic PPi concentration. The evidence consists of experiments that have searched for a correlation between Fru2,6bisP levels and a particular metabolic flux, or for a correlation between high extractable PFP activity and the predominance of a particular metabolic activity in different tissues. I argue that no single role for Fru2,6bisP or PFP has emerged, despite a considerable research effort. This does not prove that PFP has multiple roles, but it does suggest the possibility has to be considered seriously.

RELATION AMONG FRU2,6BISP, PFP, AND GLYCOLYSIS   The relation between Fru2,6bisP and respiration has been studied in several systems. Although a relation has often been found (9, 10, 83, 84, 119, 122, 144), more detailed studies have often shown it to be spurious. For example, a correlation between rising Fru2,6bisP and O_2-uptake was reported in one study of the climacteric in bananas (83), but this was less evident in a second study (10); a third study (6) showed that Fru2,6bisP actually stays constant, and that the results reported by other workers represent an improved extractability of Fru2,6bisP in ripened fruit because the tannins and phenols have decreased. Fru2,6bisP increases during wound respiration in various tubers (119, 144) or after adding ethylene (122). However, the time course of the increase did not correspond with the rise of respiration (10, 119). Fru2,6bisP also increased during short-term anaerobiosis (84), but this has not always been seen (25) and, anyway, glycolysis may actually decrease during anaerobiosis in plants because a very large component is linked to biosynthetic reactions (25). In other studies, respiration increased without any change of Fru2,6bisP, including the dramatic rise of respiration in mature *Arum* spadix (2, 3).

There is also no obvious relation between the extractable activity of PFP and the rate of O_2 uptake, either in pineapple (10), in developing *Arum* spadix (2, 3), or in different zones of pea roots (2, 3), whereas PFK did correlate with respiration. However, a dramatic increase of PFP did occur when cell suspensions became Pi limited (32). This was accompanied by an increase of a specific PEP phosphatase, suggesting glycolysis might be occurring via adenine nucleotide–independent enzymes in Pi-limited conditions, when the nucleotide pools are severely depleted.

Even if PFP activity or increased Fru2,6bisP do correlate with an increased steady-state rate of O_2 uptake, this only provides weak evidence for a causal relation. It is also necessary to show (a) that the rise of respiration is temporally related to the rise of Fru2,6bisP and (b) that glycolysis has been stimulated at the reaction catalyzed by PFP which, rigorously, requires a demonstration that PPi and Fru6P both decrease. To date, this has only been observed after adding uncoupler and hydroxyl ions simultaneously to *Chenopodium* cell suspension cultures to produce a 3-fold stimulation of dark-respiration and O_2-uptake (47c). The increased rate of glycolysis in swelling guard cells was well correlated with an increase of Fru2,6bisP and a decrease of Fru6P (48), but it is not known whether PPi changes in these cells.

GLUCONEOGENIC SUCROSE SYNTHESIS   The potential role of PFP in nonphotosynthetic gluconeogenesis has been studied in detail in two systems. In germinating castor bean endosperm, Fru2,6bisP increased when anaerobiosis or 3-mercaptopicolinate were used to interrupt the supply of 3-carbon intermediates for gluconeogenesis (66). PFP activity is also lower than Fru1,6Pase activity (71). These observations suggest that the major glycolytic flux is likely to be via Fru1,6Pase, rather than PFP. The second system is CAM plants like pineapple, which decarboxylate malate and then convert the PEP to soluble sugars in the light. PFP activity is extremely high in these plants compared to PFK or Fru1,6Pase activity (20, 36), and Fru2,6bisP rises when gluconeogenesis is occurring in the light (36). These results show that PFP is important for gluconeogenesis in these species. However, this was not the case in all CAM plants (36).

Thus, PFP may contribute to gluconeogenesis in some tissues, but even in nonphotosynthetic tissues, the cytosolic Fru1,6Pase often plays a major role. It is also clear that gluconeogenesis cannot be the major function of PFP: Appreciable PFP activity is present in members of the *Alliaceae,* which are unlikely to carry out gluconeogenesis in their nonphotosynthetic tissues because they do not store any starch or lipid (2, 3).

SYNTHESIS OF STARCH FROM INCOMING SUCROSE   On balance, available evidence indicates starch synthesis occurs via a pathway involving import of hexose units, so PFP is not directly required (see above). However, as discussed below, PFP could be involved in generating PPi to mobilize the incoming sucrose via sucrose synthase. In agreement, most available measurements suggest PFP activity is unrelated to starch accumulation in storage tissues. Indeed, the PFP:PFK ratio decreased before starch accumulation started in maize endosperm (31), pea seeds (34), and *Arum* spadix (2, 3). There is also no relation between Fru2,6bisP levels and starch accumulation in potato tubers (86), different pea genotypes (34), or *Arum* spadix. The evidence from suspension cultures is less clear. Starch accumulation correlated with high Fru2,6bisP and occurred before PFP increased in mixotrophic soybean cells (111); in contrast, in *Chenopodium* cells, starch accumulation was delayed, compared to changes of Fru2,6bisP, but did correlate with an increase of PFP (47c).

PFP AND PPi TURNOVER   The potential role of PFP in PPi turnover has received considerable attention, especially in connection with sucrose metabolism (3, 13, 14, 25, 26, 55, 139–141, 154). PPi is involved in the reversible reaction catalyzed by UDPGlc pyrophosphorylase (94, 149). During sucrose synthesis, one molecule of PPi is produced per molecule of sucrose. This PPi could be removed in a cycle between PFP and the cytosolic Fru1,6Pase (Figure 2). PPi is required when sucrose is being mobilized via sucrose synthase, and could be generated via a cycle between PFP and PFK (Figure 2).

The idea that PFP removes the PPi produced during sucrose synthesis is attractive on paper. For example, when PPi accumulates, mass action effects at UDPGlc pyrophosphorylase lead to an increase of hexose-phosphates and, thence, of Fru2,6bisP (94). A combination of rising Fru6P and Fru2,6bisP would then activate PFP, and this effect would be further enhanced if Pi were to decrease (see above). Simultaneously, the rising Fru6P and PPi and falling Pi would thermodynamically favor the forward reaction, relative to the reverse reaction. This combination of kinetic and thermodynamic effects could lead to a large stimulation of PPi removal. However, there is still no direct evidence for this proposal. Indeed, PFP activity actually decreases as sucrose synthesis increases during leaf development (13). Nevertheless, the maximum catalytic activities of PFP in mature pea, spinach, and barley leaves were 14.6 and 41 μmol mg Chl^{-1}, which is still 2–3.5-fold above the maximum rates of sucrose synthesis in these leaves (K.-P. Kraus, H. E. Neuhaus, and M. Stitt, unpublished).

The idea that PFP operates to generate PPi in sucrose-mobilizing tissues is also attractive on paper. First, PFP and sucrose synthase are both present at high activities in young, developing tissues (2, 3, 13, 14, 26, 31, 139–141). Both enzymes decrease in parallel during seed development (154) and rise in parallel during germination (154). Second, there is good evidence for rapid recycling of triose-phosphate in the cytosol of sucrose-mobilizing tissues, including developing wheat endosperm (59), maize endosperm, and potato tubers (47a). Third, the production of PPi could be regulated because the reverse reaction is strongly inhibited by PPi (12, 116). However, it is again difficult to quantify the net rate of PPi production and provide direct evidence for this hypothesis. Further, algae contain PFP, even though sucrose synthase is absent (26), suggesting that the removal of PPi cannot be the sole function of PFP.

THE ROLE OF PFP: CONCLUSIONS AND PERSPECTIVES   The available evidence shows that PFP catalyzes a near-equilibrium reaction. This reaction will therefore respond to changes in the rest of metabolism in a very flexible manner, and its precise role could vary, depending on the cell and the conditions. This flexible reaction can be turned on by increasing Fru2,6bisP or by an increase in the amount of PFP. Increased PFP activity should allow the reaction to move even closer toward its equilibrium and, hence, would allow it to respond even more sensitively to changes in the levels of the various reactants.

In two separate studies using cell suspension cultures, Fru2,6bisP changed rapidly, providing instantaneous regulation; the amount of PFP changed more slowly (47b, 111). In future more attention should probably be paid to regulation, simultaneously, at both these levels. It will also be essential to integrate studies of Fru2,6bisP and PFP with studies on the PPi-dependent H^+ pump on the tonoplast (97) if we are to understand how they interact during PPi turnover in plants. Finally, it will be rewarding to integrate studies on the turnover of PFP with studies on the expression of other enzymes, which are already known to be induced by developmental or environmental signals, including sucrose synthase and alcohol dehydrogenase (42, 104), to provide a molecular basis for the notion that PFP may act as an "adaptive" pathway, allowing plants to respond flexibly to varying conditions (14, 139, 140).

Literature Cited

1. ap Rees, T. 1985. The organisation of glycolysis and the oxidative pentose phosphate pathway in plants. In *Encylopedia of Plant Physiology* (N S), ed. R. Douce, D. A. Day, 18:390–417. Heidelberg: Springer-Verlag
2. ap Rees, T., Green, J. H., Wilson, P. M. 1985. Pyrophosphate: fructose-6-phosphate 1-phosphotransferase and glycolysis in non-photosynthetic tissues of higher plants. *Biochem. J.* 227:299–304
3. ap Rees, T., Morell, S., Edwards, J., Wilson, P. H., Green, J. H. 1985. Pyrophosphate and the glycolysis of sucrose in higher plants. In *Regulation of Carbohydrate Partitioning in Photosynthetic Tissue*, ed. R. L. Heath, J. Preiss, pp. 109–26. Baltimore: Waverley

4. Ashihara, H., Stupavska, S. 1984. Comparison of the activities and properties of pyrophosphate: fructose-6-phosphate phosphotransferase and ATP-dependent phosphofructokinases of black gram *(Phaseolus mungo)* seeds. *J. Plant Physiol.* 116:241–52
5. Badger, M. R., Sharkey, T. D., von Caemmerer, S. 1984. The relationship between steady state gas exchange of leaves and levels of carbon reduction cycle intermediates. *Planta* 160:305–13
6. Ball, K. L., ap Rees, T. 1988. Fructose-2,6-bisphosphate and the climacteric in bananas. *Eur. J. Biochem.* 177:637–41
7. Balogh, A., Wong, J. H., Wötzel, C., Soll, J., Cséke, C., Buchanan, B. B. 1984. Metabolite-mediated catalyst conversion of PFK and PFP: a mechanism

of enzyme regulation in green plants. *FEBS Lett.* 169:287–92

8. Baysdorfer, C. 1986. Purification of spinach leaf fructose-6-phosphate,2-kinase. *Plant Physiol.* 80:S204

9. Baysdorfer, C., Sicher, R. C., Kremer, D. E. 1987. Relationship between fructose-2,6-bisphosphate and carbohydrate metabolism in darkened barley primary leaves. *Plant Physiol.* 84:766–70

10. Beaudry, R. M., Paz, N., Black, C. C. Jr., Kays, S. L. 1987. Banana ripening: implications of changes in internal ethylene and CO_2 concentrations, pulp fructose-2,6-bisphosphate concentrations and glycolytic enzymes. *Plant Physiol.* 85:277–82

11. Bertagnolli, B. L., Cook, P. F. 1984. Kinetic mechanism of pyrophosphate-dependent phosphofructokinase from *Proprionibacterium freudenreichii. Biochemistry* 23:4101–7

12. Bertagnolli, B. L., Younathon, E. S., Voll, R. J., Cook, P. F. 1986. Kinetic studies on the activation of pyrophosphate-dependent phosphofructokinase from mung bean by fructose-2,6—bisphosphate and related compounds. *Biochemistry* 25:4682–87

13. Black, C. C. Jr., Carnal, C. W., Paz, N. 1985. Roles of pyrophosphate and fructose-2,6-bisphosphate in regulating plant sugar metabolism. See Ref. 3, pp. 76–92

14. Black, C. C. Jr., Mustardy, L., Kormanik, P. P., Sung, S. S., Xu, D.-P., Paz, N. 1987. Regulation and roles for alternative pathways of hexose metabolism in plants. *Physiol. Plant.* 69:387–94

15. Bligny, R., Foray, M.-F., Roby, C., Douce, R. 1989. Transport and phosphorylation of choline in higher plant cells. *J. Biol. Chem.* 264:4888–95

16. Botha, F. C., Small, J. G. C. 1987. Comparison of the activities and some properties of pyrophosphate and ATP-dependent fructose-6-phosphate 1-phosphotransferases of *Phaseolus vulgaris* seeds. *Plant Physiol* 83:772–77

17. Botha, F. C., Small, J. G. C., Burger, A. L. 1987. Characterisation of pyrophosphate dependent phosphofructokinase from germinating bean seeds. *Plant Sci.* 51:151–57

18. Botha, F. C., Small, J. G. C., de Vries, C. 1986. Isolation and characterisation of pyrophosphate: D-fructose-6-phosphate 1-phosphotransferase from cucumber seeds. *Plant Cell Physiol.* 27:1285–95

19. Buchanan, B. B. 1980. Role of light in

the regulation of chloroplasts. *Annu. Rev. Plant Physiol.* 31:341–74

20. Carnal, C. W., Black, C. C. Jr. 1983. Phosphofructokinase activities in photosynthetic organisms. *Plant Physiol.* 71:150–55

21. Cséke, C., Buchanan, B. B. 1983. An enzyme synthesising fructose-2,6-bisphosphate occurs in leaves and is regulated by metabolite effectors. *FEBS Lett.* 155:139–44

22. Cséke, C., Buchanan, B. B. 1986. Regulation of the formation and utilisation of photosynthate in leaves. *Biochim. Biophys. Acta* 853:43–63

23. Cséke, C., Stitt, M., Balogh, A., Buchanan, B. B. 1983. A product-regulated fructose-2,6-bisphosphate occurs in green leaves. *FEBS Lett.* 162:103–6

24. Cséke, C., Weeden, N. F., Buchanan, B. B., Uyeda, K. 1982. A special fructose bisphosphatase functions as a cytoplasmic regulatory metabolite in green leaves. *Proc. Natl. Acad. Sci. USA* 79:4322–26

25. Dancer, J. E., ap Rees, T. 1989. Effects of 2,4-dinitrophenol and anoxia on the inorganic pyrophosphate content of the spadix of *Arum maculatum* and the root apices of *Pisum sativum. Planta* 178:421–24

26. Dancer, J. E., ap Rees, T. 1989. Relationship between pyrophosphate: fructose-6-phosphate 1-phosphotransferase, sucrose breakdown, and respiration. *J. Plant Physiol.* In press

26a. Dancer, J., Veith, R., Fell, R., Komor, E., Stitt, M. 1990. Independent changes of inorganic pyrophosphate and the ATP/ADP or UTP/UDP ratios in plant cell suspension cultures. *Plant Sci.* In press

27. Dennis, D. T., Greyson, M. 1987. Fructose-6-phosphate metabolism in plants. *Physiol. Plant.* 69:395–404

28. Dennis, D. T., Miernyk, J. A. 1982. Compartmentation of non-photosynthetic carbohydrate metabolism. *Annu. Rev. Plant Physiol.* 33:27–50

29. Dietz, K.-J., Foyer, C. 1986. The relationship between phosphate status and photosynthesis in leaves. *Planta* 167:376–81

30. Dietz, K. J., Heber, U. 1986. Light and CO_2 limitation of photosynthesis and states of reactions regenerating ribulose-1,5-bisphosphate or reducing 3-phosphoglycerate. *Biochim. Biophys. Acta* 848:392–401

31. Doehlert, D. C., Kuo, T. M., Felker, F. C. 1988. Enzymes of sucrose and hexose metabolism in developing kernels of

two inbreds of maize. *Plant Physiol.* 86:1013–19

32. Duff, S. M. G., Moorhead, G. B. G., Lefebvre, D. D., Plaxton, W. C. 1989. Phosphate starvation inducible bypasses of adenylate and phosphate dependent glycolytic enzymes in black mustard suspension cells. *Plant Physiol.* In press

33. Echeverria, E., Boyer, C. D., Thomas, P. A., Liu, K.-C., Shannon, J. C. 1988. Enzyme activities associated with maize kernel amyloplasts. *Plant Physiol* 86:786–92

34. Edwards, J., ap Rees, T. 1986. Metabolism of UDP glucose by developing embryos of round and wrinkled varieties of *Pisum sativum. Phytochemistry* 25:2033–2039

35. Entwhistle, G., ap Rees, T. 1988. Enzymic capacities of amyloplasts from wheat endosperm. *Biochem. J.* 255: 391–96

36. Fahrendorf, T., Holtum, J. A. M., Mukherjee, U., Latzko, E. 1987. Fructose-2,6-bisphosphate, carbohydrate partitioning and crassulacean acid metabolism. *Plant Physiol.* 84:182–87

37. Fischer, E., Raschke, K., Stitt, M. 1986. Effects of abscisic acid on photosynthesis in whole leaves. *Planta* 169:535–45

38. Flügge, U.-I., Heldt, H. W. 1984. The phosphate translocator of the chloroplast. *Trends Biochem. Sci.* 9:530–33

39. Foyer, C. 1988. Feedback inhibition of photosynthesis through sink-source regulation in leaves. *Plant Physiol. Biochem.* 26:483–92

40. Foyer, C., Walker, D. A., Spencer, C., Mann, B. 1982. Observations on the phosphate status and intracellular pH of intact cells, protoplasts and chloroplasts from photosynthetic tissues using ^{31}P-NMR. *Biochem. J.* 202:429–34

41. Francois, J., Van Schaftingen, E., Hers, H.-G. 1988. Characterisation of phosphofructokinase-2 and enzymes involved in the degradation of fructose-2,6-bisphosphatase in yeast. *Eur. J. Biochem.* 171:599–608

42. Freeling, M., Bennett, D. C. 1985. Maize Adh. I. *Annu. Rev. Genet.* 19:297–323

43. Furbank, R. T., Foyer, C., Stitt, M. 1985. The localization of sucrose synthesis in maize leaves. *Planta* 164:172–78

44. Ganson, N. J., Fromm, H. J. 1982. The effect of fructose-2,6-bisphosphate on the reverse reaction kinetics of fructose-1,6-bisphosphatase from bovine liver.

Biochem. Biophys. Res. Commun. 108:233–39

45. Gerhardt R., Stitt, M., Heldt, H. W. 1987. Subcellular metabolite levels in spinach leaves. Regulation of sucrose synthesis during diurnal alterations in photosynthesis. *Plant Physiol.* 83:399–407

46. Gross, P., ap Rees, T. 1986. Alkaline inorganic pyrophosphatase and starch synthesis in amyloplasts. *Planta* 167: 140–45

47. Harris, G. C., Cheesborough, J. K., Walker, D. A. 1983. Effects of mannose on photosynthetic gas exchange in spinach leaf discs. *Plant Physiol.* 71:108–11

47a. Hatzfeld, W.-D., Stitt, M. 1990. A study of the rate of recycling of triosephosphate in heterotrophic *Chenopodium rubrum* cells, potato tubers, and maize endosperm. *Planta.* In press

47b. Hatzfeld, W.-D., Dancer, J. E., Stitt, M. 1990. Fructose-2,6-bisphosphate, metabolites and "coarse" control of pyrophosphate: fructose-6-phosphate phosphotransferase during triose phosphate recycling in heterotrophic cell suspension cultures of *Chenopodium rubrum. Planta.* In press

47c. Hatzfeld, W.-D., Dancer, J. E., Stitt, M. 1990. Direct evidence that pyrophosphate : fructose-6-phosphate phosphotransferase can act as a glycolytic enzyme in plants. *FEBS Lett.* 254:215–18

48. Hedrich, R., Raschke, K., Stitt, M. 1985. A role for fructose 2,6-bisphosphate in regulating carbohydrate metabolism in guard cells. *Plant Physiol.* 79:977–82

49. Heldt, H. W., Chon, C. J., Maronde, D., Herold, A., Stankovic, Z. S., et al. 1977. Effects of orthophosphate and other factors in the regulation of starch formation in leaves and chloroplasts. *Plant Physiol.* 59:1146–55

50. Heldt, H. W., Flügge, U.-I., Borchert, S., Bruckner, G., Ohnischi, J.-I. 1989. Phosphate translocators in plastids. In *Perspectives in Biochemical and Genetic Regulation of Photosynthesis.* New York: Liss

51. Hers, H.-G., Van Schaftingen, E. 1982. Fructose-2,6-bisphosphate two years after its discovery. *Biochem. J.* 206:1–12

52. Herzog, B., Stitt, M., Heldt, H. W. 1984. Control of photosynthetic sucrose synthesis by fructose-2,6-bisphosphate. III. Properties of the cytosolic fructose-1,6-bisphosphatase. *Plant Physiol.* 75: 561–65

53. Huber, S. C. 1986. Fructose-2,6-

bisphosphate as a regulatory metabolite in plants. *Annu. Rev. Plant Physiol.* 37:233–46

54. Huber, S. C. 1989. On the nature of the fructose-2,6-bisphosphate metabolising enzymes in plants. In *The Unique Sugar Phosphate, Fructose-2,6-Bisphosphate,* ed. S. J. Pilkis. New York: CRC. In press

55. Huber, S. C., Akazawa, T. 1986. A novel sucrose synthase pathway for sucrose degradation in cultured sycamore cells. *Plant Physiol.* 81:1008–13

56. Hue, L., Rider, M. 1987. Role of fructose-2,6-bisphosphate in the control of glycolysis in mammalian tissues. *Biochem. J.* 245:313–24

57. Journet, E.-P., Douce, R. 1985. Enzymic capacities of purified cauliflower bud plastids for lipid synthesis and carbohydrate metabolism. *Plant Physiol.* 79:458–67

58. Kacser, M., Porteous, J. W. 1987. Control of metabolism: What do we have to measure. *Trends Biochem. Sci.* 12:5–14

59. Keeling, P. L., Wood, J. R., Tyson, R. H., Bridges, I. G. 1988. Starch biosynthesis in the developing wheat grain. *Plant Physiol.* 87:311–19

60. Kerr, P. S., Huber, S. C. 1987. Coordinate control of sucrose formation in soybean leaves by sucrose phosphate synthase and fructose-2,6-bisphosphate. *Planta* 170:197–204

61. Kombrink, E., Kruger, N. J. 1984. Inhibition by metabolic intermediates of pyrophosphate : fructose-6-phosphate phosphotransferase from germinating castor bean endosperm. *Z. Pflanzenphysiol.* 114:443–53

62. Kombrink, E., Kruger, N. J., Beevers, H. 1984. Kinetic properties of pyrophosphate : fructose-6-phosphate phosphotransferase from germinating castor bean endosperm. *Plant Physiol.* 74:395–401

63. Kowalczyk, S., Januszewska, B., Cymerska, E., Maslowski, P. 1984. The occurrence of inorganic pyrophosphate : D-fructose-6-phosphate phosphotransferase in higher plants. I. Initial characterisation of partially purified enzyme from *Sanseveria trifasciata* leaves. *Physiol. Plant.* 60:31–37

64. Kruckeberg, A., Neuhaus, H. E., Feil, R., Gottlieb, L., Stitt, M. 1989. Reduced activity mutants of phosphoglucose isomerase in the cytosol and chloroplast of *Clarkia xantiana.* I. Impact on mass action ratios and fluxes to sucrose and starch. *Biochem. J.* 261:457–67

65. Kruger, N. J., Beevers, H. 1984. Effect of fructose-2,6-bisphosphate on the

kinetic properties of cytoplasmic fructose-1,6-bisphosphatase from germinating castor bean endosperm. *Plant Physiol.* 76:49–54

66. Kruger, N. J., Beevers, H. 1985. Synthesis and degradation of fructose-2,6-bisphosphate in the endosperm of castor bean seedlings. *Plant Physiol.* 77:358–64

67. Kruger, N. J., Dennis, D. T. 1985. A source of apparent pyrophosphate: fructose-6-phosphate phosphotransferase activity in rabbit muscle phosphofructokinase. *Biochem. Biophys. Res. Commun.* 126:320–26

68. Kruger, N. J., Dennis, D. T. 1985. Reassessment of an apparent hyperactive form of phosphofructokinase from plants. *Plant Physiol.* 78:645–48

69. Kruger, N. J., Dennis, D. T. 1987. Molecular properties of pyrophosphate: fructose-6-phosphate phosphotransferase from potato tuber. *Arch. Biochem. Biophys.* 256:273–79

70. Kruger, N. J., Hammond, J. B. W. 1988. Molecular comparison of pyrophosphate- and ADP-dependent fructose-6-phosphate 1-phosphotransferase from potato tuber. *Plant Physiol.* 86:645–48

71. Kruger, N. J., Kombrink, E., Beevers, H. 1983. Pyrophosphate : fructose-6-phosphate phosphotransferase in germinating castor bean seedlings. *FEBS Lett.* 153:409–12

72. Larondelle, Y., Mertens, E., Van Schaftingen, E., Hers, H.-G. 1986. Purification and properties of spinach leaf phosphofructokinase 2/fructose-2,6-bisphosphatase. *Eur. J. Biochem.* 161: 351–57

73. Larondelle, Y., Mertens, E., Van Schaftingen, E., Hers, H.-G. 1989. Fructose-2,6-bisphosphate hydrolysing enzymes in higher plants. *Plant Physiol.* 90:827–34

74. Leegood, R. C., Furbank, R. T. 1986. Stimulation of photosynthesis by 2% O_2 at low temperatures is restored by phosphate. *Planta* 168:84–93

75. Leegood, R. C., Labate, C. A., Huber, S. C., Neuhaus, H. E., Stitt, M. 1988. Phosphate sequestration by glycerol and its effects on photosynthetic carbon assimilation by leaves. *Planta* 176:117–26

76. Lehringer, A. L. 1982. *Biochemistry.* New York: Worth

77. Loughman, B. C., Rutcliffe, B. G., Southon, T. E. 1989. Observations on cytoplasmic and vacuolar orthophosphate pools in leaf tissue using in vivo

^{31}P-NMR spectroscopy. *FEBS Lett.* 242:279–84

78. MacDonald, F. D., ap Rees, T. 1983. Enzyme properties of amyloplasts from suspension cultures of soybean. *Biochim. Biophys. Acta* 755:81–89

79. MacDonald, F. D., Chou, Q., Buchanan, B. B. 1987. Ion-exchange chromatography separates activities synthesising and degrading fructose-2,6-bisphosphate from C_3 and C_4 leaves but not from rat liver. *Plant Physiol.* 85:13–16

80. MacDonald, F. D., Chou, Q., Buchanan, B. B., Stitt, M. 1989. Purification and characterisation of fructose-2,6-bisphosphatase, a substrate-specific cytosolic enzyme from leaves. *J. Biol. Chem.* 264:5540–44

81. MacDonald, F. D., Cséke, C., Chou, Q., Buchanan, B. B. 1987. Activities synthesising and degrading fructose-2,6-bisphosphate in spinach leaves reside on different proteins. *Proc. Natl. Acad. Sci. USA* 84:2742–46

82. MacDonald, F. M., Preiss, J. 1986. The subcellular location and characteristics of pyrophosphate: fructose-6-phosphate 1-phosphotransferase from suspension-cultured cells of soybean. *Planta* 167:240–45

83. Mertens, E., Marcellin, P., Van Schaftingen, E., Hers, H.-G. 1987. Effect of ethylene treatment on the concentration of fructose-2,6-bisphosphate and on the activity of phosphofructokinase 2/fructose-2,6-bisphosphatase in banana. *Eur. J. Biochem.* 167:579–83

84. Mertens, E., Van Schaftingen, E., Hers, H.-G. 1987. Fructose-2,6-bisphosphate and the control of energy charge in higher plants. *FEBS Lett.* 221:124–28

85. Mohabir, G., John, P. 1989. Effect of temperature on starch synthesis in potato tuber tissue and amyloplasts. *Plant Physiol.* 88:1222–28

86. Morell, S., ap Rees, T. 1986. Sugar metabolism in developing tubers of *Solanum tuberosum*. *Phytochemistry* 25:1579–85

87. Neuhaus, H. E., Stitt, M. 1989. Perturbation of photosynthesis in spinach leaf discs by low concentrations of methyl viologen. *Planta* 179:51–60

88. Neuhaus, H. E., Kruckeberg, A. L., Feil, R., Gottlieb, L., Stitt, M. 1989. Reduced activity mutants of phosphoglucose isomerase in the cytosol and chloroplast of *Clarkia xantiana*. II Study of the mechanisms which regulate photosynthate partitioning. *Planta* 178:110–22

89. Newsholme, E. A., Start, C. 1973. *Regulation in Metabolism.* Chichester: Wiley

90. O'Brien, W. W., Bowien, S., Wood, H. G. 1975. Isolation and characterisation of a pyrophosphate-dependent phosphofructokinase from *Proprionibacterium shermanii*. *J. Biol. Chem.* 250:8690–95

91. Pilkis, S. J., Claus, T. H., Kounitz, P. D., El-Maghrabi, M. R. 1987. Enzymes of the fructose-6-phosphate : fructose-1,6-bisphosphate substrate cycle. *Enzymes* 18:3–45

92. Preiss, J. 1982. Regulation of the biosynthesis and degradation of starch. *Annu. Rev. Plant. Physiol.* 33:431–54

93. Purwin, C., Laux, M., Holzer, H. 1987. Fructose-2-phosphate, an intermediate of the dephosphorylation of fructose-2,6-bisphosphate with a purified yeast enzyme. *Eur. J. Biochem.* 164:27

94. Quick, P., Neuhaus E., Stitt, M. 1989. Increased pyrophosphate is responsible for a restriction of sucrose synthesis after supplying fluoride to spinach leaf discs. *Biochim. Biophys. Acta* 973:263–71

95. Quick, P., Scheibe, R., Stitt, M. 1989. Use of tentoxin and nigericin to investigate the possible contribution of Δ pH to energy dissipation and control of electron transport in leaves. *Biochim. Biophys. Acta* 974:282–88

96. Quick, P., Siegl, G., Neuhaus, E., Feil, R., Stitt, M. 1989. Short-term water stress leads to a stimulation of sucrose synthesis by activating sucrose phosphate synthase. *Planta* 177:536–46

97. Rea, P. A., Sanders, D. 1987. Tonoplast energisation: two H^+ pumps, one membrane. *Physiol. Plant.* 71:131–41

98. Rebeille,, F., Bligny, R., Martin, J.-B., Douce, R. 1983. Relationship between the cytoplasm and the vacuole phosphate pool in *Acer pseudoplatanus* cells. *Arch. Biochem. Biophys.* 225:143–48

99. Roberts, J. K. M. 1984. Study of plant metabolism *in vivo* using NMR spectroscopy. *Annu. Rev. Plant Physiol.* 35:375–86

100. Roby, C., Martin, J.-B., Bligny, R., Douce, R. 1987. Biochemical changes during sucrose deprivation in higher plant cells. *J. Biol. Chem.* 262:5000–7

101. Rolleston, F. S. 1972. A theoretical background to the use of measured intermediates in the study of the control of intermediary metabolism. *Curr. Top. Cell. Regul.* 5:47–95

102. Sabularse, D. C., Anderson, R. L. 1981. Inorganic pyrophosphate: D-fructose-6-phosphate phosphotransferase in mung beans and its activation

by D-fructose-1,6-bisphosphate and D-glucose-1,6-bisphosphate. *Biochem. Biophys. Res. Commun.* 100:1423–29

103. Sabularse, D. C., Anderson, R. L. 1981. D-Fructose-2,6-bisphosphate: a naturally occurring activator for inorganic pyrophosphate: D-fructose-6-phosphate phosphotransferase in plants. *Biochim. Biophys. Res. Commun.* 103:848–54

104. Sachs, M. M., Ho, T.-H. D. 1986. Alteration of gene expression during environmental stress in plants. *Annu. Rev. Plant Physiol.* 37:363–76

105. Schäfer, G., Heber, U., Heldt, H. W. 1977. Glucose transport into spinach chloroplasts. *Plant Physiol.* 60:286–89

106. Sharkey, T. D., Stitt, M., Heineke, D., Gerhardt, R., Raschke, K., Heldt, H. W. 1986. Limitation of photosynthesis by carbon metabolism II O_2 insensitive CO_2 uptake results from limitation of triose phosphate utilisation. *Plant Physiol.* 81:1123–29

107. Sharkey, T. D., Vanderveer, P. J. 1989. Stromal phosphate concentration is low during feedback limited photosynthesis. *Plant Physiol.* In press

108. Sicher, R. C., Kremer, D. F., Harris, W. G. 1986. Control of photosynthetic sucrose synthesis in barley primary leaves. Role of fructose-2,6-bisphosphate. *Plant Physiol.* 82:15–18

109. Smyth, D. A., Black, C. C. Jr. 1984. Measurement of the pyrophosphate content of plant tissues. *Plant Physiol.* 75:862–64

110. Soll, J., Wötzel, C., Buchanan, B. B. 1984. Enzyme regulation in C_4 photosynthesis. Identification and isolation of enzymes catalysing the synthesis and hydrolysis of fructose-2,6-bisphosphate in corn leaves. *Plant Physiol.* 77:999–103

111. Spilatro, S. R., Anderson, J. M. 1989. Carbohydrate metabolism and activity of pyrophosphate : fructose-6-phosphate phosphotransferase in photosynthetic soybean (*Glycine max,* Merr.) suspension cells. *Plant Physiol.* 88:862–88

112. Steingraber, M., Outlaw, W. H. Jr., Hampp, R. 1988. Subcellular compartmentation of fructose-2,6-bisphosphate in oat mesophyll cells. *Planta* 175:204–8

113. Stitt, M. 1985. Fine control of sucrose synthesis by fructose 2,6-bisphosphate. See Ref. 3, pp. 109–26

114. Stitt, M. 1987. Fructose-2,6-bisphosphate and plant carbohydrate metabolism. *Plant Physiol.* 84:201–4

115. Stitt, M. 1989. Control analysis of

photosynthetic sucrose synthesis: assignment of elasticity coefficients and flux control coefficients to the cytosolic fructose-1,6-bisphosphatase and sucrose phosphate synthase. *Philos. Trans. R. Soc. London Ser. B* 323:327–38

116. Stitt, M. 1989. Product inhibition of potato tuber pyrophosphate: fructose-6-phosphate phosphotransferase. *Plant Physiol.* 89:628–33

117. Stitt, M. 1990. Fructose-2,6-bisphosphate. In *Methods in Plant Biochemistry,* ed. P. Lee, Vol. 7. London: Academic. In press

118. Stitt, M. 1989. Control of sucrose synthesis. Estimation of free energy charges, investigation of the contribution of equilibrium and non-equilibrium reactions, and estimation of elasticities and flux control coefficients. In *Techniques and New Developments in Photosynthetic Research,* ed. J. Barber. London: Plenum. In press

119. Stitt, M., Cséke, C. 1987. Alterations of fructose-2,6-bisphosphate during plant respiratory metabolism. In *Hungarian-USA Binational Symposium on Photosynthesis, Salve Regina College,* ed. M. Gibbs, pp. 97–104. Baltimore: Waverley

120. Stitt, M., Cséke, C., Buchanan, B. B. 1984. Regulation of fructose-2,6-bisphosphate concentration in spinach leaves. *Eur. J. Biochem.* 143:89–93

121. Stitt, M., Cséke, C., Buchanan, B. B. 1985. Occurrence of a metabolite-regulated enzyme synthesising fructose 2,6-bisphosphate in plant sinks. *Physiol. Veg.* 23:819–27

122. Stitt, M., Cséke, C., Buchanan, B. B. 1986. Ethylene-induced increase of fructose 2,6-bisphosphate in plant storage tissue. *Plant Physiol.* 80:246–49

123. Stitt, M., Foyer, C., Wirtz, W., Gerhardt, R., Heldt, H. W., et al. 1985. A comparative study of metabolite levels in plant leaf material in the dark. *Planta* 166:354–64

124. Stitt, M., Gerhardt, R., Kürzel, B., Heldt, H. W. 1983. A role for fructose-2,6-bisphosphate in the regulation of sucrose synthesis in spinach leaves. *Plant Physiol.* 72:1139–41

125. Stitt, M., Gerhardt, R., Wilke, I., Heldt, H. W. 1987. The contribution of fructose-2,6-bisphosphate to the regulation of sucrose synthesis during photosynthesis. *Physiol. Plant.* 69:377–86

126. Stitt, M., Grosse, H. 1988. Interactions between sucrose synthesis and photosynthesis. I. Slow transients during a biphasic induction of photosynthesis are

related to a delayed activation of sucrose synthesis. *J. Plant Physiol.* 133:129–37

127. Stitt, M., Grosse, H. 1988. Interaction between sucrose synthesis and photosynthesis. IV. Temperature dependent adjustment of the relation between sucrose synthesis and CO_2 fixation. *J. Plant Physiol.* 133:392–400

128. Stitt, M., Heldt, H. W. 1985. Control of photosynthetic sucrose synthesis by fructose 2,6-bisphosphate. IV. Intercellular metabolite distribution and properties of the cytosolic fructose-1,6-bisphosphatase in maize leaves. *Planta* 164:179–88

129. Stitt, M., Herzog, B., Heldt, H. W. 1984. Control of photosynthetic sucrose synthesis by fructose-2,6-bisphosphate I. Coordination of CO_2 fixation and sucrose synthesis. *Plant Physiol.* 75:548–53

130. Stitt, M., Herzog, B., Heldt, H. W. 1985. Control of photosynthetic sucrose synthesis by fructose-2,6-bisphosphate. V. Modulation of the spinach leaf cytosolic fructose-1,6-bisphosphatase in vitro by substrate, products, pH, magnesium, fructose-2,6-bisphosphate, adenosine monophosphate and dihydroxyacetone phosphate. *Plant Physiol.* 79:590–98

131. Stitt, M., Huber, S., Kerr, P. 1987. Control of photosynthetic sucrose synthesis. In *The Biochemistry of Plants*, ed. M. D. Hatch, N. K. Boardman, 10:327–409. New York: Academic

132. Stitt, M., Kürzel, B., Heldt, H. W. 1984. Control of photosynthetic sucrose synthesis by fructose-2,6-bisphosphate. II. Partitioning between sucrose and starch. *Plant Physiol.* 75:554–60

133. Stitt, M., Mieskes, G., Söling, H.-D., Grosse, H., Heldt, H. W. 1986. Diurnal changes of fructose-6-phosphate,2-kinase and fructose-2,6-bisphosphatase activities in spinach leaves. *Z. Naturforsch. Teil C* 41:291–96

134. Stitt, M., Mieskes, G., Söling, H.-D., Heldt, H. W. 1982. On a possible role of fructose-2,6-bisphosphate in regulating photosynthetic metabolism in leaves. *FEBS Lett.* 145:217–22

135. Stitt, M., Schreiber, U. 1988. Interactions between sucrose synthesis and photosynthesis. III. Response of biphasic induction kinetics and oscillations to manipulation of the relation between electron transport, the Calvin cycle, and sucrose synthesis. *J. Plant Physiol.* 133:263–71

136. Stitt, M., Steup, M. 1985. Starch and sucrose degradation. See Ref. 1, pp. 347–90

137. Stitt, M., Vasella, A. 1988. Biological action of phosphonate analogs of fructose-2,6-bisphosphate on enzymes from higher plants. *FEBS Lett.* 228:60–64

138. Stitt, M., Wilke, I., Feil, R., Heldt, H. W. 1988. Coarse control of sucrose phosphate synthase in leaves: Alterations of the kinetic properties in response to the rate of photosynthesis and the accumulation of sucrose. *Planta* 174: 217–30

139. Sung, S.-J. S., Xu, D.-P., Black, C. C. 1989. Identification of actively filling sinks. *Plant Physiol.* 89:1117–21

140. Sung, S.-J. S., Xu, D.-P., Galloway, C. M., Black, C. C. Jr. 1988. A reassessment of glycolysis and gluconeogenesis in higher plants. *Physiol. Plant.* 72:650–54

141. Sung, S. S., Xu, D.-P., Alvarez, C., Mustardy, L. A., Black, C. C. Jr. 1987. Pyrophosphate as a biosynthetic energy source and fructose-2,6-bisphosphate regulation. See Ref. 119, pp. 72–80

142. Tyson, R. H., ap Rees, T. 1988. Starch synthesis by isolated amyloplasts from wheat endosperm. *Planta* 175:33–38

143. Van Schaftingen, E. 1987. Fructose-2,6-bisphosphate. *Adv. Enzymol. Relat. Areas Mol. Biol.* 59:315–95

144. Van Schaftingen, E., Hers, H.-G. 1983. Fructose-2,6-bisphosphate in relation with the resumption of metabolic activity in slices of Jerusalem artichoke tubers. *FEBS Lett.* 164:195–200

145. Van Schaftingen, E., Lederer, B., Bartrons, R., Hers, H.-G. 1982. A kinetic study of pyrophosphate: fructose-6-phosphate phosphotransferase from potato tubers. *Eur. J. Biochem.* 129: 191–95

146. Walker, G. H., Huber, S. C. 1987. Spinach leaf 6-phosphofructo-2-kinase. Isolation of a new enzyme from spinach leaves that undergoes ATP-dependent modification. *FEBS Lett.* 213:375–80

147. Walker, J. A., Huber, S. C. 1989. Regulation of sucrose phosphate synthase activity in spinach leaves by protein level and covalent modification. *Planta* 177:116–20

148. Walker, J. A., Huber, S. C., Nielsen, T. M. 1989. Protein phosphorylation as a mechanism for regulation of spinach leaf sucrose phosphate synthase. *Arch. Biochem. Biophys.* In press

149. Weiner, H., Stitt, H., Heldt, H. W. 1987. Subcellular compartmentation of pyrophosphate and alkaline pyrophosphatase in leaves. *Biochim. Biophys. Acta* 893:13–21

150. Wong, J. H., Kang, T., Buchanan, B.

B. 1988. A novel PFP from carrot roots. *FEBS Lett.* 238:405–9

151. Wong, J. H., Yee, B. H., Buchanan, B. B. 1987. A novel type of phosphofructokinase from plants. *J. Biol. Chem.* 262:3185–91

152. Wu, M.-X., Smyth, D. A., Black, C. C. Jr. 1983. Fructose-2,6-bisphosphate and the regulation of pyrophosphate-dependent phosphofructokinase in germinating pea seeds. *Plant Physiol.* 73:188–91

153. Wu, M.-X., Smyth, D. A., Black, C. C. Jr. 1984. Regulation of pea seed pyrophosphate-dependent phosphofructokinase: evidence for interconversion of two forms as a glycolytic regulatory mechanism. *Proc. Natl. Acad. Sci. USA* 81:5051–55

154. Xu, D.-P., Sung, S.-J. S., Black, C. C. Jr. 1989. Sucrose metabolism in lima bean seeds. *Plant Physiol.* 89:1106–16

155. Yan, T.-F. J., Tao, M. 1984. Multiple forms of pyrophosphate : D-fructose-6-phosphate 1-phosphotransferase from wheat seedlings. *J. Biol. Chem.* 259: 5087–92

156. Zimmerman, G., Kelly, G. E., Latzko, E. 1978. Purification and properties of spinach leaf cytoplasmic fructose-1,6-bisphosphatase. *J. Biol. Chem.* 253: 5952–56

ACKNOWLEDGMENTS

I am grateful to S. C. Huber, T. ap Rees, B. B. Buchanan, and E. van Schaftingen, who made manuscripts available to me before publication.

Annu. Rev. Plant Physiol. Plant Mol. Biol. 1990. 41:187–223
Copyright © 1990 by Annual Reviews Inc. All rights reserved

COLD ACCLIMATION AND FREEZING STRESS TOLERANCE: ROLE OF PROTEIN METABOLISM

Charles L. Guy

Ornamental Horticulture Department, Institute of Food and Agricultural Sciences, University of Florida, Gainesville, Florida 32611

KEY WORDS: cold hardiness, cryobiology, gene expression, frost, environmental physiology

CONTENTS

INTRODUCTION

Research interest in the effects of low temperature (LT) stress on crop plant productivity has produced an avalanche of information. In 1984, Steponkus (189) noted that much of the existing knowledge had not been adequately

187

integrated, a state of affairs that still holds true for a literature that has grown by a few thousand citations. Since the only efforts to integrate completely the existing information on plants at LT required entire books (109, 111, 112, 138, 171), such a synthesis is not attempted here. Many aspects of plant responses to LT have been reviewed elsewhere (110, 114, 138, 140, 179, 194, 207, 218). Treatments in this series include an examination of frost resistance in reference to other environmental stresses (108), stresses of water redistribution during freezing and histological and molecular effects (137), physical events and mechanisms of cell injury during freezing (123), chilling injury (120), factors affecting the freezing process and supercooling (12), cold acclimation and chilling injury (53), and the influence of cold acclimation and freezing injury on the plasma membrane (189). Each represents a unique and important synthesis of this highly complex problem.

For decades, the study of plant cold-hardiness and freezing-stress injury has had two primary goals. The first was to describe the mechanism(s) during a freeze/thaw cycle that leads to cell injury and death (12, 137, 189). Of vital interest were the moment during the freeze/thaw cycle that injury occurred and the site of lethal injury. Researchers believed that once the mechanisms of injury were understood, it should be a relatively straightforward matter to determine the cellular and biochemical alterations necessary for conferring a higher level of cold tolerance. This approach was buttressed by the fact that some plants were hardier than others, and that in certain tissues of many plants hardiness varied seasonally. However, the mechanisms involved in cold injury of plants have turned out to be exceedingly complex (137), and the two most basic questions about freezing injury—when and where—remain partially unanswered. Nevertheless, researchers have developed a general understanding of the basic elements of the formation of ice in plant tissues (the freezing and thawing of cellular water) (122, 123) and identified the major probable site of lethal injury (189). Hence, the second and parallel goal, to catalog and understand the biochemical and physiological changes occurring during cold acclimation (CA), has been viewed as a viable alternative approach. Studies of the process by which plants become tolerant to the stresses imposed by freezing and thawing have gained added importance in efforts to understand plant freezing-tolerance mechanisms (52, 109, 111, 112).

In the most recent review on the subject of cold stress in this series, Steponkus (189) noted the failure of "innumerable correlative studies of biochemical changes that occur during the period of cold acclimation" to lead us to an understanding of how cold acclimation alters the tolerance of cells, tissues, and plants to freezing and thawing stresses. This failure has renewed attempts to comprehend the mechanisms of injury at the probable primary site of freeze/thaw injury, the plasmalemma or plasma membrane (189). While mechanistic approaches to the cryobehavior of the plasmalemma show great promise (50, 193), a new approach may provide insight about which biochemical and physiological changes are directly responsible for altered freezing tolerance. This approach is based on an early idea that variation in hardiness in the plant kingdom and seasonal variation within the tissues of a single species have a genetic or molecular basis (207). In this paper I briefly summarize the general aspects of the freezing process and then in specific cases suggest how a portion of a plant's freezing tolerance could be genetically controlled and regulated during cold acclimation and deacclimation in a fashion akin to a classical inducible response.

ICE FORMATION IN PLANTS

The basic structure and organization of most plant tissues and cells dictate to a large extent the site of initiation of ice crystallization, determining the course and location of cell-water freezing (12, 122, 123). The freezing process in plant tissues is affected by the following facts: (a) Plant cells are not immersed in an aqueous solution but are surrounded by a water saturated environment; (b) the amount of osmotically available extracellular water is small relative to the amount of water inside living cells (208); (c) the apoplastic solution has a much lower solute concentration than the cell (208); (d) cellular water, because of the higher solute concentration, has a greater freezing point depression (111, 112); (e) a functionally intact cell membrane is an effective barrier to the propagation of ice crystals (38, 191); (f) but liquid water can freely move across the plasma membrane in either direction (113); (g) the effectiveness of the cell membrane as a barrier to ice may vary with CA or temperature (38, 190, 191); (h) the presence of heterogeneous nucleators inside cells is minimized, excluded, or masked (12, 171:35); and (i) in many tissues a portion of the internal extracellular volume is free air space, normally saturated with water vapor. The other major factor in the freezing of plant cells is the rate of cooling (122, 123, 191).

Intracellular Freezing

Depending on the rate of tissue cooling, the crystallization of ice can occur in two markedly distinct locations within the tissues of most plants. If cooling is rapid, ice may form within the cells. Crystallization of the water inside the cell may occur by internal nucleation (113) or by penetration into the cell by an external ice crystal (123, 191). In either case such freezing, termed intracellular freezing, is considered to be universally and instantaneously lethal (12, 111, 112). The one exception may be cells and tissues that exhibit deep supercooling (4, 5, 12). For yeast cells and leaf protoplasts suspended in a liquid medium, the rate of cooling required for the formation of intracellular freezing must be, respectively, $\geq$ 10°C/min (123) and 3–16°C/min (190, 191). In contrast, some circumstantial evidence suggests cooling rates $\geq$ 3°C/hr may favor intracellular freezing in intact plants (188). Fortunately in nature, atmospheric cooling rates seldom exceed 1°C/hr (188). In situations leading to sunscald on southwest-facing tree trunks and branches, tissue cooling rates sufficient to incite intracellular freezing seem likely (207), although accurate measurements have not been reported (111). Evidence that intracellular ice formation is largely determined by factors that influence the propensity for supercooling of the cell milieu (12), and that the plasma membrane is an effective barrier to external nucleation remains insufficient (190, 191). Systematic studies on a variety of plants of the cooling rate with and without prior nucleation, of how treatments that alter freezing tolerance influence nucleation temperature, and of the effects of duration at subzero temperature and minimum temperature on nucleation must be completed with intact tissues before the contribution of intracellular ice formation in freezing injury can be fully described.

Extracellular Freezing

Ice nucleation for most plant tissues begins internally on the surface of cell walls (89), in water transporting elements (137), or on external surfaces (12). In nearly every case, ice crystals will spread from the initial foci throughout the extracellular regions of the tissue if the duration of subzero temperatures is extended (12). As long as the plasma membrane remains intact and the cooling rate is slow, ice will remain confined to regions external to the cell (111). The presence of ice exclusively in the regions of the tissue outside the cell is termed extracellular freezing. With possibly the exception of some temperate perennial plants whose xylem parenchyma cells and floral tissues deep-supercool (4, 5, 12, 171), extracellular ice formation will impose a dehydrative force on the unfrozen solution of the cell (122, 123).

The formation and presence of ice in the extracellular regions of plant tissues have been known for more than a century and a half (49), but the theoretical aspects of extracellular freezing of cells surrounded by ice or immersed in a solution in equilibrium with ice were only described by Mazur in the 1960s (122, 123). This theoretical approach illustrated two major facets of extracellular freezing that result in the dehydration of the cell. First, freezing of the extracellular solution leads to a rapid rise in the solute concentration and an equally rapid decline in vapor pressure of the unfrozen portion as a result of the freeze-concentration of solutes. Second, the vapor pressure of ice declines faster than the vapor pressure of liquid water as temperature decreases. At slow cooling rates, a water potential gradient is thus established, with liquid water moving down the gradient. During extracellular freezing in plants, liquid water will move out of the cell to the extracellular solution or ice crystal.

Much of this theory has been widely accepted as explaining the general features of extracellular freezing in plants (12, 111, 112, 189). However, Mazur's analysis assumed ideal behavior of both the surrounding solution and the cell during freezing. While his theoretical treatment was based on cells immersed in a liquid medium, the freezing of plant tissues may be slightly different. First, it can be debated whether plant cells are immersed in a solution or are simply surrounded by a water-saturated environment consisting of the cell wall and intercellular spaces. Thus, unlike the case in a cell suspended in a liquid medium, in plants an extracellular ice crystal may or may not be distant from a given cell even at subzero temperatures. Second, the ratio of extracellular water to intracellular water for many tissues may not be large—perhaps 0.1–0.2 (208). As a consequence of low solute concentration and the small volume of extracellular solution in plant tissues, there will not be massive freeze concentration of external solutes with the accompanying rise in osmotic potential. Third, Mazur's analysis excluded hydrostatic pressure differences across the cell membrane, an exclusion that, as he noted, may not be appropriate for plant cells. In spite of these potentially significant differences for plants, studies have confirmed that the freezing of water in plant tissues could follow the basic pattern of equilibrium freezing (67) and fit existing theory (123).

The importance of understanding the unique characteristics of plant-tissue freezing has prompted recent attempts to describe the physical chemical relationships of the process (67, 155, 156). Earlier work using nuclear magnetic resonance techniques to follow the course of freezing in intact tissues found that freezing of water could approximate that of an ideal solution (24, 57, 71). However, it has been shown that extracellular freezing of cellular water does not always follow ideality (2). This suggested that modification of Mazur's theoretical treatment of freezing was necessary, at least for some plant species. Rajashekar & Burke (155) employed a combination of water relations theory and the Clausius-Claperyron equation to derive a relationship that describes the water potential in megapascals (MPa) of extracellular ice and supercooled water:

$$\psi_{ice} = \frac{RT}{\bar{V}_w} \ln \frac{P_{ice}}{P_{liquid}}, \qquad\qquad 1.$$

where ψ_{ice} is the water potential of ice at a subzero temperature in MPa, R is the gas constant, T is °K, $\bar{V}_w$ is the molar volume of water, P_{ice} and P_{liquid} the vapor pressures of ice and liquid water, respectively, and ΔH_f is heat of fusion of water.

$$\psi_{ice} = \frac{\Delta H_f}{273\bar{V}_w} T(°C). \qquad\qquad 2.$$

At equilibrium during extracellular freezing, the water relations of frozen tissue can be described by Equation 3, where ψ_T(ice) is the water potential (MPa) of ice at temperature T, and π_T, P_T, and τ_T represent the osmotic, pressure, and matrix potentials at temperature T (156). Alternatively, the water potential of ice at any given freezing temperature can be determined, in MPa, by the relationship in Equation 4 where (°C) is temperature below freezing (156).

$$\psi_T(\text{ice}) = \pi_T + P_T + \tau_T.$$
3.

$$\psi_T(\text{ice}) = 1.16(°C).$$
4.

As can be seen by Equation 4, the decline in the water potential of ice with decreasing temperature is large, -1.16 MPa per °C. When coupled with the temperature dependence of osmotic pressure, cells at equilibrium with extracellular ice will be dehydrated in a strict temperature-dependent relationship. Depending on the initial osmotic potential of the unfrozen cell, at some subzero temperature all osmotically active water can be frozen out of the cell.

The water-relations approach also provides a theoretical framework to suggest negative pressure potential as an explanation of nonideal equilibrium freezing (2). Hansen & Beck (67), have also derived Equation 2 using Van't Hoff's equations for freezing-point depression and osmotic pressure and setting ψ_{ice} for $-\pi$. Experimental measurements of the water potential of ice were in good agreement with Equation 2. Further, application of the theory for two species of plants has demonstrated ideal equilibrium freezing in *Hedera helix* and non-ideal equilibrium freezing in barley. The deviation from ideality in barley was attributed to resistance of the cell wall to collapse and the generation of a negative pressure potential component as water moved out of the cell to the extracellular ice. Large negative pressure potentials would only be possible if the cell membrane adhered tightly to the cell wall. Otherwise the cell membrane could withdraw from the cell wall during freeze dehydration, a process akin to plasmolysis in hypertonic solution, known as frost plasmolysis (111). Cell collapse, not frost plasmolysis, seems to be the pattern during extracellular freezing (111). Thus negative pressure potentials may decrease cell dehydration during the course of extracellular freezing, and could function as a dehydration avoidance mechanism (2, 155, 156).

GENETICS OF FREEZING TOLERANCE

Overwintering

Freezing tolerance in temperate biennials and perennials—the phenomenon that enables them to survive winter conditions and resume growth and development in the spring—is not simply a function of the minimum survivable temperature (LT_{50}) but involves a genetically programmed, integrated process (196, 207). To survive the winter, a plant must have evolved mechanisms whereby freezing-sensitive tissues can avoid freezing or undergo a change in freezing tolerance compatible with the normal variations of the local climate, coordinate the induction of the tolerance at the appropriate time, maintain adequate tolerance during times of risk, and properly time the loss of tolerance and resumption of growth when the risk of freezing has passed. Since, overwintering or winterhardiness is an important agronomic trait in many crops, most studies have focused on cultivar and varietal trials (14, 46, 128, 153, 154, 159, 194) and selection of cold-tolerant lines (74, 134), with less attention given to the quantitative genetics (56, 80, 117, 135, 136, 195).

Breeders have exploited genetic variability present in many crops to develop cultivars with improved cold tolerance, while geneticists have used natural variation to examine inheritance. Crosses involving compatible parents of differing cold tolerance typically yield progeny exhibiting a continuous range of hardiness between the parental extremes. Since this has been the most common observation for progeny of such crosses, it has been concluded that winterhardiness is a quantitatively inherited trait (117, 135, 136, 146, 169, 212). Likewise, the absence of drastic improvement in the winterhardiness of crop species through breeding also argues in favor of the quantitative nature of the trait (153, 194). Occasionally progeny will exceed the boundaries of the parental hardiness extremes (34, 80, 135, 160). Situations where the winterhardiness of the progeny segregates outside the parental boundaries is termed transgressive segregation. Reports of transgressive segregation are consistent with the quantitative character of winterhardiness. In wheat and barley, several studies have reported evidence that winterhardiness is controlled by partially dominant and/or recessive genes (107, 167, 195), which appear to act under different levels of stress in an additive fashion (56, 167). In contrast, studies involving reciprocal crosses have not consistently indicated a maternal inheritance pattern for cold tolerance (70, 80, 132, 195, 209). More detailed cytogenetic studies using monosomic and substitution analysis in a number of laboratories suggest that in wheat at least 10 of the 21 pairs of chromosomes are involved in freezing tolerance (see 196). Chromosomes 5A and D appear to carry the major genes that influence wheat hardiness (96, 107, 113). The repeated verification of the quantitative nature of winterhardiness is not unexpected given the fact that morphology, developmental and physiological processes, and complex environmental interactions all influence cold tolerance (111).

Cold Acclimation

A significant component of winterhardiness (overwintering ability) in cold-hardy plants is the capacity to undergo CA. Virtually all temperate perennial, and many annual and biennial plants native to the regions of the world subject to subzero temperatures can alter their tissue and cellular freezing tolerance upon exposure to low nonfreezing temperature (111, 112, 171). The term cold acclimation is most often used to describe the outcome of the myriad biochemical and physiological processes associated with the increase in cold tolerance, but a more precise view of CA would include two major functions: the more universal adjustment of metabolism and basic cellular function to the biophysical constraints imposed by LT, and the induction of freezing tolerance. The first function of CA differentiates chilling-sensitive from chilling-tolerant species. The second function of CA discriminates chilling-tolerant but freezing-sensitive species from those that are freezing tolerant. At present it is not known if both functions are required for development of freezing tolerance at LT. Certainly if a plant cannot adjust cellular processes for proper function during long-term exposure to low nonfreezing temperatures, it is unlikely to become maximally freezing tolerant. However, in photoperiod-sensitive temperate woody species, a partial induction of freezing tolerance can be evoked at warmer temperatures (186, 207). Since exogenous abscisic acid (ABA) treatment of hardy species has been shown to induce freezing tolerance at normally nonacclimating temperatures (22, 25), it appears possible to separate the global metabolic adjustments to low nonfreezing temperatures from those processes involved in the induction of freezing tolerance.

Genetic studies of the cold acclimation capacity or trait are almost nonexistent. Nearly all genetic studies of cold hardiness have focused on aspects other than CA, primarily overwintering ability (13, 107, 116, 159, 167, 212), of which the CA trait is but one of many factors. Two aspects of CA that have been addressed by genetic analyses are the inheritance of the minimum survivable temperature and the timing of the photoperiodic induction of freezing tolerance. The few reports on the genetics of minimum survivable temperature associated with CA have supported a dominant/recessive pattern of inheritance of the winterhardiness character (56), which appears to be coupled in an additive-dominance system (195, 196). Chromosomal substitution with wheat has indicated additive gene effects were greater than the dominance effects (196). As freezing stress became more severe, dominant/recessive character became reversed. This is consistent with the observation that under one set of conditions genes for winterhardiness appeared to act in a dominant manner, while at other times they acted recessively (212). More specifically, at mild stress levels tolerance appeared to be a dominant trait, while under more extreme freezing stress conditions tolerance exhibited a recessive character (196, 211). This supports the hypothesis that different genes affect tolerance at different levels of stress (56). Limin & Fowler (117), examined the inheritance of minimum survivable temperature in interspecific and intergeneric hybrids of *Triticum* and related genera. The freezing tolerance of F_1 hybrids that maintained the parental chromosome number was near the midpoint between the parental extremes, indicating a quantitative character controlled by a number of additive genes.

Studies of the photoperiodic induction of freezing tolerance are equally scarce. Hummel et al (80) studied genetic control of the photoperiodic induction of freezing tolerance in F_1, F_2, and backcrosses of latitudinal ecotypes of *Cornus sericea*. While no maternal influence was observed, some indication of transgressive segregation was noted. This was not unexpected given the fact that the timing of the photoperiodic induction would be a critical factor in overall winterhardiness. Photoperiodic control of CA may be a quantitative character, as indicated indirectly by the intermediate responses of progeny with respect to the parents (80). These findings were consistent with more recent studies of *Pinus sylvestris*, where genetic analyses of the photoperiodic induction supported a polygenic inheritance model (136).

All that is clear at this stage is that the inheritance of the capacity for CA-induced freezing tolerance is polygenic. What remains unknown is how many genes are specifically involved in LT-induced CA. However, one study of the winterhardiness of pea offers a partial answer to this question (116). Analysis of segregating F_5 lines revealed a continuous range of winterhardiness between parental extremes indicative of a trait controlled by additive gene action. That lines exhibiting parental hardiness levels were obtained from a population of 50 F_5 lines suggests the inheritance is less complex than originally thought. Liesenfeld and colleagues (116) suggest that as few as three additive genes or linkage groups may control winterhardiness in pea. Since the limited genetic studies have not yet focused on the incremental change in freezing tolerance between nonacclimated and cold acclimated tissues of the same plant, the complexity of the genetic control of the inducible component of CA remains to be determined. Answers to this question may await the discovery of fertile mutants blocked in the ability to undergo CA-induced freezing tolerance.

COLD ACCLIMATION

Inducible Response

Freezing tolerance of vast numbers of temperate and alpine species is not static but varies seasonally and in response to short-term variations in temperature. For centuries it has been known that temperate perennials follow a seasonal rhythm of dormancy and freezing tolerance (see 109, 111). Temperate woody perennials rely on photoperiodic cues to signal the end of the active growth season and initiation of the developmental and physiological changes necessary to promote CA and increase freezing tolerance (202). Following initiation by photoperiodic responses, acquisition of additional freezing tolerance via CA generally requires exposure to low nonfreezing temperatures

(173, 207). Tolerance to the mildest form of freezing, frost, is typically minimal or nonexistent in spring and summer, but in fall susceptibility to even severe freezing stress declines (Figure 1A). In winter, tolerance or hardiness reaches its greatest level. This is evident by the fact that some species when fully acclimated are not injured by exposure to temperatures approaching absolute zero (O°K) (172). In spring, the warming temperatures lead to a loss of freezing tolerance (Figure 1B). This seasonal rhythm in temperate perennials derives from both developmental and physiological adjustments during CA. In contrast, for many other species (primarily but not exclusively herbaceous plants) where photoperiodic cues have been less necessary to direct seasonal rhythms of growth and hardiness, CA and induction of freezing tolerance require only LT exposure (44). Nearly 20 years ago Levitt (111) stated that "It is now standard procedure to harden off plants by exposing them for a week or two to temperatures a few degrees above the freezing point." This statement clearly demonstrated a then long known but not fully appreciated fact: CA and its associated increase in freezing tolerance were essentially inducible responses. The inducible and transient character of freezing tolerance is further demonstrated in hardy herbaceous plants that have been subjected to CA. Upon return to warm nonacclimating temperatures, freezing tolerance is lost and active growth resumes (44, 192). In woody perennials that have satisfied chilling requirements, warm temperature–induced deacclimation in spring follows a similar pattern (173).

Kinetics of Cold Acclimation and Deacclimation

Historically, cold acclimation and the induction of freezing tolerance in temperate perennials have been associated with the slowing or cessation of growth and the gradual transition from the heat of summer to the cold of winter (111). Freezing tolerance was observed to increase at a gradual rate during the early part of fall and then to accelerate in late fall to approach maximum hardiness as seen in Figure 1A (76, 111, 173, 202, 207). With the advent of controlled-environment growth rooms, most CA experiments continued to be lengthy, spanning anywhere from two weeks to several months (Figure 1C & D) (44, 64, 103, 192, 219). This generally held true even for annuals and biennials (21, 44). Time-course studies of freezing tolerance induction during CA have revealed the rate is highly temperature and species dependent (20, 44). Plants requiring photoperiodic initiation of dormancy and CA require longer periods to make the transition from sensitive to fully hardy (202), while plants strictly dependent on LT exposure may acclimate and reach maximum hardiness more rapidly. The time course for CA and deacclimation of two cultivars of spinach is shown in Figure 1E & F (44). Maximum freezing tolerance is reached in one to three weeks of LT exposure, and deacclimation is largely completed within one week. For the most part, attention has been focused on the attainment of maximum freezing tolerance,

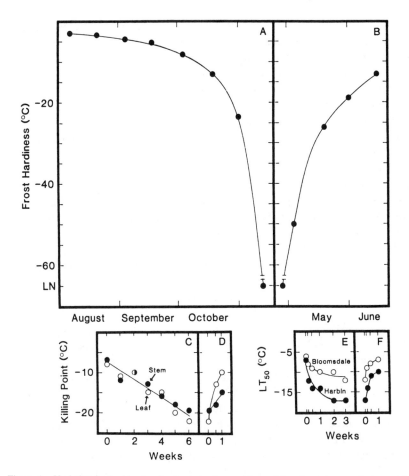

Figure 1   Variation in the kinetics for the induction and loss of freezing tolerance during cold acclimation and deacclimation of woody and herbaceous cold-hardy plants. The seasonal transition of freezing tolerance of black locust (*Robinia pseudoacacia*) cortical bark cells during cold acclimation in fall and deacclimation in spring is shown in panels *A* and *B* for trees exposed to the natural environmental conditions at Sapporo, Japan. Cold acclimation and deacclimation of ivy (*Hedera helix*) leaf and stem tissue (panels *C* and *D*), and two cultivars of spinach (*Spinacia oleracea*) (panels *E* and *F*) were carried out in controlled environments. Ivy was subjected to cold acclimation and deacclimation at 5°C and 21°C, respectively. Spinach was acclimated and deacclimated at 5°C and 21°/17°C. *A* and *B* modified after Sakai & Yoshida (173), *C* and *D* after Steponkus & Lanphear (192), *E* and *F* after Fennell & Li (44). Temperature and time scales are identical for all panels to emphasize variation in the rate of induction and relative hardiness capabilities of different plants. LN indicates liquid nitrogen.

and therefore understanding the physiology of CA rested upon this perspective. Consequently, most studies continued to utilize extended LT treatments. What was often overlooked was the fact that freezing tolerance increased during the earliest stages of CA (Figure 1A, C, E). Several studies of herbaceous species have shown freezing tolerance can be altered by as little as one day's exposure to inductive temperatures (20–22, 48, 61). When com- -pared to the most rapid temperature-inducible response known in plants, heat shock and the induction of thermotolerance (7, 95), CA and freezing tolerance appear exceedingly slow. Instead of minutes and hours as in heat shock, CA occurs over days, weeks, and months. However, the recognition that CA responses, in terms of freezing tolerance, can be detected as early as one day after exposure to LT holds promise for molecular approaches to the study of cold-tolerance mechanisms.

Low-temperature exposure is not the only means known to induce freezing tolerance. At least two methods not requiring LT exposure are known to cause the rapid induction of greater freezing tolerance of hardy species. Exogenous treatment of cell cultures, stem cultures, and seedlings with ABA at nonacclimating temperatures can alter the freezing tolerance (20, 22, 25, 94, 141, 142, 158). An interesting observation is that the level of freezing tolerance induced by ABA is in many cases comparable to that developed upon exposure to LT (20, 22, 25, 94, 141). A more important finding in the case of wheat cell cultures is that the induction of near maximal freezing tolerance by ABA is extremely rapid, occurring over a period of 2–4 days (25). The ABA concentration required to evoke freezing tolerance ranged between 10^{-4} and 10^{-5} M. This unphysiologically high concentration of ABA may have been required because of light destruction of the free ABA in the culture medium (20) or metabolism and degradation of ABA once it entered the cells. The ability of ABA to rapidly alter freezing tolerance offers two important advantages in the study of cold-tolerance mechanisms: Experiments need not be confounded by the dual aspects of CA at LT, and the rapidity of the tolerance induction greatly expands the range of experimental manipulations to include those not suitable over time intervals of weeks and months. Limited desiccation can also increase freezing tolerance. While this fact has long been known (23, 108), Canadian work in the early 1980s on desiccation-induced freezing tolerance is noteworthy (28, 29, 183). First, desiccation for as little as one day at nonacclimating temperatures induced freezing tolerance in winter rye plumules. Second, the extent of desiccation-induced freezing tolerance in epicotyls of several wheat varieties and rye correlated with the level of hardiness following four weeks of exposure to 2°C. Third, tolerance was not correlated with water content in desiccated plants but with the genetic capacity of the plant to cold-harden. These observations were consistent with the known desiccation-induced ABA accumulation (213) and suggested a potential linkage with ABA-induced freezing tolerance at nonacclimating temperatures (22, 25).

Deacclimation and loss of freezing tolerance upon return to noninductive temperatures in some species appear to be more rapid processes than cold acclimation. For example, in spinach and ivy leaf tissue, most of the freezing tolerance gained over one to six weeks of CA is lost within one week (Figure 1D & E) (44, 59, 192). In cold climates, hardy perennials can undergo partial deacclimation and lose a portion of their freezing tolerance within a few days of unseasonably warm temperatures in winter. Fortunately, with the return of seasonal temperatures partially deacclimated tissues will reharden (77).

Protein Metabolism

The metabolic responses of plants at LT have been covered in the numerous major monographs and reviews dealing with cold stress (53, 109, 111, 112, 114). In countless articles, attempts have been made to correlate metabolic and biochemical responses with cold tolerance. As Steponkus (189) has perspicaciously noted, correlative studies of biochemical changes have not yielded an understanding of how CA leads to increased freezing tolerance. Yet, elucidating cold-tolerance mechanisms may depend on analyses of biochemical responses of plants at LT. Specifically, new approaches in molecular biology and genetics have not been extensively enlisted to study cold-tolerance and injury mechanisms. When coupled with well-designed physiological, biochemical, and mechanistic studies, isolation and characterization of genes regulated by LT-responsive mechanisms will provide important new insight and a group of powerful tools with which to study cold-tolerance processes. A few studies of CA have begun to focus on some of the more rapid physiological and molecular responses in plants subjected to LTs. These studies have already revealed that plant and algal cells can rapidly begin to alter membrane lipid composition (36, 119), RNA (19), and protein content (48, 121, 214) within hours of LT exposure. These findings of rapid biochemical alterations in response to LT are in agreement with observations of the rapid induction of freezing tolerance at inductive temperatures and at noninductive temperatures by desiccation (183) and ABA (22, 25).

Two lines of evidence suggest a possible molecular basis, at minimum, for the adjustment of metabolism to low nonfreezing temperature, and perhaps for freezing tolerance. The evidence consists of repeated observations that a number of enzymes show shifts in isozymic composition upon exposure to LT, and numerous electrophoretic studies have shown both quantitative and qualitative differences in the protein content between nonacclimated and cold acclimated tissues.

ENZYME VARIATION   Studies of enzymes from plants subjected to LTs have demonstrated changes in activity, freeze stability, and isozymic variation when compared to plants that were maintained at warm temperature. In one of the earliest studies to use electrophoretic techniques to separate enzymes from nonhardened and hardened tissues, peroxidase activity was shown to be increased in hardened stems of four widely unrelated woody species (126). In three of the species, several peroxidase isozymes present in hardened tissues were not found in nonhardened tissues. In a following study, no change in peroxidases, glucose-6-phosphate, 6-phosphogluconate, and malate dehydrogenases was observed in willow stem during deacclimation (66). Differences were noted for lactate dehydrogenase, where the activity increased during deacclimation. Consistent with the above finding is the demonstration that invertase in wheat leaves at LT undergoes a shift from a lower-molecular-weight form to a higher-molecular-weight form (162). The larger form exhibits different kinetic properties and seems to functionally replace the smaller form in cold-hardened plants (163). In a series of comprehensive studies with alfalfa, Krasnuk and colleagues (97–99) observed during winter increased activity of a number of dehydrogenases associated with respiratory pathways, including glucose-6-phosphate dehydrogenase, lactate, and isocitrate dehydrogenase (98). The increases in enzyme activities paralleled increases in soluble protein content during winter, suggesting that the increase in activity may result from higher amounts of enzyme. In this sense, increased soluble protein content and enzyme activity could be part of the adjustment of metabolism to the kinetic constraints imposed by LTs. Also in winter, new isozymic variants for ATPases, esterases, acid phosphatases, leucine aminopeptidases, peroxidases, and some dehydrogenases were present (99). For many of these enzymes, freeze stability was increased during winter. However, it is not clear whether the altered stability was an intrinsic property of the enzyme, was produced by new isozymic variants, or arose from some other factor present in the tissue extracts. A more recent study of glutathione reductase from spinach has demonstrated, along with the appearance of additional isozymic forms in cold-acclimated tissue, increased activity, freezing stability, and altered kinetic behavior (58). The enzyme from hardened plants was better suited, as judged by K_m values for substrates, to function at LT, and was more stable to freeze/thaw stress. Glutathione reductase activity was decreased by freezing/thawing both in vitro and in vivo. However, the enzyme from cold-acclimated plants was less sensitive to freezing than its counterpart from nonacclimated plants. In contrast, ferredoxin NADP reductase from nonacclimated and cold-acclimated wheat showed identical kinetic parameters and freeze/thaw inactivation (161), but activity was increased during CA. These studies illustrate the potential for alterations in enzymes in response to LT exposure and the apparent selective basis on which such changes may occur.

The best-characterized enzyme relative to nonacclimated and cold-acclimated plants is ribulose bisphosphate carboxylase/oxygenase from winter rye. Early in vitro studies with purified enzyme from both non- and cold-acclimated plants demonstrated an increased stability of catalytic activity to denaturants and storage at $-25°C$ of the enzyme from cold-acclimated plants (84). Further studies provided evidence of a stable in vivo conformational change during low-temperature adaptation that was not altered by purification of the enzyme (85). The enzyme obtained from LT-treated plants exhibited altered K_m values for CO_2 with respect to temperature and pH (86). Osmotic concentration of the purified enzyme caused a greater degree of aggregation via intermolecular disulfide bond formation of the large subunit from nonacclimated plants (81). Similar studies using the enzyme purified from freezing-sensitive and -tolerant potato species demonstrated structural differences that paralleled variation in freezing tolerance much in the same way the enzyme did from acclimated and nonacclimated rye (87). What remains unclear is the rationale for the stable change in conformation, kinetic properties, and differential cryostabilities of this enzyme from cold-acclimated leaves or cold-tolerant potato species. Given that the large subunit is encoded by a single chloroplastic gene, no opportunity exists for the synthesis of an alternative cryostable large subunit from another gene. Since the small subunit in many plants is encoded by a small gene family, at LT one member of the family may be preferentially activated over others expressed at normal temperature. A change in the small subunit may have subtle effect on the cryostability and other properties of the holoenzyme. Equally possible is LT-specific posttranslational processing, although no evidence exists to support this concept.

In addition to isozymic and conformational differences of enzymes in response to LT exposure, supramolecular interactions can also be affected (55). The light-harvesting chlorophyll a/b protein of photosystem II and the chlorophyll a-protein complex of photosystem I of rye thylakoids undergo an oligomerization at LT (82). Lipid analyses have indicated that a specific decrease in $trans$-Δ^3-hexadecenoic acid of phosphatidyldiacylglycerol in thylakoid membranes paralleled the overall decline observed in leaf extracts from cold-hardened plants. Characterization of purified light-harvesting complex implicated the decrease in $trans$-Δ^3-hexadecenoic acid at LT was specifically associated with the oligomerization of the complex at LT (100). The decline in $trans$-Δ^3-hexadecenoic acid was paralleled by an increase in palmitic acid in phosphatidyldiacylglycerol. While not likely a direct factor in cold tolerance, it was interesting that the decrease in $trans$-Δ^3-hexadecenoic acid in cereals was correlated with varietal freezing tolerance (88). Not all plants have shown a decrease in $trans$-Δ^3-hexadecenoic acid at LT or the concomitant shift in the light-harvesting complex II oligomerization (88).

PROTEIN CONTENT   More than 40 years ago, accumulation of soluble protein in cold-acclimated cortical bark cells of black locust trees was first correlated with freezing tolerance (181). Study of the seasonal variation in protein content and hardiness of the bark cells demonstrated the accumulation of soluble proteins in fall, which closely paralleled the induction of freezing tolerance (182). Throughout the winter the protein concentration remained high but declined rapidly during the breaking of dormancy and the resumption of growth in the spring. The decline of soluble protein again closely matched the loss of freezing tolerance. Subsequent studies with black locust have verified these early findings (11, 151, 173, 184). Not unexpectedly, the pioneering studies by Siminovitch & Briggs prompted numerous studies of the protein content of plants during CA. Out of this cumulative effort it was established that the accumulation of soluble proteins during CA was a general response (21, 30, 47, 111, 145, 151), although not universal (59). Therefore, it is not surprising that freezing tolerance may not be explained as a simple function of the soluble protein concentration. There are many reasons why some plants might accumulate soluble proteins during CA; but with the exception of a protoplasmic augmentation hypothesis (184), no clear mechanistic rationale for conferring greater freezing tolerance has been proposed for this hardening response. In temperate deciduous perennials like black locust, export of nitrogenous materials during senescence could provide the nitrogen source for the accumulation of proteins in the cortical cells of the living bark. Therefore, in black locust the accumulation of protein in fall could be more centrally related to the storage processes necessary in deciduous species than to freezing tolerance. One group has proposed that a portion of the protein accumulated in black locust bark may be a vegetative tissue lectin storage protein (148). As such, a storage protein would seem to have little to contribute to freezing tolerance, unless it contained yet unrecognized subcellular localization, physical properties, and functions (41). Supporting a possible functional role of the increased soluble protein in cold tolerance was the fact that an evergreen, red pine, also accumulated soluble proteins during the winter (151). In evergreens there would not appear to be a need for the cortical bark cells to act in a vegetative storage capacity. However, it cannot be denied that one or more minor components of the total protein content could function in freezing-tolerance mechanisms.

In the early studies of black locust (181), not only was the soluble protein content increased during the period when the cells were most hardy, but it appeared by the primitive electrophoretic techniques available in the late 1940s and early 1950s that qualitative shifts in protein content had also occurred (182). Similar to quantitative studies of protein accumulation, scores of studies using electrophoretic separations have shown qualitative and quantitative differences in protein content between nonacclimated and cold-acclimated plants (11, 33, 34, 42, 43, 83, 91, 125, 168). Most of the studies have confirmed the notion that cold-acclimated and freezing-tolerant plants contain new protein species not present in nonacclimated plants. When compared to nonacclimated plants, the shift in protein content in cold-acclimated tissues was not dramatic but subtle, involving mostly the appearance and disappearance of minor bands in gels. Unfortunately, several major obstacles have discouraged pursuit of this line of study and obscured the significance of the cumulative findings. Consequently this approach languished until recent years.

Little information is available concerning the subcellular location of proteins that might vary in response to LT. The existing evidence includes several studies of purified plasma membranes from nonacclimated and cold-acclimated tissue. In cold-acclimated leaf plasma membranes, more than 20 proteins were found to decline or disappear, and 11 increased in concentration, while 26 proteins were new and unique to membranes from hardened tissue (200). Other alterations in protein content during CA included increased levels of high-molecular-weight glycoproteins. Similar changes were observed in mulberry bark cells and orchard grass tissue (220, 222). In another membrane system, only a single M_r 20,000 protein was accumulated during CA in the endoplasmic reticulum of *Brassica napus* (90). Cold-induced cryoprotective proteins in spinach may reside in the chloroplast (203). Rye polysomes from cold-acclimated tissues were shown to contain an acidic M_r 140,000 protein not found in polysomes from nonacclimated tissues (106). In spinach nuclei, higher-molecular-weight proteins predominate at LT, while lower-molecular-weight proteins predominate at warm temperature (61). These few studies demonstrate that qualitative and quantitative protein changes can be expected in many if not all major cell fractions one might wish to study. The one exception so far appears to be thylakoid membranes of rye (54).

PROTEIN SYNTHESIS   As early as 1970, Weiser (207) proposed that cold acclimation of temperate woody perennials may require both transcriptional activation of a set of genes normally not expressed under nonacclimating conditions and the synthesis of new proteins during the development of maximal freezing tolerance. It was clear that (*a*) presumably translation-competent polysomes could be isolated from cold-acclimated tissues (8, 10), (*b*) rRNA increased during CA (115, 176), (*c*) RNA polymerase activities were increased in response to LT (175), and (*d*) protein accumulated during the fall in woody perennial bark cells (151, 173, 184); but little direct evidence existed to support the proposal that altered gene expression and synthesis of new proteins occurred during CA. Almost simultaneously it was shown that the RNA base composition in cold-acclimated tissues differed from that in nonacclimated tissues (176), and plants at LT synthesized two unique proteins in response to CA (166). Freezing-tolerant winter wheat synthesized two hydrophilic high-molecular-weight proteins (240,000 and 115,000 M_r) at 3°C, while freezing-sensitive spring wheat was unable to make similar proteins (166). These studies provided the first evidence to support Weiser's hypothesis (207). Independent studies of wheat and/or rye subjected to CA (147, 177) or desiccation treatments have yielded analogous results (26, 27).

Numerous reports have now clearly established that plants, when exposed to LT, synthesize a new set of proteins (31, 59, 60, 63, 65, 90, 102, 106, 121, 127, 129, 144, 147, 164, 177, 199, 214, 215). Nearly all studies of protein synthesis at LT have focused on a descriptive approach of the response. Out of this collective work, which includes species that range from chilling-sensitive to freezing-tolerant, several features of the protein synthesis response at LT have emerged. Table 1 summarizes some of the documented changes in protein synthesis upon exposure to LT. Careful inspection of the available information has not revealed the uniform appearance or decline of common proteins. The set of new proteins synthesized at LT does not appear to be highly conserved as those of heat shock are (7, 95). This should be construed to indicate not that protein synthesis at LT of widely divergent species involves no similar proteins, but only that at present little or no evidence exists to support a high conservation of the response. That the set of proteins synthesized at LT appears not to be highly conserved as heat shock reflects on the diversity of cold-stress situations and adaptive responses (52, 171, 211). Also, as is not the case with heat shock, the synthesis and concentration of housekeeping proteins at LT continues, even under cold-shock treatments (48, 59, 62, 121, 147, 214, 215). Support for the idea that synthesis of housekeeping proteins at LT continues comes from the apparent increase in rRNA (106, 115, 176, 184) and protein synthesis capacity (8, 10, 106) in cold-acclimated tissues. At LT, cold-acclimated tissues can synthesize protein faster than nonacclimated tissue (63). This would seem to be consistent with the need to maintain a basal level of cellular metabolism even at LT. Therefore, the alterations in protein synthesis at LT were found to be subtler, less visually dramatic, and kinetically slower than those in response to many other environmental stresses. LT responses in protein synthesis are further distinguished from other stresses by the fact that many of the unique proteins are continuously synthesized at higher rates for extended periods. While synthesis of some of the new proteins at LT is stable, continuing for several days or weeks (59, 63), that of others is transient (19, 199). Time course studies have indicated that new proteins can appear within one day and perhaps as early as an hour following exposure to LT (59, 121). However, few studies have correlated the changes in protein synthesis with the induction and loss of freezing tolerance. Time course studies with spinach have shown the synthesis of several high-molecular-weight proteins closely correlates with the induction and loss of freezing tolerance during CA and deacclimation (59). During CA the synthesis of these proteins was upregulated, while during deacclimation synthesis was downregulated. Such time course studies provide evidence to support the notion that new proteins synthesized at LT go beyond a role in adjustment of metabolism to LT, and that some may play a central role in freezing tolerance.

At this stage, almost no information exists about the identity or function of the proteins preferentially synthesized at LT. An obvious analogy is that proteins synthesized at LT during cold stress could be related to the other temperature-responsive proteins of heat shock (7, 95). Comparison of proteins synthesized during CA and heat shock suggest that the two responses elicit the synthesis of distinctly different proteins (63, 144), although some were apparently electrophoretically similar (59, 214, 215). More specific protein blotting analyses of corn subjected to cold shock has demonstrated the accumulation of both low- and high-molecular-weight heat shock proteins (214, 215). This has been partially confirmed by the finding of increased RNA levels for the 27,000 M_r heat shock proteins in chilled soybean (104), and protein sequence homology of a 79,000 M_r protein accumulated in spinach during CA (C. L. Guy, D. Haskell, L. Green, and P. Klein, unpublished). While the heat shock response and CA may involve a common subset of "stress" proteins, it is reasonable to expect that heat and freezing tolerance involve different adaptive mechanisms that require their own unique stress proteins.

The first direct evidence to support Weiser's (207) hypothesis of altered gene expression during CA was the observation that newly translatable mRNAs were induced in spinach leaves exposed to 5°C (63). This finding also provided an explanation for the previous observation of altered base composition of RNA during CA (176). Subsequent studies have confirmed and extended this finding (19, 48, 60, 79, 90, 102, 127, 130, 178, 199). In several cases, some of the mRNAs detected in cold-acclimated tissues encode proteins that are translated and accumulated at LT (48, 59–61, 63, 102, 127). In the few studies that have examined the time course of the changes in mRNAs, it is clear that within hours (19, 48) to one day (60) of LT exposure new mRNAs begin to appear in herbaceous species. This represents one of the fastest documented biochemical responses to LT in nature and again suggests a potential role of the synthesis of new proteins in tolerance mechanisms. However, many of the responses of temperate perennial plants to LT, at the mRNA level, may parallel that found in winter flounder, where antifreeze protein mRNA synthesis is preferentially regulated at LT following priming by endogenous clock mechanisms (152).

Some mRNAs decline during exposure to LT. The mRNA level for the small subunit of ribulose bisphosphate carboxylase/oxygenase was decreased in *Brassica napus* (127), rice (65), and spinach (C. L. Guy, D. Haskell, L. Green, and P. Klein, unpublished). In rice, the decline in the mRNA level

Table 1 Listing of changes in protein synthesis and accumulation modulated at low temperature

Plant	Temperature	Protein response[a]		Comment[b]	Reference
		Increased	Decreased		
Agrostemma githago	4°,7°C	78, 28	nd	1	104
Arabidopsis thaliana	4°C	160, 47	29, 22	2, 4	48
Arabidopsis thaliana	4°C	150, 80, 69, 58, 57, 45, 33, 32, 30, 20	16	2	105
Brassica napus	5°C	c	nd	7	2
Brassica napus	0°C	80–14[d]	50–14[d]	2, 4	127
Brassica napus	4°C	20, 17	nd	1, 2, 3	90
Bromus inermis	3°C	115, 54, 48, 47, 38, 31	74, 27, 23, 22	2	164
Bromus inermis	3°C	200, 190, 165, 25	22	5, 6	165
Citrus sinensis	5°C	160, 40–16[e]	65–16[e]	6	62
Cynodon spp.	3°C	f	f	7	34
Hordeum vulgare	0°C	100, 87, 77, 67, 61, 52	56, 32	1	121
Lolium temulentum	5°C	66, 47, 38	nd	1	144
Lycopersicon esculentum	2–4°C	35	27	1, 2	31
Lycopersicon esculentum	25°/6°C	68, 62, 51, 50, 42, 40, 35, 34, 33, 32, 30, 24, 24, 19, 17, 16	nd	4	(201, pers. commun.)
Lycopersicon hiursutum	25°/6°C	69, 67, 53, 43, 42, 40, 35, 34, 32, 31, 19, 17	nd	4	(201, pers. commun.)
Medicago sativa	4°C	180, 145, 135, 90, 80, 76, 70, 60, 53, 43, 38, 27–11	nd	1, 3, 5	129, 130

Species	Temperature	Increased	Decreased	Method[b]	Apparent MW[d]
Nothofagus dombeyi	0°C	41, 35, 26	nd	5	168
Oryza sativa	4°,7°C	117, 22, 19, 17, 14	nd	1	104
Oryza sativa	11°/6°C	95, 75, 25, 21	200, 50, 42, 39, 26, 14, 10	1, 2	(65)
Secale cereale	2°C	46, 30	nd	5	26
Solanum commersonii	5°C	195, 83, 31, 27, 26, 22, 21	nd	2, 4	199
Spinacea oleracea	5°C	160, 85, 70–15[g]	50–16[g]	6	62
Spinacea oleracea	5°C	160, 117, 85, 79, 28, 26	32, 26, 25, 22, 21	2, 4	60
Thuja occidentalis	10°C	[h]	nd	7	33
Triticum aestivum	3°C	240, 115	nd	1	166
Triticum aestivun	2°C	46, 30	nd	5	26
Triticum aestivum	2°C	200, 48, 47, 42	93, 89, 80, 67, 63	6	177
Triticum aestivum	6°/2°C	200, 150, 92, 90, 80, 75, 60, 55, 48, 45, 44, 36, 31	105, 79, 75, 60, 53, 51, 49, 45, 44, 43, 42, 41, 40, 38, 35, 34	2	147
Zea mays	4°C	94–13[i]	nd	1	214, 215

[a] Responses: increased = proteins either induced, more rapidly synthesized, or accumulated at low tempertaure; decreased = proteins repressed, less rapidly synthesized, or present in lower amounts at low temperature; numbers indicate protein molecular mass in Kilodaltons; nd = not determined

[b] Method of detection: 1: fluorography of in vivo labeled proteins resolved by SDS-PAGE; 2: same as 1 except proteins resolved by 2D-PAGE; 3: fluorography of in vitro translation of mRNAs resolved by SDS-PAGE; 4: same as 3 except proteins resolved by 2D-PAGE; 5: stained proteins resolved by SDS-PAGE; 6: same as 5 except resolved by 2D-PAGE; 7: proteins separated by native PAGE.

[c] One major band at R_F 0.49.

[d] Apparent molecular weight for each protein was not listed. A total of 17 proteins or translation products were increased at low temperature. A total of 11 proteins or translation products were decreased at low temperature.

[e] Same as footnote d. At low temperature the amounts of 6 proteins were increased and of 8 were decreased.

[f] Four low-migration bands.

[g] At low temperature the amounts of 11 leaf and 15 hypocotyl proteins increased, and of 9 leaf and 7 hypocotyl proteins decreased.

[h] Three proteins.

[i] In several cultivars varying responses were observed with a total of 20 proteins that increased at low temperature.

also resulted in a lack of synthesis of the small subunit, while the large-subunit synthesis was less affected at LT (65). This decline seemed to cause a loss in the coordinate regulation of the large and small subunit syntheses. In chilling-sensitive rice, transcript levels for some organelle- and nucleus-encoded genes were also suppressed at LT (65). These included transcripts for *Sh* 1, *RbcS*, *atpE*, *coxIII*, *Cab*, *psaB*, and *rbcL* (65).

Partial cDNA clones for mRNAs accumulated during LT exposure of chilling-sensitive tomato fruit and freezing-tolerant alfalfa have been isolated. Three cDNAs were isolated from chilled tomato fruit that recognize distinct RNAs of 0.8, 1.2, and 1.9 kb (178). Time course RNA blot analyses have indicated rapid accumulation of the RNAs at temperatures below 16°C, with the greatest accumulation observed at 4°C. Sequence analysis of one of the cDNAs indicated strong homology with conserved domains of thiol proteases genes. Schaffer & Fischer (178) postulate one function of the induced thiol protease in tomato fruit might be to degrade proteins denatured at LT. If they are correct, the appearance of this enzyme at LT in chilling-sensitive tissues might be an injury response to cold stress instead of an adaptive response that increases tolerance. In contrast, RNA blot analysis of rice showed the transcript of ubiquitin to decrease during LT exposure (65). The biological role of the other two genes remains unknown. The same situation holds true for the CA-specific cDNAs isolated from alfalfa (131). The three cDNAs hybridize to transcripts ranging from 0.9 to 1.4 kb. RNA blot analyses have indicated that the transcripts' steady-state levels increase rapidly during CA but do not increase in response to ABA administration, wounding, water stress, or heat shock. The relative levels of expression of the three genes seem to correspond to cultivar hardiness levels during CA. Since no one knows how many genes are activated at LT, the two studies discussed above may only have scratched the surface. As more genes are cloned and characterized, a better understanding of both adjustment of metabolism to LT and cold tolerance mechanisms should emerge.

INDUCTION BY DESICCATION AND ABSCISIC ACID   Some desiccation prior to freezing enhances survival in a number of species (108). The demonstration of (*a*) ABA induction of freezing tolerance at room temperature in cell and stem cultures (20, 22, 25) and recently in seedlings (105), and (*b*) rapid induction of freezing tolerance by desiccation (183) suggests a possible linkage among desiccation, ABA, and CA. It is well established that desiccation results in an accumulation of ABA (213), and so apparently does CA (22, 60). Since exogenous application of ABA can substitute for LT exposure (22, 25), the parallel between desiccation and CA may go beyond simply changes in ABA accumulation. As a consequence of the similarity between extracellular freezing and desiccation (i.e. the removal of water from the cell), plants may have developed responses to both stresses that involve common adaptive mechanisms. Cloutier (27) has demonstrated that a 46,000 M_r protein accumulated in response to CA and/or desiccation of wheat. Further comparisons of the proteins synthesized or accumulated during induction of freezing tolerance at LT or with ABA at room temperature have revealed a common subset of proteins (90, 105, 164, 165, 199). Recent protein blot analyses in our laboratory have shown that two high-molecular-weight proteins accumulated in spinach during CA were more actively accumulated during desiccation (C. L. Guy, D. Haskell, L. Green, and P. Klein, unpublished). Along similar lines, Mundy & Chua (133) have isolated a gene for a hydrophilic protein from rice, a gene also induced by ABA or desiccation. Rice seedlings exposed to LT accumulate large quantities of the *Rab* transcript (65). A demonstration that the *Rab* gene is induced at LT in freezing-tolerant species would provide additional impetus for studies linking freezing tolerance with desiccation responses.

Candidate Genes for Upregulation during Cold Acclimation

At this early stage in the characterization of the molecular-genetic basis of cold tolerance, several cDNAs for genes upregulated at LT have been isolated (131, 178). In both cases, the cDNAs were isolated by taking advantage of the upregulation and accumulation of the corresponding mRNAs in plant tissues exposed to LT. This has been and continues to be a standard approach in the isolation of stress-responsive genes (69, 95, 133). Given the fact that CA encompasses two processes, adjustment of metabolism and induction of freezing tolerance, there is no way of knowing in which capacity a gene cloned by this approach may function. In keeping with the genetic evidence of the quantitative character of cold tolerance, information obtained from in vivo labeling studies and in vitro translations suggests the number of changes in protein synthesis at LT (induction/repression, up- or downregulated) in spinach could well exceed 300, most of which are beyond the sensitivity of current electrophoretic techniques (C. L. Guy, D. Haskell, unpublished). At the present pace, it seems likely that a great deal of time and research will be required to work through this large set of genes.

A second approach to the molecular biology of cold tolerance seems to have been largely ignored. More is known about the biochemistry, physiology, and metabolism of cold stress than about any other stress, including some aspects that are closely associated with cold stress tolerance and injury (111, 112). It is here that an opportunity exists to target a given response and explore its role in cold-tolerance mechanisms. The major opportunities include characterization of (a) genes that encode cold-labile enzymes and proteins, (b) enzymes in the biosynthesis of low molecular weight cryoprotectants, (c) lipid metabolism, and (d) key regulatory enzymes of respiratory pathways, to list just a few.

COLD-LABILE ENZYMES   Plant enzymes and proteins can be divided into two classes—those whose activity or function is sensitive to low nonfreezing temperatures, and those inactivated by freezing/thaw cycles. Enzymes exhibiting cold lability to nonfreezing temperatures have mostly been identified in yeast or chilling-sensitive plants (Table 2). To this reviewer's knowledge there exists no report that describes the in vivo inactivation at nonfreezing temperatures of an enzyme from chilling-tolerant plants. The closest parallel in a chilling-tolerant plant is the depolymerization of root tip microtubules at 0°C (18). As in the depolymerization of onion microtubules, the most frequent feature of cold inactivation of enzymes is loss of quaternary structure caused by dissociation into individual subunits (9). In contrast, freeze/thaw cycles can lead to the inactivation of enzymes not labile at nonfreezing temperatures in chilling-tolerant plants, although the mechanism may differ (Table 2). Enzyme inactivation at LT may result from dissociation or aggregation (9, 110). Notable in this respect was the work of Gorke (51) more than 80 years ago that demonstrated freeze precipitation of proteins in plant extracts. Heber (72) showed soluble proteins to be largely unaffected by freezing; most of the precipitated protein had a membranous origin. The mechanism of inactivation by freeze/thaw cycles of plant enzymes and proteins has not been thoroughly examined but could stem from the loss of tertiary or quaternary structure brought about by extreme shifts in pH and ionic strength (9) during extracellular freeze dehydration of the cell (111). Freeze/thaw mediation of membrane enzyme–complex dissociation has been demonstrated by the release of proteins from thylakoid membranes (73, 179, 204). Alternatively, aggregation through intermolecular associations, such as the formation of disulfide linkages, has been proposed (110) and indirectly supported experimentally (54, 58, 85).

Several of the cold- and freezing-labile enzymes belong to respiratory pathways (Table 2). Since a number of respiratory dehydrogenase enzymes possess seemingly differential freezing stabilities and higher activities during winter (98), it would be reasonable to expect a need for upregulation (beyond that afforded by mass action, allosteric regulation, or activation/deactivation by covalent modification) in an attempt to overcome LT inactivation. This could be accomplished by making more enzyme or through the activation of a homologous gene(s) that encodes a more cryotolerant form of the enzyme (162).

Two cold-labile enzymes appear to be at key regulatory steps in glycolysis: phosphofructokinase and pyruvate kinase. At the phosphofructokinase major control point, plants also have a pyrophosphate-dependent phosphotransferase

Table 2 Chilling-labile or freezing-sensitive enzymes

Enzyme	Organism	Temperature	Molecular Consequence	Reference
Pyruvate Phosphate	corn	0–10°C	dissociation	96, 180, 197
Dikinase	barnyardgrass	0°C	nd[a]	185
NADP-malate Dehydrogenase	sorghum, maize		inactivation	197
NADP-malic enzyme	barnyardgrass	0°C	nd	185
PEP carboxylase	corn	0–4°C	dissociation	205
	Cynodon dactylon	0°C	dissociation	3
	barnyardgrass	0°C	nd	185
Ribulose bisphosphate carboxylase/oxygenase	tobacco, rye	0°C	conformational change	85, 93
	spinach	−20°C	altered redox state	198
Catalase	sorghum, cucumber	10°C	nd	197
	maize	5°C	nd	139
Ascorbate peroxidase	poplar	−10°C	nd	170
Glutathione reductase	spinach,	freeze/thaw,	nd	58, 170
	poplar	−10°C	nd	
Polyphenol oxidase	poplar	−10°C	nd	170
Starch synthetase	maize, avocado, sweet potato	3–12°C	protein-lipid perturbation	39
Phosphofructokinase	potato	12°C	dissociation	37
Fructose-1,6-bisphosphatase	spinach, pea	0°C	nd	206
Glucose-6-phosphate dehydrogenase	poplar	−10°C	nd	170
6-Phosphogluconate dehydrogenase	poplar	−10°C	nd	170
Glyceraldehyde-3-phosphate dehydrogenase	yeast	0–10°C	dissociation	187
Pyruvate kinase dehydrogenase	yeast	0–10°C	dissociation	101
Aspartic β-semialdehyde dehydrogenase	yeast	0°C	dissociation	75
Glutamate dehydrogenase	*Neurospora*	21°C	no change	45
Tonoplast H⁺-ATPase	mung bean, rice	0–10°C	nd	92, 221
Chloroplast Ca²⁺ ATPase	spinach	0°C	dissociation	124
CF₁	spinach	freeze/thaw	dissociation	204
NADP reductase	spinach	freeze/thaw	dissociation	204
Plastocyanin	spinach	freeze/thaw	dissociation	204

[a] Not determined

that can function as a kinase on fructose-6-phosphate or as a bisphosphatase on fructose-1,6-bisphosphate (16). The catalytic direction of this enzyme is highly regulated by the allosteric effector fructose-2,6-bisphosphate (78). Since glycolysis/gluconeogenesis processes must continue at LT to provide carbon flux for respiratory needs and synthesis of sugars, these respiratory enzymes may be upregulated during CA (Table 3) as part of an adjustment of metabolism and indirectly function in tolerance mechanisms by the biosynthesis of sugar cryoprotectants in photosynthetic tissues. In peppers and *Lolium*, fructose-2,6-bisphosphate declines at LT (149, 150), which may favor gluconeogenesis and sugar accumulation. Since, fructose-2,6-bisphosphatase is responsible for degrading fructose-2,6-bisphosphate to fructose-6-phosphates, the enzyme may be more active at LT or a candidate for upregulation (Table 3).

CRYOPROTECTANTS   In many plants, exposure to LT causes the accumulation of low-molecular-weight compounds with demonstrated cryoprotectant activity (21, 64, 103, 109, 111, 112, 145, 151, 173, 174, 184, 192, 217, 219). Known cryoprotectants include disaccharide and trisaccharide sugars; the polyol, sorbitol; a quaternary ammonium compound, glycinebetaine; proline; and polyamines. All of these compounds are compatible solutes (216). They may function as cryoprotectants by helping to sustain the ordered vicinal water around proteins by decreasing protein-solvent interactions in such a way that subunit association and native conformation are maintained at low water activities (216). Many of these compounds may also stabilize membranes through interactions with the polar head groups of phospholipids, while others may form hydrophobic interactions with the membrane (1).

The carbohydrates sucrose (111), raffinose (145), and sorbitol (111) are the primary cryoprotectants present in plants. Of these, sucrose is the most widely found in freezing-tolerant plants; its levels can increase 10-fold during exposure to LT (174). The importance of sugar accumulation to freezing tolerance is demonstrated by the fact that the ability to become freezing tolerant at LT is lost if sugar accumulation is blocked (64, 192). While starch-sucrose interconversions have been a well-known feature of LT exposure, and numerous studies have documented the accumulation of sugars in hardened plants (111), no studies have yet focused on the enzymology of carbohydrate metabolism during CA. The activities of sucrose synthase in wheat (15) and sucrose phosphate synthase in *Chlorella* (174) were recently shown to be increased at LT. One possible cause among many for this increased activity is increased synthesis of the enzyme at LT. Given the central role of sucrose accumulation as an osmolyte (216) and cryoprotectant (1, 17), enzymes of starch-sucrose interconversions would seem to be possible targets for upregulation at LT. Similar arguments also hold true for the accumulation of proline (96, 103, 217), polyamines (103), antioxidants like ascorbate and glutathione (58, 111), and possibly glycinebetaine in the Chenopodiaceae and grasses (32, 68).

MEMBRANE LIPIDS   Another well-known feature of cold acclimation is the biochemical (36, 100, 118, 119, 200, 220, 222) and physical restructuring (50, 189–192) of cell membranes. Restructuring, presumably within the lipid

Table 3 Abbreviated listing of potential targets for upregulated expression at low temperature

Class	Function	Enzyme/Protein
Respiratory processes[a]		
	glycolysis/gluconeogenesis	
		phosphofructokinase
		pyrophosphate: fructose 6-phosphate phosphotransferase
		fructose bisphosphatase
		pyruvate kinase
	Krebs cycle	
		pyruvate dehydrogenase complex
		citrate synthase
	oxidative pentose pathway	
		glucose-6-phosphate dehydrogenase
Regulatory/processes		
	glycolysis/gluconeogenesis	
		fructose-2,6-bisphosphatase
Cryoprotectants		
	starch-sucrose interconversion	
		starch phosphorylase
		starch hydrolysis enzymes
		sucrose phosphate synthase
		invertase
	proline	
		glutamate kinase
		ornithine-oxo-acid aminotransferase
		pyrroline 5-carboxylate reductase
	putrescine	
		arginine decarboxylase
		ornithine decarboxylase
		agmatine deiminase
	betaine	
		choline monooxygenase
		betaine aldehyde dehydrogenase
Antioxidants		
	glutathione	
		cysteine synthase
		glutathione synthase
		glutathione reductase
	ascorbate	
		dehydroascorbate reductase
Membrane lipids		
	fatty acids	
		acetyl-CoA carboxylase
		desaturases
Nonenzymatic proteins		
	thermal hysteresis	
	ice nucleation	
	cryoprotective	
	LEA/WSPs[b]	

[a] This list represents only a small portion of possible proteins and enzymes that may be upregulated at either transcription or translation upon exposure to low temperature.
[b] Late Embryogenesis Abundant/Water Stress Proteins (41).

components, has been shown to alter the cryostability of the membrane in cold-hardened rye protoplasts (192). With all of the modifications in lipid metabolism occurring during CA, it is probable that some of the biosynthetic enzymes of the various lipid classes are upregulated as well. Likely candidates include, but may not be limited to, the initial and regulatory step of fatty acid biosynthesis, acetyl-CoA carboxylase, and of course the desaturases responsible for increasing membrane lipid unsaturation at LT. As lipid metabolic pathways become better understood through enzyme characterization or the isolation of mutants, great opportunity for the study of LT regulation of membrane biogenesis will unfold.

NONENZYMATIC PROTEINS   In recent years it has become clear that organisms have proteins that alter the freezing of water (157). These consist of two types: (a) insect thermal hysteresis proteins (40) or fish antifreeze proteins (35), which retard freezing; and (b) ice nucleation proteins, which promote freezing (143, 210). While neither of these types has been reported in plants, the precedent in nature exists; such proteins could be present in some plants. If plants contain such proteins, their synthesis likely varies seasonally. More probable in plants are cryoprotective proteins, which have been detected in cold-acclimated spinach leaf tissue (203). These proteins appear not to be present in nonacclimated leaves and therefore seem to fit the model for LT modulation of expression. However, if plants contain cryoprotectant proteins that lack any other enzymatic function, then they will probably first be detected by the nonspecific molecular cloning approach. Along this line, a new class of plant proteins has been found to be inducible by ABA (69, 133), desiccation (133), and during developmentally programmed seed dry-down (6). A role in protecting cell structures during times of low water activity has been proposed for this related group of proteins (41). Analysis of the group reveals a repeating 11-amino-acid sequence that may form an amphiphilic α-helical structure that serves as a foundation for a higher-order structure functional in stabilization of desiccation-sensitive sites in cells. In fact, RNA blot analysis has already demonstrated the steady-state level of the *Rab* 21 gene (133) transcript to be greatly increased in rice seedlings exposed to LT (65). Because extracellular freezing and drought stress are similar, and given the fact that ABA and/or desiccation can induce freezing tolerance at noninductive temperatures, hardy plants may express and derive a portion of their tolerance from a subset of proteins related to the Late Embryogenesis Abundant proteins (41).

CONCLUDING REMARKS

The scatter of plant life over the globe from warm equatorial regions to frigid polar regions during the course of evolutionary time (211) has resulted in an array of developmental, biochemical, and physiological adaptations to cope with the adversities of decreasingly lower atmospheric and soil temperatures (52, 109, 111, 112, 171). This has become increasingly apparent during the last ten millennia of agricultural development, because of people's desire and expanding need to grow crop plants well beyond their native ranges. What is not generally understood is the multitude of LT constraints that must be balanced with overall competitiveness (against biotic and abiotic forces) or fitness. Thus, a vast diversity of adaptive and avoidance strategies have evolved to cope with the myriad variations of LT stress (171). Many higher plants can withstand the freezing of tissue water that few other higher eukaryotes can tolerate. Molecular approaches to the study of the responses of plants to LT will, in all likelihood, reveal higher orders of complexity not yet apparent. When utilized within a biochemical or physiological approach, molecular genetics offers great potential for understanding how plants adjust metabolism to function at low nonfreezing temperatures and how they tolerate freezing.

ACKNOWLEDGMENTS

I thank Dale Haskell, Lisa Neven, and Mike Kane for helpful comments during preparation of this manuscript, and colleagues who provided manuscripts prior to their publication. This article was based on literature available prior to August 14, 1989, from which it was possible to cite only a fraction of the important works.

This work was supported by the Institute of Food and Agricultural Sciences, and the Interdisciplinary Center for Biotechnology Research, The University of Florida, and grants from the US Department of Agriculture. Florida Agricultural Experiment Station Journal Series R-00091.

Literature Cited

1. Anchordoguy, T. J., Rudolph, A. S., Carpenter, J. F., Crowe, J. H. 1987. Modes of interaction of cryoprotectants with membrane phospholipids during freezing. *Cryobiology* 24:324–31
2. Anderson, J. A., Gusta, L. V., Buchanan, D. W., Burke, M. J. 1983. Freezing of water in citrus leaves. *J. Am. Soc. Hort. Sci.* 108:397–400
3. Angelopoulos, K., Gavalas, N. A. 1988. Reversible cold inactivation of C₄-phosphoenolpyruvate carboxylase: factors affecting reactivation and stability. *J. Plant Physiol.* 132:714–19
4. Ashworth, E. N. 1982. Properties of peach flower buds which facilitate supercooling. *Plant Physiol.* 70:1475–79
5. Ashworth, E. N. 1984. Xylem development in *Prunus* flower buds and the relationship to deep supercooling. *Plant Physiol.* 74:862–65
6. Baker, J., Steele, C., Dure, L. III, 1988. Sequence and characterization of 6 *Lea* proteins and their genes from cotton. *Plant Mol. Biol.* 11:277–91
7. Barnett, T., Altschuler, M., McDaniel, C. N., Mascarenhas, J. P. 1980. Heat shock induced proteins in plant cells. *Dev. Genet.* 1:331–40
8. Bixby, J. A., Brown, G. N. 1975. Ribosomal changes during induction of cold hardiness in black locust seedlings. *Plant Physiol.* 56:617–21
9. Bock, P. E., Frieden, C. 1978. Another look at the cold lability of enzymes. *Trends Biochem. Sci.* 3:100–3

10. Brown, G. N. 1972. Changes in ribosomal patterns and a related membrane fraction during induction of cold hardiness in *Mimosa* epicotyl tissues. *Plant Cell Physiol.* 13:345–51

11. Brown, G. N., Bixby, J. A. 1975. Soluble and insoluble protein patterns during induction of freezing tolerance in black locust seedlings. *Physiol. Plant.* 34:187–91

12. Burke, M. J., Gusta, L. V., Quamme, H. A., Weiser, C. J., Li, P. H. 1976. Freezing and injury in plants. *Annu. Rev. Plant Physiol.* 27:507–28

13. Cahalan, C., Law, C. N. 1979. The genetical control of cold resistance and vernalisation requirement in wheat. *Heredity* 42:125–32

14. Cain, D. W., Andersen, R. L. 1980. Inheritance of wood hardiness among hybrids of commercial and wild Asian peach genotypes. *J. Am. Soc. Hort. Sci.* 105:349–54

15. Calderon, P., Pontis, H. G. 1985. Increase of sucrose synthase activity in wheat plants after a chilling shock. *Plant Sci.* 42:173–76

16. Carnal, N. W., Black, C. C. 1983. Phosphofructokinase activities in photosynthetic organisms. The occurrence of pyrophosphate-dependent 6-phosphofructokinase in plants and algae. *Plant Physiol.* 71:150–55

17. Carpenter, J. F., Hand, S. C., Crowe, L. M., Crowe, J. H. 1986. Cryoprotection of phosphofructokinase with organic solutes: characterization of enhanced protection in the presence of divalent cations. *Arch. Biochem. Biophys.* 250:505–12

18. Carter, J. V., Wick, S. M. 1984. Irreversible microtubule depolymerization associated with freezing injury in *Allium cepa* root tip cells. *Cryo-Lett.* 5:373–82

19. Cattivelli, L., Bartels, D. 1989. Cold-induced mRNAs accumulate with different kinetics in barley coleoptiles. *Planta* 178:184–88

20. Chen, H. H., Gavinlertvatana, P., Li, P. H. 1979. Cold acclimation of stem-cultured plants and leaf callus of solanum species. *Bot. Gaz.* 140:142–47

21. Chen, H. H., Li, P. H. 1980. Biochemical changes in tuber-bearing *Solanum* species in relation to frost hardiness during cold acclimation. *Plant Physiol.* 66:414–21

22. Chen, H. H., Li, P. H., Brenner, M. L. 1983. Involvement of abscisic acid in potato cold acclimation. *Plant Physiol.* 71:362–65

23. Chen, P., Li, P. H., Weiser, C. J. 1975. Induction of frost hardiness in red osier dogwood stems by water stress. *HortScience* 10:372–74

24. Chen, P. M., Burke, M. J., Li, P. H. 1976. The frost hardiness of several solanum species in relation to the freezing of water, melting point depression, and tissue water content. *Bot. Gaz.* 137:313–17

25. Chen, T. H. H., Gusta, L. V. 1983. Abscisic acid-induced freezing resistance in cultured plant cells. *Plant Physiol.* 73:71–75

26. Cloutier, Y. 1983. Changes in the electrophoretic patterns of the soluble proteins of winter wheat and rye following cold acclimation and desiccation stress. *Plant Physiol.* 71:400–3

27. Cloutier, Y. 1984. Changes of protein patterns in winter rye following cold acclimation and desiccation stress. *Can. J. Bot.* 62:366–71

28. Cloutier, Y., Andrews, C. J. 1984. Efficiency of cold hardiness induction by desiccation stress in four winter cereals. *Plant Physiol.* 76:595–98

29. Cloutier, Y., Siminovitch, D. 1982. Correlation between cold- and drought-induced frost hardiness in winter wheat and rye varieties. *Plant Physiol.* 69:256–58

30. Coleman, E. A., Bula, R. J., Davis, R. L. 1966. Electrophoretic and immunological comparisons of soluble root proteins of *Medicago sativa* L. Genotypes in the cold hardened and non-hardened condition. *Plant Physiol.* 41:1681–85

31. Cooper, P., Ort, D. R. 1988. Changes in protein synthesis induced in tomato by chilling. *Plant Physiol.* 88:454–61

32. Coughlan, S. J., Heber, U. 1982. The role of glycinebetaine in the protection of spinach thylakoids against freezing stress. *Planta* 156:62–69

33. Craker, L. E., Gusta, L. V., Weiser, C. J. 1969. Soluble proteins and cold hardiness of two woody species. *Can. J. Plant Sci.* 49:279–86

34. Davis, D. L., Gilbert, W. B. 1970. Winter hardiness and changes in soluble protein fraction of bermudagrass. *Crop Sci.* 10:7–9

35. DeVries, A. L. 1983. Antifreeze peptides and glycopeptides in cold-water fishes. *Annu. Rev. Physiol.* 45:245–60

36. Dickens, B. F., Thompson, G. A. 1981. Rapid membrane response during low temperature acclimation. Correlation of early changes in the physical properties and lipid composition of *Tetrahymena* microsomal membranes. *Biochim. Biophys. Acta* 644:211–18

37. Dixon, W. L., Franks, F., Ap Rees, T. 1981. Cold-lability of phosphofructokinase from potato tubers. *Phytochemistry* 20:969–72

38. Dowgert, M. F., Steponkus, P. L. 1984. Behavior of the plasma membrane of isolated protoplasts during a freeze-thaw cycle. *Plant Physiol.* 75:1139–51

39. Downton, W. J., Hawker, J. S. 1985. Evidence for lipid-enzyme interaction in starch synthesis in chilling sensitive plants. *Phytochemistry* 14:1259–63

40. Duman, J. G. 1979. Thermal-hysteresis-factors in overwintering insects. *J. Insect Physiol.* 25:805–10

41. Dure, L. III, Crouch, M., Harada, J., Ho, T.-H. D., Mundy, J., et al. 1989. Common amino acid sequence domains among the LEA proteins of higher plants. *Plant Mol. Biol.* 12:475–86

42. Faw, W. F., Jung, G. A. 1972. Electrophoretic protein patterns in relation to low temperature tolerance and growth regulation of alfalfa. *Cryobiology* 9:548–55

43. Faw, W. F., Shih, S. C., Jung, G. A. 1976. Extractant influence on the relationship between extractable proteins and cold tolerance of alfalfa. *Plant Physiol.* 57:720–23

44. Fennell, A., Li, P. H. 1985. Rapid cold acclimation and deacclimation in winter spinach. *Acta Hortic.* 168:179–83

45. Fincham, J. R. S., Garner, H. R. 1967. Effects of pH and temperature on the wildtype and a mutant form of *Neurospora* glutamate dehydrogenase. *Biochem. J.* 103:705–8

46. Fowler, D. B., Gusta, L. V. 1979. Selection for winterhardiness in wheat. I. Identification of genotypic variability. *Crop. Sci.* 19:769–72

47. Gerloff, E. D., Stahmann, M. A., Smith, D. 1967. Soluble proteins in alfalfa roots as related to cold hardiness. *Plant Physiol.* 42:895–99

48. Gilmour, S. J., Hajela, R. K., Thomashow, M. F. 1988. Cold acclimation in *Arabidopsis thaliana*. *Plant Physiol.* 87:745–50

49. Goppert, H. R. 1830. Über die Wärme-Entwicklung in den Pflanzen, deren Gefrieren und die Schützmittel gegen dasselbe. Berlin: Max and Comp. 273 pp.

50. Gordon-Kamm, W. J., Steponkus, P. L. 1984. Lamellar-to-hexagonal$_{II}$ phase transitions in the plasma membrane of isolated protoplasts after freeze-induced dehydration. *Proc. Natl. Acad. Sci. USA* 81:6373–77

51. Gorke, H. 1906. Über chemische Vorgängel beim Erfrieren der Pflanzen. *Landw. Vers. Sta.* 65:149–60

52. Grace, J. 1987. Climatic tolerance and the distribution of plants. *New Phytol.* 106 (Suppl.):113–30

53. Graham, D., Patterson, B. D. 1982. Responses of plants to low, nonfreezing temperatures: proteins, metabolism, and acclimation. *Annu. Rev. Plant Physiol.* 33:347–72

54. Griffith, M., Brown, G. N., Huner, N. P. A. 1982. Structural changes in thylakoid proteins during cold acclimation and freezing of winter rye (*Secale cereale* L. cv. Puma). *Plant Physiol.* 70:418–23

55. Griffith, M., Huner, N. P. A., Kyle, D. J. 1984. Fluorescence properties indicate that photosystem II reaction centers and light-harvesting complex are modified by low temperature growth in winter rye. *Plant Physiol.* 76:381–85

56. Gullord, M., Olien, C. R., Everson, E. H. 1975. Evaluation of freezing hardiness in winter wheat. *Crop. Sci.* 15:153–57

57. Gusta, L. V., Burke, M. J., Kapoor, A. C. 1975. Determination of unfrozen water in winter cereals at subfreezing temperatures. *Plant Physiol.* 56:707–9

58. Guy, C. L., Carter, J. V. 1984. Characterization of partially purified glutathione reductase from cold-hardened and nonhardened spinach leaf tissue. *Cryobiology* 21:454–64

59. Guy, C. L., Haskell, D. 1987. Induction of freezing tolerance in spinach is associated with the synthesis of cold acclimation induced proteins. *Plant Physiol.* 84:872–78

60. Guy, C. L., Haskell, D. 1988. Detection of polypeptides associated with the cold acclimation process in spinach. *Electrophoresis* 9:787–96

61. Guy, C. L., Haskell, D. 1989. Preliminary characterization of high molecular mass proteins associated with cold acclimation in spinach. *Plant Physiol. Biochem.* 27: In press

62. Guy, C. L., Haskell, D., Yelenosky, G. 1988. Changes in freezing tolerance and polypeptide content of spinach and citrus at 5° C. *Cryobiology* 25:264–71

63. Guy, C. L., Niemi, K. J., Brambl, R. 1985. Altered gene expression during cold acclimation of spinach. *Proc. Natl. Acad. Sci. USA* 81:3673–77

64. Guy, C. L., Yelenosky, G., Sweet, H. C. 1980. Light exposure and soluble sugars in citrus frost hardiness. *Fla. Sci.* 43:268–73

65. Hahn, M., Walbot, V. 1989. Effects of cold-treatment on protein synthesis and mRNA levels in rice leaves.

Plant Physiol. 91:930–38

66. Hall, T. C., McLeester, R. C., McCown, B. H., Beck, G. E. 1970. Enzyme changes during deacclimation of willow stem. *Cryobiology* 7:130–35

67. Hansen, J., Beck, E. 1988. Evidence for ideal and non-ideal equilibrium freezing of leaf water in frosthardy ivy *(Hedera helix)* and winter barley *(Hordeum vulgare). Bot. Acta* 101:76–82

68. Hanson, A. D., Hitz, W. D. 1982. Metabolic responses of mesophytes to plant water deficits. *Annu. Rev. Plant Physiol.* 33:163–203

69. Harada, J. J., DeLisle, A. J., Baden, C. S., Crouch, M. L. 1989. Unusual sequence of an abscisic acid-inducible mRNA which accumulates late in *Brassica napus* seed development. *Plant Mol. Biol.* 12:395–401

70. Harris, R. E. 1965. The hardiness of progeny from reciprocal *Malus* crosses. *Can. J. Plant Sci.* 45:159–61

71. Harrison, L. C., Weiser, C. J., Burke, M. J. 1978. Freezing of water in red-osier dogwood stems in relation to cold hardiness. *Plant Physiol.* 62:899–901

72. Heber, U. 1959. Ursachen der Frostresistenz bei Winterweizen. III. Mitteilung die Bedeutung von Proteinen für die Frostresistenz. *Planta* 54:34–67

73. Hincha, D. K., Heber, U., Schmitt, J. M. 1985. Antibodies against individual thylakoid membrane proteins as molecular probes to study chemical and mechanical freezing damage in vitro. *Biochim. Biophys. Acta* 809:337–44

74. Hoard, K. G., Crosbie, T. M. 1985. S$_1$-line recurrent selection for cold tolerance in two maize populations. *Crop. Sci.* 25:1041–45

75. Holland, M. J., Westhead, E. W. 1973. Adenosine 5'-triphosphate induced cold inactivation of yeast aspartic β-semialdehyde dehydrogenase. *Biochemistry* 12:2270–75

76. Howell, G. S., Weiser, C. J. 1970. The environmental control of cold acclimation in apple. *Plant Physiol.* 45:390–94

77. Howell, G. S., Weiser, C. J. 1970. Fluctuations in cold resistance of apple twigs during spring dehardening. *J. Am. Soc. Hort. Sci.* 95:190–92

78. Huber, S. C. 1986. Fructose 2,6-bisphosphate as a regulatory metabolite in plants. *Annu. Rev. Plant Physiol.* 37:233–46

79. Hughes, M. A., Pearce, R. S. 1988. Low temperature treatment of barley plants causes altered gene expression in shoot meristems. *J. Exp. Bot.* 39:1461–67

80. Hummel, R. L., Ascher, P. D., Pellet, H. M. 1982. Inheritance of the photoperiodically induced cold acclimation response in *Cornus sericea* L., red-osier dogwood. *Theor. Appl. Genet.* 62:385–94

81. Huner, N. P. A., Carter, J. V. 1982. Differential subunit aggregation of a purified protein from cold-hardened and unhardened puma rye. *Z. Pflanzenphysiol.* 106:179–84

82. Huner, N. P. A., Krol, M., Williams, J. P., Maissan, E., Low, P. S., et al. 1987. Low temperature development induces a specific decrease in trans-Δ^3-hexadecenoic acid content which influences LHCII organization. *Plant Physiol.* 84:12–18

83. Huner, N. P. A., Macdowall, F. D. H. 1976. Chloroplastic proteins of wheat and rye grown at warm and cold-hardening temperatures. *Can. J. Biochem.* 54:848–53

84. Huner, N. P. A., Macdowall, F. D. H. 1976. Effect of cold adaptation of puma rye on properties of RUDB carboxylase. *Biochem. Biophys. Res. Commun.* 73:411–20

85. Huner, N. P. A., Macdowall, F. D. H. 1978. Evidence for an in vivo conformational change in ribulose bisphosphate carboxylase-oxygenase from puma rye during cold adaptation. *Can. J. Biochem.* 56:1154–61

86. Huner, N. P. A., Macdowall, F. D. H. 1979. The effects of low temperature acclimation of winter rye on catalytic properties of its ribulose bisphosphate carboxylase-oxygenase. *Can. J. Biochem.* 57:1036–41

87. Huner, N. P. A., Palta, J. P., Li, P.-H., Carter, J. V. 1981. Comparison of the structure and function of ribulose bisphosphate carboxylase-oxygenase from a cold-hardy and nonhardy potato species. *Can. J. Biochem.* 59:280–89

88. Huner, N. P. A., Williams, J. P., Maissan, E. E., Myscich, E. G., Krol, M., et al. 1989. Low temperature–induced decrease in trans-Δ^3-hexadecenoic acid content is correlated with freezing tolerance in cereals. *Plant Physiol.* 89:144–50

89. Jeffree, C. E., Read, N. D., Smith, J. A. C., Dale, J. E. 1987. Water droplets and ice deposits in leaf intercellular spaces: redistribution of water during cryofixation for scanning electron microscopy. *Planta* 172:20–37

90. Johnson-Flanagan, A. M., Singh, J. 1987. Alteration of gene expression during the induction of freezing tolerance in

Brassica napus suspension cultures. *Plant Physiol.* 85:699–705

91. Kacperska-Palacz, A., Dlugokecka, E., Breitenwald, J., Wcislinska, B. 1977. Physiological mechanisms of frost tolerance: possible role of protein in plant adaptation to cold. *Biol. Plant.* 19:10–17

92. Kasamo, K. 1988. Response of tonoplast and plasma membrane ATPases in chilling-sensitive and -insensitive rice (*Oryza sativa* L.) culture cells to low temperature. *Plant Cell Physiol.* 29:1085–94

93. Kawashima, N., Singh, S., Wildman, S. G. 1971. Reversible cold inactivation and heat reactivation of RuDP carboxylase activity of crystallized tobacco fraction I protein. *Biochem. Biophys. Res. Commun.* 42:664–68

94. Keith, C. N., McKersie, B. D. 1986. The effect of abscisic acid on the freezing tolerance of callus cultures of *Lotus corniculatus* L. *Plant Physiol.* 80:766–70

95. Key, J. L., Lin, C. Y., Chen, Y. M. 1981. Heat shock proteins of higher plants. *Proc. Natl. Acad. Sci. USA* 78:3526–30

96. Krall, J. P., Edwards, G. E., Andreo, C. S. 1989. Protection of pyruvate, Pi dikinase from maize against cold lability by compatible solutes. *Plant Physiol.* 89:280–85

97. Krasnuk, M., Jung, G. A., Witham, F. H. 1975. Electrophoretic studies of the relationship of peroxidases, polyphenol oxidase, and indoleacetic acid oxidase to cold tolerance of alfalfa. *Cryobiology* 12:62–80

98. Krasnuk, M., Jung, G. A., Witham, F. H. 1976. Electrophoretic studies of several dehydrogenases in relation to cold tolerance of alfalfa. *Cryobiology* 13:375–93

99. Krasnuk, M., Witham, F. H., Jung, G. A. 1976. Electrophoretic studies of several hydrolytic enzymes in relation to cold tolerance of alfalfa. *Cryobiology* 13:225–42

100. Krupa, Z., Huner, N. P. A., Williams, J. P., Maissan, E., James, D. R. 1987. Development at cold-hardening temperatures. *Plant Physiol.* 84:19–24

101. Kuczenski, R. T., Suelter, C. H. 1970. Effect of temperature and effectors on the conformations of yeast pyruvate kinase. *Biochemistry* 9:939–45

102. Kurkela, S., Franck, M., Heino, P., Lang, V., Palva, E. T. 1988. Cold induced gene expression in *Arabidopsis thaliana* L. *Plant Cell Rep.* 7:495–98

103. Kushad, M. M., Yelenosky, G. 1987. Evaluation of polyamine and proline levels during low temperature acclimation of citrus. *Plant Physiol.* 84:692–95

104. Kuznetsov, V. V., Kimpel, J. A., Goekjian, G., Key, J. L. 1987. Elements of nonspecificity in responses on the plant genome to chilling and heat stress. *Sov. Plant Physiol.* 34:685–93

105. Lang, V., Heino, P., Palva, E. T. 1989. Low temperature acclimation and treatment with exogenous abscisic acid induce common polypeptides in *Arabidopsis thaliana* (L.) Heynh. *Theor. Appl. Genet.* 77:729–34

106. Laroche, A., Hopkins, W. G. 1987. Polysomes from winter rye seedlings grown at low temperature. I. Size class distribution, composition, and stability. *Plant Physiol.* 85:648–54

107. Law, C. N., Jenkins, G. 1970. A genetic study of cold resistance in wheat. *Genet. Res.* 15:197–208

108. Levitt, J. 1951. Frost, drought, and heat resistance. *Annu. Rev. Plant Physiol.* 2:245–68

109. Levitt, J. 1956. *The Hardiness of Plants.* New York: Academic. 278 pp.

110. Levitt, J. 1962. A sulfhydryl-disulfide hypothesis of frost injury and resistance in plants. *J. Theor. Biol.* 3:355–91

111. Levitt, J. 1972. *Responses of Plants to Environmental Stresses.* New York: Academic. 697 pp.

112. Levitt, J. 1980. *Responses of Plants to Environmental Stresses: Vol. 1. Chilling, Freezing and High Temperature Stresses.* New York: Academic. 497 pp. 2nd ed.

113. Levitt, J., Scarth, G. W. 1936. Frost-hardening studies with living cells. II. Permeability in relation to frost resistance and the seasonal cycle. *Can. J. Res. Ser. C* 14:285–305

114. Li, P. H. 1984. Subzero temperature stress physiology of herbaceous plants. *Hortic. Rev.* 6:373–416

115. Li, P. H., Wesier, C. J. 1969. Metabolism of nucleic acids in one-year-old apple twig during cold hardening and de-hardening. *Plant Cell Physiol.* 10:21–30

116. Liesenfeld, D. R., Auld, D. L., Murray, G. A., Swensen, J. B. 1986. Transmittance of winterhardiness in segregated populations of peas. *Crop Sci.* 26:49–54

117. Limin, A. E., Fowler, D. B. 1988. Cold hardiness expression in interspecific hybrids and amphiploids of the Triticeae. *Genome* 30:361–65

118. Lynch, D. V., Steponkus, P. L. 1987. Plasma membrane lipid alterations associated with cold acclimation of win-

ter rye seedlings *(Secale cereale* L. cv Puma). *Plant Physiol.* 83:761–67
119. Lynch, D. V., Thompson, G. A. 1984. Microsomal phospholipid molecular species alterations during low temperature acclimation in *Dunaliella. Plant Physiol.* 74:193–97
120. Lyons, J. M. 1973. Chilling injury in plants. *Annu. Rev. Plant Physiol.* 24: 445–66
121. Marmiroli, N., Terzi, V., Odoardi Stanca, M., Lorenzoni, C., Stanca, A. M. 1986. Protein synthesis during cold shock in barley tissues. *Theor. Appl. Genet.* 73:190–96
122. Mazur, P. 1963. Kinetics of water loss from cells at subzero temperatures and the likelihood of intracellular freezing. *J. Gen. Physiol.* 47:347–69
123. Mazur, P. 1969. Freezing injury in plants. *Annu. Rev. Plant Physiol.* 20: 419–48
124. McCarty, R. E., Racker, E. 1966. Effect of a coupling factor and its antiserum on photophosphorylation and hydrogen ion transport. *Brookhaven Symp. Biol.* 19:202–12
125. McCown, B. H., Beck, G. E., Hall, T. C. 1968. Plant leaf and stem proteins. I. Extraction and electrophoretic separation of basic, water-soluble fraction. *Plant Physiol.* 43:578–82
126. McCown, B. H., McLeester, R. C., Beck, G. E., Hall, T. C. 1969. Environment-induced changes in peroxidase zymograms in the stems of deciduous and evergreen plants. *Cryobiology* 5:410–12
127. Meza-Basso, L., Alberdi, M., Raynal, M., Ferrero-Cadinanos, M. L., Delseny, M. 1986. Changes in protein synthesis in rapeseed *(Brassica napus)* seedlings during a low temperature treatment. *Plant Physiol.* 82:733–38
128. Mock, J. J., Eberhart, S. A. 1972. Cold tolerance in adapted maize populations. *Crop Sci.* 12:466–69
129. Mohapatra, S. S., Poole, R. J., Dhindsa, R. S. 1987. Cold acclimation, freezing resistance and protein synthesis in alfalfa *(Medicago sativa* L. cv. Saranac). *J. Exp. Bot.* 38:1697–1703
130. Mohapatra, S. S., Poole, R. J., Dhindsa, R. S. 1987. Changes in protein patterns and translatable messenger RNA populations during cold acclimation of alfalfa. *Plant Physiol.* 84:1172–76
131. Mohapatra, S., Wolfraim, L., Poole, R. J., Dhindsa, R. S. 1989. Molecular cloning and relationship to freezing tolerance of cold-acclimation-specific genes of alfalfa. *Plant Physiol.* 89:375–80

132. Muehlbauer, F. J., Marshall, H. G., Hill, R. R. Jr. 1970. Winter hardiness in oat populations derived from reciprocal crosses. *Crop. Sci.* 10:646–49
133. Mundy, J., Chua, N. H. 1988. Abscisic acid and water-stress induce the expression of a novel rice gene. *EMBO J.* 8:2279–86
134. Nilsson, J. E., Andersson, B. 1987. Performance in freezing tests and field experiments of full-sib families of *Pinus sylvestris* (L.). *Can. J. For. Res.* 17:1340–47
135. Nilsson-Ehle, H. 1912. Zur Kenntnis der Erblichkeitsverhältnisse der Eigenschäft Winterfestigkeit beim Weizen. *Z. Pflanzenzuecht.* 1:3–12
136. Norell, L., Eriksson, G., Ekberg, I., Dormling, I. 1986. Inheritance of autumn frost hardiness in *Pinus sylvestris* L. seedlings. *Theor. Appl. Genet.* 72: 440–48
137. Olien, C. R. 1967. Freezing stresses and survival. *Annu. Rev. Plant Physiol.* 18:387–408
138. Olien, C. R., Smith, M. B. 1981. *Analysis and Improvement of Plant Cold Hardiness.* Boca Raton: CRC Press. 215 pp.
139. Omran, R. G. 1980. Peroxide levels and the activities of catalase, peroxidase, and indoleacetic acid oxidase during and after chilling cucumber seedlings. *Plant Physiol.* 65:407–8
140. Oquist, G., Martin, B. 1986. Cold climates. In *Photosynthesis in Contrasting Environments,* ed. N. R. Baker, S. P. Long, pp. 238–93. New York: Elsevier
141. Orr, W., Keller, W. A., Singh, J. 1986. Induction of freezing tolerance in an embryogenic cell suspension culture of *Brassica napus* by abscisic acid at room temperature. *J. Plant Physiol.* 126:23–32
142. Orr, W., Singh, J., Brown, D. C. W. 1985. Induction of freezing tolerance in alfalfa cell suspension cultures. *Plant Cell Rep.* 4:15–18
143. Orser, C., Staskawicz, B. J., Panopoulus, N. J., Dahlbeck, D., Lindow, S. E. 1985. Cloning and expression of bacterial ice nucleation genes in *Escherichia coli. J. Bacteriol.* 164:359–66
144. Ougham, H. J. 1987. Gene expression during leaf development in *Lolium temulentum:* patterns of protein synthesis in response to heat-shock and cold-shock. *Physiol. Plant.* 70:479–84
145. Parker, J. 1962. Relationships among cold hardiness, water-soluble protein, anthocyanins, & free sugars in *Hedera helix* L. *Plant Physiol.* 37:809–13
146. Parodi, P. C., Nyquist, W. E., Patter-

son, F. L., Hodges, H. F. 1983. Traditional combining-ability and Gardner-Eberhart analyses of a diallel for cold resistance in winter wheat. *Crop. Sci.* 23:314–18

147. Perras, M., Sarhan, F. 1989. Synthesis of freezing tolerance proteins in leaves, crown, and roots during cold acclimation of wheat. *Plant Physiol.* 89:577–85

148. Peumans, W. J., Nsimba-Lubaki, M., Broekaert, W. F., Van Damme, E. J. M. 1986. Are bark lectins of elderberry *(Sambucus nigra)* and black locust *(Robinia pseudoacacia)* storage proteins? In *Molecular Biology of Seed Storage Proteins and Lectins,* ed. L. M. Shannon, M. J. Chrispeels, pp. 53–63. Baltimore: Waverly. 239 pp.

149. Phelps, D. C., Mcdonald, R. E. 1989. Changes in fructose 2,6-bisphosphate levels in green pepper *(Capsicum annuum* L.) fruit in response to temperature. *Plant Physiol.* 90:458–62

150. Pollock, C. J., Cairns, A. J., Collis, B. E., Walker, R. P. 1989. Direct effects of low temperature upon components of fructan metabolism in leaves of *Lolium temulentum* L. *J. Plant Physiol.* 134: 203–8

151. Pomeroy, M. K., Siminovitch, D., Wrightman, F. 1970. Seasonal biochemical changes in the living bark and needles of red pine *(Pinus resinosa)* in relation to adaptation to freezing. *Can. J. Bot.* 48:953–67

152. Price, J. L., Gourlie, B. B., Lin, Y., Huang, R. C. C. 1986. Induction of winter flounder antifreeze protein messenger RNA at 4°C in vivo and in vitro. *Physiol. Zool.* 59:679–95

153. Quamme, H. A. 1978. Breeding and selecting temperate fruit crops for cold hardiness. In *Plant Cold Hardiness and Freezing Stress,* ed. P. H. Li, A. Sakai, 1:313–32. New York: Academic. 416 pp.

154. Quamme, H. A., Stushnoff, C., Weiser, C. J. 1972. Winter hardiness of several blueberry species and cultivars in Minnesota. *HortScience* 7:500–2

155. Rajashekar, C., Burke, M. J. 1982. Liquid water during slow freezing based on cell water relations and limited experimental testing. In *Plant Cold Hardiness and Freezing Stress,* ed. P. H. Li, A. Sakai, 2:211–20. New York: Academic. 694 pp.

156. Rajashekar, C. B., Burke, M. J., Li, P. H., Carter, J. V. 1989. Freezing characteristics of rigid plant tissues: experimental data and analyses involving negative pressure potential. *Plant Physiol.* In press

157. Raymond, J. A., Wilson, P., DeVries, A. L. 1989. Inhibition of growth of nonbasal planes in ice by fish antifreezes. *Proc. Natl. Acad. Sci. USA* 86:881–85

158. Reaney, M. J. T., Gusta, L. V. 1987. Factors influencing the induction of freezing tolerance by abscisic acid in cell suspension cultures of *Bromus inermis* Leyss and *Medicago sativa* L. *Plant Physiol.* 83:423–27

159. Rehfeldt, G. E. 1977. Growth and cold hardiness of intervarietal hybrids of douglas fir. *Theor. Appl. Genet.* 50:3–15

160. Reid, D. A. 1965. Winter hardiness of progenies from winter × spring barley *(Hordeum vulgare,* L. emend. Lam.) crosses. *Crop Sci.* 5:263–66

161. Riov, J., Brown, G. N. 1976. Comparative studies of activity and properties of ferredoxin-NADP$^+$ reductase during cold hardening of wheat. *Can. J. Bot.* 54:1896–1902

162. Roberts, D. W. A. 1974. The invertase complement of cold-hardy and cold-sensitive wheat leaves. *Can. J. Bot.* 53:1333–37

163. Roberts, D. W. A. 1978. Changes in the proportions of two forms of invertase associated with the cold acclimation of wheat. *Can. J. Bot.* 57:413–19

164. Robertson, A. J., Gusta, L. V., Reaney, M. J. T., Ishikawa, M. 1987. Protein synthesis in bromegrass *(Bromus inermis* Leyss) cultured cells during the induction of frost tolerance by abscisic acid or low temperature. *Plant Physiol.* 84:1331–36

165. Robertson, A. J., Gusta, L. V., Reaney, M. J. T., Ishikawa, M. 1988. Identification of proteins correlated with increased freezing tolerance in bromegrass *(Bromus inermis* Leyss. cv Manchar) cell cultures. *Plant Physiol.* 86:344–47

166. Rochat, E., Therrien, H. P. 1975. Etude des protéines des blés résistant, Kharkov, et sensible, Selkirk, au cours de l'endurcissement au froid. I. Protéines solubles. *Can. J. Bot.* 53:2411–16

167. Rohde, C. R., Pulham, C. F. 1960. Heritability estimates of winter hardiness in winter barley determined by the standard unit method of regression analysis. *Agron. J.* 52:584–86

168. Rosas, A., Alberdi, M., Delseny, M., Meza-Basso, L. 1986. A cryoprotective polypeptide isolated from *Nothofagus dombeyi* seedlings. *Phytochemistry* 25:2497–2500

169. Rudolph, T. D., Nienstaedt, H. 1962. Polygenic inheritance of resistance to winter injury in jack pine-lodgepole pine hybrids. *J. For.* 60:138–39

170. Sagisaka, S. 1985. Injuries of cold acclimatized poplar twigs resulting from enzyme inactivation and substrate depression during frozen storage at ambient temperatures for a long period. *Plant Cell Physiol.* 26:1135–45

171. Sakai, A., Larcher, W. 1987. *Frost Survival of Plants.* Berlin: Springer-Verlag. 321 pp.

172. Sakai, A., Sugawara, Y. 1973. Survival of popular callus at super-low temperatures after cold acclimation. *Plant Cell Physiol.* 14:1201–4

173. Sakai, A., Yoshida, S. 1968. The role of sugar and related compounds in variations of freezing resistance. *Cryobiology* 5:160–74

174. Salerno, G. L., Pontis, H. G. 1989. Raffinose synthesis in *Chlorella* vulgaris cultures after a cold shock. *Plant Physiol.* 89:648–51

175. Sarhan, F., Chevrier, N. 1985. Regulation of RNA synthesis by DNA-dependent RNA polymerases and RNases during cold acclimation in winter and spring wheat. *Plant Physiol.* 78:250–55

176. Sarhan, F., D'Aoust, M. J. 1975. RNA synthesis in spring and winter wheat during cold acclimation. *Physiol. Plant.* 35:62–65

177. Sarhan, F., Perras, M. 1987. Accumulation of a high molecular weight protein during cold hardening of wheat (*Triticum aestivum* L.). *Plant Cell Physiol.* 28:1173–79

178. Schaffer, M. A., Fischer, R. L. 1988. Analysis of mRNAs that accumulate in response to low temperature identifies a thiol protease gene in tomato. *Plant Physiol.* 87:431–36

179. Schmitt, J. M., Schramm, M. J., Pfanz, H., Coughlan, S., Heber, U. 1985. Damage to chloroplast membranes during dehydration and freezing. *Cryobiology* 22:93–104

180. Shirahashi, K., Hayakawa, S., Sugiyama, T. 1978. Cold lability of pyruvate, orthophosphate dikinase in the maize leaf. *Plant Physiol.* 62:826–30

181. Siminovitch, D., Briggs, D. R. 1949. The chemistry of the living bark of the black locust tree in relation to frost hardiness. I. Seasonal variations in protein content. *Arch. Biochem.* 23:8–17

182. Siminovitch, D., Briggs, D. R. 1953. Studies on the chemistry of the living bark of the black locust tree in relation to frost hardiness. IV. Effects of ringing on translocation, protein synthesis, and the development of hardiness. *Plant Physiol.* 28:177–200

183. Siminovitch, D., Cloutier, Y. 1982. Twenty-four-hour induction of freezing and drought tolerance in plumules of winter rye seedlings by desiccation stress at room temperature in the dark. *Plant Physiol.* 69:250–55

184. Siminovitch, D., Rheaume, B., Pomeroy, K., Lepage, M. 1968. Phospholipid, protein, and nucleic acid increases in protoplasm and membrane structures associated with development of extreme freezing resistance in black locust tree cells. *Cryobiology* 5:202–25

185. Simon, J. P. 1987. Differential effects of chilling on the activity of C_4 enzymes in two ecotypes of *Echinochloa crusgalli* from sites of contrasting climates. *Physiol. Plant.* 69:205–10

186. Smithberg, M. H., Weiser, C. J. 1968. Patterns of variation among climatic races of red-osier dogwood. *Ecology* 49:495–505

187. Stancel, G. M., Deal, W. C. Jr. 1969. Reversible dissociation of yeast glyceraldehyde 3-phosphate dehydrogenase by adenosine triphosphate. *Biochemistry* 8:4005–11

188. Steffen, K. L., Arora, R., Palta, J. P. 1989. Relative sensitivity of photosynthesis and respiration to freeze-thaw stress in herbaceous species. *Plant Physiol.* 89:1372–79

189. Steponkus, P. L. 1984. Role of the plasma membrane in freezing injury and cold acclimation. *Annu. Rev. Plant Physiol.* 35:543–84

190. Steponkus, P. L., Dowgert, M. F., Evans, R. Y., Gordon-Kamm, W. 1982. Cryobiology of isolated protoplasts. See Ref. 155, pp. 459–74

191. Steponkus, P. L., Dowgert, M. F., Gordon-Kamm, W. J. 1983. Destabilization of the plasma membrane of isolated plant protoplasts during a freeze-thaw cycle: the influence of cold acclimation. *Cryobiology* 20:448–65

192. Steponkus, P. L., Lanphear, F. O. 1968. The relationship of carbohydrates to cold acclimation of *Hedera helix* L. cv. Thorndale. *Physiol. Plant.* 21:777–91

193. Steponkus, P. L., Uemura, M., Balsamo, R. A., Arvinte, T., Lynch, D. V. 1988. Transformation of the cryobehavior of rye protoplasts by modification of the plasma membrane lipid composition. *Proc. Natl. Acad. Sci. USA* 85:9026–30

194. Stushnoff, C. 1972. Breeding and selection methods for cold hardiness in deciduous fruit crops. *HortScience* 7:10–13

195. Sutka, J. 1981. Genetic studies of frost resistance in wheat. *Theor. Appl. Genet.* 59:145–52

196. Sutka, J., Veisz, O. 1988. Reversal of dominance in a gene on chromosome 5A controlling frost resistance in wheat. *Genome* 30:313–17

197. Taylor, A. O., Slack, C. R., McPherson, H. G. 1974. Plants under climatic stress. VI. Chilling and light effects on photosynthetic enzymes of sorghum and maize. *Plant Physiol.* 54:696–701

198. Tenaud, M., Jacquot, J. P. 1987. In vitro thiol-dependent redox regulation of purified ribulose-1,5-bisphosphate carboxylase. *J. Plant Physiol.* 130:315–26

199. Tseng, M. J., Li, P. H. 1987. Changes in nucleic acid and protein synthesis during induction of cold hardiness. In *Plant Cold Hardiness*, ed. P. H. Li, A. Sakai, pp. 1–27. New York: Liss

200. Uemura, M., Yoshida, S. 1984. Involvement of plasma membrane alterations in cold acclimation of winter rye seedlings (*Secale cereale* L. cv Puma). *Plant Physiol.* 75:818–26

201. Vallejos, C. E. 1989. Low night temperatures have a differential effect on the diurnal cycling of gene expression in cold sensitive and tolerant tomatoes. (In review)

202. Van Huystee, R. B., Weiser, C. J., Li, P. H. 1967. Cold acclimation in *Cronus stolonifera* under natural and controlled photoperiod and temperature. *Bot. Gaz.* 128:200–5

203. Volger, H. G., Heber, U. 1975. Cryoprotective leaf proteins. *Biochim. Biophys. Acta* 412:335–49

204. Volger, H., Heber, U., Berzborn, R. J. 1978. Loss of function of biomembranes and solubilization of membrane proteins during freezing. *Biochim. Biophys. Acta* 511:455–69

205. Walker, G. H., Ku, M. S. H., Edwards, G. E. 1986. Catalytic activity of maize of phosphoenolpyruvate carboxylase in relation to oligomerization. *Plant Physiol.* 80:848–55

206. Weeden, N. F., Buchanan, B. B. 1983. Leaf cytosolic fructose-1,6-bisphosphatase. *Plant Physiol.* 72:259–61

207. Weiser, C. J. 1970. Cold resistance and injury in woody plants. *Science* 169:1269–78

208. Wenkert, W. 1980. Measurement of tissue osmotic pressure. *Plant Physiol.* 65:614–17

209. Wilner, J. 1965. The influence of maternal parent on frost-hardiness of apple progenies. *Can. J. Plant Sci.* 45:67–71

210. Wolber, P. K., Deininger, C. A., Southworth, M. W., Vandekerckhove, J., Van Montagu, M., Warren, G. J. 1986. Identification and purification of a bacterial ice-nucleation protein. *Proc. Natl. Acad. Sci. USA* 83:7256–60

211. Woodward, F. I., Williams, B. G. 1987. Climate and plant distribution at global and local scales. *Vegetatio* 69:189–97

212. Worzella, W. W. 1935. Inheritance of cold resistance in winter wheat, with preliminary studies on the technique of the artifical freezing test. *J. Agric. Res.* 50:625–35

213. Wright, S. T. C. 1977. The relationship between leaf water potential (ψleaf) and the levels of abscisic acid and ethylene in excised wheat leaves. *Planta* 134:183–89

214. Yacoob, R. K., Filion, W. G. 1986. The effects of cold-temperature stress on gene expression in maize. *Biochem. Cell Biol.* 65:112–19

215. Yacoob, R. K., Filion, W. G. 1986. Temperature-stress response in maize: a comparison of several cultivars. *Can. J. Genet. Cytol.* 28:1125–35

216. Yancey, P. H., Clark, M. E., Hand, S. C., Bowlus, R. D., Somero, G. N. 1982. Living with water stress: evolution of osmolyte systems. *Science* 217:1214–22

217. Yelenosky, G. 1979. Accumulation of free proline in citrus leaves during cold hardening of young trees in controlled temperature regimes. *Plant Physiol.* 64:425–27

218. Yelenosky, G. 1985. Cold hardiness in citrus. *Horic Rev.* 7:201–38

219. Yelenosky, G., Guy, C. L. 1977. Carbohydrate accumulation in leaves and stems of 'valencia' orange at progressively colder temperatures. *Bot. Gaz.* 138:13–17

220. Yoshida, S. 1984. Chemical and biophysical changes in the plasma membrane during cold acclimation of mulberry bark cells (*Morus bombycis* Koidz. cv Goroji). *Plant Physiol.* 76:257–65

221. Yoshida, S., Matsuura, C., Etani, S. 1989. Impairment of tonoplast H^+-ATPase as an initial physiological response of cells to chilling in mung bean (*Vigna radiata* [L.] Wilczek). *Plant Physiol.* 89:634–42

222. Yoshida, S., Uemura, M. 1984. Protein and lipid compositions of isolated plasma membranes from orchard grass (*Dactylis glomerata* L.) and changes during cold acclimation. *Plant Physiol.* 75:31–37

Annu. Rev. Plant Physiol. Plant Mol. Biol. 1990. 41:225–53
Copyright © 1990 by Annual Reviews Inc. All rights reserved

ASSIMILATORY NITRATE REDUCTASE: FUNCTIONAL PROPERTIES AND REGULATION[1]

Larry P. Solomonson and Michael J. Barber

Department of Biochemistry and Molecular Biology, University of South Florida College of Medicine, Tampa, Florida 33612

KEY WORDS: nitrate assimilation, Mo-pterin enzymes, nitrogen metabolism, b-type cytochrome, nitrate reductase structure

CONTENTS

[1]Abreviations used: NR, nitrate reductase; MV, reduced methyl viologen cation; CD, circular dichroism; EPR, electron paramagnetic resonance; FH, reduced flavin.

1040-2519/90/0601-0225$02.00

INTRODUCTION

Inorganic nitrogen in the biosphere is converted to a biologically useful form (organic nitrogen) by either the "fixation" of molecular nitrogen (N_2) or the "assimilation" of nitrate. Nitrogen fixation occurs in certain prokaryotes and in some higher plants that have a symbiotic association with nitrogen-fixing bacteria. Considerable effort has been directed towards the incorporation of nitrogen fixation genes into higher plants, an effort so far unrewarded. Thus nitrate assimilation remains the major pathway by which inorganic nitrogen is converted to organic form. As noted in an earlier review by Guerrero et al (50), nitrate assimilation is estimated to produce in excess of 2×10^4 megatons of organic nitrogen per year compared to 2×10^2 megatons for nitrogen fixation. As much as 25% of the energy of photosynthesis is consumed in driving nitrate assimilation. Thus the process of nitrate assimilation is of fundamental biological importance. It occurs in a wide variety of organisms including bacteria, yeast, fungi, algae, and higher plants. These organisms in turn supply the nutritional requirements for nitrogen in other forms of life.

The conversion of nitrate to ammonia is an 8-electron reduction process that occurs in two steps. The first step is a 2-electron reduction of nitrate to nitrite, catalyzed by the enzyme nitrate reductase (NR). NAD(P)H serves as the physiological electron donor for this reaction in eukaryotes. The second step is a 6-electron reduction of nitrite to ammonia, catalyzed by the enzyme nitrite reductase. This step is coupled to photosynthetic electron transport in algae and higher plants via reduced ferredoxin, a product of the "light" reactions of photosynthesis, which serves as the physiological electron donor for nitrite reductase.

The rate-limiting and regulated step of nitrate assimilation appears to be the initial reaction, catalyzed by NR (16). This enzyme is considered to be a limiting factor for growth, development, and protein production in plants and other nitrate-assimilating organisms. Nitrate reductase has therefore been intensively studied in order to delineate properties related to its catalytic efficiency and regulation.

Here we focus on the properties and regulation of assimilatory NR associated with green algae and higher plants with only occasional reference to the enzyme from other eukaryotic sources. Several recent reviews cover various aspects of assimilatory nitrate reduction in cyanobacteria (49), yeast (58), and fungi (119).

FUNCTIONAL PROPERTIES

Nitrate reductase is a complex enzyme containing several different redox-active prosthetic groups. Nitrate reductases isolated from eukaryotic sources have been shown to be multimeric proteins containing FAD, heme (cytochrome b_{557}) and Mo-pterin prosthetic groups in a 1:1:1 stoichiometry per subunit. In addition to the full, NAD(P)H:nitrate reductase (NADH:NR) activity, the enzyme catalyzes a variety of partial activities that use alternate artificial electron donors and acceptors that involve one or more of the enzyme's prosthetic groups. These partial activities have been divided into two classes referred to as either "diaphorase" or "nitrate-reducing" activities, respectively. Diaphorase activities use NAD(P)H as the electron donor and include NADH:ferricyanide reductase (NADH:FR), NADH:cytochrome c reductase (NADH:CR), and NADH:dichlorophenolindophenol reductase (NADH:DR) activities. Nitrate-reducing activities include reduced flavin:nitrate reductase (FH:NR), reduced methyl viologen:nitrate reductase (MV:NR), and reduced bromophenol blue:nitrate reductase (BB:NR) activities. Limited proteolysis studies have established that NADH:FR activity requires FAD, whereas NADH:CR and NADH:DR activities require both FAD and heme prosthetic groups. Reduced flavin:nitrate reductase activity requires heme and Mo-pterin whereas MV:NR and BB:NR activities require only Mo-pterin. Diaphorase activities are inhibited by sulfhydryl reagents such as N-ethylmaleimide (NEM) whereas nitrate-reducing activities are inhibited by cyanide. Figure 1 is a schematic diagram indicating the prosthetic groups, and the intra- and intermolecular electron transfer reactions comprising these various activities.

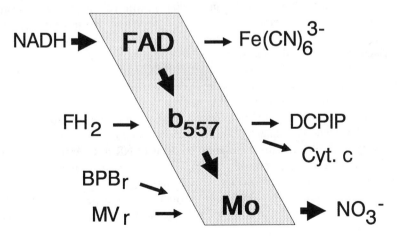

Figure 1   Interaction of electron donors and acceptors with nitrate reductase. The various inter- and intraelectron transfer reactions that comprise the full and partial activities associated with nitrate reductase are shown together with the individual prosthetic groups required for these activities.

In this section we review what is presently known about various molecular and physical properties of NR important to the catalytic functions of the enzyme.

Functional Size and Quaternary Structure

There is wide variation in reported molecular masses of assimilatory NR isolated from different sources ranging from 200 kd for the spinach enzyme (57) to 500 kd for the enzyme from *Ankistrodesmus braunii* (31). A similar variation has been reported for the size and number of the polypeptide subunits, resulting in several complex structural models. Recent work has clarified our understanding and enables us to account for some of the previously reported variations.

Through a combination of gel chromatography, sedimentation equilibrium, gel electrophoresis, cross-linking, and peptide-mapping studies, Howard & Solomonson (60) obtained conclusive evidence that *Chlorella* NR is a homotetramer with dihedral symmetry ("dimer of dimers"). The mode of association between subunits was isologous (head to head) rather than heterologous (head to tail). It was also shown that the active tetramer dissociates to active dimers at low enzyme concentrations without significant loss of activity, suggesting that association/dissociation plays no role in regulation of enzyme activity (60). The NR homotetramer exhibits a rather high partial specific volume at neutral pH, which may have led to erroneous conclusions about the molecular mass and quaternary structure of the native enzyme. No lipid or carbohydrate was found associated with the enzyme, suggesting that the high value for the partial specific volume may arise from preferential solvent interactions (60). The observation that the enzyme can exist as an active dimer or tetramer may explain some of the reported variations in observed molecular masses of the native enzyme.

The subunit molecular mass of *Chlorella* NR, determined by SDS polyacrylamide gel electrophoresis, was approximately 100 kd, present as a single band (109). Similar results were obtained for *Neurospora* and barley NR (75, 93). Reports of multiple, lower-molecular-mass subunits for NR from spinach and barley were subsequently shown to be the consequence of proteolytic "nicking" (19, 39). In vitro translation products of NR mRNA also appear to be in this molecular mass range (23, 26, 27). These results suggest a similar subunit structure for eukaryotic nitrate reductases. A possible exception is the enzyme from *Ankistrodesmus braunii*, which has been purified in a high-activity state and appears to be a homooctomer with a subunit molecular mass of 59 kd (31).

Radiation inactivation analysis has been utilized to determine the functional size of *Chlorella* NR at high and low concentrations of enzyme, where the principal physical species would be either tetrameric or dimeric, respectively.

In both cases, target sizes corresponding to approximately 100 kd were obtained, suggesting that each subunit in the tetramer or dimer functions independently (110, 111). These results confirmed earlier studies that indicated independent, identical subunits containing a full complement of prosthetic groups.

Functional Domains

Radiation inactivation analysis and limited proteolysis have been applied to NR to yield structural information and to determine whether or not the protein is folded in a manner consistent with active "domains" comprising the individual prosthetic groups. Target size analysis, using radiation inactivation, of *Chlorella* NADH:NR activity has indicated the "functional unit" to be the 100-kd subunit. In contrast, target sizes for the partial activities NADH:FR, NADH:CR, and MV:NR (60, 78 and 40 kd, respectively) were in each case different and significantly smaller than the minimum polypeptide mass, suggesting that these activities reside on functionally independent domains (110, 111). The apparent functionl size for NADH:CR activity was significantly larger than for NADH:FR activity, suggesting that these alternate electron acceptors do not act at the same site (111). Because Mo-pterin is not required for either activity, this result may suggest that ferricyanide accepts electrons via the FAD center while cytochrome *c* accepts electrons via the heme moiety of NR. Preliminary results for spinach NR (M. J. Barber, B. A. Notton, L. P. Solomonson, and M. J. McCreery, unpublished) have indicated target sizes for the various activities similar to those obtained for *Chlorella* NR.

Incubation of native *Chlorella* NR with either trypsin, *Staphylococcus aureus* V8 protease, or a natural inactivator protease (CIP) from corn resulted in loss of NADH:NR and NADH:CR activities but retention of nitrate-reducing activities (107, 108). Incubation of NR with V8 protease or CIP resulted in two different products, each of which retained a different partial activity (107). Nitrate-reducing activity was associated with a homotetrameric fragment of about 260 kd that contained heme and Mo-pterin but no FAD, whereas NADH:FR activity was associated with a monomeric species of approximately 30 kd that contained FAD and the NADH-binding site. The heme associated with the nitrate-reducing fragment could not be reduced by NADH, indicating that FAD was essential for the transfer of electrons between NADH and heme. Circular dichroism, visible, and EPR spectral analysis of the limited proteolysis fragments compared to the native enzyme indicated no significant perturbation of the protein environment surrounding the prosthetic groups as a result of limited proteolysis (69, 107).

Trypsin and V8 proteolysis has also been applied to spinach NR (74, 88). Incubation with trypsin resulted in the generation of two fragments of 59 and

45 kd, the latter retaining NADH:FR and NADH:CR partial activities and containing FAD and heme prosthetic groups. The larger fragment was devoid of MV:NR activity but was presumed to contain Mo-pterin. In contrast, incubation of spinach NR with V8 protease yielded two fragments, a monomeric 30-kd fragment that retained NADH:FR and contained FAD, and a dimeric 170-kd fragment that retained FH:NR and MV:NR activities and contained heme and Mo-pterin. Further, the 45-kd fragment obtained by trypsin treatment could be further digested with V8 protease to yield two fragments of 28 and 14 kd, respectively. The former retained NADH:FR and contained FAD while the latter was devoid of any partial activity but contained heme. These results suggested that there are two exposed proteolysis sites within spinach NR and that the FAD, heme, and Mo-pterin groups may be found in discrete domains within the overall subunit structure. A structural model consistent with the results of the above limited proteolysis studies of NR is shown in Figure 2.

Immunological Properties

Immunological methods have been widely used to analyze structural similarities, in terms of conserved antigenic recognition sites, between nitrate reductases isolated from different sources. Polyclonal antibodies have been raised against a variety of nitrate reductases including those from spinach (47), squash (103), barley (116), and *Chlorella vulgaris* (35). In several cases, immunodetection methods such as inhibition of enzyme activity, Ouchterlony double diffusion, and rocket immunoelectrophoresis were used to compare the cross-reactivity of polyclonal antibodies elicited against one form of NR with that isolated from other sources. The extent of cross-reactivity is apparently dependent on the phylogenetic distances between the various organisms. All nitrate reductases have been found to have some common determinants. For example, polyclonal antibodies elicited against *Chlorella* NR cross-react and inhibit the purified spinach enzyme and vice versa (89).

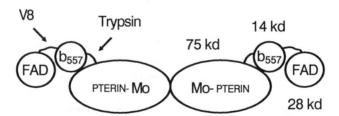

Figure 2 Limited proteolysis models of spinach and *Chlorella* nitrate reductase. The sizes of the major domains obtained following limited proteolysis of spinach and *Chlorella* nitrate reductase are indicated together with their associated prosthetic groups.

Monoclonal antibodies have been developed against the purified NR enzymes from spinach (89) and corn (25) and have been characterized with respect to their inhibition of the various partial activities or binding to selective domains. Spinach monoclonal antibodies have been shown to cross-react with a variety of other higher-plant NR from both monocotyledon and dicotyledon species, though with varying affinities. These monoclonal antibodies have also been demonstrated to cross-react with the *Chlorella* enzyme, though with decreased affinity (87). The production of monoclonal antibodies has resulted in the development of a rapid and efficient immunoaffinity chromatography purification procedure for the spinach enzyme (37).

Molecular Biology and Derived Primary Structures

Recent progress in the molecular biology of NR has been rapid and promises to have a major impact on both our understanding of various structural aspects of the enzyme and its regulation at the genetic level. The first cloning of cDNA for a eukaryotic NR was reported in 1986 by Cheng et al (23) for barley NR. There followed in rapid succession reports of cDNA clones for squash (27), tobacco (18), *Arabidopsis thaliana* (28), and genomic clones for the NR structural gene from *Neurospora* (42), tomato (43), tobacco (123), rice (53), and *A. thaliana* (24). The cDNAs from several higher plant species all hybridize to mRNA that is 3.2–3.5 kb long and whose concentration is responsive to nitrate (18, 27, 28).

Deduced amino acid sequences for NR have been derived from clones to tobacco (18, 122, 123), *Arabidopsis* (28), tomato (45), and corn (46). The complete sequence corresponded to single polypeptides of 904–917 residues that exhibited significant (70–80%) sequence similarity, as shown in Figure 3. Limited sequence information from *Chlorella* NR, obtained by direct amino acid sequencing (A. C. Cannons, P. J. Neame, M. J. Barber, and L. P. Solomonson, unpublished), demonstrated less sequence similarity to *Arabidopsis* NR than to the other higher plant enzymes.

Analysis of the amino acid sequence of nitrate reductase suggested the presence of three "domains," corresponding to the regions forming the Mo-pterin-, cytochrome b_5-, and flavin-binding domains (28, 45). These domains were indicated by the extensive conservation of residues in these regions with similar cofactor-binding domains in other proteins. Thus, the N-terminal region of NR (residues 1–542) binds Mo-pterin and exhibits significant sequence similarity to the Mo-pterin domain of chicken liver sulfite oxidase (28, 85), whereas the heme-binding domain (residues 543–638) exhibited significant sequence similarity to the b_5-binding domains of sulfite oxidase and microsomal cytochrome b_5. The flavin-binding C-terminal region of NR (residues 639–917) was very similar to the FAD-binding domain of microsomal cytochrome b_5 reductase.

```
ATNR    MAASVDNRQYARLEPGLNGVVRSYKPPVPGRSDSPKAHQNQTTNQTVFLK    50
TOM     .....E....SH.....S..G.TF..-----.P...VRGC.FPPSSNHE.P
TOB S   .....E...FSHI.A...---.SF..-----.S...VRGC.FP-SPNSTN-
TOB T   .....E...FSHI.A...---.SF..-----.S...VRGCHFPP-PNSTN-
CORN
CVNR

ATNR    PAKVHDDDEDVS-SEDENETHNSNAVYYKEMIRKSNAELEPSVLDPRDEYT   100
TOM     FQ.KQNPPIFLDY.SS.D.DDDDEKNE.VQ..K.GKT.....IH.T...G.
TOB S   FQ.KPNSTIFLDY.SS.DDDDDDEKNE.LQ..K.GNS.....VH.T...G.
TOB T   FQ.KPNSTIFLDY.SS.DDDDDDEKNE.LQ..K.GNS.....VH.S...G.
CORN
CVNR

ATNR    ADSWIERNPSMVRLTGKHPFNSEAPLNRLMHHGFITPVPLHYVRNHGHVP   150
TOM     ..N.....F.LI...........P..S....................P..
TOB S   ..N.....FMLI..........P.NS....................P..
TOB T   ..N.....F.LI..........P.NS....................P..
CORN
CVNR

ATNR    KAQWAEWTVEVTGFVKRPMKFTMDQLVSEFAYREFAATLVCAGNRRKEQN   200
TOM     ..S.SD.......L..............N..PS...PV............
TOB S   GTSDSD.......L..............N..CS.L.PV............
TOB T   GTSDSD.......L..............N..PS.L.PV............
CORN
CVNR                            PSVDVTC..T........

ATNR    MVKKSKGFNWGSAGVSTSVWRGVPLCDVLRRCGIFSRKGGALNVCFEGSE   250
TOM     ...QTI.....A.A...T.......RAL.K...VQ.K.K..........D
TOB S   ...QTI.....A.A..IT.......RAL.K...FVQNK.K.......A..D
TOB T   ...QTI.....A.A...T.......RIL.K..FVQNK.K.......A..D
CORN
CVNR                                        NRA.

ATNR    DLPGGAG-TAGSKYGTSIKKEYAMDPSRDIILAYMQNGEYLTPDHGFPVRI  300
TOM     V....----G............F..........V......M.S........M
TOB S   V....----G.........F.S..A....V.V.....K.A.S........M
TOB T   V....----G.........F.S..A....V.....K.A.S........M
CORN                                                  .....V
CVNR    .VEAA.AA.VAPPPAPAGA.SFT-..AS.V....K...RL......Y.P.L

ATNR    IIPGFIGGRMVKWLKRIIVTTKESDNFYHFKDNRVLPSLVDAELADEEGW   350
TOM     .................V...Q..ESY..Y......PH......NA.A.
TOB S   ...........I...............Y.........PH......NT.A.
TOB T   ...........I...............Y.........PH......NT.A.
CORN    ....C.............PA....Y.........H.......NA.A.
CVNR    R.

ATNR    WYKPEYIINELNINSVITTPCHEEILPINAFTTQRPYTLKGYAYSGGGKK   400
TOM     ............................W........R..........
TOB S   ............................W........R..S.......
TOB T   ............................W...P....R..S.......
CORN    ...........................D...................
CVNR

ATNR    VTRVEVTVDGGETWNVCALDHQEKPNKYGKFWCWCFWSLEVEVLDLLSAK   450
TOM     .......L......S..T...P...T....Y.................
TOB S   .......L......Q.ST..P...T...Y..................
TOB T   .......L......Q..T...P...T...Y..................
CORN    ......L......L..T-..P...T....Y..................
CVNR
```

Figure 3 Comparison of the amino acid sequences of higher plant nitrate reductases. The *Arabidopsis thaliana* (ATNR) (28) sequence was used as the base sequence with which sequences for tomato (TOM) (45), two forms of tobacco (TOBS and TOBT) (122), and the partial sequence for corn (CORN) (46) were aligned using the software Genepro (version 4.1). Also shown are partial sequences for *Chlorella* (CVNR) NR obtained from direct amino acid sequencing of peptide fragments derived from the Mo-domain of the enzyme (A. C. Cannons, P. J. Neame, M. J. Barber, and L. P. Solomonson, unpublished). Periods indicate identical residues while dashes indicate spaces added to maximize the alignment.

```
ATNR    EIAVRAWDETLNTQPEKMIWNLMGMMNNCWFRVKTNVCKPHKGEIGIVFE    500
TOM     ......T..........L...V.............M.............
TOB S   ......T..........L...V.............M.............
TOB T   ......T..........L...V.............M.............
CORN    .........S.......L...V.........................D
CVNR

ATNR    HPTLPGNESGGWMAKERHLEKSADAPPSLKKSVSTPFMNTT--AKMYSMSEV  550
TOM     ...Q...Q.............I..V...T....I.......A--S........
TOB S   ...Q...Q.............I..E..QT....I.......A--S........
TOB T   ...Q...Q.............I..E...T....I.......A--S.....C..
CORN    ................K...TAEA.A.GY.R.T........DVG.EFT....
CVNR

ATNR    KKHNSADSCWIIVHGHIYDCTRFLMDHPGGSDSILINAGTDCTEEFEAIH    600
TOM     R....S..A..........AS..K.....V................D...
TOB S   R....S..A..........A....K.....T..............D...
TOB T   R....S..A..........A....K.....................D...
CORN    R..A.QE.A..V....V....K..K.....A...............A...
CVNR

ATNR    SDKAKKMLEDYRIGELITTGYSSDSSSPNNSVHGSSAVFSLLAPIGE--ATP  650
TOM     ......L...F..........T..-....S......SIS.F....K.LVQ..
TOB S   ......L...F..........T....-.GN......SFS.F....K.LV-PA
TOB T   ......L..EF..........T....-.GN......SFS.F....K.LV-PA
CORN    .....AL.DT..........---TGY.SD.....G.-.L.H....----RRA
CVNR

ATNR    VRNLALVNPRAKVPVQLVEKTSISHDVRKFRFALPVEDMVLGLPVGKHIF    700
TOM     T.SV..I-..E.I.CK..D.Q..........K....S..Q...........
TOB S   Q.SV..I..E.I.CK.ID.Q..............S..Q...........
TOB T   Q.SV..I..E.I.CK.ID.Q...P..........S..Q...........
CORN    ..AP..SD..E.IHCR..G.KEL.R...L...S..SPTE.....I.....
CVNR

ATNR    LCATINDKLCLRAYTPSSTVDVVGYFELVVKIYFGGVHPRFPNGGLMSQY    750
TOM     ....VD....M.....T....E..F...........K....K.....Q...H
TOB S   ...V.D....M.....T..I.E...........K.I..K.....Q...Y
TOB T   ...V.D....M.....T..I.E...........K.I..K.....Q...Y
CORN    V..RLEG..M.R-.RT.M..EI.H.D.L..V..KNE..K.......T..
CVNR

ATNR    LDSLPIGSTLEIKGPLGHVEYLGKGSFTVHGKPKFADKLAMLAGGTGITP    800
TOM     ......AF.DV......I..Q...N.L....Q...K....I........
TOB S   ...M.L..F.DV......I..Q...N.L....Q...K....I........
TOB T   ...MQL..F.DV......I..Q...N.L....Q...K....I........
CORN    .....V.GYIDV.........T.R...VIN..QRH.SR...IC..S....
CVNR

ATNR    VYQIIQAILKD-PEDETEMYVIYANRTEEDILLREELDGWAEQYPDRLKVW  850
TOM     ...VM.S....-...D.....V......D....KD...A....V.N.V...
TOB S   ...VM......-...D.....V......D....K....S...KI.E.V...
TOB T   ...VM......-...D.....V......D....K....S...KI.E.V...
CORN    MT.....VVR.Q...H...HLV......D.....D...R..AE..Y.....
CVNR

ATNR    YVV-ESAKEGWAYSTGFISEAIMREHIPDGLDGSALAMACGPPPMIQFAVQ  900
TOM     ...Q..ITQ..K.....VT.S.L.....EP-SHTT..L............IN
TOB S   ...QD......K.I..IT...L.....EP-SHTT..L.............N
TOB T   ...QD......K.I..I....L.....EP-SHTT..L.............N
CORN    ..I-DQV....K..V..VT..VL...V.E.-GDDT..L............IS
CVNR

ATNR    PNLEKMQYNIKEDFLIF                                    917
TOM     ......G.D...EL.V.
TOB S   ......G.D..DSL.V.
TOB T   ......G.D..DSL.V.
CORN    ......K.DMANS.VV.
CVNR
```

Active Site Residues

Chemical modification studies have suggested that a number of specific amino acid residues are required for both diaphorase and nitrate-reducing activities of NR. The possible participation of sulfhydryl groups in the transfer of reducing equivalents between NAD(P)H and FAD was initially proposed based on labeling studies of *Neurospora* NR (1). More recent studies on the role of the essential sulfhydryl group of *Chlorella* NR have combined spin-labeling and inhibition experiments (13). Incubation of the enzyme with either NEM or spin-labeled derivatives of NEM resulted in a time-dependent inactivation of NADH:NR and NADH:CR activities with no effect on nitrate-reducing activity. This inactivation was prevented by NADH and involved the modification of a single sulfhydryl group. The EPR spectrum of the spin-labeled enzyme demonstrated the presence of a single species, with the nitroxide retaining substantial motional freedom. The absence of any change in the nitrogen hyperfine coupling constant of the bound label compared with that of the free nitroxide in buffer and its facile reduction by ascorbate indicated that the sulfhydryl group is probably located on the external surface of the enzyme close to the NADH-binding site. Cleavage of the spin-labeled enzyme using CIP and separation into its FAD and heme/Mo domains followed by EPR spectroscopy indicated that the modified sulfhydryl group is associated with the heme/Mo fragment. These domains may therefore interact closely in the region of the nucleotide-binding site. These results suggest that there is one essential sulfhydryl group per subunit that is required for NADH-binding and catalytic activity but is probably not directly involved in electron transfer between prosthetic groups since it reacts with SH-specific reagents when the enzyme is in the fully oxidized state. In addition, the results of microcoulometry experiments using *Chlorella* NR have demonstrated that a maximum of five reducing equivalents are required for total reduction of the enzyme, consistent with FAD, heme, and Mo being the only redox-active prosthetic groups (117).

Arginine residues have also been implicated in the function of *Amaranthus* NR (6). Inactivation of both diaphorase and nitrate-reducing activities, although at different rates, has been shown following incubation with the arginine-specific reagents phenylglyoxal and butanedione, suggesting that arginine residues are required for the catalytic function of both the FAD and Mo-pterin centers. Nitrate-utilizing activities were inactivated more rapidly than the diaphorase activities, suggesting that the arginine residue(s) in the vicinity of the flavin site is less reactive. NADH protected against inactivation of the diaphorase activities, whereas nitrate failed to protect the nitrate-reducing activity. Inactivation of *Chlorella* NR by phenylglyoxal has also been reported (105).

Modification of histidine using diethylpyrocarbonate has been shown to inhibit diaphorase but not nitrate-reducing activities of *Amaranthus* NR (5), suggesting a role for a histidyl residue(s) in the NADH-binding domain. The presence of NADH resulted in protection from inactivation by diethylpyrocarbonate.

Spectroscopic Properties

Spectroscopy—visible, CD, and EPR—has been used extensively to define the properties of the FAD, cytochrome b_{557}, and Mo-pterin prosthetic groups. The visible spectrum of the oxidized enzyme is dominated by the heme absorbance, with Soret and α bands at 413 nm and 550 nm, respectively, which effectively obscure any contributions from flavin and Mo-pterin. Utilization of the small fragment obtained following limited proteolysis of the *Chlorella* enzyme, which contains only the FAD prosthetic group, has enabled the flavin absorbance bands at 460 nm and 487 nm to be resolved (69). Reduction with NADH or dithionite resulted in a bleaching of the flavin absorbance. In contrast to the visible spectrum, contributions from FAD could be discerned in the CD spectrum of the oxidized holoenzyme at 311 nm and 375 nm (positive CD) and at 460 nm and 487 nm (negative CD); both contributions are lost following reduction. The formation of an anionic flavin semiquinone radical (FAD$^{\cdot-}$) following partial reduction of the *Chlorella* enzyme with NADH has been demonstrated by low-temperature EPR spectroscopy (106).

CD and EPR spectroscopies have been used to characterize the Mo-pterin center of both *Chlorella* and spinach nitrate reductases, with changes at 333 nm in the CD spectrum ascribed to reduction of this prosthetic group (12, 69). The formation of a number of different Mo(V) species following partial reduction of both *Chlorella* (68, 106) and spinach (11, 51) enzymes has been demonstrated by EPR spectroscopy. At low pH, the Mo(V) EPR spectrum was a composite of two paramagnetic species of identical g-value ($g_{av} = 1.977$) but distinguishable by the presence or absence of superhyperfine interaction ($A_{av}^{H} = 1.35$ mT) as a result of coupling of a single exchangeable proton. The pK_a of this exchangeable proton was observed to be influenced by the presence of anions such as chloride or nitrite that were found to increase the proportion of the signal exhibiting superhyperfine interaction by increasing the pK_a. Phosphate did not alter the relative signal proportions. Enzyme samples poised at alkaline pH (pH 9) have indicated the formation of a single Mo(V) "high-pH" species of axial symmetry ($g_{av} = 1.961$) that showed no superhyperfine interaction (106). Recent studies (C. J. Kay, M. J. Barber, and L. P. Solomonson, unpublished) have indicated that this may result from the nature of the buffer utilized and may not represent a distinct "high-pH" species.

Thermodynamic Properties

Oxidation-reduction midpoint potentials have been determined for the FAD, b_{557}, and Mo-pterin prosthetic groups of *Chlorella* and spinach nitrate reductases using various techniques including visible, CD, and EPR potentiometric titrations and microcoulometry (14, 69, 70, 106, 117). The visible spectrum of the heme chromophore dominates the optical spectrum and masks absorbance changes from either FAD or Mo-pterin, effectively precluding optical titration of these prosthetic groups in the holoenzyme. However, optical titrations of the proteolytically generated FAD domain showed a reversible $n=2$ redox process with a midpoint potential of -288 mV that was shifted approximately 60 mV more positive in the presence of NAD^+. This value was confirmed by CD spectroscopy (-272 mV) and microcoulometry (-283 mV) (69, 117). Utilizing estimates of the amount of $FAD^{\cdot-}$ formed during the titrations ($<1\%$), the values for the individual flavin couples were calculated to be -372 mV for $FAD/FAD^{\cdot-}$ and -172 mV for $FAD^{\cdot-}/FADH_2$. Similar behavior has been shown for the flavin center of spinach NR, CD titrations yielding values of -380 mV and -180 mV for the $FAD/FAD^{\cdot-}$ and $FAD^{\cdot-}/FADH_2$ couples, respectively (9). Preliminary work on the pH dependence of the midpoint potentials of the flavin center (C. J. Kay, M. J. Barber, and L. P. Solomonson, unpublished) indicate that over the pH range 6–10, the potential of the $FAD/FADH_2$ couple decreases by approximately -30 mV/pH unit.

The midpoint potential for the heme prosthetic group of *Chlorella* NR has been shown to be -164 mV and to exhibit a pH dependence of approximately -20 mV/pH unit within the range 5.5–7. This behavior indicated the presence of a single, redox-associated, ionizable functional group in the protein with $pK_{ox}=5.8$ and $pK_{red}=6.2$. At pH 7 and within the range 12–38°C, the midpoint potential of the heme decreased by approximately 1 mV per degree. Values for $\Delta S°$ and $\Delta H°$ were calculated to be -25.6 eu and -4.0 kcal/mol (70). The midpoint potential for the heme in spinach NR has been determined to be -123 mV, significantly lower than the value originally reported (9, 38) but comparable to that obtained for the *Chlorella* enzyme. Data obtained for the midpoint potential of the heme prosthetic group of *Candida nitratophila* NR yielded a value of -174 mV, similar to that of the *Chlorella* enzyme (9, 58, 70). However, the midpoint potential of the heme prosthetic group of *Ankistrodesmus* NR has been determined to be -73 mV, suggesting possible differences in the structure of this center in enzymes from different sources (31).

Both room temperature and low temperature EPR potentiometric tritrations have been utilized to determine the midpoint potentials of the Mo-pterin prosthetic group for *Chlorella* and spinach nitrate reductases (9, 69). Room temperature EPR potentiometry (pH 7.0, 25°C) yielded midpoint potentials of

+15 mV and −25 mV for the Mo(VI)/Mo(V) and Mo(V)/Mo(IV) redox couples, respectively, and +2 mV and −6 mV for the spinach couples. The similarity in the redox potentials of the Mo centers for the algal and plant enzymes suggests similar structure for the Mo-pterin prosthetic groups. Similar values for the Mo(VI)/Mo(V) and Mo(V)/Mo(IV) redox couples of the *Chlorella* enzyme have been obtained from CD potentiometric titrations (+26 mV and −40 mV, respectively) and microcoulometry (+16 mV and −27 mV, respectively), confirming these measurements. However, these values are slightly higher than the Mo-pterin midpoint potentials first obtained using low-temperature EPR potentiometric titrations (Mo(VI)/Mo(V)=−34 mV and Mo(V)/Mo(IV)=−54 mV) for the *Chlorella* enzyme, which indicates that there is some redistribution of reducing equivalents within the enzyme subunits upon freezing that particularly affects the measured potential of the Mo(VI)/Mo(V) couple. Initial pH-dependence studies of the potentials of the Mo-pterin center in *Chlorella* NR indicate that the Mo(VI)/Mo(V) and Mo(V)/Mo(IV) couples shift approximately −60 mV/pH unit over the pH range 6–9 (C. J. Kay, M. J. Barber, and L. P. Solomonson, unpublished).

Kinetic Properties

Various steady-state kinetic mechanisms have been reported for nitrate reductases from different sources. Herrero et al (56) reported an iso ping-pong bi-bi kinetic mechanism for *Ankistrodesmus braunii* NR based on initial velocity and product inhibition patterns. Similarly, Campbell & Smarrelli (20) proposed a two-site ping-pong mechanism for squash and corn nitrate reductases. In contrast, McDonald & Coddington (83) proposed a random-order rapid-equilibrium mechanism for the enzyme from *Aspergillus nidulans*. Howard & Solomonson (59) also reported a rapid-equilibrium random bi-bi kinetic mechanism for *Chlorella* NR based on initial velocity, product inhibition, and dead-end inhibitor patterns. These reports may suggest fundamental differences in kinetic mechanisms for nitrate reductases from different sources. However, there are considerations that might explain these apparent differences. It is noteworthy that, where studied, the product inhibition patterns for all eukaryotic nitrate reductases have been the same. The apparent differences have been in the initial velocity patterns, which have in some cases been parallel (20, 33, 59), suggesting ping-pong kinetics, and in other cases have been intersecting (59, 83), indicating sequential kinetics. Parallel initial velocity patterns suggest, but are not sufficient evidence for, ping-pong mechanism. The apparent patterns will be dictated by the relative magnitudes of the individual dissociation constants comprising the kinetic rate expression. Intersecting initial velocity patterns, on the other hand, are not consistent with a ping-pong mechanism but rather indicate a sequential mechanism. In addition, NR catalyzes an essentially irreversible reaction ($K_{eq} = 10^{40}$) so the products (nitrite and NAD^+) should behave as dead-end inhibitors, and product inhibition patterns established for reversible bi-substrate reactions would therefore not be applicable. Nonproduct dead-end inhibitors could be used to distinguish between possible kinetic mechanisms for NR but have been used only for the *Chlorella* enzyme (59).

Kay & Barber (67) have determined the kinetic parameters for the various partial activities of *Chlorella* NR and the effect of ionic strength on these parameters. While FH:NR and NADH:NR had the same V_{max}, all other partial activities had significantly higher V_{max} values. In addition, FH:NR and NADH:NR were the only activities stimulated by increased ionic strength. These results, together with the effect of ionic strength on the steady-state reduction of the heme prosthetic group of NR during turnover, suggested that the transfer of electrons from heme to Mo is the rate-limiting step in the transfer of electrons between NADH and nitrate in the *Chlorella* enzyme.

Similar steady-state studies of the effects of ionic strength and pH on the full and partial activities of spinach NR (10) have indicated a number of differences between the *Chlorella* and spinach enzymes. For the spinach enzyme, all partial activities were faster than the full NADH:NR activity. In addition, increased ionic strength inhibited NADH:NR activity. These results have indicated that, in contrast to the *Chlorella* enzyme, intramolecular electron transfer is not rate limiting for spinach NR.

Phosphate has long been known to enhance NR activity (86), although a part of this stimulatory effect may have resulted from ionic strength augmentation in cases where ionic strength was not adequately controlled (71, 86, 92). Phosphate stimulation is specific for nitrate-reducing activities and is likely the result of interactions of phosphate with the Mo-pterin center (59, 67, 106). However, EPR analysis of the direct binding of phosphate to Mo, using ^{17}O-substituted phosphate, has failed to detect any ^{17}O superhyperfine interaction. In addition, phosphate did not alter the redox potentials of the Mo(VI)/Mo(V) or Mo(V)/Mo(IV) couples, suggesting that phosphate binds with equal affinity to all three redox states of Mo. Steady-state kinetic analysis has indicated that phosphate functions as a weak non-essential activator effecting stimulation by a weakening of nitrate-binding (68).

Halides have also been shown to influence NR activity for both the *Chlorella* (68) and spinach (11) enzymes. Under controlled conditions of ionic strength, the presence of halides resulted in inhibition of NADH:NR and MV:NR activity by formation of ternary dead-end complexes with NR. This inhibitory effect is also associated with the nitrate-reducing moiety.

REGULATION

The regulation of nitrate assimilation has been the focus of intense research activity because of the potential for improving the efficiency of the process and enhancing agricultural productivity through intervention in this growth-limiting process. Substantial understanding has been achieved of how nitrate assimilation, and particularly NR, can be regulated. However, a complete understanding of how NR is regulated under various environmental conditions is still lacking. It is likely that there is no single mode of regulation of the enzyme; rather, several different modes of regulation may act simultaneously or sequentially in response to different signals or changing environmental conditions. Some of the modes of regulation that serve as potential controls of NR activity are illustrated in Figure 4. Cellular activity of NR can be regulated at the levels of enzyme synthesis and degradation, reversible inactivation, concentration of substrates and effectors, and intracellular sequestration of enzyme. Factors that can influence activity directly or serve as signals for the various modes of regulation include light, oxygen, carbon dioxide, nitrate, fixed nitrogen, and other metabolites.

We will attempt to summarize our current understanding of various aspects

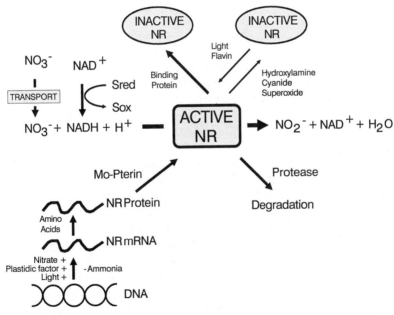

Figure 4 Modes of regulation of assimilatory nitrate reductase. The various potential models of regulation of assimilatory nitrate reductase indicated include regulation of enzyme synthesis by positive effectors (such as nitrate, plastidic factor, and light) and by a negative effector derived from ammonia; incorporation of Mo-pterin as a rate-limiting step for the full expression of NR activity; reversible inactivation of NR by combination with hydroxylamine, cyanide, or superoxide and reversal of this inactivation by blue light in the presence of flavin; inactivation by combination with specific binding proteins or by limited proteolysis; and availability of the substrate nitrate via transport and the substrate NADH formed via intracellular metabolism.

of the regulation of NR, including localization of the enzyme, induction/repression, and enzyme activity modulation.

Localization

An important consideration in the regulation of NR is the intracellular location of the enzyme relative to organelles or enzymes that may be involved in its processing or regulation. Surprisingly, the intracellular location of NR has not yet been firmly established some 35+ years after its initial description and characterization by Nason and coworkers (36). Several features of the enzyme and the reactions it catalyzes should be taken into account in considering a "functional location" for the enzyme. First, the physiological electron donor for the major form of NR in green (i.e. photosynthetic) cells is NADH, produced in the cytosol, and not NADPH, which is produced in the chloroplast. Second, the product, nitrite, is toxic at high levels, which necessitates a mechanism for efficient transfer to the second enzyme in the pathway, nitrite reductase, which is clearly located in the chloroplast. Third, there is an apparent coupling or close association between nitrate uptake and nitrate reduction (62).

Early localization studies involved cell disruption followed by density gradient centrifugation to separate the various organelles and membrane fractions. Most of these earlier studies suggested a cytosolic location for NR; nitrite reductase was localized in the chloroplast. Some reports, however, using the same general approach, suggested that NR is associated with the chloroplast or with microbodies (49, 77). Under the conditions used for these studies it is possible that a loosely bound enzyme could be dissociated from its original site and/or associate with another organelle. The availability of monospecific antibodies to NR has made possible the localization of NR by immunoelectron microscopy techniques that avoid possible artifacts arising from cell disruption and fractionation techniques. This technique, however, yielded conflicting results, with some reporting a cytosolic location, some a chloroplastic location, and others a plasmalemma location. Our own experience (V. Kriho, A. Michaels, and L. P. Solomonson, unpublished) suggested that results obtained by this technique could vary widely depending on the immunolocalization conditions used and the nature of the antibody. Despite these reservations, this should be the method of choice provided adequate controls are utilized, the antibody preparation is of high titer and specificity, and the results are reproducible.

Immunocytochemical localization studies using conventional techniques and utilizing immunopurified, and apparently highly specific, anti-NR preparations have given widely disparate results. Vaughn et al (125) used immunofluorescence to localize NR in soybean cotyledons and found intense fluorescence in the cytoplasm and weaker reactions associated with organelles tentatively identified as plastids. Roldan et al (100) found by immunoelectron microscopy that NR in mycelial cells of *Neurospora crassa* is located in the cell wall–plasmalemma region and in the tonoplast membranes. Lopez-Ruiz et al (78) used similar techniques and reported that the NR of the green alga *Monoraphidium braunii* is specifically located in the pyrenoid region of the chloroplast, a region generally associated with starch synthesis. Immuno-gold labeling has been employed to localize NR in leaves of two different higher-plant species and has produced two different results. Kamachi et al (64) reported that NR of spinach leaves is specifically located in the chloroplast, while Vaughn & Campbell (124) reported that NR in corn leaves is located exclusively in the cytoplasm of mesophyll cells. It is unlikely that the intracellular location of NR varies in different species, although these different results may reflect alternate locations under different physiological states or different isoforms of the enzyme. However, NR may also share common epitopes with more abundant proteins, which could produce erroneous conclusions. For example, Gowri & Campbell (46) have recently reported that NR and chloroplast $NAD(P)^+$:glyceraldehyde-3-phosphate dehydrogenase may have an epitope in common, because their anti-NR preparation cross-reacts with the latter enzyme. In addition, sequence information for NR has indicated significant sequence similarities in the FAD, heme, and Mo-pterin domains with other enzymes that contain analogous domains (28, 45). As noted by Oaks et al (90), Western blotting is insufficient to demonstrate monospecificity of anti-NR preparations because the subunit size of a major protein in green cells, phosphoenolpyruvate carboxylase, also has a subunit molecular mass of about 100 kd. Fischer & Klein (40) have recently carried out a careful study of the localization of nitrogen-assimilating enzymes in the unicellular green alga *Chlamydomonas reinhardtii* and could detect no NR activity in chloroplasts, concluding that NR is a cytoplasmic enzyme. Photo-oxidative treatment of the plastids of carotenoid-free mustard cotyledons confirmed the cytosolic location of NR, but a plastidic signal appeared to be required for the synthesis of NR (91). These conclusions agree with the majority of investigations on the compartmentation of NR in plant cells.

Although most evidence points to a cytoplasmic location for NR, recent evidence from Huffaker's and Tischner's laboratories suggests there is a plasma membrane–bound form of NR that is involved in nitrate uptake. Ward et al (129, 130) reported that a latent NR is associated with the plasma membrane of corn roots and barley seedlings. Anti-*Chlorella* NR IgG fragments inhibited both the soluble NR activity and the plasma membrane–bound NR activity of barley roots (130). The anti-NR IgG fragments also specifically inhibited nitrate transport, raising the interesting possibility that this plasma membrane–bound form of NR may function as a nitrate transporter. Tischner et al (118) reported similar results for the plasma membrane–bound NR of

Chlorella sorokiniana and further showed that the subunit sizes of the plasma membrane and soluble NR were 60 and 95 kd, respectively. The plasma membrane–bound form of NR accounts for only 1–4% of the total NR activity under normal conditions, which could possibly be attributed to contamination of the membrane fractions with soluble enzyme. In addition to having a different subunit size, the plasma membrane–bound form also appears to be under a different regulatory control than the soluble form. For example, it is detected in *Chlorella* cells grown in the presence of ammonia, while the soluble form was completely repressed. These initial findings will require further study and confirmation, but they may well lead to an improved understanding of the molecular properties of nitrate uptake and its possible relationship to nitrate reduction.

Induction/Repression

Nitrate reductase and associated partial activities are largely absent in cells that utilize ammonia rather than nitrate as the source of nitrogen. This absence of NR activities was assumed to be paralleled by loss of NR protein, but this could not be demonstrated conclusively until recently using immunochemical techniques.

Nitrate has been considered to be an inducer of NR synthesis while ammonia, or a product of ammonia metabolism, was considered to act as a repressor of NR synthesis. In several species, however, nitrate does not appear to be required for the synthesis of NR protein; rather, derepression, or removal of ammonia, is sufficient for NR synthesis. This appears to be the case for green algae and yeast (21, 22, 132, 133). The case is somewhat more complex in higher plants owing to the presence of storage vacuoles that accumulate nitrate, but enzyme synthesis occurs at approximately equal rates in the presence of nitrate alone or of equimolar concentrations of ammonia and nitrate, which suggests a true "induction" by nitrate in higher plants (76, 81, 90, 99). Oaks et al (90) showed, however, that the induction of inactive NR protein in corn occurs in response to light in the presence of very low levels of nitrate, while full expression of active enzyme required higher levels of nitrate. Studies in which the kinetics of the appearance of NR protein and the expression of NR activity were followed during "induction" showed that NR protein generally appears more rapidly than NR activity. This has been shown for corn leaf NR (90) and for *Chlorella* NR (133). It was also shown that the partial activity, NADH:CR, associated with *Chlorella* NR was "induced" faster than the nitrate-reducing activity, suggesting that incorporation of the Mo-pterin cofactor may be rate limiting for the expression of full NR activity (132, 133). Consistent with this hypothesis was the accompanying 4–6-fold increase in Mo-pterin levels during "induction" (132). Zeiler (132) showed that nitrate-reducing activity could be increased to near normal levels if samples taken during the initial lag period are exposed to exogenous, biologically active Mo-pterin cofactor. Thus, availability of Mo-pterin cofactor rather than Mo alone, or a protein factor, appears to be the limiting factor for the expression of full NR activity.

Exposure of fully "induced" *Chlorella* cells to ammonia resulted in an apparently more rapid loss of nitrate-reducing activities than NADH-dependent partial activities and immunoreactive protein resulting from the conversion of NR to the reversibly inactive form (133). When activities were measured following full activation, all activities and immunoreactive protein decayed at the same rate (133). Remmler & Campbell (99) observed a similar reversible drop in NR activity after transfer of fully induced corn plants to the dark. After this initial activity drop, NR activity and protein decayed at the same rate.

Recent reports by Mohr and coworkers (91, 102) indicate that a "plastidic signal" is required for nitrate-induced appearance of nitrate and nitrite reductases. Phytochrome appears to mediate a stimulation of NR expression by light (84, 98). Oelmuller et al (91) concluded that the action of nitrate and phytochrome on enzyme appearance depends on the plastidic factor. Although nitrate reductase synthesis is generally repressed in darkness, when the green alga *Monoraphidium braunii* was grown under heterotrophic conditions with glucose as the carbon and energy source, synthesis of NR was maintained in the dark at near-normal levels. NR protein, in fact, appeared to be elevated under these conditions (32).

Regulation of NR synthesis by plant hormones has not been extensively investigated. Synthesis of higher plant NR has been reported to be stimulated by cytokinins (51). Schmerder & Borriss (101) showed that NR activity in *Agrostemma githago* was also induced by ethylene and that induction by cytokinins only occurred in the presence of ethylene. They speculated that cytokinins and nitrate regulate NR activity independently.

The bulk of NR activity in photosynthetic tissue appears to be under metabolic control—that is, induced in response to nitrate or absence of ammonia and repressed by ammonia or ammonia metabolites. Three distinct isoforms have been identified in soybeans; a bispecific NAD(P)H:NR with a pH optimum of 6.5, an NADH:NR with a pH optimum of 6.5, and a more typical higher-plant NR that is an inducible NADH:NR with maximal activity at pH 7.5. The pH 6.5–linked activities were less sensitive to nitrate flux than the pH 7.5–linked activity (29). The physiological role(s) of these apparently constitutive forms is not known, but Dean & Harper (30) reported that the constitutive NAD(P)H:NR is responsible for the nitric oxide and nitrogen dioxide evolution activity extracted from soybean leaflets. Constitutive NR isoforms have also been found in corn, barley, wheat, oats, and mustard (54, 55, 102).

The availability of cDNAs for several nitrate reductases has permitted an examination of regulation at the transcriptional level. Cheng et al (23) initially demonstrated that nitrate induction of barley seedlings increased NR protein and translatable mRNA as well as the steady-state level of NR mRNA. Crawford et al (27) showed that NR mRNA was 120 times as abundant in nitrate-induced squash cotyledons than in uninduced tissue. Nitrate-reductase activity and protein were also detected only in trace amounts in uninduced squash tissue compared to induced tissue, suggesting that nitrate induction of NR occurs primarily at the RNA level.

Other workers have carried out more-detailed studies of the effects of nitrate and light on NR mRNA accumulation and expression of NR activity. It has been shown for tomato, tobacco, barley, and rice that NR mRNA accumulates rapidly followed by a decline during which time NR activity increases (43, 53, 84). Galangau et al (43) showed that during a light-dark cycle, both tomato and tobacco NR mRNA increased rapidly before the end of the dark period and reached a maximum at the beginning of the light period, while NR protein peaked 2–4 hr after mRNA peaked. Melzer et al (84) studied the effects of nitrate and light on NR mRNA accumulation in barley roots and leaves. NR mRNA, which was low or not detected in the absence of nitrate in both roots and leaves, accumulated at a faster rate in roots than in leaves and also declined more rapidly to a steady-state level. In both roots and leaves the decline in NR mRNA occurred while NR activity was increasing. Light enhanced NR mRNA accumulation. Green seedlings did not respond to red, far-red, or blue light; but accumulation of NR mRNA was stimulated by white light. In contrast, NR mRNA in etiolated seedlings increased 20-fold in response to red and blue light, indicating that phytochrome or another photo-receptor may facilitate induction of NR transcription by nitrate in etiolated seedlings but not in green leaves. Gowri & Campbell (46) also showed that nitrate induces NR mRNA formation in the dark in both etiolated and green corn leaves. The maximum levels reached in the dark were about 40% and 60% of the maximum levels reached in the light for etiolated and green leaves, respectively. NR activity was also expressed under these conditions but at lower levels. Maximum activity levels reached in the dark were about 25% and 45% of the maximum levels reached in the light for etiolated and green leaves, respectively. Thus, neither light nor functional chloroplasts are required for expression of NR activity, but light and functional chloroplasts markedly enhance the nitrate-induced expression of NR.

Activity Modulation

Nitrate reductase activity may be modulated by several potential effectors in response to particular environmental conditions leading in some cases to a reversible inactivation of the enzyme when it reacts with molecules such as cyanide, hydroxylamine, or superoxide radicals under reducing conditions. Reactivation may be mediated by exposure to an oxidant system such as blue light, flavin, or oxygen. Other potential effectors include inactivator proteins that appear to bind specifically to NR and proteases that appear to have a high specificity for NR. The possible physiological importance of these various modulators of NR activity has not been established, but it is likely that one or more of these modes of regulation may be important for the short-term regulation of NR activity. Other products or intermediates of nitrate assimilation such as nitrate, ammonia, or amino acids are unlikely to play a direct role in the short-term regulation of NR. Nitrite is a competitive inhibitor of NR; but the K_i is high relative to the K_m for nitrate, so it is unlikely to play a significant role as an effector in vivo. However, Kramer et al (73) have reported that the chloroplast envelope contains a nitrite-binding site that may regulate the availability of nitrite within the chloroplast, thereby serving as an additional regulatory site in nitrate assimilation. Ammonia and amino acids have little or no effect on NR in vitro.

The availability of the substrates NADH and nitrate would be an important determinant of the rate of nitrate assimilation. Reactions that generate NADH are important in this regard, as are the processes that maintain intracellular levels of nitrate, such as the nitrate uptake system and mobilization of nitrate from storage vacuoles. Although substrate availability is of potential importance to the overall regulation of nitrate assimilation, the processes that govern the maintenance of celluar nitrate and NADH levels are not considered here.

REVERSIBLE INACTIVATION   Losada et al (80) initially observed a reversible inactivation of NR in *Chlorella* cells in response to addition of ammonia to the algal culture. A rapid and complete reactivation occurred on subsequent removal of ammonia. Vennesland & Jetchmann (126) also observed an inactive form of NR in crude extracts of *Chlorella* that was slowly converted to an active form by incubation with nitrate. In this and other respects, the inactive form of the enzyme resembled enzyme that had been inactivated by treatment with NADH and low levels of cyanide in vitro (115). The inactive enzyme could be rapidly activated in vivo by exposure to nitrate or in vitro by treatment with low concentrations of ferricyanide (63, 104, 113). The inactive NR-cyanide complex is quite stable in the absence of NADH and cyanide; it has a dissociation constant of about 10^{-10} M, which is in the same range as that of hormone-receptor complexes (79). Only the Mo-dependent nitrate-reducing activities are affected (104). NADH-dependent partial activities remain fully active following inactivation of the enzyme with NADH and cyanide. Solomonson (104) suggested that cyanide forms a complex with a reduced form of molybdenum, thereby preventing further catalytic turnover of the enzyme. EPR evidence also suggested that cyanide reacts with the Mo-pterin center (106). After conversion of NR to the inactive form in vivo by brief treatment of *Chlorella* cells with ammonia, the inactive form of NR was isolated, purified, and shown to contain a stoichiometric amount of bound cyanide. This cyanide was released upon conversion of NR to the active form, suggesting its possible physiological role (79). Pistorius et al (95) showed that light and oxygen were required for the efficient conversion of NR to the inactive form, which suggested a connection between cyanide generation in vivo and photorespiration. Several reports have suggested the occurrence of an inactive form of NR in other photosynthetic organisms and higer-plant sources (3, 17, 34, 120, 128). To date, however, an inactive NR-cyanide complex has not been isolated from a higher-plant source. The presence of nitrate stores in higher-plant cells, which could be released upon cell disrup-tion and cause an activation of the enzyme, may preclude such a demonstra-tion.

The amount of free cyanide present in *Chlorella* cells and in a variety of higher plants would be adequate for the inactivation of NADH-reduced NR (44). There are a number of potential sources for the generation of cyanide in vivo other than the cyanogenic glycosides. Vennesland and coworkers (96) showed that cyanide can be produced from amino acids such as D- or L-histidine through the action of D- or L-amino acid oxidase, respectively. Cyanide is also a product of the last step of the ethylene biosynthetic pathway and would therefore be formed concurrently with ethylene (94). Solomonson & Spehar (112–114) proposed a model for the regulation of Nr by cyanide, hypothesizing an enzyme that catalyzed the formation of cyanide from glyoxylate oxime. This enzyme was subsequently found in *Chlorella* and spinach (112–114). It has a molecular mass of about 40 kd, requires Mn^{2+}, and has an absolute requirement for ADP or a combination of ADP and ATP (113, 114). Hucklesby et al (61) confirmed these results and showed that the enzyme is also present in corn and barley leaves.

Other potential effectors, with activities similar to cyanide's, include super-oxide radicals and hydroxylamine (7, 8, 121). The inactivation of NR with these reagents is not as readily reversible as for the inactivation by cyanide. Blue light is an effective activating agent both in vitro and in vivo and may play an important role in the photoregulation of nitrate utilization in green algae and higher plants (2, 4, 65, 66). Aparicio et al (2) were the first to observe a photoactivation of NR by blue light. This photoregulation of NR activity may be of physiological importance and may account for the higher protein content in plants grown under blue light (4, 72, 127). In vitro, blue light reactivates NR inactivated by cyanide, hydroxylamine, or superoxide (8, 34, 41, 82). This photoreactivation was enhanced by flavins. The effect of flavins was greater under anaerobic conditions (41). Fritz & Ninneman (41) showed that triplet flavin is the species responsible for the reactivation. Oxygen would act as a quencher of triplet flavin, generating singlet oxygen in the process, which could lead to a photodestruction of NR. An irreversible photoinactivation of NR occurs upon irradiation of the enzyme with blue light in the presence of FMN and oxygen (121). Singlet oxygen appeared to be the primary agent involved in the photoinactivation of the enzyme.

In summary, it is likely that reversible inactivation of NR plays an important role in its regulation and that blue light stimulates the process of nitrate assimilation through a rapid activation of the inactive form of NR. In contrast, the absorption of red light by phytochrome would stimulate nitrate assimilation through the enhancement of the rate of NR synthesis (84, 98). The nature of the inactive form of NR has not been fully elucidated, although an inactive NR-cyanide complex has been isolated from *Chlorella* cells. However, hydroxylamine, superoxide, or possibly other metabolites could also play an important role in the formation of the reversible inactive form of NR in vivo.

INACTIVATOR PROTEINS   Inactivator proteins that act through either binding or limited proteolysis, could also play an important role in the regulation of NR. Endopeptidases that specifically inactivate and are correlated with reduced tissue levels of NR have been isolated from corn roots and from barley leaves (50, 52). Corn root endopeptidase cleaves NR at only one site, leading to loss of NADH-associated activities, and was initially thought to act only on NR (97, 108). It was subsequently shown, however, that the specificity for NR is not absolute (15). Therefore, the contribution of this enzyme to the regulation of NR levels is not certain. Protein-type inhibitors that do not appear to degrade NR have been isolated from rice seedlings, soybean, wheat, and spinach leaves (51, 131). The inactivator proteins isolated from rice and from spinach have been shown to bind stoichiometrically and with high affinity to NR, a finding consistent with a role for these proteins as specific inactivators of NR. Although NADH protected NR against inactivation by either the rice or the spinach inactivator proteins, efforts to reverse the inactivation were unsuccessful (108, 131).

CONCLUSIONS AND FUTURE DIRECTIONS

The general features of NR and its regulation are well understood. For example, the quaternary structure, prosthetic group composition, midpoint potentials, steady-state kinetics, and identity of effectors that act at the transcriptional level or as activity modulators have been established. Significant gaps still remain in our knowledge regarding intracellular location, linkages between transport and reduction, coupling between nitrate and nitrite reduction, details of the catalytic mechanism, functional groups involved in binding and catalysis, determinants of enzyme stability, and the three-dimensional structure of the enzyme. Although our understanding is still far from complete, we are moving with increasing speed to a stage where we can consider modifying regulatory and catalytic properties of the enzyme through genetic engineering. The application of these and other molecular biological techniques should lead to exciting new discoveries concerning NR structure, function, and regulation that may provide a means for improving the efficiency of nitrate utilization and hence agricultural productivity.

ACKNOWLEDGMENTS

We thank Ms. Shirley Lundy for her skillful assistance in the preparation of this review and gratefully acknowledge the participation of current and past associates in work from the authors' laboratories described here, especially Dr. C. J. Kay, and fruitful collaborations with other investigators, particularly Drs. D. C. Eichler, M. J. McCreery, P. J., Neame, B. A. Notton, A. Oaks, K. V. Rajagopalan, and J. T. Spence. Research from the authors' laboratories was supported by grants from NIH(GM32696), NSF(DMB8214001 and DCB8615836), USDA(GAM8400528 and 88-37120-3871), and NATO(86-0015).

Literature Cited

1. Amy, N. K., Garrett, R. H., Anderson, B. M. 1977. Reaction of the *Neurospora crassa* nitrate reductase with NAD(P) analogs. *Biochim. Biophys. Acta.* 480:83–95

2. Aparicio, P. J., Roldan, J. M., Calero, F. 1976. Blue light photoreactivation of nitrate reductase from green algae and higher plants. *Biochem. Biophys. Res. Commun.* 70:1071–77

3. Aryan, A. P., Batt, R. G., Wallace, W. 1983. Reversible inactivation by NADH and the occurrence of partially inactive enzyme in the wheat leaf. *Plant Physiol.* 71:582–87

4. Azuara, M. P., Aparicio, P. J. 1983. *In Vivo* blue-light activation of *Chlamydomonas reinhardii* nitrate reductase. *Plant Physiol.* 71:86–90

5. Baijal, M., Sane, P. V. 1987. Chemical modification of nitrate reductase from *Amaranthus. Ind. J. Biochem. Biophys.* 24:75–79

6. Baijal, M., Sane, P. V. 1988. Arginine residue(s) at the active site(s) of the nitrate reductase complex from *Amaranthus. Phytochemistry* 27:1969–72

7. Balandin, T., Aparicio, P. J. 1987. Hydroxylamine metabolism in *Monor-*

aphidium braunii. II. Its interference with the utilization of NO_3^-. *New Phytol.* 107:523–30

8. Balandin, T., Fernandez, V. M., Aparicio, P. J. 1986. Characterization of the reversible inactivation of *Ankistrodesmus braunii* nitrate reductase by hydroxylamine. *Plant Physiol.* 82:65–70

9. Barber, M. J., Kay, C. J., Notton, B. A., Solomonson, L. P. 1989. Oxidation-reduction midpoint potentials of the flavin, heme and Mo-pterin centers in spinach nitrate reductase. *Biochem. J.* 263:285–87

10. Barber, M. J., Notton, B. A. 1989. Spinach nitrate reductase: effect of ionic strength and pH on the full and partial enzyme activities. *Plant Physiol.* In press

11. Barber, M. J., Notton, B. A., Kay, C. J., Solomonson, L. P. 1989. Chloride inhibition of spinach nitrate reductase. *Plant Physiol.* 90:70–74

12. Barber, M. J., Notton, B. A., Solomonson, L. P. 1987. Oxidation-reduction midpoint potentials of the molybdenum center in spinach NADH:nitrate reductase. *FEBS Lett.* 213:372–74

13. Barber, M. J., Solomonson, L. P. 1986. The role of the essential sulfhydryl

group in assimilatory NADH:nitrate reductase of *Chlorella*. *J. Biol. Chem.* 261:4562–67

14. Barber, M. J., Solomonson, L. P. 1986. Properties of the molybdenum domain of nitrate reductase. *Polyhedron* 5:577–80

15. Batt, R., Wallace, W. 1989. Characteristics of the active site and substrate specificity of a maize root endopeptidase. *Biochim. Biophys. Acta* 990:109–12

16. Beevers, L., Hageman, R. H. 1969. Nitrate reduction in higher plants. *Annu. Rev. Plant Physiol.* 20:495–522

17. Benzioni, A., Heimer, Y. M. 1977. Temperature effect on nitrate reductase activity *in vivo. Plant Sci. Lett.* 9:225–31

18. Calza, R., Huttner, E., Vincentz, M., Rouze, P., Galangau, F., et al. 1987. Cloning of DNA fragments complementary to tobacco nitrate reductase mRNA and encoding epitopes common to the nitrate reductases from higher plants. *Mol. Gen. Genet.* 209:552–62

19. Campbell, J. McA., Wray, J. L. 1983. Purification of barley nitrate reductase and demonstration of nicked subunits. *Phytochemistry* 22:2375–82

20. Campbell, W. H., Smarrelli, J. Jr. 1978. Purification and kinetics of higher plant NADH:nitrate reductase. *Plant Physiol.* 61:611–16

21. Cannons, A. C., Ali, A. H., Hipkin, C. R. 1986. Regulation of nitrate reductase in the yeast *Candida nitratophila*. *J. Gen. Microbiol.* 132:2005–11

22. Cannons, A. C., Hipkin, C. R. 1987. Evidence for the transcriptional control of nitrate reductase in *Candida nitratophila* from *in vitro* translation studies. *Eur. J. Biochem.* 164:383–87

23. Cheng, C.-L., Dewdney, J., Kleinhofs, A., Goodman, H. M. 1986. Cloning and nitrate induction of nitrate reductase mRNA. *Proc. Natl. Acad. Sci. USA* 83:6825–28

24. Cheng, C., Dewdney, J., Nam, H., deBoer, B. G. W., Goodman, H. M. 1988. A new locus (NIA 1) in *Arabidopsis thalina* encoding nitrate reductase. *EMBO J.* 7:3309–14

25. Cherel, I., Marion-Poll, A., Meyer, C., Rouze, P. 1986. Immunological comparisons of nitrate reductases of different plant species using monoclonal antibodies. *Plant Physiol.* 81:376–78

26. Commere, B., Cherel, I., Kronenberger, J., Galangan, F., Caboche, M. 1986. In vitro translation of nitrate reductase messenger RNA from maize and tobacco and detection with an antibody directed against the enzyme of maize. *Plant Science* 44:191–203

27. Crawford, N. M., Campbell, W. H., Davis, R. W. 1986. Nitrate reductase from squash: cDNA cloning and nitrate regulation. *Proc. Natl. Acad. Sci. USA* 83:8073–76

28. Crawford, N. M., Smith, M., Bellissimo, D., Davis, R. W. 1988. Sequencing and nitrate regulation of the *Arabidopsis thaliana* mRNA encoding nitrate reductase, a metalloflavoprotein with three functional domains. *Proc. Natl. Acad. Sci. USA* 85:5006–10

29. Curtis, L. T., Smarrelli, J. Jr. 1986. Metabolic control of nitrate reductase activity in cultured soybean cells. *J. Plant Physiol.* 127:31–39

30. Dean, J. V., Harper, J. E. 1988. The conversion of nitrite to nitrogen oxide(s) by the constitutive NAD(P)H-nitrate reductase enzyme from soybean. *Plant Physiol.* 88:389–95

31. De la Rosa, M. A., Vega, J. M., Zumft, W. G. 1981. Composition and structure of assimilatory nitrate reductase from *Ankistrodesmus braunii. J. Biol. Chem.* 256:5814–19

32. Diez, J., Lopez-Ruiz, A. 1989. Immunological approach to the regulation of nitrate reductase in *Monoraphidium braunii. Arch Biochem. Biophys.* 268:707–15

33. Eaglesham, A. R. J., Hewitt, E. J. 1975. Inhibition of nitrate reductase from spinach leaf by adenosine nucleotides. *Plant Cell Physiol.* 16:1137–49

34. Echevarria, C., Maurino, S. G., Maldonado, J. M. 1984. Reversible inactivation of maize leaf nitrate reductase. *Phytochemistry* 23:2155–58

35. Eichler, D. C., Barber, M. J., Solomonson, L. P. 1985. Anti-nitroxide immunoglobulin G: analysis of antibody specificity and their application as probes for spin-labeled proteins. *Biochemistry* 24:1181–86

36. Evans, H. J., Nason, A. 1953. Pyridine nucleotide nitrate reductase from extracts of higher plants. *Plant Physiol.* 28:233–54

37. Fido, R. J. 1987. Purification of nitrate reductase from spinach *(Spinacea oleracea)* by immunoaffinity chromatography using a monoclonal antibody. *Plant Science* 50:111–15

38. Fido, R. J., Hewitt, E. J., Notton, B. A., Jones, O. T. G., Nasrulhaq-Boyce, A. 1979. Heme of spinach nitrate reductase: low temperature spectrum and midpoint potential. *FEBS Lett.* 99:180–82

39. Fido, R. J., Notton, B. A. 1984. Spinach nitrate reductase: further purifica-

tion and removal of 'nicked' subunits by affinity chromatography. *Plant Sci. Lett.* 37:87–91

40. Fischer, P., Klein, U. 1988. Localization of nitrogen assimilating enzymes in the chloroplast of *Chlamydomonas reinhardtii. Plant Physiol.* 88:947–52

41. Fritz, B., Ninnemann, H. 1985. Photoreactivation by triplet flavin and photoinactivation by singlet oxygen of *Neurospora crassa* nitrate reductase. *Photochem. Photobiol.* 41:39–45

42. Fu, Y.-H., Marzluf, G. A. 1987. Molecular cloning and analysis of the regulation of Nit-3, structural gene for nitrate reductase in *Neurospora crassa. Proc. Natl. Acad. Sci. USA* 84:8243–47

43. Galangau, F., Daniel-Vedele, F., Maureaux, T., Dorbe, M.-F., Leydecker, M.-T., Caboche, M. 1988. Expression of nitrate reductase genes from tomato in relation to light-dark regimes and nitrate supply. *Plant Physiol.* 88:383–88

44. Gewitz, H. S., Lorimer, G. H., Solomonson, L. P., Vennesland, B. 1974. The presence of HCN in *Chlorella vulgaris* and its possible role in controlling the reduction of nitrate. *Nature* 249:79–81

45. Gonneau, M., Cherel, I., Deng, M., Kavanagh, M., Maron-Poll, A., et al. 1988. Biochemistry and genetics of nitrate reductase in three Solanaceae. In *Proceedings of the 8th International Biotechnology Symposium, Vol. II,* ed. G. Durand, L., Bobichon, J. Florent. Paris: Soc. Franc. Microbiol.

46. Gowri, G., Campbell, W. H. 1989. cDNA clones for corn leaf NADH: nitrate reductase and chloroplast NAD(P)$^+$:glyceraldehyde-3-phosphate dehydrogenase. *Plant Physiol.* 90:792–98

47. Graf, L., Notton, B. A., Hewitt, E. J. 1975. Serological estimation of spinach nitrate reductase. *Phytochemistry* 14:1241–44

48. Grant, B. R., Atkins, C. A., Canvin, D. T. 1970. Intracellular location of nitrate reductase and nitrite reductase in spinach and sunflower leaves. *Planta* 134:195–200

49. Guerrero, M. G., Lara, C. 1987. Assimilation of inorganic nitrogen. In *The Cyanobacteria,* ed. P. Fay, C. Van Baalen, pp. 163–86. Amsterdam: Elsevier

50. Guerrero, M. G., Vega, J. M., Losada, M. 1981. The assimilatory nitrate-reducing system and its regulation. *Annu. Rev. Plant Physiol.* 32:169–204

51. Gutteridge, S., Bray, R. C., Notton, B.

A., Fido, R. J., Hewitt, E. J. 1983. Studies by electron paramagnetic resonance spectroscopy of the molybdenum center of spinach nitrate reductase. *Biochem. J.* 213:137–42

52. Hamano, T., Oji, Y., Okamoto, S., Mitsuhashi, Y., Matsuki, Y. 1985. Inverse correlation of thiol proteinase with nitrate reductase activities in barley leaves. *Plant Physiol.* 76:353–58

53. Hamat, H., Kleinhofs, A., Warner, R. L. 1989. Nitrate reductase induction and molecular characterization in rice. *Mol. Gen. Genet.* 218:93–98

54. Harker, A. R., Narayanan, K. R., Warner, R. L., Kleinhofs, A. 1986. NAD(P)H bispecific nitrate reductase in barley leaves: partial purification and characterization. *Phytochemistry* 25:1275–79

55. Heath-Pagliuso, S., Huffaker, R. C., Allard, R. W. 1984. Inheritance of nitrate reductase and regulation of nitrate reductase, nitrite reductase and glutamine synthetase isozymes. *Plant Physiol.* 76:353–58

56. Herrero, A., De la Rosa, M. A., Diez, J., Vega, J. M. 1980. Catalytic properties of *Ankistrodesmus braunii* nitrate reductase. *Plant Sci. Lett.* 17:409–15

57. Hewitt, E. J. 1975. Assimilatory nitrate-nitrite reduction. *Annu. Rev. Plant Physiol.* 26:73–100

58. Hipkin, C. R. 1989. Nitrate assimilation in yeast. In *Molecular and Genetic Aspects of Nitrate Assimilation,* ed. J. L. Wray, J. R. Kinghorn, pp. 51–68. London: Oxford Univ. Press

59. Howard, W. H., Solomonson, L. P. 1981. Kinetic mechanism of assimilatory NADH:nitrate reductase from *Chlorella. J. Biol. Chem.* 256:12725–30

60. Howard, W. D., Solomonson, L. P. 1982. Quaternary structure of assimilatory NADH:nitrate reductase from *Chlorella. J. Biol. Chem.* 257:10243–50

61. Hucklesby, D. P., Dowling, M. J., Hewitt, E. J. 1982. Cyanide formation from glyoxylate and hydroxylamine catalyzed by extracts of higher plant leaves. *Planta* 156:487–91

62. Ingemarsson, B. 1987. Nitrogen utilization in *Lemma.* I. Relations between net nitrate flux, nitrate reduction, and *in vitro* activity and stability of nitrate reductase. *Plant Physiol.* 85:856–59

63. Jetschmann, K., Solomonson, L. P., Vennesland, B. 1972. Activation of nitrate reductase by oxidation. *Biochim. Biophys. Acta* 75:276–78

64. Kamachi, K., Amemiya, Y., Ogura, N., Nakagawa, H. 1987. Immuno-gold localization of nitrate reductase in spin-

ach *(Spinacia oleracea)* leaves. *Plant Cell Physiol.* 28:333–38

65. Kamiya, A. 1988. Blue light-induced *in vivo* absorbance changes and *in vitro* activation of nitrate reductase in a nitrate-starved *Chlorella* mutant. *Plant Cell Physiol.* 29:489–96

66. Kamiya, A. 1989. Effects of blue light and ammonia on nitrogen metabolism in a colorless mutant of *Chlorella*. *Plant Cell Physiol.* 30:513–21

67. Kay, C. J., Barber, M. J. 1986. Assimilatory nitrate reductase from *Chlorella:* effect of ionic strength and pH on catalytic activity. *J. Biol. Chem.* 261:14125–29

68. Kay, C. J., Barber, M. J. 1989. EPR and kinetic analysis of the interaction of halides and phosphate with nitrate reductase. *Biochemistry* 28:5750–58

69. Kay, C. J., Barber, M. J., Solomonson, L. P. 1988. CD and potentiometry of FAD, heme and Mo-pterin prosthetic groups of assimilatory nitrate reductase. *Biochemistry* 27:6142–49

70. Kay, C. J., Solomonson, L. P., Barber, M. J. 1986. Thermodynamic properties of the heme prosthetic group in assimilatory nitrate reductase. *J. Biol. Chem.* 261:5799–5802

71. Kimsky, S. C., McElroy, W. D. 1958. *Neurospora* nitrate reductase: the role of phosphate flavine and cytochrome *c* reductase. *Arch. Biochem. Biophys.* 73:466–83

72. Kowalik, W. 1982. Blue light effects on respiration. *Annu. Rev. Plant Physiol.* 33:51–72

73. Kramer, E., Tischner, R., Schmidt, A. 1988. Regulation of assimilatory nitrate reduction at the level of nitrite in *Chlorella fusca*. *Planta* 176:28–35

74. Kubo, Y., Ogura N., Nakagawa, H. 1988. Limited proteolysis of the nitrate reductase from spinach leaves. *J. Biol. Chem.* 263:19684–89

75. Kuo, T., Kleinhofs, A., Warner, R. L. 1980. Purification and partial characterization of nitrate reductase from barley leaves. *Plant Sci. Lett.* 17:371–81

76. Langendorfer, R. L., Walters, M. T., Smarrelli, J. 1988. Metabolite control of squash nitrate reductase. *Plant Science* 57:119–25

77. Lips, S. H., Avissar, Y. 1972. Plant-leaf microbodies as the intracellular site of nitrate reductase and nitrite reductase. *Eur. J. Biochem.* 29:20–24

78. Lopez-Ruiz, A., Roldan, J. M., Verbelen, J. P., Diez, J. 1985. Nitrate reductase from *Monoraphidium braunii:* immunocytochemical localization and immunological characterization. *Plant Physiol.* 78:614–18

79. Lorimer, G. H., Gewitz, H. S., Volker, W., Solomonson, L. P., Vennesland, B. 1974. The presence of bound cyanide in the naturally inactivated form of nitrate reductase of *Chlorella*. *J. Biol. Chem.* 249:6074–79

80. Losada, M., Paneque, A., Aparicio, P. J., Vega, J. M., Cárdenas, J., Herrera, J. 1970. Inactivation and repression by ammonium of the nitrate reducing system in *Chlorella*. *Biochem. Biophys. Res. Commun.* 38:1009–15

81. Martino, S. J., Smarrelli, J. 1989. Nitrate reductase synthesis in squash cotyledons. *Plant Sci.* 61:61–67

82. Maurino, S. G., Vargas, M. A., Aparicio, P. J., Maldonado, J. M. 1983. Blue-light reactivation of spinach nitrate reductase inactivated by acetylene or cyanide. Effects of flavins and oxygen. *Physiol. Plant* 57:411–16

83. McDonald, D. W., Coddington, A. 1974. Properties of the assimilatory nitrate reductase from *Aspergillus nidulans*. *Eur. J. Biochem.* 46:169–78

84. Melzer, J. M., Kleinhofs, A., Warner, R. L. 1989. Nitrate reductase regulation: effects of nitrate and light on nitrate reductase mRNA accumulation. *Mol. Gen. Genet.* 217:341–46

85. Neame, P. J., Barber, M. J. 1990. Conserved domains in Mo-hydroxylase enzymes: determination of the amino acid sequence of sulfite oxidase. *J. Biol. Chem.* In press

86. Nicholas, D. J. D., Scawin, J. H. 1956. A phosphate requirement for nitrate reductase from *Neurospora crassa*. *Nature* 178:1474–75

87. Notton, B. A., Barber, M. J., Fido, R. J., Whitford, P. N., Solomonson, L. P. 1988. Immunological comparison of spinach and *Chlorella* nitrate reductase. *Phytochemistry* 27:1965–68

88. Notton, B. A., Barber, M. J., Fido, R. J., Whitford, P. N., Solomonson, L. P. 1989. Limited proteolysis of spinach nitrate reductase. *J. Cell Biol.* 107:846 (Abstr.)

89. Notton, B. A., Fido, R. J., Galfre, G. 1985. Monoclonal antibody to higher plant nitrate reductase: differential inhibition of enzyme activities. *Planta* 165:114–19

90. Oaks, A., Poulle, M., Goodfellow, V. J., Class, L. A., Deising, H. 1988. The role of nitrate and ammonium ions and light on the induction of nitrate reductase in maize leaves. *Plant Physiol.* 88:1067–72

91. Oelmuller, R., Schuster, C., Mohr, H.

1988. Physiological characterization of a plastidic signal required for nitrate-induced appearance of nitrate and nitrite reductases. *Planta* 174:75–83

92. Oji, Y., Ryoma, Y., Wakinchi, N., Okamoto, S. 1987. Effect of inorganic orthophosphate on in vitro activity of NADH-nitrate reductase isolated from 2-row barley leaves. *Plant Physiol.* 83:472–74

93. Pan, S.-S., Nason, A. 1978. Purification and characterization of homogeneous assimilatory reduced nicotinamide adenine dinucleotide phosphate-nitrate reductase from *Neurospora crassa*. *Biochim. Acta* 523:297–313

94. Peiser, G. D., Wang, T.-T., Hoffman, N. E., Yang, S. F., Liu, H., Walsh, C. T. 1984. Formation of cyanide from carbon 1 of 1-aminocyclopropane-1-carboxylic acid during its conversion to ethylene. *Proc. Natl. Acad. Sci. USA* 81:3059–63

95. Pistorius, E. K., Gewitz, H.-S., Voss, H., Vennesland, B. 1976. Reversible inactivation of nitrate reductase in *Chlorella vulgaris in vivo*. *Planta* 128:73–80

96. Pistorius, E. K., Gewitz, H.-S., Voss, H., Vennesland, B. 1977. Cyanide formation from histidine in *Chlorella*. A general reaction of aromatic amino acids catalyzed by amino acid oxidase systems. *Biochem. Biophys. Acta* 481:384–94

97. Poulle, M., Oaks, A., Bzonek, P., Goodfellow, V. J., Solomonson, L. P. 1987. Characterization of nitrate reductase from corn leaves and *Chlorella vulgaris*. *Planta Physiol.* 85:375–78

98. Rajasekar, V. K., Gowri, G., Campbell, W. H. 1988. Phytochrome-mediated light regulation of nitrate reductase expression in squash cotyledons. *Plant Physiol.* 88:242–44

99. Remmler, J. L., Campbell, W. H. 1986. Regulation of corn leaf nitrate reductase. II. Synthesis and turnover of the enzyme's activity and protein. *Plant Physiol.* 80:442–47

100. Roldan, J. M., Verbelen, J. P., Butler, W., Tokuyasu, K. 1982. Intracellular localization of nitrate reductase in *Neurospora crassa*. *Plant Physiol.* 70:872–74

101. Schmerder, B., Borriss, H. 1986. Induction of nitrate reductase by cytokinin and ethylene in *Agrostemma githago* L. embryos. *Planta* 169:539–93

102. Schuster, C., Schmidt, S., Mohr, H. 1989. Effect of nitrate, ammonium, light and plastidic factor on the appearance of multiple forms of nitrate reductase in mustard cotyledons. *Planta* 177:74–83

103. Smarrelli, J. Jr., Campbell, W. H. 1981. Immunological approach to comparisons of assimilatory nitrate reductase. *Plant Physiol.* 68:1226–30

104. Solomonson, L. P. 1974. Regulation of nitrate reductase activity by NADH and cyanide. *Biochim. Biophys. Acta* 334:297–308

105. Solomonson, L. P., Barber, M. J. 1989. Essential arginyl groups in molybdoenzymes. *J. Cell Biol.* 107:844a

106. Solomonson, L. P., Barber, M. J., Howard, W. D., Johnson, J. L., Rajagopalan, K. V. 1984. Electron paramagnetic resonance studies on the molybdenum center of assimilatory NADH:nitrate reductase from *Chlorella vulgaris*. *J. Biol. Chem.* 259:849–53

107. Solomonson, L. P., Barber, M. J., Robbins, A. P., Oaks, A. 1986. Functional domains of assimilatory NADH:nitrate reductase from *Chlorella*. *J. Biol. Chem.* 261:11290–94

108. Solomonson, L. P., Howard, W. D., Yamaya, T., Oaks, A. 1984. Mode of action of natural inactivator proteins from corn and rice on a purified assimilatory nitrate reductase. *Arch. Biochem. Biophys.* 233:469–74

109. Solomonson, L. P., Lorimer, G. H., Hall, R. H., Borchers, R., Bailey, J. L. 1975. Reduced nicotinamide adenine dinucleotide-nitrate reductase of *Chlorella vulgaris*: purification, prosthetic groups, and molecular properties. *J. Biol. Chem.* 250:4120–27

110. Solomonson, L. P., McCreery, M. J. 1986. Radiation inactivation of assimilatory NADH:nitrate reductase from *Chlorella*: catalytic and physical sizes of functional units. *J. Biol. Chem.* 261:806–10

111. Solomonson, L. P., McCreery, M. J., Kay, C. J., Barber, M. J. 1987. Radiation inactivation analysis of assimilatory NADH:nitrate reductase: apparent functional sizes of partial activities associated with intact and proteolytically modified enzyme. *J. Biol. Chem.* 262:8934–39

112. Solomonson, L. P., Spehar, A. M. 1977. Model for the regulation of nitrate assimilation. *Nature* 265:373–75

113. Solomonson, L. P., Spehar, A. M. 1979. Stimulation of cyanide formation by ADP and its possible role in the regulation of nitrate reductase. *J. Biol. Chem.* 254:2176–79

114. Solomonson, L. P., Spehar, A. M. 1981. Glyoxylate and cyanide forma-

tion. In *Cyanide in Biology*, ed. B. Vennesland, E. E. Conn, C. J. Knowles, J. Westley, F. Wissing, pp. 363–70. New York: Academic
115. Solomonson, L. P., Vennesland, B. 1972. Properties of nitrate reductase of *Chlorella*. *Biochem. Biophys. Acta* 75:276–78
116. Somers, D. A., Kuo, T.-M., Kleinhofs, A., Warner, R. L. 1983. Nitrate reductase-deficient mutants in barley. Immunoelectrophoretic characterization. *Plant Physiol.* 71:145–49
117. Spence, J. T., Barber, M. J., Solomonson, L. P. 1988. Stoichiometry of electron uptake and oxidation-reduction potentials of NADH:nitrate reductase. *Biochem. J.* 250:921–23
118. Tischner, R., Ward, M. R., Huffaker, R. C. 1989. Evidence for a plasma-membrane-bound nitrate reductase involved in nitrate uptake of *Chlorella sorokiniana*. *Planta* 178:19–24
119. Tomsett, A. B. 1989. The genetics and biochemistry of nitrate assimilation ascomycete fungi. *Microbiol. Rev.* 2:31–55
120. Trinity, P. M., Filner, P. 1979. Activation and inhibition of nitrate reductase extracted from cultured tobacco cells. *Plant Physiol.* 63:133
121. Vargas, M. A., Maurino, S. G., Maldonado, J. M. 1987. Flavin-mediated photoinactivation of spinach leaf nitrate reductase involving superoxide radical and activating effect of hydrogen peroxide. *J. Photochem. Photobiol. B. Biol.* 1:195–201
122. Vauchert, H., Kronenberger, J., Rouze, P., Caboche, M. 1989. Complete nucleotide sequence of the two homeologous tobacco nitrate reductase genes. *Plant Mol. Biol.* 12:597–600
123. Vauchert, H., Vincentz, M., Kronenberger, J., Caboche, M., Rouze, P. 1989. Molecular cloning and characterization of the two homeologous genes coding for nitrate reductase in tobacco. *Mol. Gen. Genet.* 216:10–15
124. Vaughn, K., Campbell, W. H. 1988. Immuno-gold localization of nitrate re-

ductase in maize leaves. *Plant Physiol.* 88:1354–57
125. Vaughn, K. C., Duke, S. O., Funkhauser, E. R. 1984. Immunochemical characterization and localization of nitrate reductase in norflurazon-treated soybean cotyledons. *Physiol. Plant.* 62:481–84
126. Vennesland, B., Jetschmann, K. 1971. The nitrate reductase of *Chlorella vulgaris*. *Biochim. Biophys. Acta* 227:544–57
127. Voskresenskaya, N. P. 1979. Effect on light quality on carbon metabolism. In *Encyclopedia of Plant Physiology*, ed. M. Gibbs, E. Latzko, 6:174–180. Berlin: Springer-Verlag
128. Wallace, W., Aryan, A. P. 1988. The role of oxidation and reduction of leaf nitrate reductase in modulating its activity. In *Advances in Frontier Areas of Plant Biochemistry*, ed. R. Singh, S. K. Sawhney, pp. 214–28. New Delhi: Prentice-Hall
129. Ward, M. R., Grimes, H. D., Huffaker, R. C. 1989. Latent nitrate reductase activity is associated with the plasma membrane of corn root. *Planta* 177:470–75
130. Ward, M. R., Tischner, R., Huffaker, R. C. 1988. Inhibition of nitrate transport by antinitrate reductase IgG fragments and the identification of plasma membrane associated nitrate reductase in roots of barley seedlings. *Plant Physiol.* 88:1141–45
131. Yamagishi, K., Sato, T., Ogura, N., Nakgawa, H. 1988. Isolation and some properties of a 115-kilodalton nitrate reductase-inactivator protein from *Spinacia oleracea*. *Plant Cell Physiol.* 29:371–76
132. Zeiler, K. Z. 1989. *Regulation of* Chlorella *nitrate reductase by nitrogen source*. PhD thesis, Univ. South Florida
133. Zeiler, K. G., Solomonson, L. P. 1989. Regulation of *Chlorella* nitrate reductase: control of enzyme activity and immunoreactive protein levels by ammonia. *Arch. Biochem. Biophys.* 269:46–54

Annu. Rev. Plant Physiol. Plant Mol. Biol. 1990. 41:255–76

PHOTOSYSTEM II AND THE OXYGEN-EVOLVING COMPLEX

Demetrios F. Ghanotakis

Department of Chemistry, University of Crete, Iraklion, Crete, Greece

Charles F. Yocum

Departments of Biology and Chemistry, The University of Michigan, Ann Arbor, Michigan 48109–1048

KEY WORDS: oxygen evolution, calcium, manganese, chloride

CONTENTS

INTRODUCTION

Remarkable progress has occurred in deducing the structure and function of photosystem II (PSII) and its associated O_2-evolving reaction. Once characterized as the "inner sanctum" of photosynthesis, PS II is now among the best-characterized multi-subunit membrane protein complexes. A variety of techniques are providing increasingly refined information on the structure of

255

the photoreaction and the function of the water oxidation reaction. Significant guideposts in the characterization of PSII have been the pioneering observations of Joliot & Kok (80), showing that O_2 evolution is a linear 4-electron oxidation reaction, and Kok's model for the process (consisting of discrete oxidation states, S_n, where n = 0, 1, 2, 3, or 4):

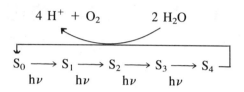

Among the S-states only S_1 exhibits long-term dark stability; in the dark, S_2 and S_3 decay to S_1 in minutes, and S_0 is oxidized slowly in the dark to S_1; the reaction sequence requires Mn, Ca^{2+}, and Cl^- ligated to an increasingly well-defined ensemble of polypeptides. Although much is now known about the polypeptide and inorganic ion cofactor requirements for O_2 evolution, elucidation of the mechanism of the O_2-evolving reaction and the precise organization of its constituent parts still presents a formidable challenge to investigators working in the field of photosynthesis.

In this review we address selected recent advances in analyses of the polypeptide and inorganic components of PSII, as well as findings relevant to the mechanism of H_2O oxidation. Interested readers will find alternative views (8, 26, 46, 64, 133) to be useful, as well as the discussion by Andréasson & Vänngård (7) of the photosynthetic electron transfer chain, which may be found in Volume 39 of this series.

POLYPEPTIDE CONSTITUENTS OF PHOTOSYSTEM II

Research prior to 1981 showed that partial separations of PSII and PSI activities could be obtained by treating thylakoid membranes with non-ionic detergents (3). Coupled with work by Izawa, Good, and their coworkers (114) defining the properties of efficient PSII-specific electron acceptors (*p*-benzoquinones, *p*-phenylenediamines), the groundwork was laid for the isolation of highly active PSII preparations (15, 82, 140) which, along with everted vesicles prepared by mechanical shear techniques (2), have served as the raw material for the discovery of polypeptides required for the O_2-evolving reaction. Crude detergent preparations consist of membrane "sheets" containing 15–20 polypeptides (67); purer preparations retain O_2-evolution activity but with a much smaller number (7–10) of polypeptides (58, 65, 74, 139). Table 1 summarizes the molecular weights and selected properties of a number of PSII polypeptides. As can be seen from the table, both intrinsic and extrinsic species are present, and several of the intrinsic proteins provide binding sites for chlorophylls (47, 43, 34, 32, and 25–29 kDa) and Pheo a (34 and/or 32 kDa).

Nanba & Satoh (98) have shown that the reaction center chlorophyll (P680) and the intermediate acceptor Pheo a are located on the 34- and 32-kDa intrinsic species (also called "D2" and "D1," respectively). As a consequence of these findings, the current model for the PSII reaction center assigns binding sites for P680, Pheo a, and the quinones Q_A and Q_B to D1 and D2. The model is attractive owing to similarities in amino acid sequences between D1 and D2, and between D1 and D2 and the L and M subunits of the reaction centers of photosynthetic bacteria; the crystal structures of the latter have revealed that the L and M polypeptides bind the BChls, Bpheos, quinones and non-heme iron reaction center components (1, 37, 142). The D1 polypeptide has been established as the locus of herbicide binding in PSII (110) and has also been proposed to participate in Mn binding (89, 90). It has been shown that species Z, which transfers electrons from the oxygen-evolving complex to P680$^+$, is Tyr 161 in the D1 amino acid sequence of $Synechocystis$ 6803 (36). The dark-stable radical known as D$^{\cdot+}$ (or Signal II slow) has been identified as Tyr 160 in D2 of the cyanobacterium (10, 35). Thus, current evidence points to the location of key components of PSII photochemistry and electron transfer on the D1 and D2 proteins. Cytochrome b559 copurifies with D1 and D2 and consists of polypeptides with molecular masses of 4.5 and 9 kDa. Since hydropathy plots show one membrane-spanning region for each subunit of the cytochrome (34, 72) and EPR spectroscopy has demonstrated that two histidines are axial ligands of the heme iron (9), it has been proposed that the heme of b559 crosslinks separate 4.5- and 9-kDa polypeptides, each of which contains a single histidine, to form the holoprotein. In spite of this structural information on b559, its role in PSII activity remains obscure, and its presence along with D1 and D2 points to a major difference between PSII and bacterial reaction centers.

The most highly purified PSII preparations capable of O$_2$-evolution activity consist of the D1, D2, and cytochrome b559 species discussed above, augmented only by species with apparent molecular masses of 47, 43, and 33 kDa (58, 74). (These proteins, along with D1, D2, and b559, are ubiquitous constituents of all oxygen-evolving PSII preparations examined so far. As discussed below, the other polypeptides listed in Table 1, and found in less highly refined preparations, serve regulatory functions.) The intrinsic 47- and 43-kDa proteins bind chlorophyll a and are viewed as antenna pigment binding sites. This probably represents an oversimplification of the role(s) of these proteins in PSII. The 43-kDa species is more loosely bound to the photochemical reaction center than is the 47-kDa protein, and is the first intrinsic protein to be removed in the purification of D1-D2-b559; harsher

Table 1 Polypeptides of the PS II O_2 evolving reaction center complex

Molecular mass[a] (kDa)	Intrinsic or extrinsic?	Bound chromophores	Proposed function	Coding[b] site
47	intrinsic	Chl a	antenna	C(psbB)
43	intrinsic	Chl a	antenna	C(psbC)
33	extrinsic	none	regulation of Mn stability	N
34 (D2)	intrinsic	$D^{\cdot+}$, Chl a, Pheo a, Q_A (?)	photochemical reaction	C(psbD)
32 (D1)	intrinsic	$Z^{\cdot+}$, Chl a, Pheo a,Q_B	center	C(psbA)
28	intrinsic	Chl a	regulation of Q_A/Q_B e^- transfer	N?
23	extrinsic	none	regulation of Ca^{2+} concentration	N
22	intrinsic	none	?	N?
17	extrinsic	none	regulation of Cl^- concentration	N
9	intrinsic	cyt b559	? (copurifies with D1/ D2 photochemical	C(psbE)
4.5	intrinsic	cyt b559	reaction center)	C(psbF)
<4.5	?	?	?	C(psb I-K)

[a] Values determined by SDS-PAGE, and commonly found in the literature; the actual values (from DNA-sequence-derived amino acid sequences) may vary considerably from those given here.
[b] C: chloroplast DNA (psb = PSII gene, followed by alphabetical identifier according to Ref. 69); N: nuclear DNA.

conditions (use of chaotropes or extended washing with Triton X-100) are required for the removal of the 47-kDa species (60, 98). It is possible that removal of these proteins either singly or together affects the ability of the reaction center species to bind the plastoquinone acceptors that are missing in D1-D2-b559. There is clearly an effect on electron transfer within PSII after removal of the 47- and 43-kDa proteins, as evidenced by the inability to detect either the stable ($D^{\cdot+}$) or transient ($Z^{\cdot+}$) tyrosine radicals (60) and the formation on illumination of a spin-polarized triplet species (100). It is therefore likely that the 47- and 43-kDa proteins, in addition to binding chlorophyll, are important structural elements of the intact oxygen-evolving system of PSII, and roles for these proteins in the regulation of quinone and Mn binding to PSII cannot be excluded.

Beginning with the results of experiments on everted vesicles (2) and subsequent investigations on detergent-isolated PSII membranes (95, 115, 140), substantial experimentation has been directed towards elucidating the roles of the extrinsic 33-, 23-, and 17-kDa polypeptides associated with PSII. These proteins are amenable to extraction/reconstitution procedures, so it has been possible to induce a lesion in PSII activity by polypeptide removal, and then to characterize the lesion and repair it by reconstitution techniques. The 33-kDa extrinsic protein has a pI of 5 but contains substantial amounts of Lys (83). The protein is tenaciously bound to PSII; its removal requires either treatments that in some cases also liberate functional Mn (29, 51, 82, 115, 140) or exposure of PSII preparations to 1 M solutions of $CaCl_2$ or $MgCl_2$ (101). The former observation might be taken as evidence that the 33-kDa protein is required for Mn binding to PSII, but exposure to 1 M $CaCl_2$ does not release Mn; addition of the protein reactivates O_2-evolution activity (75, 96, 102). We speculate that release of the 33-kDa protein along with Mn is a coincidental phenomenon related to ionic strength (for example, high concentrations of protonated Tris buffer). In the absence of the 33-kDa protein, both advancement to the S_2 state and low rates of O_2 evolution have been reported (103, 104, 123, 132). These observations have been challenged on the grounds that $CaCl_2$-washed preparations retain small amounts of the 33-kDa protein, and that only the intact centers are competent in electron transfer (27). Complete amino acid sequence data are available for the protein (99); the proposal that the spinach protein contains a Mn-binding sequence similar to that for a Mn superoxide dismutase has not been supported by sequence data from other species.

The 23-kDa protein (pI = 6.5) is removed from PSII preparations by treatments utilizing elevated ionic strength, most commonly NaCl at concentrations up to 2 M (2, 96, 115). This treatment, which also removes the 17-kDa extrinsic protein (see below), results in a loss of as much as 90% of O_2-evolving capacity and a slowing of the rereduction of $Z^{\bullet+}$ (53, 56); but addition of Ca^{2+} (56) and Cl^- (4) to assay mixtures can partially restore activity. Highly purified O_2-evolving preparations that lose the 23-kDa species during isolation routinely require elevated concentrations of Ca^{2+} and Cl^- for assays of activity. The 17-kDa protein is basic (pI = 9.2), and like the 23-kDa species, can be extracted by elevated ionic strength, although much lower salt conditions are required for its removal (136). Many functions of the normal O_2-evolving apparatus can be observed in the presence of the 33- and 23-kDa species alone, and Mn is resistant to extraction by bulky reductants (see below); Miyao & Murata (92) have shown that the 17-kDa protein is involved in retention of Cl^- at the active site of water oxidation.

Several results provide information on the structure and topology of the extrinsic proteins. Since removal of the 23-kDa species exposes functional Mn to bulky reductants (hydroquinone, TMPD) (63, 126, 129), the topological position of this protein is exterior to the site of O_2 evolution. An estimate of the size of the structure of all three extrinsic proteins is based on work by Innes & Brudvig (76), who used Dy^{3+} complexes to estimate the distance between the lumenal surface of PSII and the stable $D^{\bullet+}$ radical; minimal distances of about 42 Å (extrinsic polypeptides present) and 27 Å (after Tris washing) give an estimated width of approximately 14 Å for the entire ensemble of extrinsic proteins. On the basis of electron microscopy, Dekker et al (38) suggest that the 33-kDa protein has a width of about 8 Å. Evidence from crosslinking studies on intact PSII centers indicate that the 33-kDa protein binds to the 47-kDa intrinsic protein of the PSII reaction center (24, 48) and perhaps also to D1-D2-b559 (22). In the intact oxygen-evolving complex, it seems likely that the 23-kDa protein binds to the 33-kDa species, and the 17-kDa species in turn binds to the 23 kDa polypeptide (84, 95), although no data as yet identify the amino acid sequences in these proteins that are involved in the binding interactions.

Polypeptides of 28, 22, and 10 kDa apparent molecular masses are removed from PSII preparations either by Tris [10 kDa (87)] or by detergents used to prepare O_2-evolving reaction centers (59). Although information on these proteins is still emerging, a principal effect of extraction of this group of polypeptides is a shift in reducing-side acceptor preference, with a diminution of p-benzoquinone-catalyzed activity, an increase in ferricyanide-catalyzed activity, and a reduced sensitivity of substrate reduction to inhibition by DCMU (59, 74). Preliminary studies point to removal of the 28-kDa protein as the cause of the shift in acceptor specificity and inhibitor sensitivity (23). The 22-kDa protein has been proposed as a binding site for extrinsic polypeptides (88).

The stoichiometries of the polypeptides comprising PSII have been examined (28, 97, 141), with results that support either one or two copies of the extrinsic species per reaction center. In view of the measurements on the size of the extrinsic protein ensemble mentioned above, single copies of each of the extrinsic species would seem to be the most likely stoichiometry. There are as yet no data on the stoichiometries of the 28-, 22-, and 10-kDa species; but for stoichiometries of the intrinsic polypeptides of the reaction center, De Vitry et al (45) show data indicating the presence of one copy each of CP47, CP43, D1, and D2, and 1 or 2 copies of the 9-kDa subunit of b559. Little information is available on the stoichiometries or possible functions, if any, of low-molecular-mass polypeptides (< 5 kDa) encoded by chloroplast DNA (94) that copurify with O_2-evolution activity (71).

INORGANIC COFACTORS OF THE OXYGEN-EVOLVING REACTION

Involvement of Manganese in the Oxygen-Evolving Complex

Manganese is a critical cofactor in O_2 evolution, and there is general agreement that activity requires 4 metal atoms per center. The sites of Mn ligation and the structure of the 4-atom assembly are not clearly understood. Experiments on Mn extraction from PSII indicate heterogeneity in Mn binding to PSII; inhibitory treatment with Tris at high concentrations releases 2 of the 4 Mn atoms, but NH_2OH releases 3 atoms (143). An extensive exploration of inhibitory treatments by Murata and his associates (95) provides even more evidence for heterogeneity in the Mn ensemble. The origin of this heterogeneity is unknown. One can imagine either that the ligands to the metal are heterogeneous or that some of the Mn atoms exist in higher oxidation states than others, in which case inhibitory extraction of the metal would be affected if the inhibitor is also a reducing agent, such as NH_2OH. The inhibitory extraction of Mn from PSII preparations by NH_2OH can be reversed by readdition of the metal, as shown by Cheniae's group (127), who observed substantial recovery of oxygen-evolution activity. Although Ca^{2+} is required for completion of the reactivation process, it is not needed for the light-induced ligation of Mn, nor can Mn-depleted PSII centers bind Ca^{2+} (128).

Attempts to probe the function of Mn in H_2O oxidation have been hampered by the fact that the metal is not easily probed by spectroscopic techniques. Encouraging progress has occurred, however, in characterizing an EPR signal detected at cryogenic temperatures, centered at approximately g = 2.0, which is comprised of at least 16 lines spanning 1500 G. Discovered by Dismukes & Siderer (47), the signal has been subjected to intense scrutiny by a number of investigators, leading to a consensus that the signal originates from Mn in the S_2 state. Although the multiline EPR signal was the first direct probe of Mn in PSII, a major limitation to the use of EPR spectroscopy for studying the oxygen-evolving complex is the fact that only the S_2 state has been detected.

Studies by EPR of the S_2 state have revealed another signal that exhibits a turning point at g = 4.1 and that is the precursor to the g = 2 multiline signal (30, 144, 145). Speculations about the molecular origins of these signals have resulted in two hypotheses. According to the first hypothesis, one binuclear and two mononuclear metal centers are involved in a structural arrangement where the binuclear center gives rise to the multiline EPR signal and one of the other two Mn ions gives rise to the g = 4.1 signal; Hansson et al (70) ascribe the g = 4.1 signal to a mononuclear Mn (IV) in redox equilibrium with a binuclear Mn species. An alternate proposal by de Paula et al (42, 43) suggests that both the multiline and g = 4.1 signals arise from the same tetranuclear Mn complex, where the conversion of the g = 4.1 signal into the multiline EPR signal upon incubation at 200 K in the dark can be explained by a temperature-dependent structural change of the Mn complex that alters the exchange couplings between the Mn ions.

Evidence for changes in the oxidation state of Mn in PSII has also been found by detection of optical absorption changes in the UV which are best resolved around 305 nm. Dekker et al (39–41) found equivalent absorbance changes on $S_0 \rightarrow S_1$, $S_1 \rightarrow S_2$, and $S_2 \rightarrow S_3$ transitions; a comparison of the absorbance changes to difference spectra of Mn-containing model compounds led to the conclusion that a Mn(III) is oxidized to Mn(IV) in each of the first 3 S-state transitions. Work by other investigators has complicated the interpretation of Dekker et al (85, 86, 117). Arguments concerning extinction coefficients for the $S_0 \rightarrow S_1$ change obscure the fact that there has been no serious challenge to the hypothesis that these changes arise from Mn oxidation-state advancements. High-sensitivity measurements of H_2O proton relaxation rates by NMR have also revealed changes that correlate with S-state advancement (120, 121). In contrast to the UV absorption change measurements, however, the NMR changes do not occur on the $S_2 \rightarrow S_3$ transition. While the NMR relaxation changes correlate with S-state advancement, the molecular identity of the relaxing species in these experiments is not known.

A study of the structural organization of the Mn complex has been carried out by X-ray absorption spectroscopy and a series of extended X-ray absorption fine-structure (EXAFS) experiments. Investigations by Goodin et al (66) found that the Mn K-edge, a measure of the coordination charge for the Mn ions, is shifted to a higher energy in the S_2 state than in the S_1 state, an observation that provides evidence for Mn oxidation during the $S_1 \rightarrow S_2$ transition. The EXAFS study of the S_1 state indicated that each Mn ion is six coordinate and ligated to oxygen and/or nitrogen. The same study also suggests that the Mn ions in the S_1 state are organized as bridged dimers; the Mn–Mn distance detected—about 2.7 Å—is the same as that in model di-u-oxo-bridged Mn dimer complexes. Later EXAFS studies by George et al (55) and Penner-Hahn et al (109) with improved signal to noise have refined that early work; an Mn–Mn distance of 3.3 Å has been discovered, and the Mn–O(N) distance has been shown to be longer (1.9 Å) than first reported. These refined metal–ligand distances have been interpreted to suggest a structure for the Mn ensemble comprised of a trinuclear Mn cluster plus a single Mn atom; an X-ray scatterer detected at 4.2 Å may arise from Ca^{2+}.

The chemical identity of the Mn ligands in PSII has been addressed by Britt et al (25), who probed the S_2 multiline signal using the electron spin-echo envelope modulation technique, and by Andréasson (5) using low-temperature EPR. The results of these spectroscopic investigations suggest that the ligands bound to the Mn atoms that produce the multiline signal are oxygen rather than nitrogen atoms. At the same time, these results do not eliminate nitrogen as a potential ligand to any Mn atoms that are not detectable by EPR in the S_2 state.

The Roles of Calcium and Chloride as Cofactors for Oxygen Evolution

The involvement of Cl^- as a cofactor in the O_2-evolving reaction was first examined in detail by Arnon (21). Izawa et al (79) revived an interest in the role of the anion in PSII activity that continues to the present. After Cl^- depletion, restoration of activity can be achieved by adding anions whose effectiveness follows the order $Cl^- > Br^- > I^- > NO_3^-$; F^- and ^-OH are inhibitory (81). The stoichiometry of Cl^- in the oxygen-evolving complex has not been extensively probed; Theg & Homann (130) report 4–5 Cl^-/PSII in thylakoids. Proposals about the role of activating anions in the O_2-evolving reaction are of two kinds. One holds that the anions activate O_2-evolving activity by acting as charge-neutralizing counterions in the environment of the active site of the O_2-evolving reaction (33, 73). The alternative hypothesis is that the anions bind to Mn, acting to facilitate electron transfer among the metal atoms (116), or that the anion occupies a site close to, but not on Mn (26); the chemical nature of the latter site has not been specified.

Formation of the $g = 2$ multiline signal is suppressed in the absence of Cl^- in favor of the $g = 4.1$ species, but the multiline signal forms on subsequent addition of Cl^- in the dark (107). The possibility of a direct binding of Cl^- to the Mn complexes has been investigated by EXAFS and EPR spectroscopy. No changes in the superhyperfine structure of the multiline signal have been reported in sytems in which Cl^- was replaced by Br^- (68); moreover, Yachandra et al (138) failed to observe, by EXAFS, evidence of Cl^- in the first coordination sphere of the Mn complex. More recent EXAFS studies (109) indicate that much higher sample concentrations will be required if EXAFS is to successfully resolve the question.

Calcium ions are required for O_2 evolution activity as evidenced by the need to include the ion in assays of PSII preparations deprived of the extrinsic 23- and 17-kDa proteins (56, 62, 91). Of the two extrinsic species removed by salt-washing, it is the 23-kDa protein that most strongly influences Ca^{2+} binding (93, 136). The actual number of Ca^{2+} atoms minimally required for activity is 2–3 per reaction center in highly active PSII preparations from wheat (28); a lower number, 1, has been obtained in other work (119), but a large number of inactive centers may account for the lower stoichiometry. Similarly, estimates of the K_m for the Ca^{2+} binding site in PSII vary. In spinach preparations, a low-affinity site is seen (56), whereas in wheat, a high-affinity binding site has been detected (28). At least part of the discrepancy has been resolved by the discovery that Na^+ and other monovalent cations compete for the Ca^{2+} binding site (137); the monovalent cations, especially Na^+, are common constituents of assay media used for the assay of O_2 evolution activity, and would increase values estimated for the Ca^{2+} K_m.

The discovery of inhibitory effects of lanthanides on Ca^{2+}-stimulated activity led Ghanotakis et al (57) to suggest a Concanavalin A–type structure in which Ca^{2+} binds in close proximity to the Mn complex to affect its structure. Such a structural arrangement is supported by recent data from EPR experiments with samples in which Ca^{2+} is replaced by Sr^{2+}; a modified multiline signal is obtained with narrowed lines, suggesting a change in the structure of the Mn complex upon replacement of Ca^{2+} with the larger ion (20). The Concanavalin A analogy is also attractive on account of the requirement that Mn bind to Concanavalin A to facilitate Ca^{2+} binding; at the same time, K_d values for Ca^{2+} binding in the two systems differ by an order of magnitude, with a higher value for Concanavalin A. In PSII, higher oxidation states of Mn are present, which could influence Ca^{2+} binding. It has been suggested that a calmodulin-type binding site exists in PSII (33), but, as just noted, the binding properties of Ca^{2+} in its role in activation of PSII in preparations lacking the 23- and 17-kDa proteins do not resemble the properties reported for Ca^{2+} binding to calmodulin. The principal difference resides in the substantially higher affinity of calmodulins for Ca^{2+}, which no doubt reflects the very different function of Ca^{2+} in the oxygen-evolving complex as contrasted with the biological functions of calmodulin. Although at this point a purely structural role for Ca^{2+} in the oxygen-evolving complex seems most likely, a more direct involvement of the ion in the O_2-evolution mechanism cannot be excluded.

As regards the localization of the site of Ca^{2+} binding in PSII, the observation of a Ca^{2+} requirement for O_2 evolution activity in the most highly purified PSII preparations (74) indicates that the binding site(s) for the ion must reside on intrinsic polypeptides of the reaction center and/or the 33-kDa extrinsic species. There is no information available on the ligands that bind Ca^{2+} in PSII, nor have the responsible polypeptides been identified. At the same time, it is reasonable to assume that the site is located on the interior of the structure provided by the extrinsic polypeptides, since reactivation of oxygen-evolution activity by Ca^{2+} with the polypeptides present requires long-term incubation (62, 105), in contrast to the situation in polypeptide-depleted preparations, where reactivation occurs immediately upon addition of the ion to assay mixtures (56).

INHIBITORY AMINES AS PROBES OF MANGANESE IN THE OXYGEN-EVOLVING COMPLEX

Much of the information regarding the properties of the oxygen-evolving complex has resulted from the use of various amines as inhibitors of activity. Izawa et al (79) showed that amines reversibly inhibit O_2 evolution and proposed the existence of an inhibition site near the site in PSII where Cl^- stimulated activity. Velthuys (134) provided the first evidence that S_2 and S_3 were the sites of NH_3 and CH_3NH_2 binding, and Cheniae's group (31, 54) subsequently showed that the inhibitory binding of Tris that leads to Mn extraction is also S_2 specific. Ghanotakis et al (61) showed that the strength of the binding of amines to PSII was proportional to pKa, and proposed that amines bind as Lewis bases to sites on one or more of the Mn atoms of the oxygen-evolving complex. It was subsequently shown that primary amines inhibit O_2 evolution by competing with Cl^- for the Cl^- binding site, and Sandusky & Yocum (116) found that there are two distinct sites on the oxygen-evolving complex where amines bind. In contrast to the first site, the second site is NH_3-specific but Cl^- insensitive and highly selective for small ligands; this site has been proposed to be the site at which substrate water ligates to the Mn complex.

Investigations of the NH_3-specific site with EPR spectroscopy by Beck et al (13) showed that NH_3 binds only after formation of the S_2 state and induces an altered multiline EPR signal; later investigations by Andréasson et al (6) have provided evidence for weaker NH_3 binding in the S_1 state. The observation that amines larger than NH_3 do not alter the S_2 multiline signal supports the proposal that the NH_3 binding site is sterically selective for small molecules that ligate to the first coordination sphere of the Mn. Further investigations of the effect of primary amines on the oxygen-evolving complex using EPR spectroscopy have confirmed the existence of the second, Cl^--sensitive binding site for amines in both the S_1 and S_2 states (6, 11). Binding of the amines to the S_1 state causes a reduction in the yield of the S_2 multiline signal with a concomitant increase in the yield of the S_2 g = 4.1 signal. Since exchange of ligands at this particular site affects the stability of the g = 4.1 signal, and therefore the structure of the Mn complex, the Cl^- binding site must be located close to Mn.

Amine reductants (hydroxylamines and hydrazines) have been shown to inhibit O_2 evolution by two separate mechanisms. At low concentrations, NH_2OH delays the release of O_2 by two charge separations (17), while at higher concentrations, extraction of Mn is observed (32). The first models regarding the effect of hydroxylamines and hydrazines suggested that they acted as H_2O analogs that bind to Mn and are subsequently oxidized in the oxygen-evolving complex (49, 50). It was later shown, however, that hydroxylamine binding and reduction of Mn are regulated by Cl^- (12), although the interaction between NH_2OH and the anion was not shown to be competitive. Thus the current model for amine inhibitions postulates that they act as ligands that interact with Mn in the oxygen-evolving complex by coordinating at or near the Cl^- binding site. It has been proposed recently (14) that primary and secondary amines can inactivate the Mn complex in the S_2 state by a reductive mechanism, leading to the extraction of Mn(II) and a concomitant loss of O_2-evolution activity.

THE MECHANISM OF PHOTOSYNTHETIC H_2O OXIDATION

The period-4 oscillation in O_2 yield from PSII indicates the existence of a linear oxidation reaction within the oxygen-evolving complex. In the original hypothesis of Kok, the O_2 yield optimum on the third flash, with a lower yield on the fourth flash, was proposed to result from an unequal distribution of S-states (75% S_1/25% S_0) in the resting, dark-adapted system. Further examinations of O_2 yield patterns have shown that imperfections in electron transfer reactions within PSII influence the apparent distribution of S-states, as deduced from O_2 yield measurements. It has been suggested that a slowly reduced species (ascribed to the $D^{\cdot+} \rightarrow D$ transition) may be present in 15–25% of PSII centers, creating an added electron donor outside the S-state system that would retard S-state advancement by one oxidizing equivalent to produce the higher yields observed on the fourth flash (135). This suggestion is supported by EPR examinations of the interaction between D^+ and the S_0 state which show that centers in S_0 are slowly converted to S_1 by reduction of $D^{\cdot+}$ (124). It seems reasonable to conclude that the original estimates of S_0 concentrations were high, and that in fact nearly all PSII centers in material subjected to long-term dark adaptation reside in the S_1 oxidation state.

An important question concerning the S-state ensemble relates to the oxidation states of Mn, the function of the metal in storage of oxidizing equivalents, and what sort of synthetic metal cluster structures may have relevant similarities to the natural system. Here, there are a number of speculations and a number of promising Mn-containing synthetic models but little firm evidence as yet to link any of these models to the biological system (108). Substantial effort has gone into modeling the paramagnetic S_2 state, where the $S = 1/2$ multiline signal might indicate the overall oxidation state of the Mn atoms in PSII (118). A variety of compounds (di-, tri-, and tetranuclear Mn complexes) are EPR active, producing multiline-type signals; the minimum requirement is a pair of Mn atoms in a mixed-valence state (II-III or III-IV). Based on such models for S_2 and on the fact that the S_1 state is diamagnetic, it has been suggested that S_1 may consist of all Mn (III), whose 1-electron oxidation to S_2 would produce a paramagnetic oxidation state [Mn(III) Mn(III)]-[Mn(III) Mn(IV)]. On the other hand, XANES data (109) indicate the possibility of a Mn(II), 2 Mn(III), Mn(IV) distribution of oxidation states in S_1; this finding does not rule out the generation of a paramagnetically active Mn(III)/Mn(IV) species in the S_2 state. It is likely that more experimental work both on the natural system and on synthetic Mn-containing clusters will bring a clearer understanding of the structure of the metals in the oxygen-evolving complex.

The appearance of O_2 in a "gush" following the photochemical removal of 4 electrons from Mn in the oxygen-evolving complex of PSII has led to speculations about the mechanism by which this reaction occurs. A sequential, 4-electron oxidation of H_2O (invoking formation of hydroxyl radical) is unlikely to be part of the mechanism on thermodynamic grounds. (Estimates of the potential available at the oxidizing side of PSII are inadequate to permit hydroxyl radical formation.) Therefore, mechanisms that permit a bypass of the 1-electron oxidation of H_2O must either propose a concerted 2-electron oxidation reaction leading to the formation of a hypothetical "peroxyl" intermediate, or alternatively propose a concerted 4-electron oxidation of two H_2O molecules. The latter is the prevalent model, as shown below:

$$S_0^0 \xrightarrow{h\nu} S_1^{1+} \xrightarrow{h\nu} S_2^{2+} \xrightarrow{h\nu} S_3^{3+} \xrightarrow{h\nu} S_4^{4+} \xrightarrow{\text{DARK}} S_0^0 + O_2 + 4H^+$$

$$\uparrow$$

$$2\ H_2O$$

Evidence supporting this model was provided by Radmer & Ollinger (112), who used mass spectroscopy to examine the isotopic composition of O_2 released from PSII, and showed that the isotopic form of oxygen (^{16}O or ^{18}O) present in the H_2O to which the S_3 state was exposed appeared in the O_2 released on the next flash, which catalyzed the $S_3 \rightarrow S_4 \rightarrow S_0$ transition. Barring the possibility of rapid oxygen isotope exchange with bulk H_2O in either the S_3 or S_4 state, where a bound intermediate oxidation state of H_2O (such as a peroxyl species) might exist, the isotope experiments point to the accumulation of stable oxidizing equivalents, followed by the concerted oxidation of 2 H_2O as the mechanism of photosynthetic O_2 evolution.

Support for a concerted mechanism of H_2O oxidation can be found in the measurements of UV absorption changes, attributed to oxidation-state advancements in Mn (39), discussed above. The UV absorbance changes corresponding to the formation of S_0, S_1, S_2, and S_3 are dark stable, with lifetimes approximating those of the corresponding S-states; but the flash that catalyzes the $S_3 \rightarrow S_4 \rightarrow S_0$ transition produces a rapid collapse of the absorption, which is interpreted to indicate the reduction of the accumulated higher oxidation states of Mn accompanying the oxidation of H_2O to O_2. An interesting feature of the oxygen-evolving reaction has been described by Plijter et al (111), namely that the time required for O_2 release from PSII may be longer than previously thought (>30 msec rather than 1.2 msec). If this is correct, then the time required for O_2 release from the oxygen-evolving complex, rather than the time required for H_2O oxidation, is the rate-limiting step.

An understanding of the exact mechanism of H_2O oxidation will require a complete delineation of the roles of Ca^{2+} and Cl^-, since both cofactors are required for O_2 production. The findings reviewed above suggest to us that the effects of Ca^{2+} and Cl^- are not related to Mn oxidation per se; rather, these cofactors are part of the structure in which the accumulated oxidizing equivalents on the Mn can be coupled to H_2O oxidation. For example, Cl^- displacement from the oxygen-evolving complex does not affect Mn oxidation, but the oxidation step produces the g = 4.1 signal rather than the multiline species. Delayed luminescence and fluorescence measurements of the effects of Cl^- on the accumulation of oxidizing equivalents in the oxygen-evolving complex are consistent with the EPR observations; oxidizing equivalents, presumably from sites including Mn, are photochemically removed in the absence of Cl^- (78, 131) to form modified higher S-states (106).

Studies by EPR on Ca^{2+} function in the oxygen-evolving complex have produced opposing results on the requirement for the ion in multiline-signal formation (20, 44). A major concern in such experiments is whether Ca^{2+} has been effectively removed from PSII preparations, especially the dense suspensions required for EPR spectroscopy. We view it as likely, based on experiments with highly purified PSII preparations, that both Ca^{2+} and Cl^- are required for formation of the S_2 multiline signal (58). That a chelator such as EGTA, used to deplete Ca^{2+} from PSII, might interfere with formation of the multiline signal by ligating Mn seems unlikely, since these chelators have no discernible effect on steady-state rates of O_2 evolution. Delayed luminescence measurements again provide alternative evidence that Ca^{2+} depletion allows for Mn oxidation-state advancements—i.e. 2 electrons may be removed from the oxidizing side of PSII in the absence of Ca^{2+} (18), a result similar to that observed in Cl^--depleted preparations. An intriguing observation, reported by Boussac & Rutherford (19), is that the S_3 state binds Ca^{2+} less strongly than do the lower S-states. This observation would be consistent with a model in which Ca^{2+} either shares ligands with, or is in very close proximity to, Mn. Oxidation of Mn during S-state advancement would increase the affinity of shared ligands for Mn and would at the same time decrease the affinity of these ligands for Ca^{2+}. Transition-metal-catalyzed reactions rarely require facilitation by a nontransition metal such as Ca^{2+}; Cu-Zn superoxide dismutase may be a pertinent example, where the Zn is present at a binding site near the Cu to regulate the redox potential of the transition metal (113). At present it seems most likely to us that Ca^{2+} plays a structural role in PSII, but we cannot exclude the possibility that the metal could provide ligand sites for substrate H_2O, depending on the number and type of other ligands bound to the metal.

As regards the role of Cl^-, spectroscopy has proven useful in confirming the results of the steady-state kinetic experiments demonstrating a competition between activating anions and a wide range of inhibitory Lewis base ligands. At the same time, spectroscopic examinations of Mn have not detected a ligation of Cl^- to the metal. Recent estimates are that sample concentrations some 10 times as great as the highest so far used in EXAFS experiments (109) would be required to detect Cl^-, presenting a formidable barrier to the use of this technique to resolve the question. Similarly, EPR does not detect super-hyperfine phenomena from Mn-bound halides, but without a detailed knowledge of either the structure of the Mn ensemble or the oxidation states of the metal, and without information on the nature of other ligands binding the metals, negative spectroscopic results cannot be viewed as definitive. The available evidence from all assay techniques points to a critical involvement of Cl^- in water oxidation, at the level of Mn itself. The point of greatest controversy remains the actual molecular identity of the site or sites at which Cl^- exerts its activating effect on O_2 evolution.

CONCLUSIONS

Table 2 summarizes our current opinion regarding the stoichiometries and functions of essential components of PSII and the oxygen-evolving complex. The organic and inorganic catalysts of photochemical O_2 evolution by PSII are ligated to a template comprised of several intrinsic polypeptides accompanied by a single 33-kDa extrinsic species. The O_2-evolving process is one in which at least 3, and perhaps 4, oxidizing equivalents are stored as oxidation-state advancements in Mn; the fourth oxidation-state advancement within the oxygen-evolving complex leads to the 4-electron oxidation of H_2O. Although Ca^{2+} and Cl^- are not required for Mn oxidation, omission of either of these essential cofactors blocks H_2O oxidation, so Ca^{2+} and Cl^- are necessary elements in maintaining the active structure of the Mn ensemble. The presence of these cofactors may contribute to the behavior of the O_2-evolution reaction. Oxidation of substrate appears not to proceed through intermediate steps, which means that the Mn atoms must be ligated in such a way as to prevent H_2O oxidation from occurring except in one discrete oxidation state. This implies changes in the ligand properties of the Mn

Table 2   Essential components of the oxygen-evolving complex and PSII

Cofactor	Stoichiometry	Function	Polypeptide(s) involved in cofactor ligation[a]
Mn	4	H_2O oxidation	? (47, 43, 33, D1, D2)
Ca^{2+}	2–3	regulation of Mn function in H_2O oxidation	? (47, 43, 33, D1, D2, 23)
Cl^-	4–5(?)		
PQ	2	electron acceptors (Q_A and Q_B)	34(D2), 32(D1)
Chl a	50	photochemistry/antenna	47, 43, D1, D2
Pheophytin a	2	charge separation	34(D2), 32(D1)
$D^{\cdot +}$	1	oxidation of S_0 to S_1	34(D2)
$Z^{\cdot +}$	1	oxidation of Mn/reduction of $P680^+$	32(D1)
Non-heme Fe	1	regulation of Q_A/Q_B electron transfer	D1, D2
Heme Fe	2($b559$)	?	9, 4.5 kDa

[a] The numbers given are the estimated molecular masses in kDa

ensemble in its highest oxidation state that permit access of H_2O to the site(s) on Mn responsible for formation of the O–O bonds.

The current state of knowledge about PSII and H_2O oxidation defines the obvious challenges that must be overcome to arrive at a complete understanding of the photosystem and its most important function. The molecular identity and structure of cofactor ligation sites must be uncovered and the interactions among essential polypeptides must be elucidated. The contributions, if any, of low-molecular-weight polypeptides to activity are as yet unknown, and the way in which polypeptide structure regulates oxygen evolution activity remains a mystery. Perhaps the greatest surprise to photosynthesis researchers in this past decade was the finding that PSII can be manipulated like any other enzyme; the surprises of the next decade should be interesting indeed.

ACKNOWLEDGMENTS

We thank a number of colleagues who shared reprints and preprints with us, and are especially grateful to Lars-Erik Andréasson and Tore Vänngård, who provided us with an advance copy of their article for Volume 39 of the *Annual Review of Plant Physiology and Plant Molecular Biology*. We thank Drs. Bridgette Barry, R. J. Debus, J. P. Dekker, Julio de Paula, and Prof. G. T. Babcock for their helpful comments. The Metabolic Biology (now Cellular Biochemistry) Program of the National Science Foundation and the Competitive Research Grants Office of the United States Department of Agriculture are gratefully acknowledged for their continuing support to C.F.Y.

Literature Cited

1. Allen, J. P., Feher, G., Yeates, T. O., Komiya, H., Rees, D. C. 1988. Structure of the reaction center from *Rhodobacter Sphaeroides* R-26: protein-cofactor (quinones and Fe^{2+}) interactions. *Proc. Natl. Acad. Sci. USA* 85:8487–91
2. Akerlund, H.-E., Jansson, C., Andersson, B. 1982. Reconstitution of photosynthetic water splitting in inside-out thylakoid vesicles and identification of a participating polypeptide. *Biochim. Biophys. Acta* 681:1–10
3. Anderson, J., Boardman, N. 1966. Fractionation of the photochemical systems of photosynthesis. I. Chlorophyll contents and photochemical activities of particles isolated from spinach chloroplasts. *Biochim. Biophys. Acta* 112:403–21
4. Andersson, B., Critchley, C., Ryrie, I. J., Jansson, C., Larsson, C., et al. 1984. Modification of the chloride requirement for photosynthetic O_2 evolution. The

role of the 23 kDa polypeptide. *FEBS Lett.* 168:113–17
5. Andréasson, L.-E. 1989. Is nitrogen liganded to manganese in the photosynthetic oxygen-evolving system? EPR studies after isotopic replacement with ^{15}N. *Biochim. Biophys. Acta* 973:465–67
6. Andréasson, L.-E., Hansson, O., Von Schenck, K. 1988. The interaction of ammonia with the photosynthetic oxygen-evolving system. *Biochim. Biophys. Acta* 936:351–60
7. Andréasson, L. E., Vänngård, T. 1988. Electron transport in photosystems I and II. *Annu. Rev. Plant Physiol. Plant Mol. Biol.* 39:379–411
8. Babcock, G. T. 1987. The photosynthetic oxygen-evolving process. In *New Comprehensive Biochemistry*, Vol. 15, *Photosynthesis*, ed. J. Amesz, pp. 125–58. Amsterdam: Elsevier
9. Babcock, G. T., Widger, W. R., Cramer, W. A., Oertling, W. A., Metz, J. G.

1985. Axial ligands of chloroplast cytochrome b559: identification and requirement for a heme-cross-linked polypeptide structure. *Biochemistry* 24: 3639–45

10. Barry, B. A., Babcock, G. T. 1987. Tyrosine radicals are involved in the photosynthetic oxygen-evolving system. *Proc. Natl. Acad. Sci. USA* 84:7009–7103

11. Beck, W. F., Brudvig, G. W. 1986. Binding of amines to the O_2-evolving center of photosystem II. *Biochemistry* 25:6479–86

12. Beck, W. F., Brudvig, G. W. 1987. Reactions of hydroxylamine with the electron-donor side of photosystem II. *Biochemistry* 26:8285–95

13. Beck, W. F., de Paula, J. C., Brudvig, G. W. 1986. Ammonia binds to the manganese site of the O_2-evolving complex of photosystem II in the S_2 state. *J. Am. Chem. Soc.* 108:4018–22

14. Beck, W. F., Sears, J., Brudvig, G. W., Kuwaliec, R. J., Crabtree, R. H. 1989. Oxidation of exogenous substrates by the O_2-evolving center of photosystem II and related catalytic air oxidation of secondary alcohols via a tetranuclear manganese (IV) complex. *Tetrahedron Lett.* 45:4903–11

15. Berthold, D. A., Babcock, G. T., Yocum, C. F. 1981. A highly resolved, oxygen-evolving photosystem II preparation from spinach thylakoid membranes: EPR and electron transport properties. *FEBS Lett.* 134:231–34

16. Biggins, J., ed. 1987. *Progress in Photosynthesis Research* (Proc. 7th Int. Congr. Photosynth.). Dordrecht: Nijhoff

17. Bouges-Bocquet, B. 1973. Limiting steps in photosystem II and water decomposition in *Chlorella* and spinach chloroplasts. *Biochim. Biophys. Acta* 292:772–85

18. Boussac, A., Maison-Peteri, B., Etienne, A.-L., Vernotte, C. 1985. Reactivation of oxygen evolution of NaCl-washed photosystem-II particles by Ca^{2+} and/or the 24 kDa protein. *Biochim. Biophys. Acta* 808:231–34

19. Boussac, A., Rutherford, A. W. 1988. Ca^{2+} binding to the oxygen evolving enzyme varies with the redox state of the Mn cluster. *FEBS Lett.* 236:432–36

20. Boussac, A., Rutherford, A. W. 1988. The nature of the inhibition of the oxygen evolving enzyme of photosystem II which is induced by NaCl washing and reversed by the addition of Ca^{2+} or Sr^{2+}. *Biochemistry* 27:3476–83

21. Bove, J. M., Bove, C., Whatley, F. R., Arnon, D. I. 1963. Chloride require-

ment for oxygen evolution in photosynthesis. *Z. Naturforsch.* 18b:683–88

22. Bowlby, N. R., Frasch, W. D. 1986. Isolation of a manganese containing protein complex from photosystem II preparations of spinach. *Biochemistry* 25:1402–7

23. Bowlby, N. R., Ghanotakis, D. F., Yocum, C. F., Petersen, J., Babcock, G. T. The functional unit of oxygen evolving activity: implications for the structure of photosystem II. See Ref. 122, pp. 215–26

24. Bricker, T. M., Frankel, L. K. 1987. Use of a monoclonal antibody in structural investigations of the 49-kDa polypeptide of photosystem II. *Arch. Biochem. Biophys.* 256:295–301

25. Britt, R. D., Zimmermann, J.-L., Sauer, K., Klein, M. P. 1989. Ammonia binds to the catalytic Mn of the oxygen evolving complex of photosystem II: evidence by electron spin echo envelope modulation spectroscopy. *J. Am. Chem. Soc.* 111:3522–32

26. Brudvig, G. W., Beck, W. F., de Paula, J. C. 1989. Mechanism of photosynthetic water oxidation. *Annu. Rev. Biophys. Chem.* 18:25–46

27. Camm, E. L., Green, B. R., Allred, D. R., Staehelin, L. A. 1987. Association of the 33 kDa extrinsic polypeptides (water splitting) with PSII particles: immunochemical quantification of residual polypeptide after membrane extraction. *Photosynth. Res.* 13:69–80

28. Cammarata, K., Cheniae, G. 1987. Studies on 17, 24 kD depleted photosystem II membranes. I. Evidence for high and low affinity calcium sites in 17, 24 kD depleted PSII membranes from wheat versus spinach. *Plant Physiol.* 84:8577–95

29. Cammarata, K., Tamura, N., Sayre, R., Cheniae, G. 1986. Identification of polypeptides essential for oxygen evolution by extraction and mutational analyses. See Ref. 125, pp. 311–20

30. Casey, J. L., Sauer, K. 1984. EPR detection of a cryogenically photogenerated intermediate in photosynthetic oxygen evolution. *Biochim. Biophys. Acta* 767:21–28

31. Cheniae, G. M., Martin, I. F. 1978. Studies on the mechanism of Tris-induced inactivation of oxygen evolution. *Biochim. Biophys. Acta* 502:321–44

32. Cheniae, G. M., Martin, I. F. 1971. Effects of hydroxylamine on photosystem II. 1. Factors affecting the decay of O_2 evolution. *Plant Physiol.* 47:568–75

33. Coleman, W. J., Govindjee. 1987. A model for the mechanism of chloride

activation of oxygen evolution in photosystem II. *Photosynth. Res.* 13:199–223

34. Cramer, W. A., Theg, S. M., Widger, W. R. 1986. On the structure and function of cytochrome *b-559*. *Photosynth. Res.* 10:393–403

35. Debus, R. J., Barry, B. A., Babcock, G. T., McIntosh, L. 1988. Site-directed mutagenesis identifies a tyrosine radical involved in the photosynthetic oxygen-evolving system. *Proc. Natl. Acad. Sci. USA* 85:427–30

36. Debus, R. J., Barry, B. S., Sithole, I., Babcock, G. T., McIntosh, L. 1988. Directed mutagenesis indicates that the donor to P680$^+$ in photosystem II is *tyr*-161 of the D1 polypeptide. *Biochemistry* 27:9071–74

37. Deisenhofer, J., Epp, O., Miki, K., Huber, R., Michel, H. 1984. X-ray structure analysis of a membrane protein complex: electron density map at 3 Å resolution and a model of the chromophores of the photosynthetic reaction center from *Rhodopseudomonas viridis*. *J. Mol. Biol.* 180:385–98

38. Dekker, J. P., Boekema, E. J., Witt, H. T., Rogner, M. 1988. Refined purification and further characterization of oxygen-evolving and Tris-treated photosystem II particles from the thermophilic cyanobacterium *Synechococcus* sp. *Biochim. Biophys. Acta* 936:307–18

39. Dekker, J. P., Plijter, J. J., Ouwehand, L., van Gorkom, H. J. 1984. Kinetics of manganese redox transitions in the oxygen-evolving apparatus of photosynthesis. *Biochim. Biophys. Acta* 767:176–79

40. Dekker, J. P., van Gorkom, H. J., Brok, M., Ouwehand, L. 1984. Optical characterization of photosystem II electron donors. *Biochim. Biophys. Acta* 764:301–9

41. Dekker, J. P., van Gorkom, H. J., Wensink, J., Ouwehand, L. 1984. Absorbance difference spectra of the successive redox states of the oxygen-evolving apparatus of photosynthesis. *Biochim. Biophys. ACta* 767:176–79

42. de Paula, J. C., Beck, W. F., Brudvig, G. W. 1986. Magnetic properties of manganese in the photosynthetic O$_2$-evolving complex. 2. Evidence for a manganese tetramer. *J. Am. Chem. Soc.* 108:4002–9

43. de Paula, J. C., Brudvig, G. W. 1985. Magnetic properties of manganese in the photosynthetic O$_2$-evolving complex. *J. Am. Chem. Soc.* 107:2643–48

44. de Paula, J. C., Li, P. M., Miller, A.-F., Wu, B. W., Brudvig, G. W. 1986. Effect of the 17- and 23-kilodalton

polypeptides, calcium and chloride on electron transfer in photosystem II. *Biochemistry* 25:6487–94

45. De Vitry, C., Diner, B. A., Lemoine, Y. 1987. Chemical composition of photosystem II reaction centers (PSII): phosphorylation of PSII polypeptides. See Ref. 16, pp. 105–8

46. Dismukes, G. C. 1986. The metal centers of the photosynthetic oxygen-evolving complex. *Photochem. Photobiol.* 43:99–115

47. Dismukes, G. C., Siderer, Y. 1980. EPR spectroscopic observation of a manganese center associated with water oxidation in spinach chloroplasts. *FEBS Lett.* 121:78–80

48. Enami, I., Miyaoka, T., Mochizuki, Y., Shen, J.-R., Satoh, K., Katoh, S. 1989. Nearest neighbor relationships among constituent proteins of oxygen-evolving photosystem II membranes: binding and function of the extrinsic 33 kDa protein. *Biochim. Biophys. Acta* 973:35–40

49. Forster, V., Junge, W. 1985. Cooperative and reversible action of three of four hydroxylamine molecules on the water-oxidizing complex. *FEBS Lett.* 186:153–57

50. Forster, V., Junge, W. 1986. On the action of hydroxylamine, hydrazine and their derivatives on the water-oxidizing complex. *Photosynth. Res.* 9:197–210

51. Franzen, L.-G., Andreasson, L.-E. 1984. Studies on manganese binding by selective solubilization of photosystem-II polypeptides. *Biochim. Biophys. Acta* 765:166–70

52. Deleted in proof

53. Franzen, L.-G., Hansson, O., Andreasson, L.-E. 1985. The roles of the extrinsic subunits in photosystem II as revealed by EPR. *Biochim. Biophys. Acta* 808:171–79

54. Frasch, W. D., Cheniae, G. M. 1980. Flash inactivation of oxygen evolution. Identification of S$_2$ as the target of inactivation by Tris. *Plant Physiol.* 65:735–45

55. George, G. N., Prince, R. C., Cramer, S. P. 1989. The Mn site of the photosynthetic water-splitting enzyme. *Science* 243:789–91

56. Ghanotakis, D. F., Babcock, G. T., Yocum, C. F. 1984. Calcium reconstitutes high rates of oxygen evolution in polypeptide depleted photosystem II preparations. *FEBS Lett.* 167:127–30

57. Ghanotakis, D. F., Babcock, G. T., Yocum, C. F. 1985. Structure of the oxygen-evolving complex of photosystem II: calcium and lanthanum compete for sites on the oxidizing side of photo-

system II which control the binding of water-soluble polypeptides and regulate the activity of the manganese complex. *Biochim. Biophys. Acta* 809:173–80

58. Ghanotakis, D. F., Demetriou, D. M., Yocum, C. F. 1987. Isolation and characterization of an oxygen-evolving photosystem II reaction center core preparation and a 28 kDa Chl-a-binding protein. *Biochim. Biophys. Acta* 891:15–21

59. Ghanotakis, D. F., Demetriou, D. M., Yocum, C. F. 1987. Purification of an oxygen evolving photosystem II reaction center core preparation. See Ref. 16, pp. 681–84

60. Ghanotakis, D. F., de Paula, J. C., Demetriou, D. M., Bowlby, N. R., Petersen, J., et al. 1989. Isolation and characterization of the 47 kDa protein and the D1-D2-cytochrome *b*559 complex. *Biochim. Biophys. Acta* 974:44–53

61. Ghanotakis, D. F., O'Malley, P. J., Babcock, G. T., Yocum, C. F. 1983. Structure and inhibition of components on the oxidizing side of photosystem II. See Ref. 77, pp. 91–102

62. Ghanotakis, D. F., Topper, J. N., Babcock, G. T., Yocum, C. F. 1984. Water-soluble 17 and 23 kDa polypeptides restore oxygen evolution activity by creating a high-affinity binding site for Ca^{2+} on the oxidizing side of photosystem II. *FEBS Lett.* 170:169–73

63. Ghanotakis, D. F., Topper, J. N., Yocum, C. F. 1984. Structural organization of the oxidizing side of photosystem II. Exogenous reductants reduce and destroy the Mn-complex in photosystem II membranes depleted of the 17 and 23 kDa polypeptides. *Biochim. Biophys. Acta* 767:524–31

64. Ghanotakis, D. F., Yocum, C. F. 1985. Polypeptides of photosystem II and their role in oxygen evolution. *Photosynth. Res.* 7:97–114

65. Ghanotakis, D. F., Yocum, C. F. 1986. Purification and properties of an oxygen-evolving reaction center complex from photosystem II membranes: a simple procedure utilizing a non-ionic detergent and elevated ionic strength. *FEBS Lett.* 197:244–48

66. Goodin, D. B., Yachandra, V. K., Britt, R. D., Sauer, K., Klein, M. P. 1984. The state of manganese in the photosynthetic apparatus. 3. Light-induced changes in X-ray absorption (K-edge) energies of manganese in photosynthetic membranes. *Biochim. Biophys. Acta* 767:209–16

67. Goodman Dunahay, T., Staehelin, L. A., Seibert, M., Ogilvie, P. D., Berg, S. P. 1984. Structural, biochemical and

biophysical characterization of four oxygen-evolving photosystem II preparations from spinach. *Biochim. Biophys. Acta* 764:179–93

68. Haddy, A., Aasa, R., Andreasson, L.-E. 1989. S-band studies of the S_2-state multiline signal from the photosynthetic oxygen evolving complex. *Biochemistry* 28:6954–59

69. Hallick, R. B., Bottomley, W. 1983. Proposals for the naming of chloroplast Genes. *Mol. Biol. Rep.* 4:38–43

70. Hansson, O., Aasa, R., Vänngård, T. 1987. The origin of the multiline and g = 4.1 electron paramagnetic resonance signals from the oxygen-evolving system of photosystem II. *Biophys. J.* 51:825–32

71. Henrysson, T., Ljungberg, U., Franzen, L.-G., Andersson, B., Akerlund, B. 1987. Low molecular weight polypeptides in photosystem II and protein dependent acceptor requirement for photosystem II. See Ref. 16, pp. 125–28

72. Herrmann, R. G., Alt, J., Schiller, B., Widger, W. R., Cramer, W. A. 1984. Nucleotide sequence of the gene for apocytochrome *b*-559 on the spinach plastid chromosome: implications for the structure of the membrane protein. *FEBS Lett.* 176:239–44

73. Homann, P. H. 1988. The chloride and calcium requirement of photosynthetic water oxidation: effects of pH. *Biochim. Biophys. Acta* 934:1–13

74. Ikeuchi, M., Yuasa, M., Inoue, Y. 1985. Simple and discrete isolation of an O_2-evolving PS II reaction center complex retaining Mn and the extrinsic 33 kDa protein. *FEBS Lett.* 185:316–22

75. Imaoka, A., Yanagi, M., Akabori, K., Toyoshima, Y. 1984. Reconstitution of photosynthetic charge accumulation and oxygen evolution in $CaCl_2$-treated PSII particles. I: Establishment of a high recovery of O_2 evolution and examination of the effect of Cl^- on O_2 evolution. *FEBS Lett.* 176:341–45

76. Innes, J. B., Brudvig, G. W. 1989. Location and magnetic relaxation properties of the stable tyrosine radical in photosystem II. *Biochemistry* 28:1116–25

77. Inoue, Y., Crofts, A. R., Govindjee, Murata, N., Renger, G., Satoh, K., eds. 1983. *The Oxygen Evolving System of Photosynthesis.* Tokyo: Academic. 459 pp.

78. Itoh, S., Yerkes, C. T., Koike, H., Robinson, H. H., Crofts, A. R. 1984. Effects of chloride depletion on electron donation from the water-oxidizing com-

plex to the photosystem II reaction center as measured by the microsecond rise of chlorophyll fluorescence in isolated pea chloroplasts. *Biochim. Biophys. Acta* 766:612–22

79. Izawa, S., Heath, R. L., Hind, G. 1969. The role of chloride ion in photosynthesis. III. The effect of artificial electron donors upon electron transport. *Biochim. Biophys. Acta* 180:388–98

80. Joliot, P., Kok, B. 1975. Oxygen evolution in photosynthesis. In *Bioenergetics of Photosynthesis*, ed. Govindjee, pp. 387–411. London/New York: Academic

81. Kelley, P. M., Izawa, S. 1978. The role of chloride ion in photosystem II. I. Effects of chloride ion on photosystem II electron transport and on hydroxylamine inhibition. *Biochim. Biophys. Acta* 502:198–210

82. Kuwabara, T., Murata, N. 1982. Quantitative analysis of the inactivation of photosynthetic oxygen evolution and the release of polypeptides and manganese in the photosystem II particles of spinach chloroplasts. *Plant Cell Physiol.* 24:741–47

83. Kuwabara, T., Murata, N. 1982. An improved method and further characterization of the 33-kilodalton protein of spinach chloroplasts. *Biochim. Biophys. Acta* 680:210–15

84. Kuwabara, T., Murata, T., Miyao, M., Murata, N. 1986. Partial degradation of the 18-kDa protein of the photosynthetic oxygen-evolving complex: a study of a binding site. *Biochim. Biophys. Acta* 850:146–55

85. Lavergne, J. 1986. Stoichiometry of the redox changes of manganese during the photosynthetic water oxidation cycle. *Photochem. Photobiol.* 43:311–17

86. Lavergne, J. 1987. Optical-difference spectra of the S-state transitions in the photosynthetic oxygen-evolving complex. *Biochim. Biophys. Acta* 894:91–107

87. Ljungberg, U., Akerlund, H.-E., Andersson, B. 1984. The release of a 10-kDa polypeptide from everted photosystem I thylakoid membranes by alkaline tris. *FEBS Lett.* 175:255–58

88. Ljungberg, U., Akerlund, H.-E., Larsson, C., Andersson, B. 1984. Identification of polypeptides associated with the 23 and 33 kDa proteins of photosynthetic oxygen evolution. *Biochim. Biophys. Acta* 767:145–52

89. Metz, J. G., Pakrasi, H. B., Seibert, M., Arntzen, C. J. 1986. Evidence for a dual function of the herbicide-binding D1 protein in photosystem II. *FEBS Lett.* 205:269–74

90. Metz, J. G., Seibert, M. 1984. Presence in photosystem II core complexes of a 34-kilodalton polypeptide required for water photolysis. *Plant Physiol.* 76:829–32

91. Miyao, M., Murata, N. 1984. Calcium ions can be substituted for the 24-kDa polypeptide in photosynthetic oxygen evolution. *FEBS Lett.* 168:118–20

92. Miyao, M., Murata, N. 1985. The Cl- effect on photosynthetic oxygen evolution: interaction of Cl- with 18-kDa, 24-kDa and 33-kDa proteins. *FEBS Lett.* 180:303–8

93. Miyao, M., Murata, N. 1986. Light-dependent inactivation of photosynthetic oxygen evolution during NaCl treatment of photosystem II particles: the role of the 24-kDa protein. *Photosynth. Res.* 10:489–96

94. Murata, N., Miyao, M., Hayashida, N., Hidaka, T., Sugiura, M. 1988. Identification of a new gene in the chloroplast genome encoding a low-molecular-mass polypeptide of photosystem II complex. *FEBS Lett.* 235:283–88

95. Murata, N., Miyao, M., Kuwawbara, T. 1983. Organization of the photosynthetic oxygen evolution system. See Ref. 77, pp. 213–22

96. Miyao, M., Murata, N., Lavorel, J., Maison-Peteri, B., Boussac, A., et al. 1987. Effect of the 33-kDa protein on the S-state transitions in photosynthetic oxygen evolution. *Biochim. Biophys. Acta* 890:151–59

97. Murata, N., Miyao, M., Omata, T., Matsunami, H., Kuwabara, T. 1984. Stoichiometry of components in the photosynthetic oxygen evolution system of photosystem II particles prepared with Triton X-100 from spinach chloroplasts. *Biochim. Biophys. Acta* 765:363–69

98. Nanba, O., Satoh, K. 1987. Isolation of a photosystem II reaction center consisting of D-1 and D-2 polypeptides and cytochrome b-559. *Proc. Natl. Acad. Sci. USA* 84:109–12

99. Oh-oka, H., Tanaka, S., Wada, K., Kuwabara, T., Murata, N. 1986. Complete amino acid sequence of 33 kDa protein isolated from spinach photosystem II particles. *FEBS Lett.* 197:63–66

100. Okamura, M. Y., Satoh, K., Isaacson, R. A., Feher, G. 1987. Evidence of the primary charge separation in the D_1D_2 complex of photosystem II from spinach: EPR of the triplet state. See Ref. 16, pp. 379–81

101. Ono, T.-A., Inoue, Y. 1983. Mn-preserving extraction of 33-, 24- and 16-kDa proteins from O_2-evolving PSII particles by divalent salt-washing. *FEBS Lett.* 164:255–59

102. Ono, T., Inoue, Y. 1984. Reconstitution

of photosynthetic oxygen evolving activity by rebinding of 33 kDa protein to $CaCl_2$-extracted PS II particles. *FEBS Lett.* 166:381–84

103. Ono, T., Inoue, Y. 1985. S-state turnover in the O_2-evolving system of $CaCl_2$-washed photosystem II particles depleted of three peripheral proteins as measured by thermoluminescence. Removal of 33 kDa protein inhibits S_3 to S_4 transition. *Biochim. Biophys. Acta* 806:331–40

104. Ono, T., Inoue, Y. 1986. Effects of removal and reconstitution of the extrinsic 33, 24 and 16 kDa proteins on flash oxygen yield in photosystem II particles. *Biochim. Biophys. Acta* 850:380–89

105. Ono, T., Inoue, Y. 1988. Discrete extraction of the Ca atom functional for O_2 evolution in higher plant photosystem II by a simple low pH treatment. *FEBS Lett.* 227:147–52

106. Ono, T., Nakayama, H., Gleiter, H., Inoue, Y., Kawamori, A. 1987. Modification of the properties of S_2 state in photosynthetic O_2-evolving center by replacement of chloride with other anions. *Arch. Biochem. Biophys.* 256:618–24

107. Ono, T., Zimmermann, J. L., Inoue, Y., Rutherford, A. W. 1986. EPR evidence for a modified S-state transition in chloride-depleted photosystem II. *Biochim. Biophys. Acta* 851:193–201

108. Pecoraro, V. L. 1988. Structural proposals for the manganese centers of the oxygen evolving complex: an inorganic chemist's perspective. *Photochem. Photobiol.* 48:249–64

109. Penner-Hahn, J. E., Fronko, R. M., Pecoraro, V. L., Yocum, C. F., Betts, S. D., et al. 1990. Structural characterization of the Mn sites in the photosynthetic oxygen evolving complex using X-ray absorption spectroscopy. *J. Am. Chem. Soc.* In press

110. Pfister, K., Steinback, K. E., Gardner, G., Arntzen, C. J. 1981. Photoaffinity labeling of an herbicide receptor protein in chloroplast membranes. *Proc. Natl. Acad. Sci. USA* 78:981–85

111. Plijter, J. J., Aalbers, S. E., Barends, J.-P. F., Vos, M. H., Van Gorkom, H. J. 1988. Oxygen release may limit the rate of photosynthetic electron transport: the use of a weakly polarized oxygen cathode. *Biochim. Biophys. Acta* 935:299–311

112. Radmer, R., Ollinger, O. 1986. Do the higher oxidation states of the photosynthetic O_2-evolving system contain bound H_2O? *FEBS Lett.* 195:285–89

113. Richardson, J. R., Thomas, K. A., Rubin, B. H., Richardson, D. C. 1975.

Crystal structure of bovine Cu/Zn superoxide dismutase at 3 A resolution: chain tracing and metal ligands. *Proc. Natl. Acad. Sci. USA* 72:1349–53

114. Saha, S., Ouitrakul, R., Izawa, S., Good, N. E. 1971. Electron transport and photophosphorylation as a function of the electron acceptor. *J. Biol. Chem.* 246:3204–9

115. Sandusky, P. O., Selvius DeRoo, C., Hicks, D. B., Yocum, C. F., Ghanotakis, D. F., Babcock, G. T. 1983. Electron transport activity and polypeptide composition of the isolated photosystem II complex. See Ref. 77, pp. 189–99

116. Sandusky, P. O., Yocum, C. F. 1984. The chloride requirement for photosynthetic oxygen evolution. Analysis of the effects of chloride and other anions on the amine inhibition of the oxygen-evolving complex. *Biochim. Biophys. Acta* 766:603–11

117. Saygin, O., Witt, H. T. 1987. Optical characterization of intermediates in the water-splitting enzyme system of photosynthesis—possible states and configurations of manganese and water. *Biochim. Biophys. Acta* 893:452–69

118. Sheats, J. E., Czernuszewicz, R. S., Dismukes, G. C., Rheingold, A. L., Petrouleas, V., et al. 1987. Binuclear manganese (III) complexes of potential biological significance. *J. Am. Chem. Soc.* 109:1435–44

119. Shen, J.-R., Satoh, K., Katoh, S. 1988. Isolation of an oxygen-evolving photosystem II preparation containing only one tightly bound calcium atom from a chlorophyll *b*-deficient mutant of rice. *Biochim. Biophys. Acta* 386:386–94

120. Srinivasan, A. N., Sharp, R. R. 1986. Flash-induced enhancements in the proton NMR relaxation rate of photosystem II particles. *Biochim. Biophys. Acta* 850:211–17

121. Srinivasan, A. N., Sharp, R. R. 1986. Flash-induced enhancements in the proton NMR relaxation rate of photosystem II particles: response to flash trains of 1–5 flashes. *Biochim. Biophys. Acta* 851:369–76

122. Stevens, S. E., Bryrant, D., eds. 1988. *Light Energy Transduction in Photosynthesis: Higher Plant and Bacterial Models.* Rockville: Am. Soc. Plant Physiol. 388 pp.

123. Styring, S., Miyao, M., Rutherford, A. W. 1987. Formation and flash-dependent oscillation of the S_2-state multiline EPR-signal in an oxygen evolving photosystem-II preparation lacking the three extrinsic proteins in the oxygen-evolving system. *Biochim. Biophys. Acta* 890:32–38

124. Styring, S., Rutherford, A. W. 1987. In the oxygen-evolving complex of photosystem II the S_0-state is oxidized to the S_1-state by D^+ (signal II_{slow}). *Biochemistry* 26:2401–5

125. Sybesma, C. 1984. *Advances in Photosynthesis Research*. The Hague: Nijhoff/Junk

126. Tamura, N., Cheniae, G. 1985. Effects of photosystem II extrinsic proteins on microstructure of the oxygen-evolving complex and its reactivity to water analogs. *Biochim. Biophys. Acta* 809:245–59

127. Tamura, N., Cheniae, G. 1987. Photoactivation of the water-oxidizing complex in photosystem II membranes depleted of Mn and extrinsic proteins. I. Biochemical and kinetic characterization. *Biochim. Biophys. Acta* 890:179–94

128. Tamura, N., Cheniae, G. 1988. Photoactivation of the water oxidizing complex: the mechanisms and general consequences to photosystem 2. See Ref. 122, pp. 227–42

129. Tamura, N., Radmer, R., Lantz, S., Cammarata, K., Cheniae, G. 1986. Depletion of photosystem II-extrinsic proteins. II. Analysis of the PS II/water-oxidizing complex by measurements of N,N,N',N',-tetramethyl-p-phenylenediamine oxidation following an actinic flash. *Biochim. Biophys. Acta* 850:369–79

130. Theg, S. M., Homann, P. H. 1982. Light-, pH-, and uncoupler-dependent association of chloride with chloroplast thylakoids. *Biochim. Biophys. Acta* 679:221–34

131. Theg, S. M., Jursinic, P. A., Homann, P. H. 1984. Studies on the mechanism of chloride action on photosynthetic water oxidation. *Biochim. Biophys. Acta* 766:636–46

132. Toyoshima, Y., Akabori, K., Imaoka, A., Nakayama, H., Ohkoushi, N., et al. 1984. Reconstitution of photosynthetic charge accumulation and oxygen evolution in $CaCl_2$-treated PSII particles. II: EPR evidence for reactivation of the $S_1 \rightarrow S_2$ transition in $CaCl_2$-treated PSII particles with the 17-, 23-, and 34-kDa proteins. *FEBS Lett.* 176:346–50

133. van Gorkom, H. J. 1985. Electron transfer in photosystem II. *Photosynth. Res.* 6:97–112

134. Velthuys, B. R. 1975. Binding of the inhibitor NH_3 to the oxygen-evolving apparatus of spinach chloroplasts. *Biochim. Biophys. Acta* 396:392–401

135. Vermaas, W. F. J., Renger, G., Dohnt, G. 1984. The reduction of the oxygen-evolving system in chloroplasts by thylakoid components. *Biochim. Biophys. Acta* 764:194–202

136. Waggoner, C. M., Yocum, C. F. 1987. Selective depletion of water-soluble polypeptides associated with photosystem II. See Ref. 16, pp. 685–88

137. Waggoner, C. M., Pecoraro, V. L., Yocum, C. F. 1989. Monovalent cations (Na^+, K^+, Cs^+) inhibit calcium activation of photosynthetic oxygen evolution. *FEBS Lett.* 244:237–40

138. Yachandra, V. K., Guiles, R. D., Sauer, K., Klein, M. P. 1986. The state of manganese in the photosynthetic apparatus. 5. The chloride effect in photosynthetic oxygen evolution. Is halide coordinated to the EPR-active manganese in the O_2-evolving complex? Studies of the substructure of the low-temperature multiline EPR signal. *Biochim. Biophys. Acta* 850:333–42

139. Yamada, Y., Tang, X.-S., Itoh, S., Satoh, K. 1987. Purification and properties of an oxygen-evolving photosystem II reaction-center complex from spinach. *Biochim. Biophys. Acta* 891:129–37

140. Yamamoto, Y., Doi, M., Tamura, N., Nishimura, N. 1981. Release of polypeptides from highly active O_2-evolving photosystem-2 preparation by tris treatment. *FEBS Lett.* 133:265–68

141. Yamamoto, Y., Tabata, K., Isogai, Y., Nishimura, M., Okayama, S., et al. 1984. Quantitative analysis of membrane components in a highly active O_2-evolving photosystem II preparation from spinach chloroplasts. *Biochim. Biophys. Acta* 767:493–500

142. Yeates, T. O., Komiya, H., Chirino, A., Rees, D. C., Allen, J. P., Feher, G. 1988. Structure of the reaction center from *Rhodobacter sphaeroides* R-26 and 2.4.2: Protein-cofactor (bacteriochlorophyll, bacteriopheophytin, and carotenoid) interaction. *Proc. Natl. Acad. Sci. USA* 85:7993–97

143. Yocum, C. F., Yerkes, C. T., Blankenship, R. E., Sharp, R. R., Babcock, G. T. 1981. Stoichiometry, inhibitor sensitivity, and organization of manganese associated with photosynthetic oxygen evolution. *Proc. Natl. Acad. Sci. USA.* 78:7507–11

144. Zimmermann, J. L., Rutherford, A. W. 1984. EPR studies of the oxygen-evolving enzyme of photosystem II. *Biochim. Biophys. Acta* 767:169–67

145. Zimmermann, J.-L., Rutherford, A. W. 1986. Electron paramagnetic resonance properties of the S_2 state of the oxygen-evolving complex of photosystem II. *Biochemistry* 25:4609–15

Annu. Rev. Plant Physiol. Plant Mol. Biol. 1990. 41:277–315

THE STRUCTURE AND FUNCTION OF THE MITOTIC SPINDLE IN FLOWERING PLANTS

T. I. Baskin and W. Z. Cande

Department of Molecular and Cell Biology, University of California, Berkeley, California 94720

KEY WORDS: centrosome, cytoskeleton, kinetochore, microtubules, mitosis

CONTENTS

1040-2519/90/0601-0277$02.00

INTRODUCTION

Many reviews have been written about the structure and function of the mitotic spindle (13, 22, 91, 124, 134, 163, 204). These usually describe examples from all eukaryotes, making implicit the assumption that mitosis, having evolved so early and being so basic a process, can tolerate only minor changes. However, the very importance of mitosis for survival has led to its being provisioned with redundant ways to achieve the same goals (151, 203). This redundancy provides an avenue for evolutionary change, but it is not known how far various taxa may have diverged (77, 102). Therefore, it is important to consider mitosis among groups of related organisms. In this review, by limiting our attention to the structure and function of the mitotic spindle in flowering plants, we hope to provide a reference from which features of mitosis specific to plant cells may be distinguished from those that are widely shared or universal among eukaryotic cells.

Recent experiments have profoundly altered the way many scientists view mitosis (124, 134). Two findings have been crucial in causing this change. First, microtubules of the mitotic spindle have been found to turn over faster than expected, having half-lives on the order of 10 s (162). Second, the kinetochore, the structure that attaches the chromosome to the spindle, has been shown to capture and to translocate microtubules (27, 124) and to be the site where force is produced for chromosome movements in anaphase (67, 135). These experiments have all been performed on animal cells, and so it is not known whether the results also apply to plants. In this review, we pay close attention to indirect evidence that suggests that plant spindle microtubules and kinetochores share these newfound capacities with their animal counterparts.

Another reason motivating our preparation of a review on mitosis limited to flowering plants is to focus on the plant cytoskeleton. This subject has been receiving attention among plant scientists, and general reviews have been published (112, 113). Mitosis offers the plant biologist an outstanding problem of cytoskeletal function. By comparing what is known about the mitotic cytoskeleton between plant and animal cells, we hope to draw attention to areas where the plant cytoskeleton is not understood and to guide future research.

This review starts with a description of the components of the mitotic spindle; this is followed by an account of the formation of the spindle at prophase and of both its form and function at metaphase. Chromosome separation at anaphase is then described, both structurally and mechanistically. Finally, the dissolution of the spindle and the formation of the cytokinetic organelle, the phragmoplast, are briefly discussed. The emphasis of this review is on structural information; a recent review on the regulation of spindle function, with good coverage material, has already appeared (204).

There are several aspects of mitosis that are conceptually linked to the workings of the mitotic apparatus but are not discussed in this review owing to insufficient space. Passage through the cell cycle is controlled (87), and some knowledge about this control in plant cells has been gained (33, 64, 85). Chromosomes undergo special movements at prophase (157); may (142) or may not (156) rest in specified locations on the metaphase plate; must pair homologously at meiosis (132); and sometimes undergo cycles of replication uncoupled from cell division, leading to amplified genomes (133). Finally, morphogenesis requires plants to control the placement of new cross-walls (70, 112, 136, 174).

COMPONENTS OF THE MITOTIC APPARATUS

Microtubules

The most extensively studied component of the mitotic apparatus is the microtubule. Animal microtubules are better understood than those of plants because of the abundance of microtubule protein in neural tissue. Nevertheless, plant tubulin has been purified and characterized biochemically (129, 130), and genes coding for plant tubulin have been cloned (172). Plants, like animals, produce several isoforms of α- and β-tubulin, and microtubules comprise mixtures of these (88). Tubulins are similar among kingdoms, with the α-tubulin family tending to be more divergent than the β-tubulin family (130). The amount of tubulin present in a cell varies during the cell cycle; the regulation of tubulin synthesis in plants has also begun to be studied (56, 101).

Microtubules are similar across kingdoms biochemically (47) and ultrastructurally (180); the greatest differences seem to lie in sensitivities to various drugs with depolymerizing action (129). The biochemistry of microtubule assembly and disassembly is complex. Microtubules are dynamic polymers, and relatively minor changes in the state of their surroundings can drive assembly or disassembly (91, 163). They are also polar polymers, with each end having different biochemical properties. One end of the microtubule (the "plus" end) grows faster than the other (the "minus" end). It has been recently discovered that microtubule ends can alternate between growing and shrinking states, a behavior termed dynamic instability (124). It is probably the hydrolysis of GTP during polymerization that gives rise to the kinetic differences between the two ends (124). In cells, the behavior of microtubules is modified by microtubule-associated proteins. Several classes of these proteins have been characterized from animal tissues but, with rare exception, are unidentified in plant tissues (46). Different repertoires of microtubule-associated proteins in plant and animal cells would be expected to result in different microtubule behavior in the two cell types.

The mitotic spindle comprises microtubules organized into two half-spindles. Each half-spindle contains a pole and many hundreds or thousands of microtubules that radiate away from it. In animal cells, the pole is a tight focus from which microtubules radiate in all directions, whereas in plant cells, the pole is usually diffuse and microtubules radiate predominantly toward the opposite half-spindle. The polarity of microtubules is highly uniform, in both plant and animal spindles, with the plus ends being distal to the pole (54). Microtubules that terminate at a kinetochore are called kinetochore microtubules. These are responsible for attaching the chromosome to the spindle and may play a role in providing the force for chromosome movements. Other microtubules are called nonkinetochore microtubules. These may be confined to the polar regions (polar microtubules) or may run into the opposite half-spindle (interzonal microtubules).

Kinetochores

The kinetochore is a specialized region of the mitotic chromosome that has long been known to attach chromosomes to the mitotic apparatus (27). This knowledge rests on results of microsurgical experiments, which show a mechanical connection between chromosome and spindle at the kinetochore, and on the observation that fragments of chromosomes lacking kinetochores fail to segregate during mitosis (134). Recently the converse has been demonstrated for animal cells: fragments of chromosomes containing only their kinetochoric regions carry out the same movements during mitosis as do intact chromosomes (28). The rest of the chromosome seems to have no part to play in the process by which it is moved.

Kinetochores are embedded in a special region of the chromosome, the centromere. They contain DNA, protein, and possibly RNA (27, 53, 134). As seen by electron microscopy, kinetochores appear in vertebrates as trilamellate structures, with one layer usually more electron dense than the surrounding chromatin (27). In contrast, plant kinetochores are more variable in structure, may not be layered, and are usually less electron dense than the chromatin (26, 160). Furthermore, they may appear either as an inclusion set into the surface of the chromosome, the so-called ball-in-cup configuration (84, 201), or, more commonly, as a protuberance from the chromosome surface (53, 160). Some investigators claim that the protuberance takes on a special geometry, such as annular (121), cylindrical (72), or a double protrusion (11). In both plants (97) and animals (27), microtubules penetrate deeply (several micrometers) into the kinetochore material. Kinetochores first become distinct from the mass of condensing chromatin at prophase (9, 201).

Spindle Poles

The pole of each half-spindle in animal cells is characterized by a well-defined structure called a centrosome. A centrosome comprises, with few exceptions, a centriole surrounded by an osmiophilic cloud of material, termed pericentriolar material (118). Centrosomes nucleate microtubules very effectively, and they are important in establishing the bipolar organization of the mitotic spindle (119a, 124). There are no centrioles at the poles of higher-plant mitotic spindles (13, 118, 151); and it is not certain what takes on the centrosomal function in plants. The prevailing view states that the pericentriolar material is responsible for microtubule organization; in plant spindles, this material is presumed to be present in a diffuse zone at the poles, and thus gives rise to the broad poles of the plant metaphase spindle (151). Mazia has termed this a flexible centrosome (118). He has found examples in animal cells in which a spindle with broad poles (i.e. plantlike) is organized following the dispersal of centrosomal material away from the centrioles (118).

The flexible-centrosome model predicts that at least some of the components of pericentriolar material of animal centrosomes are present throughout the diffuse poles of plant spindles. In ultrastructural observations, osmiophilic material is sometimes (151), but not always (50, 97), associated with the polar ends of spindle microtubules. In immunofluorescence studies, antisera that recognize unidentified pericentriolar components in animal cells have been reported to stain the poles of plant spindles in root tips (42, 195, 196) but not in endosperm cells (11). However, these antisera are unfractionated; a recent study suggests that different antigens are in fact being recognized in the plant and animal cells and that the plant staining probably corresponds to a membranous component that is present at the poles during mitosis (75). Several protein constituents of pericentriolar material in animal cells have been identified, and antibodies have been raised against them (e.g. 137). To confirm the flexible-centrosome model for plant spindle poles, antibodies such as these will have to be tested on plant material, and both positive and negative results published.

Other Spindle Components

Microtubules and kinetochores are essential components of the mitotic spindle: their disruption is inevitably associated with disruption of function at mitosis. Many other components have been observed to be associated with mitotic structures; however, in no case have they been demonstrated to have an essential function at mitosis.

MEMBRANES   The mitotic apparatus of both plants and animals contains a large quantity of membranous material, including endoplasmic reticulum (ER) and remnants of the nuclear envelope (80, 154). Membranes are found not only surrounding the mitotic spindle, but also penetrating the spindle, usually parallel and near to kinetochore microtubules (e.g. 76, 79). During anaphase, they may undergo a net movement toward the poles of the spindle

(76). Very little is known about the biochemical composition of these membranes. Researchers have suggested that they play a number of roles in mitosis, including sequestering calcium (80, 204), anchoring the spindle microtubules (37), and providing sites for spindle microtubule nucleation (202). Their role in mitosis has been reviewed (80), but certain specific observations about membranes during plant mitosis are considered below.

ACTIN MICROFILAMENTS   The presence and significance of actin microfilaments, i.e. F-actin, in the mitotic spindle has long been a subject of controversy for both plant and animal material. The controversy is fueled by the lack of a detectable signal from F-actin in living material (in contrast to the birefringence from microtubules) and the particular difficulties of faithfully preserving F-actin in plant material. This problem has been reviewed in general for plants (111, 176). In root tips, F-actin is localized to the spindle at late anaphase, but not usually before then (71, 146). In *Haemanthus katherinae* endosperm, microfilaments are present as a cage around the mitotic spindle, with some F-actin within the spindle, running parallel to the long axis of the spindle (127, 128, 165, 166). In suspension culture cells (98, 114, 170, 182) and microsporocytes (171, 181, 184), F-actin is associated with the spindle in both metaphase and anaphase, roughly colocalizing with microtubules. The protocol for some of these studies includes a permeabilization and extraction pretreatment, whose effect on microfilament structure is unknown. Other protocols omit fixation, which could allow the introduction of time-dependent changes in structure during staining and observation. Despite potential technical problems, the results from the different methods concur in finding spindle-associated F-actin. The reported presence of F-actin in the plant spindle is therefore probably real; its significance has yet to be learned (111, 176).

Many studies have tested the effects of microfilament inhibitors on mitosis, in both plants and animals; these inhibitors generally fail to disrupt chromosome separation (91, 111, 143, 166, 167, 176). These compounds do inhibit actin-based motility in plant cells, such as cytoplasmic streaming (111, 176). Microfilament inhibitors such as cytochalasins bind to a similar site on F-actin as does rhodamine-phalloidin, a commonly used probe for F-actin; the large number of reports claiming a drug-induced disappearance of F-actin during mitosis must therefore be viewed with caution (178). Recently, an antibody against actin was microinjected into mitotic cells of *Haemanthus* endosperm (127). This treatment prevents the specific rearrangement of the actin cytoskeleton that occurs in this cell type during mitosis (128, 165, 166), but has no effect on spindle formation or on chromosome movement (127). Therefore, even if the actin cytoskeleton responds to the presence of the mitotic spindle by becoming colocalized with it, there is little to indicate that F-actin plays a functional role during mitosis.

OTHER PROTEINS LOCALIZED TO THE SPINDLE   The distribution during mitosis of several antigens has been studied in plants. In some cases, localization of the antigen to spindle or phragmoplast is reported. The problems of adequate fixation of plant cells are notorious (111, 113), and one study has shown how different fixation protocols give rise to different patterns of antigen localization (198). Less well known is the tendency of the physical organization of the spindle per se to give rise to the appearance of specific spindle association (93). To control for artifacts related to volume exclusion and other physical effects, the distribution during mitosis of an inert probe molecule, such as fluorescent dextran, must be examined (93).

Antigens related to the intermediate filament class of proteins have recently been detected in plant cells. They have been localized to microtubule-containing structures, including the mitotic spindle, throughout the cell cycle. This contrasts with the localization observed in animal cells, where intermediate filaments dissociate from microtubules during mitosis (65, 74, 150). Testing the function of the intermediate filament-like proteins is hampered by the lack of any inhibitor specific for them; perhaps the injection of antibodies could be used to probe for a requirement for intermediate filaments during mitosis.

The distribution of the calcium-binding protein calmodulin during the plant cell cycle has also been studied. Its distribution in mitosis is roughly parallel to the distribution of membrane, being greatest at the spindle poles and along kinetochore microtubules (71, 185, 196, 198, 199). This distribution also parallels that of calcium, observed by various techniques in the spindle of *H. katherinae* endosperm cells (80, 100, 185, 204). Treatment of root tips with inhibitors of calmodulin function leads to abnormalities in calmodulin localization, but not necessarily to mitotic abnormalities (199). A role for calmodulin in mitosis has not been demonstrated in studies of animal cells (204). Calmodulin may therefore be a membrane protein without any specific mitotic function, or it may be one of several calcium-binding proteins that funtion in parallel during mitosis.

Unconfirmed reports of the localization of other proteins to the plant spindle or phragmoplast have appeared, but without any accompanying data about the function of the localized molecule. These proteins include myosin (149), troponin T (110), phosphorylated antigens (183), and several unidentified cytoskeletal proteins (117).

FORMATION OF THE MITOTIC APPARATUS: PROPHASE

In this section, we describe two processes that occur during prophase: establishment of the plane of cell division, and construction of the mitotic spindle. The plane of future division is marked in many plant cell types by the elaboration, starting before prophase, of two specialized cytoplasmic structures, the phragmosome and the cortical division site (70). For convenience, this review treats them separately, but they are probably not independent structures. The mitotic spindle forms in a process that occurs gradually during prophase and then rapidly during prometaphase. Until the end of prophase, the spindle has focused poles; it is only after the reorganization of spindle microtubules in prometaphase that the diffuse poles, typically associated with the plant spindle, usually develop. This review refers to the "prophase spindle" as though it were a structure distinct from the metaphase spindle, but the two structures are stages in a continuous process of spindle formation.

Establishment and Structure of the Division Plane

PHRAGMOSOME AND CORTICAL DIVISION SITE   A detailed description of the establishment of the division plane is given by Gunning (70). In general, hours before division (8–20 h), the nucleus moves to the center of the plane of the future cross-wall. Evidence exists that this migration is microtubule (59) and microfilament (189) dependent. Many thin transvacuolar strands connect migrating nuclei to the cortical cytoplasm. Once the nucleus has attained its position, strands that lie in the plane of eventual division ramify into a more or less contiguous sheet of cytoplasm. This was termed the "phragmosome" by Sinnott & Bloch, who showed it to be present in the vacuolated cells of a large number of plant species (174). The phragmosome forms after nuclear migration and several hours before spindle formation (174, 187, 189). It is present in some form throughout cell division, eventually to become cortical cytoplasm at the end walls of the daughter cells (70, 71). A probable function of the phragmosome is to maintain the centered position of the nucleus (70, 189); other functions in cytokinesis are likely but undocumented.

A system for observing phragmosome development in living cells has been developed that takes advantage of periclinal divisions induced in epidermal cells peeled from young leaves. The transvacuolar strands are seen to fuse and form the phragmosome (187). In these large vacuolated cells, both microtubules and microfilaments are present in the phragmosome (19, 66). The phragmosome is also seen in suspension culture cells (24), where it contains microtubules (120, 192) and F-actin (98, 114, 170, 182). Visualizing the presence of a phragmosome in cells with small vacuoles is problematic. In stamen hair cells a phragmosome is usually not apparent, but centrifugation of these cells to displace the nucleus laterally does reveal a plane of cytoplasm linking the displaced nucleus to the opposite wall (141). In root tip cells, although the nuclei do migrate to specific positions before division, the nuclear diameter is only a little smaller than the cell diameter, and a typical phragmosomal organization of the cytoplasm is not observed. Microtubules are sometimes observed to lie in the plane of future division, between nucleus and cortex (19, 153), but they are also observed to extend from the nucleus to many areas of the cytoplasm (44, 69, 71).

The plane of future cell division is also marked by cytoplasmic specialization in the cortex where that plane intersects the parent wall. This ring surrounding the cell is called the "division site" and is usually found in all cell types that make a phragmosome and in meristematic cells (70). The most prominent feature of the division site is the massive band of microtubules that forms there, called the "preprophase band" (153). No function for these microtubules is known, but because of their conspicuous marking of the division site, they have often been observed (19, 31, 44, 52, 62, 68, 117), and stages in their formation have been studied (45, 123, 192, 197). Comparison of the above studies shows that the timing of formation of the preprophase band with respect to the formation of the prophase spindle is variable. Finally, when cells being prepared for immunofluorescence are broken open and their contents dispersed, the preprophase band often stays attached to the nucleus, which implies that the two are somehow connected (71).

Recently, another marker has been detected, along with the preprophase band microtubules: cortical actin microfilaments are aligned in parallel with the preprophase band microtubules. These microfilaments have been seen in suspension culture cells (98, 114) as well as in root tips (119, 145). Both microtubule (e.g. 44) and actin (e.g. 114) bands disappear from the cortex at some time before metaphase spindle formation is complete.

FUNCTION OF THE CORTICAL DIVISION SITE   The only established function of the division site is to guide the expanding phragmoplast to the parent wall, after the separation of chromosomes (70). The means whereby the division site exerts its attraction for the growing phragmoplast is not known, but several lines of evidence show that the cortical site interacts with the growing phragmoplast over only a short distance (2, 59–61, 71, 144). Because cytochalasin is known to inhibit the reorientation of oblique cell plates (70, 143), these steering interactions may be mediated by actin microfilaments that link the expanding phragmoplast with the cortex (66, 98, 114, 182).

The cortical division site is apparently not required for the formation of a functional mitotic spindle. The usual positional relationship between nucleus and division site can be altered without impairing the ability of the spindle subsequently to segregate chromosomes (70, 71). The nucleus can be displaced from the division site naturally (60) or experimentally by drugs (114, 189) or centrifugation (61, 141), and functional spindles are formed. Also, several types of cells form spindles without any detectable division site, namely, endosperm cells (49, 168), microsporocytes (29, 86, 181), megasporocytes (23), and cells of the male gametophyte (35, 36, 83, 148, 179). Finally, in multinucleate cells, division site(s) may form that are not strictly related to nuclear position (2, 193). However, the cortical division site may play a role in orienting the forming spindle appropriately. The prophase spindle normally forms with its long axis perpendicular to the plane of the phragmosome and division site (31, 69). As mentioned above, cytoplasmic links between nucleus and cortical division site are commonly observed; it is not known how such links would exert an orienting influence upon the forming spindle. The placement of the division site and orientation of the spindle probably are parallel responses to the same fundamental polarizing influence within the cell.

Formation of the Prophase Spindle

CLEAR ZONE   The formation of the prophase spindle begins with the elaboration of a special region surrounding the nucleus. This homogeneous, organelle-devoid region is called the clear zone (13, 190). This zone has been observed in vivo in many plant cell types, including endosperm (3, 6), stamen hair cells (190), and microsporocytes (115). The first clue to its composition came from observations of endosperm cells by polarized light: the clear zone was found to be birefringent and was predicted on that basis to comprise the same subunits as did the spindle and phragmoplast (92). Both electron microscopy (34, 52, 161) and immunofluorescence light microscopy (e.g. 49) have confirmed the extensive presence of microtubules in the clear zone. These microtubules appear at first to be randomly arranged and tangential to the nuclear surface (31, 44, 120, 197), often running parallel to it for long distances (9, 97). The clear zone may be more or less extensive in different species or cell types (6, 34, 115).

DEVELOPMENT OF THE POLES   The next stage of formation of the prophase spindle is the emergence of a bipolar organization among the microtubules. On-line observations show the development of fusiform poles, one on each side of the nucleus, which have been called polar caps (13, 158, 190). Polar caps create the appearance of a spindle, hence leading to the term prophase spindle. The long axis of the prophase spindle is roughly the same as that of the subsequently formed metaphase spindle. In endosperm, sometimes three (or rarely more) polar caps form, which are not resolved into two poles until later in metaphase (9, 13, 49, 51, 168, 169). Formation of well-organized multiple poles that persist into metaphase was reported in early observations on meiocytes (108, 131, 139). More recent examinations of the microtubule cytoskeleton in microsporocytes have revealed some instances of multipolar prophase spindle organization (29, 86, 175, 181), but these studies have not emphasized the problem of spindle formation, and multipolar intermediates may have been missed.

During the formation of the prophase spindle, observations on root tips and on *H. katherinae* endosperm are in accord. In general, microtubules around the nucleus increase in number and in alignment, while their numbers decrease in other cellular locations (44, 49, 97, 168, 185, 197). Organization of these clear-zone microtubules is first seen as strands that emanate from focal centers on the nuclear envelope surface (68, 69, 71, 86, 103, 120, 122, 173, 192, 196). The foci sometimes appear annular in pole view (117, 197). In nonendosperm tissue, there are usually only two such centers; even at early stages they are at opposite sides of the nucleus, on an axis perpendicular to the division plane. These foci become the spindle poles. As the number of microtubules in the spindle increases, the poles move away from the surface of the nuclear envelope (31, 34, 52, 153, 161, 194, 197). Microtubules come to a tight focus at these poles, radiating chiefly toward the opposite pole. Occasionally, microtubules also radiate in other directions, appearing similar to the asters of animal cells (e.g. 71, 120); however, evidence from immunocytochemistry shows that these "asters" are probably remnants of an interphase microtubule array associated with the nuclear envelope (31, 44, 69, 113). Ultrastructural observations have revealed the presence of smooth ER closely associated with the prophase spindle poles (34, 37, 52, 76, 153).

PROBLEMS OF MICROTUBULE ORGANIZATION   In animal cells, spindle organization is mediated by the centrosome; however, in plant cells, in the absence of a conspicuous centrosome, the means whereby microtubules become organized into a bipolar spindle are not certain (70, 113). In insect spermatocytes, a well-organized spindle can form without the presence of a centrosome or detectable pericentriolar material at one pole (177), which implies that modes of spindle organization without typical centrosomes can also occur in animal cells. The flexible centrosomes, believed to organize plant spindle poles (118, 151), must somehow become distributed to two localized areas on opposite sides of the nucleus. When there is more than one nucleus (or micronucleus) per cell, each is able to form a spindle (or microspindle) (30, 41, 175, 191, 193); therefore, the organizing material must be flexible enough to become distributed over more than one nucleus.

The nuclear envelope has been suggested to organize the microtubules of the prophase spindle (104); several questions are prompted by this suggestion. The prophase array of microtubules forms tangentially around the nucleus; however, at other times, for example at telophase, a radial array of microtubules forms. How is this difference in organization achieved? What limits the formation of stable microtubule foci to two sites, on opposite sides of the nuclear envelope? How do these foci, once formed, become lifted up and away from the surface of the nuclear envelope? Although the nuclear envelope may participate in ordering microtubules at prophase, other components seem to be needed to fully organize the prophase spindle. One such component could be the sheets of smooth ER that are present at prophase behind the spindle poles (153, 202). However, the smooth ER occupies an expanse on either side of the nucleus far wider in extent than that adjacent to the prophase spindle poles (76, 154), implying the restriction of nucleation capacity to small zones of the ER.

Other components that could contribute to spindle organization are microtubule-associated proteins that specify lateral interactions between microtubules. Microtubule-associated proteins could allow parallel microtubules to associate and grow and antiparallel ones to become stabilized and not grow. If so, preferential microtubule nucleation at two antipodal sites on the nucleus could result in a stable, bipolar organization of microtubules. Indirect evidence for the presence of lateral interactions between the microtubules of the polar caps in endosperm has been found: polar caps are stable to cell-crushing pressures (51) and are more stable than other microtubules to cell lysis in calcium-containing buffers (169); also, when microtubule polymerization is inhibited, each cap is observed to form a tight, rod-shaped bundle (16). In diatoms, overlapping antiparallel microtubules of the spindle remain stable to isolation (38); in plants, they apparently become stabilized in the phragmoplast (54, 70). The functions that the animal centrosome carries out are nucleation and orientation of microtubules; both of these functions must be fulfilled in mitotic plant cells, but not necessarily by the same structure or set of proteins. Much more must be learned about the flexible centrosome and microtubule-associated proteins before the basis of the organization of the plant mitotic spindle is known.

CONTROL OF PROPHASE SPINDLE ORIENTATION WITHIN THE TISSUE   The prophase spindle normally forms perpendicular to the future division plane (31, 69), even when the subsequent metaphase or anaphase spindle is not perpendicular to that plane (40, 45, 122). Little is known about how spindle orientation is controlled. The long axes of the cell and the spindle are often parallel in microsporocytes (175); this implies that whatever polarizes the growth of plant cells can also affect the establishment of the spindle. However, there are many instances in which the spindle axis is perpendicular to the long axis of the cell, for example in the division of cambial initials (66) or guard cell mother cells (45) or in the production of new cell files in a meristem; and there are instances in which the division plane has no evident relation to cell geometry (152, 174, 187). Therefore, the polarizing influence acting on the spindle is not obligately coupled to cell growth or shape.

Insight into the means whereby the spindle direction is fixed can be gained from consideration of two specialized divisions that are asymmetric. The first example is from the development of stomatal complexes in the leaves of grasses. A large epidermal cell cuts off a small lens-shaped region, which becomes a subsidiary cell. This division is preceded by migration of the nucleus to a site just opposite a neighboring guard mother cell (152). A variety of lines of evidence suggest that proximity to the guard mother cell fixes the location of the proximal spindle pole (2, 45, 59). For example, divisions are observed where the nucleus of a cell that is not a subsidiary mother cell is nevertheless drawn toward a neighboring guard mother cell, and a spindle is formed with one pole adjacent to the guard mother cell (60). The implication of these studies is that proximity to a guard mother cell is responsible not only for nuclear migration, but also for providing the influence that stabilizes at least one pole of the forming prophase spindle.

Another example of an asymmetric division is the first microspore division, which happens after meiosis to produce the generative and vegetative nuclei of the pollen grain (116). In many species, division is preceded by migration of the nucleus to the outer wall of the microspore and a concomitant stratification of cytoplasm into generative and vegetative components. After mitosis, a hemispherical wall curves around the outer daughter nucleus (116). This division has several unique features: first, each half-spindle at metaphase has a different type of pole, one broad and the other focused (32, 116); and second, despite the curving path of the forming cell plate, no specialized cortical site (such as a preprophase band) has been found (36, 83, 186). However, its most unusual feature is the presence of a polar organizing center, seen ultrastructurally as an electron-dense mass, several micrometers across, from which spindle microtubules radiate. This structure is prominent at the generative pole (i.e. within the future generative cell) (83), but is also evident at the other pole (36). The presence of an active spindle organizer in this division is further hinted at by observations of multinucleate microspores: at metaphase, a single spindle was observed, even though multiple spindles had formed at metaphase of the preceding (meiotic) division (41). These results, taken together, suggest that an active microtubule-organizing region at the generative pole of the first microspore division is involved in orienting the spindle in this asymmetric division.

The means whereby the spindle is oriented in the above-mentioned asymmetric divisions may represent an elaboration of the same mechanism that is present in the more usual type of ordered division in plant cells, for example in the production of regular files of cells. As we argue above, the essential feature of the mechanism may be to endow the location of the future spindle poles (or of at least one pole) with the ability to promote the nucleation, or the minus-end stability, of microtubules. Interactions among microtubules may then continue the process responsible for the appearance of the prophase spindle.

Prophase in Plant Cells Compared with Animal Cells

The events in prophase in plant cells stand in contrast to those in animal cells. The phragmosome and cortical division site, formed even before prophase, have no counterparts in most animal cells. They function in precisely control-ling the geometry of cell division (111, 113). Animal cells do have a specialized structure in the cortical cytoplasm, the contractile ring, that functions in cytokinesis, but this structure forms after the spindle and usually depends on the spindle for its placement. The formation of the spindle is also different in the two kingdoms. In animal cells the centrosome duplicates, and a bipolar array of microtubules grows between the separating centrosomes (119a). In plant cells a clear zone full of microtubules and devoid of organelles forms around the nucleus; this is followed by the appearance of microtubule foci that are already situated at opposite sides of the nuclei. An understanding of the significance of these differences in spindle organization can come only from learning more about the mechanistic basis for spindle organization in the cells of both animals and plants.

THE MITOTIC APPARATUS AT METAPHASE

Formation of the Metaphase Spindle

The formation of the metaphase spindle begins with the breakdown of the nuclear envelope, marking the start of prometaphase. This is followed by complex changes in microtubule deployment, as microtubules become associ-ated with chromosomes, and by movements of the chromosomes that result in their congression at the equator of the metaphase spindle. Once this equilibri-um position has been reached, metaphase begins; it lasts until the start of chromosome separation. The principles by which the metaphase spindle becomes organized are not known. Prior construction of a prophase spindle seems to be required, because spindles are not re-formed during recovery from microtubule depolymerization induced after prophase (44, 55).

NUCLEAR ENVELOPE BREAKDOWN   In higher plants, as in animals, the nuclear envelope breaks down for mitosis. In plants this process appears to be a fragmentation rather than a complete dissolution (80). Ultrastructural observations show that at or about the time of breakdown, the nuclear envelope may become sharply undulated (35, 36, 160) or may expand with ripples (described as boiling) prominent at the poles (9, 13). The breakdown products of the envelope apparently become mixed with ER (mostly smooth), forming a membrane complex, perhaps with specialized functions in mitosis (80). Fragments of nuclear envelope sometimes show remnants of nuclear-pore complexes; this makes a windowed structure, which has been likened to the sarcoplasmic reticulum of muscle cells (76, 79). In animal cells, regula-tion of breakdown involves a phosphorylation cascade, involving structural proteins (lamins) of the nuclear envelope (63), but nothing is known about the analogous regulation in plants.

After breakdown, membranes accumulate at the spindle poles by the end of prometaphase, and they remain there throughout mitosis (52, 104, 153, 154, 161). In one study, in which the area occupied by membranes was measured from serial sections, the amount of membranous material throughout the spindle doubled between interphase and late prometaphase (160). In early prometaphase, membranous elements are found throughout the forming spindle structure (76, 154). As spindle formation proceeds, the internal elements become fewer and are aligned along microtubules, typically around kinetochore microtubules (52, 79, 95, 97, 153, 160).

MICROTUBULE REARRANGEMENTS   Extensive and rapid microtubule rearrangements occur throughout prometaphase. The fusiform structure of the prophase spindle is lost as microtubules invade the nucleoplasm. The first sign of the new metaphase structure is the appearance of microtubules organized into bundles and sheets (44, 52, 90, 92, 96, 97). Some of these become associated with kinetochores, whereas others become nonkinetochore microtubules. An overall bipolar organization remains, but with broad rather than focused poles. Observations of the above process have been made by polarized light for microsporocytes (90) and endosperm (57, 92), and by immunofluorescence and electron microscopy for a number of different kinds of cells (44, 49, 86, 103, 107, 153). Because prometaphase is much shorter than prophase, the immunofluorescence observations of populations of cells have not been very successful in identifying intermediate stages in prometaphase microtubule rearrangements. The most complete observations of prometaphase have been made with *H. katherinae* endosperm, for which ultrastructural observations are made on cells that had been previously observed in vivo. Many kinds of structural arrangements between kinetochore microtubules and nonkinetochore microtubules have been found and are inferred to be dynamic (9, 72, 96, 97, 107). Dynamic kinetochore–spindle associations at prometaphase also occur in animal cells (134).

KINETOCHORE–MICROTUBULE ASSOCIATION   During prometaphase, microtubules become associated with kinetochores. This occurs asynchronously among kinetochores over a 10–20-min time span after nuclear envelope breakdown (9, 11). Two possible mechanisms could cause this association: the kinetochore may capture extant microtubules that reach its vicinity, or, alternatively, it may act as a nucleation site from which microtubules grow. In animal cells, labile microtubules constantly grow out from the centrosome, radiating in all directions. Those that reach a kinetochore are captured and stabilized, whereas those that do not are short-lived (124, 134). Although direct evidence from plants is lacking for kinetochore capture, the same process is likely to occur, with the polar regions serving as diffuse but active sites for microtubule nucleation.

The strongest objection to a leading role for kinetochore nucleation in the development of the spindle arises from studies of the polarity of its microtubules. These studies show that in both plant and animal spindles, kinetochore microtubules (and other microtubules in the same half-spindle as well) are nearly all of the same polarity, such that the plus ends are at the kinetochore, and the minus ends at the poles (54). Because all known microtubule-nucleating structures nucleate the minus end of the polymer, nucleation at the kinetochore is expected to produce microtubules of opposite polarity from those started at the pole, and few of these are observed. The centrosome, whether localized as in animal cells or flexible as in plant cells, is therefore assumed to nucleate almost all of the microtubules in the spindle (118).

Evidence consistent with kinetochore capture of microtubules comes from counts of microtubules made on *H. katherinae* endosperm cells, fixed at known times following nuclear envelope breakdown. Kinetochores are never observed with less than 12 microtubules (97); microtubules are more easily believed simultaneously captured than nucleated by a kinetochore. These groups of microtubules may have previously been part of the prophase spindle. The average number of microtubules per kinetochore rises steadily with time, reaching 80 by the end of metaphase (97); this could result from nucleation or capture of single microtubules. However, short microtubules ($<$ 2 μm) are rarely associated with kinetochores at any stage, except at telophase when they are common (97). Short microtubules would be expected if the kinetochore were nucleating microtubules, given a typical microtubule growth rate of a few micrometers per minute. Finally, in cross-section, ends of assembling (or disassembling) microtubules may appear "c"-shaped, rather than with the usual "o" shape. Examination of the microtubule ends terminating at the kinetochore shows that the percentage of c-shaped termini is greatest at prometaphase (20%) and zero at anaphase (97). These c-shaped termini probably indicate assembling microtubules that have been captured by the kinetochore.

However, other observations suggest that the kinetochore can nucleate microtubules (11). Short tufts of microtubules form at kinetochores following recovery from complete drug-induced depolymerization of microtubules (44, 55, 129); it is not known whether these have the same polarity as is normally found for kinetochore microtubules. One study found what appeared to be submicrometer-length stubs of microtubules embedded in the kinetochore and resistant to depolymerization (44); these stubs may serve as the basis for regrowth of microtubules in the recovery experiments. Other evidence suggests that kinetochore nucleation may happen normally during prometaphase. When counted in serial sections, more microtubules are present in the general vicinity of the chromosomes than at the poles (96). Although numbers of microtubules fall in general with proximity to the pole, the largest number of microtubules in the kinetochore bundle occurs a few micrometers distal to the kinetochore, as nonkinetochore microtubules mix with the true kinetochore ones (97). Therefore, it may be that the flexibile centrosome is not only spread in the plane perpendicular to the spindle axis but it also spread along that axis, extending some way toward the kinetochores.

Because kinetochores have the capacity to support microtubule growth or shrinkage while the plus end remains bound to the kinetochore (27, 124, 134), we suggest that if kinetochores were able to form a microtubule seed, subunits might add to its bound plus end while its minus end translocated away from the kinetochore. This suggestion allows microtubules originating at the kinetochore to have the same polarity as those nucleated at the poles. Kinetochore-generated microtubules may be important in circumstances when the amount of microtubule nucleation at the poles is small. It remains for future investigators to show whether plant kinetochores share the microtubule capping and translocating properties of their animal counterparts.

Chromosome Movements at Prometaphase

The chromosomes come to lie at the metaphase plate in a process called prometaphase congression. At prophase, chromosomes usually condense near the periphery of the nucleus, appressed to the nuclear membrane (139, 157). With the breakdown of the nuclear membrane, the chromosomes may suddenly move a small distance toward the center of the nucleus, in what has been termed prometaphase collapse (200). This movement has been seen in endosperm (4) and has been inferred in root tips from fixed and stained material (200), but has not been reported in other instances (13). This collapse does not result in any noticeable order among the chromosomes. Prometaphase chromosome movements then occur to bring the chromosomes to their equilibrium positions at the spindle equator. These movements have not been extensively studied in plants, but certain common features have been reported. First, chromosomes move independently, with velocities changing frequently in absolute value and also in sign. In endosperm cells (6), microsporocytes (159), and stamen hair cells (20), the typical velocity of these chromosome movements is 0.5 μm min^{-1}. This is significantly slower than that in animal cells, where movements, especially in early prometaphase, may be at least 20 times as fast (134). In endosperm cells, for which the most detailed observations have been made, the paths taken by the chromosomes are unpredictable and complex (7, 13, 107).

Once metaphase has been reached, the chromosomes may remain stably anchored at the equator (57) or may occasionally move to one pole and then move back to the equator again (7). Duration of metaphase is variable, ranging from a few minutes to an hour (13). When chromosomes are large, as they often are in plants (the largest one in *H. katherinae* cells measures 30 μm at metaphase), they have the appearance of lying in an irregular heap; however, ultrastructural observations of metaphase plates in root tip cells have shown that the kinetochore regions are precisely arrayed in a plane (84).

The mechanism whereby chromosomes are brought to and held at the equator is unknown. From observations in plant cells, Östergren proposed that microtubules pulled on kinetochores with a force proportional to their length (138), which gives rise to a stable equilibrium position for the chromosomes at the equator of the spindle. Further evidence for this model comes from measurements of kinetochore fiber lengths of trivalents at metaphase in animal cells (124, 134). Such a length-dependent force would arise if subunits were lost from over the whole microtubule length, as first proposed by the dynamic equilibrium theory (91); however, this would require the number of microtubules to be exactly the same at each sister kinetochore. Although this has in fact been observed to within counting error (i.e. ±2 microtubules) in *H. katherinae* cells (97), no mechanism to ensure such constancy has been proposed. An alternative to generating a length-dependent force is to suppose that a force-transducing protein on the surface of the kinetochore fiber engages some element of the surroundings (or vice versa, i.e. with the transducer bound in the matrix engaging the kinetochore microtubules) and exerts a force that moves the chromosome (134). This is at present no more than a theoretical possibility.

It has also been suggested that kinetochore-based motility may drive prometaphase movements in animal cells (124, 134); however, it is difficult to see how kinetochore-based motility alone would result in a stable equilibrium position for the chromosomes at the spindle equator. Indirect evidence, from endosperm, that the kinetochore alone does not drive prometaphase motility comes from irradiations with an ultraviolet (UV) microbeam. When kinetochores are irradiated at anaphase, chromosome movement is drastically reduced; however, when kinetochores are irradiated at any time before anaphase, including early prometaphase, no effect at all on congression is observed (8) (although subsequent anaphase separation of the irradiated kinetochores is inhibited). Clearly, more must be learned about the plant kinetochore before its role at prometaphase can be evaluated.

Microtubule Dynamics at Metaphase

Microtubules in the metaphase spindle of animal cells show dynamic instability, i.e. they turnover rapidly throughout the spindle (124, 162). Half-lives of microtubules in plant spindles have not been measured. An indication that plant microtubules might rapidly turn over comes from experiments in which microtubule polymerization is inhibited. Polar and interzonal microtubules are lost within 2–15 min of application of inhibitor (16, 129, 167); this could indicate that ongoing microtubule disassembly is no longer balanced by assembly. However, it is also possible that the inhibitors themselves induce disassembly of the polar and interzonal microtubules.

In addition to the rapid turnover of whole microtubules in the spindle of animal cells, Mitchison has recently shown that a flux of microtubule subunits moves poleward in one population of metaphase microtubules, evidently the kinetochore microtubules (125). The rate of movement was about one-third of the typical velocity of anaphase chromosome movement. This poleward flux probably results from addition of tubulin to the plus ends of kinetochore microtubules and concomitant loss from the minus ends. A poleward flux of unidentified material has been seen in *H. katherinae* endosperm; high-resolution Nomarski images of metaphase spindles show a movement of unresolved objects along kinetochore microtubules toward the poles (73). This movement occurs at a uniform velocity, about that of anaphase chromosome movement. Observations of spindle inclusions have shown that these also move to the pole at metaphase, with approximately the velocity of later chromosome movement to the pole (13, 17, 18). It is not known, however, whether these objects are stuck to the microtubule lattice and hence reflect a flux of subunits through the kinetochore microtubules, similar to what Mitchison observed. Alternatively, the flux of material may reflect the activity of some other transport system within the spindle (13, 17). Further experiments to test the dynamics of plant spindle microtubules in the spindle are required.

THE MITOTIC APPARATUS AT ANAPHASE

Anaphase chromosome movement involves two distinct processes: one in which chromosomes move toward the pole, called anaphase A, and one in which the poles are moved farther apart, called anaphase B (91). In plants both of these occur. In this section we first consider matters pertaining to anaphase in general, then those pertaining to anaphase A, and, finally, those pertaining to anaphase B.

General Anaphase Events

DISJUNCTION—THE START OF ANAPHASE   When chromosomes replicate, the daughter chromatids are held together in a tight complex. The start of anaphase is marked by the dissolution of this complex and the consequent "popping apart" of the chromatids. In both plants and animals, this popping, likened to an electrostatic repulsion, occurs simultaneously over all chromosomes and does not require microtubules. Synchronized popping occurs for acentric fragments, as well as for cells in which all traces of microtubules have been removed with inhibitors (104).

CHANGES IN STRUCTURE OF THE POLES   The poles of plant spindles have a physical coherence, which may increase during anaphase. Cohesion of the poles of plant spindles is indicated by the formation of monopolar spindles, with microtubules radiating in all directions, after treatment with any of a variety of compounds, e.g. taxol (13, 194). Chromosomes remain attached to the monopolar spindles and apparently can move toward the monopole (13, 81). Structural integrity of the poles is probably required to prevent anaphase chromosomes from dispersing too widely to be encompassed within a single reforming nucleus. A mutant of maize is known in which the meiotic spindle poles become divergent in metaphase, instead of convergent, and, as a result, numerous micronuclei form in telophase (41, 175). In the endosperm of a variety of plants, the poles become more focused during anaphase than they were in metaphase, which may indicate an increased polar cohesion (5, 57, 185). Images of nonendosperm cells also sometimes suggest a tightening of the poles in anaphase spindles compared wih metaphase spindles (32, 40, 147, 153). In microsporocytes, the spindle poles become focused by metaphase and remain so during anaphase (90, 131). The polar focusing in anaphase (or in metaphase for microsporocytes) may reflect a change in the composition or organization of the flexible centrosome; an anaphase change in the flexible centrosome is also indicated by the increased microtubule polymerization in the polar regions that occurs in *H. katherinae* endosperm (15).

ENERGETICS OF ANAPHASE   Nicklas has aptly said that mitosis is a question of finesse, not power (134). Given a reasonable estimate of cytoplasmic viscosity, the energy required to move the chromosomes is very low (134). A variety of evidence suggests that in animal cells ATP is not needed for chromosome-to-pole movement (anaphase A) but is needed for spindle elongation (anaphase B) (38, 124). In contrast, respiration inhibitors (i.e. azide and dinitrophenol) rapidly and reversibly inhibit anaphase A in guard mother cells and stamen hair cells (82). This inhibition was taken to mean that respiratory energy, if not ATP, is continuously needed during anaphase in plant cells. However, Amoore showed that the average rate of mitosis in a population of root tip cells was half-maximal at an oxygen tension low enough to totally inhibit respiration and to reduce the ATP level in the cell to 1% of its interphase value (1). Furthermore, on the basis of differences in sensitivity to inhibitors between mitosis and respiration, he inferred the involvement in plant mitosis of a nonrespiratory ferrous complex (1). This complex could well be sensitive to azide and dinitrophenol; therefore, until the effects of low oxygen levels on the rate of anaphase are tested in a variety of plant cell types, it is premature to conclude that respiration is required during anaphase A in plant cells.

Anaphase A

KINETICS OF CHROMOSOME-TO-POLE MOVEMENT The kinetics of chromosome-to-pole movements are generally similar between plants and animals. The velocity at which chromosomes move poleward in a variety of plant cell types has been measured (13, 20, 82, 159). An unexplained correlation exists in endosperm cells between cell size and chromosome velocity: the smaller the cell, the higher the speed (6). The average magnitude of anaphase A is usually 1–2 μm min^{-1}, with slower and faster examples known. Chromosomes usually move with a more or less stable rate from equator to pole (6, 20, 82, 141, 190), but they may show complex time-dependent changes in rate (57). Also, all the chromosomes of a half-spindle may move at identical rates (57) or may move somewhat independently (6). Separated sister chromatids may move at identical speeds (159). Chromosome movement in animals has similar rates and variations (39). Further evidence of similarity of anaphase A between plants and animals comes from a comparison of the effect of temperature on the rate of chromosome movement. Increasing temperature leads to an exponential rise in velocity, and the curves for several plant and animal cell types are the same (no normalization) (58). Although this accord could be fortuitous, it could imply the existence in plant and animal cells of a common mechanism driving chromosome movement.

CHANGES IN THE ORGANIZATION OF MICROTUBULES Considerable effort has been expended in attempts to understand the structural changes that occur in the microtubules of the spindle during anaphase. Perhaps as much is known about these changes in *H. katherinae* endosperm as is known about them in any animal cell type. Kinetochore microtubules shorten and diverge during anaphase, whereas polar microtubules increase in number and interact with the kinetochore microtubules. Detailed observations are discussed below.

Kinetochore microtubules are long (10–20 μm), running from the kinetochore to the polar region (97). Bundles of 8–20 microtubules are common in the interzone between kinetochores, and these often become associated with chromosomes and intermingle with one or more kinetochore fibers (96, 97, 105). Many other microtubules are found poleward of the kinetochores and are not particularly bundled. The disposition of these with respect to the kinetochore microtubules has been described as analogous to that of the boughs on a fir tree, with the kinetochore microtubules being the trunk and the kinetochore being the ground (15, 16, 126). This fir tree arrangement is particularly clear in preparations of *H. katherinae* endosperm made by using immunogold visualization of microtubules (15) or by using scanning electron microscopy (78). Aggregations of membranes are associated with the polar regions of these structures (95). Immunofluorescence images of metaphase or anaphase spindles in other cell types have small links between neighboring bright bundles of what are presumably kinetochore microtubules, and these have also been called fir trees (e.g. 31, 147). However, whether the spindle organization in these cell types truly resembles that of endosperm will require ultrastructural examination.

The significance of fir tree organization for spindle function is not clear. Not all ultrastructural examinations show fir tree–like structures. For example, electron-microscopic serial reconstruction of a meiotic spindle found a large central shaft of nonkinetochore microtubules; chromosomes arrayed on the outside of the central shaft had more or less uninterrupted kinetochore fibers, and few oblique, polar microtubules (i.e. fir trees) were observed (50). Furthermore, in cytoplasts, membrane-enclosed fragments of cytoplasm without a nucleus, which occur naturally in *H. katherinae* endosperm preparations, microtubules take on the fir tree organization spontaneously, without the presence of chromosomes (14, 15). Fir tree forms among groups of *H. katherinae* microtubules may thus reflect an inherent organizational tendency in these cells and need not reflect an essential element of spindle function.

As anaphase proceeds in *H. katherinae* endosperm, several kinds of changes happen to kinetochore microtubules. On average, they shorten continuously (15). They also increase their angle of divergence (i.e. the angle included by the most divergent members of the kinetochore fiber). This angle goes from about $10°$ at metaphase to nearly $60°$ at late anaphase (97). Kinetochore microtubules are also seen to diverge in on-line observations; soon after the start of anaphase, divergence of the kinetochore fibers begins at the poles and moves in a wave toward the kinetochore itself (73). To document further the changes in kinetochore fibers, Jensen counted kinetochore microtubules in serial sections at different stages of mitosis and found that the number of true kinetochore microtubules drops precipitously from 80 to 70 at the moment of chromatid popping (97); the significance of this result remains to be shown. Following that stage, kinetochore microtubules are lost at a constant rate, until few of these microtubules remain at telophase; furthermore, microtubules are removed only from the periphery of the kinetochore fiber (97). Jensen hypothesized that the lost kinetochore microtubules are broken off at the kinetochore and become oblique polar microtubules, i.e. fir tree boughs.

Changes during anaphase have also been found for the nonkinetochore microtubules in the spindle of *H. katherinae* endosperm cells. New microtubule polymerization occurs in the spindle. This is seen as an increase in quantity of microtubules in the polar regions of the cells and an increase in length of microtubules penetrating the interzone (14, 15, 126). Although new microtubule polymerization in many cell types is thought to occur in late anaphase or telophase, as a microtubule array radiates out from the reforming nuclear surface (e.g. 195), so far polymerization of new microtubules in early and mid-anaphase has been documented only for *H. katherinae* cells.

EFFECTS OF MICROTUBULE INHIBITORS   A pronounced difference between anaphase in plant and animal cells is the observed response to microtubule inhibitors. In animal cells, drugs such as colchicine will accelerate chromosome movement to the spindle poles (91). However, in plant cells, treatment with colchicine (or other compounds that specifically affect plant microtubules) freezes chromosome motion instantly (16, 129, 167), and, thereafter, chromosomes may slowly become randomized in the cell center (82). In animal cells, inhibitors prevent the polymerization of new microtubules and promote the depolymerization of extant ones (163). The mechanistic basis for the action of plant microtubule inhibitors on anaphase is not clear. Their overall effect seems to result from permanently preventing the kinetochore microtubules from depolymerizing; however, in vitro, these compounds have been shown to enhance the depolymerization of plant microtubules (129). We suggest that herbicides bind to plant kinetochores and inhibit the loss of microtubule subunits at the kinetochore. This suggestion implies that in plant cells, as in animal cells, subunits of kinetochore microtubules are lost during chromosome-to-pole movement at the kinetochore (27, 67, 124).

Only a few studies have observed the effects of microtubule inhibitors on anaphase on-line and then correlated them with structural observations of microtubules. These have provided information about the stability of different classes of spindle microtubules, as well as information about the mechanism of inhibitor action. Studies of endosperm cells are partly contradictory. In one, changes in microtubule organization or abundance were not detected until 15 min after chromosome motion had been frozen by APM (a dinitroaniline herbicide specific for plant microtubules), when there were fewer interzonal microtubules and oblique polar ones. Kinetochore fibers became increasingly parallel with time in the drug, and they persisted for many hours (16, 167). However, another study found that at a paralyzing concentration of oryzalin (a compound related to APM), all microtubules vanished within 2 min, except some kinetochore microtubules (129). However, this study did report that at a 10-fold-lower drug concentration, when anaphase slowed appreciably from control, no disturbance of the microtubule cytoskeleton could be detected. In neither study could the drug effects be reversed. Immunofluorescence observations of root tips have also shown that polar and interzonal microtubules are more sensitive than are kinetochore microtubules to depolymerization by oryzalin (44).

MICROTUBULE-BASED MODELS FOR FORCE PRODUCTION   Several models for anaphase A are based on properties of microtubules. The model called zipping, proposed by Bajer, posits that peripheral kinetochore microtubules interact laterally with other microtubules in the spindle, especially the oblique polar ones that are seen as fir tree boughs (10). This model has kinetochore microtubules zipping together with oblique ones, which will exert a poleward force on the chromosome. Lateral associations between microtubules are common in fixed preparations (10, 97), and labile lateral associations between microtubules have also been seen on-line in high-resolution video microscopy of anaphase in *H. katherinae* cells (94). Furthermore, in cells shifted to near-freezing temperatures, chromosome motion stops as kinetochore microtubules converge, and with a return to room temperature, chromosome motion resumes as kinetochore microtubules diverge (106). The conspicuous fir tree morphology of endosperm kinetochore bundles has been taken as evidence to support the importance of zipping in anaphase A (15).

Although lateral interactions between microtubules are likely to be real, their contribution to chromosome movement is far from clear. We suggest alternatively that in the absence of any polar organizing structure, lateral zipping interactions between some kinetochore and nonkinetochore microtubules may help anchor the polar regions so that they remain fixed while the chromosomes move. Evidence has been obtained that supports a role for lateral interactions between microtubules in providing rigidity to the polar regions of the spindle in insect spermatocytes. Nicklas severed metaphase spindles between pole and kinetochore, thus removing about one-third of the spindle, including the pole; subsequently, chromosomes moved poleward and reached within 1 μm of the cut edge (135). For this to have happened, microtubule ends near the cut edge must be held together strongly enough to resist the force moving the chromosomes.

Another microtubule-based model for anaphase A invokes microtubule disassembly as the source of energy for chromosome movement. Inoué observed that in a variety of animal material, chromosome velocity, whether induced or natural, is always correlated with the rate of loss of birefringence of kinetochore microtubules (91). He proposed a model in which loss of subunits from the microtubule drives chromosomes poleward (91). In an important study of the endosperm of *Tilia americana*, Fuseler measured the birefringence over time of a spindle region just poleward of the kinetochores and found that birefringence decayed exponentially with time and that the rate constant for this decay was proportional to the velocity of chromosome movement (58). The absolute value of birefringence was not related to chromosome velocity. Comparing anaphase *T. americana* with that of an echinoderm egg, the constant of proportionality between chromosome velocity and rate of birefringence decay is not the same: a given rate of birefringence decay produces faster chromosome movement in the plant spindle than in that of the animal. However, both extrapolate to the same (positive) rate constant of birefringence decay at zero chromosome velocity, which could be coincidence or could indicate an underlying invariant in the force producing systems of the two species. Furthermore, Fuseler's data extend to plants the range of material in which microtubule disassembly is closely linked to chromosome movement (58, 91, 163).

Further support for microtubule disassembly's being important in anaphase A comes from studies in which microtubule assembly is promoted. The compound taxol is known to promote microtubule assembly in both plant and animal material (129, 180, 194). When taxol is added to endosperm cells at anaphase, a slowing and even a reversal of chromosome motion occurs (12), although some chromosomes eventually move poleward. When endosperm cells are lysed in microtubule-stabilizing buffers, chromosome movement slows (17). These examples demonstrate that in plant cells, as has been shown for animal cells, microtubule assembly (disassembly) has the ability to move chromosomes. Whether and how this ability is used in undisturbed cells remain a matter of debate.

KINETOCHORE-BASED MODELS FOR FORCE PRODUCTION   In animal cells, the kinetochore is now thought to be the site where force is generated to move the chromosomes poleward (27, 119a, 124, 134). Force may be produced by microtubule depolymerization at the kinetochore or from the presence there of a microtubule-dependent translocator protein. For plants, the most that may be said is that some observations are consistent with a role for the kinetochore in chromosome movement other than attaching the chromosome to the spindle. Irradiation of kinetochores with a UV microbeam at any time during anaphase in *H. katherinae* cells inhibits the poleward movement of only the irradiated chromosome and results instead in prometaphase-like oscillatory movement toward the equator. Because the irradiation did not sever the kinetochore microtubules, the implication from the UV-induced loss of anaphase chromosome movement is that force transduction at the kinetochore was inhibited (8). Finally, as we argue above, the simplest explanation for the herbicide effects on anaphase A is that herbicides inhibit the force-generating process at the kinetochore. The structure and function of the plant kinetochore remain important subjects for future research.

Anaphase B

Spindle elongation, the separation of half-spindles, or anaphase B, has long been known to occur in plants (190). The contribution made by spindle elongation to the total separation of chromosomes may be substantial or minor (6, 57). In some cells, especially those in which the spindle poles at metaphase abut the end walls of the cell, but in others as well, spindle elongation may be entirely absent (82, 159). In maize (51) and *T. americana* (57) endosperm there is substantial anaphase B, but in other species there is scant elongation of the spindle (6). Anaphase B appears substantial in published micrographs of the generative cell division in pollen tubes (116, 140, 179), but neither the extent nor the rate of spindle elongation in this cell type has been quantified. It is not known whether spindle elongation occurs in meristematic cells, such as root tips. Spindle elongation may precede chromosome-to-pole movement (57), but the two are usually more or less concurrent (6). Similar variability of the observed timing and extent of anaphase B is also observed in animal cells (39). The rate of elongation is similar to the rate of chromosome movement, about $1 \mu m$ min^{-1} (6, 57), which is also about the same as that for animal cells (39). In general, quantification of the separation of the spindle poles in plants is not as straightforward as in animals, because there is usually not a well-defined, pointlike pole in plants; moreover changes in the polar organization occur during anaphase in *H. katherinae* endosperm cells (15).

The mechanism of anaphase B in diatoms is an ATP-dependent process, which is driven by interactions between antiparallel microtubule arrays in the zone of overlap between half-spindles (38). Microtubule polymerization does not drive anaphase B in diatoms; tubulin polymerization provides polymer that can slide through the zone of overlap as the spindle elongates, and thus determines the total extent of elongation (38). Evidence suggests that the same may be true for diverse animal cell types, such as mammalian cells, insect spermatocytes, and echinoderm eggs (38). There is relatively little evidence bearing directly on the mechanism of anaphase B in plants. On the one hand, in images of fixed cells, abundant interzonal microtubules are seen [e.g. root tips (31, 147, 195), maize endosperm (51), and wheat meristems (153)], which could be involved in spindle elongation. However, the images are often of stages late in anaphase, and so these interzonal microtubules could instead be establishing the phragmoplast. In *H. katherinae* endosperm, in which a modest degree of spindle elongation usually occurs (6), the number of microtubules in the interzone in mid-anaphase often becomes small (49, 105, 167, 185) but their alignment becomes greater (73, 78, 96). Anaphase B of these cells has been said to result from a kind of autonomous swimming apart of the half-spindles (126), pushing off, perhaps, against the cytomatrix. Because new microtubules invade the interzone during the middle and late stages of anaphase, interactions between antiparallel microtubules have also been proposed to drive spindle elongation (14, 15). The latter possibility is similar to what occurs in the diatom spindle during anaphase B (38).

Finally, many studies have shown no effect of actin microfilament poisons on chromosome-to-pole movement in plants (176), but these have not considered spindle elongation. In one exception, treatment with cytochalasin inhibited anaphase B (167); however, the treatment also inhibited the stretching of the F-actin cage surrounding the spindle, which may have physically prevented spindle elongation from occurring at all. In conclusion, studies of systems with extensive anaphase B, such as *T. americana* endosperm or perhaps generative cell division, are required to gather a better understanding of this facet of plant mitosis.

FORMATION AND FUNCTION OF THE PHRAGMOPLAST

The phragmoplast is a cytoskeletal structure of great complexity that forms after chromosome separation is mostly complete. It functions to construct a new wall between daughter cells; however, the mechanism whereby the phragmoplast helps control the deposition of wall material is not understood. The formation of the phragmoplast involves the new polymerization of microtubules from the polar regions of the spindle and from the reforming nuclear envelope and involves the stabilization of interdigitated, antiparallel microtubules at the site of the future wall. Filamentous actin also appears at early stages in phragmoplast formation. The phragmoplast forms centrally in the cell, between nuclei, and then expands until the parent walls have been reached (70). The cortical division site and the phragmosome, formed at preprophase, are instrumental in guiding the expanding phragmoplast to the correct site where the new wall joins the old; this is discussed above in the section on prophase.

The phragmoplast should be considered a new structure and not merely a derivative of the mitotic spindle. Phragmoplasts commonly form between any adjacent nuclei, not just between sisters where spindles would have been previously. Phragmoplasts form between non-sister nuclei during the cellularization of endosperm (70) and of megagametophyte tissue (25); they form between adjacent micronuclei that occur naturally (30, 175, 191) or artificially (for example, when multipolar spindles are induced by spindle-disrupting agents) (43, 71, 114).

The organizing centers for the phragmoplast microtubules are not known, but evidence suggests that the nuclear envelope plays a role in the organization. Ultrastructural observations reveal that the earliest sign of cell plate formation is clusters of interdigitating microtubules, pointed toward each spindle pole, with vesicles and osmiophilic fuzz gathered in the region of microtubule overlap (80a, 70). It was once assumed that this material was the microtubule-organizing center for the phragmoplast. However, later studies show that the polarity of phragmoplast microtubules, in early and late stages, is overwhelmingly parallel in each half-spindle, with plus ends being in the zone of overlap, suggesting a site of nucleation toward the spindle poles (54). The nuclear envelope itself may play this role. In root tips (and also in other cell types), the phragmoplast microtubules are seen first as long, parallel fibers filling the space from one daughter nucleus to the other (44, 69, 117, 195), with a region of overlap halfway between nuclei. At the same time, microtubules, albeit in lesser numbers, also radiate from all sides of the daughter nuclei into the cytoplasm (31, 69, 120).

However, observations in endosperm, where temporal stages can be related to ultrastructure, suggest that phragmoplast-organizing material may be present, at least initially, in diffuse form at the spindle poles. Increased polymerization of microtubules at the polar regions and increased microtubule presence in the interzone begin before the nuclear envelope has reformed, while the chromosomes are still moving poleward (14, 15, 49, 126, 165, 186). The birefringence of the forming phragmoplast rises while that of the kinetochore fibers falls in *T. americana* (57). Once the nuclear envelope forms in endosperm, microtubules do radiate from its surface, and many of these are involved in the phragmoplast (15, 129, 168). Hence, the diffuse polar material involved in spindle organization may also participate in phragmoplast organization, possibly coalescing on the surface of the nuclear envelope once it has re-formed.

The electron-dense material present between interdigitating microtubules at the earliest stage of phragmoplast formation (36, 70, 80a) may cross-link and stabilize antiparallel microtubules and mark the beginning of new wall formation. If microtubules grow at equal rates away from each spindle pole or daughter nucleus, the zone of antiparallel interactions will be halfway between them, which is where the cell plate forms. When there are micronuclei of different sizes, the cell plate forms closer to the smaller one, suggesting that the zone of antiparallel interactions is displaced toward the smaller nucleus because of its smaller production of microtubules (e.g. 30). The importance of microtubule interactions in organizing the phragmoplast is further emphasized by observations of cytoplasts, formed naturally from *H. katherinae* endosperm, in which apparently functional phragmoplasts form in the absence of any detectable chromatin (14).

Once established, the phragmoplast expands centrifugally until it meets the parent walls. During expansion, the microtubules in the center disappear, so that an annular structure is formed. Microtubules in these outer regions are usually shorter than the original ones and no longer have obvious links to the nuclear surface (e.g. 44, 71). The truly annular shape of the growing phragmoplast has been demonstrated in on-line observations of periclinally dividing epidermal cells (188). The velocity of growth in diameter was measured as 0.2 μm min^{-1}. Given the large area of the phragmoplast and its velocity of expansion, Gunning has pointed out that a rapid rate of turnover of component microtubules is required (70); nevertheless, microtubules in the phragmoplast are resistant to drug- and cold-induced depolymerization (14, 44, 129, 167).

A challenge for the future is to discover the basis for this dynamic *stability* of phragmoplast microtubules.

Actin microfilaments are detected in phragmoplasts (e.g. 146, 181); in mature stages, they appear as a ring of short filaments, colocalizing with phragmoplast microtubules (98, 111, 176). Some studies report that cytochalasin inhibits cytokinesis (167, 181), whereas others report that it does not (127, 143, 148). Recently, phragmoplasts were isolated from synchronized tobacco tissue culture cells (99). Microfilaments run mostly parallel to, and sometimes laterally associated with, the microtubules. Actin polarity is appropriate to support streaming toward the forming cell plate. Part of the inconsistency in the cytochalasin studies may result from redundant mechanisms for delivery of material to the forming cell plate, with greater and lesser contributions from actin-based motility.

Contradictory observations concerning the role of actin in the early stages of phragmoplast formation have been made in *H. katherinae* endosperm. There is agreement that early in this process, short, random microfilaments are seen at the equator (127, 128, 165–167). However, on the one hand, the microfilaments are claimed to precede the increased microtubule presence at the equator and therefore to be active participants in localizing cell plate formation (166); on the other hand, they are said to appear along with the microtubule buildup at the equator and therefore to be fragments of the actin cytoskeleton passively gathered by the same process that also gathers bits of broken-off chromosomes and other particles (127, 128). A gathering of equatorial actin late in anaphase has also been seen in other cell types (e.g. 71, 181), which implies that F-actin is generally present at early stages of phragmoplast formation; studies of these cell types have not been made with sufficient temporal resolution to compare the timing of appearance of equatorial F-actin and microtubules. Further study is needed to understand the roles of both microfilaments and microtubules in the process of wall construction carried out by the phragmoplast.

DIVISION OF THE GENERATIVE CELL IN POLLEN TUBES

A division that offers us an opportunity to learn more about spindle organization and function in plants occurs when the generative cell nucleus divides while within the pollen tube (116). In some species (or under some pollen tube growth conditions), a typical metaphase plate is not produced; instead, the chromatid pairs form a line parallel to the direction of separation (116, 140, 164). Interestingly, in one species, the onset of chromosome movement in these linear arrangements of chromosomes begins at one end of the tube and sweeps toward the other end in a rapid wave (109). A predictable spatial pattern to the onset of anaphase may be unique and may provide a good experimental system to study the trigger for anaphase onset.

The organization of the spindle microtubules in this division is unusual. This may reflect the normal pattern of spindle formation confined in the narrows of the pollen tube or, alternatively, may reflect the induction of systems that function specifically in this division. When a normal metaphase plate forms, the spindle resembles that formed in other plant cells (155), but when a metaphase plate does not form, the spindle appears to be different (35, 148, 179). The particular differences reported were not all shared, which may be explained by the use of different species in each investigation. In one report, the spindle comprised two half-spindles, separated from one another by the row of chromatid pairs positioned between them. Anaphase appeared to result from a sliding apart of the two half-spindles (179). In another report, coherent half-spindle architecture was not apparent. Instead, many microtubule bundles ran longitudinally in the generative cell cytoplasm; some bundles terminated at kinetochores and others continued uninterrupted. The chromosomes did not appear in any obvious arrangement (148). Further on-line and ultrastructural observations are needed to clarify the process of generative cell division and to learn whether it uses unique components.

OTHER MOTILITY DURING MITOSIS

Movements of chromosomes are studied more often than those of other cell components at mitosis, but other movements within the cell occur and are of possible relevance for understanding mitosis. Cytoplasmic streaming and even Brownian motion slow in prophase, stop entirely at metaphase, and return vigorously by telophase (20, 141). This loss of motility presumably reflects an underlying reorganization of the cytoskeleton in dividing cells, especially of actomyosin (111, 176). In thin strands of cytoplasm connecting isolated endosperm cells, rapid organelle movements (1 μm s^{-1}) occur (18), but their ultrastructural basis has not been shown.

Spindle-associated motility of elements other than chromosomes in endosperm has been studied extensively by Bajer et al (7, 13). Recently, they have monitored the movements of injected 40-nm gold particles or of X-ray-induced chromosome fragments (17, 18). At prometaphase, a uniform poleward flow was seen, at about the same velocity as chromosomes would move at anaphase; then, by mid- to late anaphase, the movement became equatorial and nonuniform, two- to fivefold faster than chromosome velocity; finally, in telophase, on the phragmoplast, a slow poleward movement returned, again at about the same velocity as that of anaphase chromosomes. This poleward flow of particles has been seen on the phragmoplast-like structures that form in fragments of endosperm cytoplasm that lack nuclei (14). However, during anaphase, an equatorial flow of material at the spindle surface has been reported for stamen hair cells (141). These observations are intriguing, but their mechanistic basis and their relationship to the mechanisms responsible for chromosome movements are not known.

CONCLUSION

In this review, we have attempted to build a picture of the structure and function of the plant mitotic spindle. Overall, the evidence indicates that mitotic processes in plants and animals have a great deal in common. Indirect evidence suggests that similar mechanisms of prometaphase chromosome congression, anaphase chromosome movement, and spindle elongation are found in both kingdoms. The biggest differences seem to lie in establishing the spindle, when plant but not animal cells organize a bipolar prophase spindle out of a proliferation of microtubules around the nucleus, and in cytokinesis, when plant cells form a phragmoplast and animal cells form a contractile ring. Differences in the organization of the spindle poles, which are often heralded as important distinctions between animal and plant mitotic spindles, probably do not reflect any fundamental difference in spindle organization or function.

More must be learned about the mechanism of mitosis in the cells of higher plants before these conclusions can be confirmed. We can offer some suggestions for future experimentation. On-line observations must be made of a variety of cell types beyond those few kinds usually observed; the need is especially acute for ultrastructural observations following observation of living cells in cell types other than *H. katherinae* endosperm. Microtubule dynamics must be studied directly in living plant cells by monitoring the incorporation of derivatized tubulin. Cell models, involving either lysed cells or isolated spindles, must be developed for plant material. Spindle-associated proteins must be characterized biochemically, with close attention to proteins that mediate microtubule-microtubule interactions or cause nucleotide-dependent microtubule motility. Finally, genetic and molecular approaches are required that lead to the analysis of the genes required for mitosis. Meeting all these needs will be challenging; but this is necessary to an understanding of the structure and function of the mitotic spindle in flowering plants.

ACKNOWLEDGMENTS

We thank Chris Staiger for helpful discussions and Anne Sylvester and Chris Hogan for critical reading of the typescript. This work was supported by a National Science Foundation Postdoctoral Research Fellowship in Plant Biology, awarded to T.I.B. in 1986, and by Public Health Service grant GM 23238 from the National Institutes of Health and U.S. Department of Agriculture grant 8901117 to W.Z.C.

Literature Cited

1. Amoore, J. E. 1963. Non-identical mechanisms of mitotic arrest by respiratory inhibitors in pea root tips and sea urchin eggs. *J. Cell Biol.* 18:555–66
2. Apostolakos, P., Galatis, B. 1987. Induction, polarity and spatial control of cytokinesis in some abnormal subsidiary cell mother cells of *Zea mays*. *Protoplasma* 140:26–42
3. Bajer, A. 1957. Ciné-micrographic studies on mitosis in endosperm. III. The origin of the mitotic spindle. *Exp. Cell Res.* 13:493–502
4. Bajer, A. 1958. Ciné-micrographic studies on mitosis in endosperm. IV. The mitotic contraction stage. *Exp. Cell Res.* 14:245–56
5. Bajer, A. 1965. Behavior of chromosomal spindle fibers in living cells. *Chromosoma* 16:381–90
6. Bajer, A., Molè-Bajer, J. 1954. Endosperm, material for study on the physiology of cell division. *Acta Soc. Bot. Polon.* 23:69–98
7. Bajer, A., Molè-Bajer, J. 1956. Ciné-micrographic studies on mitosis in endosperm. II. Chromosome, cytoplasmic and brownian movements. *Chromosoma* 7:558–607
8. Bajer, A., Molè-Bajer, J. 1961. Ultraviolet microbeam irradiation of chromosomes during mitosis in endosperm. *Exp. Cell Res.* 25:251–67
9. Bajer, A., Molè-Bajer, J. 1969. Formation of spindle fibers, kinetochore orientation and behavior of the nuclear envelope during mitosis in endosperm. *Chromosoma* 27:448–84
10. Bajer, A. S. 1973. Interaction of microtubules and the mechanism of chromosome movement (zipper hypothesis). I. General principle. *Cytobios* 8:139–60
11. Bajer, A. S. 1987. Substructure of the kinetochore and reorganization of kinetochore microtubules during early prometaphase in *Haemanthus* endosperm. *Eur. J. Cell Biol.* 43:23–34
12. Bajer, A. S., Cypher, C., Molè-Bajer, J., Howard, H. M. 1982. Taxol-induced anaphase reversal: evidence that elongating microtubules can exert a pushing force in living cells. *Proc. Natl. Acad. Sci. USA* 79:6569–73
13. Bajer, A. S. Molè-Bajer, J. 1972. Spindle dynamics and chromosome movements. *Int. Rev. Cytol. Supl.* 3:1–273
14. Bajer, A. S., Molè-Bajer, J. 1982. Asters, poles, and transport properties within spindlelike microtubule arrays. *Cold Spring Harbor Symp. Quant. Biol.* 46:263–83
15. Bajer, A. S., Molè-Bajer, J. 1986. Reorganization of microtubules in endosperm cells and cell fragments of the higher plant *Haemanthus*. *J. Cell Biol.* 102:263–81
16. Bajer, A. S., Molè-Bajer, J. 1986. Drugs with colchicine-line effects that specifically disassemble plant but not animal microtubules. *Ann. NY Acad. Sci.* 466:767–84
17. Bajer, A. S., Vantard, M. 1988. Microtubule dynamics determine chromosome lagging and transport of acentric fragments. *Mut. Res.* 201:271–81
18. Bajer, A. S., Vantard, M., Molè-Bajer, J. 1987. Multiple mitotic transports expressed by chromosome and particle movements. *Fortschr. Zool.* 34:171–86
19. Bakhuizen, R., Van Spronsen, P. C., Sluiman-den Hertog, F. A. J., Venverloo, C. J., Goosen-de Roo, L. 1985. Nuclear envelope radiating microtubules in plant cells during interphase mitosis transition. *Protoplasma* 128:43–51
20. Barber, H. N. 1939. The rate of movement of chromosomes on the spindle. *Chromosoma* 1:33–50
21. Baskin, T. I., Cande, W. Z. 1989. Kinetic analysis of anaphase B *in vitro*. *J. Cell Biol.* In press
22. Becker, W. A. 1938. Recent investigations in vivo on the division in plant cells. *Bot. Rev.* 4:446–72
23. Bednara, J., Van Lammeren, A. M. M., Willemse, M. T. M. 1988. Microtubular configuration during meiosis and megasporogenesis in *Gasteria verrucosa* and *Chamaenerion angustifolium*. *Sex. Plant Reprod.* 1:164–72
24. Bergmann, L. 1960. Growth and division of single cells of higher plants in vitro. *J. Gen. Physiol.* 43:841–51
25. Bhandari, N. N., Chitralekha, P. 1989. Cellularization of the female gametophyte in *Rannunculus scerleratus*. *Can. J. Bot.* 67:1244–53
26. Braselton, J. P., Bowen, C. C. 1971. The ultrastructure of the kinetochores of *Lilium longiflorum* during the first meiotic division. *Caryologia* 24:49–60
27. Brinkley, B. R., Valdivia, M. M., Tousson, A., Balczon, R. D. 1989. See Ref. 89, pp. 76–118
28. Brinkley, B. R., Zinkowski, R. P., Mollow, W. L., Davis, F. M., Pisegna, M. A., et al. 1988. Movement and segregation of kinetochores experimentally detached from mammalian chromosomes. *Nature* 336:251–54
29. Brown, R. C., Lemmon, B. E. 1988. Microtubules associated with simulta-

neous cytokinesis of coenocytic microsporocytes. *Am. J. Bot.* 75:1848–56

30. Brown, R. C., Lemmon, B. E. 1989. Minispindles and cytoplasmic domains in microsporogenesis of orchids. *Protoplasma* 148:26–32

31. Brown, R. C., Lemmon, B. E., Mullinax, J. E. 1989. Immunofluorescent staining of microtubules in plant tissues: improved embedding and sectioning techniques using polyethylene glycol (peg) and Steedman's wax. *Bot. Acta* 102:54–61

32. Brumfield, R. T. 1941. Asymmetrical spindles in the first microspore division of certain angiosperms. *Am. J. Bot.* 41:713–22

33. Bryant, J. A., Francis, D., eds. 1985. *The Cell Division Cycle in Plants*. Cambridge, UK: Cambridge Univ. Press

34. Burgess, J. 1969. Two cytoplasmic features of prophase in wheat root cells. *Planta* 87:259–70

35. Burgess, J. 1970. Cell shape and mitotic spindle formation in the generative cell of *Endymion non-scriptus*. *Planta* 95:72–85

36. Burgess, J. 1970. Microtubules and cell division in the microspore of *Dactylorchis fuschii*. *Protoplasma* 69:253–64

37. Burgess, J., Northcote, D. H. 1968. The relationship between the endoplasmic reticulum and microtubular aggregation and disaggregation. *Planta* 80:1–14

38. Cande, W. Z., Hogan, C. J. 1989. The mechanism of anaphase spindle elongation. *BioEssays* 11:5–9

39. Carlson, J. G. 1977. Anaphase chromosome movement in the unequally dividing grasshopper neuroblast and its relation to anaphases of other cells. *Chromosoma* 64:191–206

40. Cho, S.-O., Wick, S. M. 1989. Microtubule orientation during stomatal differentiation in grasses. *J. Cell Sci.* 92:581–94

41. Clark, F. J. 1940. Cytogenetic studies of divergent meiotic spindle formation in *Zea mays*. *Am. J. Bot.* 27:547–59

42. Clayton, L., Black, C. M., Lloyd, C. W. 1985. Microtubule nucleating sites in higher plant cells identified by an auto-antibody against pericentriolar material. *J. Cell Biol.* 101:319–24

43. Clayton, L., Lloyd, C. W. 1984. The relationship between the division plane and spindle geometry in *Allium* cells treated with CIPC or griseofulvin: an anti-tubulin study. *Eur. J. Cell Biol.* 34:248–53

44. Cleary, A. L., Hardham, A. R. 1988. Depolymerization of microtubule arrays in root tip cells by oryzalin and their recovery with modified nucleation patterns. *Can. J. Bot.* 66:2353–66

45. Cleary, A. L., Hardham, A. R. 1989. Microtubule organization during development of stomatal complexes in *Lolim rigidum*. *Protoplasma* 149:67–81

46. Cyr, R. J., Palevitz, B. A. 1989. Microtubule binding proteins from carrot. I. Initial characterization and microtubule bundling. *Planta* 177:245–60

47. Dawson, P. J., Lloyd, C. W. 1987. A comparison of plant and animal tubulins. In *The Biochemistry of Plants: A Comprehensive Treatise*, Vol. 9: *Metabolism*, ed. D. D. Davies. London: Academic

48. DeBrabander, M., De Mey, J., eds. 1985. *Microtubules and Microtubule Inhibitors, 1985*. Amsterdam: Elsevier Science

49. De Mey, J., Lambert, A.-M., Bajer, A. S., Moeremans, M., De Brabander, M. 1982. Visualization of microtubules in interphase and mitotic plant cells of *Haemanthus* endosperm with the immunogold staining method. *Proc. Natl. Acad. Sci. USA* 79:1898–1902

50. Dietrich, J. 1979. Reconstructions tridimensionnelles de l'appareil mitotique à partir de coupes sériées longitudinales de méiocytes polliniques. *Biol. Cell.* 34:77–82

51. Duncan, R. E., Persidsky, M. D. 1958. The achromatic figure during mitosis in maize endosperm. *Am. J. Bot.* 45:719–29

52. Esau, K., Gill, R. H. 1969. Structural relations between nucleus and cytoplasm during mitosis in *Nicotiana tabacum* mesophyll. *Can. J. Bot.* 47:581–91

53. Esponda, P. 1978. Cytochemistry of kinetochores under electron microscopy. *Exp. Cell Res.* 114:247–52

54. Euteneuer, U., Jackson, W. T., McIntosh, J. R. 1982. Polarity of spindle microtubules in *Haemanthus* endosperm. *J. Cell Biol.* 94:644–53

55. Falconer, M. M., Donaldson, G., Seagull, R. W. 1988. MTOC's in higher plant cells: an immunofluorescent study of microtubule assembly sites following depolymerization by APM. *Protoplasma* 144:46–55

56. Fukuda, H. 1989. Regulation of tubulin degradation in isolated *Zinnia* mesophyll cells in culture. *Plant Cell Physiol.* 30:243–52

57. Fuseler, J. W. 1975. Mitosis in *Tilia americana* endosperm. *J. Cell Biol.* 64:159–71

58. Fuseler, J. W. 1975. Temperature dependence of anaphase chromosome

velocity and microtubule depolymerization. *J. Cell Biol.* 67:789–800

59. Galatis, B., Apostolakos, P., Katsaros, C. 1983. Synchronous organization of two preprophase microtubule bands and final cell plate arrangement in subsidiary cell mother cells of some *Triticum* species. *Protoplasma* 117:24–39

60. Galatis, B., Apostolakos, P., Katsaros, C. 1984. Positional inconsistency between preprophase microtubule band and final cell plate arrangement during triangular subsidiary cell and atypical hair cell formation in two *Triticum* species. *Can. J. Bot.* 62:343–59

61. Galatis, B., Apostolakos, P., Katsaros, C. 1984. Experimental studies on the function of the cortical cytoplasmic zone of the preprophase microtubule band. *Protoplasma* 122:11–26

62. Galatis, B., Apostolakos, P., Katsaros, C., Loukari, H. 1982. Pre-prophase microtubule band and local wall thickening in guard mother cells of some leguminosae. *Ann. Bot.* 50:779–91

63. Gerace, L., Barhe, B. 1988. Functional organization of the nuclear envelope. *Annu. Rev. Cell Biol.* 4:335–74

64. González-Fernández, A., Selman, A. M., Giménez-Martín, G., De la Torre, C. 1989. Protein synthesis in late mitosis and in G_1 required for cell cycle progression in synchronous onion root meristem cells. *Eur. J. Cell Biol.* 49:87–91

65. Goodbody, K. C., Hargreaves, A. J., Lloyd, C. W. 1989. On the distribution of microtubule-associated intermediate filament antigens in plant suspension cells. *J. Cell Sci.* 93:427–38

66. Goosen-de Roo, L., Bakhuizen, R., Van Spronsen, P. C., Libbenga, K. R. 1984. The presence of extended phragmosomes containing cytoskeletal elements in fusiform cambial cells of *Fraxinus excelsior* L. *Protoplasma* 122:145–52

67. Gorbsky, G. J., Sammak, P. J., Borisy, G. G. 1987. Chromosomes move poleward in anaphase along stationary microtubules that coordinately disassemble from their kinetochore ends. *J. Cell Biol.* 104:9–18

68. Gorst, J., Wernicke, W., Gunning, B. E. S. 1986. Is the preprophase band of microtubules a marker of organization in suspension cultures? *Protoplasma* 134:130–40

69. Gubler, F. 1989. Immunofluorescence localization of microtubules in plant root tips embedded in butyl-methyl methacrylate. *Cell Biol. Int. Rep.* 13:137–45

70. Gunning, B. E. S. 1982. See Ref. 112, pp. 229–92

71. Gunning, B. E. S., Wick, S. M. 1985. Preprophase bands, phragmoplasts and spatial control of cytokinesis. *J. Cell Sci. Suppl.* 2:157–79

72. Hanaoka, A. 1981. Kinetochore and kinetochore fiber. I. Electron microscopic studies on the mitotic endosperm protoplast. *Cytologia* 46:331–42

73. Hard, R., Allen, R. D. 1977. Behavior of kinetochore fibers in *Haemanthus katherinae* during anaphase movements of chromosomes. *J. Cell Sci.* 27:47–56

74. Hargreaves, A. J., Dawson, P. J., Butcher, G. W., Larkins, A., Goodbody, K. C., Lloyd, C. W. 1989. A monoclonal antibody raised against cytoplasmic fibrillar bundles from carrot cells, and its cross-reaction with animal intermediate filaments. *J. Cell Sci.* 92:371–78

75. Harper, J. D. I., Mitchison, J. M., Williamson, R. E., John, P. C. L. 1989. Does the autoimmune serum 5051 specifically recognize microtubule organizing centers in plant cells? *Cell Biol. Int. Rep.* 13:471–83

76. Hawes, C. R., Juniper, B. E., Horne, J. C. 1981. Low and high voltage electron microscopy of mitosis and cytokinesis in maize roots. *Planta* 152:397–407

77. Heath, I. B. 1980. Variant mitoses in lower eukaryotes: indicators of the evolution of mitosis? *Int. Rev. Cytol.* 64:1–80

78. Heneen, W. K. 1981. Mitosis as described in the scanning electron microscope. *Eur. J. Cell Biol.* 25:242–47

79. Hepler, P. K. 1980. Membranes in the mitotic apparatus of barley cells. *J. Cell Biol.* 86:490–99

80. Hepler, P. K. 1989. See Ref. 89, pp. 241–71

80a. Hepler, P. K., Jackson, W. T. 1968. Microtubules and early stages of cell plate formation in the endosperm of *Haemanthus katherinae* Baker. *J. Cell Biol.* 38:437–46

81. Hepler, P. K., Jackson, W. T. 1969. Isopropyl N-phenylcarbamate affects spindle microtubule orientation in dividing endosperm cells of *Haemanthus katherinae* Baker, *J. Cell Sci.* 5:727–43

82. Hepler, P. K., Palevitz, B. A. 1986. Metabolic inhibitors block anaphase A in vivo. *J. Cell Biol.* 102:1995–2005

83. Heslop-Harrison, J. 1968. Synchronous pollen mitosis and the formation of the generative cell in massulate orchids. *J. Cell Sci.* 3:457–66

84. Heslop-Harrison, J. S., Bennett, M. D. 1983. The positions of centromeres on

the somatic metaphase plate of grasses. *J. Cell Sci.* 64:163–77

85. Hess, F. D. 1987. Herbicide effects on the cell cycle of meristematic plant cells. *Rev. Weed Sci.* 3:183–203

86. Hogan, C. J. 1987. Microtubule patterns during meiosis in two higher plant species. *Protoplasma* 138:126–36

87. Hunt, T. 1989. Maturation promoting factor, cyclin and the control of M-phase. *Curr. Opin. Cell Biol.* 1:268–74

88. Hussey, P. J., Traas, J. A., Gull, K., Lloyd, C. W. 1987. Isolation of cytoskeletons from synchronized plant cells: the interphase array utilizes multiple tubulin isotypes. *J. Cell Sci.* 88:225–30

89. Hyams, J. S., Brinkley, B. R., eds. 1989. *Mitosis: Molecules and Mechanisms.* London: Academic

90. Inoué, S. 1953. Polarization optical studies of the mitotic spindle. I. The demonstration of spindle fibers in living cells. *Chromosoma* 5:487–500

91. Inoué, S. 1981. Cell division and the mitotic spindle. *J. Cell Biol.* 91:131S–47S

92. Inoué, S., Bajer, A. 1961. Birefringence in endosperm mitosis. *Chromosoma* 12:48–63

93. Inoué, S., Kiehart, D. P., Mabuchi, I., Ellis, G. W. 1979. Molecular mechanism of mitotic chromosome movement. In *Motility in Cell Function,* ed. F. Pepe, J. Sanger, V. Nachmias, pp. 301–11. New York: Academic

94. Inoué, S., Molè-Bajer, J., Bajer, A. S. 1985. See Ref. 48, pp. 269–76

95. Jackson, W. T., Doyle, B. G. 1982. Membrane distribution in dividing endosperm of *Haemanthus. J. Cell Biol.* 94:637–43

96. Jensen, C., Bajer, A. 1973. Spindle dynamics and arrangement of microtubules. *Chromosoma* 44:73–89

97. Jensen, C. G. 1982. Dynamics of spindle microtubule organization: kinetochore fiber microtubules of plant endosperm. *J. Cell Biol.* 92:540–58

98. Kakimoto, T., Shibaoka, H. 1987. Actin filaments and microtubules in the preprophase band and phragmoplast of tobacco cells. *Protoplasma* 140:151–56

99. Kakimoto, T., Shibaoka, H. 1988. Cytoskeletal ultrastructure of phtramoplast-nuclei complexes isolated from cultured tobacco cells. *Protoplasma,* Suppl. 2:95–103

100. Keith, C. H., Ratan, R., Maxfield, F. R., Bajer, A., Shelanski, M. L. 1985. Local cytoplasmic calcium gradients in living mitotic cells. *Nature* 316:848–50

101. Kloth, R. H. 1989. Changes in the level of tubulin subunits during development

of cotton *(Gossypium hirsutum)* fiber. *Physiol. Plant.* 76:37–41

102. Kubai, D. F. 1975. The evolution of the mitotic spindle. *Int. Rev. Cytol.* 43:167–227

103. Kubiak, J., De Brabander, M., De Mey, J., Tarkowska, J. A. 1986. Origin of the mitotic spindle in onion root cells. *Protoplasma* 130:51–56

104. Lambert, A.-M. 1980. The role of chromosomes in anaphase trigger and nuclear envelope activity in spindle formation. *Chromosoma* 76:295–308

105. Lambert, A.-M., Bajer, A. 1972. Dynamics of spindle fibers and microtubules during anaphase and phragmoplast formation. *Chromosoma* 39:101–44

106. Lambert, A.-M., Bajer, A. 1977. Microtubule distribution and reversible arrest of chromosome movements induced by low temperature. *Cytobiologie* 15:1–23

107. Lambert, A.-M., Bajer, A. S. 1975. Fine structure dynamics of the prometaphase spindle. *J. Micro. Biol. Cell.* 23:181–95

108. Lawson, A. A. 1900. Origin of the cones of the multipolar spindle in *Gladiolus. Bot. Gaz.* 30:145–53

109. Lewandowska, E., Charzyñska, M. 1977. *Tradescantia bracteata* pollen in vitro: pollen tube development and mitosis. *Acta Soc. Bot. Pol.* 47:587–97

110. Lim, S. S., Hering, G. E., Borisy, G. G. 1986. Widespread occurrence of anti-troponin-T crossreactive components in non-muscle cells. *J. Cell Sci.* 85:1–19

111. Lloyd, C. 1988. Actin in plants. *J. Cell Sci.* 90:185–88

112. Lloyd, C. W., ed. 1982. *The Cytoskeleton in Plant Growth and Development.* London: Academic

113. Lloyd, C. W. 1987. The plant cytoskeleton: the impact of fluorescence microscopy. *Annu. Rev. Plant Physiol.* 38:119–39

114. Lloyd, C. W., Traas, J. A. 1988. The role of F-actin in determining the division plane of carrot suspension cells. Drug studies. *Development* 102:211–21

115. Longwell, A., Mota, M. 1960. The distribution of cellular matter during meiosis. *Endeavour* 19:100–7

116. Maheshwari, P. 1950. *An Introduction to the Embryology of Angiosperms,* pp. 154–81. New York, NY:McGraw-Hill Book Co. Inc. 453 pp.

117. Marc, J., Gunning, B. E. S. 1988. Monoclonal antibodies to a fern spermatozoid detect novel components of the mitotic and cytokinetic apparatus in higher plant cells. *Protoplasma* 142:15–24

118. Mazia, D. 1987. The chromosome cycle and the centrosome cycle in the mitotic cycle. *Int. Rev. Cytol.* 100:49–92

119. McCurdy, D. W., Sammut, M., Gunning, B. E. S. 1988. Immunofluorescent visualization of arrays of transverse cortical actin microfilaments in wheat root-tip cells. *Protoplasma* 147:204–6

119a. McIntosh, J. R., Koonce, M. P. 1989. Mitosis. *Science* 246:622–28

120. Meijer, E. G. M., Simmonds, D. H. 1988. Microtubule organization during the development of the mitotic apparatus in cultured mesophyll protoplasts of higher plants—an immunofluorescence microscopy study. *Physiol. Plant.* 74:225–32

121. Mesquita, J. F. 1970. Sur l'ultrastructure de la zone d'insertion des fibres chromosomyiques dans les cellules méristématiques des racines d'*Allium cepa* L. *Z. Pflanzenphysiol.* 63:276–80

122. Mineyuki, Y., Marc, J., Palevitz, B. A. 1988. Formation of the oblique spindle in dividing guard mother cells of *Allium*. *Protoplasma* 147:200–3

123. Mineyuki, Y., Marc, J., Palevitz, B. A. 1989. Development of the preprophase band from random cytoplasmic microtubules in guard mother cells of *Allium cepa* L. *Planta* 178:291–96

124. Mitchison, T. J. 1989. Mitosis: basic concepts. *Curr. Opin. Cell Biol.* 1:67–74

125. Mitchison, T. J. 1989. Polewards microtubule flux in the mitotic spindle: evidence from photoactivation of fluorescence. *J. Cell Biol.* 109:637–52

126. Molè-Bajer, J., Bajer, A. S. 1983. Action of taxol on mitosis: modification of microtubule arrangements and function of the mitotic spindle in *Haemanthus* endosperm. *J. Cell Biol.* 96:527–40

127. Molè-Bajer, J., Bajer, A. S. 1988. Relation of F-actin organization to microtubules in drug treated *Haemanthus* mitosis. *Protoplasma,* Suppl. 1:91–112

128. Molè-Bajer, J., Bajer, A. S., Inoué, S. 1988. Three dimensional localization and redistribution of F-actin in higher plant mitosis and cell plate formation. *Cell Motil. Cytoskel.* 10:217–28

129. Morejohn, L. C., Burean, T. E., Molè-Bajer, J., Bajer, A. S., Fosket, D. E. 1987. Oryzalin, a dinitroaniline herbicide, binds to plant tubulin and inhibits microtubule polymerization in vitro. *Planta* 172:252–64

130. Morejohn, L. C., Fosket, D. E. 1982. Higher plant tubulin identified by self-assembly in microtubules in vitro. *Nature* 297:426–28

131. Mottier, D. M. 1903. The behavior of the chromosomes in the spore mother-cells of higher plants and the homology of the pollen- and embryo-sac mother cells. *Bot. Gaz.* 35:250–80

132. Murray, A. W., Szostak, J. W. 1985. Chromosome segregation in mitosis and meiosis. *Annu. Rev. Cell Biol.* 1:289–315

133. Nagl, W., Pohl, J., Radler, A. 1985. See Ref. 33, pp. 216–32

134. Nicklas, R. B. 1988. The forces that move chromosomes in mitosis. *Annu. Rev. Biophys. Biophys. Chem.* 17:431–39

135. Nicklas, R. B., 1989. The motor for polar chromosome movement in anaphase is at or near the kinetochore. *J. Cell Biol.* 109:2245–55

136. Niklas, K. J. 1989. The cellular mechanics of plants. *Am. Sci.* 77:344–49

137. Ohta, K., Toriyama, M., Endo, S., Sakai, H. 1988. Localization of mitotic-apparatus-associated 51 kD protein in unfertilized and fertilized sea urchin eggs. *Cell. Motil. Cytoskel.* 10:496–505

138. Östergren, G. 1951. The mechanism of co-orientation. Bivalents, multivalents and the theory of orientation by pulling. *Hereditas* 37:85–156

139. Osterhout, W. J. V. 1902. Cell studies. I. Spindle formation in *Agave*. *Proc. Calif. Acad. Sci. Third Ser. Bot.* 2:255–84

140. Ôta, T. 1957. Division of the generative cell in the pollen tube. *Cytologia* 22:15–27

141. Ôta, T. 1961. The role of the cytoplasm in cytokinesis of plant cells. *Cytologia* 26:428–47

142. Oud, J. L., Mans, A., Brakenhoff, G. J., Van der Voort, H. T. M., Van Spronsen, E. A., Nanninga, N. 1989. Three dimensional chromosome arrangement of *Crepis capillaris* in mitotic prophase and anaphase as studied by confocal scanning laser microscopy. *J. Cell Sci.* 92:329–39

143. Palevitz, B. A. 1980. Comparative effects of phalloidin and cytochalasin B on motility and morphogenesis in *Allium*. *Can. J. Bot.* 58:773–85

144. Palevitz, B. A. 1986. Division plane determination in guard mother cells of *Allium:* video time-lapse analysis of nuclear movements and phragmoplast rotation in the cortex. *Dev. Biol.* 117:644–54

145. Palevitz, B. A. 1987. Actin in the preprophase band of *Allium cepa*. *J. Cell Biol.* 104:1515–19

146. Palevitz, B. A. 1987. Accumulation of F-actin during cytokinesis in *Allium*.

Correlation with microtubule distribution and the effects of drugs. *Protoplasma* 141:24–32

147. Palevitz, B. A. 1988. Microtubular firtrees in mitotic spindles of onion roots. *Protoplasma* 142:74–78

148. Palevitz, B. A., Cresti, M. 1989. Cytoskeletal changes during generative cell division and sperm formation in *Tradescantia virginiana*. *Protoplasma* 150:54–71

149. Parke, J., Miller, C., Anderton, B. H. 1986. Higher plant myosin heavy-chain identified using a monoclonal antibody. *Eur. J. Cell Biol.* 41:9–13

150. Parke, J. M., Miller, C. C. J., Cowell, I., Dodson, A., Dowding, A., et al. 1987. Monoclonal antibodies against plant proteins recognize animal intermediate filaments. *Cell Motil. Cytoskel.* 8:312–23

151. Pickett-Heaps, J. D. 1969. The evolution of the mitotic apparatus: an attempt at comparative ultrastructural cytology in dividing plant cells. *Cytobios* 3:257–80

152. Pickett-Heaps, J. D. 1969. Preprophase microtubules and stomatal differentiation; some effects of centrifugation on symmetrical and asymmetrical division. *J. Ultrastr. Res.* 27:24–44

153. Pickett-Heaps, J. D., Northcote, D. H. 1966. Organization of microtubules and endoplasmic reticulum during mitosis and cytokinesis in wheat meristems. *J. Cell Sci.* 1:109–20

154. Porter, K. R., Machado, P. D. 1960. Studies on the endoplasmic reticulum. IV. Its form and distribution during mitosis in cells of onion root tip. *J. Biophys. Biochem. Cytol.* 7:167–80

155. Raudaskoski, M., Aström, H., Perttila, K., Virtanen, I., Louhelainen, J. 1987. Role of the microtubule cytoskeleton in pollen tubes: an immunocytochemical and ultrastructural approach. *Biol. Cell* 61:177–88

156. Rawlins, D. J., Shaw, P. J. 1988. Three-dimensional organization of chromosomes of *Crepis capsilaris* by optical tomography. *J. Cell Sci.* 91:401–14

157. Rickards, G. K. 1981. Chromosome movements within prophase nuclei. In *Mitosis/Cytokinesis*, ed. A. Zimmerman, A. Forer, pp. 103–31. New York: Academic

158. Robyns, W. 1929. La figure achromatique sur matériel frais dans les divisions somatiques des phanérogames. *La Cellule* 39:85–117

159. Ryan, K. G. 1983. Prometaphase and anaphase chromosome movements in

living pollen mother cells. *Protoplasma* 116:24–33

160. Ryan, K. G. 1984. Membranes in the spindle of iris pollen mother cells during the second division of meiosis. *Protoplasma* 122:56–67

161. Sakai, A. 1969. Electron microscopy of dividing cells. II. Microtubules and formation of the spindle in root tip cells of higher plants. *Cytologia* 34:57–70

162. Salmon, E. D., Leslie, R. J., Saxton, W. M., Karaw, M. L., McIntosh, J. R. 1984. Spindle microtubule dynamics in sea urchin embryos. *J. Cell Biol.* 99:2165–74

163. Sato, H., Kobayashi, A., Itoh, T. J. 1989. Molecular basis of physical and chemical probes for spindle assembly. *Cell Struc. Func.* 14:1–34

164. Sax, K., O'Mara, J. G. 1941. Mechanism of mitosis in pollen tubes. *Bot. Gaz.* 102:629–36

165. Schmit, A.-C., Lambert, A.-M. 1985. See Ref. 48, pp. 243–52

166. Schmit, A.-C., Lambert, A.-M. 1987. Characterization and dynamics of cytoplasmic F-actin in higher plant endosperm cells during interphase, mitosis and cytokinesis. *J. Cell Biol.* 105:2157–66

167. Schmit, A.-C., Lambert, A.-M. 1988. Plant actin filament and microtubule interactions during anaphase-telophase transition: effects of antagonist drugs. *Biol. Cell* 64:309–19

168. Schmit, A.-C., Vantard, M., De Mey, J., Lambert, A.-M. 1983. Aster-like microtubule centers establish spindle polarity during interphase-mitosis transition in higher plant cells. *Plant Cell Rep.* 2:285–88

169. Schmit, A.-C., Vantard, M., Lambert, A.-M. 1985. Microtubule and F-actin rearrangement during the initiation of mitosis in acentriolar plant cells. In *Cell Motility: Mechanism and Regulation*, ed. H. Ishikawa, S. Hatano, H. Sato, pp. 415–33. Tokyo: Univ. Tokyo Press

170. Seagull, R. W., Falconer, M. M., Weerdenburg, C. A. 1987. Microfilaments: dynamic arrays in plant cells. *J. Cell Biol.* 104:995–1004

171. Sheldon, J. M., Hawes, C. 1988. The actin cytoskeleton during male meiosis in *Lilium*. *Cell Biol. Int. Rep.* 12:471–76

172. Silflow, C. D., Oppenheimer, D. G., Kopczak, S. D., Ploense, S. E., Ludwig, S. R., et al. 1987. Plant tubulin genes: structure and differential expression during development. *Dev. Genet.* 8:435–60

173. Simmonds, D. H. 1986. Prophase bands of microtubules occur in protoplast cul-

tures of *Vicia hajastana* Grossh. *Planta* 167:469–72

174. Sinnott, E. W., Bloch, R. 1940. Cytoplasmic behavior during division of vacuolate plant cells. *Proc. Natl. Acad. Sci. USA* 26:223–27

175. Staiger, C. J., Cande, W. Z., 1989. Microtubule distribution in *dv*, a maize meiotic mutant defective in the prophase to metaphase transition. *Dev. Biol.* In press

176. Staiger, C. J., Schliwa, M. 1987. Actin localization and function in higher plants. *Protoplasma* 141:1–12

177. Steffen, W., Fuge, H., Dietz, R., Bast-meyer, M., Müller, G. 1986. Aster-free spindle poles in insect spermatocytes: evidence for chromosome-induced spindle formation? *J. Cell Biol.* 102:1679–87

178. Tang, X., Lancelle, S. A., Hepler, P. K., 1989. Fluorescence microscopic localization of actin in pollen tubes: comparison of actin antibody and phalloidin staining. *Cell Motil. Cytoskel.* 12:216–24

179. Terasaka, O., Niitsu, T. 1989. Peculiar spindle configuration in the pollen tube revealed by the anti-tubulin immunofluorescence method. *Bot. Mag. Tokyo* 102:143–47

180. Tiezzi, A., Moscatelli, A., Cresti, M. 1988. Taxol-induced microtubules from different sources: an ultrastructural comparison. *J. Submicro. Cytol. Pathol.* 20:613–17

181. Traas, J. A., Burgain, S., De Vaulx, R. D. 1989. The organization of the cytoskeleton during meiosis in eggplant (*Solanum melongena* L.): microtubules and F-actin are both necessary for coordinated meiotic division. *J. Cell Sci.* 92:541–50

182. Traas, J. A., Doonan, J. H., Rawlins, D. J., Shaw, P. J., Watts, J., Lloyd, C. W. 1987. An actin network is present in cytoplasm throughout the cell cycle of carrot cells and associates with the dividing nucleus. *J. Cell Biol.* 105:387–95

183. Vandre, D. D., Davis, F. M., Rao, P. N., Borisy, G. G. 1986. Distribution of cytoskeletal proteins showing a conserved phosphorylated epitope. *Euro. J. Cell Biol.* 41:72–81

184. Van Lammeren, A. M. M., Bednara, J., Willemse, M. T. M. 1989. Organization of the actin cytoskeleton during pollen development in *Gasteria verrucosa* (Mill.) H. Duval visualized with rhodamine-phalloidin. *Planta* 178:531–39

185. Vantard, M., Lambert, A.-M., De Mey, J., Picquot, P., Van Eldic, L. J. 1985. Characterization and immunocytochemi-

cal distribution of calmodulin in higher plant endosperm cells: localization in the mitotic apparatus. *J. Cell Biol.* 101:488–99

186. Van Went, J., Cresti, M. 1988. Cytokinesis in microspore mother cells of *Impatiens sultani*. *Sex. Plant Reprod.* 1:228–33

187. Venverloo, C. J., Hovenkamp, P. H., Weeda, A. J., Libbenga, K. R. 1980. Cell division in *Nautilocalyx* explants. I. Phragmosome, preprophase band and plane of division. *Z. Pflanzenphysiol.* 100:161–74

188. Venverloo, C. J., Libbenga, K. R. 1981. Cell division in *Nautilocalyx* explants. II. Duration of cytokinesis and velocity of cell plate growth in large, highly vacuolated cells. *Z. Pflanzenphysiol.* 102:389–95

189. Venverloo, C. J., Libbenga, K. R. 1987. Regulation of the plane of cell division in vacuolated cells. I. The function of nuclear positioning and phragmosome formation. *J. Plant Physiol.* 131:267–84

190. Wada, B. 1965. Analysis of mitosis. *Cytologia* 30(Suppl.):1–158

191. Walters, M. S. 1958. Aberrant chromosome movement and spindle formation in meiosis of *Bromus* hybrids: an interpretation of spindle organization. *Am. J. Bot.* 45:271–89

192. Wang, H., Cutler, A. J., Fowke, L. C. 1989. High frequencies of preprophase bands in soybean protoplast cultures. *J. Cell Sci.* 92:575–80

193. Wang, H., Cutler, A. J., Fowke, L. C. 1989. Preprophase bands in cultured multinucleate soybean protoplasts. *Protoplasma* 150:110–16

194. Weerdenburg, C., Falconer, M. M., Setterfield, G., Seagull, R. W. 1986. Effects of taxol on microtubule arrays in cultured higher plant cells. *Cell Motil. Cytoskel.* 6:469–78

195. Wick, S. M. 1985. Immunofluorescence microscopy of tubulin and microtubule arrays in plant cells. III. Transition between mitotic/cytokinetic and interphase microtubule arrays. *Cell Biol. Int. Rep.* 9:357–71

196. Wick, S. M. 1985. The higher plant mitotic apparatus: redistribution of microtubules, calmodulin and microtubule initiation material during its establishment. *Cytobios* 43:285–94

197. Wick, S. M., Duniec, J. 1984. Immunofluorescence microscopy of tubulin and microtubule arrays in plant cells. II. Transition between the pre-prophase band and the mitotic spindle. *Protoplasma* 122:45–55

198. Wick, S. M., Duniec, J. 1986. Effectiveness of various fixatives on the reactivity of plant cell tubulin and calmodulin in immunofluorescence microscopy. *Protoplasma* 133:1–18

199. Wick, S. M., Muto, S., Duniec, J. 1985. Double immunofluorescence labeling of calmodulin and tubulin in dividing plant cells. *Protoplasma* 126: 198–206

200. Wilson, G. B., Hyypio, P. A. 1955. Some factors concerned in the mechanism of mitosis. *Cytologia* 20:177–84

201. Wilson, H. J. 1968. The fine structure of the kinetochore in meiotic cells of *Tradescantia. Planta* 78:379–85

202. Wilson, H. J. 1970. Endoplasmic reticulum and microtubule formation in dividing cells of higher plants—a postulate. *Planta* 94:184–90

203. Wise, D. 1988. The diversity of mitosis: the value of evolutionary experiments. *Biochem. Cell Biol.* 66:515–29

204. Wolniak, S. M. 1988. The regulation of mitotic spindle function. *Biochem. Cell Biol.* 66:490–514

Annu. Rev. Plant Physiol. Plant Mol. Biol. 1990. 41:317–38

GENE ACTIVITY DURING POLLEN DEVELOPMENT

Joseph P. Mascarenhas

Department of Biological Sciences, State University of New York at Albany, Albany, New York 12222

KEY WORDS: pollen-specific genes, pollen promoters, haploid transcription, sperm

CONTENTS

INTRODUCTION

The male gametophyte (pollen grain) of flowering plants is a microscopic structure that completes its early development within the sporophyte tissue of the anther. The role of the male gametophyte is the production of two sperm cells and their transport within the pollen tube through the tissues of the style and into the embryo sac in the ovule. In the double fertilization that follows, fusion of one sperm with the egg and of the second sperm with the central cell

317

results in the formation of the zygote and primary endosperm cell, respectively.

The extremely reduced haploid male plant (gametophyte) has a number of specialized functions to perform. The critical steps in the transformation of the pollen mother cells into haploid microspores are two meiotic divisions that result in each pollen mother cell's producing a tetrad of microspores. Each microspore within the tetrad and the entire tetrad are surrounded by a callose (1,3-β-glucan) wall. Upon release from the tetrad, the microspores increase rapidly in volume and also change in shape. An extended interphase period terminates with a very unequal division (3) of the microspore (microspore mitosis), forming a large vegetative cell and a small generative cell; both cells are included within the wall of the original microspore. The structure is now termed a pollen grain. The generative cell, which inherits a small amount of cytoplasm, lies within the vegetative cell. The immature pollen grain progresses through a number of identifiable developmental changes prior to release from the anther. In several plants, such as maize and *Brassica* species, the generative cell undergoes a mitotic division within the pollen grain, forming two sperm cells. In most plants, however, the generative cell completes its division during the growth of the pollen tube in the style (12). When fully mature, the male gametophyte consists of three cells, the vegetative cell and the two sperm cells.

For a short while following release from the anther, the pollen grain exists as a free organism until it is carried by wind, insects, or other agents to the stigma of an appropriate or compatible flower, where the right conditions are present for the germination of the pollen grain. The field of self-incompatibility has been reviewed recently (9, 19, 21, 28) and is not covered here. On the stigma, the pollen grain begins a new phase of its life and development. It germinates by the extrusion of a tube through a germ pore in the pollen wall. The tube grows down into the style. The vegetative nucleus and (depending on the plant species) the generative cell or the sperm cells move out of the pollen grain and into the pollen tube. In most plants germination and pollen tube growth are relatively rapid events, the period from pollination to fertilization ranging from 1 to 48 hr. Pollen tubes of different species show different rates of growth; rates as high as 35 mm/hr have been reported (56). The pollen tube grows through the style, enters the ovule through the micropyle, and reaches the embryo sac. It penetrates one of the synergids (cells flanking the egg), normally the one that has begun to degenerate. The tube stops growing, and the sperm cells together with some of the other tube contents are discharged into the synergid. One of the sperm cells fuses with the egg cell to form the zygote. The second sperm fuses with the central cell of the embryo sac to give rise to the primary endosperm cell, thus completing the process of double fertilization.

In this review I cover primarily topics concerned with the regulation of gene activity during pollen development that I consider important and likely to further our understanding of pollen development in the foreseeable future. Several reviews can be consulted for other areas of study of the male gametophyte and the anther (23, 24, 36, 57–59, 75, 91, 130).

THE REGULATION AND EXPRESSION OF GENES IN THE DEVELOPING MALE GAMETOPHYTE

How Important is the Tapetum in Pollen Development?

The tapetum, which is the tissue in the anther in closest contact with the pollen mother cells and microspores, is generally believed to be involved with the nutrition of the developing pollen grains. Here I treat the role of the tapetum only briefly; the reader is referred to a recent review (1) for further information. The importance of the tapetum in pollen development is supported by some cases of cytoplasmic male sterility, where the initial lesions in anther development are found in the tapetal tissue and not in the sporogenous cells (10, 44, 53; see reviews in 40, 51, 54).

The tapetum is active in the synthesis of all the major classes of compounds such as proteins, lipids, carbohydrates, etc, and the levels of their synthesis are stage specific (1, 79, 93, 104). The available evidence seems to suggest that the tapetum is the site of synthesis of the proteins that are contained within the cavities of the exine of the pollen grain wall (reviewed in 58). The tapetal cells produce and secrete material utilized in the formation of the sporopollenin of pollen walls (29, 96). The tapetum is also involved in the production of the *pollenkitt,* a lipoidal, often pigmented layer, and the *tryphine,* a complex mixture of hydrophobic and hydrophilic substances, which coat mature pollen grains (reviewed in 58). Flavonoids and other phenylpropanoids are present in large amounts in the cavities of the exine of many plant species. In tulip anthers the synthesis of phenylpropanes, chalcones, flavonols, and anthocyanins occurs at different periods during postmeiotic pollen development (133). There is evidence to suggest that several key enzymes of phenylpropanoid metabolism, such as phenylalanine ammonia lyase, cinnamic acid-4-hydroxylase, etc, are produced by the tapetal cells, secreted into the anther loculus, and are active in phenylpropanoid metabolism either in the loculus or in the exine cavities of the developing microspores (7, 42, 133).

A large number of genes are active in the anther tissues. The total complexity of tobacco anther messenger RNA (mRNA) has been estimated to be 3.23×10^7 nucleotides, equivalent to about 26,000 different genes. Of these about 15,500 mRNAs were similar to those found in the leaf (48). Recombinant cDNA libraries have been constructed to mRNAs from tobacco anthers (37).

In situ hybridizations using some of these clones as probes have identified several tapetum-specific mRNAs. The levels of these mRNAs were correlated with the establishment and degeneration of the tapetum. Other clones from the library represent mRNAs present in other parts of the anther (37, 38). cDNA clones have also been made to mRNAs from tomato anthers, and the tissue localization and developmental time course of synthesis of the corresponding mRNAs have been studied by in situ hybridizations (34). Whether any of these cloned genes plays a role in pollen development, as opposed to tapetal development, remains to be determined.

Because isolated microsporocytes and early microspores continue their development in tissue culture without the tapetum and other anther tissues, it is unlikely that these supportive tissues provide any unique factors essential for pollen development. Isolated microsporocytes of lily and *Trillium* when cultured at the leptotene stage, completed meiosis in a defined culture medium (119). Moreover, uninucleate microspores of lily and tulip at the G_1 phase of the cell cycle could be successfully cultured to produce normal pollen grains that germinated and produced normal pollen tubes (120, 121). These results suggest that the tapetum and other anther tissues might be concerned with the mobilization and production of simple nutrients and the provision of the necessary physical conditions for development, rather than the supply of any critical morphogenetic components essential for the progress of pollen differentiation. The tapetum, as discussed earlier, is essential for the deposition of sporopollenin and other coatings on the pollen wall, and there is evidence for its role in breaking down the callose that surrounds the microspores when they are first formed after meiosis (reviewed in 57).

Is the Differentiation of the Male Gametophyte Dependent on Genes Transcribed Prior to Meiosis in the Sporophyte, or Is It Dependent on Transcription from the Haploid Genome?

Transcription of specific genes occurs in the haploid microspore after its formation following meiosis, and during its subsequent differentiation. Genetic analysis has provided the first evidence for this point (see review in 59). The ribosomal RNA (rRNA) genes are actively transcribed in the microspore. Both in lily and in *Tradescantia* large amounts of rRNA are transcribed prior to microspore mitosis. Following microspore mitosis there is a sharp decrease in rRNA synthesis, and the ribosomal genes become transcriptionally inactive during the terminal stages of pollen maturation and during pollen tube growth (57, 61, 84, 114). In addition, no ribosomal proteins are synthesized in lily pollen tubes (115). Further support for the observation that rRNA is not synthesized in the mature pollen grain and pollen tube is the fact that both in the pollen grain at anthesis and in the pollen tube of most species the nucleoli (sites of synthesis of rRNA) are either absent or greatly reduced in size. In tobacco, based on the isolation of ribosomes from pollen at different stages of development, the maximum rate of increase in ribosome content was found after microspore mitosis (125). The transfer RNA genes seem to follow a pattern of activity similar to that of the ribosomal genes, with greater synthesis prior to microspore mitosis and a subsequent decrease until the genes are transcriptionally inactive during the terminal stages of pollen maturation (84).

No new ribosomes or tRNAs are synthesized in the mature pollen grain and during pollen tube growth, but the mature pollen grain is equipped with a large store of ribosomes and tRNAs. These ribosomes and tRNAs are utilized in the protein synthesis that occurs during germination and that is required for pollen tube growth and the completion of generative cell division in those pollens released in the bicellular condition (reviewed in 58).

At anthesis, the ungerminated pollen grain contains a large quantity of presynthesized mRNAs that are translated early during germination. The evidence for the presence and functions of these mRNAs has been obtained from studies of the effects of inhibitors of RNA and protein synthesis on germination and tube growth (early work reviewed in 57). More direct evidence for the presence of mRNAs in the ungerminated pollen grains of *Tradescantia* and maize has been obtained by their isolation and translation in cell-free systems into polypeptides, many of which show similarity to proteins made during germination (31, 64). These mRNAs are capped at their 5'-termini with a guanosine 5' phosphate moiety that is methylated (31). Each pollen grain of *Tradescantia paludosa* contains 196 pg of total RNA, of which 5.1 pg is poly(A)RNA. This is equivalent to 6×10^6 molecules of poly(A)-RNA per pollen grain (63). Each pollen grain of tobacco contains about 230 pg of total RNA, of which 6.2 pg is poly(A)RNA (124). Individual maize pollen grains contain 352–705 pg of total RNA and 8.9–17.8 pg of poly(A)R-NA (64). From studies in the literature it would appear that mature pollen grains contain a store of stable mRNAs that are translated early during germination and that play a greater or lesser role in pollen germination and tube growth depending on the plant species (reviewed in 57; 118).

Numbers of Genes Expressed in the Male Gametophyte

It is of interest to know how many different kinds of mRNAs are present in the pollen grain, since this information would provide an estimate of the number of different genes active in the haploid male plant. Such studies have been carried out with pollen of *Tradescantia* and maize. From an analysis of the kinetics of hybridization of ^{3}H-cDNA with poly(A)RNA in excess, the mRNAs in mature pollen grains of *Tradescantia* have been found to have a total complexity of 2.3×10^7 nucleotides, which corresponds to about 20,000 different sequences (134). The mRNAs are present in three abundance classes with complexities of 5.2×10^4, 1.6×10^6 and 2.1×10^7 nucleotides.

Approximately 15% of the mRNAs are very abundant and comprise about 40 different sequences, each present on average in 26,000 copies per pollen grain. The major fraction of the mRNA (60%) consists of 1,400 different sequences, each sequence present in about 3,400 copies per pollen grain. The least abundant fraction is a relatively small fraction (24%) of the mRNA and contains 18,000 different sequences, each present in about 100 copies per grain. In contrast to pollen mRNA sequences, *Tradescantia* shoot mRNAs have a total complexity of 3.4×10^7 nucleotides, which is equivalent to about 30,000 different sequences (134). Moreover, in pollen about 75% of the mRNA sequences occur in the two more abundant frequency classes, whereas only 35% of the shoot mRNAs are abundant. Even the least-abundant fraction in pollen contains sequences that are much more abundant (100 copies) than in the corresponding fraction in shoots (5–10 copies).

Similar estimates of different mRNA sequences have been obtained with maize pollen (135). The total complexity of maize pollen poly(A)RNA is 2.4×10^7 nucleotides or about 24,000 different mRNAs. The most abundant fraction (35%) of maize pollen mRNA consists of about 240 sequences, each present in about 32,000 copies per pollen grain. The middle abundance class (49%) comprises about 6,000 different sequences, each present in 1,700 copies per grain. The third fraction (15%) contains about 17,000 diverse sequences, each present in about 200 copies per pollen grain. In contrast, maize shoot mRNAs show a total complexity of 4.0×10^7 nucleotides or about 31,000 different sequences (135). As in *Tradescantia,* in maize the mRNAs are much more abundant in pollen than in the shoot.

These studies show that a large number of genes are transcribed in the male gametophyte. About 60% as many genes are expressed in the morphologically simple male gametophyte as are expressed in shoots that consist of several different cell types. The 20,000–24,000 different mRNAs are the sequences found in mature pollen and represent genes primarily activated late in pollen development (see the section below on pattern of transcription of pollen mRNAs). Additional genes are likely active soon after meiosis and during the early stages of microspore development. We do not at present have any estimates of the numbers of these early genes, or of their similarity to the genes expressed in later pollen development. It is clear, however, that in terms of the numbers of genes required male gametophyte development is complex.

Are specific genes transcribed during the development of the male gametophyte, or is a subset of genes active in the sporophyte expressed in pollen? From a comparison of isozyme profiles in several vegetative tissues and pollen of tomato, 60% of the isozymes (i.e. genes coding for these isozymes) in vegetative tissues were also expressed in pollen, whereas 95% of pollen expressed isozymes were present in one or more vegetative tissues (122). In maize a similar analysis indicated that 72% of the isozymes studied were expressed in both pollen and sporophyte tissues, while only 6% of the isozymes were pollen specific (102). Extensive overlap has also been found between isozyme profiles of pollen and sporophyte tissues of three *Populus* species (90). In barley, similarly, 60% of the isozymes studied were expressed in both sporophyte tissues and the male gametophyte; 30% were sporophyte specific and 10% were male gametophyte specific (85). An extensive overlap of gene expression as measured by isozymes has also been reported between the endosperm and the male gametophyte (80).

Heterologous hybridizations of pollen cDNA with shoot poly(A)RNA and shoot cDNA to pollen poly(A)RNA indicate that in *Tradescantia* no more than 60% of the shoot mRNAs are represented in pollen, whereas a minimum of 64% of the pollen mRNA mass is present in shoot RNA (134). Likewise in maize, a minimum of 65% of the mRNAs in pollen are similar to those found in shoots. Because of various constraints in this type of analysis, the estimate of pollen mRNA sequences in common with the sporophyte could be in excess of 90% (135). From colony hybridizations with cDNA libraries made to pollen poly(A)RNA and hybridized with ^{32}P-cDNAs from pollen and vegetative tissues, it has been estimated that about 10% and 20% of the total sequences expressed in maize and *Tradescantia* pollen, respectively, might be pollen specific (116). It is thus apparent that the genetic program expressed during pollen development is extensive and that a substantial overlap occurs between genes active in the male gametophyte and in the sporophyte.

Although most of the isozymes expressed in the male gametophyte appear to be expressed in the sporophyte also, certain enzymes in pollen are coded for by genes different from those coding for the same enzyme activity in sporophytic tissues—i.e. the genes code for nonallelic isozymes. Nonallelic genes may code for ADP-glucose pyrophosphorylases in pollen and in the endosperm of maize (15). β-Glucosidase in maize appears to be another enzyme encoded by different genes in pollen and sporophyte (33).

Six well-defined β-tubulin isotypes characterized by immunoblotting of two-dimensional gels were found to be expressed differentially within the carrot plant (45). The β_5- and β_6-tubulin isotypes were absent from reproductive organs. β_4-Tubulin was found in stamens and was the most abundant β-tubulin in the mature pollen grain. The β_1- and β_3-tubulins were expressed in all the organs of the floret except the pollen grain. These results suggest that different β-tubulin genes may be expressed in pollen and vegetative tissues of carrot. The possibility that posttranslational modifications could produce the multiple β-tubulins has, however, not yet been ruled out. Using gene-specific probes with RNA blot hybridizations, α_1- and α_3-tubulin gene expression has been studied in various tissues of *Arabidopsis thaliana* (55). The α_3-tubulin mRNA was found in roots, leaves, and flowers. The gene-specific probe for α_1-tubulin hybrizided very weakly to RNA from roots and leaves, but strongly to RNA from flowers. Of several stages in flower development, the α_1-transcript was most abundant in flowers actively shedding pollen. The results suggest a preferential expression of the α_1-tubulin gene in pollen. Unfortunately, in this study the levels of the α_1-transcripts were not determined in pollen.

Mulcahy (73, 75) has proposed that genetic selection could operate during male gametophyte growth in the style, and this selection could have a positive effect on the resulting sporophyte generation. This pollen selection by competition has also been implicated as an important evolutionary process in the rise and success of the angiosperms (73, 75). The rapidity of germination and the rate of pollen tube growth are the primary factors that govern the competition between pollen tubes in reaching and effecting fertilization in a limited number of ovules. For these male gametophyte traits to have selective value in the sporophyte, it is necessary that pollen germination and tube growth be regulated by a large number of genes that are expressed and affect basic functions in both the gametophyte and the sporophyte. As discussed earlier, the genetic program expressed during pollen development is extensive, and there is a substantial overlap between genes active in the gametophyte and those active in the sporophyte. It is thus reasonable to expect that selecting for genes in the male haploid phase could increase the success of the sporophyte. Positive correlations between pollen tube growth rate and various sporophytic traits have been reported for several plants (69, 71, 72, 74, 77, 81, 82). The competitive ability of the haploid male gametophyte in the style and the selective advantages of this competition have been successfully utilized in screening under selective conditions for low-temperature tolerance in tomato hybrids (140, 141), for tolerance to high salt in hybrid progeny from a *Lycopersicon esculentum* (salt intolerant) × *Solanum pennellii* (salt tolerant) cross (100), copper or zinc tolerance in *Silene dioica* and *Mimulus guttatus* (106, 107), and for herbicide tolerance in maize (103). A good correlation has been found between seedling tolerance of individual lines of sugarbeet to the herbicide ethofumesate and the tolerance of pollen germination of the same lines to the herbicide, indicating that the genes for herbicide tolerance are active both in the sporophyte and in pollen (110).

There are several advantages to the use of haploid pollen to screen for desirable genes (75, 139). The most important is that extremely large numbers of pollen grains can be screened within a short time on the styles of a relatively few plants in a very small area, whereas a similar screening of diploid plants might require several months and many acres of land. In theory, one should be able to use pollen selection for any agronomic trait for which a suitable selection protocol can be devised. The screening for desirable traits can utilize stages of pollen development other than tube growth, as has been done to identify chilling resistant genotypes in tomato (83) and maize (6). This is a method with much potential, but which has thus far, been underutilized as an approach in plant breeding.

The Pattern of Transcription of Specific Pollen-Expressed mRNAs

The mature pollen grain contains a store of presynthesized mRNAs. When during pollen development are these mRNAs transcribed? To answer this question, clones from cDNA libraries made to mature pollen mRNA have been used as probes in RNA blot hybridizations. Both in *Tradescantia* and in maize, specific mRNAs are first detectable in the young pollen grain after microspore mitosis. These mRNAs continue to accumulate thereafter, reaching their maximum concentrations in the pollen grain just before anthesis (62, 116). In contrast, using an actin clone as a probe, actin mRNA was detectable in *Tradescantia* microspores soon after release from the tetrads. The mRNA accumulated thereafter reached a maximum concentration at late pollen interphase and decreased substantially in the mature pollen grain (116).

These results suggest that at least two sets of genes are activated at different times during pollen development. Genes in the first set (the "early" genes, such as that for actin) become active soon after meiosis is completed. The appearance and pattern of increase of alcohol dehydrogenase activity in maize (117) and of β-galactosidase in *Brassica campestris* (109) are what would be expected if their mRNAs were synthesized in a manner similar to that of actin in *Tradescantia*. Similarly, the maize glutamic-oxalacetic transaminase and β-glucosidase genes are activated early in pollen development (33). Genes in the second group (the "late" genes) become active after microspore mitosis, and the mRNAs increase in abundance up to maturity. All the clones from the late cDNA libraries in both maize and *Tradescantia* thus far analyzed belong to the second group.

cDNA libraries to poly(A)RNA from mature pollen of tomato (126, 127) and of *Oenothera organensis* (14; S. M. G. Brown and M. L. Crouch, manuscript in preparation) have recently been constructed. Five of the tomato clones appear to correspond to "late" genes, their transcripts being first detectable around the time of microspore mitosis and increasing progressively in concentration until anthesis. All five of the tomato clones are expressed in anther tissues in addition to being expressed in pollen (126, 127). In *Oenothera* the mRNAs for three pollen-specific clones (P1, P2, and P3) follow a similar accumulation pattern. The mRNA for one other pollen-specific clone, P6, is not detectable until fairly late in pollen development, whereas those for two other clones, P4 and P5, which are expressed in pollen and in leaves and ovaries, are present at all stages of pollen development (14).

The pattern and amount of accumulation of the "late" mRNAs would suggest a major function for these mRNAs during the latter part of pollen maturation and/or during pollen germination and tube growth. Are these mRNAs translated before anthesis or are they stored and utilized for protein synthesis during germination? The enzymes for the synthesis of both neutral and polar lipids necessary for cell membrane formation are already present in the mature ungerminated *Tradescantia* pollen grain and are functionally stable for several hours in the pollen tube (132).

Similarly, a large number of enzymes have been reported present in pollen grains of various species (reviewed in 57). All the enzymes studied have been found in ungerminated pollen; and although increases in the activity of a few enzymes have been reported during pollen tube growth, there is (with the exceptions discussed below) no good evidence for the new synthesis, on newly made or presynthesized mRNAs, of any enzyme not present in the mature pollen grain prior to germination. The enzyme phytase is synthesized during germination of *Petunia* pollen on mRNAs presynthesized in the developing pollen grain prior to anthesis (46). Three phytases with different pH optima have been detected in germinating lily pollen (5, 52). One is already present in mature ungerminated pollen. A second is newly synthesized during germination from preexisting mRNA (52). A 65-kD protein of unknown function has been reported to be newly synthesized during tobacco pollen tube growth. It is not present prior to germination (16) and appears to be synthesized on mRNA present in the pollen grain prior to anthesis (17).

Antibodies prepared to a portion of the protein coding region of the *Oenothera* P2 cDNA clone were used in immunoblots to determine the timing of synthesis and accumulation of the protein during pollen development. The P2 protein was present in pollen at later stages of development and in germinating pollen tubes. The P2 mRNA is thus translated in the pollen grain before anthesis, not merely stored for translation later, during germination (14). Whether the protein product has functions during pollen maturation or is made for use during germination and tube growth has still to be determined.

The "late" mRNAs in the pollen grain appear to be products of the vegetative cell. In situ hybridizations show that Zmc13 mRNA is located in the cytoplasm of the vegetative cell of the maize pollen grain and, after germination, is distributed throughout the pollen tube cytoplasm (39). Similar in situ hybridizations with the tomato clones as probes also show localization of the mRNAs in the vegetative cell cytoplasm (127). Conclusive experiments have not yet been done to determine whether the mRNAs are present in the generative cell or sperm cells, in addition to being present in the vegetative cell. It seems likely, however, that the mRNAs are not present in the sperm cells in maize (39).

Mature pollen grains do not appear to synthesize heat shock proteins (HSPs) during germination or tube growth. Pollen tubes of *Tradescantia* do not synthesize HSPs at either 37° or 40°C or after a gradual increase in temperature from 29° to 40°C (2, 60). Nor does the ungerminated *Tradescantia* pollen grain at the time it is released from the anther appear to contain hsps that might have been synthesized during maturation (138). No hsps were found in germinating maize or lily pollen (20) or in pollen tubes of petunia (105). Subsets of the HSPs are, however, synthesized in immature maize pollen during a heat shock (33). A chimeric gene under the control of the *hsp70* promoter of *Drosophila* is not activated after a heat shock in germinating pollen of transgenic tobacco plants, although it is heat activated in roots, stems, and leaves (112). This is good supporting evidence for the inactivity of the *hsp* genes in pollen.

Sequence Characterization and Functions of Pollen-Expressed Genes

Several cDNA clones from the libraries prepared to pollen poly(A)RNA from *Tradescantia*, maize, tomato, and *Oenothera* have been sequenced. In addition a few of the corresponding genomic clones have been isolated and sequenced. A pollen-specific cDNA clone from maize, Zmc13, which is a full length copy of the mRNA, is 929 nucleotides long. It codes for a predicted polypeptide of 170 amino acid residues with a molecular mass of 18.3 kD (39). The hydropathy profile of the polypeptide suggests a possible signal sequence at the amino terminus. The mRNA contains a 5'-untranslated region of 127 nucleotides and a 3'-untranslated region of 292 nucleotides to the polyadenylation site. An unusual feature of Zmc13 is the great distance between the putative polyadenylation signal and the actual site of poly(A) addition. The consensus AATAAA polyadenylation signal motif is located 180 nucleotides upstream from the site of poly(A) addition, and 110 bases downstream from the presumptive stop codon (39). Both the alcohol dehydrogenase gene *(Adh1)* from maize, which is expressed in pollen although it is not pollen specific (101), and a pollen-specific clone from *Oenothera* (P22) (14) also exhibit an extended region between the consensus AATAAA signal and the actual polyadenylation sites. Another pollen-specific clone (P1), however, does not have this characteristic (14).

The *lat52* cDNA clone from tomato contains an open reading frame that codes for a putative protein of 17.8 kD that has an amino terminal hydrophobic region with characteristics of a signal sequence (126). It is an interesting coincidence that *lat52* and the pollen-specific sequence from maize, *Zmc13* (39), exhibit substantial amino acid homology in their putative amino acid sequences. *Lat52* shows 32% amino acid identity to the predicted polypeptide sequence of *Zmc13*, including the presence of six conserved cysteine residues. Neither of the two clones exhibits significant sequence homology with any currently known protein in DNA or protein data banks.

Two of the cDNA clones from the P1/P2 family of *Oenothera* pollen cDNAs have been sequenced, and their putative protein sequences have been determined (14). There is substantial homology in the amino acid sequence between the two clones and a polygalacturonase from tomato fruit (14). Polygalacturonases have been found in fairly high concentrations in pollen of 12 monocotyledonous species examined (89). Approximately 200 times more polygalacturonase is found in maize pollen (89) than in maize seedlings (88).

The proteins corresponding to two cDNA clones (*lat56* and *lat59*) expressed in tomato anthers and pollen are similar in amino acid sequence to pectate lyases from the plant bacterial pathogen *Erwinia* (68). A pollen-specific cDNA clone from maize, *Zmc58*, also has sequences homologous to those of pectate lyases (D. Hamilton, D. Bashe, and J. P. Mascarenhas, unpublished results).

The "late" pollen-expressed genes tentatively identified with respect to their function are enzymes involved in the degradation of the middle lamella region of cell walls. It is thus likely that a major function of these genes is to provide the enzymes necessary to facilitate the growth of the pollen tubes through the tissues of the style.

It is also possible that these enzymes function in germination and pollen tube extension, and some evidence in the literature, although it is not particularly strong, supports such a role. Several hydrolytic enzymes have been found in the pollen walls of many species examined (50). The enzyme activity is associated principally with the intine, with a concentration of activity in or around the apertural intine, the potential sites of emergence of the pollen tube. Pollen of *Pyrus communis* (pear) releases β-1,4-glucanase and pectinase immediately after being placed in a germination medium (113). Low concentrations of β-1,3-glucanase and pectinase added to the growth medium had a stimulating effect on pollen tube growth (94). The effects were specific since other enzymes, such as pectin esterase, acid phosphatase, and α-amylase, were without tube-promoting activity.

It will be of interest to identify the functions of the several other pollen-specific genes that have been isolated. The primary synthetic reactions that occur during pollen tube growth are concerned with the synthesis of the pollen tube wall and the synthesis of the cell membrane of the elongating pollen tube. One might thus expect many of the presynthesized mRNAs and proteins present in the pollen grain at anthesis to be involved in these two processes. Unfortunately, our knowledge of the enzymology of plant cell wall synthesis, including that of the pollen tube, is not profound. This serious deficiency in our knowledge will make it difficult to identify the many gene products present in the pollen grain that probably function in pollen tube wall synthesis.

Mutants of *Arabidopsis thaliana* deficient in adenine phosphoribosyltransferase activity have been identified (70). The mutants are male sterile, with pollen abortion occurring after release of the microspores from the tetrads. The tapetum in the mutant anthers is morphologically normal. This is to my knowledge the first known nuclear male sterile mutant in which the biochemical basis of the sterility has been determined. It has been suggested that the defect may be due to a change in the cytokinin metabolism (70).

Southern hybridizations indicate that the pollen-expressed genes that have been analyzed are present in one or a very few copies in the maize genome (116). Restriction fragment polymorphism mapping indicates that the maize pollen-specific gene $Zm13$, is located near the centromere on the short arm of chromosome 10 (38a). It is interesting that two male sterile mutants, $ms10$ and $ms11$, map very near $Zm13$ on chromosome 10 (43). The $LAT52$ gene appears to be present in one copy in the tomato genome (126). The six analyzed pollen-expressed cDNA clones in *Oenothera* include single copy genes and members of relatively small families (showing 3 to 12 bands on the DNA hybridization blots) (14).

Isolation of Genomic Clones and Identification of Pollen Promoters

A few pollen-expressed but not pollen-specific genes have been sequenced. The *Adh1* gene from maize is expressed in pollen and in many other tissues of the plant (30, 32). The complete sequence of the gene including extensive 5'- and 3'-flanking regions has been determined. It includes nine introns (101). The *waxy (wx)* locus of maize is also genetically well characterized. It encodes the UDP-glucose:starch glycosyltransferase that is bound to starch granules and is expressed in pollen (22), in the embryo sac (13), and in the endosperm (131). About 900 bp of 5'-flanking sequences of the *waxy* gene, 3718 bp of the coding region, which is composed of 14 exons and 13 small introns, and several hundred nucleotides of 3'-flanking sequences are available (49). The *lat52* gene from tomato is expressed in abundance in anthers and pollen, and weakly in petals (126). It contains a single intron with a highly repetitive sequence. The pollen-specific gene from maize, $Zm13$, including extensive 5'- and 3'-flanking regions, has been sequenced (38a). The $Zm13$ gene does not contain any introns although it shares homology at the amino acid level with the $LAT52$ gene product.

There is currently no information about *cis*-acting regulatory elements that control gene transcription in the male gametophyte. The sequences in the presumptive promoter regions of the four genes discussed in the preceding paragraph have been compared to determine if specific elements are common to all these genes that could potentially be the portion of the promoter responsible for pollen specificity (38a). In the comparison of sequence homology, the promoter was considered to include the entire 5'-nontranslated region of each sequence. About 20 sequence elements 10 or 11 nucleotides long were common to all the pollen-expressed genes (38a). Characterization of DNA sequences that exhibit enhancer activity has shown that the "enhancer effect" is produced by a combination of various sequence motifs or modules, all relatively small and each contributing to the overall activity of the enhancer. Such DNA sequence motifs that play a significant role in transcriptional regulation can lie a considerable distance upstream of the transcriptional start site (reviewed in 4, 27). Whether any of the common sequence elements found in the various pollen-transcribed genes regulates pollen expression of these genes remains to be determined by mutational and other types of analyses currently in progress in several laboratories. Preliminary analysis indicates that 1.4 kb of 5'-flanking DNA of *lat59* is sufficient to direct pollen expression of the *E. coli* β-glucuronidase (GUS) reporter gene in tobacco and tomato (68).

Genomic and cDNA clones have been isolated that code for the flavonoid biosynthetic enzyme chalcone flavonone isomerase (CHI) in *Petunia* (128, 129). There are two CHI genes. Chi gene A *(Chi-A)* is expressed in the corolla and tube of the flower and can be induced by UV light. *Chi* gene B *(Chi-B)* is expressed only in immature anthers and is not UV inducible. Mature anthers have a larger CHI transcript 437 nucleotides longer than the *CHI-A* transcript in the corolla. The *CHI-A* gene apparently has two promoters, one of which is active in the corolla, whereas the second promoter which gives rise to a larger transcript is active in late stages of anther development and in older pollen grains (129). Comparison of the 5'-flanking sequences of the *Chi-B* gene with those of two other flavonoid genes expressed in immature anthers (chalcone synthase A and J genes) resulted in the identification of a sequence element conserved in all the genes. This sequence motif was designated an "anther box" (129). Until more definitive evidence is obtained for the involvement of the motif in the regulation of genes in the anther, the designation "anther box" seems premature.

The enzyme 5-enolpyruvylshikimate-3-phosphate (EPSP) synthase catalyzes a reaction in the biosynthetic pathway of aromatic amino acids. This gene is expressed in anthers of both petunia and tomato in addition to other tissues (35). The petunia EPSP synthase promoter region from −1800 to −285 was inserted in front of a cauliflower mosaic virus (CaMV) 35S promoter region (−90 to +8) fused to the GUS coding sequence (8). This construct was sufficient to obtain pollen expression of the reporter gene in both petunia and tobacco pollen in transgenic plants. The 35S promoter region alone without the EPSP synthase promoter gave low levels of expression only in roots. In these experiments wild-type tobacco showed no GUS histochemical activity in pollen or floral tissues, whereas wild-type petunia showed light staining in the stigma and in anthers and pollen of immature flowers. The entire CaMV 35S upstream region (-941 to $+8$) when fused to a GUS coding sequence in the absence of the EPSP synthase promoter induced GUS activity in petunia pollen (8).

The CaMV 35S promoter alone was not active in inducing transcription in pollen of either tobacco or tomato (68). In contrast the promoter of the nopaline synthase *(Nos)* gene of the *Agrobacterium tumefaciens* Ti plasmid appears to be active in pollen. The *Nos* promoter coupled to a kanamycin resistance gene confers kanamycin tolerance to pollen of transgenic tomato plants (11).

The *E. coli* GUS gene has been considered a very satisfactory reporter gene for use with plants, because of the lack of endogenous β-glucuronidase activity in all plants assayed (47). There has been a recent report, however, that substantial endogenous GUS activity is present after meiosis, both in the tapetum and in pollen grains of potato, tomato, and tobacco (87). In addition, endogenous GUS activity has been found in pollen of several other bicellular pollen species. The four tricellular pollens examined (*Brassica oleracea,* maize, *Tagetes erecta, Bellis perennis*) exhibited no endogenous GUS activity (89). These results disagree with other reports in the literature. In both tomato and tobacco no endogenous GUS activity was found in mature pollen (68). Similarly, no GUS activity has been reported in pollen of wild-type tobacco (8). The contradictions among these results require explanation.

SPERM CELL BIOLOGY

We know little about the biochemistry and molecular biology of sperm cells of flowering plants. During germination in vitro sperm cells in the pollen grain of tricellular rye *(Secale cereale)* have been shown by ^{3}H-uridine labeling and autoradiography to synthesize RNA (41). In contrast, sperm cells of *Hyoscyamus niger,* in which generative cell division is completed during pollen tube growth, do not synthesize RNA after short pulses of ^{3}H-uridine and autoradiography (92).

Methods for the isolation of sperm cells have recently been described for several plants, including *Plumbago* (99), *Brassica oleracea* (65), *Triticum aestivum* (65), maize (18, 26, 65), *Rhodendron* and *Gladiolus* (108), *Gerbera* (111), and spinach (123).

Isolated sperm from maize were frozen at $-75°C$ for two weeks, and a sizeable fraction of the cells were viable after this storage (18). Monoclonal antibodies have been produced to a pollen fraction enriched in sperm from *Plumbago* with the long-term goal of identifying the sperm cell surface molecules responsible for the recognition of the two female cells involved in double fertilization (86). With these technical advances one might expect rapid increases in our knowledge of the biochemistry and molecular biology of sperm cells in the next few years.

Are the two sperm cells different or is the fusion of one with the egg and the other with the central cell simply a matter of chance? The two sperm cells of *P. zeylanica* differ significantly with respect to morphology and organelle content (97). The larger of the sperm cells is intimately associated with the vegetative nucleus. The two sperm cells and the vegetative nucleus travel as a linked unit within the pollen tube, which has now been termed the "male germ unit" (25). One sperm is practically devoid of plastids, whereas the other contains a large number of these organelles. The plastid-rich, mitochondrion-poor sperm cell preferentially fuses with the egg (98). This is an important finding because it indicates that in this species the two sperm cells have differentiated, that there is a specificity in gamete recognition, and that the process of double fertilization is not a random fusion event.

A similar dimorphism has been described for the sperm of *Brassica campestris* and *B. oleracea* (67), maize (66), *Euphorbia dulcis* (78), and spinach (137). Genetic evidence for preferential fertilization in maize was obtained forty years ago (95). The preferential fusion of the two sperm cells with the female cells may thus be a general phenomenon.

CONCLUDING REMARKS AND SUMMARY

Several important aspects of pollen development have not been considered in this review because of limitations of time and space. One example is the cytoskeleton and its role in pollen tube function, including cytoplasmic streaming, and the mechanism(s) of pollen tube wall synthesis. Another burgeoning area concerns second messengers, such as Ca^{2+}, and their involvement in the regulation of pollen germination and tube growth. The exlusion of male organelle DNAs from transmission to the zygote is a third area of study important to our understanding of sperm development and function.

The mature male gametophyte of flowering plants, although morphologically simple, undergoes a series of discrete differentiation events. A relatively large number of genes are required to program its entire development. A small fraction of these genes are pollen specific while the large majority are expressed both in the sporophyte and in the male gametophyte. Pollen-expressed genes have recently been isolated from several plant species. It is now essential to determine the functions of these genes in pollen development. Does any of these genes have a regulatory role? The mRNAs tentatively identified with respect to function code for enzymes that may be involved in the growth of the pollen tube through the style. Several of the pollen-expressed genes, including pollen-specific genes, are being analyzed with respect to promoter elements required for pollen expression. It would be interesting to see if the promoter structure of genes expressed in the haploid generation are different from that of genes expressed in the diploid sporophyte.

ACKNOWLEDGMENTS

Work in the author's laboratory has been supported by grants from the National Science Foundation.

Literature Cited

1. Albertini, L., Souvre, A., Audran, J. C. 1987. Le tapis de l'anthere et ses relations avec les microsporocytes et les grains de pollen. *Rev. Cytol. Biol. Veget.-Bot.* 10:211–42
2. Altschuler, M., Mascarenhas, J. P. 1982. The synthesis of heat shock and normal proteins at high temperatures and their possible roles in survival under heat stress. In *Heat Shock: From Bacteria to Man*, ed. M. J. Schlessinger, M. Ashburner, M. A. Tissieres, pp. 291–97. Cold Spring Harbor, New York: Cold Spring Harbor Lab.
3. Angold, R. E. 1968. The formation of the generative cell in the pollen grain of *Endymion non-scriptus* (L). *J. Cell Sci.* 3:573–78
4. Atchison, M. L. 1988. Enhancers: mechanisms of action and cell specificity. *Annu. Rev. Cell Biol.* 4:127–53
5. Baldi, B. G., Scott, J. J., Everard, J. D., Loewus, F. A. 1988. Localization of constitutive phytases in lily pollen and properties of the pH 8 form. *Plant Sci.* 56:137–47
6. Barnabas, B., Kovacs, G. 1988. Perspectives of pollen and male gamete selection in cereals. See Ref. 137. pp. 137–47
7. Beerhues, L., Forkmann, G., Schopker, H., Stotz, G., Wiermann, R. 1989. Flavanone 3-hydroxylase and dihydroflavonol oxygenase activities in anthers of *Tulipa*. The significance of the tapetum fraction in flavonoid metabolism. *J. Plant Physiol.* 133:743–46
8. Benfey, P. N., Chua, N.-H. 1989. Regulated genes in transgenic plants. *Science* 244:174–81
9. Bernatzky, R., Anderson, M. A., Clarke, E. 1988. Molecular genetics of self-incompatibility in flowering plants. *Dev. Genet.* 9:1–12
10. Bino, R. J. 1985. Histological effects of microsporogenesis in fertile, cytoplasmic male sterile and restored fertile *Petunia hybrida. Theor. Appl. Genet.* 69:425–28
11. Bino, R. J., Hille, J., Franken, J. 1987. Kanamycin resistance during in vitro development of pollen from transgenic tomato plants. *Plant Cell Rep.* 6:333–36
12. Brewbaker, J. L. 1967. The distribution and phylogenetic significance of binucleate and trinucleate pollen grains in the angiosperms. *Am. J. Bot.* 54:1069–83
13. Brink, R. A. 1925. Mendelian ratios and the gametophyte generation in angiosperms. *Genetics* 10:359–88
14. Brown, S. M. G. 1988. *Molecular analysis of gene expression during pollen development in* Oenothera organensis. PhD thesis. Indiana Univ., Bloomington
15. Bryce, W. H., Nelson, O. E. 1979. Starch synthesizing enzymes in the endosperm and pollen of maize. *Plant Physiol.* 63:312–17
16. Capkova, V., Hrabetova, E., Tupy, J. 1987. Protein changes in tobacco pollen culture; a newly synthesized protein related to pollen tube growth. *J. Plant Physiol.* 130:307–14
17. Capkova, V., Hrabetova, E., Tupy, J. 1988. Protein synthesis in pollen tubes: preferential formation of new species independent of transcription. *Sex. Plant Reprod.* 1:150–55
18. Cass, D. D., Fabi, G. C. 1988. Structure and properties of sperm cells isolated from the pollen of *Zea mays. Can. J. Bot.* 66:819–25
19. Clarke, A. E., Anderson, M. A., Bernatsky, R., Cornish, E. C., Mau, S.-L. 1989. Molecular aspects of self-incompatibility. In *The Molecular Basis of Plant Development*, ed. R. B. Goldberg, pp. 87–98. New York: Liss
20. Cooper, P., Ho, T.-H. D., Hauptman, R. M. 1984. Tissue specificity of the heat shock response in maize. *Plant Physiol.* 75:431–41

21. Cornish, E. C., Anderson, M. A., Clarke, A. E. 1988. Molecular aspects of fertilization in flowering plants. *Annu. Rev. Cell Biol.* 4:209–28

22. Demerec, M. 1924. A case of pollen dimorphism in maize. *Am. J. Bot.* 11:461–64

23. Dickinson, H. G. 1987. The physiology and biochemistry of meiosis in the anther. *Int. Rev. Cytol.* 107:79–109

24. Dickinson, H. G. 1987. Nucleocytoplasmic interaction. *Ann. Bot.* 60 (Suppl. 4):61–73

25. Dumas, C., Knox, R. B., McConchie, C. A., Russell, S. D. 1984. Emerging physiological concepts on sexual reproduction in angiosperms. *What's New in Plant Physiol.* 15:17–20

26. Dupuis, I., Roeckel, P., Matthys-Rochon, E., Dumas, C. 1987. Procedure to isolate viable sperm cells from corn *(Zea mays)* pollen grains. *Plant Physiol.* 85:876–78

27. Dynan, W. S. 1989. Modularity in promoters and enhancers. *Cell* 58:1–4

28. Ebert, P. R., Anderson, M. A., Bernatzky, R., Altschuler, M., Clarke, A. E. 1989. Genetic polymorphism of self-incompatibility in flowering plants. *Cell* 56:255–62

29. Echlin, P. 1971. The role of the tapetum during microsporogenesis of angiosperms. In *Pollen: Development and Physiology,* ed. J. Heslop-Harrison, pp. 41–61. London: Butterworths

30. Felder, M. R., Scandalios, J. R., Liu, E. H. 1973. Purification and partial characterization of two genetically defined alcohol dehydrogenase isozymes in maize. *Biochim. Biophys. Acta* 317:149–59

31. Frankis, R. C., Mascarenhas, J. P. 1980. Messenger RNA in the ungerminated pollen grain: a direct demonstration of its presence. *Ann. Bot.* 45:595–99

32. Freeling, M., Schwartz, D. 1973. Genetic relationships between the multiple alcohol dehydrogenases of maize. *Biochem. Genet.* 8:27–36

33. Frova, C., Binelli, G., Ottaviano, E. 1987. Isozyme and *hsp* gene expression during male gametophyte development in maize. In *Isozymes: Current Topics in Biological and Medical Research,* Vol. 15, *Genetics, Development and Evolution,* ed. M. C. Rattazzi, J. G. Scandalios, pp. 97–120. New York: Liss

34. Gasser, C. S., Smith, A. G., Budelier, K. A., Hinchee, M. A., McCormick, S., et al. 1988. Isolation of differentially expressed genes from tomato flowers. In *Temporal and Spatial Regulation of*

Plant Genes, ed. D. P. S. Verma, R. B. Goldberg, pp. 83–96. New York: Springer-Verlag

35. Gasser, C. S., Winter, J. A., Hironaka, C. M., Shah, D. M. 1988. Structure, expression, and evolution of the 5-enolpyruvylshikimate-3-phosphate synthase genes of petunia and tomato. *J. Biol. Chem.* 263:4280–89

36. Giles, K. L., Prakash, J., eds. 1987. *Pollen: Cytology and Development, International Review of Cytology,* Vol. 107. New York: Academic

37. Goldberg, R. B. 1987. Emerging patterns of plant development. *Cell* 49:298–300

38. Goldberg, R. B. 1988. Plants: novel developmental processes. *Science* 240:1460–67

38a. Hamilton, D. A., Bashe, D. M., Stinson, J. R., Mascarenhas, J. P. 1989. Characterization of a pollen-specific genomic clone from maize. *Sex. Plant. Reprod.* 2:208–12

39. Hanson, D. D., Hamilton, D. A., Travis, J. L., Bashe, D. M., Mascarenhas, J. P. 1989. Characterization of a pollen-specific cDNA clone from *Zea mays* and its expression. *Plant Cell* 1:173–79

40. Hanson, M. R., Conde, M. F. 1985. Functioning and variation of cytoplasmic genomes: lessons from cytoplasmic-nuclear interactions affecting male fertility in plants. *Int. Rev. Cytol.* 94:213–67

41. Haskell, D. W., Rogers, O. M. 1985. RNA synthesis by vegetative and sperm nuclei of trinucleate pollen. *Cytologia* 50:805–9

42. Herdt, E., Sutfield, R., Wiermann, R. 1978. The occurrence of enzymes involved in phenylpropanoid metabolism in the tapetum fraction of anthers. *Cytobiologie* 17:433–41

43. Hoisington, D. 1989. Working linkage maps. *Maize Genet. Coop. News Lett.* 63:141–51

44. Horner, H. T., Rogers, M. A. 1974. A comparative light and electron microscopic study of microsporogenesis in male-fertile and cytoplasmic male-sterile pepper *(Capsicum annuum).* *Can. J. Bot.* 52:435–41

45. Hussey, P. J., Lloyd, C. W., Gull, K. 1988. Differential and developmental expression of β-tubulins in a higher plant. *J. Biol. Chem.* 263:5474–79

46. Jackson, J. F., Linskens, H. F. 1982. Phytic acid in *Petunia hybrida* pollen is hydrolyzed during germination by a phytase. *Acta Bot. Neerl.* 31:441–47

47. Jefferson, R. A., Kavanagh, T. A., Bevan, M. W. 1987. GUS fusions: β-glucuronidase as a sensitive and versatile

gene fusion marker in higher plants. *EMBO J.* 6:3901–7

48. Kamalay, J. C., Goldberg, R. B. 1980. Regulation of structural gene expression in tobacco. *Cell* 19:934–46

49. Klosgen, R. B., Gierl, A., Schwarz-Sommer, Z., Saedler, H. 1986. Molecular analysis of the *waxy* locus of *Zea mays. Mol. Gen. Genet.* 203:237–44

50. Knox, R. B., Heslop-Harrison, J. 1970. Pollen wall proteins: Localization and enzymatic activity. *J. Cell Sci.* 6:1–27

51. Laughnan, J. R., Gabay-Laughnan, S. 1983. Cytoplasmic male sterility in maize. *Annu. Rev. Genet.* 117:27–48

52. Lin, J.-J., Dickinson, D. B., Ho, T.-H. D. 1987. Phytic acid metabolism in lily (*Lilium longiflorum* Thunb.) pollen. *Plant Physiol.* 83:408–13

53. Liu, X. C., Jones, K., Dickinson, H. G. 1987. DNA synthesis and cytoplasmic differentiation in tapetal cells of normal and cytoplasmically male sterile lines of *Petunia hybrida. Theor. Appl. Genet.* 74:846–51

54. Lonsdale, D. M. 1987. Cytoplasmic male sterility: a molecular perspective. *Plant Physiol. Biochem.* 25:265–71

55. Ludwig, S. R., Oppenheimer, D. G., Silflow, C. D., Snustad, D. P. 1988. The α_1-tubulin gene of *Arabidopsis thaliana:* primary structure and preferential expression in flowers. *Plant Mol. Biol.* 10:311–21

56. Maheshwari, P. 1950. *An Introduction to the Embryology of Angiosperms.* New York: McGraw-Hill

57. Mascarenhas, J. P. 1975. The biochemistry of angiosperm pollen development. *Bot. Rev.* 41:259–314

58. Mascarenhas, J. P. 1988. Anther- and pollen-expressed genes. In *Temporal and Spatial Regulation of Plant Genes,* ed. D. P. S. Verma, R. B. Goldberg, pp. 97–115. New York: Springer-Verlag

59. Mascarenhas, J. P. 1989. The male gametophyte of flowering plants. *The Plant Cell* 1:657–64

60. Mascarenhas, J. P., Altschuler, M. 1983. The response of pollen to high temperatures and its potential applications. See Ref. 76, pp. 3–8

61. Mascarenhas, J. P., Bell, E. 1970. RNA synthesis during development of the male gametophyte of *Tradescantia. Dev. Biol.* 21:475–90

62. Mascarenhas, J. P., Eisenberg, A., Stinson, J. R., Willing, R. P., Pe, M. E. 1985. Genes expressed during pollen development. In *Plant Cell/Cell Interactions,* ed. I. Sussex, A. Ellingboe, M. Crouch, E. Malmberg, pp. 19–23. Cold

Spring Harbor, New York: Cold Spring Harbor Lab.

63. Mascarenhas, J. P., Mermelstein, J. 1981. Messenger RNAs: their utilization and degradation during pollen germination and tube growth. *Acta Soc. Bot. Polon.* 50:13–20

64. Mascarenhas, N. T., Bashe, D., Eisenberg, A., Willing, R. P., Xiao, C. M., Mascarenhas, J. P. 1984. Messenger RNAs in corn pollen and protein synthesis during germination and pollen tube growth. *Theor. Appl. Genet.* 68:323–26

65. Matthys-Rochon, E., Vergne, P., Detchepare, S., Dumas, C. 1987. Male germ unit isolation from three tricellular pollen species: *Brassica oleracea, Zea mays* and *Triticum aestivum. Plant Physiol.* 83:464–66

66. McConchie, C. A., Hough, T., Knox, R. B. 1987. Ultrastructural analysis of the sperm cells of mature pollen of maize, *Zea mays. Protoplasma* 139:9–19

67. McConchie, C. A., Russell, S. D., Dumas, C., Knox, R. B. 1987. Quantitative cytology of the mature sperm cells of *Brassica campestris* and *B. oleracea. Planta* 170:446–52

68. McCormick, S., Twell, D., Wing, R., Ursin, V., Yamaguchi, J., Larabell, S. 1989. Anther-specific genes: molecular characterization and promoter analysis in transgenic plants. In *Plant Reproduction: From Floral Induction to Pollination,* ed. E. Lord, G. Bernier, pp. 128–35. Rockville, MD: Am. Soc. Plant Physiol.

69. McKenna, M., Mulcahy, D. L. 1983. Ecological aspects of gametophytic competition in *Dianthus chinensis.* See Ref. 76, pp. 419–24

70. Moffatt, B., Somerville, C. 1988. Positive selection for male sterile mutants of *Arabidopsis* lacking adenine phosphoribosyl transferase activity. *Plant Physiol.* 86:1150–54

71. Mulcahy, D. L. 1971. A correlation between gametophytic and sporophytic characteristics in *Zea mays* L. *Science* 171:1155–56

72. Mulcahy, D. L. 1974. Correlation between speed of pollen tube growth and seedling height in *Zea mays* L. *Nature* 249:491–92

73. Mulcahy, D. L. 1979. The rise of the angiosperms: a genecological factor. *Science* 206:20–23

74. Mulcahy, D. L., Mulcahy, G. B. 1975. The influence of gametophytic competition on sporophytic quality in *Dianthus chinensis. Theor. Appl. Genet.* 46:277–80

75. Mulcahy, D. L., Mulcahy, G. B. 1987. The effects of pollen competition. *Am. Sci.* 75:44–50

76. Mulcahy, D. L., Ottaviano, E., ed. 1983. *Pollen: Biology and Implications for Plant Breeding.* New York: Elsevier

77. Mulinix, C. A., Iezzoni, A. F. 1988. Microgametophyte selection in two alfalfa (*Medicago sativa* L.) clones. *Theor. Appl. Genet.* 75:917–22

78. Murgia, M., Wilms, H. J. 1988. Three-dimensional image and mitochondrial distribution in sperm cells of *Euphorbia dulcis.* See Ref. 136, pp. 75–79

79. Nave, E. B., Sawhney, V. K. 1986. Enzymatic changes in post-meiotic anther development in *Petunia hybrida.* I. Anther ontogeny and isozyme analyses. *J. Plant Physiol.* 125:451–65

80. Ottaviano, E., Petroni, D., Pe, M. E. 1988. Gametophytic expression of genes controlling endosperm development in maize. *Theor. Appl. Genet.* 74:252–58

81. Ottaviano, E., Sari-Gorla, M. 1979. Genetic variability of male gametophyte in maize: pollen genotype and pollen-style interaction. *Monographie in Genetica Agraria,* 4:89–106

82. Ottaviano, E., Sari-Gorla, M., Mulcahy, D. L. 1980. Pollen tube growth rates in *Zea mays:* implications for genetic improvement of crops. *Science* 210:437–38

83. Patterson, B. D., Mutton, L., Paull, R. E., Nguyen, V. Q. 1987. Tomato pollen development: stages sensitive to chilling and a natural environment for the selection of resistant genotypes. *Plant Cell Environ.* 10:363–68

84. Peddada, L., Mascarenhas, J. P. 1975. 5S ribosomal RNA synthesis during pollen development. *Dev. Growth Diff.* 17:1–8

85. Pedersen, S., Simonsen, V., Loeschcke, V. 1987. Overlap of gametophytic and sporophytic gene expression in barley. *Theor. Appl. Genet.* 75:200–6

86. Pennell, R. I., Geltz, N. R., Koren, E., Russell, S. D. 1987. Production and partial characterization of hybridoma antibodies elicited to the sperm of *Plumbago zeylanica. Bot. Gaz.* 148:401–6

87. Plegt, L., Bino, R. J. 1989. β-Glucuronidase activity during development of the male gametophyte from transgenic and non-transgenic plants. *Mol. Gen. Genet.* 216:321–27

88. Pressey, R., Avants, J. K. 1977. Occurrence and properties of polygalacturonase in *Avena* and other plants. *Plant Physiol.* 60:548–53

89. Pressey, R., Reger, B. J. 1989. Polyga-lacturonase in pollen from corn and other grasses. *Plant Sci.* 59:57–62

90. Raghavan, V. 1987. Developmental strategies of the angiosperm pollen: a biochemical perspective. *Cell Differ.* 21:213–26

91. Rajora, O. P., Zsuffa, L. 1986. Sporophytic and gametophytic gene expression in *Populus deltoides* Marsh, *P. nigra* L., and *P. maximowiczii* Henry. *Can. J. Genet. Cytol.* 28:476–82

92. Reynolds, T. L., Raghavan, V. 1982. An autoradiographic study of RNA synthesis during maturation and germination of pollen grains of *Hyoscyamus niger. Protoplasma* 111:177–82

93. Reznickova, S. A. 1978. Histochemical study of reserve nutrient substances in anther of *Lilium candidum. C. R. Acad. Bulg. Sci.* 31:1067–70

94. Roggen, H. P., Stanley, R. G. 1969. Cell wall hydrolyzing enzymes in wall formation as measured by pollen tube extension. *Planta* 84:295–303

95. Roman, H. 1948. Directed fertilization in maize. *Proc. Natl. Acad. Sci. USA* 34:36–42

96. Rowley, J. R., Walles, B. 1987. Origin and structure of Ubisch bodies in *Pinus sylvestris. Acta Soc. Bot. Polon.* 56: 215–27

97. Russell, S. D. 1984. Ultrastructure of the sperm of *Plumbago zeylanica.* 2. Quantitative cytology and three-dimensional organization. *Planta* 162:385–91

98. Russell, S. D. 1985. Preferential fertilization in *Plumbago:* ultrastructural evidence for gamete-level recognition in an angiosperm. *Proc. Natl. Acad. Sci. USA* 82:6129–34

99. Russell, S. D. 1986. Isolation of sperm cells from the pollen of *Plumbago zeylanica. Plant Physiol.* 81:317–19

100. Sacher, R., Mulcahy, D. L., Staples, R. 1983. Developmental selection for salt tolerance during self-pollination of *Lycopersicon* × *Solanum* F1 for salt tolerance of F2. See Ref. 76, pp. 329–34

101. Sachs, M. M., Dennis, E. S., Gerlach, W. L., Peacock, W. J. 1986. Two alleles of maize alcohol dehydrogenase 1 have 3' structural and poly(A) addition polymorphisms. *Genetics* 113:449–67

102. Sari-Gorla, M., Frova, C., Binelli, G., Ottaviano, E. 1986. The extent of gametophytic-sporophytic gene expression in maize. *Theor. Appl. Genet.* 72:42–47

103. Sari-Gorla, M., Ottaviano, E., Frascaroli, E., Landi, P. 1989. Herbicide-tolerant corn by pollen selection. *Sex. Plant Reprod.* 2:65–69

104. Sawhney, V. K., Nave, E. B. 1986. Enzymatic changes in post-meiotic anther development in *Petunia hybrida*. II. Histochemical localization of esterase, peroxidase, malate- and alcohol dehydrogenase. *J. Plant Physiol.* 125:467–73

105. Schrauwen, J. A. M., Reijnen, W. H., DeLeeuw, H. C. G. M., van Herpen, M. M. A. 1986. Response of pollen to heat stress. *Acta Bot. Neerl.* 35:321–27

106. Searcy, K. B., Mulcahy, D. L. 1985. Pollen tube competition and selection for metal tolerance in *Silene dioica* (Caryophyllaceae) and *Mimulus guttatus* (Scrophulariaceae). *Am. J. Bot.* 72: 1695–99

107. Searcy, K. B., Mulcahy, D. L. 1985. Pollen selection and the gametophytic expression of metal tolerance in *Silene dioica* (Caryophyllaceae) and *Mimulus guttatus* (Scrophulariaceae). *Am. J. Bot.* 72:1700–6

108. Shivanna, K. R., Xu, H., Taylor, P., Knox, R. B. 1988. Isolation of sperms from the pollen tubes of flowering plants during fertilization. *Plant Physiol.* 87:647–50

109. Singh, M. B., O'Neill, P., Knox, R. B. 1985. Initiation of postmeiotic β-galactosidase synthesis during microsporogenesis in oilseed rape. *Plant Physiol.* 77:225–28

110. Smith, G. A., Moser, H. S. 1985. Sporophytic-gametophytic herbicide tolerance in sugarbeet. *Theor. Appl. Genet.* 71:231–37

111. Southworth, D., Knox, R. B. 1989. Flowering plant sperm cells: isolation from pollen of *Gerbera jamesonii* (Asteraceae). *Plant Sci.* 60:273–77

112. Spena, A., Schell, J. 1987. The expression of a heat-inducible chimeric gene in transgenic tobacco plants. *Mol. Gen. Genet.* 206:436–40

113. Stanley, R. G., Thomas, D. S. 1967. Pollen enzymes and growth. *Proc. Assoc. Soc. Agri. Workers* 13:265

114. Steffensen, D. M. 1966. Synthesis of ribosomal RNA during growth and division in *Lilium. Exp. Cell Res.* 44:1–12

115. Steffensen, D. M. 1971. Ribosome synthesis compared during pollen and pollen tube development. See Ref. 29, pp. 223–29

116. Stinson, J. R., Eisenberg, A. J., Willing, R. P., Pe, M. E., Hanson, D. D., Mascarenhas, J. P. 1987. Genes expressed in the male gametophyte of flowering plants and their isolation. *Plant Physiol.* 83:442–47

117. Stinson, J., Mascarenhas, J. P. 1985.

118. Suss, J., Tupy, J. 1979. Poly(A)RNA synthesis in germinating pollen of *Nicotiana tabacum* L. *Biol. Plant.* 21:365–71

119. Takegami, M. H., Yoshioka, M., Tanaka, I., Ito, M. 1981. Characteristics of isolated microsporocytes from liliaceous plants for studies of the meiotic cell cycle in vitro. *Plant Cell Physiol.* 22:1–10

120. Tanaka, I., Ito, M. 1980. Induction of typical cell division in isolated microspores of *Lilium longiflorum. Plant Sci. Lett.* 17:279–85

121. Tanaka, I., Ito, M. 1981. Studies on microspore development in liliaceous plants. II. Pollen tube development in lily pollens cultured from the uninucleate microspore stage. *Plant Cell Physiol.* 22:149–53

122. Tanksley, S. D., Zamir, D., Rick, C. M. 1981. Evidence for extensive overlap of sporophytic and gametophytic gene expression in *Lycopersicon esculentum. Science* 213:454–55

123. Theunis, C. H., Van Went, J. L. 1989. Isolation of sperm cells from mature pollen grains of *Spinacia oleracea* L. *Sex. Plant Reprod.* 2:97–102

124. Tupy, J. 1982. Alterations in polyadenylated RNA during pollen maturation and germination. *Biol. Plant.* 24:331–40

125. Tupy, J., Suss, J., Hrabetova, E., Rihova, L. 1983. Developmental changes in gene expression during pollen differentiation and maturation in *Nicotiana tabacum* L. *Biol. Plant.* 25:231–37

126. Twell, D., Wing, R., Yamaguchi, J., McCormick, S. 1989. Isolation and expression of an anther-specific gene from tomato. *Mol. Gen. Genet.* 247:240–45

127. Ursin, V. M., Yamaguchi, J., McCormick, S. 1989. Gametophytic and sporophytic expression of anther-specific genes in developing tomato anthers. *The Plant Cell* 1:727–36

128. van Tunen, A. J., Hartman, S. A., Mur, L. A., Mol, J. N. M. 1989. Regulation of chalcone flavanone isomerase (CHI) gene expression in *Petunia hybrida*: the use of alternative promoters in corolla, anthers and pollen. *Plant Mol. Biol.* 12:539–51

129. van Tunen, A. J., Koes, R. E., Spelt, C. E., van der Krol, A. R., Stuitze, A. R., Mol, J. N. M. 1988. Cloning of the two chalcone flavanone isomerase genes from *Petunia hybrida*: coordinate, light-regulated and differential expression of flavonoid genes. *EMBO J.* 7:1257–63

130. van Went, J. L., Willemse, M. T. M.

Onset of alcohol dehydrogenase synthesis during microsporogenesis in maize. *Plant Physiol.* 77:222–24

1984. Fertilization. In *Embryology of Angiosperms*, ed. B. M. Johri, pp. 273–317. Berlin/New York: Springer-Verlag

131. Weatherwax, P. 1922. A rare carbohydrate in waxy maize. *Genetics* 7:568–72

132. Whipple, A. P., Mascarenhas, J. P. 1978. Lipid synthesis in germinating *Tradescantia* pollen. *Phytochem.* 17: 1273–74

133. Wiermann, R. 1979. Stage-specific phenylpropanoid metabolism during pollen development. In *Regulation of Secondary Product and Plant Hormone Metabolism*, ed. M. Luckner, K. Schreiber, pp. 231–39. Oxford: Pergamon

134. Willing, R. P., Bashe, D., Mascarenhas, J. P. 1988. An analysis of the quantity and diversity of messenger RNAs from pollen and shoots of *Zea mays*. *Theor. Appl. Genet.* 75:751–53

135. Willing, R. P., Mascarenhas, J. P. 1984. Analysis of the complexity and diversity of mRNAs from pollen and shoots of *Tradescantia*. *Plant. Physiol.* 75:865–68

136. Wilms, H. J., Keijer, C. J., ed. 1988.
Plant Sperm Cells as Tools for Biotechnology. Wageningen: Pudoc

137. Wilms, H. J., Van Aelst, A. C. 1983. Ultrastructure of spinach sperm cells in mature pollen. In *Fertilization and Embryogenesis in Ovulated Plants*, ed. O. Erdelska, pp. 105–12. Bratislava: VEDA

138. Xiao, C. M., Mascarenhas, J. P. 1985. High temperature induced thermotolerance in pollen tubes of *Tradescantia* and heat shock proteins. *Plant Physiol.* 78:887–90

139. Zamir, D. 1983. Pollen gene expression and selection: applications in plant breeding. In *Isozymes in Plant Genetics and Breeding*, ed. S. D. Tanksley, T. J. Orton, pp. 313–30. Amsterdam: Elsevier

140. Zamir, D., Tanksley, S. D., Jones, R. A. 1982. Haploid selection for low temperature tolerance of tomato pollen. *Genetics* 101:129–37

141. Zamir, D., Vallejos, E. C. 1983. Temperature effects on haploid selection of tomato microspores and pollen grains. See Ref. 76, pp. 335–42

Annu. Rev. Plant Physiol. Plant Mol. Biol. 1990. 41:339–67

MOLECULAR COMMUNICATION IN INTERACTIONS BETWEEN PLANTS AND MICROBIAL PATHOGENS

Richard A. Dixon

Plant Biology Division, Samuel Roberts Noble Foundation, P.O. Box 2180, Ardmore, Oklahoma 73402

Christopher J. Lamb

Plant Biology Laboratory, Salk Institute for Biological Studies, 10010 North Torrey Pines Road, La Jolla, California 92037

KEY WORDS: avirulence genes, defense genes, disease resistance genes, elicitors, pathogenicity genes

CONTENTS

339

1040-2519/90/0601-0339$02.00

INTRODUCTION

Fungal and bacterial pathogens express sets of genes involved in establishing infection while novel genes are expressed in the host as part of its response. An understanding of the molecular communication that underlies the temporal and spatial control of these gene expression events is now within reach, as more sophisticated techniques of molecular and genetic analysis are applied to plant-pathogen interactions. Various aspects of the biochemistry and physiology of induced defense in hosts and of the molecular genetic basis of avirulence and virulence in bacterial phytopathogens have already been reviewed (21, 22, 26, 43, 56, 67). Here we review recent results shedding light on the nature of the molecular signals that determine or modulate host-pathogen recognition, specificity, and induced defense.

SIGNALS FOR THE ESTABLISHMENT OF INFECTION

Successful infection by microbial pathogens requires surface attachment, degradation of host chemical and physical barriers, production of toxins, and inactivation of plant defenses (67). Overall, up to 100 genes, including those involved in fitness, may be needed for bacterial pathogenicity (21), and the differentiation of elaborate infection structures by fungal pathogens suggests a correspondingly greater complexity.

Regulation of microbial pathogenicity genes involves a complex interplay of signals between host and pathogen. For example, in *Fusarium solani* f. sp. *pisi*, a fungal pathogen of pea, transcripts encoding the extracellular cutinase involved in penetration of the plant cuticle rapidly accumulate after addition of cutin monomers to fungal cultures (63). Cutinase transcription can be induced in isolated *Fusarium* nuclei by cutin monomers in the presence of a soluble protein factor from the fungal extract (90). In vivo induction of cutinase by polymeric cutin is greatly enhanced by small amounts of the active enzyme. This feed-forward regulation suggests that the fungal spore senses contact with the host by responding to cutin monomers released by a low-level constitutive cutinase activity. Cutinase is then induced to levels required for host penetration.

Pectin-degrading enzymes are likewise induced in response to polymer degradation products. The sequential appearance of different classes of cell wall–degrading enzymes during fungal growth on plant cell walls suggests

that specific degradative enzymes may be induced by different signal molecules as the wall is progressively degraded (63). A reiterative signal hierarchy
could allow the pathogen to effect rapid entry with minimal degradation of the
host cell wall. Such a strategy may be critical for successful infection because
certain plant cell wall fragments act as elicitors of host defense responses.

In the tumor-inducing bacterium *Agrobacterium tumefaciens*, virulence
genes involved in the transfer of T-DNA into the plant cells are induced by
factors from the host (104). Small molecules involved in this process were
discovered as follows. Bacteria carrying reporter genes with the promoter of
an inducible virulence gene led to the identification of acetosyringone and
related phenylpropanoid products as potent and specific inducers. Accumulation of these metabolites is stimulated by wounding of plant tissue, and hence
the cue for expression of virulence functions is provided by components of the
stress response machinery of the host. Perception and transduction of this
signal involve two bacterial genes, *virA* and *virG,* which are homologous to
other two-component bacterial regulatory systems—e.g. the *ntrB* and *ntrC*
genes of *Rhizobium* species (93). Similar regulatory systems may also be
involved in the induction of other bacterial pathogenicity genes, such as those
encoding wall-degrading enzymes and components of their secretion pathway
(42). Several approaches have been developed recently to isolate bacterial
genes whose induction requires the presence of host-derived factors. These
methods may be of value in dissecting other signal systems in the determinative stages of plant–pathogen interactions (21).

Specific signals may also be involved in regulating the expression of
microbial virulence genes involved in disarming inducible plant defenses. For
example, accumulation of the phytoalexin pisatin in peas infected with *Nectria haematococca* causes induction of the phytoalexin-detoxifying enzyme
pisatin demethylase (PDA). PDA is a cytochrome P-450–dependent enzyme
that converts the pterocarpan pisatin to the less fungitoxic product 3,6a-
dihydroxy-8,9-methylenedioxypterocarpan. This enzyme is encoded by three
nonallelic genes in *N. haematococca* with one locus, *Pda1,* conferring high
demethylating activity. A PDA gene was cloned by expression in *Aspergillus
nidulans* (116). PDA transcripts are induced by pisatin, and genetic analysis
showed a close correlation between the ability to demethylate pisatin and the
virulence of *N. haematococca* isolates. Although phytoalexin detoxification
may not be a major determinative factor in most interactions, this analysis
indicates that specific signals may be deployed even at relatively late stages in
pathogenesis.

MICROBIAL DETERMINANTS OF AVIRULENCE

Detailed genetic analysis of a relatively small number of plant–pathogen
interactions has shown a "gene-for-gene" interaction. That is, resistance or

susceptibility in host cultivars to distinct physiological races of a pathogen is determined by pairs of corresponding genes in the host and pathogen (37). In such "gene-for-gene" interactions, a resistance gene in a particular host cultivar confers resistance against physiological races that express the matching avirulence gene; this leads to genetic incompatibility characterized by the hypersensitive response (HR), which involves rapid death of the first infected cell and elaboration of a number of inducible defenses (see below).

In the last several years, a number of bacterial avirulence genes have been isolated by a shotgun cloning strategy involving broad-host-range plasmids such as pLAFR1. Identification of these avirulence genes was based on the ability of specific transconjugated cosmid clones to confer cultivar-specific patterns of avirulence on recipient virulent races (56, 57). Cloned bacterial avirulence genes identified to date do not impart the cultivar-specific virulence phenotype of the donor in this way; this confirms the dominance of the avirulence genotype. Furthermore, in cases where congenic host lines have been analyzed, as in cotton, it has been possible to show that the cloned avirulence genes match specifically, on a gene-for-gene basis, with individual resistance genes.

Expression of avirulence genes may also underlie nonhost resistance. For example, *Xanthomonas campestris* pv. *vesicatoria,* the agent of leaf spot disease of pepper and tomato, is unable to cause disease on nonhost plants such as bean, soybean, and cotton. Instead it induces an HR in all cultivars. A cosmid clone from *X. campestris* pv. *vesicatoria* tomato race 1 converted *X. campestris* pv. *phaseoli* to avirulence by inducing an HR on the bean cultivar Sprite but not on the cultivar Bush Blue Lake (117). This nonhost avirulence gene, designated *avrRxv,* also inhibited disease production by several other *X. campestris* pathovars on their normally susceptible hosts—e.g. *X. campestris* pv. *glycines* on soybean and *X. campestris* pv. *malvacearum* on cotton. Hypersensitive resistance in bean to *X. campestris* pv. *phaseoli* carrying the *avrRxv* gene segregated as a single, incompletely dominant gene. A parallel analysis showed that when the *Pseudomonas syringae* pv. *tomato* gene *avrD* is introduced into *P. syringae* pv *glycinea,* it elicits a unique pattern of cultivar-specific HR in soybean (61). This pattern had not previously been observed with any known *P. syringae* pv. *glycinea* race, or any avirulence gene cloned to date from the soybean pathogen.

These studies indicate that avirulence genes may contribute to the restriction of host range at higher taxonomic levels than race-cultivar specificity and suggest that genetic control of nonhost resistance is similar to that of host resistance. Mutating *avrRxv* did not allow *X. campestris* pv. *vesicatoria* to become pathogenic on bean or other species, even though the wild-type *avrRxv,* when mobilized into appropriate pathovars, induced the HR in the corresponding plant host species (117). It will be of great interest to determine whether *X. campestris* pv. *vesicatoria* lacks positive gene functions required for pathogenicity on nonhost species, such as bean, or whether avirulence genes in addition to *avrRxv* condition resistance reactions to *X. campestris* pv. *vesicatoria* in bean and other species. *avrD*, *avrRxv*, and their corresponding resistance genes were recognized only after mobilization of the bacterial genes. These data raise the possibility that nonhost plants may contain batteries of resistance genes that correspond to avirulence genes unique to each potential pathogen and that such resistance genes have additional functions in the plants.

A major question is how the avirulence gene products confer the avirulent phenotype. Several avirulence genes from *P. syringae* pv. *glycinea* and *P. syringae* pv. *tomato* have been sequenced and shown to encode single protein products ranging from 34 to 100 kDa, that exhibit no homology to other known protein sequences (55, 56). The protein products of the *P. syringae* pv. *glycinea* avrB and *avrC* genes share considerable sequence homology, except in the central region, which was shown by analysis of the phenotypic effects of *avrB/C* recombinants to be involved in conferring HR specificity (109). The proteins encoded by the bacterial avirulence genes sequenced to date are hydrophilic and contain no readily identifiable signal sequences for extracellular transport (55, 56). Therefore, it is not clear how these proteins could interact directly with a plant recognition factor such as the product of a disease-resistance gene. The AvrB and AvrC gene products have been produced in large amounts in *Escherichia coli* cells, where they are deposited in cytoplasmic inclusion bodies. Following solubilization with urea, these proteins did not elicit a HR when infiltrated into soybean leaves (56).

An alternative hypothesis is that avirulence gene products act indirectly, either controlling functions related to the surface architecture of the bacterium or participating in the production of diffusible cultivar-specific elicitors. Support for the involvement of an elicitor comes from recent studies of the *P. syringae* pv. *tomato avrD* gene, which establishes a unique pattern of race-cultivar specificity in *P. syringae* pv. *glycinea*–soybean interactions when mobilized into *P. syringae* pv. *glycinea* race 4 (see above). *E. coli* overexpressing the *avrD* gene can induce the HR in the same set of soybean cultivars (106). Expression of *avrD* apparently leads to the synthesis of a low-molecular-weight race-cultivar-specific elicitor of the HR; and because the gene is functional in *E. coli*, the elicitor must be derived from a common bacterial metabolite.

Bacterial avirulence gene expression does not seem to depend on highly specific plant signals. This is in contrast to the induction of *nod* genes in *Rhizobium* species or *vir* genes in *Agrobacterium tumefaciens*. Thus, induction of the *P. syringae* pv. *glycinea avrB* gene in planta requires only simple nutritional factors, is under catabolite repression, and occurs in both compatible and incompatible host cultivars. If *avrB* is expressed, no further RNA or protein synthesis by the bacterium is necessary for HR induction (49). *avrB* is, however, under the control of a 2-kb region within the *hrp* locus. This locus is a 22-kb region containing a cluster of pathogenicity genes; mutations in the locus abolish induction of the HR on resistant cultivars and render the bacterium nonpathogenic on susceptible cultivars (73). Both physical expression and biological activity of *avrB* are therefore controlled by other bacterial genes.

Most avirulence genes are chromosomal, but they do not appear to be closely linked; this contrasts with many pathogenicity genes—e.g. the *hrp* locus, or a number of plant disease-resistance genes. Retention of an avirulence gene appears to be a selective disadvantage on cultivars containing the complementary resistance gene, but surprisingly little attention has been given to other possible functions for avirulence genes in addition to their role in HR induction. Southern hybridization detected alleles of avirulence genes in some, but not all *P. syringae* pv. *glycinea* pathovars; e.g. *P. syringae* pv. *glycinea* contains sequences homologous to the *P. syringae* pv. *tomato avrD* gene that are presumably inactive, at least in relation to HR induction (55, 61).

In bacterial spot disease of pepper, the stability of resistance is a function of the mutation frequency of *X. campestris* pv. *vesicatoria* avirulence genes. Mutation rates are substantially higher in genetic backgrounds containing a 1.2-kb transposable element IS476, which can account for all natural mutations in the *avrBs₁* gene. *avrBs₁* is on a plasmid that contains three copies of IS476 (at least one of which is active) and a gene for copper resistance (54). The plasmid is transmitted through the population under copper selection, and mutation to virulence may unwittingly have been increased by the use of copper sprays as a chemical defense against the bacterium. The evolution and population genetics of avirulence genes as well as the elucidation of their other potential functions are topics of considerable interest for future research.

Similarly, recent advances in genetic manipulation in several phytopathogenic fungi and development of appropriate transformation vectors with positive selective markers for filamentous fungi (86, 113, 115) opens the way for molecular analysis of the determinants of fungal pathogenicity and avirulence in a number of plant-pathogen interactions.

PLANT DISEASE-RESISTANCE GENES

Shotgun cloning by function, analogous to the strategy used for bacterial avirulence genes, is not yet feasible for plant disease-resistance genes because of the large sizes of plant genomes and the lack of scorable phenotypes in simplified experimental systems such as isolated protoplasts. Current strategies for cloning these genes involve (*a*) transposon tagging (*b*) physical isolation relying on genetic maps based on restriction-fragment length polymorphisms and other markers for linkage analysis, followed by "chromosome walking" (84). Considerable effort has been devoted to the isolation of disease-resistance genes from maize. The transposable elements of maize have been well characterized and several loci that condition race-cultivar specificity have been analyzed in detail (91). Although more than a dozen genes in maize have been isolated by transposon tagging (39), for disease-resistance genes this approach has been slowed by two factors: (*a*) the high copy number of certain transposable elements, such as *Mutator,* used in some of the tagging studies; and (*b*) in the case of the complex *Rp1* locus, a high rate of spontaneous mutation, even in genetic backgrounds lacking known active transposable elements (6). Genetic and mutational analysis indicates that in *Rp1* and many other plant disease-resistance genes new specificities can be generated by rearrangement or recombination events within the locus (91). Such genetic instability makes it difficult to recognize potential transposon-induced alleles.

The second strategy may benefit from the observation that a number of disease-resistance genes map close to other loci for which cloned hybridization probes are already available or may be generated by standard approaches. When a high-density map is available, "chromosome walking" is greatly facilitated. For example, nematode pathogens exhibit race-cultivar-specific interactions analogous to those observed with microbial pathogens, and a specific nematode-resistance locus in tomato maps adjacent to an acid phosphatase locus (92a). Development of yeast artificial chromosomes to clone large fragments of plant genomic DNA will greatly assist efforts to clone disease-resistance genes by physical mapping. An exciting opportunity for cloning a disease-resistance gene is afforded by the observation that a single dominant locus, *Pc2,* not only conditions the sensitivity of oat plants to the host-specific toxin victorin (produced by *Helminthosporium victoriae*) but also is a disease-resistance gene against the fungus *Puccinia coronata,* conditioning a HR to avirulent races. Victorin is a basic peptide containing unusual amino acid residues, and its structure has been completely determined. Recently, a 100-kDa polypeptide was shown to be a specific binding site for victorin (121). It should now be possible to isolate this putative toxin receptor and test whether it is the product of the *Pc2* gene.

The small genome size of *Arabidopsis thaliana* and the molecular and genetic tools available for study of this plant [including extensive physical and genetic mapping and development of a library of ordered contiguous cosmid clones (84)] make it an attractive organism in which to clone both disease-resistance genes and other genes involved in stress signal transduction and response. However, because it is a weed of no agricultural significance, little is known about its potential pathogens or biology of infection. In the absence of intensive breeding programs, highly specialized interactions equivalent to the race-cultivar-specific ones in established crops may not have developed in *A. thaliana*. It will be interesting to see whether race-cultivar-specific pathogens of other crucifers exhibit differential interactions with *A. thaliana* biotypes.

ELICITORS AND ELICITOR RECEPTORS

Race-Specific Elicitors

A major effort has been made to characterize microbial preparations active in various bioassays based on the plant HR. Biotic elicitors are molecules of either pathogen or host origin that can induce defense responses (such as phytoalexin accumulation) in plant tissue. Many elicitors have been described, including various polysaccharides, oligosaccharide fragments, proteins, glycoproteins, and fatty acids (3, 22, 45, 110). Although this diverse array of bioactive molecules may reflect the complexity of plant-pathogen signal systems, it may also indicate the inherent limitations of a bioassay approach in the absence of molecular genetic tools for the dissection of such complex interactions. Race-specific elicitors would induce a response only in host cultivars on which a particular race of pathogen is avirulent and would be direct or indirect products of avirulence genes. The low-molecular-weight elicitor isolated from culture fluids of *E. coli* expressing the *P. syringae* pv. *tomato avrD* gene to high levels (see above) may be the first race-specific elicitor whose role in determining avirulence has direct support.

Only two reports provide good evidence for the isolation of race-specific elicitors from fungi. A partially purified galactose- and mannose-rich glycoprotein from the α race of *Colletotrichum lindemuthianum* induces phytoalexin accumulation in a bean cultivar resistant to the α race but not in a susceptible cultivar (112). Unfortunately, the structural basis for this specificity has not been established. The only macromolecular race-specific elicitor so far characterized as a discrete molecular species is a peptide from the tomato pathogen *Cladosporium fulvum* (98). A necrosis-inducing peptide was isolated from the intracellular space of tomato leaves infected with a compatible race of *Cladosporium fulvum*. In a test panel of tomato cultivars, this peptide exhibited appropriate race-cultivar specificity with respect to the *A9* resistance gene. Growth in planta appears necessary for the formation of this race-specific elicitor. The 27-amino-acid peptide (M_r = 3 kDa) has been sequenced and is cysteine rich. By using oligonucleotide probes deduced from the amino acid sequence, mRNA hybridization was detected in the compatible interaction but has yet to be detected in the incompatible interaction or in axenically grown fungus. The potential coding capacity of the transcript detected by Northern hybridization is much greater than required for a 3-kDa product. The peptide may be a specific degradation product.

Despite the identification of the *C. fulvum* peptides, most elicitors do not show appropriate race-cultivar specificity. This absence of specificity may reflect both the use of axenic cultures as source material and the employment of harsh treatments such as acid hydrolysis or heat solubilization for elicitor purification (67). Nonetheless, study of such elicitors defined by activity in various bioassays has provided useful information about the properties of signal transduction mechanisms associated with induction of defense responses.

Structure–Function Relationships

From analysis of a glucan elicitor isolated from cell walls of *Phytophthora megasperma* f. sp. *glycinea* it was determined that a specific branched β-3,6-heptaglucoside fragment is the minimal elicitor-active structure in this cell wall preparation (101). Other structural isomers co-isolated from hydrolyzed *P. megasperma* f. sp. *glycinea* walls were inactive, an analysis confirmed by chemical synthesis. One implication of this exquisite structural specificity is that the active elicitor is likely to exert its effects by interaction with a cognate plant receptor. In contrast, studies of how degree of polymerization (DP) and extent of N-acetylation affect the activity of chitosan as an elicitor of callose synthesis in *Catharanthus roseus* cell cultures suggest that this elicitor interacts primarily with regularly spaced negative charges on the plant plasma membrane—e.g. with arrays of charged phospholipid headgroups—rather than with a discrete macromolecular receptor (53).

Elicitor-Binding Sites

A *P. megasperma* f. sp. *glycinea* cell wall β-3,6-glucan elicitor with an average DP of 22 has been used to identify high-affinity binding sites on soybean cell membranes (18). Binding was reversible, and labeled glucan elicitor could be displaced by unlabeled elicitative derivatives but not by inactive glucans—observations consistent with the existence of a receptor. The affinity was greatest for glucans with a DP > 12, and the relationship between these binding sites and the as-yet-unidentified receptor for the elicitor-active heptaglucoside remains to be established. As with the heptaglucoside (see above), modification of the reducing end did not inhibit elicitor activity. Synthesis of high-specific-activity [^{125}I] glucan with elicitor activity was thus possible, and this was used to demonstrate high-affinity binding ($K_d = 37$ nM) to soybean protoplasts in vivo. The most binding was observed in a plasma membrane–enriched fraction, and conditions for solubilization with retention of binding activity have been determined as a first step in the purification of the putative receptor.

Multiple Signal Functions

A second, well-characterized class of elicitors comprises oligogalacturonide fragments obtained from host-plant cell walls (94). The linear oligomers of α-1,4-linked galacturonic acid induce the synthesis of phytoalexins in legume cell suspension cultures, lignin in cucurbits, and proteinase inhibitors in plants that respond systemically to insect attacks. Oligogalacturonides with a DP of 9–12 are optimal for induction of phytoalexin and lignin production, whereas oligogalacturonides with a DP as low as 2 can induce proteinase inhibitors (94). Recent data indicate that oligogalacturonide fragments stimulate floral development when applied to tobacco stem thin-layer cultures maintained on appropriate concentrations of auxin and cytokinin (35). These studies suggest that the plant may utilize specific signal systems very flexibly, depending on the biological context. The release, modification, and turnover of elicitor-active molecules may be crucial in determining signal strength and duration.

Phytophthora cryptogea and *P. capsici,* two pathogens causing systemic leaf necrosis on tobacco, produce two closely related 98-amino-acid peptides called cryptogein and capsicein, with M_rs of 10,323 and 10,155 respectively. These molecules are sufficient to cause necrosis. Application of these peptides can protect tobacco against *P. nicotianae,* a pathogen unable to produce such an elicitor (92). Cryptogein causes visible necrosis at a dose of 1 μg plant^{-1}, whereas 50 times as much capsicein is required to produce a similar reaction; capsicein induces protection in the near absence of necrosis. These data imply that the protective and necrosis-inducing activities are distinct. The peptides contain extensive internal repetitive sequences and are identical except for heterologous carboxy and amino terminal regions. These elicitors will reward structure–function studies and are potentially exciting targets for biotechnological manipulation.

Elicitor Release, Modification, and Turnover

Elicitor-active oligogalacturonides can be released from pectic material or isolated plant cell walls in vitro by microbial polygalacturonases or pectic lyases, which are themselves elicitative in vivo (94). The products of polygalacturonic acid degradation by lyases are oligogalacturonides with the nonreducing terminus composed of Δ-4,5-galacturonic acid; these products are as active as the corresponding saturated oligomers produced by hydrolysis with polygalacturonases.

Although production of these elicitors does not appear to occur in a race-specific manner, it may be important for the plant to regulate the rate of appearance and size of these endogenous elicitor fragments. Many plants appear to contain polygalacturonase-inhibiting proteins (PGIPs), some of which have been purified to homogeneity (45, 118). Recent data indicate that the bean PGIP retards polygalacturonase-mediated hydrolysis of polypectate, resulting in the formation of larger oligomers (DP > 4); this suggests that PGIPs may function in vivo to increase the potential biological activity of the fragments released as a result of degradation of the host cell wall (13). Release of these fragments does not appear to occur in a race-specific manner, however; consequently there is no direct correlation between the levels of PGIPs and resistance.

Plants contain enzymes capable of releasing elicitor-active fragments from fungal cell wall polymers in situ. Thus, in addition to their antimicrobial lytic activities, chitinase and glucanase may be involved in the amplification of elicitor signals in the initial stages of microbial attack (26, 45). Bacterial chitinases often exhibit exo-chitinase activity, whereas the plant enzymes, for which there are no obvious endogenous substrates, appear to be endo-chitinases that catalyze the formation of relatively low-molecular-weight chitin oligomers. Basic forms of chitinase and glucanase accumulate in the vacuole—a location inconsistent with a putative role in signal generation, at least in the initial stages of infection prior to host cell penetration. Studies noting that several pathogenesis-related (PR) proteins that appear in intercellular fluids of infected leaves are acidic forms of chitinase and glucanase (52, 64, 69) did not establish whether these forms were actively secreted or were released into the intercellular space because of cell death in the infected tissue.

Detailed immunocytochemical and biochemical fractionation studies of bean cultivars detected a strong induction of chitinase and glucanase activity in the absence of tissue damage, leaf yellowing, or abscission; although both chitinase and glucanase were present at high levels in vacuoles, only the glucanase was present in the cell wall (77). On the basis of these studies it was proposed that the cell wall–localized glucanase is indeed involved in recognition processes, releasing defense-activating signal molecules from the walls of invading pathogens. It will be interesting to determine whether regulation of the plant cell wall glucanase system resembles that of fungal cutinase (see above). That is, a low-level constitutive activity is involved in the initial release of elicitor-active fragments to further induce glucanase, chitinase, and other defenses.

Signal Interactions

Emerging evidence indicates considerable complexity in interactions among various elicitors and between elicitors and other modulatory factors. For example, with cultivar-specific preparations from the α race of *Colletotrichum lindemuthianum*, the patterns of phenolic synthesis and induced mRNA activities in the plant varied with the purity of the extracellular elicitor fraction (111). Similarly, various glycoprotein fractions differed in their ability to induce enzymes of phytoalexin synthesis in cultured bean cells, such that two fractions rich in galactose and mannose could induce only a subset of the enzymes of phytoalexin biosynthesis (46). These fractions also induced different patterns of total mRNA translation products in the cells. Hence, induction of the overall phytoalexin defense response may involve several distinct elicitor activities and activation of more than one signal transduction pathway. One example of the complexity of the interactions is that the *Phytophthora megasperma* f. sp. *glycinea* glucan elicitor of soybean phytoalexins is inactive in parsley cells, possibly because of rapid metabolism; the protein component of a *P. megasperma* f. sp. *glycinea* glycoprotein is the elicitor in parsley (85).

Quantitative synergism between different factors has also been observed in a number of cases, including interactions between arachidonic acid and glucan elicitors in potato, between glucan elicitor and the agrichemical probenazole in rice (C. A. West, personal communication), and between glucan elicitor and oligogalacturonide fragments (33, 45). The last example is of particular interest, because this represents an interaction between microbial and endogenous plant signals. Thus, in the presence of low concentrations of polygalacturonic acid lyase and heptaglucoside, an approximately 12-fold stimulation in phytoalexin accumulation above the calculated additive response was observed (24). A similar synergism was observed between glucan elicitor and pectic fragments in parsley cell suspension cultures (23) but not in bean cell cultures exposed to elicitor from the cell walls of *C. lindemuthianum*. Such synergistic interactions might sensitize the plant to low levels of fungal elicitors when host cell wall fragments are released by mechanical damage prior to infection and thus might provide a mechanism for integration of responses to wounding and attempted infection.

A number of macromolecules with the ability to suppress elicitation have been isolated from microbial culture fluids (33, 45). Such suppressors include water-soluble, anionic, and neutral glucans from *Phytophthora infestans* (29), a peptide from the chickpea pathogen *Ascochyta rabiei* (60), small glycopeptides from the pea pathogen *Mycosphaerella pinodes* (122), and unidentified heat-labile components (possibly enzymes) in intercellular fluids of tomato leaves infected with *Cladosporium fulvum* (87). It is not known whether suppressors act by competition for elicitor-binding sites or by the induction of a separate response that overcomes the consequences of elicitor action. The *A. rabiei* suppressor inhibited the synthesis of both elicitor-induced and constitutive isoflavonoids in chickpea, suggesting the existence of sites of action at least partly distinct from those of the elicitor (60). In the pea–*M. pinodes* interaction, the suppressor caused a delay in the accumulation of phenylalanine ammonia-lyase (PAL) and chalcone synthase (CHS) transcripts in response to glucan/glycoprotein elicitors but did not affect their final levels (122). A similar delay was observed if reduced glutathione (GSH) was used as elicitor (see below), which suggests that the suppressor does not compete for specific elicitor receptors but acts at a later stage in the signal pathway. Whether delay or inhibition of defense responses as a result of suppressor activity affects compatibility remains to be established. In the *C. fulvum*–tomato interaction, suppression of nonspecific elicitor activity showed no race-cultivar specificity (87), and race-cultivar-specific suppression of nonspecific elicitors would not account for the dominance of the avirulent phenotype (37).

DEFENSE GENE REGULATION: IMPLICATIONS FOR SIGNALING

Defense Mechanisms

Elicitor treatment, mechanical damage, and microbial attack induce a number of host defense responses, including (*a*) phytoalexin synthesis; (*b*) cell wall reinforcement by deposition of callose and lignin, and accumulation of hydroxyproline-rich glycoproteins (HRGPs); and (*c*) production of proteinase inhibitors and lytic enzymes such as chitinase and glucanase (26, 67). Recent data have delineated a number of novel systems that may serve in defense against pathogens, including a novel lysozyme/chitinase activity (82). As discussed above, a number of PR proteins have been identified as acidic forms of chitinase and glucanase, distinct from the well-characterized basic vacuolar forms. Novel wound-inducible genes have been characterized that encode proteins with significant homology to the carbohydrate-binding sites of chitinase and certain lectins (105). A pollen allergen has also been shown to have considerable homology to the product of an infection-induced gene (11). Finally, a group of novel sulfur-rich vacuolar and cell wall proteins, called thionins, have been shown to have strong antimicrobial activity (8). These studies reinforce the conclusion that plants have evolved a diverse array of inducible defenses.

Induction of these defenses is observed in the early stages of attempted infection by a nonpathogen or by an avirulent race of a pathogen associated with the HR. The close correlation between phytoalexin detoxification and pathogenicity in *Nectria haematococca* (see above) and the ability of aminooxyphenylpropionic acid [a specific inhibitor of phenylalanine ammonia-lyase (PAL), the first enzyme in lignin and isoflavonoid phytoalexin synthesis] to make soybean seedlings susceptible to normally avirulent races of *P. megasperma* f. sp. *glycinea* provide direct evidence for the role of inducible defense responses in the expression of disease resistance (67). Defense responses are also often induced in the later stages of infection by a virulent race, at the onset of lesion formation; under appropriate physiological conditions they can restrict further development of infection. Defenses may also be induced in uninfected tissue at a distance from the initial site of microbial attack associated with the establishment and/or subsequent expression of induced systemic resistance. The molecular cloning of specific defense genes will allow the use of gene-transfer techniques to distinguish the specific contributions of particular defenses in the various forms of resistance.

Defense Gene Activation: Signal Complexity

Callose production involves stimulation of preexisting callose synthase at the plasma membrane (62). Rapid deposition of a proline-rich cell wall protein appears to involve elicitor-stimulated insolubilization of preexisting soluble precursors (D. J. Bradley, unpublished data). With these exceptions, the known induction of defense responses involve transcriptional activation of the corresponding defense genes as part of a massive switch in host gene expression (26, 67). Thus, in suspension-cultured bean cells treated with fungal elicitor, defense genes encoding chitinase and enzymes of phenylpropanoid biosynthesis involved in phytoalexin and lignin production are activated within 2–3 min. These kinetics imply that the signal transduction system is in place prior to elicitor treatment and that there are few intervening steps between elicitor binding to a receptor and activation of these genes. In contrast, elicitor activation of cell wall HRGP transcription is only observed after a 1-hr lag. These markedly different induction kinetics imply the existence either of more than one stimulus or of a single stimulus leading to sequential effects or divergent pathways.

Studies of the responses of intact plant tissue to microbial infection have shown that the temporal and spatial patterns of defense gene activation differ markedly in incompatible and compatible interactions (67). Tissue dissection and in situ hybridization have shown that certain defense genes are activated at a distance from the site of localized infection or wounding. Only a weak response of the phenylpropanoid genes is found at a distance; in contrast, transcripts encoding HRGPs, proteinase inhibitors, and certain PR proteins accumulate to high levels in tissue distant from, as well as adjacent to, the site of damage or infection. In some cases induction occurs throughout the plant (94).

In elicitor-treated bean cells, the action at a distance and the lag for activation of genes encoding HRGP may therefore reflect the generation and transmission of intercellular signals, whereas phenylpropanoid and chitinase genes are activated directly following intracellular transduction of the initial external signal (67). Stress-induced cell wall HRGPs are encoded by a family of genes, the transcripts of which exhibit markedly different patterns of accumulation during different stress conditions. This selective activation of individual members of a gene family indicates that different signal transduction systems operate in the early stages of an incompatible interaction, in the later stages of a compatible interaction, and in wounded tissue. Utilization of a complex array of cues for defense-gene activation may facilitate the flexible deployment of the same battery of defense responses in several rather different biological circumstances.

Developmental Regulation of Defense Genes

In addition to the roles of isoflavonoids and furanocoumarins as phytoalexins, and of lignin as a structural barrier in the cell walls of peripheral tissues, phenylpropanoid products derived from phenylalanine have a number of diverse functions in plant development. Lignin is a major structural polymer in the walls of water-conducting xylem elements, and flavonoids are pigments and UV protectants in epidermal cells. Moreover, phenylpropanoid products function as signals not only in the rhizosphere (31, 104) but also internally as modulators of hormone transport or action (7, 50) and potentially as metabolic regulators of phenylpropanoid-biosynthetic enzyme levels (9).

The regulation of phenylpropanoid-biosynthetic genes is correspondingly complex. For example, bean and parsley each contain families of three PAL genes (19, 75). These genes encode distinct polypeptide isoforms in bean (70). The corresponding transcripts exhibit markedly different patterns of accumulation, leading to the selective synthesis of functional variants of the enzyme in different biological situations. In particular, the isoforms of the native enzyme exhibit different K_ms; forms with low K_ms are preferentially induced by wounding or by the elicitor, thereby exerting a metabolic priority for phenylpropanoid synthesis in the cellular economy of phenylalanine specifically under conditions of stress. The complex patterns of PAL gene regulation and the apparent biochemical specialization of the encoded isopolypeptides may be related to the highly diverse biological functions of phenylpropanoid natural products.

A PAL2 promoter–β-glucuronidase (GUS) reporter gene fusion in transgenic tobacco plants is specifically induced at an early stage of vascular development at the inception of xylem differentiation, as well as in floral tissue associated with pigmentation (71). In addition, the promoter is induced by light and wounding, which evoke specific changes in the spatial pattern of GUS activity in stems, including induction in the epidermis. The PAL2 promoter apparently transduces a complex set of developmental and environmental cues into an integrated spatial and temporal program of gene expression that regulates the synthesis of an array of natural products.

Other classes of defense genes are also developmentally regulated in healthy plants. Thus, chitinase and glucanase are under hormonal control, and the gene products accumulate to high levels in roots and lower leaves (81, 102). This may be correlated with a buildup of general, nonspecific resistance as a plant matures. Emerging data indicate that plant development involves precise morphogenetic control of cell wall architecture mediated by the selective transcription of genes encoding specific glycine-rich proteins (GRPs) and HRGPs, some of which are also stress inducible (57, 58). For example, histochemical analysis of the pattern of GUS activity in transgenic plants containing an HRGP4.1–GUS gene fusion showed that this stress-inducible bean promoter also exhibited a specific pattern of expression during normal development, being particularly active in root and shoot tips and in stem nodes (T. Powell & K. Wycoff, unpublished data). The spatial patterns of wound induction of the PAL2, HRGP4.1, and GRP genes are clearly different and contain features reminiscent of the respective developmental patterns. Thus wounding may reactivate transiently expressed or cryptic developmental programs (57, 71).

Many defense genes are strongly expressed during floral development. As expected, PAL and other phenylpropanoid genes involved in pigment formation are highly active in petals, but strong expression of PAL and CHS, the first enzyme of the flavonoid branch pathway, has recently been observed in other floral organs (71; J. Schmid & P. Doerner, unpublished data). More-over, chitinase and glucanase activities are present at high levels in styles, and a number of flower-induced transcripts identified by subtractive cloning have been identified as PR protein homologs (79). HRGPs accumulate in styles; in HRGP4.1-GUS transgenic plants high levels of expression occur during floral development, particularly in styles (Powell & Wycoff, unpublished data). These observations are of interest in view of the report that oligogalacturonide fragments not only exhibit elicitor activity but also stimulate floral development in a tissue culture system (see above). It remains to be established whether the defense-gene products are involved in protection against microbial attack or have other functions in reproductive organs. Also it is of interest that stylar and other floral tissues are richly endowed with differentiating vascular tissue.

The possible roles of chitinase and glucanase in the generation and/or amplification of elicitor signals have been discussed above. These and other defense-gene products may likewise be involved in other forms of molecular communication. For example, a surprising feature of the PAL2 promoter is the high level of activity in the zone of cell proliferation immediately adjacent to the root apical meristem and in the actual shoot apical meristem (71). That the latter observation might reflect PAL gene expression associated with the synthesis of flavonoids for protection of meristematic cells against UV damage does not account for the strong expression in root tips. Recent data indicate that flavonoids may be natural regulators of polar auxin transport (50) and that phenylpropanoid derivatives such as dehydrodiconiferyl glucosides have cytokinin-like activities (7). Hence, PAL2 expression in juvenile cells may be related to the generation of morphogenetic signals. Consistent with this hypothesis is the observation that inappropriate expression of the PAL2 gene in transgenic tobacco can cause abnormal development (Y. Elkind, unpublished data).

Defense Gene Regulatory Elements

Functional dissection of defense-gene promoters will identify the *cis*-acting elements required for the developmental and environmental induction of these genes. 5'-Deletion analysis of a bean CHS promoter in elicitor-responsive soybean protoplasts detected an upstream silencer element and an elicitor-inducible activator in the TATA-proximal region of the promoter (32). Coelectroporation of the silencer element in *trans* markedly stimulated promoter activity; the presumed mechanism is competition for the binding of transcription factors that function as repressors in these protoplasts (M. A. Lawton, unpublished data). The activator element contains three versions of a sequence motif that is also found in the promoters of the bean PAL2 gene and several stress-inducible genes in parsley and other species. Analysis of sites in the CHS promoter hypersensitive to digestion by DNase I treatment of isolated nuclei indicated that transcriptional activation is accompanied by changes in chromatin structure in the proximal region of the promoter. Two hypersensitive sites, which are markedly induced upon elicitation, map to conserved activator sequence elements and appear to reflect the binding of specific transcription factors (68a).

Specific binding of nuclear factors to the activator and silencer regions has been detected in vitro; this assay may allow the isolation and molecular cloning of transcription factors involved in the regulation of CHS, PAL, and other defense genes in bean. In the course of these studies, gel retardation and DNase I footprinting experiments demonstrated binding in vitro of one specific nuclear protein to three discrete *cis*-acting elements within the silencer of the CHS promoter (M. A. Lawton, unpublished data). Competition experiments indicate that this factor is identical with or closely related to GT-1, which binds to *cis*-acting elements involved in the light regulation of genes encoding the small subunit of ribulose bisphosphate carboxylase (41). Analysis of the function of this factor in diverse programs of gene regulation may give an insight into the molecular basis of developmental plasticity in higher plants.

Elicitor-responsive protoplasts have also been obtained from parsley cell suspension cultures (20). In this system the short period of contact with fungal elicitor needed for stimulation of phytoalexin synthesis is not sufficient for cell wall regeneration, suggesting that the wall is not required for signal perception. In principle, receptor activity could be functionally assayed by the fluorescence of furanocoumarins excreted into the medium following induction of these parsley phytoalexins by elicitor. Functional analysis of elicitor-responsive promoters in this preparation has not yet been reported, but in vivo genomic footprinting has shown elicitor-induced binding of factors to conserved activator elements of a parsley PAL promoter (75). Detailed functional analysis of a parsley CHS promoter shows that one of these elements is also involved in UV-induction (100). This CHS promoter is not induced by elicitor, in line with the synthesis of furanocoumarin rather than isoflavonoid phytoalexins in parsley. It will be of interest to determine the molecular basis for the selective induction of PAL but not CHS in this plant.

The properties of promoters relating to tissue- or cell-type-specificity and the interactions between environmental signal pathways and developmental programs of regulation cannot currently be addressed in protoplast transient-assay systems using suspension cultures; these must be examined in protoplasts derived from plant tissues or in intact plants. In addition to the PAL2-GUS gene fusion discussed above, the appropriate regulation of a number of other defense genes or defense-gene promoter-reporter gene fusions has been observed in transgenic plants. These include ethylene induction of a chitinase-GUS gene fusion (12); wound induction of *wun1* gene fusions (74); wound induction of the HRGP4.1 gene and of an HRGP4.1-GUS gene fusion (D. R. Corbin and T. Powell, unpublished data); wound induction of the proteinase inhibitor II gene (2); and developmental regulation and stress induction of CHS-GUS gene fusions (J. Schmid and P. Doerner, unpublished data). In the last study, high levels of GUS were observed immediately adjacent to local HR lesions induced by *Pseudomonas syringae*, together with significant induction throughout the rest of the leaf (B. Stermer, unpublished data). These data suggest that the CHS promoter responds to both microbial and plant elicitation signals.

Functional analysis of the bean CHS promoter in transgenic tobacco plants confirmed conclusions drawn from transient assays in electroporated protoplasts about the presence of elicitor-responsive elements in the proximal region of the promoter. This analysis has so far failed to separate the *cis*-acting elements for floral expression and elicitor induction, which suggests that these apparently diverse modes of expression may be mechanistically related (J. Kooter, unpublished data). Further characterization of these regulatory elements and the corresponding transcription factors may shed light on the putative pleiotropic functions of oligogalacturonides and other factors in stress and floral signaling.

SIGNAL TRANSDUCTION MECHANISMS

Many factors have been considered as possible components of stress-signal pathways. Here we assess the evidence for a proposed component's function in vivo.

Membrane Potentials and Ion Fluxes

Infection or exposure of plant cells to fungal elicitors causes rapid changes in membrane potential and proton transport (3). The causes of such changes are often difficult to determine, and the effects of elicitors on membrane potential in vitro are not necessarily reproduced in the intact host–pathogen system (114). Modulation of plasma membrane ATPase activity has been implicated in some of these effects (3, 78, 107); the recent cloning of the plasma membrane H^+-ATPase gene in A. *thaliana* (47) will enable its putative function in the internalization of elicitor signals to be examined directly by gene transfer techniques.

Perturbations in Oxidative Metabolism

The roles of activated oxygen species and/or perturbations of the redox state as early signals have been the focus of a number of studies. The initial work of Doke and colleagues implicated an NADPH-dependent generation of superoxide anion as an important factor in the HR of potato tuber tissue to *Phytophthora infestans* (28). Cells undergoing the HR suffer lipid peroxidation associated with superoxide production, and the HR can be delayed by inhibition of lipid peroxidation—e.g. by GSH treatment (1). A Tn5 mutant of *Pseudomonas syringae* pv. *syringae,* inactive in HR induction in tobacco, also fails to induce superoxide production (1).

These studies establish a link with the HR, but superoxide production may be a metabolic symptom of cell death rather than an early inductive signal. Thus, singlet-oxygen quenchers fail to prevent the HR in tobacco cells exposed to *P. syringae* pv. *pisi* (95), and the production of cytochrome c–reducing equivalents (the basis for the measurement of superoxide production) is similar whether potato tuber tissues are treated with eliciting (arachidonic) or noneliciting (oleic, linolenic) fatty acids (83). Transcripts encoding a manganese superoxide dismutase are induced by infection of *Nicotiana plumbaginifolia* with an HR-inducing strain of *P. syringae* (10). If superoxide is a signal molecule, induction of dismutase activity could act to limit the spread of the signal.

Elicitors from *Verticillium dahliae* induce striking decreases in the fluorescence of membrane potential–sensitive and pH-sensitive dyes in cotton, tobacco, and soybean cells, and this was shown to result from fluorescence quenching by the rapid production of H_2O_2 (4). The ability of various elicitors to induce these changes within 5 min closely correlates with their relative activities as inducers of phytoalexin accumulation measured 60 hr later. H_2O_2 is thought to be generated in the cell wall by the action of an elicitor-stimulated vectorial NADH oxidase at the plasma membrane (4). Exogenous H_2O_2 is itself an elicitor, and phytoalexin induction by fungal elicitor is inhibited by exogenous catalase, which suggests a causal role for H_2O_2. The elicitor-stimulated insolubulization of a proline-rich cell wall protein can be mimicked by exogenous H_2O_2 (D. J. Bradley, unpublished data). Thus stress-induced extracellular H_2O_2 may have a dual function as a substrate for initial ultrarapid defense responses in situ and as a component of the signal pathway for induction of responses dependent on defense-gene activation.

GSH induces a rapid, massive, and selective appearance of defense-gene transcripts in cultured bean cells (119). Although GSH is a powerful cellular reductant, in animal cells it can act as a substrate in a complex oxygenase reaction to generate H_2O_2 (79a, 120). This mechanism is consistent with (a) the very low concentration of GSH (1 μM) needed to insolubulize the bean proline-rich wall protein and (b) the inhibition of both this effect and defense-gene induction at high GSH concentrations. GSH is not active as an elicitor in all systems. Thus, a bean CHS promoter in transgenic tobacco plants is readily activated by fungal elicitor, $HgCl_2$, or oxalic acid, but not by either GSH or its monoethyl ester (a form more readily taken up by plants) (B. Stermer and R. Edwards, unpublished data). In parsley cell suspension cultures GSH itself was inactive, but it increased the sensitivity of the cells to elicitation by chitosan in the presence of conditioned medium (17). In cells transferred to fresh medium, GSH inhibited chitosan-stimulated increases in coumarin and callose formation. GSH does not elicit isoflavonoid phytoalexin accumulation or induction of phytoalexin-biosynthetic enzymes in alfalfa cell suspension cultures or in protoplasts derived from them (A. Choudhary and R. Edwards, unpublished data). Comparison of the uptake and metabolism of labeled GSH in bean and alfalfa cells indicates that GSH activity is correlated with uptake and a fungal elicitor–mediated increase in thiol levels (R. Edwards and J. Blount, unpublished data). It is not clear whether the site of GSH action in bean is intra- or extracellular.

Calcium- and Protein Kinase–Mediated Pathways

Chitosan stimulation of callose synthesis apparently involves Ca^{2+}-mediated activation of callose synthase at the plasma membrane. This finding led to the hypothesis that internalization of signals for defense-gene activation may likewise be mediated by Ca^{2+} (62). Experimental determination of changes in Ca^{2+} flux in plant cells is hindered by the presence of the cell wall, which is itself a rich store of Ca^{2+}. Evidence of a role of changes in Ca^{2+} flux or compartmentation in stress signaling has been indirect.

In pea pod tissues infected with *Fusarium solani* or treated with chitosan, no effects on phytoalexin accumulation and/or leakage of UV-absorbing material were observed when high levels of exogenous Ca^{2+}, Ca^{2+} channel blockers, or calmodulin inhibitors were added (59). In contrast, a role for plasma membrane Ca^{2+} fluxes in the elicitation of phytoalexin synthesis by *Phytophthora megasperma* f. sp. *glycinea* elicitor in soybean cell suspension cultures has been proposed for two reasons: Elicitation is inhibited by external Ca^{2+} depletion and by either La^{2+} or the Ca^{2+} channel blocker verapamil; and phytoalexin synthesis is stimulated by the Ca^{2+} ionophore A23187 (103). Similarly, external Ca^{2+} was necessary for elicitor induction of transcripts encoding enzymes of furanocoumarin phytoalexin biosynthesis in parsley cell suspension cultures but was not required for UV induction of CHS transcripts (16). A23187 induced the accumulation of the phytoalexin 6-methoxymellein in carrot cell suspension cultures, and induction by a crude elicitor preparation from carrot tissues was inhibited by the addition of verapamil within 30 min of elicitation (66). The development of patch clamp techniques for plant research should now allow the direct measurement of Ca^{2+} fluxes following addition of elicitor (99).

Calmodulin antagonists strongly inhibit the accumulation of terpenoid phytoalexins in tobacco cell suspension cultures if added prior to the elicitor, and experiments with structural analogs suggest that these effects are calmodulin specific (U. Vögeli and J. Chappell, unpublished data). Measurement of cyclic AMP (cAMP) levels in soybean and parsley cell cultures did not detect changes in response to elicitors (16, 44). In contrast, dibutyryl cAMP was reported to induce 6-methoxymellein accumulation in carrot cultures, and elicitor induced a rapid but transient increase in endogenous cAMP levels (66). The recent cloning of plant genes that encode protein kinases with catalytic domains that resemble those in animal cyclic nucleotide-dependent protein kinases (68), and plant genes encoding transcription factors that resemble the animal cAMP-responsive element–binding proteins (51), now provides the opportunity to delineate the possible roles of cyclic nucleotides as components in stress and other signal pathways in higher plants. Any role for cyclic nucleotides has been a controversial subject (91a).

An elicitor-mediated increase in inositol triphosphate preceding phytoalexin accumulation has been reported in cultured carrot cells, which suggests the involvement of a phosphatidylinositol turnover–mediated signal pathway (65). Elicitation of phytoalexin accumulation in parsley cells was not correlated with changes in the levels of polyphosphoinositides involved in signal transduction in animals (108). However, although the key enzyme, phosphoinositide-specific phospholipase C, appears to be present in plant membrane fractions, the level of its substrate phosphatidylinositol 4,5-bisphosphate, as a proportion of the total phosphoinositide pool is at least 10 times lower in plant than in animal cells (80, 89). Moreover, whereas diacylglycerol is an activator of protein kinase C in animal cells, plants contain novel phospholipids similar to mammalian platelet-activating factor, which can stimulate protein kinases and the plasma membrane H^+-ATPase (97). Thus, if a phosphoinositide cycle does operate in plants, it may differ in several crucial respects from the animal cycle. Molecular cloning of plant homologs of G proteins and other components of this signal pathway should help define its role in stress signaling (76).

Operation of one or more of these signal pathways would imply the involvement of protein kinases. In vivo labeling of soybean cells has shown changes in the phosphorylation state of specific proteins in response to glucan elicitors (34); oligogalacturonide and noncarbohydrate elicitors of proteinase inhibitors cause the phosphorylation of specific plasma membrane proteins in vitro (38). The functional significance of these events remains to be established. However, the phosphorylation in vitro of a specific plasma membrane protein in response to addition of a pure oligogalacturonide may represent ligand-dependent receptor autophosphorylation (38), and hence further characterization of this protein will be of considerable interest.

Signals for Induction at a Distance

Three distinct spatial patterns of defense-gene activation have been observed: (a) localized to the cells immediately adjacent to the perturbation; (b) throughout the organ; and (c) throughout the plant. Overlapping but distinct sets of defense genes can be defined on the basis of the distance of signal transduction. As discussed above, these data imply the operation of a number of distinct intercellular signaling systems.

True systemic responses are seen in the induction of proteinase inhibitors in solanaceous species by wounding or insect attack (94) and in the induction of systemic resistance in a number of species, including cucurbits and bean, by pre-inoculation with a fungal or viral pathogen (15, 25). The application of oligogalacturonides to cut plant surfaces causes systemic induction of proteinase inhibitor genes (94), but the oligogalacturonides do not themselves move through the plant (5). The oligogalacturonides (PIIF) are now believed to act as a local signal, potentiating or transducing the systemic signal. A more potent, small noncarbohydrate molecule or group of molecules termed "super PIIF" has recently been reported (38) that could act as a systemically mobile signal.

Recent evidence indicates that the hormone abscisic acid (ABA) has a key function in the systemic induction of proteinase inhibitor genes (88). ABA itself induces systemic proteinase inhibitor synthesis when sprayed on potato leaves. This was also shown with leaves of mutants of potato and tomato deficient in ABA, in which proteinase inhibitor synthesis could not be induced by wounding alone. Endogenous ABA levels increased both locally and systemically when wild-type plants were wounded. Although direct transport of ABA from the wound site remains to be demonstrated, it seems clear that this hormone is intimately related to systemic signal transduction in response to mechanical damage. In a separate study, the induction of proteinase inhibitors in tomato plants by wounding or by applying pectic fragments or chitosan was reversibly inhibited by prior treatment with aspirin (acetylsalicylic acid) and related hydroxybenzoic acids (27). The molecular basis of this effect is unclear, although it is interesting that similar structural specificities among the hydroxybenzoic acids were seen for the inhibition of proteinase inhibitor accumulation in tomato and the appearance of PR proteins in tobacco.

The signals involved in induced systemic resistance, in which the plant becomes sensitized toward subsequent attempts at infection, are not known. This type of resistance may depend not on a specific signal from the pathogen but on the persistence of a low level of metabolic perturbation (40). The induction of systemic resistance by applied phosphates (40) or oxalic acid (30) suggests that sequestration of Ca^{2+} ions may be important.

Ethylene release is often associated with the response of plants to physical or microbial stress. Ethylene induces defense-gene transcription (36), but there is little evidence that it plays a causal role in the effects either of microbial elicitors or of infection. Ethylene formation appears to be more a symptom than a signal (14). In bean leaves, ethylene and PAL activity may be induced by different mechanisms, but ethylene may stimulate PAL appearance at suboptimal concentrations of fungal elicitor (48). The observation that ethylene modulates gene expression at the posttranscriptional level (72) suggests a role as a downstream modulator of responses initiated by elicitor-induced transcriptional activation. Many studies of ethylene function have involved the use of chemical inhibitors and stimulators of ethylene synthesis. Analysis of stress responses in ethylene-insensitive mutants of *Arabidopsis thaliana* (84) and gene transfer experiments that exploit the recent cloning of aminocyclopropane carboxylate synthase (96) provide new and potentially powerful approaches to determining the role of ethylene in stress signaling.

CONCLUDING REMARKS

Mechanistic analysis is the cornerstone of modern biology. Significant advances have recently been made in our understanding of the exceedingly complex molecular communication between plant hosts and pathogens. Further progress will require the concerted application of genetic, molecular, and biochemical approaches to the elucidation of selected plant-microbe interactions.

ACKNOWLEDGMENTS

We thank Cindy Doane and Scotty McGill for assembling the manuscript, and colleagues for permission to quote unpublished data. We are deeply grateful to the Samuel Roberts Noble Foundation for its support and encouragement.

Literature Cited

1. Adam, A., Farkas, T., Somlyai, G., Hevesi, M., Kiraly, Z. 1989. Consequence of O_2^- generation during a bacterially induced hypersensitive reaction in tobacco: deterioration of membrane lipids. *Physiol. Mol. Plant Pathol.* 34:13–26

2. An, G., Mitra, A., Choi, H. K., Costa, M. A., An, K., et al. 1989. Functional analysis of the 3' control region of the potato wound-inducible proteinase inhibitor II gene. *The Plant Cell* 1:115–22

3. Anderson, A. J. 1989. The biology of glycoproteins as elicitors. In *Plant:Microbe Interactions: Molecular and Genetic Perspectives,* ed. T. Kosuge, E. W. Nester, 3:87–130. New York: McGraw-Hill

4. Apostol, I., Heinstein, P. F., Low, P. S. 1989. Rapid stimulation of an oxidative burst during elicitation of cultured plant cells. *Plant Physiol.* 90:109–16

5. Baydoun, E. A.-H., Fry, S. C. 1985. The immobility of pectic substances in injured tomato leaves and its bearing on the identity of the wound hormone. *Planta* 165:269–76

6. Bennetzen, J. L., Qin, M.-M., Ingels, S., Ellingboe, A. H. 1989. Allele-specific and Mutator-associated instability at the *Rp1* disease-resistance locus of maize. *Nature* 332:369–70

7. Binns, A. N., Chen, R. H., Wood, H. N., Lynn, D. G. 1987. Cell division promoting activity of naturally occurring dehydrodiconiferyl glucosides: Do cell wall components control cell division? *Proc. Natl. Acad. Sci. USA* 84:980–84

8. Bohlmann, H., Clausen, S., Behuke, S., Giese, H., Hiller, C., et al. 1988. Leaf-specific thionins of barley—a novel class of cell wall proteins toxic to plant-pathogenic fungi and possibly involved in the defense mechanism of plants. *EMBO J.* 7:1559–65

9. Bolwell, G. P., Mavandad, M., Millar, D. J., Edwards, K. H., Schuch, W., Dixon, R. A. 1989. Inhibition of mRNA levels and activities by *trans*-cinnamic acid in elicitor-induced bean cells. *Phytochemistry* 27:2109–17

10. Bowler, C., Alliotte, T., DeLoose, M., Van Montagu, M., Inze, D. 1989. The induction of manganese superoxide dismutase in response to stress in *Nicotiana plumbaginifolia. EMBO J.* 8:31–38

11. Breitender, H., Pettenburger, K., Bito, A., Valenta, R., Kraft, D., et al. 1989. The gene coding for the major birch pollen allergen *BetvI* is highly homologous to a pea disease resistance response gene. *EMBO J.* 8: 1935–38

12. Broglie, K. E., Biddle, P., Cressman, R., Broglie, R. 1989. Functional analysis of DNA sequences responsible for ethylene regulation of a bean chitinase gene in transgenic tobacco. *The Plant Cell* 1:599–607

13. Cervone, F. C., Hahn, M. G., De Lorenzo, G., Darvill, A., Albersheim, P. 1989. Host-pathogen interactions. XXXIII. A plant protein converts a fungal pathogenesis factor into an elicitor of plant defense responses. *Plant Physiol.* 90:542–48

14. Chappell, J., Hahlbrock, K., Boller, T. 1984. Rapid induction of ethylene biosynthesis in cultured parsley cells by fungal elicitor and its relationship to the induction of phenylalanine ammonia-lyase. *Planta* 161:475–80

15. Cloud, A. M. E., Deverall, B. J. 1987. Induction and expression of systemic resistance to the anthracnose disease of bean. *Plant Pathol.* 36:551–57

16. Colling, C., Hahlbrock, K., Scheel, D. 1989. Studies on signal transduction in plant defense gene activations. *NATO Int. Symp. Signal Perception and Transduction in Higher Plants, Toulouse, July 9–13, 1989* (Abstr.)

17. Conrath, U., Domard, A., Kauss, H. 1989. Chitosan-elicited synthesis of callose and of coumarin derivatives in parsley cell suspension cultures. *Plant Cell Rep.* 8:152–55

18. Cosio, E. G., Popperl, H., Schmidt, W. E., Ebel, J. 1988. High-affinity binding of fungal glucan fragments to soybean (*Glycine max* L.) microsomal fractions and protoplasts. *Eur. J. Biochem.* 175:309–15

19. Cramer, C. L., Edwards, K., Dron, M., Liang, X., Dildine, S. L., et al. 1989. Phenylalanine ammonia-lyase gene organization and structure. *Plant Mol. Biol.* 12:367–83

20. Dangl, J. L., Hanfle, K. D., Lipphardt, S., Hahlbrock, K., Scheel, D. 1987. Parsley protoplasts retain differential responsiveness to u.v. light and fungal elicitor. *EMBO J.* 6:2551–56

21. Daniels, M. J., Dow, J. M., Osbourn, A. E. 1988. Molecular genetics of pathogenicity in phytopathogenic bacteria. *Annu. Rev. Phytopathol.* 26:285–312

22. Darvill, A. G., Albersheim, P. 1984. Phytoalexins and their elicitors—a defense against microbial infection in

plants. *Annu. Rev. Plant. Physiol.* 35:
243–75

23. Davis, K. R., Hahlbrock, K. 1987. Induction of defense responses in cultured parsley cells by plant cell wall fragments. *Plant Physiol.* 85:1286–90

24. Davis, K. R., Darvill, A. G., Albersheim, P. 1986. Host-pathogen interactions. XXXI. Several biotic and abiotic elicitors act synergistically in the induction of phytoalexin accumulation in soybean. *Plant Mol. Biol.* 6:23–32

25. Dean, R. A., Kuc', J. 1986. Induced systemic protection in cucumbers: the source of the "signal". *Physiol. Mol. Plant Pathol.* 28:227–33

26. Dixon, R. A., Harrison, M. J. 1990. Activation, structure and organization of genes involved in microbial defense in plants. *Adv. Genet.* In press

27. Doherty, H. M., Selvendran, R. R., Bowles, D. J. 1988. The wound response of tomato plants can be inhibited by aspirin and related hydroxy-benzoic acids. *Physiol. Mol. Plant Pathol.* 33:377–84

28. Doke, N., Chai, H. B. 1985. Activation of superoxide generation and enhancement of resistance against compatible races of *Phytophthora infestans* in potato plants treated with digitonin. *Physiol. Plant Pathol.* 27:323–34

29. Doke, N., Garas, N. A., Kuc', J. 1979. Partial characterization and aspects of the mode of action of a hypersensitivity-inhibiting factor (HIF) isolated from *Phytophthora infestans*. *Physiol. Plant Pathol.* 15:127–40

30. Doubrava, N. S., Dean, R. A., Kuc', J. 1988. Induction of systemic resistance to anthracnose caused by *Colletotrichum lagenarium* in cucumber by oxalate and extracts from spinach and rhubarb leaves. *Physiol. Mol. Plant Pathol.* 33:69–79

31. Downie, J. A., Johnston, A. W. B. 1986. Nodulation of legumes by *Rhizobium*: The recognized root? *Cell* 47:153–54

32. Dron, M., Clouse, S. D., Lawton, M. A., Dixon, R. A., Lamb, C. J. 1988. Glutathione and fungal elicitor regulation of a plant defense gene promoter in electroporated protoplasts. *Proc. Natl. Acad. Sci. USA* 85:6738–42

33. Ebel, J. 1986. Phytoalexin synthesis: the biochemical analysis of the induction process. *Annu. Rev. Phytopathol.* 24:235–64

34. Ebel, J. 1989. Elicitor-binding proteins and signal transduction in elicitor action. *J. Cell Biochem.* 13D (Suppl.):248

35. Eberhard, S., Doubrava, N., Marfa, B., Mohnen, D., Southwick, A., et al. 1989. Pectic cell wall fragments regulate tobacco thin-cell-layer explant morphogenesis. *The Plant Cell* 1:747–55

36. Ecker, J. R., Davis, R. W. 1987. Plant defense genes are regulated by ethylene. *Proc. Natl. Acad. Sci. USA* 84:5202–6

37. Ellingboe, A. H. 1981. Changing concepts in host-pathogen genetics. *Annu. Rev. Phytopathol.* 19:125–43

38. Farmer, E. E., Pearce, G., Ryan, C. A. 1989. *In vitro* phosphorylation of plant plasma membrane proteins in response to the proteinase inhibitor inducing factor. *Proc. Natl. Acad. Sci. USA* 86:1539–42

39. Fedoroff, N. V. 1989. About maize transposable elements and development. *Cell* 56:181–91

40. Gottstein, H. D., Kuc', J. 1989. Induction of systemic resistance to anthracnose in cucumber by phosphates. *Phytopathology* 79:176–79

41. Green, P. J., Yong, M.-H., Cuozzo, M., Kano-Murakami, Y., Silverstein, P., Chua, N.-H. 1988. Binding site requirements for pea nuclear protein factor GT-1 correlate with sequences required for light-dependent transcriptional activation of the *rbcS-3A* gene. *EMBO J.* 7:4035–44

42. Grimm, C., Panopoulos, N. J. 1989. The predicted protein product of a pathogenicity locus from *Pseudomonas syringae* pv. *phaseolicola* is homologous to a highly conserved domain of several procaryotic regulatory proteins. *J. Bacteriol.* 171:5031–38

43. Hahlbrock, K., Scheel, D. 1989. Physiology and molecular biology of phenylpropanoid metabolism. *Annu. Rev. Plant Physiol. Plant Mol. Biol.* 40:347–64

44. Hahn, M. G., Grisebach, H. 1983. Cyclic AMP is not involved as a second messenger in the response of soybean to infection by *Phytophthora megasperma* f. sp. *glycinea*. *Z. Naturforsch.* 38C: 578–82

45. Hahn, M. G., Bucheli, P., Cervone, F., Doares, S. H., O'Neill, R. A., et al. 1989. Roles of cell wall constituents in plant-pathogen interactions. See Ref. 3, pp. 131–81

46. Hamdan, M. A. M. S., Dixon, R. A. 1987. Differential patterns of protein synthesis in bean cells exposed to elicitor fractions from *Colletotrichum lindemuthianum*. *Physiol. Mol. Plant Pathol.* 31:105–21

47. Harper, J. F., Surowy, T. K., Sussman,

M. R. 1989. Molecular cloning and sequence of cDNA encoding the plasma membrane proton pump (H^+-ATPase) of *Arabidopsis thaliana*. *Proc. Natl. Acad. Sci. USA* 86:1234–38

48. Hughes, R. K., Dickerson, A. G. 1989. The effect of ethylene on phenylalanine ammonia-lyase (PAL) induction by a fungal elicitor in *Phaseolus vulgaris*. *Physiol. Mol. Plant Pathol.* 34:361–78

49. Huynh, T. V., Dahlbeck, D., Staskawicz, B. J. 1989. Bacterial blight of soybean: regulation of a pathogen gene determining host cultivar specificity. *Science* 245:1374–77

50. Jacobs, M., Rubery, P. H. 1988. Naturally occurring auxin transport regulators. *Science* 241:346–49

51. Katagiri, F., Lam, E., Chua, N.-H. 1989. Two tobacco DNA-binding proteins with homology to the nuclear factor CREB. *Nature* 340:727–29

52. Kaufmann, S., Legrand, M., Geoffroy, P., Fritig, B. 1987. Biological functions of "pathogenesis-related" proteins: Four PR proteins of tobacco have 1,3-β-glucanase activity. *EMBO J.* 6:3209–12

53. Kauss, H., Jeblick, W., Domard, A. 1989. The degrees of polymerization and *N*-acetylation of chitosan determine its ability to elicit callose formation in suspension cells and protoplasts of *Catharanthus roseus*. *Planta* 178:385–92

54. Kearney, B., Ronald, P. C., Dahlbeck, D., Staskawicz, B. J. 1988. Molecular basis for evasion of plant host defense in bacterial spot disease of pepper. *Nature* 332:541–43

55. Keen, N. T., Staskawicz, B. 1988. Host range determinants in plant pathogens and symbionts. *Annu. Rev. Microbiol.* 42:421–40

56. Keen, N. T., Kobayashi, D., Tamaki, S., Thordal-Christensen, H., Masaghi, I., et al. 1988. In *Molecular Genetics of Plant-Microbe Interactions*, ed. R. Palacios, D. P. S. Verma, pp 15–19. Minneapolis: APS Press

57. Keller, B., Schmid, J., Lamb, C. J. 1989. Vascular expression of a bean cell wall glycine-rich protein-β-glucuronidase gene fusion in transgenic tobacco. *EMBO J.* 8:1309–14

58. Keller, B., Lamb, C. J. 1989. Specific expression of a novel cell wall hydroxyproline-rich glycoprotein gene in lateral root initiation. *Genes Devel.* 3:1639–46

59. Kendra, D. F., Hadwiger, L. A. 1987. Calcium and calmodulin may not regulate the disease resistance and pisatin formation responses of *Pisum sativum* to chitosan or *Fusarium solani*. *Physiol. Mol. Plant Pathol.* 31:337–48

60. Kessmann, H., Barz, W. 1986. Elicitation and suppression of phytoalexin and isoflavone accumulation in cotyledons of *Cicer arietinum* L. as caused by wounding and by polymeric components from the fungus *Ascochyta rabiei*. *J. Phytopathol.* 117:321–35

61. Kobayashi, D. Y., Tamaki, S. J., Keen, N. T. 1989. Cloned avirulence genes from the tomato pathogen *Pseudomonas syringae* pv. *tomato* confer cultivar specificity on soybean. *Proc. Natl. Acad. Sci. USA* 86:157–61

62. Köhle, H., Jeblick, W., Poten, F., Blaschek, W., Kauss, H. 1985. Chitosan-elicited callose synthesis in soybean cells as a Ca^{+2}-dependent process. *Plant Physiol.* 77:544–S1

63. Kolattukudy, P. E., Podila, G. K., Mohan, R. 1989. Molecular basis of the early events in plant-fungus interaction. *Genome* 31:342–49

64. Kombrink, E., Schröder, M., Hahlbrock, K. 1988. Several "pathogenesis-related" proteins in potato are 1,3-β-glucanases and chitinases. *Proc. Natl. Acad. Sci. USA* 85:782–86

65. Kurosaki, F., Tsurukawa, Y., Nishi, A. 1987. Breakdown of phosphatidyl inositol during the elicitation of phytoalexin production in cultured carrot cells. *Plant Physiol.* 85:601–4

66. Kurosaki, F., Tsurukawa, Y., Nishi, A. 1987. The elicitation of phytoalexins by Ca^{2+} and cyclic AMP in carrot cells. *Phytochemistry* 26:1919–23

67. Lamb, C. J., Lawton, M. A., Dron, M., Dixon, R. A. 1989. Signals and transduction mechanisms for activation of plant defenses against microbial attack. *Cell* 56:215–24

68. Lawton, M. A., Yamamoto, R. T., Hanks, S. K., Lamb, C. J. 1989. Molecular cloning of plant transcripts encoding protein kinase homologs. *Proc. Natl. Acad. Sci. USA* 86:3140–44

68a. Lawton, M. A., Clouse, S. D., Lamb, C. J. 1990. Glutathione-elicited changes in chromatin configuration in the promoter of the plant defense gene encoding chalcone synthase. *Plant Cell Rep.* In press

69. Legrand, M., Kauffmann, S., Geoffroy, P., Fritig, B. 1987. Biological function of pathogenesis-related proteins: four tobacco pathogenesis-related proteins are chitinases. *Proc. Natl. Acad. Sci. USA* 84:6750–54

70. Liang, X., Dron, M., Cramer, C. L., Dixon, R. A., Lamb, C. J. 1989. Differ-

ential regulation of phenylalanine ammonia-lyase genes during plant development and by environmental cues. *J. Biol. Chem.* 264:14486–92

71. Liang, X., Dron, M., Schmid, J., Dixon, R. A., Lamb, C. J. 1989. Developmental and environmental regulation of a phenylanine ammonia-lyase: β-glucuronidase gene fusion in transgenic tobacco plants. *Proc. Natl. Acad. Sci. USA* 86:9284–88

72. Lincoln, J. E., Fischer, R. L. 1988. Diverse mechanisms for the regulation of ethylene-inducible gene expression. *Mol. Gen. Genet.* 212:71–75

73. Lindgren, P. B., Panopoulos, N. J., Staskawicz, B. J., Dahlbeck, D. 1988. Genes required for pathogenicity and hypersensitivity are conserved and interchangeable among pathovars of *Pseudomonas syringae*. *Mol. Gen. Genet.* 211:499–506

74. Logemann, J., Lipphardt, S., Lörz, H., Häuser, I., Willmitzer, L., Schell, J. 1989. 5' Upstream sequences from the *wun1* gene are responsible for gene activation by wounding in transgenic plants. *The Plant Cell* 1:151–58

75. Lois, R., Dietrich, A., Hahlbrock, K., Schulz, W. 1989. A phenylalanine ammonia-lyase gene from parsley: structure, regulation and identification of elicitor and light responsive *cis*-acting elements. *EMBO J.* 8:1641–48

76. Matsui, M., Sasamoto, S., Kunieda, T., Nomura, N., Ishizaki, R. 1989. Cloning of *ara*, a putative *Arabidopsis thaliana* gene homologous to the *ras*-related gene family. *Gene* 76:313–19

77. Mauch, F., Staehlin, L. A. 1989. Functional implications of the subcellular localization of ethylene-induced chitinase and β-1,3-glucanase in bean leaves. *The Plant Cell* 1:447–57

78. Mayer, M. G., Ziegler, E. 1988. An elicitor from *Phytophthora megasperma* f. sp. *glycinea* influences the membrane potential of soybean cotyledonary cells. *Physiol. Mol. Plant Pathol.* 33:397–407

79. Meeks-Wagner, D. R., Dennis, E. S., Thanh Van, K. T., Peacock, W. J. 1989. Tobacco genes expressed during *in vitro* floral initiation and their expression during normal development. *The Plant Cell* 1:25–35

79a. Meister, A., Anderson, M. E. 1983. Glutathione. *Annu. Rev. Biochem.* 52:711–60

80. Melin, P.-M., Sommarin, M., Sandelius, A. S., Jergil, B. 1987. Identification of Ca²⁺-stimulated polyphosphoinositide phospholipase C in isolated

plant plasma membranes. *FEBS Lett.* 223:87–91

81. Memelink, J., Hoge, J. H. C., Schilperoort, R. A. 1987. Cytokinin stress changes the developmental regulation of several defense-related genes in tobacco. *EMBO J.* 6:3579–84

82. Metraux, J. P., Burkhart, W., Moyer, M., Dincher, S., Middlesteadt, W., et al. 1989. Isolation of a complementary cDNA encoding a chitinase with a structural homology to a bifunctional lysozyme/chitinase. *Proc. Natl. Acad. Sci. USA* 86:896–900

83. Moreau, R. A., Osman, S. F. 1989. The properties of reducing agents released by treatment of *Solanum tuberosum* with elicitors from *Phytophthora infestans*. *Physiol. Mol. Plant Pathol.* 35:1–10

84. Myerowitz, E. M. 1989. *Arabidopsis*, a useful weed. *Cell* 56:263–69

85. Parker, J. E., Hahlbrock, K., Scheel, D. 1988. Different cell wall components from *Phytophthora megasperma* f. sp. *glycinea* elicit phytoalexin production in soybean and parsley. *Planta* 176:75–82

86. Parsons, K. A., Chumley, F. G., Valent, B. 1987. Genetic transformation of the fungal pathogen responsible for rice blast disease. *Proc. Natl. Acad. Sci. USA* 84:4161–65

87. Peever, T. L., Higgins, V. J. 1989. Suppression of the activity of nonspecific elicitor from *Cladosporium fulvum* by intercellular fluids from tomato leaves. *Physiol. Mol. Plant Pathol.* 34:471–82

88. Pena-Cortes, H., Prat, S., Sanchez-Serrano, J. J., Willmitzer, L. 1989. The wound-induced expression of the proteinase inhibitor II gene in potato and tomato plants is mediated by abscisic acid. *NATO Int. Symp. Signal Perception and Transduction in Higher Plants, Toulouse, July 9–13, 1989* (Abstr.)

89. Pfaffmann, H., Hartmann, E., Brightman, A. O., Morré, D. J. 1987. Phosphatidylinositol specific phospholipase C of plant stems. *Plant Physiol.* 85:1151–55

90. Podila, G. P., Dickman, M. B., Kolattukudy, P. E. 1988. Transcriptional activation of a cutinase gene in isolated fungal nuclei by plant cutin monomers. *Science* 242:922–25

91. Pryor, A. 1987. The origin and structure of fungal disease resistance genes in plants. *Trends Genet.* 3:157–61

91a. Ranjeva, R., Boudet, A. M. 1987. Phosphorylation of proteins in plants: Regulatory effects and potential involvement in stimulus/response coupling. *Annu. Rev. Plant Physiol.* 38:73–93

92. Ricci, P., Bonnet, P., Huet, J.-C., Sallantin, M., Beauvais-Cante, F., et al. 1989. Structure and activity of proteins from pathogenic fungi *Phytophthora* eliciting necrosis and acquired resistance in tobacco. *Eur. J. Biochem.* 183:555–63

92a. Rick, C. M. 1982. Linkage map of the tomato *(Lycopersicon esculentum)*. *Genet. Maps* 2:360–65

93. Ronson, C. W., Nixon, B. T., Ausubel, F. M. 1987. Conserved domains in bacterial regulatory proteins that respond to environmental stimuli. *Cell* 49:579–81

94. Ryan, C. A. 1988. Oligosaccharides as recognition signals for the expression of defensive genes in plants. *Biochemistry* 27:8879–83

95. Salzwedel, J. L., Daub, M. E., Huag, J. 1989. Effects of singlet oxygen quenchers and pH on the bacterially induced hypersensitive reaction in tobacco suspension cell cultures. *Plant Physiol.* 90:25–28

96. Sato, T., Theologis, A. 1989. Cloning the mRNA encoding 1-aminocyclopropane-1-carboxylate synthase, the key enzyme for ethylene biosynthesis in plants. *Proc. Natl. Acad. Sci. USA* 86:6621–25

97. Scherer, G. F. E., Martiny-Baron, G., Stoffel, B. 1988. A new set of regulatory molecules in plants: A plant phospholipid similar to platelet-activating factor stimulates protein kinase and proton-translocating ATPase in membrane vesicles. *Planta* 175:241–53

98. Schottens-Toma, I. M. J., de Wit, P. J. G. M. 1988. Purification and primary structure of a necrosis-inducing peptide from the apoplastic fluids of tomato infected with *Cladosporium fulvum* (sym. *Fulvia fulva*). *Physiol. Mol. Plant Pathol.* 33:59–67

99. Schroeder, J. I., Hagiwara, S. 1989. Cytosolic calcium regulates ion channels in the plasma membrane of *Vicia faba* guard cells. *Nature* 338:427–30

100. Schulze-Lefert, P., Becker-Andre, M., Schulz, W., Hahlbrock, K., Dangl, J. L. 1989. Functional architecture of the light-responsive chalcone synthase promoter from parsley. *The Plant Cell* 1:707–14

101. Sharp, J. K., Albersheim, P., Ossowski, P., Pilotti, A., Garegg, P., Lindberg, B. 1984. Comparison of the structures and elicitor activities of a synthetic and a mycelial-wall-derived hexa-(β-D-glu-

copyranosyl)-D-glucitol. *J. Biol. Chem.* 259:11341–45

102. Sinshi, H., Mohnen, D., Meins, F. Jr. 1987. Regulation of a plant pathogenesis-related enzyme: Inhibition of chitinase and chitinase mRNA accumulation in cultured tobacco tissues by auxin and cytokinin. *Proc. Natl. Acad. Sci. USA* 84:89–92

103. Stab, M. R., Ebel, J. 1987. Effects of Ca^{2+} on phytoalexin induction by fungal elicitor in soybean cells. *Arch. Biochem. Biophys.* 257:416–23

104. Stachel, S. E., Zambryski, P. C. 1986. *Agrobacterium tumefaciens* and the susceptible plant cell: a novel adaptation of extracellular recognition and DNA conjugation. *Cell* 47:155–57

105. Stanford, A., Bevan, M., Northcote, D. 1989. Differential expression within a family of novel wound-induced genes in potato. *Mol. Gen. Genet.* 215:200–8

106. Stayton, M., Tamaki, S., Kobayashi, D., Keen, N. 1989. A cultivar-specific elicitor of the hypersensitive response in soybean has been identified and may be the signal molecule that interacts with the plant disease resistance gene product to trigger host defense. *J. Cell Biochem.* Suppl. 13D:326

107. Steffens, M., Ettle, F., Kranz, D., Kindl, H. 1989. Vanadate mimics effects of fungal cell wall in eliciting gene activation in plant cell cultures. *Planta* 177:160–68

108. Strasser, H., Hoffmann, C., Grisebach, H., Matern, U. 1986. Are polyphosphoinositides involved in signal transduction of elicitor-induced phytoalexin synthesis in cultured plant cells? *Z. Naturforsch.* 41C:717–24

109. Tamaki, S., Dahlbeck, D., Staskawicz, B., Keen, N. T. 1988. Characterization and expression of two avirulence genes cloned from *Pseudomonas syringae* pv. *glycinea. J. Bacteriol.* 170:4846–54

110. Templeton, M. D., Lamb, C. J. 1988. Elicitors and defense gene activation. *Plant, Cell and Environ.* 11:395–401

111. Tepper, C. S., Albert, F. G., Anderson, A. J. 1989. Differential mRNA accumulation in three cultivars of bean in response to elicitors from *Colletotrichum lindemuthianum. Physiol. Mol. Plant Pathol.* 34:85–98

112. Tepper, C. S., Anderson, A. J. 1986. Two cultivars of bean display a differential response to extracellular components from *Colletotrichum lindemuthianum. Physiol. Mol. Plant Pathol.* 29:411–20

113. Timberlake, W. E., Marshall, M. A.

1989. Genetic engineering of filamentous fungi. *Science* 244:1313–17

114. Tomiyama, K., Okamoto, H., Katou, K. 1983. Effect of infection by *Phytophthora infestans* on the membrane potential of potato cells. *Physiol. Plant Pathol.* 22:233–43

115. Wang, J., Holden, D. W., Leong, S. A. 1988. Gene transfer systems for the phytopathogenic fungus *Ustilago maydis. Proc. Natl. Acad. Sci. USA* 85:865–69

116. Weltring, K.-M., Turgeon, B. G., Yoder, O. C., VanEtten, H. D. 1988. Isolation of a phytoalexin-detoxification gene from the plant pathogenic fungus *Nectria haematococca* by detecting its expression in *Aspergillus nidulans. Gene* 68:335–44

117. Whalen, M. C., Stall, R. E., Staskawicz, B. J. 1988. Characterization of a gene from a tomato pathogen determining hypersensitive resistance in non-host species and genetic analysis of this resistance in bean. *Proc. Natl. Acad. Sci. USA* 85:6743–47

118. Wijesundera, R. L. C., Bailey, J. A., Byrde, R. J. W., Fielding, A. H.

1989. Cell wall degrading enzymes of *Colletotrichum lindemuthianum:* their role in the development of bean anthracnose. *Physiol. Mol. Plant Pathol.* 34:403–13

119. Wingate, V. P. M., Lawton, M. A., Lamb, C. J. 1988. Glutathione causes a massive and selective induction of plant defense genes. *Plant Physiol.* 87:206–10

120. Winterbourn, C. C. 1989. Inhibition of autoxidation of divicine and isouramil by the combination of superoxide dismutase and reduced glutathione. *Arch. Biochem. Biophys.* 271:447–55

121. Wolpert, T. J., Macko, V. 1989. Specific binding of victorin to a 100 kDa protein from oats. *Proc. Natl. Acad. Sci. USA* 86:4092–96

122. Yamada, T., Hashimoto, H., Shiraishi, T., Oku, H. 1989. Suppression of pisatin, phenylalanine ammonia-lyase mRNA, and chalcone synthase mRNA accumulation by a putative pathogenicity factor from the fungus *Mycosphaerella pinodes. Mol. Plant Microbe Int.* 2:256–61

Annu. Rev. Plant Physiol. Plant Mol. Biol. 1990. 41:369–419

PLASMODESMATA

A. W. Robards

Institute for Applied Biology, University of York, York YO1 5DD, United Kingdom

W. J. Lucas

Department of Botany, University of California, Davis, California 95616

KEY WORDS: intercellular communication, virus-plasmodesma interaction, symplasmic transport, molecular size exclusion limits, regulation of cell-to-cell transport

CONTENTS

INTRODUCTION

Plasmodesmata are cytoplasmic connections between adjacent plant cells, constituting the structural and functional analogs of gap junctions found between the cells of animals; as such, they create an intercellular continuum—the symplasm. Although plasmodesmata were first described by Tangl (233) more than 100 years ago, it is only within the past 2 decades that investigations have provided significant information on both the structure and function of these minute cellular connections. The pace of progress has accelerated within the past few years (there have been more than 150 papers referring to plasmodesmata within the past 5 years, although very few of these were concerned with plasmodesmata as the major topic of study) so that now, for the first time, it is possible to make realistic assessments of the capability of plasmodesmata not only to facilitate, but also to regulate, the transport of water and solutes through the plant symplasm.

The previous review of plasmodesmata in this series was by Robards (188) in 1975. Subsequently, the role of plasmodesmata in plant intercellular communication was comprehensively reviewed in the book edited by Gunning & Robards and published in 1976 (92). This volume covered all aspects of plasmodesmatal studies, from their structure and distribution to their capacity to transport solutes and their ability to allow the transmission of viruses from cell to cell; it has served as an information base, summarizing the literature accumulated over almost 100 years, and we make no attempt to incorporate that material into the present review. Since 1976 there have been several further reviews (178, 190, 191) and a number of major articles relating to specific areas of plasmodesmatal structure and function. Nevertheless, the relatively few groups of investigators around the world who are actively carrying out studies on plasmodesmata per se is a matter of both interest and concern.

Functional studies, rather than structural investigations, have increasingly pointed to the physiological analogies between plasmodesmata and their animal counterparts—gap junctions. This resulted in the organization of a NATO Advanced Research Workshop in 1989, the invited papers of which will be published in the volume edited by Robards et al (194). Significant areas of commonality have now been found between plasmodesmata and gap junctions, many of which are referred to below.

The purpose of the present review is, therefore, to provide a contemporary statement on progress in studies on plasmodesmata against the background of the well-referenced earlier work. Space precludes reference to the many citations of plasmodesmatal frequency and distribution reported as minor elements of many published papers.

STRUCTURE

The Structure of Higher-Plant Plasmodesmata

As has been well emphasised by many authors (174, 178, 189), the technicalities of viewing ultrathin sections of plasmodesmata in the electron microscope severely limit the resolution of the structural information that can be obtained. In short, the extremely small size of a plasmodesma means that the image of any part of it will always be superimposed on other information in the final micrograph. The plant scientist does not have the opportunity to form two-dimensional arrays of plasmodesmata as is possible with gap junctions. In consequence, the quality of information from electron microscopy has improved little over the past 20 years. Nevertheless, significant advances in the interpretation of plasmodesmatal structure certainly have been made, the most significant being reported by Overall et al (178).

For the purposes of discussion, a generalised model of a simple plasmodesma is illustrated in Figure 1. The points of general agreement are that the pore through the wall is lined by the plasmalemma; that an axial component passes through the plasmalemma-lined tube and that this axial strand often connects in some way with the endoplasmic reticulum (ER) in the neighboring cells; and that there is a space, the "cytoplasmic sleeve," between the axial strand and the plasmalemma. This cytoplasmic sleeve may be more or less occluded in the neck region of the plasmodesmata, where the axial strand and the plasmalemma are in close contact with each other. The basic structure of a plasmodesma is therefore rather simple, although the dimensions and relative dispositions at the molecular level—so important to understanding how plasmodesmata actually work—remain inadequately resolved.

Earlier studies had produced a number of models for plasmodesmata (reviewed in 189), the most generally accepted of which was that proposed by Robards in 1968 (186). This envisaged the axial strand to be a tubule—the desmotubule—running from cell to cell and connecting at each end with the ER. The model arose from the consideration that a membrane in the configuration of a bimolecular leaflet could not be curved around the small radius (<10 nm) measured for this axial strand (196). The desmotubule was considered to be composed of protein subunits and, in that sense, would be rather similar to a microtubule. Work on membrane structure and chemistry during the 1970s changed views on the arrangement of lipids in membranes, and this, together with other considerations, led to the proposal of the model of Overall et al (178). This model no longer assumed the presence of a proteinaceous desmotubule with a patent channel through its center, but considered the desmotubule to be an extension of the ER membrane from cell to cell, with

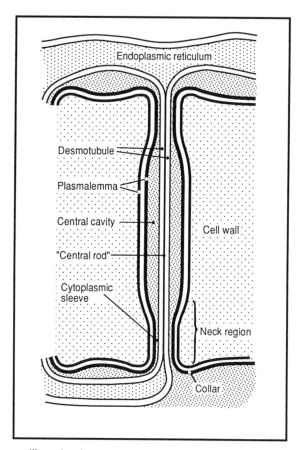

Figure 1 Diagram illustrating the component parts of a simple plasmodesma. [Adapted from (174) with permission.]

the central axis of the structure almost, if not completely, occluded by the lipidic head groups. This interpretation was made possible only by a detailed and rigorous analysis of lipid packing in membranes (178) although it provided a model that was essentially the same as that proposed earlier by López-Sáez et al (137). Both structural (172–174) and physiological (e.g. 234) studies have subsequently tended to provide general support for this model. However, details at the molecular level remain incomplete, and it is important to observe that some features, such as the precise number of subunits around the desmotubule, are by no means fully determined.

The connection of the desmotubule with the ER has frequently been observed and is more clearly shown by the use of tannic acid (158, 171) or tannic acid plus ferric chloride (178). Potassium ferrocyanide–containing fixatives give a selective deposition of electron-opaque stain in the lumen of ER cisternae, which does not, however, appear to penetrate inside the desmo-tubules (100). Hawes and co-workers (97, 225) used another ER-selective stain, zinc iodide, which allowed the visualization of very thin filaments (10–30 nm) of positively stained membranelike material extending through cell walls and connecting ER cisternae on either side. This work is the best current evidence that there may still be a very small pore through the centre of the desmotubule, although, as Overall et al (178) observed, this would be large enough to accommodate only a few water molecules at most.

The different interpretations of desmotubular structure have been well reviewed by Gunning & Overall (88). The model of Overall et al (178) draws on data from Olesen (171) for *Salsola* leaves and their own work on *Azolla* roots with tannic acid or tannic acid plus ferric chloride used for contrast enhancement. They concluded that the inner translucent ring of the desmo-tubule is equivalent to the central layer of a membrane bilayer derived from membranes of ER tubules connected to cortical ER cisternae. Here, the ER membranes merge in the desmotubule to form a more or less solid cylinder, where the hydrophilic (and electron-opaque) parts of the inner leaflet are close-packed to form the dark central rod.

The neck region is an almost universal feature of simple plasmodesmata; i.e. at each side the plasmodesmatal canal narrows into a more or less intimate contact between the axial desmotubule and the plasma membrane tube (61, 171, 189): the neck constriction. A model incorporating many of the above features has been developed by Olesen (174) and is illustrated in Figure 2.

The overall diameter of the plasmodesmatal canal is significantly larger and somewhat more variable within the wall than at the neck (171); neither the plasma membrane nor the desmotubule shows the parallel-sided appearance seen in the neck regions. A special case seems to exist in the bundle sheath/mesophyll interface in C_4 grasses, for which a number of authors have reported that a median suberin lamella in the cell walls imposes a similar constriction on the plasmodesmatal canal (15, 61; P. Olesen, unpublished data).

In principle, there are two pathways for volumetric flow through plas-modesmata (90, 91, 170, 171, 187, 189), one via the pore through the desmotubule and the other through the cytoplasmic sleeve between the two cytosolic compartments. With the introduction of the model of Overall et al (178) and the subsequent discussions of Gunning & Overall (88) on the relative efficacy of the two pathways, the cytoplasmic sleeve route, and its possible analogy with gap junctions in animal cells, has received much more attention as the potentially major pathway for intercellular communication and symplasmic transport between plant cells.

Robards (186) originally suggested 11 subunits for the desmotubule in

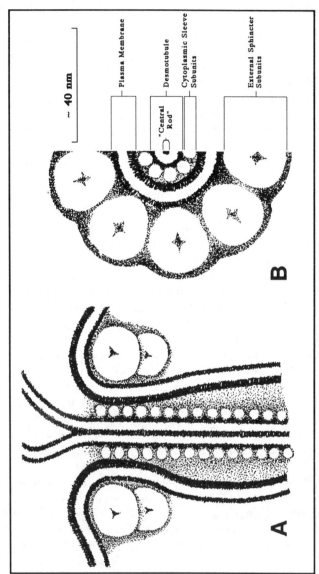

Figure 2 Diagrammatic representation of a plasmodesma drawn approximately to scale, based on electron micrographs of plasmodesmata from the mesophyll/bundle sheath interface in *Salsola kali* leaves (174). Components are shown in longitudinal (A) and transverse (B) sections. The relationships between ER, desmotubule, and the 5-nm subunits of the cytoplasmic sleeve are fully compatible with the model of Overall et al [Figures 19 and 20 in (178)]. Electron micrographs depicting such views of plasmodesmata must take account of four structural components. A central dark dot (about 1.0 nm in diameter) is surrounded by a pale ring about 12 nm in diameter. This pale zone is covered on its outer surface by an electron-opaque layer a few nanometers thick. The area between this structure—the desmotubule—and the inner face of the plasma membrane has the form of a sleeve or tube, whose appearance varies according to the fixation and staining conditions. (Adapted from reference 174 with permission.)

Labels in figure: Plasma Membrane; Desmotubule; "Central Rod"; Cytoplasmic Sleeve Subunits; External Sphincter Subunits; ~ 40 nm; A; B

Salix, whereas Zee (272) showed some evidence for 14 subunits in plasmodesmata of *Vicia* phloem. Later, tannic acid–mediated contrast enhancement facilitated the more precise determination of 9 lucent subunits in the cytoplasmic sleeve of plasmodesmata in *Salsola* leaves and *Epilobium* roots (171), as well as in the grass *Sporobolus* (Olesen, unpublished). Overall et al (178) suggested that there might be 9 subunits within the cytoplasmic sleeve in their model, although they emphasised that their evidence on this specific point was weak. Subsequent authors, such as Terry & Robards (234), have shown that a model containing 9 subunits would be compatible with experimental determinations of intercellular transport through the cytoplasmic sleeve, but it should be stressed that such results do not, per se, give added support to the suggestion that there is a specific number of subunits. Thomson & Platt-Aloia (236), in a freeze-fracture study of chemically fixed and cryoprotected *Tamarix* gland cells, obtained some evidence for a cluster of 6 subunits in the desmotubule at the point where the E-fracture face breaks off at the ER–desmotubule connection. However, this observation cannot necessarily be considered representative for the internal structure of the cytoplasmic sleeve inside the neck constriction and, in any case, must be regarded with caution in view of the fixed and cryoprotected nature of the material. Freeze-fracture studies on plasmodesmata have, in general, been extremely disappointing and have yielded little new information (193, 236, 264). This is because the fracture plane usually passes uninterestingly across the collars of the plasmodesmata without revealing internal details (193). Given the very high rates of transport through plasmodesmata (192), it is surprising that rapid-freezing techniques have not played a greater part in the structural studies of plasmodesmata, because chemical immobilization of such a dynamic system cannot possibly be expected to retain in vivo structure at the molecular level.

Estimation of accurate dimensions within the cytoplasmic sleeve is very difficult because of the relatively ill-defined surfaces of the outer side of the desmotubule and the inner side of the plasma membrane and because of uncertainties in delineating the precise location of the surfaces of particles. However, Overall et al (178) and Terry & Robards (234) have suggested about 5–6 nm for the width of the cytoplasmic sleeve, 4.5–5 nm for the diameter of each of the nine lucent particles, and, accordingly, 1.5–3 nm for the radial gaps between the particles.

Very little is known about the anchorage of the subunit particles, which, theoretically at least, could be bound either to the desmotubule or to the plasma membrane in the neck constriction. In *Salsola* plasmodesmata, which display a very conspicuous neck constriction (171), cross-sections located more deeply in the cell wall (i.e. below the neck constriction) show an increasing separation between the particle ring and the plasma membrane, although there is a constant association between the desmotubule and the particles. Here, although the diameter of the desmotubule expands significantly, electron-translucent subunits can still be seen at its outer surface: a situation also clearly illustrated in Figures 11–18 of Overall et al (178). It therefore seems probable that the subunit particles are bound to the desmotubule surface. Nothing is known about possible changes in their number outside the neck constriction; i.e. whether the enlarged desmotubule would be covered by more than nine particles.

In most transverse sections of plasmodesmata, the plasma membrane appears with high contrast and clarity, mainly because of its parallel-sided nature normal to the section thickness. In neck regions, the traditional asymmetry of outer and inner leaflet opacity in the plasma membrane, with the thicker (or more opaque) leaflet seen on the extracellular side, often appears reversed (171). However, this is not always the case (178), and such variation is most probably related to difficulties in defining the outer surface of the inner leaflet, which more or less forms a continuum with the opaque material separating the subunits in the cytoplasmic sleeve. Whether the appearance of closely packed, more or less globular subunits in the plasma membrane tube in the neck region (171) indicates a different structure and composition relative to other parts of the plasma membrane inside and outside plasmodesmata or whether it reflects simply a pronounced vertical superposition in the parallel-sided neck region is unknown. Interestingly, however, the monoclonal antibody MAC 207, which specifically recognizes an arabinogalactan-rich epitope in all higher plant plasma membranes (183), was shown by immunogold cytochemistry to be excluded from plasmodesmata in frozen ultrathin sections. If, as suggested by Gunning & Overall (88), the neck constriction can be seen as a well-defined macromolecular construction, a different structure and composition of the plasma membrane in this area would by no means be surprising. Such a concept receives support from observations on newly formed plant protoplasts, where remnants of plasmodesmata can be observed at the surface, apparently still locked into the plasma membrane (Olesen, unpublished). Here the complete neck constriction structure remained intact, indicating a very tightly integrated structure. The appearance of a ninefold radial symmetry, both in the cytoplasmic sleeve and in the outer sphincter (see below), supports the view that the sphincter has a defined, stable, and relatively invariable structure.

The Sphincter Concept

The conclusion to be drawn from the Overall et al model for plasmodesmatal structure (88, 178) is that intercellular transport must occur via the cytoplasmic sleeve. More specifically, the channel(s) for communication would be through the spaces between the ring of 5-nm subunits that partly occlude this sleeve (Fig. 2). The cross-sectional area of each of these spaces has been calculated to be 2.6–3.6 nm^2 for plasmodesmata in *Azolla* roots (178). This is remarkably similar to the 3.1-nm^2 cross-sectional area of the pores in the macromolecular components of animal gap junctions, the connexons. If this is so, even small dimensional modulations of the cytoplasmic sleeve might significantly influence the cross-sectional area available for transport through this area (Figure 3).

It has long been realized that the cytoplasmic sleeve might serve as the conducting pathway through plasmodesmata (90, 91, 170), and since then the question of the possible regulation of intercellular solute transport has led to a search for ultrastructural sphincters or valves that could control symplastic transport by dimensional modulations in the neck region (61, 90, 171, 264). Freeze-fracture images of nonfixed and noncryoprotected plasmodesmata (193, 264) have unequivocally shown that around each plasmodesmatal en-

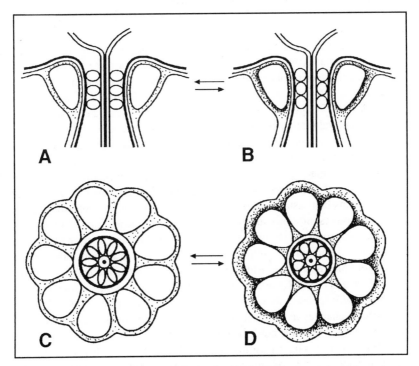

Figure 3 Simplified version of a model for the physiological sphincter activity of plasmodesmata (174). (A and C) Longitudinal and transverse views, respectively, through the neck region of open plasmodesmata. As the result of an appropriate stimulus (53), the sphincter action may tighten on the subunits in the cytoplasmic sleeve, so impeding transport through the plasmodesma (B and D). In the model of Olesen & Robards (174), the stimulus could be high cytosolic Ca^{2+} leading to the UDP-glucose-stimulated synthesis of callose and hence to the closing of the sphincter.

trance or "mouth," the plasma membrane appears raised above its normal level to form a distinct collar. Olesen (171) demonstrated, by using high-contrast ultrathin sections following tannic acid fixation, that an extracellular ring of large particles can be found just below the plasma membrane at this same location: i.e. in the collar surrounding the outer part of the neck constriction (Figures 2 and 3). Both the location and the dimensions of this external ring structure correspond exactly to the collar seen by freeze-fracture. Furthermore, previous studies on enzyme cytochemistry and ion localization (171 and references therein) have shown this area to be relatively more active in terms of enzyme activity and/or the presence of strongly reducing substances than other parts of plasmodesmata and plasma membrane. [It might be noted here that there are many citations of enzyme activity associated with plasmodesmata, although not necessarily with the neck region alone (113, 191, 253). These include α-amylase activity located around plasmodesmata in barley aleurone cell walls (e.g. 84) and other enzymes such as ATPase (e.g. 42, 273), peroxidase (213) and 5'-nucleotidase (167) actually within the plasmodesmata themselves.]

If we accept the evidence that the cytoplasmic sleeve contains nine subunit particles in radial symmetry (171), it is most interesting that rotational image enhancement experiments in the same study also indicated a ninefold symmetry of the external ring structure (Figure 2; cf Figures 9–11 in reference 171). Olesen (171) put forward the hypothesis that these ring structures could be ultrastructural equivalents of hypothetical sphincters involved in the regulation and direction of symplastic transport of solutes through plasmodesmata.

This outer-ring structure, suggested by Olesen (171) to act as a sphincter, can be seen in many published electron micrographs even when tannic acid fixation has not been used (see references in 171). However, in most such cases it seems that visualization has been facilitated by the contrasting action of endogenous tanniferous substances themselves. The addition of exogenous tannic acid to fixatives strongly emphasises the sphincter structure (158, 171, 173, 189). Although the sphincter structure was not demonstrated in the *Azolla* root plasmodesmata which were the subject of the meticulous study by Overall et al (178), Robards (189) published a micrograph from roots of another *Azolla* species, showing distinct sphincter structures. It could be that this reflects changes in sphincter structure in relation to the stage of development as, judged from the published micrographs, the tissue used by Overall et al (178) appears younger than that used by Robards (189). It is also possible that sphincters are not present in all types of tissue but reflect a specialization in physiological conditions where a very high capacity for symplasmic transport is needed (61). This suggestion would be supported by the demonstration of very distinct sphincters in locations such as the mesophyll–bundle sheath interface of *Salsola* leaves (171), which is a site of very rapid cell-to-cell transport occurring during C_4 photosynthesis and where sphincters regulating flow through the cytoplasmic sleeve could well be envisaged to be necessary.

There thus appears to be a growing consensus that plasmodesmatal regulation may take place at the level of the neck region, where particulate subunits are closely packed between the plasmalemma and the desmotubule. Such a model is consistent with the views of a number of authors (171, 178, 187, 234, 268). It is relatively simple, then, to envisage a mechanism for using such a structure to control transport, even if the precise signaling and transduction components remain to be elucidated. One model for plasmodesmatal regulation at the sphincter has been put forward by Olesen & Robards (174) (Fig. 3). On an extremely speculative note, Xu et al (268) have hypothesized that the effective intercellular junction in plasmodesmata may be connexin-like proteins packed into the cytoplasmic sleeve within the plasmodesmatal neck.

Future studies will undoubtedly require simple systems, such as plasmodesmata between fused cells in protoplast cultures, and will need to use techniques that are as yet incompletely perfected, such as isolation of plasmodesmata from their surrounding cell wall: an approach presently being pioneered by Epel et al (B. Epel, personal communication).

Evidence has been presented to suggest that some polypeptides in plant cells are immunologically related to rat liver-type connexin, the animal gap junctional protein (151, 268). Immunolocalization studies suggest that the polypeptide is widely distributed in plants and their tissues and is localized at the cell periphery (152). It is too early to comment on the generality or significance of this result, but it is clearly an area open for further study (150). In a somewhat similar intriguing study, Conkling et al (28) have reported that *Arabidopsis thaliana* and *Nicotiana tabacum* each contain a gene coding for a 26-kDa polypeptide that is approximately 30–40% related in amino acid sequence to membrane intrinsic protein (MIP) 26 isolated from animal eye lens. MIP 26 has been suggested to be a gap junction protein, whereas the plant 26-kDa protein appears to be confined exclusively to roots.

Although the Overall et al (178) model of plasmodesmata now seems broadly consistent with the observations of authors studying plasmodesmata in higher plants (88, 171, 173, 178, 234, 268), one or two entirely different suggestions have been made. For example, Rowley (200, 201) has interpreted the structure of plasmodesmata-like processes at the tapetal cell surface of *Epilobium angustifolium,* but further work is required before such findings can be correlated with results such as those described above.

Plasmodesmata in Lower Plants

As observed in an earlier review (188), the structure of plasmodesmata and plasmodesmata-like intercellular connections in algae and fungi remains relatively poorly studied. Over the past decade or so, there have been few publications concerned primarily with plasmodesmata in such organisms and fewer still that have concerned themselves with specific details of structure or function. We have come little further forward since the detailed review of Marchant (146), in which he stated that "The variety of intercellular connections that is to be found among the algae and fungi strongly suggests that there is considerable selective advantage for cellular continuity and that these intercellular bridges have diverse evolutionary origins." Indeed, it has been suggested (145) that pit connections or plasmodesmata might well be used as morphogenetic differentiators in the neothalli of the Chloro-, Rhodo-, and Phaeophyta. It is a pity that the little high-resolution work that has been carried out on plasmodesmata has been undertaken largely on those of higher plants, whereas many of the most informative experiments concerning intercellular transport and its regulation have been performed with the characean algae.

Having said this, a number of relevant papers have been published by a relatively small group of authors. The physiologically intensively studied genus *Chara* is not the easiest subject for electron-microscopic examination but has been studied, among others, by Kwiatowska et al (125–127) and Godlewski (80). Electrophysiological work conducted by Côté et al (32) led to the conclusion that the plasmodesmata could act as valves when a pressure difference was induced across their ends. Godlewski (80) was able to demonstrate that *Chara* exhibits a traumatic callose synthesis response, whereas Trebacz et al (239) concluded that transnodal transport, including that via the plasmodesmata, is at least partly active and requires metabolic energy to sustain it. These findings, correlated with those previously reported (146), provide an overall functional picture very similar to that in higher plants: the unresolved question remains the molecular structure of the plasmodesmata themselves and the sphincter mechanism, if any. The antheridium of *Chara vulgaris* has been shown to be connected by plasmodesmata to the thallus via a basal cell (125), these connections being broken prior to the initiation of spermatozoid differentiation and thus showing parallels with higher-plant symplasmic isolation during gamete formation; the same authors had earlier (126, 127) documented detailed changes in the form and plugging-unplugging of plasmodesmata during antheridial development. Open plasmodesmata occurred between "differentiation-synchronized" cells, whereas cells that were not at similar stages of development had plugged plasmodesmata. The large phaeophytan alga *Dictyopteris membrancea* has been shown (116) to have numerous plasmodesmata in its end walls; the authors compare these structures to the sieve elements of the Laminariales. There are numerous other citations of plasmodesmatal presence in the large algae, including in the walls between hairs and underlying filaments of *Hydroclathrus clathratus* (168) and a detailed survey of plasmodesmata in two *Laminaria* species (212).

Chapman et al have contributed a number of publications on *Cephaleuros* and other chroolepidacean algae (22–24). These algae have simple plasmodesmata, 40–50 nm in diameter, which occupy a central pit in the crosswalls (22, 25); the authors contrast this pit with pit connections in Rhodophyta and Cyanophyta. In a study of cytokinesis in the apical cells of *Cephaleuros* (24), it was noted that bundles of microtubules in the higher-plant-like phragmoplast appear to mark the sites of plasmodesmata but ER is not directly involved in their formation.

In a study of the parasitic relationship between *Stephanoma phaeospora* on its host, *Fusarium* sp., Hoch (102) described plasmodesmata 24–28 nm in diameter, which must presumably be of secondary origin, traversing the interface between the two fungi. Giddings & Staehelin (78), using freeze-fracture methods, described plasmodesma-like connections, which they referred to as microplasmodesmata, between vegetative cells of different filamentous cyanobacteria, including *Anabaena, Nostoc, Phormidium,* and *Plectonema*.

PLASMODESMATA FORMATION

It has long been known and accepted that most plasmodesmata are formed from the entrapment of strands of ER in the developing cell plate (185). However, this leaves open questions such as how plasmodesmata come to be arranged in regular patterns (e.g. pit fields) within the cell wall and how specific cell types consistently insert more or less the same frequency of plasmodesmata across a specific interface, this frequency sometimes varying in relation to the age and/or number of divisions of the parent cell (87, 103). Hepler (100) has provided a detailed electron-microscopic study of plasmodesmatal formation during cell plate deposition by using ER-impregnating and contrasting stains which appears to confirm the formation of plasmodesmata by the entrapment of ER but does not support the idea that the desmotubular core is in open continuity from one cell to another. Recent publications have also referred to the continuity, not only of ER from cell to cell via the plasmodesmata but also to an intercellular continuum from nuclear envelope to ER and thence to plasmodesmata (99, 205).

Although most plasmodesmata in mature cells have arisen from the process described above, it is now beyond argument that plants also have the capacity, under appropriate circumstances, to form plasmodesmata secondarily in nondivision walls (see 110 for a review of the literature until 1976). Plasmodesmata have more recently been found in nondivision walls such as those between the two components of chimeras (12), between host and parasite cells (36, 230), and in grafts (109, 119, 120, 270). Secondary cell connections have also been reported to develop between normally growing cells in intact plants, such as between pollen mother cells (26) and their presence has been deduced from maintenance of plasmodesmatal frequencies in elongating cells (213, 216). Plasmodesmata that penetrate only halfway through the cell wall have often been considered to be indicative of the plant's attempt at secondary plasmodesmatal formation, and these structures have been fairly widely found in a number of situations, including callus cultures (159). The distribution of plasmodesmata in cambial initials and their xylem derivatives is of particular interest (7) because although there are frequent anticlinal divisions, the relative paucity of periclinal divisions would lead to extreme plasmodesmatal dilution in the radial walls of, e.g., ray cells. Barnett (8) showed that the plasmodesmata in *Sorbus aucuparia* cambial pit fields are widely separated, although they are later found close-packed during secondary wall formation. He was not able to distinguish whether this arrangement arose from a relocation of existing plasmodesmata or the de novo synthesis of more connections.

A recent review of this topic (159) concentrated on the secondary formation of plasmodesmata in cultured cells. Monzer's observations led to the suggestion of a model for secondary plasmodesmata formation. Tubular cisternae of ER are constricted to the size of a desmotubule. Golgi vesicles are involved in the enclosure, and exclusion, of protoplasmic strands containing the constricted ER cisternae. This model unexpectedly raises striking similarities to normal plasmodesmatal formation, in which the ER strands are enmeshed in a mass of Golgi vesicles. The earlier view of secondary plasmodesmata formation was that progress through the cell wall was achieved by enzymatic digestion in front of the inpushing cytoplasmic strand. In fact, there has so far been no demonstration of the necessary enzymes at the appropriate time and place. If secondary plasmodesmata are produced by the temporally and spatially synchronized growth toward each other of "half-plasmodesmata," a number of problems remain, not the least of which is how the adjacent cells can communicate and control such processes and how progress is maintained through the intervening cell wall. This appears to be less of a problem in Monzer's culture system, in which wall development is minimal and it may be, as he points out, that an entirely different mechanism exists in this experimental culture situation (159).

In general, secondary plasmodesmatal formation appears to arise in relatively undifferentiated cells. Those forming connections between grafts, for example, are dedifferentiated cells (109, 120, 228). Secondary formation of plasmodesmata remains one of the more intriguing aspects of this topic, and much remains to be done to establish how the process takes place and whether there is a single unifying mechanism.

PLASMODESMATA DISTRIBUTION AND FREQUENCY

Plasmodesmata are found between all young, living cells of higher plants. Their frequencies are broadly within the range of $0.1–10.0$ μm^{-2}, although exceptions can be found on either side of this range. Plasmodesmatal frequencies from many plants and tissues have been comprehensively cited (92, 188, 191).

Fisher (66) has recently reviewed the methods for citing the distribution of plasmodesmatal frequencies. His work has been specifically related to the minor veins of leaves, but can be applied more widely. Plasmodesmatal frequencies have been cited in a number of different forms: (a) plasmodesmata per square micrometer of interface per thin section (62, 64, 67, 202), (b) plasmodesmata per square micrometer of interface (64, 73, 76, 122), (c) percentage of the total number of plasmodesmata that appears in each interface (15, 89, 115), (d) plasmodesmata at each interface per millimeter of vein (251), and (e) the surface:volume ratio of the cells in a given compartment multiplied by the number of plasmodesmata per micrometer connecting it to the next compartment (44). A trend has developed to illustrate plasmodesmatal frequencies diagrammatically by using, for example, the "plasmodesmograms" of van Bel et al (251) (Figure 4), but care must be taken if these snapshots of intercellular communication are not to give a misleading impression of the actual quantitative level of communication (66). Accurate collection of plasmodesmatal frequency data is important, and correction for section thickness must be made (189). Robins & Juniper (197) criticized the method suggested by Gunning (189), who subsequently responded and, in combination with Robins & Juniper themselves (93), concluded that the original recommendation was valid. It is, indeed, now widely accepted and used.

Fisher (66) points out that the use of plasmodesmatal frequency to encompass a citation of symplasmic transport capacity is complex because of factors such as variation in cell shape, presence of intercellular spaces, etc. In the species studied and/or cited by him (*Coleus, Cananga, Populus,* and *Beta*), frequencies between mesophyll cells were within the range $0.09–0.62$ plasmodesmata μm^{-2} of wall interface. He concludes that, at present, the simplest and probably the best measure is the number of plasmodesmata per unit of vein length for interfaces at the mesophyll/bundle sheath boundary and within, and plasmodesmata per micrometer or per square micrometer for interfaces in the mesophyll.

Since then, a number of publications have been concerned with the distribution and frequency of plasmodesmata, and there has been a movement toward the consideration of these data in relation to the transport capacity of the intercellular connections.

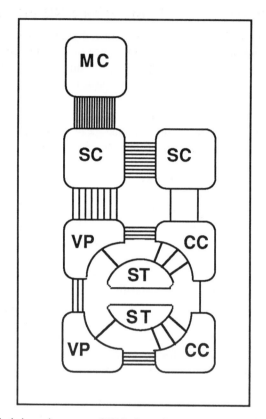

Figure 4   A typical plasmodesmogram (249) [redrawn from (251)]. Plasmodesmograms attempt to quantify the plasmodesmatal frequency between different cell types. The absolute number of plasmodesmata, as well as the quantitative expression of the number of contacts, varies with both species and authors. Therefore, to enable comparisons of several species [as in (251)], the number of stripes represents the percent contribution to the total number of plasmodesmatal units in a particular cell interface. The plasmodesmogram here gives a quantitative impression of the cytoplasmic connections between different cells associated with phloem loading in *Amaranthus retroflexus* (67). Abbreviations: MC, mesophyll cell; SC, sheath cell (mestome or bundle); VP, vascular parenchyma; CC, companion cell; ST, sieve tube.

Improvement in the ability to measure plasmodesmatal frequencies with high accuracy and then to relate these to observed rates of intercellular transport has led to a number of publications concerned with the quantitative correlation of plasmodesmatal structure, frequency, and transport capacity. The demands of this approach are that it is necessary to obtain high-grade ultrastructural data from a system that is also amenable to the experimental determination of rates of water and solute movement. This approach was reported in papers such as those by Robards & Clarkson (192), in which measured rates of movement across the root endodermis were interpreted in relation to plasmodesmatal frequencies and pore dimensions. Later, attention was transferred to the hypodermal/epidermal interface in maize roots (27), where, as in the endodermis, plasmodesmata were present in sufficient numbers to accommodate the measured fluxes and flows and were not occluded when suberin lamellae were deposited in the walls. In addition to root endodermis and hypodermis, some attention has been given to plasmodesmata in the walls of root hairs, including those of *Trianea bogotensis* (124), where the distribution of plasmodesmata was observed to be related to the anticipated requirements for higher axial than radial transport. In all cases studied so far, it has been found that the observed numbers of plasmodesmata, assuming reasonable dimensions for the patent pore channels, are adequate to cope with the transport capacity required across a specific interface. However, plasmodesmatal frequencies in roots are relatively low compared with some other high transport activity sites in plants, such as in glands. It is for this reason, as much as any other, that secretory glands have been studied in great detail in an attempt to elucidate the relationships between structure and function (reviewed in 50).

Probably the best-studied gland, in the present context, is the nectary of *Abutilon*, which has been used by Gunning and colleagues (86, 103) and, later, by Robards and colleagues (121, 195). The closely related gland of *Hibiscus* has been studied by Sawidis and colleagues (209–211). Among the many merits of these glands as model systems, one of the most important is that the prenectar is moved into a nectary hair across an interface, where it is considered that apoplasmic transport is blocked and therefore that all movement must be via the numerous plasmodesmata. This is a closely analagous situation to that seen in the endodermis (with its Casparian band) and is recapitulated in a number of different situations in plants. Although the rate of flow through the cells of such glandular trichomes is high, calculations demonstrate that the plasmodesmata can always cope with the anticipated demand. Indeed, it is interesting, and probably physiologically significant, that the plasmodesmatal frequency at all stages of development is seen to change from about 12 μm^{-2} at the base of the trichome to around 4 μm^{-2} at its tip (121), broadly comparable to the earlier data of Hughes (103) and also to those for *Gossypium* obtained by Eleftheriou & Hall (51, 52) and for *Hibiscus* by Sawidis et al (210), apparently closely correlating supply (of plasmodesmatal intercellular continuity) with demand (for a declining rate of intercellular transport as nectar is unloaded from cells along the trichome).

A number of different nectaries and other glands have been investigated, usually in less detail than those of the Malvaceae mentioned above. These include *Gossypium hirsutum,* which was investigated by Wergin et al (262) and, later, by Eleftheriou & Hall (51, 52); the salt glands of *Tamarix,* for which Thomson & Platt-Aloia (236) have made a detailed electron-microscopic study of the plasmodesmata; *Cressa* glandular trichomes (259); *Ricinus* (164), *Vigna* (123, 182), and *Acacia* (147) extrafloral nectaries; and the carnivorous bromeliad *Brocchinia reducta* (179), in which no apoplasmic barrier was detected at the base of the hairs. In such situations it is then much more difficult to distinguish between the relative contributions of the symplastic and apoplasmic pathways.

The mestome sheath in C_4 grasses (170) is another system involving active transport which has also been used beneficially in studying the interrelationships between plasmodesmatal structure and function (see also below), as are the minor veins and phloem of leaves; for example, Russin & Evert (202) have carried out a correlative investigation on the leaves of *Populus deltoides* and concluded that plasmodesmata are present in all the walls between the palisade parenchyma cells and the sieve tubes of the minor veins and that they exist in sufficient numbers to allow diffusion along a concentration gradient from the palisade layer to the companion cells. The transport of metabolites along the rays of woody plants has long been understood to represent an interesting site for the study of intercellular communication. This system has been studied in detail in *Populus canadensis* by Sauter & Kloth (208), who obtained data on both fluxes and flows as well as plasmodesmatal frequencies. The maximum frequency within pit fields was 39 μm^{-2}, although this reduced to 8.0 μm^{-2} for the wall as a whole. As in the glandular systems referred to above, it was concluded that the fluxes were sufficiently high that they could not proceed by a transmembrane mechanism but that there were enough plasmodesmata to accommodate the observed rates of translocation.

In addition to the above, there are numerous incidental observations on the occurrence and frequency of plasmodesmata from a wide variety of tissues, but these add little to our general understanding and are not recorded here.

PLASMODESMATA IN DEVELOPMENT

Although it now seems accepted that the secondary formation of plasmodesmata may have a role to play in the normal development of some relatively undifferentiated plant cells, the number of plasmodesmata that are inserted into a given cell plate and what subsequent losses and/or dilution there may subsequently be during development are important aspects of the regulation of cell growth and development. Different numbers of plasmodesmata are inserted into the cell plates of different cell types. The few detailed experiments carried out in this area suggest the frequency of plasmodesmata between cells may be a determinant of the ability to maintain cell division (e.g. 85). Gunning concluded that although there was no secondary formation of plasmodesmata during the growth and elongation of the roots of *Azolla,* there could be loss from the walls of both xylem and phloem elements as well as through cell separation. Eventually, the apical cell of this small aquatic fern (which alone produces all the cells of the mature root) fails to maintain the same number of plasmodesmata in each successive cell plate. Consequently, the apical cell becomes more and more isolated symplasmically: a feature that could ultimately be responsible for the determinant growth pattern of the root.

The ultimate extent of plasmodesmatal dilution is, of course, total symplasmic isolation (113). This occurs, for example, in stomatal guard cells (e.g. 72, 113, 181, 263, 265).

BIOPHYSICAL STUDIES ON PLASMODESMATA

Theoretical and Experimental Studies

Electrophysiological experiments have been conducted on a range of plant cell types to determine whether plasmodesmata truly represent a low-resistance pathway for the movement of ions and solutes. Some of the pioneering studies were performed on the giant characean cells, as their shape and size facilitated the unambiguous location of the tips of the microelectrodes. Additionally, considerable information was available with respect to the electrical properties of the plasmalemma of these cells. The experimental system consisted of two adjoining internodal cells that were connected by the nodal complex from which the associated branch, or whorl cells, had been removed. Voltage-recording microelectrodes were placed within the vacuoles of each cell, and these were used to monitor the voltage changes elicited in both cells when current was injected into one of the cells via a separate current-injecting electrode. With such a system, Spanswick & Costerton (221) were able to show that the specific resistance of the node was approximately 50 times lower than the specific resistance of the plasma membranes of the two internodal cells.

Ultrastructural studies conducted on the same experimental tissue revealed that numerous plasmodesmata were present within the cell walls that connected the internodal cells with the cells of the nodal complex (221). The desmotubule is absent from the plasmodesmata of characean cells (69, 221) and so Spanswick & Costerton (221) were able to model this low-electrical-resistance pathway as being represented by simple pores filled with 100 mM KCl. From this analysis they deduced that the measured resistance should have been even lower than they actually measured. Thus, these electrophysiological and ultrastructural results were consistent with the hypothesis that plasmodesmata constitute a low-resistance pathway. However, the discrepancy between experimental and theoretically determined plasmodesma resistance led Spanswick & Costerton (221) to conclude that there must be some restriction within the plasmodesma that limits the free diffusion of ions.

Tyree (247) used the principles of irreversible thermodynamics to evaluate various transport processes that may function at the level of the plasmodesma. It is unfortunate that Tyree elected to restrict his theoretical modeling of the plasmodesma to the Robards model (186, 187), as we now know that this almost certainly means that most of the physical parameters used in Tyree's study overestimate the dimensions of the cytoplasmic sleeve. In any event, Tyree also concluded that plasmodesmata constitute the pathway of least resistance for the cell-to-cell movement of all small solutes. Furthermore, he concluded that whereas complex transport phenomena involving active mechanisms cannot be completely discounted, diffusion will be the predominant mechanism of transport through plasmodesmata.

Since this pioneering attempt by Tyree, there have been only limited attempts to develop biophysical models of plasmodesmata. Blake (13) used Newtonian and non-Newtonian fluid mechanics to model the hydrodynamics of plasmodesmata. Unfortunately, the models and conditions that he employed are now known to be either unrealistic or at variance with our present model of a plasmodesma. Information provided on the overall influence of the neck constriction region, in terms of restricting fluxes (flow) through the plasmodesma, may still be applicable.

The complexity of the cellular arrangement found in most plant tissues has made it very difficult to interpret electrical-coupling experiments performed on higher-plant tissues. However, the development of an appropriate electrical-circuit analog for the simple leaf cell arrangement of *Elodea canadensis* allowed Spanswick (220) to establish that the junctional resistance between neighboring cells is much lower than the resistance of the plasma membrane. Electrical coupling between *Elodea* cells was such that 72% of the injected current passed through their plasmodesmata to the neighboring cytoplasm. Similar electrical-coupling experiments have now been conducted on a range of tissues, including oat coleoptiles (48, 49), *Azolla* (177) and *Trianea bogotensis* (255) roots and the trap lobes of *Aldrovanda vesiculosa* (108); coupling values range from 9% up to 79%. Using submersed trichomes of *Salvinia*, Lyalin et al (140) found rapid changes in intercellular resistance, which they attributed to the possible rapid opening and closing of plasmodesmata.

It is worth noting that Drake (48) used this biophysical approach to investigate possible energy requirements for plasmodesmatal function. In the presence of metabolic inhibitors such as azide and cyanide, the electrical coupling between parenchyma cells in the oat coleoptile was almost completely eliminated. Although some part of this reduction could be attributed to the deposition of callose, this could not account for the loss of electrical coupling in the presence of azide. Whether these results can be explained on the basis of an inhibition of the histochemically identified ATPase activity within plasmodesmata (see 69 and references therein) remains to be established. The perplexing aspect to this work is that Tucker (241) used high levels of cyanide and azide to inhibit cytoplasmic streaming in staminal hairs of *Setcreasea purpurea,* and under these conditions the plasmodesmata appeared normal, at least in terms of dye coupling (see the section, below, on molecular size exclusion limits).

Tomiyama et al (238) used electrical coupling experiments to investigate the response of potato cells to infection by *Phytophthora infestans.* They found that early in the infection process, electrical coupling was completely eliminated. Further studies of this nature may provide valuable information on pathogen-induced changes in plasmodesmata and the host hypersensitivity response.

The structural and biophysical studies performed on the plasmodesmata of *Azolla* roots by Overall et al (177, 178) is perhaps the best example available of an attempt to integrate ultrastructural details into a biophysical model to test the predicted characteristic of plasmodesmata against experimental data. The *Azolla* root provides an interesting experimental system for electrical-coupling studies because it is relatively simple in structure and, as it ages, its apical cell lays down walls with progressively fewer plasmodesmata. An additional cytological feature of importance, in terms of analyzing the electrical-coupling data, is that young cells near the apex have very small vacuoles, thus making it easy to insert microelectrodes into the cytoplasm.

From the combined electrical, anatomical, and ultrastructural data obtained on the *Azolla* root system, Overall & Gunning (177) established that a quantitative relationship does exist between electrical coupling and the number of plasmodesmata in the cell wall across which the injected current flows. The relationship between cell age and single plasmodesma resistance was also investigated; although the values ranged from 2×10^{11} to 8×10^{11} Ω, with the average being 4.4×10^{11} Ω, there was no correlation with age. Finally, while recognizing the gross oversimplification of their approach, these workers used their elegant ultrastructural data to compute the resistance of a single plasmodesma. Their calculations were based on the following: (*a*) only the cytoplasmic annulus is available for the passage of current, i.e. the desmotubule exists as a solid lipidic cylinder; (*b*) the medium within the cytoplasmic annulus has a conductivity equivalent to a 100 mM KCl solution; (*c*) the cross-sectional area available for current flow can be computed by assuming that the particles observed in the cytoplasmic annulus form orderly files; and (*d*) the physical dimensions of a plasmodesma, obtained from electron microscopy of fixed material, represent the in vivo situation.

With these criteria, the computed resistance of a single plasmodesma was approximately 2.5×10^9 Ω, which is some 170 times lower than the experimentally derived value. Overall & Gunning (177) attributed this discrepancy to an impediment factor, which might be a tortuosity factor if the particles do not form files but, rather, are helically arranged, or a callose-mediated constriction of the plasmodesma as a result of wounding of the cell upon electrode insertion. Since the 3.0-nm^2 cross-sectional area of the interparticle spaces is close to the known dimensions of animal gap junctions [3.1 nm^2 (135, 222)], and as there is close agreement between the experimental and theoretical resistances of the gap junction system, we used these equations to further investigate this discrepancy. The relevant plasmodesmatal parameters that we used in the gap junction equations (222) were as follows: resistivity of pore fluid, 91 Ω-cm; pore radius, 8×10^{-8} cm; pore length, 1.47×10^{-5} cm; geometric packing of nine particles within the cytoplasmic annulus. Based on these calculations, the theoretical resistance of a single plasmodesma would be approximately 6.7×10^9 Ω (or a conductance of 149 pS), which is very close to the value obtained by Overall & Gunning (177). In that the two theoretical values have approximately the same value and are much lower than the experimental resistance, it would seem that the most likely basis for the discrepancy resides with the spatial arrangement of the particles within the cytoplasmic annulus. However, it is worth noting that changing the resistivity of the pore from 91 to 200 Ω-cm raised the single-plasmodesma resistance to 1.5×10^{10}; this reduced the discrepancy to less than a factor of 30.

Future studies aimed at establishing the unitary conductance of plasmodesmata should be performed on the simplest cellular systems available; cells in callus or suspension culture would seem ideal. If a two-celled system can be developed, the voltage clamp approach used for gap junction characterizations (111, 222) could be employed. Every care must be taken to reduce the complications associated with wound-induced callose formation around the plasmodesmata. In addition, since there is preliminary evidence that plasmodesmata may be pressure sensitive (32), care must be taken to ensure that osmotic perturbations are also avoided. With unambiguous measurements of single plasmodesmal conductance, it should be possible to further resolve the roles played by the particles of the cytoplasmic annulus and the region of the neck constriction in terms of establishing the biophysical properties of plasmodesmata.

Molecular Size Exclusion Limits

Electrical-coupling experiments indicated that ions could pass through plasmodesmata, with the most likely route being the cytoplasmic sleeve. The molecular properties of this pathway were further probed by the use of fluorescent dyes. Tyree and co-workers (5, 248) were the first to conduct a semiquantitative analysis on the diffusion of fluorescein through staminal hairs of *Tradescantia virginia* and the trichomes of *Lycopersicon esculentum*. Although dye movement was found to be proportional to the square root of time, which would be expected if movement were diffusion limited, these studies were of limited value because the plasmodesmata appeared to be blocked. This was probably caused by the rather crude wounding method used to introduce the fluorescein into the cells, which would have induced callose formation around the wound. [This is an appropriate point to mention that electron-microscopic studies of callose synthesis and localization are made more difficult by the glutaraldehyde-induced deposition of this glucan (104).]

Microinjection studies in which fluorescent dye molecules were introduced into the cytoplasm, or the nucleoplasm of the staminal hair cells of *Setcreasea purpurea*, indicated that the plasmodesmata that interconnect these cells constitute a rather complex molecular diffusion barrier (240). Using an image intensification system to detect the time-dependent cell-to-cell movement of carboxyfluorescein in the staminal hair system, Tucker et al were able to develop a numerical analysis model that allowed them to compute the junctional (intercellular) diffusion coefficient for carboxyfluorescein (243, 244). Although dye transfer to the vacuole had previously been considered unimportant, Tucker et al (244) had to incorporate this transfer coefficient in their numerical model to obtain reasonable agreement between simulated and experimental data.

Computation of the plasmodesma-specific diffusion coefficient from this junctional coefficient required information on plasmodesmatal frequency. Such information was accessible to these workers, in that Tucker (240) had investigated the ultrastructure of the plasmodesmata in this *Setcreasea* system, so that representative dimensions for diameter, length, etc, were available. However, it is unfortunate that plasmodesmatal frequency data were not available, and although such data obtained on a close relative, i.e. *Tradescantia* (254), were used, such an extrapolation must place limits on our confidence in the computed plasmodesma-specific diffusion coefficient for carboxyfluorescein. In any event, the value calculated by Tucker et al (244) was 5.37×10^{-8} cm^2 s^{-1}, which is approximately 80 times lower than the value of 4.5×10^{-6} cm^2 s^{-1} obtained for the diffusion of carboxyfluorescein in water (248). This result provides the first experimental support for the hypothesis that plasmodesmata represent a significant barrier to the diffusion of small molecules.

The adaptation of fluorescent dye-coupling techniques used to probe gap junctions (68, 215, 219, 226) has recently provided a direct means for evaluating the functional status of plasmodesmata. Goodwin (82) and Tucker (240) established that the molecular size exclusion limits of plasmodesmata were from 700 to 874 Da, which is very close to the limits observed for gap junctions (219). In initial experiments, the highly vacuolated nature of mature plant cells placed some restrictions on the type of tissue or cells that could be used for dye-coupling experiments. Goodwin (82) performed his experiments on the simple leaf of the aquatic plant *Elodea canadensis,* and conjugates of fluorescein isothiocyanate (FITC) and various amino acids were iontophoresed into the cell via a microelectrode that was most probably located in the vacuole; dye presumably leaked from the vacuole to the cytoplasm during electrode withdrawal. Tucker (240) overcame the problem of delivering the dye to the cytoplasm by placing the tip of the microelectrode in the nucleoplasm of staminal hair cells of *Setcreasea.* In these experiments, a very high concentration of dye within the tip of the micropipette was used to drive the diffusion of the fluorescent probe into the nucleus and hence, via the nuclear pores, to the cytoplasm. Both systems provided comparable size exclusion limits, but the actual rate of dye movement was higher in the staminal hair cell system.

The relationship between the molecular size exclusion limit, expressed in terms of molecular weight, and the actual physical dimensions of the fluorescent dyes has been addressed by several groups. Goodwin (82) used a simple molecular modeling approach to predict that the physical size limits of plasmodesmata in *E. canadensis* would be within the range 3.0–5.0 nm. Tucker (240), on the other hand, relied on the earlier gap junction studies of Loewenstein (133) to predict that in *S. purpurea* this physical limit would be somewhat greater than 1.4 nm. Perhaps the most extensive and rigorous evaluation of the physical properties responsible for establishing the size limits of plasmodesmata was performed by Terry & Robards (234). An extensive array of FITC–peptide conjugates was evaluated by using space-filling models to determine the most energetically favorable conformation and, thus, the most probable molecular dimensions. These probes were then used in dye-coupling experiments performed on the nectary trichome cells of *Abutilon striatum* to determine the molecular size exclusion limits for plasmodesmata in this system. Terry & Robards (234) then adapted the permeability equations, developed for permeation through the pores of the nuclear envelope (180), to their plasmodesmatal system to explore the relationship between the radius of the permeant dye and the rate of diffusion. From these studies they determined that, at least for the *Abutilon* system, the mobility of a probe is determined solely by its effective Stokes radius. In addition, the rates of diffusion observed for the different probes used in their study can best be explained if the cytoplasmic annulus is subdivided into a number of discrete channels (from 9 to 20), each having a diameter of approximately 3 nm.

Regulation of Plasmodesmatal Permeability

An extension of the concept that plasmodesmata are involved in regulating intercellular transport of solutes on the basis of size (or effective Stokes radius) is that their actual size exclusion limits are dynamic, being under some form of cellular control. Support for this concept was provided by dye-coupling experiments in which cytosolic Ca^{2+} levels were raised either by including Ca^{2+} in the injection pipette or by using A23187 (Ca ionophore)-mediated delivery. Erwee & Goodwin (53) found that pretreatment of *Egeria densa* (synonym *Elodea densa*) cells with Ca^{2+} caused a dramatic reduction in the subsequent movement of fluorescent dye molecules; similar results were also obtained with Mg^{2+} and Sr^{2+}. However, if they waited for 30 min after Ca^{2+} had been iontophoresed into the cell, normal kinetics for cell-to-cell dye movement were obtained. As callose synthesis is strongly influenced by Ca^{2+}, these effects could have been caused by callose deposition in the neck constriction region of the plasmodesmata. In addition, the elevated Ca^{2+} levels caused an inhibition of cyclosis, which, in itself, could have influenced the intercellular movement of the dyes. The latter possibility was eliminated on the basis of cytochalasin experiments in which complete inhibition of cytoplasmic streaming had no effect on dye transfer (see also 241). Further-more, aniline blue staining for callose did not reveal a marked increase in fluorescence, and this finding, along with the rapid recovery of the dye-coupling capability of these cells, supports the contention that plasmodesmata are regulated by a Ca^{2+}-mediated processes other than callose formation (53). This conclusion is further supported by recent fluorescein photobleaching studies performed on soybean suspension culture cells (9).

Interestingly, this regulation by Ca^{2+} was eliminated if the tissue was first plasmolyzed and then deplasmolyzed (54). Microinjection of dye molecules into plasmolyzed cells established that all cytoplasmic bridges had been ruptured. Within 10 min of returning previously plasmolyzed *Egeria* cells to normal solutions, the authors found that 75% of injected cells displayed dye coupling. [Note that this time course for the recovery of dye coupling is considerably shorter than that reported by Drake & Carr (49) for electrical coupling in oat coleoptile cells.] The size exclusion limits in these de-plasmolyzed cells had undergone a substantial increase from approximately 670 to 1678 Da, and, as mentioned above, elevated cytoplasmic Ca^{2+} no longer had an effect on dye movement. Deplasmolyzed cells also allowed the cell-to-cell movement of FITC conjugates of aromatic amino acids which, under control conditions, were unable to move through the plasmodesmata of staminal hair cells of *S. purpurea* (240) or leaf cells in *E. densa* (54). These findings indicate that plasmolysis can disrupt a normal regulatory mech-anism(s), the function of which appears to be the control of plasmodesmatal size exclusion limits.

The inositol trisphosphate-diacylglycerol (IP_3-DG) second-messenger sys-tem (11) may form part of this regulatory pathway. Tucker (242) investigated the effects of various D-*myo*-inositol metabolites on the movement of car-boxyfluorescein in the *Setcreasea* staminal hair system. Elevated cytoplasmic (or nucleoplasmic) levels of either D-*myo*-inositol 1,4-bisphosphate or D-*myo*-

inositol 1,4,5-trisphosphate caused a complete blockage of the symplasmic pathway. This down-regulation of permeability may involve phosphorylation of a plasmodesmatal protein, via a Ca^{2+}-activated kinase (see 184, 214, 235). Interpretation of these results must remain equivocal, because extremely high inositol concentrations were used in the pipette tip, and such levels may elicit a general wound response. Baron-Epel et al (9) used fluorescence photobleaching to study the effect of 12-O-tetradecanoyl-phorbol-13-acetate (TPA) on fluorescein and carboxyfluorescein movement between adjacent soybean suspension cultured cells. Inasmuch as phorbol esters are thought to bypass diacylglycerol to activate protein kinase C through the mobilization of cytosolic Ca^{2+}, TPA inhibition of dye coupling in soybean cells is consistent with the hypothesis that the IP_3-DG second-messenger system is involved in regulating plasmodesmata.

Parallels Between the Regulation of Cell-to-Cell Transport in Plasmodesmata and Gap Junctions

Gap junction permeability appears to be regulated in an analogous manner to that found for plasmodesmata (for recent reviews of this literature, see 101 and 194). The molecular size exclusion limits for gap junctions of different animal and insect tissues range from as low as 464 Da to as high as 1830 Da, with the norm being in the 800-Da range. The effect of increasing intracellular levels of Ca^{2+} on the permeability of gap junctions has been studied in several laboratories. In general, the results support the hypothesis that either the entry of Ca^{2+} into the cytoplasm or the release of intracellular stores of Ca^{2+} causes the down regulation of electrical coupling (39, 133–135) and decreases the junctional conductance (223). However, Spray (222) cautions that because very high intracellular levels of Ca^{2+} are required to achieve these effects, Ca^{2+} could play a regulatory role during cell injury but may not be involved under normal physiological conditions. (As we cannot dismiss the possibility of wound effects on plant cells, the same conclusions may apply to the regulation of plasmodesmata.)

Experimental evidence has also been obtained for the involvement of DG in the control of gap junction permeability (71, 269). Pretreatment of cultured epithelial cells with DG resulted in the down regulation of gap junction permeability to Lucifer Yellow CH in a manner similar to that found for Ca^{2+}. Treatment of these cells with TPA gave similar results. Furthermore, the effects of DG could be blocked by the addition of 8-N,N-(diethyl-amino)octyl-3,4,5-trimethoxybenzoate, a blocker of internal Ca^{2+} mobilization.

Other parameters that have been shown to influence gap junctional conductance are cytosolic pH and transjunctional voltage (222). Many types of gap junctions have now been shown to be closed by acidification of the cytoplasm. This phenomenon is thought to involve the titration of charged groups on the junctional proteins that could bring about a conformational change which results in channel closure. Voltage gating has received considerable attention, especially in terms of its effects during the early events of embryogenesis (20, 199, 222). At present, neither the effect of cytoplasmic pH nor the influence of junctional voltage has been investigated in terms of plasmodesmatal properties. Clearly, such experiments must be conducted to ascertain the full extent of the common control features used by animal and plant cells to regulate their intercellular communication pathways.

Role of Symplasmic Domains in Development

Analysis of dye movement has been used as a qualitative assessment of the extent of symplasmic continuity between cells of the same tissue and those of neighboring tissues. The concept of symplasmic domains within a particular plant received experimental support from the work of Erwee & Goodwin (55). They used molecular dyes to establish the size exclusion limits at various locations on plants of *Egeria densa*. Plasmodesmata within the apex of this plant had the highest size exclusion limits (749 Da), with epidermal cells of the leaf being next (674 Da), followed by epidermal and cortical cells of the stem and root (376 Da). Although plasmodesmata connected the cells of the epidermis and cortex, there was a barrier to dye movement between these cell types. Similar barriers were discovered at the nodes and were found to be correlated with development.

The onset of synchronous cell divisions in the apex of *Silene coeli-rosa* that can be induced by a long-day photoperiod has also been correlated with a down regulation of plasmodesmatal size exclusion limits (83). A direct correlation between this down regulation of the symplasmic pathway and the developmental events that give rise to flowering in *S. coeli-rosa* could not be established (207). However, the mitotic index of shoot apices from long-day-induced plants was significantly increased over that of control (short-day) plants, and this change was highly correlated with the down regulation of the symplasmic pathway within these apices. Although these studies are quite limited in terms of the scope of plant material investigated, they suggest that dye-coupling experiments may provide important insights into the cellular processes that underlie development.

SYMPLASMIC TRANSPORT AND PLASMODESMATA

Phloem Loading

A fundamental issue surrounding the symplasmic concept is whether the plasmodesmata within the cells of a mature leaf provide a continuous symplasmic route from the mesophyll to the lumen of the sieve element. It has been argued that as phloem translocation is driven by a pressure gradient, the presence of functional plasmodesmata between the sieve element–companion cell complex and the surrounding phloem parenchyma or bundle sheath cells would constitute a low-resistance shunt. Thus, it may be that the plasmodesmata that are present within the walls of these cells are vestigial, in that they were necessary for the development of the phloem cells but became down regulated following the sink-to-source transition (76). Consequently, loading of sugars into the phloem would involve an apoplasmic step. Several recent reviews on this specific topic are available (38, 73, 138, 143, 245, 250, 251), and so we will restrict our attention to the question of whether the plasmodesmata are down regulated.

The technical difficulties associated with placing a micropipette into the cytoplasm of a mesophyll cell located with a mature, intact, functional leaf were overcome by utilizing the large size of the vacuole. Madore et al (144) encapsidated the fluorescent, membrane-impermeant dye Lucifer Yellow CH into 0.2-μm phosphatidylserine liposomes, which were then loaded into a micropipette (diameter, 0.5 μm). A small portion of the lower epidermis was removed to allow the micropipette to be inserted into the vacuole of a mesophyll cell, and then pressure pulses were applied to eject the liposomes into the vacuole. The low pH and high Ca^{2+} present in the vacuole in combination with the composition of the liposome permitted fusion between the tonoplast and the liposome, thereby delivering the Lucifer Yellow CH to the cytosol.

Fluorescence microscopy performed on source leaves of *Ipomoea tricolor* indicated that Lucifer Yellow moved symplasmically from the site of injection to the cells of the minor veins. An ultrastructural study conducted on the same leaves indicated that plasmodesmata interconnect all cell types within the leaf; however, attempts to locate the dye within the sieve elements of sectioned material proved to be difficult. Consequently, although Madore et al (144) established that molecular dyes could move to the vascular bundle and that subsequent movement took place preferentially along the minor vein (see also 56), the question of symplasmic continuity with the sieve elements remained conjectural. A further point established by these studies was that apoplasmic pH values above 8.0 prevented Lucifer Yellow from moving into or along the minor veins. The corollary to this experiment was that such pH treatments also prevented the transfer of photosynthate from the mesophyll to the minor veins (142).

Ultrastructural studies on the mature leaf of *Coleus blumei* established that plasmodesmata were distributed among all cell types, but were especially abundant at the bundle sheath/intermediary (companion) cell interface (64). In view of these findings, Fisher (65) also conducted Lucifer Yellow injection studies to determine whether dye could move from the mesophyll to the intermediary cells. Again, although in vivo light microscopy established that Lucifer Yellow could move up to and along the veins, an examination of fixed and embedded tissue failed to confirm that dye had penetrated beyond the bundle sheath. Since Wang & Canny (257) have established that recently fixed carbon can be translocated over considerable distances (up to 500 μm) by cells other than the sieve tubes, extreme caution must be taken when interpreting fluorescence profiles.

Microsurgery on the upper and lower epidermis of *Commelina benghalensis*, followed by a wall-loosening treatment with pectinase allowed Van Kesteren et al (252) to produce "free-lying viable minor veins with small lumps of mesophyll cells attached." Fluorescein was iontophoresed into either mesophyll or mestome sheath cells, and continuous monitoring of dye movement allowed these workers to identify the exact pathway; the dye appeared to move freely between mesophyll cells as well as through the mestome sheath into the sieve tubes. These results are supported by a subsequent study in which fluorescent dyes were iontophoresed into intermediary cells of the minor veins of *Cucurbita pepo* leaves. In these experiments, Turgeon & Hepler (246) removed a small portion of the lower epidermis and then abraded the mesophyll tissue to gain access to the cells of the minor vein. Lucifer Yellow or carboxyfluorescein was able to move into neighboring intermediary cells as well as to the bundle sheath and mesophyll cells.

Based on current dye-coupling experiments performed on intact and surgically modified leaf tissues, it would appear that plasmodesmata establish a symplasmic pathway that interconnects the cytosol of the mesophyll to the long-distance pathway of the phloem. Clearly, these dye-coupling results do not establish that the process(es) involved in phloem loading occur via this symplasmic pathway. However, it is obvious that future modeling of the phloem system must incorporate the finding that plasmodesmata within the vascular bundle are not vestigial.

Plasmodesmata and C_4 Photosynthesis

The special spatial attributes associated with the C_4 carbon fixation pathway implicate the involvement of plasmodesmata as the most likely pathway for the intercellular trafficking of metabolites such as malate, aspartate, alanine, pyruvate, phosphoenolpyruvate, 3-phosphoglycerate, and triose phosphate (for a review of this concept, see 96). Ultrastructural studies of C_4 plants have established that numerous plasmodesmata are located in the adjoining walls of the mesophyll and bundle sheath cells (61, 67). Several strategies have been used in an attempt to evaluate the role of plasmodesmata as an appropriate diffusion pathway for the supply of these substrates. Rapid-fractionation techniques that permit the separation of mesophyll and bundle sheath cells (130) have established that significant concentration gradients exist between these cell types (130, 227). In *Zea mays*, a 10–15 mM diffusion gradient for triose phosphate is directed from the mesophyll to the bundle sheath cells, where starch synthesis takes place (70). Such gradients provide evidence that plasmodesmata constitute the most likely pathway for the intercellular movement of metabolites (96).

A more direct method for the study of bundle sheath plasmodesmata was recently developed by Burnell & Hatch (17, 18). It has long been recognized that isolated bundle sheath cells are capable of rapid response to added biochemical substrates (21, 95), yet protoplasts isolated from either the mesophyll (35) or bundle sheath cells (260) are relatively impermeable to the same molecules. Photosynthetic studies performed on bundle sheath cell strands isolated from the C$_4$ plant *Urochloa panicoides* established that biochemical substrates as large as nucleotides (ATP or ADP) were rapidly transferred between the suspension medium and the cytoplasm (17, 18). Burnell & Hatch (17) suggested that during the physical isolation of the bundle sheath cell strands the plasmodesmata remain functional.

Burnell (16) further investigated this possibility by using an ingenious approach in which the sensitivity of bundle sheath cytosolic enzymes to inhibitors was investigated. The basis of these experiments was the knowledge that in C$_4$ plants alanine aminotransferase is located in the cytosol of bundle sheath cells, and that Reactive Yellow 2 is a potent inhibitor of this enzyme. An increasing molecular weight series of Reactive Yellow 2 dye derivatives was used to establish the molecular size exclusion limits of these "opened" plasmodesmata (Figure 5). The correlation between this biochemically established size exclusion limit of 870 Da and the 800–900-Da range established by dye-coupling experiments offers strong support for the conclusion that plasmodesmata of isolated bundle sheath cells provide a diffusion pathway for metabolite transfer. As Burnell stressed (16), such a system provides an important alternative method for studying plasmodesmata (see also 260).

The intriguing question that remains is why does the bundle sheath plasmodesma remain open once these cells have been physically separated from their neighboring mesophyll cells? Obviously $\beta(1{\to}3)$-D-glucan synthase activity must be either down-regulated or absent from the neck constriction region of these plasmodesmata, or such physical wounding would cause callose deposition. During plasmolysis the plasma membrane seals at both ends of the plasmodesmata, and one would have expected this to have taken place on the mesophyll side of the wound. Clearly, something must have kept the plasma membrane and the desmotubule in place at the orifice of the plasmodesma so that membrane fusion could not take place. A possible explanation can be found from work on protoplasts, where with certain species half-plasmodesmata have been observed projecting out from the protoplast surface (P. Olesen, personal communication). In such a situation,

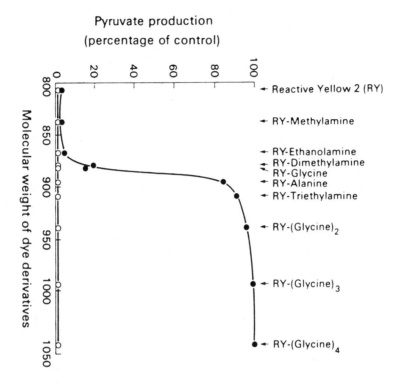

Figure 5   Inhibition of aniline aminotransferase in isolated bundle sheath cells of *U. panicoides* by Reactive Yellow 2 dye derivatives of different molecular weights. Pyruvate production was measured both in bundle sheath cells (●) and in extracts of bundle sheath cells (○). [From (16), with permission of the publisher.]

electron-opaque material is observed at the orifice of the plasmodesmata and appears to form a scaffolding between the plasma membrane and the ER. If such structures were present at the ends of the plasmodesmata that interconnect C_4 mesophyll and bundle sheath cells, the added stability may allow the maintenance of a functional diffusion pathway, as the plasma membrane could not seal across this substructural component. An electron microscopy study of this system may provide valuable supporting information.

VIRUS–PLASMODESMA INTERACTION

Ultrastructural Studies

Numerous cytological studies have been conducted on plant tissues that have been infected by plant viruses and, on the basis of these investigations, it is generally accepted that the systemic spread of virus infection occurs through plasmodesmata (57, 106, 156, 271). As the particles of most viruses are either globular (icosahedral with a diameter of 18–80 nm), helical, or filamentous rods (rigid or flexuous; diameters ranging from 10 to 25 nm and lengths up to 2.5 μm), with a physical size that is considerably greater than the associated molecular size exclusion limit(s) of the cytoplasmic annulus, viruses must interact with plasmodesmata to gain access to neighboring cells. In some cases, as with dahlia mosaic virus (DMV)-infected tissue, some of the plasmodesmata undergo significant substructural modification, in that the desmotubule (axial component) is removed and the cross-sectional diameter is increased to approximately 60–80 nm. In addition, tubular extensions of the cell wall develop at the cytoplasmic boundaries of individual plasmodesma, thereby increasing the physical length of the modified cytoplasmic sleeve. These structural changes allow the isometric DMV particles (diameter, 50 nm) to pass from cell to cell through plasmodesmata (118).

Systemic infection by cauliflower mosaic virus (CaMV) on host plants such as turnip (132) or Chinese cabbage (29) results in a DMV-like modification to the plasmodesmata interconnecting mesophyll cells (see also 203). However, unlike infection with DMV, all of the plasmodesmata in CaMV-infected turnip mesophyll cells are modified (132). The cell wall projections or plasmodesmatal protuberances induced by CaMV infection are present on one or both sides of the modified plasmodesma and may extend up to 800 nm into the cytoplasm. Other structural modifications have been observed, in conjunction with infection by viruses that cause the production of wall protuberances, include branching of plasmodesmata and the formation of a large central (median) cavity (1, 98). It is interesting that virus particles have always been observed within these structurally modified plasmodesmata. This correlation led to the general concept that removal of the desmotubule was an essential prerequisite for systemic infection (59).

An additional cytological feature of interest, with respect to plasmodesma-mediated viral movement, relates to the production of inclusions bodies, or cylinders, that become closely associated with the neck region of the plasmodesma. These inclusions are encoded by the genome of the infecting virus (47), and one generally finds virus particles in close association with both these structures and the neighboring plasmodesma (128, 129, 149), but this is not always the case (19). It has been suggested that these cylindrical inclusions may function to align elongated rod or flexuous virions with the cytoplasmic annulus of the plasmodesma to permit penetration to a neighboring cell (128). If this hypothesis were correct, it would mean that these virally encoded structures contain a molecular recognition sequence that identifies a constituent of the plasmodesmatal orifice or neck constriction region (174). Future studies aimed at further characterizing the viral gene product responsible for these inclusions may potentiate the immunolocalization and subsequent purification of the putative plant-encoded proteins of the neck constriction/cytoplasm interface. Information of this nature would be of immense value in advancing current models of plasmodesmatal structure.

Substructural modifications to plasmodesmata, mediated by viruses, should significantly increase the molecular size exclusion limit of the molecules moving between infected cells. Such modifications to molecular traffic may well be responsible for some of the symptoms that develop during infection. In addition, a perturbation of the cell–cell communication system of the infected plant may serve, in itself, as a signal to indicate wounding and thereby limit the ability of the virus to modify plasmodesmata of neighboring cells. Thus, from a viewpoint of systemic infectivity, one may speculate that the optimal virus–plasmodesma interaction would be one in which the virus genome passes through the plasmodesma with a minimal influence on the substructure and consequent molecular size exclusion limit of the plasmodesma. An example of a virus that may well be approaching this goal would be the tobacco mosaic virus (TMV) group.

Early ultrastructural studies conducted on TMV-infected tissue failed to identify TMV particles within the plasmodesmata. Esau (57) dismissed this apparent contradiction to the general concept that rod-type viruses pass through modified plasmodesmata on the grounds that insufficient material had been studied. However, Weintraub et al (261) later pointed out that more than 100 ultrastructural studies were conducted on TMV-infected tissue from 1950 to 1975; all studies failed to provide evidence that the TMV–plasmodesma interaction results in removal of the desmotubule or any modification to the ER–plasmodesma association. Weintraub et al (261) did, however, report that in TMV-infected tobacco, viral particles were observed in plasmodesmata where the desmotubule was apparently displaced, but its association with the ER was not affected. As they pointed out, the location of these TMV particles could have been an artifact of tissue preparation. The possibility that such observations represent situations in which an infected cell is dying should not be ignored. Subsequent studies on TMV have confirmed that in newly infected tissues no major substructural changes to the plasmodesmata are detected (218; W. J. Lucas & T. C. Pesacreta, unpublished results). The method by which genomic material of TMV may move through plasmodesmata is addressed in the next section.

A further illustration of the complexity of the interaction between viruses and plasmodesmata can be found by examining studies conducted on viruses that have been shown to be restricted to the tissues of the phloem. Such viruses are usually insect transmitted (by aphids or white flies), and upon delivery into the sieve element they are carried along with the translocation stream. Esau et al (58) have shown that viruses of this type (e.g. the beet yellows virus) can move into and replicate in phloem parenchyma cells. These viruses also cause similar substructural modifications to the plasmodesmata of the sieve elements and phloem parenchyma, but the extent of these modifications is restricted to the phloem tissues (33). Thus, the plasmodesmata between adjoining phloem parenchyma and bundle sheath cells are not modified, whereas the plasmodesmata on the inner tangential walls of the same phloem parenchyma cell have lost their desmotubules and their close association with the ER.

These findings suggest that there is tissue specificity with respect to the cellular control processes that the plant uses to regulate plasmodesmatal properties. Such an interpretation is supported by virus complementation studies in which a helper virus, so called because it can move systemically in the host plant, potentiates the systemic spread of an otherwise phloem-limited virus. Barker (6) recently described such a situation in *Nicotiana clevelandii* in which potato Y potyvirus acted as the helper virus for potato leafroll luteovirus. Although the mechanism(s) by which these symplasmic barriers to specific viral movement has not been fully resolved, the involvement of two specific virally encoded proteins has been implicated. As we will discuss shortly for TMV, one protein is thought to mediate in the short-distance cell-to-cell movement of the virus, while the other protein, which in some cases appears to be a special region on the coat protein (106), functions in mediating in the long-distance spread via the phloem.

One concept that has been clearly established by these ultrastructural studies performed on virally infected cells (tissues) is that plasmodesmata are not static entities but, rather, are dynamic structures that are capable of undergoing considerable morphological transformation. In general, the virus genome is rather simple, encoding only a few proteins, and so it should be possible to use molecular cloning to explore the basis for the virally induced changes to plasmodesmata. Possible sites of molecular interaction include modulation or deletion of the proteins within the cytoplasmic annulus and activation of cellulase enzymes that would mediate in reducing cell wall material to give rise to the observed increase in plasmodesmatal diameter.

As much of the cell wall material present in the plasmodesmata–cell wall protuberances appears to be callose (1, 43), the virus must be able to exert spatial and temporal control over $\beta(1\rightarrow3)$-D-glucan synthase and $(1\rightarrow3)$-β-D-glucanase enzymes. As these enzymes have been isolated and reconstituted (37, 160, 161), the effects of virally encoded proteins on the in vitro synthesis and degradation of callose can now be investigated.

There is presently no information on the mechanism by which the physical continuity of the ER and the desmotubule might be disrupted during virus infection. It is possible that the region in which the phospholipids of the ER bilayer undergo the transition to the tightly appressed cylindrical arrangement of the desmotubule represents a zone of membrane lipid instability. Interaction with a virally encoded product may allow the lipids in this zone to relax into a less compressed state, which could introduce a discontinuity into the membrane.

Molecular Studies

Details on the molecular mechanism(s) by which a certain virus effects its cell-to-cell and long-distance systemic spread, in a particular plant host, are presently the center of intense investigation. Although there was considerable speculation on likely modes of interaction, the pioneering studies conducted by Nishiguchi et al (165, 166) provided valuable insight into the location of the viral cistron that encodes viral capacity for cell-to-cell movement. These workers isolated a temperature-sensitive mutant of TMV (Ls1), which, under nonpermissive temperatures (32°C), replicated and assembled normally in both epidermal cells and protoplasts but did not move from cell to cell in inoculated tobacco leaves. Under permissive temperatures (22°C) the Ls1 mutant could spread systemically in inoculated tobacco plants. The mutation was thought to be located in a 30-kDa protein (P30) encoded by the TMV genome, and Atabekov & Morozov (4) speculated that this protein may be responsible for cell-to-cell movement of TMV; they called it the 30-kDa transport protein.

Experimental support for this hypothesis was provided by Leonard & Zaitlin (131). They found that two-dimensional tryptic peptide maps of in vitro translation products from the 30,000, 130,000, and 165,000 M_r TMV RNA of Ls1 and the closely related normal TMV strain (L) differed only slightly in the 30-kDa transport protein. After this study, several groups cloned either the entire TMV RNA sequence (81) or the 30-kDa protein cistron of various strains of TMV (154, 155, 175, 231). A comparison of the deduced amino acid sequences of the 30-kDa proteins of the tomato strain (L) and the common strain (OM) revealed that the N-terminal regions of the two proteins were highly conserved, but that they became highly divergent in the regions near the C terminus. Ohno et al (169) then determined the nucleic acid sequence of the 30-kDa protein cistron of Ls1 and compared it with the L strain 30-kDa protein: only one amino acid substitution was found between the two sequences; at amino acid residue 154, substitution of proline for serine results in the Ls1 phenotype.

Complementation studies in which a helper virus was used to overcome the cell-to-cell movement defect in Ls1 further supported the hypothesis that the 30-kDa protein was responsible for the spread of infection. When Ls1 was coinoculated with a common strain (like *T. vulgare* or U2), Ls1 spread through the symplasm. Of equal interest, coinoculation of Ls1 with an evolutionarily very different virus (potato virus X) also allowed Ls1 to move under nonpermissive temperatures (3, 232). This may indicate that the underlying mechanism of cell-to-cell movement may be common to many groups of viruses (157).

Antibodies directed against a synthetic peptide corresponding to the C-terminal amino acid sequence of the TMV 30-kDa protein cistron were used to immunoprecipitate proteins extracted from TMV-infected tobacco plants (112, 117, 176). Only a single protein with an M_r of 30,000 was precipitated, and in tobacco protoplasts and inoculated leaves this protein was synthesized early in the infection process. Protein levels began to increase approximately 8–10 h after infection, rising to a maximum at about 24 h and then declining over the next 48 h. The relationship between the infection process and the transient expression of the TMV 30-dDa transport protein suggests that changes to the plasmodesmata may be temporary. It is possible that for TMV, once the infection front has passed, the plasmodesmata return to normal before the viral genome undergoes encapsidation by the coat protein. This would explain why TMV particles are rarely, if ever, seen in plasmodesmata.

The cellular location of the putative transport protein has been investigated by using antibodies directed against various regions on the C terminus of the 30-kDa TMV protein. Watanabe et al (258) found that in tobacco protoplasts inoculated with TMV, the protein accumulated in the nucleus. However, this experimental system is not really appropriate, as it has no potential for systemic movement. Tomenius et al (237) used immunogold cytochemistry to localize the 30-kDa protein in infected tobacco leaves, and they reported antigenic reactions located over the plasmodesmata. Furthermore, they found that their immunogold-labeling patterns paralleled the time course of 30-kDa protein synthesis and turnover.

To further explore the mechanism whereby TMV gains access to the symplasm, Deom et al (40) constructed a chimeric gene encoding the TMV 30-kDa protein, and this construct was introduced into tobacco plants by using a modified Ti plasmid in *Agrobacterium tumefaciens*. Expression of the 30-kDa protein in transgenic plants allowed the function of this gene product to be explored in the absence of the expression of other TMV genes. Complementation studies performed on transgenic tobacco plants, maintained at the nonpermissive temperature for Ls1, revealed that the Ls1 mutant could move from cell to cell in both the inoculated and upper systemic leaves. Furthermore, the 30-kDa protein was found to be associated with cell wall fractions that were obtained from either transgenic (40) or virus-infected (162) plants, findings that are consistent with the immunolocalization studies of Tomenius et al (237). In a related study, Meshi et al (157) constructed a range of TMV mutants by using various frame shifts in the 30-kDa cistron of the TMV genome. All such mutants were replication competent, but none were capable of establishing a systemic infection when inoculated onto tobacco plants. Collectively, these results provide unequivocal support for the hypothesis that the TMV 30-kDa protein is an essential component for cell-to-cell movement.

Since the expression of the TMV 30-kDa protein in transgenic tobacco plants potentiated cell-to-cell movement of Ls1, Wolf et al (266) studied the influence of this protein on the molecular size exclusion limit of plasmodesmata in these plants. These experiments were performed by using the liposome delivery system of Madore et al (144) to introduce fluorescent molecules into mesophyll cells within intact mature tobacco leaves. The size exclusion limit for cell-to-cell mobility in control tobacco plants was found to be in the 750-Da range; this value is consistent with the limits established for a range of plants (53, 54, 234, 240, 243). Wolf et al (266) found that in the transgenic tobacco plants, large fluorescein-labeled dextran molecules could move from cell to cell; the exclusion limit was greater than 9,400 Da but less than 17,200 Da. Studies performed on developing leaves of these transgenic plants revealed that the increase in the plasmodesmatal size exclusion limit does not occur until a leaf achieves a critical size (41).

To assess the importance of this 10-fold increase in the plasmodesmatal size exclusion limit, in terms of potentiating cell-to-cell viral movement, requires a comparison between the molecular dimensions of the dextran probes and TMV. Wolf et al (266) assumed that the dextrans used in their experiments existed in the cytoplasm as random coils, and from this they deduced that the size exclusion limit of plasmodesmata in control plants would be equivalent to molecules having an effective Stokes radius of 0.73 nm, whereas this value would be in the range from 2.4 to 3.1 nm in the 30-kDa transgenic plants. TMV particles are 18 nm in diameter and 300 nm long, and so the observed change in the plasmodesmatal properties would be insufficient to allow particles to move from cell to cell, a finding consistent with earlier ultrastructural studies conducted on TMV-infected tissues. The molecular dimensions of TMV-RNA have yet to be established, but the size may be within the 10-nm range (77). If this value were correct, the increase in plasmodesmatal size exclusion limit caused by the 30-kDa protein would be insufficient to allow TMV-RNA to pass from cell to cell by diffusion.

Dorokhov et al (45) reported isolating a new informosomelike ribonucleoprotein particle from TMV-infected tobacco plants (vRNP), and they proposed that these filamentous particles may be the infective form that moves from cell to cell. Studies with temperature-sensitive TMV mutants supported this contention in that systemic spread and detection of vRNP were highly correlated (46). Furthermore, recently constructed mutants of TMV in which the coat protein cistron was either deleted or rendered nonfunctional were still capable of cell-to-cell movement (34, 157), thereby establishing that encapsidation is not essential for cell-to-cell movement. Only when the effective Stokes radius of these vRNP filaments becomes available will we be able to establish whether a mechanism more complex than diffusion is involved in TMV movement through plasmodesmata (2, 3). In any event, considering the information currently available, the proposal that the 30-kDa transport protein forms a special transport complex with virus-specific RNAs appears to be worthy of close attention (232). Finally, in view of the present state of our knowledge in this area, the term transport protein can be used interchangeably with movement protein, although the latter more accurately reflects the known function of the TMV-encoded 30-kDa protein.

Several putative movement proteins in other viral systems, including CaMV and alfalfa mosaic virus (AlMV), have now been studied (for a review of this literature, see 106). It is interesting that immunolocalization studies in these systems have identified different regions of the cell wall as likely target sites for these virally-encoded proteins. Linstead et al (132) found that in turnip leaves infected with CaMV, the gene 1 product [P1 protein proposed to be the systemic spread factor (see also 107)] was located in the regions around the modified plasmodesmata of mesophyll and phloem parenchyma cells. According to their interpretation of the immunolocalization results, the P1 protein appeared to be located in the extracellular space. Here we should stress that plant fractionation studies conducted on CaMV-infected plants revealed that antibodies raised against the P1 cistron reacted with at least five infection-specific proteins (148). It will be very interesting to see whether any of these proteins has cellulase or glucanase activity, as one of the features of CaMV-modified plasmodesmata is an increase in overall diameter.

Godefroy-Colburn et al (79) and Stussi-Garaud et al (229), on the other hand, reported that during the early events of AlMV infection in tobacco, a 32-kDa nonstructural protein (P3) that is thought to be involved in systemic infection (163) was not detected in the vicinity of the plasmodesmata. Rather, the protein accumulated in the middle lamella and remained in the wall throughout infection (10). Clearly, these results further illustrate the complexities of the potential interactions between the virally encoded proteins that permit systemic spread and the constituents of the plant cell that are involved in regulating plasmodesmata.

The fundamental question that remains to be answered is what is the exact molecular mechanism by which these putative movement proteins modify the infected plant cell to allow the viral genome to pass to neighboring cells? The main focus has always been on plasmodesmata, but Atabekov and Dorokhov (3) point out that an alternative mechanism may involve the suppression of a plant defense response. Virally encoded proteins that accumulate in the nucleus may represent an example of this second strategy (141). Although we can envisage a mechanism whereby a viral protein could act to regulate gene transcription in the host plant, we have difficulty in seeing how repression of the host resistance response could give rise to viral movement without in some way modifying plasmodesmata.

Clearly, different viruses have evolved alternative mechanisms by which to interact with and modify plasmodesmata for cell-to-cell movement. As a corollary, plants have evolved a range of defense mechanisms that enable them to prevent, or limit, the ability of a particular virus to modify plasmodesmata. One way to elucidate the exact cellular processes involved in these various viral strategies would be to examine the genetic sequences directly involved in host–pathogen interactions. We will restrict our attention to the putative movement proteins, owing to limitations of space, but we stress that resolving the complexities that underlie resistance (198) will also provide important information on the cellular processes involved in controlling plasmodesmata. In this regard, there is considerable experimental evidence linking callose formation to the local lesion or hypersensitive response (43, 63, 204, 267).

Putative movement protein genes for a number of evolutionarily related and divergent viruses have now been cloned and sequenced (14, 30, 31, 81, 136, 154, 155, 206, 231). Unfortunately, no secondary or tertiary structural information is presently available for any of these proteins. Although a high degree of similarity has been reported between the amino acid sequences of some of the nonstructural virally encoded proteins, these specific proteins are thought to be involved in replication rather than cell-to-cell movement (94).

In general, little similarity has been found between the amino acid sequences of most of the known movement proteins. Of the four tobamovirus movement proteins that have been studied, a high degree of homology has been identified only between closely related strains (206). Lucas et al (139) reported 79% amino acid homology between the TMV common strain and the TMV tomato strain (L), but the level of homology between TMV and the related sunnhemp mosaic virus and cucumber green mosaic virus fell to 34–41%. Two regions located in the N-terminal four-fifths of the tobamovirus movement protein, denoted I and II by Saito et al (206), appear to be highly conserved and consequently may be important for movement function. Region II contains the mutation sites for temperature-sensitive movement in tobacco plants. Saito et al (206) also pointed out that although the C-terminal region of the TMV movement proteins reflected a low degree of amino acid homology, this region could be divided into three subregions on the basis of charged amino acid residues. The importance of these subregions has yet to be established (e.g. 31), but it may be of interest that this region of the protein is highly hydrophilic, while the rest of the protein is very hydrophobic (139).

Comparative studies performed on the predicted amino acid sequences of AlMV, brome mosaic virus (BMV), and TMV movement proteins indicated that no significant homology could be detected (30). This is interesting, as the predicted molecular masses of these proteins are very similar (32, 32, and 30 kDa for AlMV, BMV, and TMV, respectively). Furthermore, AlMV has a temperature sensitivity mutation in its movement protein that is comparable with the TMV Ls1 (105). This lack of amino acid homology may indicate that numerous sites are available by which plasmodesmatal function can be modified to allow the passage of viral genomic material. Alternatively, these movement proteins may all elicit a common response, in terms of causing a minor perturbation to the molecular size exclusion limits of plasmodesmata, but may do so by interacting at very different sites on the cellular pathway that controls the turnover of plasmodesmatal constituents. It has long been known that the active site for the host hypersensitive response in tobacco maps to the movement protein cistron (114, 274). Meshi et al (153) recently established that this is also true for the tomato Tm-2 resistance gene. Elucidating the action of this plant gene should aid considerably in the development of a model to explain the mode of action of the viral movement proteins.

DYNAMICS OF PLASMODESMATA

The current literature provides ample indication that plasmodesmata are dynamic entities. Substructural details vary markedly and depend very much upon the specific tissue being examined, as well as the developmental and physiological status of the cells therein. The extent of the plant's genetic plasticity, in terms of controlling frequency, as well as events associated with dedifferentiation and the construction of secondary plasmodesmata, are only now becoming apparent. This new framework, along with the concept that the plasma membrane is a dynamic structure that is constantly being turned over (224), enables us to better envisage the ways by which the plant may modify its plasmodesmata.

At present, little is known about the nature of the putative proteins that are thought to regulate the size exclusion limits of the cytoplasmic sleeve. Progress will be attendant upon the development of a protocol that will enable the isolation and purification of these substructural entities. In that the absolute amount of material available from an individual cell will be incredibly small, an immunopurification scheme may be the best way to approach this problem. Since genetic and molecular biology studies on certain viruses have established that specific proteins potentiate the movement of a virus through plasmodesmata of the host, these proteins may have domains that interact with the substructural elements of either the orifice or the neck constriction region of the plasmodesmata. Identification of these putative sites would potentiate the preparation of antibodies that could then be used to immunoprecipitate the plant material. There is an equally important need to obtain information on the dynamics of the membranous material that forms the desmotubule. More extensive application of the fluorescence photobleaching technique, as pioneered by Schindler and co-workers (9, 256), should provide important information on the dynamics of the ER–desmotubule association. Ideally, these photobleaching studies should use fluorescent molecules that can be inserted, preferentially, into the ER or plasma membrane from the cytosol.

Finally, the concept that callose is involved in establishing dynamic control over the cytoplasmic annulus within the neck constriction region warrants further investigation.

CONCLUDING REMARKS

Over the past decade considerable progress has been made in terms of elucidating structural and functional aspects of plasmodesmata. Although there is now a substantial body of information on plasmodesmatal formation, structure, and tissue-specific variation in substructure, there is a pressing need to obtain information on the dynamic aspects of plasmodesmata. Such details, together with a further refinement of our understanding of the physiological roles played by plasmodesmata, will establish the necessary framework for future studies on the role of cell-to-cell communication in orchestrating plant development.

Although plasmodesmata and animal gap junctions exhibit certain parallels in terms of their physiological size exclusion limits, they possess no structural similarities. However, it appears highly likely that these two divergent cell-to-cell communication systems have evolved, by completely separate pathways, to permit the molecular and biochemical coordination of cells within specialized tissues. It will be of great interest to determine the cellular mechanisms used by these two systems to control the size exclusion limits of their cell-to-cell junctions) and, in particular, to identify common attributes. Studies of this kind may well establish a new operational framework for comparative analysis of developmental systems in plants and animals.

ACKNOWLEDGMENTS

Research in our laboratories has been supported by grants from the AFRC and SERC (to A.W.R.) and the National Science Foundation (grants PCM 83-15408 and DMB-87-03624 to W.J.L.). We also thank all those colleagues who kindly provided us with unpublished information and preprints.

Literature Cited

. 1. Allison, A. V., Shalla, T. A. 1974. The ultrastructure of local lesions induced by potato virus: a sequence of cytological events in the course of infection. *Phytopathology* 64:784–93
2. Atabekov, J. G. 1977. Defective and satellite plant viruses. *Comp. Virol.* 11:143–200
3. Atabekov, J. G., Dorokhov, Yu. L. 1984. Plant virus-specific transport function and resistance of plants to viruses. *Adv. Virus Res.* 29:313–64

4. Atabekov, J. G., Morozov, S. Yu. 1979. Translation of plant virus messenger RNAs. *Adv. Virus Res.* 25:1–91
5. Barclay, G. F., Peterson, C. A., Tyree, M. T. 1982. Transport of fluorescein in trichomes of *Lycopersicon esculentum*. *Can. J. Bot.* 60:397–402
6. Barker, H. 1987. Invasion of non-phloem tissue in *Nicotiana clevelandii* by potato leafroll luteovirus is enhanced in plants also infected with potato Y potyvirus. *J. Gen. Virol.* 68:1223–27

7. Barnett, J. R. 1982. Plasmodesmata and pit development in secondary xylem elements. *Planta* 155:251–60

8. Barnett, J. R. 1987. Changes in the distribution of plasmodesmata in developing fibre-tracheid pit membranes of *Sorbus aucuparia* L. *Ann. Bot.* 59:269–79

9. Baron-Epel, O., Hernandez, D., Jiang, L. -W., Meiners, S., Schindler, M. 1988. Dynamic continuity of cytoplasmic and membrane compartments between plant cells. *J. Cell Biol.* 106:715–21

10. Berna, A., Briand, J. -P., Stussi-Garaud, C., Godefroy-Colburn, T. 1986. Kinetics of accumulation of the three non-structural proteins of alfalfa mosaic virus in tobacco plants. *J. Gen. Virology* 67:1135–47

11. Berridge, M. J. 1987. Inositol trisphosphate and diacylglycerol: two interacting second messengers. *Annu. Rev. Biochem.* 56:159–228

12. Binding, H., Witt, D., Monzer, J., Mordhorst, G., Kollmann, R. 1987. Plant cell graft chimeras obtained by coculture of isolated protoplasts. *Protoplasma* 141:64–73

13. Blake, J. R. 1978. On the hydrodynamics of plasmodesmata. *J. Theor. Biol.* 74:33–47

14. Boccara, M., Hamilton, W. D. O., Baulocombe, D. C. 1986. The organisation and interviral homologies of genes at the 3' end of tobacco rattle virus RNA1. *EMBO J.* 5:223–29

15. Botha, C. E. J., Evert, R. F. 1988. Plasmodesmatal distribution and frequency in vascular bundles and contiguous tissues of the leaf of *Themeda triandra*. *Planta* 173:433–41

16. Burnell, J. N. 1988. An enzymic method for measuring the molecular weight exclusion limit of plasmodesmata of bundle sheath cells of C_4 plants. *J. Exp. Bot.* 39:1575–80

17. Burnell, J. N., Hatch, M. D. 1988a. Photosynthesis in phosphoenolpyruvate carboxykinase-type C_4 plants: photosynthetic activities of isolated bundle sheath cells from *Urochloa panicoides*. *Arch. Biochem. Biophys.* 260:177–86

18. Burnell, J. N., Hatch, M. D. 1988b. Photosynthesis in phosphoenolpyruvate carboxykinase-type C_4 plants: pathways of C_4 acid decarboxylation in bundle sheath cells of *Urochloa panicoides*. *Arch. Biochem. Biophys.* 260:187–99

19. Castellano, M. A., Di Franco, A., Martelli, G. P. 1987. Electron microscopy of two olive viruses in host tissues. *J. Submicrosc. Cytol.* 19:495–508

20. Caveney, S. 1990. Patterns of junctional permeability in developing insect tissues. See Ref. 194

21. Chapman, K. S. R., Berry, J. A., Hatch, M. D. 1980. Photosynthetic metabolism in the bundle sheath cells of the C_4 species *Zea mays*. *Arch. Biochem. Biophys.* 202:330–41

22. Chapman, R. L., Good, B. H. 1978. Ultrastructure of plasmodesmata and cross walls in *Cephaleuros, Phycopeltis* and *Trentepholia* (Chroolepidaceae; Chlorophyta). *Br. Phycol. J.* 13:241–46

23. Chapman, R. L., Henk, M. C. 1985. Observations on the habit, morphology and ultrastructure of *Cephaleuros parasiticus*, Chlorophyta, and a comparison with *Cephaleuros virescens*. *J. Phycol.* 21:513–22

24. Chapman, R. L., Henk, M. C. 1986. Phragmoplasts in cytokinesis of *Cephaleuros parasiticus*, Chlorophyta, vegetative cells. *J. Phycol.* 22:83–88

25. Chappell, D. F., Stewart, K. D., Mattox, K. R. 1978. On pits and plasmodesmata of Trentepholiacean algae, Chlorophyta. *Trans. Am. Microsc. Soc.* 97:88–94

26. Cheng, K. C., Nie, X. W., Chen, S. W., Jian, L. C., Sun, L. H., Sun, D. L. 1987. Studies on the secondary formation of plasmodesmata between the pollen mother cells of lily before cytomixis. *Acta Biol. Exp. Sinica* 20:1–11

27. Clarkson, D. T., Robards, A. W., Stephens, J. E., Stark, M. 1987. Suberin lamellae in the hypodermis of *Zea mays* root development and factors affecting the permeability of hypodermal layers. *Plant Cell Environ.* 10:83–94

28. Conkling, M. A., Yamamoto, Y. T., Acedo, G. N. 1990. A tobacco root-specific gene homologous to the mammalian lens major intrinsic in soybean nodulin26 proteins. *Science.* Submitted

29. Conti, G. G., Vegetii, G., Bassi, M., Favali, M. A. 1972. Some ultrastructural and cytochemical observations on Chinese cabbage leaves infected with cauliflower mosaic virus. *Virology* 47:694–700

30. Cornelissen, B. J. C., Bol, J. F. 1984. Homology between the proteins encoded by tobacco mosaic virus and two tricornviruses. *Plant Mol. Biol.* 3:379–84

31. Cornelissen, B. J. C., Linthorst, H. J. M., Brederode, F. T., Bol, J. F. 1986. Analysis of the genome structure of tobacco rattle virus strain PSG. *Nucleic Acids Res.* 14:2157–69

32. Côté, R., Thain, J. F., Fenson, D. S. 1987. Increase in electrical resistance of

plasmodesmata of *Chara* induced by an applied pressure gradient across nodes. *Can. J. Bot.* 65:509–11

33. D'Arcy, C. J., de Zoeten, G. A. 1979. Beet western yellows virus in phloem tissue of *Thlaspi arvense*. *Phytopathology* 69:1194–98

34. Dawson, W. O., Bubrick, P., Grantham, G. L. 1988. Modifications of the tobacco mosaic virus coat protein gene affecting replication, movement and symptomatology. *Phytopathology* 78:783–89

35. Day, D. A., Jenkins, C. L. D., Hatch, M. D. 1981. Isolation and properties of functional mesophyll protoplasts and chloroplasts from *Zea mays*. *Aust. J. Plant Physiol.* 8:21–30

36. Dell, B., Kuo, J., Burbidge, H. H. 1982. Anatomy of *Pilostyles hamiltonii* (Rafflesiaceae) on stems of *Daviesia*. *Aust. J. Bot.* 30:1–9

37. Delmer, D. P. 1987. Cellulose biosynthesis. *Annu. Rev. Plant. Physiol.* 38:259–90

38. Delrot, S. 1987. Phloem loading: apoplastic or symplastic? *Plant Physiol. Biochem.* 25:667–76

39. DeMello, W. C. 1984. Modulation of junctional permeability. *Fed. Proc.* 43:2692–96

40. Deom, C. M., Oliver, M. J., Beachy, R. N. 1987. The 30-kilodalton gene product of tobacco mosaic virus potentiates virus movement. *Science* 337:389–94

41. Deom, C. M., Schubert, K., Wolf, S., Holt, C., Lucas, W. J., Beachy, R. N. 1990. Molecular characterization and biological function of the movement protein of tobacco mosaic virus in transgenic plants. *Proc. Natl. Acad. Sci. USA.* In press

42. Didhevar, F., Baker, D. A. 1986. Localization of ATPase in sink tissues of *Ricinus communis*. *Ann. Bot.* 57:823–28

43. Di Franco, A., Martelli, G. P., Russo, M. 1983. An ultrastructural study of olive latent ringspot virus in *Gomphrena globosa*. *J. Submicrosc. Cytol.* 15:539–48

44. Ding, B., Parthasarathy, M. V., Niklas, K., Turgeon, R. 1988. A morphometric analysis of the phloem-unloading pathway in developing tobacco leaves. *Planta* 176:307–18

45. Dorokhov, Y. L., Alexandrova, N. M., Miroshnichenko, N. A., Atabekov, J. G. 1983. Isolation and analysis of virusspecific ribonucleoprotein of tobacco mosaic virus-infected tobacco. *Virology* 127:237–52

46. Dorokhov, Y. L., Alexandrova, N. M.,

Miroshnichenko, N. A., Atabekov, J. G. 1984. The informosome-like virusspecific ribonucleoprotein (vRNP) may be involved in the transport of tobacco mosaic virus infection. *Virology* 137:127–34

47. Dougherty, W. G., Carrington, J. C. 1988. Expression and function of potyviral gene products. *Annu. Rev. Phytopathol.* 26:123–43

48. Drake, G. 1979. Electrical coupling, potentials, and resistances in oat coleoptiles: effect of azide and cyanide. *J. Exp. Bot.* 30:719–25

49. Drake, G., Carr, D. J. 1978. Plasmodesmata, tropisms, and auxin transport. *J. Exp. Bot.* 29:1309–18

50. Eleftheriou, E. P. 1990. Plasmodesmatal structure and function in nectaries. See Ref. 194

51. Eleftheriou, E. P., Hall, J. L. 1983a. The extrafloral nectaries of cotton. I. Fine structure of the secretory papillae. *J. Exp. Bot.* 34:103–19

52. Eleftheriou, E. P., Hall, J. L. 1983b. The extrafloral nectaries of cotton. II. ATPase activity, Ca^{2+} binding sites and selective osmium impregnation. *J. Exp. Bot.* 34:1066–79

53. Erwee, M. G., Goodwin, P. B. 1983. Characterisation of the *Egeria densa* Planch. leaf symplast: inhibition of the intercellular movement of fluorescent probes by group II ions. *Planta* 158:320–28

54. Erwee, M. G., Goodwin, P. B. 1984. Characterization of the *Egeria densa* leaf symplast: response to plasmolysis, deplasmolysis and to aromatic amino acids. *Protoplasma* 22:162–68

55. Erwee, M. G., Goodwin, P. B. 1985. Symplastic domains in extrastelar tissues of *Egeria densa* Planch. *Planta* 163:9–19

56. Erwee, M. G., Goodwin, P. B., Van Bel, A. J. E. 1985. Cell-cell communication in the leaves of *Commelina cyanea* and other plants. *Plant Cell Env.* 8:173–78

57. Esau, K. 1968. *Viruses in Plant Hosts*. Madison, WI: Univ. Wisconsin Press. 225 pp.

58. Esau, K., Cronshaw, J., Hoefert, L. L. 1967. Relation of beet yellows virus to the phloem and to movement in the sieve tube. *J. Cell Biol.* 32:71–87

59. Esau, K., Hoefert, L. L. 1972. Ultrastructure of sugar beet leaves infected with beet western yellows virus. *J. Ultrastruct. Res.* 40:556–71

60. Esau, K., Thorsch, J. 1985. Sieve plate pores and plasmodesmata, the communication channels of the symplast: ul-

trastructural aspects and developmental relations. *Am. J. Bot.* 72:1641–53

61. Evert, R. F., Eschrich, W., Heyser, W. 1977. Distribution and structure of plasmodesmata in mesophyll and bundle-sheath cells of *Zea mays* L. *Planta* 136:77–89

62. Evert, R. F., Mierzwa, R. J. 1986. Pathway(s) of assimilate movement from mesophyll cells to sieve tubes in the *Beta vulgaris* leaf. In *Phloem Transport,* ed. J. Cronshaw, W. J. Lucas, R. T. Giaquinta, pp. 419–32. New York: Alan Liss

63. Favali, M. A., Conti, G. G., Bassi, M. 1978. Modifications of the vascular bundle ultrastructure in the "Resistant Zone" around necrotic lesions induced by tobacco mosaic virus. *Physiol. Plant Pathol.* 13:247–51

64. Fisher, D. G. 1986. Ultrastructure, plasmodesmatal frequency, and solute concentration in green areas of variegated *Coleus blumei* Benth. leaves. *Planta* 169:141–52

65. Fisher, D. G. 1988. Movement of lucifer yellow in leaves of *Coleus blumei* Benth. *Plant Cell Env.* 11:639–44

66. Fisher, D. G. 1990. Distribution of plasmodesmata in leaves. A comparison of *Cananga odorata* with other species using different measures of plasmodesmatal frequency. See Ref. 194

67. Fisher, D. G., Evert, R. F. 1982. Studies on the leaf of *Amaranthus retroflexus* (Amaranthaceae): ultrastructure, plasmodesmatal frequency, and solute concentration in relation to phloem loading. *Planta* 155:377–87

68. Flagg-Newton, J., Simpson, I., Loewenstein, W. R. 1979. Permeability of the cell-to-cell membrane channels in mammalian cell junction. *Science* 205:404–7

69. Franceschi, V. R., Lucas, W. J. 1982. The relationship of the charasome to chloride uptake in *Chara corallina:* physiological and histochemical investigations. *Planta* 154:525–37

70. Furbank, R. T., Stitt, M., Foyer, C. H. 1985. Intercellular compartmentation of sucrose synthesis in leaves of *Zea mays* L. *Planta* 164:172–78

71. Gainer, H. St. C., Murray, A. W. 1985. Diacylglycerol inhibits gap junctional communication in cultured epidermal cells: evidence for a role of protein kinase C. *Biochem. Biophys. Res. Commun.* 126:1109–13

72. Galatis, B., Mitrakos, K. 1980. The ultrastructural cytology of the differentiating guard cells of *Vigna sinensis. Am. J. Bot.* 67:1243–61

73. Gamalei, Y. V. 1986. Characteristics of phloem loading in woody and herbaceous plants. *Sov. Plant Physiol.* 32:656–65

74. Gamalei, Y. V. 1988. The structural and functional evolution of minor veins of the leaf. *Bot. Zh.* 73:1513–22

75. Gamalei, Y. V., Pakhomova, M. V. 1981. Distribution of plasmodesmata and parenchyma transport of assimilates in the leaves of several dicots. *Soviet Plant Physiol.* 28:649–61

76. Giaquinta, R. T. 1983. Phloem loading of sucrose. *Annu. Rev. Plant Physiol.* 34:347–87

77. Gibbs, A. 1976. Viruses and plasmodesmata. In *Intercellular Communication in Plants: Studies on Plasmodesmata,* ed. B. E. S. Gunning, A. W. Robards, pp. 149–63. Berlin: Springer-Verlag

78. Giddings, T. H. Jr., Staehelin, L. A. 1981. Observation of microplasmodesmata in both heterocyst-forming and non-heterocyst forming filamentous cyanobacteria by freeze-fracture electron microscopy. *Arch. Microbiol.* 129:295–98

79. Godefroy-Colburn, T., Gagey, M. -J., Berna, A., Stussi-Garaud, G. 1986. A non-structural protein of alfalfa mosaic virus in the walls of infected tobacco cells. *J. Gen. Virol.* 67:2233–39

80. Godlewski, M. 1988. Callose formation in injured cells of the vegetative and generative thallus of *Chara vulgaris* L. Absence of callose in the process of cytodifferentiation. *Acta Soc. Bot. Pol.* 57:21–30

81. Goelet, P., Lomonossoff, G. P., Butler, P. J. G., Akam, M. E., Gait, M. J., Karn, J. 1982. Nucleotide sequence of tobacco mosaic virus RNA. *Proc. Natl. Acad. Sci. USA* 79:5818–22

82. Goodwin, P. B. 1983. Molecular size limit for movement in the symplast of the *Elodea* leaf. *Planta* 157:124–30

83. Goodwin, P. B., Lyndon, R. F. 1983. Synchronisation of cell division during transition to flowering in *Silene apices* not due to increased symplast permeability. *Protoplasma* 116:219–22

84. Gubler, F., Ashford, A. E., Jacobsen, J. V. 1987. The release of alpha-amylase through gibberellin-treated barley aleurone cell walls: an immunocytochemical study with Lowicryl K4M. *Planta* 172:155–61

85. Gunning, B. E. S. 1978. Age-related and origin-related control of the numbers of plasmodesmata in cell walls of developing *Azolla* roots. *Planta* 143:181–90

86. Gunning, B. E. S., Hughes, J. E. 1976. Quantitative assessment of symplastic transport of pre-nectar into trichomes of *Abutilon* nectaries. *Aust. J. Plant Physiol.* 3:619–37

87. Gunning, B. E. S., Hughes, J. E., Hardham, A. R. 1978. Formative and proliferative divisions, cell differentiation and developmental changes in the meristem of *Azolla* roots. *Planta* 143:121–44

88. Gunning, B. E. S., Overall, R. L. 1983. Plasmodesmata and cell-to-cell transport in plants. *Bioscience* 33:260–65

89. Gunning, B. E. S., Pate, J. S., Minchin, F. R., Marks, I. 1974. Quantitative aspects of transfer cell structure in relation to vein loading in leaves and solute transport in legume nodules. *Symp. Soc. Exp. Biol.* 28:87–126

90. Gunning, B. E. S., Robards, A. W. 1976a. Plasmodesmata: current knowledge and outstanding problems. See Ref. 92, pp. 297–311

91. Gunning, B. E. S., Robards, A. W. 1976b. Plasmodesmata and symplastic transport. In *Transport and Transfer Processes in Plants,* ed. I. F. Wardlow, J. B. Passioura, pp. 15–41, New York and London: Academic

92. Gunning, B. E. S., Robards, A. W., eds. 1976c. *Intercellular Communication in Plants: Studies on Plasmodesmata.* Heidelberg: Springer-Verlag

93. Gunning, B. E. S., Robins, R. J., Juniper, B. E. 1981. Considerations of the estimation of plasmodesmatal frequencies: emendations to a previous paper by Robins and Juniper. *J. Theor. Biol.* 89:711–3

94. Haseloff, J., Goelet, P., Zimmern, D., Ahlquist, P., Dasgupta, R., Kaesberg, P. 1984. Striking similarities in amino acid sequence among nonstructural proteins encoded by RNA viruses that have dissimilar genome organization. *Proc. Natl. Acad. Sci. USA* 81:4358–62

95. Hatch, M. D., Kagawa, T. 1976. Photosynthetic activities of isolated bundle sheath cells in relation to differing mechanisms of C_4 photosynthesis. *Arch. Biochem. Biophys.* 175:39–53

96. Hatch, M. D., Osmond, C. B. 1976. Compartmentation and transport in C_4 photosynthesis. In *Encyclopedia of Plant Physiology. New Series, Transport in Plants,* ed. C. R. Stocking, U. Heber, 3: pp. 144–84. Berlin: Springer-Verlag

97. Hawes, C. R., Juniper, B. E., Horne, J. C. 1981. Low and high voltage electron microscopy of mitosis and cytokinesis in maize roots. *Planta* 152:397–407

98. Hearon, S. S., Lawson, R. H. 1981.

Effects of light intensity, photoperiod, and temperature on symptom expression and host and virus ultra-structure in *Saponaria vaccaria* infected with carnation etched ring virus. *Phytopathology* 71:645–52

99. Hensel, W. 1987. Caffeine-induced alterations in the generation and cytodifferentiation of root cap cells from cress—*Lepidium sativum,* L. *Eur. J. Cell Biol.* 43:208–14

100. Hepler, P. K. 1982. Endoplasmic reticulum in the formation of the cell plate and plasmodesmata. *Protoplasma* 111:121–33

101. Hertzberg, E. L., Johnson, R. G. 1988. *Gap Junctions.* New York: Alan R. Liss. In press

102. Hoch, H. C. 1978. Mycoparasitic relationships. IV. *Stephanoma phaeospora* parasitic on a species of *Fusarium. Mycologia* 70:370–79

103. Hughes, J. E. 1977. *Aspects of ultrastructure and function in* abutilon *nectaries.* M. Sc. Thesis, Australian Natl. Univ., Canberra

104. Hughes, J. E., Gunning, B. E. S. 1980. Glutaraldehyde-induced deposition of callose. *Can. J. Bot.* 58:250–58

105. Huisman, M. J., Sarachu, A. N., Ablas, F., Broxterman, H. J. G., Van Vloten-Doting, L., Bol, J. F. 1986. Alfalfa mosaic virus temperature-sensitive mutants. III. Mutants with a putative defect in cell-to-cell transport. *Virology* 154:401–4

106. Hull, R. 1989. The movement of viruses in plants. *Annu. Rev. Phytopathol.* 24:213–40

107. Hull, R., Covey, S. N. 1985. Cauliflower mosaic virus: pathways to infection. *BioEssays* 3:160–63

108. Iijima, T., Sibaoka, T. 1982. Propagation of action potential over the traplobes of *Aldrovanda vesiculosa. Plant Cell Physiol.* 23:679–88

109. Jeffree, C. E., Yeoman, M. M. 1983. Development of intercellular connections between opposing cells in a graft union. *New Phytol.* 93:491–509

110. Jones, M. G. K. 1976. The origin and development of plasmodesmata. See Ref. 92, pp. 81–105

111. Jongsma, H. J., Rook, M. B. 1990. Cardiac gap junctions: gating properties of single channels. See Ref. 194

112. Joshi, S., Pleij, C. W. A., Haenni, A. L., Chapeville, F., Bosch, L. 1983. Properties of the tobacco mosaic virus intermediate length RNA-2 and its translation. *Virology* 127:100–11

113. Juniper, B. E. 1977. Some speculations on the possible roles of the plas-

modesmata in the control of differentiation. *J. Theor. Biol.* 66:583–92

114. Kado, C. I., Knight, C. A. 1966. Location of a local lesion gene in tobacco mosaic virus RNA. *Proc. Natl. Acad. Sci. USA* 55:1276–83

115. Kaneko, M., Chonan, N., Matsuda, T., Kawahara, H. 1980. Ultrastructure of the small vascular bundles and transfer pathways for photosynthate in the leaves of rice plant. *Jpn. J. Crop Sci.* 49:42–50

116. Katsaros, C., Galatis, B. 1988. Thallus development in *Dictyopteris membranacea* Phaeopyta Dictyotales. *Br. Phycol. J.* 23:71–88

117. Kiberstis, P. A., Pessi, A., Atherton, E., Jackson, R., Hunter, T., Zimmern, D. 1983. Analysis of in vitro and in vivo products of the TMV 30 kDa open reading frame using antisera raised against a synthetic peptide. *FEBS Lett.* 164:355–60

118. Kitajima, E. W., Lauritis, J. A. 1969. Plant virions in plasmodesmata. *Virology* 37:681–85

119. Kollmann, R., Glockmann, C. 1985. Studies on graft unions. I. Plasmodesmata between cells of plants belonging to different unrelated taxa. *Protoplasma* 124:224–35

120. Kollmann, R., Yang, S., Glockmann, C. 1985. Studies on graft unions. II. Continuous and half plasmodesmata in different regions of the graft interface. *Protoplasma* 126:19–29

121. Kronestedt, E. C., Robards, A. W., Stark, M., Olesen, P. 1986. Development of trichomes in the *Abutilon* nectary gland. *Nord. J. Bot.* 6:627–40

122. Kuo, J., O'Brien, T. P., Canny, M. J. 1974. Pit-field distribution, plasmodesmatal frequency, and assimilate flux in the mestome sheath cells of wheat leaves. *Planta* 121:97–118

123. Kuo, J., Pate, J. S. 1985. The extrafloral nectaries of cowpea (*Vigna Xunguiculata* (L.) Walp): I. Morphology, anatomy and fine structure. *Planta* 166:15–27

124. Kurkova, E. B., Vakhmistrov, D. B. 1984. Distribution of plasmodesmata in the rhizodermis along the root axis in *Trianea bogotensis. Soviet Plan Physiol.* 31:115–19

125. Kwiatkowska, M. 1988. Symplasmic isolation of *Chara vulgaris* antheridium and mechanisms regulating the process of spermatogenesis. *Protoplasma* 142:137–46

126. Kwiatkowska, M., Maszewski, J. 1985. Changes in ultrastructure of plasmodesmata during spermatogenesis in *Chara vulgaris* L. *Planta* 166:46–50

127. Kwiatkowska, M., Maszewski, J. 1986. Changes in the occurrence and ultrastructure of plasmodesmata in antheridia of *Chara vulgaris* L. during different stages of spermatogenesis. *Protoplasma* 132:179–88

128. Langenberg, W. G. 1986. Virus protein association with cylindrical inclusions of two viruses that infect wheat. *J. Gen. Virol.* 67:1161–68

129. Lawson, R. H., Hearon, S. S. 1971. The association of pinwheel inclusions with plasmodesmata. *Virology* 44:454–56

130. Leegood, R. C. 1985. The intercellular compartmentation of metabolites in the leaves of *Zea mays* L. *Planta* 164:163–71

131. Leonard, D. A., Zaitlin, M. 1982. A temperature-sensitive strain of tobacco mosaic virus defective in cell-to-cell movement generates an altered viral-coded protein. *Virology* 117:416–24

132. Linstead, P. J., Hills, G. J., Plaskitt, K. A., Wilson, I. G., Harker, C. L., Maule, A. J. 1988. The subcellular location of the gene 1 product of cauliflower mosaic virus is consistent with a function associated with virus spread. *J. Gen. Virol.* 69:1809–18

133. Loewenstein, W. R. 1979. Junctional intercellular communication and the control of growth. *Biochim. Biophys. Acta* 560:1–65

134. Loewenstein, W. R. 1981. Junctional intercellular communication: the cell to cell membrane channel. *Physiol. Rev.* 61:809–913

135. Loewenstein, W. R., Rose, B. 1978. Calcium in (junctional) intercellular communication and a thought on its behavior in intracellular communication. *Ann. NY Acad. Sci.* 307:285–307

136. Lommel, S. A., Weston-Fina, M., Lomonossoff, G. P. 1988. The nucleotide sequence and gene organization of red clover necrotic mosaic virus RNA-2. *Nucleic Acids Res.* 16:8587–602

137. López-Sáez, J. F., Giménéz-Martin, G., Risueño, M. C. 1966. Fine structure of the plasmodesm. *Protoplasma* 61:81–84

138. Lucas, W. J., Madore, M. A. 1988. Recent advances in sugar transport. In *The Biochemistry of Plants*, ed. J. Preiss, 14:35–84. New York: Academic

139. Lucas, W. J., Wolf, S., Deom, C. M., Kishore, G. M., Beachy, R. N. 1990. Viruses and plasmodesmata. See Ref. 194

140. Lyalin, O. O., Ktitorova, I. N., Barmicheva, E. M., Akhmedov, N. I. 1986. Intercellular connections in sub-

mersed trichomes of *Salvinia. Fiziol. Rast. (Mosc.)* 33:432–46

141. Mackenzie, D. J., Tremaine, J. H. 1988. Ultrastructural location of nonstructural protein 3A of cucumber mosaic virus in infected tissue using monoclonal antibodies to a clones chimeric fusion protein. *J. Gen. Virol.* 69: 2387–95

142. Madore, M. A., Lucas, W. J. 1987. Control of photoassimilate movement in source-leaf tissues of *Ipomoea tricolor* Cav. *Planta* 171:197–204

143. Madore, M. A., Lucas, W. J. 1989. Transport of photoassimilates between leaf cells. In *Transport of Photoassimilates,* ed. D. A. Baker, J. Milburn, pp. 49–78. London: Longman

144. Madore, M. A., Oross, J. W., Lucas, W. J. 1986. Symplastic transport in *Ipomoea tricolor* source leaves. *Plant Physiol.* 82:432–42

145. Magne, F. 1988. Archethallus and nematothallus concepts. *Crypto. Algol.* 9:267–72

146. Marchant, H. J. 1976. Plasmodesmata in algae and fungi. See Ref. 92, pp. 59–80

147. Marginson, R., Sedgley, M., Knox, R. B. 1985. Structure and histochemistry of the extrafloral nectary of *Acacia terminalis* (Leguminosae, Mimosidae). *Protoplasma* 127:21–30

148. Maule, A. J., Harker, C. L., Wilson, I. G. 1989. The pattern of accumulation of cauliflower mosaic virus-specific products in infected turnips. *Virology* 169: 436–46

149. McMullen, C. R., Gardner, W. S. 1980. Cytoplasmic inclusions induced by wheat streak mosaic virus. *J. Ultra. Res.* 72:65–75

150. Meiners, S., Baron-Epel, O., Schindler, M. 1988. Intercellular communication—filling the gaps. *Plant Physiol.* 88:791–93

151. Meiners, S., Schindler, M. 1987. Immunological evidence for gap junction polypeptide in plant cells. *J. Biol. Chem.* 262:951–53

152. Meiners, S., Schindler, M. 1989. Characterization of a connexin homologue in cultured soybean cells and diverse plant organs. *Planta* 179:148–55

153. Meshi, T., Motoyoshi, F., Maeda, T., Yoshiwoka, S., Watanabe, H., Okada, Y. 1989. Mutations in the tobacco mosaic virus 30-kD protein gene overcome Tm-2 resistance in tomato. *The Plant Cell* 1:515–22

154. Meshi, T., Ohno, T., Okada, Y. 1982a. Nucleotide sequence of the 30K protein cistron of cowpea strain of tobacco

mosaic virus. *Nucleic Acids Res.* 10: 6111–17

155. Meshi, T., Ohno, Y., Okada, Y. 1982b. Nucleotide sequence and its character of cistron coding for the 30K protein of tobacco mosaic virus (OM strain). *J. Biochem.* 91:1441–44

156. Meshi, T., Okada, Y. 1987. Systemic movement of viruses. In *Plant-Microbe-Interactions: Molecular and Genetic Perspectives,* ed. T. Kosuge, E. W. Nester, 2:285–304. New York: Macmillan

157. Meshi, T., Watanabe, Y., Saito, T., Sugimoto, A., Maeda, T., Okada, Y. 1987. Function of the 30 kd protein of tobacco mosaic virus: involvement in cell-to-cell movement and dispensability for replication. *EMBO J.* 6:2557–63

158. Mollenhauer, H., Morré, J. 1987. Some unusual staining properties of tannic acid in plants. *Histochemistry* 88:17–22

159. Monzer, J. 1990. Secondary formation of plasmodesmata in cultured cells. See Ref. 194

160. Morrow, D. L., Lucas, W. J. 1986. (1-3)-β-D-Glucan synthase from sugar beet. I. Isolation and solubilization. *Plant Physiol.* 81:171–76

161. Morrow, D. L., Lucas, W. J. 1987. (1-3)-β-D-Glucan synthase from sugar beet. II. Product inhibition by UDP. *Plant Physiol.* 84:565–67

162. Moser, O., Gagey, M. -J., Godefroy-Colburn, T., Stussi-Garaud, C., Ellwart-Tschurtz, M., Nitschko, H., Mundry, K. -W. 1988. The fate of the transport protein of tobacco mosaic virus in systemic and hypersensitive tobacco hosts. *J. Gen. Virol.* 69:1367–73

163. Nassuth, A., Bol, J. F. 1983. Altered balance of the synthesis of plus and minus strand RNAs induced by RNA 1 and 2 of alfalfa mosaic virus in absence of RNA 3. *Virology* 124:75–85

164. Nichol, P., Hall, J. L. 1988. Characteristics of nectar secretion by the extrafloral nectaries of *Ricinus communis. J. Exp. Bot.* 39:573–86

165. Nishiguchi, M., Motoyoshi, F., Oshima, N. 1978. Behaviour of a temperature sensitive strain of tobacco mosaic virus in tomato leaves and protoplasts. *J. Gen. Virol.* 39:53–61

166. Nishiguchi, M., Motoyoshi, F., Oshima, N. 1980. Further investigation of a temperature-sensitive strain of tobacco mosaic virus: its behaviour in tomato leaf epidermis. *J. Gen. Virol.* 46:497–500

167. Nougarede, A., Landre, P., Rembur, J., Hernandez, M. N. 1985. Are variations in the activities of 5'nucleotidase and

416 ROBARDS & LUCAS

adenylate cyclase components in the release of inhibition in the pea cotyledonary bud? *Can. J. Bot.* 63:309–23

168. Oates, B. R., Cole, K. M. 1987. Ultrastructure and ontogeny of hair filaments in the subtidal saccate alga *Hydroclathrus clathratus* Phaeophyta Scytosiphonales. *Can. J. Bot.* 65:1687–93

169. Ohno, T., Takamatsu, N., Meshi, T., Okada, Y., Nishiguchi, M., Kiho, Y. 1983. Single amino acid substitution in 30K protein of TMV defective in virus transport function. *Virology* 131:255–58

170. Olesen, P. 1975. Plasmodesmata between mesophyll and bundle sheath cells in relation to the exchange of C_4-acids. *Planta* 123:199–202

171. Olesen, P. 1979. The neck constriction in plasmodesmata evidence for a peripheral sphincter-like structure revealed by fixation with tannic-acid. *Planta* 144:349–58

172. Olesen, P. 1980. A model of a possible sphincter associated with plasmodesmatal neck regions. *Eur. J. Cell Biol.* 22:250

173. Olesen, P. 1986. Interactions between cell wall and plasmodesmata: model of a possible sphincter mechanism. In *Cell Walls '86*, ed. B. Vian, D, Reis, R. Goldberg. Paris: Univ. P. and M. Curie

174. Olesen, P., Robards, A. W. 1990. The neck region of plasmodesmata: general architecture and some functional aspects. In Ref. 194

175. Oliver, J., Deom, C. M., De, B. K., Beachy, R. N. 1986. In vitro transcription and translation of cloned cDNAs encoding the 30-kDa protein gene of TMV. *Virology* 155:277–83

176. Ooshika, I., Watanabe, Y., Meshi, T., Okada, Y., Igano, K., Inouye, K., Yoshida, N. 1984. Identification of the 30K protein of TMV by immunoprecipitation with antibodies directed against a synthetic peptide. *Virology* 132:71–78

177. Overall, R. L., Gunning, B. E. S. 1982. Intercellular communication in *Azolla* roots: II. Electrical coupling. *Protoplasma* 111:151–60

178. Overall, R. L., Wolfe, J., Gunning, B. E. S. 1982. Intercellular communication in *Azolla* roots: I. Ultrastructure of plasmodesmata. *Protoplasma* 111:134–50

179. Owen, T. P., Benzing, D. H., Thomson, W. W. 1988. Apoplastic and ultrastructural characterisations of the trichomes from the carnivorous bromeliad *Brocchinia reducta. Can. J. Bot.* 66:941–48

180. Paine, P. L., Moore, L. C., Horowitz, S. B. 1975. Nuclear envelope permeability. *Nature* 254:109–14

181. Palevitz, B. A., Hepler, P. K. 1985. Changes in dye coupling of stomatal cells of *Allium* and *Commelina* demonstrated by microinjection of Lucifer yellow. *Planta* 164:473–79

182. Pate, J. S., Peoples, M. B., Storer, P. J., Atkins, C. A. 1985. The extrafloral nectaries of cowpea (*Vigna unguiculata* (L.) Walp.). II. Nectar composition, origin of nectar solutes, and nectary functioning. *Planta* 166:28–38

183. Pennell, R. I., Knox, J. P., Scofield, G. N., Selvendran, R. R., Roberts, K. 1989. A family of abundant plasmamembrane-associated glycoproteins related to the arabinogalactan proteins is unique to flowering plants. *J. Cell Biol.* 108:1967–77

184. Poovaiah, B. W., Reddy, A. S. N., McFadden, J. J. 1987. Calcium messenger system: role of protein phosphorylation and inositol bisphospholipids. *Physiol. Plant.* 69:569–73

185. Porter, K. R., Machado, R. D. 1960. Studies on the endoplasmic reticulum. IV. Its form and distribution during mitosis in cells of onion root tip. *J. Biophys. Biochem. Cytol.* 7:167–80

186. Robards, A. W. 1968. A new interpretation of plasmodesmatal ultrastructure. *Planta* 82:200–10

187. Robards, A. W. 1971. The ultrastructure of plasmodesmata. *Protoplasma* 72:315–23

188. Robards, A. W. 1975. Plasmodesmata. *Annu. Rev. Plant Physiol.* 26:13–29

189. Robards, A. W. 1976. Plasmodesmata in higher plants. See Ref. 92, pp. 15–57

190. Robards, A. W. 1980. General and molecular cytology. *Prog Bot.* 42:1–15

191. Robards, A. W. 1982. Cell interactions in plants—a comparative survey. In *The Functional Integration of Cells in Animal Tissues*, ed. J. D. Pitts, M. E. Finbow, pp. 57–79. London: Cambridge Univ. Press

192. Robards, A. W., Clarkson, D. T. 1976. The role of plasmodesmata in the transport of water and nutrients across roots. See Ref. 92, pp. 181–201

193. Robards, A. W., Clarkson, D. T. 1984. Effects of chilling temperatures on root cell membranes as viewed by freeze-fracture electron microscopy. *Protoplasma* 122:75–85

194. Robards, A. W., Jongsma, H., Lucas, W. J., Pitts, J., Spray, D., eds. 1990. *Parallels in Cell to Cell Junctions in Plants and Animals.* Heidelberg: Springer-Verlag. In press

195. Robards, A. W., Stark, M. 1988. Nectar secretion in *Abutilon:* a new model. *Protoplasma* 142:79–91
196. Robertson, J. D. 1964. Unit membranes: a review with recent new studies of experimental alterations and a new subunit structure in synaptic membranes. In *Cellular Membranes in Development,* ed. M. Locke, pp. 1–81. New York and London: Academic
197. Robins, R. J., Juniper, B. E. 1980. Considerations of the estimation of plasmodesmatal frequencies. *J. Theor. Biol.* 83:405–9
198. Roggero, P., Pennazia, S. 1988. Effects of salicylate on systemic invasion of tobacco plants by various viruses. *J. Phytopathol.* 123:207–16
199. Rose, B., Socolar, S. J., Obaid, A. L. 1984. Cell-to-cell channels with two independent gates in series regulated by membrane potentials, by pCa_i and by pH_i. *Biophys. J.* 45:64–66
200. Rowley, J. R. 1986. A model for plasmodesmata. In *Biology of Reproduction and Cell Motility in Plants and Animals,* ed. M. Cresti, R. Dallai, pp. 175–80. Sienna: Univ. Sienna
201. Rowley, J. R. 1987. Plasmodesmata-like processes in tapetal cells. *Cellule* 74:227–42
202. Russin, W. A., Evert, R. F. 1985. Studies on the leaf of *Populus deltoides* (Salicaceae): ultrastructure, plasmodesmatal frequency, and solute concentrations. *Am. J. Bot.* 72:1232–47
203. Russo, M., Castellano, M. A., Martelli, G. P. 1982. The ultrastructure of broad bean stain and broad bean true mosaic virus infections. *J. Submicrosc. Cytol.* 14:149–60
204. Russo, M., Martelli, G. P., Di Franco, A. 1981. The fine structure of local lesions of beet necrotic yellow vein virus in *Chenopodium amaranticolor*. *Physiol. Plant Pathol.* 19:237–42
205. Sack, F. D., Kiss, J. Z. 1989. Root-cap structure in wild type and in a starchless mutant of *Arabidopsis*. *Am. J. Bot.* 76:454–64
206. Saito, T., Imai, Y., Meshi, T., Okada, Y. 1988. Interviral homologies of the 30K proteins of tobamoviruses. *Virology* 167:653–56
207. Santiago, J. F., Goodwin, P. B. 1988. Restricted cell-cell communication in the shoot of *Silene coeli* Rosa during the transition to flowering is associated with a high mitotic index rather than with evocation. *Protoplasma* 146:52–60
208. Sauter, J. J., Kloth, S. 1986. Plasmodesmatal frequency and radial translocation rates in ray cells of poplar

(*Populus canadensis* Moench "robusta"). *Planta* 168:377–80
209. Sawidis, T., Eleftheriou, E. P., Tsekos, I. 1987a. The floral nectaries of *Hibiscus rosa-sinensis* L. I. Development of secretory hairs. *Ann. Bot.* 59:643–52
210. Sawidis, T., Eleftheriou, E. P., Tsekos, I. 1987b. The floral nectaries of *Hibiscus rosa-sinensis* L. II. Plasmodesmatal frequencies. *Phyton (Austria)* 27:155–64
211. Sawidis, T., Eleftheriou, E. P., Tsekos, I. 1989. The floral nectaries of *Hibiscus rosa-sinensis* L. III. A morphometric and ultrastructural approach. *Nord. J. Bot.* In press
212. Schmitz, K., Kuhn, R. 1982. Fine structure, distribution and frequency of plasmodesmata and pits in the cortex of *Laminaria hyperborea* and *L. saccharina*. *Planta* 154:385–92
213. Schnepf, E., Sych, A. 1983. Distribution of plasmodesmata in developing *Sphagnum* leaflets. *Protoplasma* 116: 51–56
214. Schumaker, K. S., Sze, H. 1987. Inositol 1,4,5-trisphosphate releases Ca^{2+} from vacuolar membrane-vesicles of oat root. *J. Biol. Chem.* 262:3944–46
215. Schwartzmann, G., Wiegandt, H., Rose, B., Zimmerman, A., Ben-Haim, D., Loewenstein, W. R. 1981. Diameter of the cell-to-cell junctional membrane channels as probed with neutral molecules. *Science* 213:551–53
216. Seagull, R. W. 1983. Differences in the frequency and disposition of plasmodesmata resulting from root cell elongation. *Planta* 159:497–504
217. Serras, F., van den Biggelaar, J. A. M. 1990. Progressive restrictions in gap junctional communication during development. See Ref. 94
218. Shalla, T. A., Petersen, L. J., Zaitlin, M. 1982. Restricted movement of a temperature-sensitive virus in tobacco leaves is associated with a reduction in numbers of plasmodesmata. *J. Gen. Virol.* 60:355–58
219. Simpson, I., Rose, B., Loewenstein, W. R. 1977. Size limits of molecules permeating the junctional membrane channels. *Science* 197:294–96
220. Spanswick, R. M. 1972. Electrical coupling between cells of higher plants: a direct demonstration of intercellular communication. *Planta* 102:215–27
221. Spanswick, R. M., Costerton, J. W. F. 1967. Plasmodesmata in *Nitella translucens:* structure and electrical resistance. *J. Cell Sci.* 2:451–64
222. Spray, D. C. 1990. Electrophysiological properties of gap junction channels. See Ref. 94

223. Spray, D. C., Stern, J. H., Harris, A. L., Bennett, M. V. L. 1982. Comparison of sensitivities of gap junctional conductance to H and Ca ions. *Proc. Natl. Acad. Sci. USA* 79:441–45

224. Steer, M. W. 1988. Plasma membrane turnover in plant cells. *J. Exp. Bot.* 39:987–96

225. Stephenson, J. L. M., Hawes, C. R. 1986. Stereology and stereometry of endoplasmic reticulum during differentiation in the maize root cap. *Protoplasma* 131:32–46

226. Stewart, W. W. 1981. Lucifer dyes—highly fluorescent dyes for biological tracing. *Nature* 292:17–21

227. Stitt, M., Heldt, H. W. 1985. Generation and maintenance of concentration gradients between the mesophyll and bundle sheath in maize leaves. *Biochem. Biophys. Acta* 808:400–14

228. Stoddard, F. L., McCully, M. E. 1979. Histology of the development of the graft union in pea roots. *Can. J. Bot.* 57:1486–1501

229. Stussi-Garaud, C., Garaud, J. -C., Berna, A., Godefroy-Colburn, T. 1987. In situ location of an alfalfa mosaic virus non-structural protein in plant cell walls: correlation with virus transport. *J. Gen. Virol.* 68:1779–84

230. Tainter, F. H. 1971. The ultrastructure of *Arceuthobium pusillum*. *Can. J. Bot.* 49:1615–22

231. Takamatsu, N., Ohno, T., Meshi, T., Okada, Y. 1983. Molecular cloning and nucleotide sequence of the 30K and the coat protein cistron of TMV (tomato strain) genome. *Nucleic Acids Res.* 11: 3767–78

232. Taliansky, M. E., Malyshenko, S. I., Pshennikova, E. S., Kaplan, I. B., Ulanova, E. F., Atabekov, J. G. 1982. Plant virus-specific transport function. I. Virus genetic control required for systemic spread. *Virology* 122:318–26

233. Tangl, E. 1879. Ueber offene Communicationen zwischen den Zellen des Endosperms einiger Samen. *Jb. Wiss Bot.* 12:170–90

234. Terry, B. R., Robards, A. W. 1987. Hydrodynamic radius alone governs the mobility of molecules through plasmodesmata. *Planta* 171:145–57

235. Teuliers, C., Alibert, G., Ranjeva, R. 1985. Reversible phosphorylation of tonoplast proteins involves tonoplast-bound calcium-calmodulin-dependent protein kinase(s) and protein phosphatase(s). *Plant Cell Rep.* 4:199–201

236. Thomson, W. W., Platt-Aloia, K. 1985. The ultrastructure of the plasmodesmata of the salt glands of *Tamarix-aphylla* as

revealed by transmission electron microscopy and freeze-fracture electron microscopy. *Protoplasma* 125:13–23

237. Tomenius, K., Clapham, D., Meshi, T. 1987. Localization by immunogold cytochemistry of the virus-coded 30K protein in plasmodesmata of leaves infected with tobacco mosaic virus. *Virology* 160:363–71

238. Tomiyama, K., Okamoto, H., Katou, K. 1987. Membrane potential change induced by infection by *Phytophthora infestans* of potato cells. *Ann. Phytopath. Soc. Jpn.* 53:310–22

239. Trębacz, K., Fensom, D. S., Harris, A., Zawadzki, T. 1988. Transnodal transport of carbon-14 in *Nitella flexilis*. III. Further studies on dissolved inorganic carbon movements in tandem cells. *J. Exp. Bot.* 39:1561–74

240. Tucker, E. B. 1982. Translocation in the staminal hairs of *Setcreasea purpurea*. I. Study of cell ultrastructure and cell-to-cell passage of molecular probes. *Protoplasma* 113:193–201

241. Tucker, E. B. 1987. Cytoplasmic streaming does not drive intercellular passage in staminal hairs of *Setcreasea purpurea*. *Protoplasma* 137:140–44

242. Tucker, E. B. 1988. Inositol bisphosphate and inositol trisphosphate inhibit cell-to-cell passage of carboxyfluorescein in staminal hairs of *Setcreasea purpurea*. *Planta* 174:358–63

243. Tucker, E. B., Spanswick, R. M. 1985. Translocation in the staminal hairs of *Setcreasea purpurea*. II. Kinetics of intercellular transport. *Protoplasma* 128: 167–72

244. Tucker, J. E., Mauzerall, D., Tucker, E. B. 1989. Symplastic transport of carboxyfluorescein in staminal hairs of *Setcreasea purpurea* is diffusive and includes loss to the vacuole. *Plant Physiol.* 90:1143–47

245. Turgeon, R. 1989. The sink-source transition in leaves. *Annu. Rev. Plant Physiol. Plant Mol. Biol.* 40:119–38

246. Turgeon, R., Hepler, P. K. 1989. Symplastic continuity between mesophyll and companion cells in minor veins of mature *Cucurbita pepo* L. leaves. *Planta* 179:24–31

247. Tyree, M. T. 1970. The symplast concept. A general theory of symplastic transport according to the thermodynamics of irreversible processes. *J. Theor. Biol.* 26:181–214

248. Tyree, M. T., Tammes, P. M. L. 1975. Translocation of uranin in the symplasm of staminal hairs of *Tradescantia*. *Can. J. Bot.* 53:2038–46

249. Van Bel, A. J. E. 1986. Amino acid

loading by minor veins of *Commelina benghalensis:* an integration of structural and physiological aspects. In *Fundamental, Ecological and Agricultural Aspects of Nitrogen Metabolism in Higher Plants,* ed. H. Lambers, J. J. Neeteson, I. Stulen, pp. 111–14. Dordrecht/Boston/Lancaster: Martinus Nijhoff

250. Van Bel, A. J. E. 1987. The apoplast concept of phloem loading has no universal validity. *Plant Physiol. Biochem.* 25:677–86

251. Van Bel, A. J. E., Van Kesteren, W. J. P., Papenhuijzen, C. 1988. Ultrastructural indications for coexistence of symplastic and apoplastic phloem loading in *Commelina benghalensis* leaves. *Planta* 176:159–72

252. Van Kesteren, W. J. P., Van Der Schjoot, C., Van Bel, A. J. E. 1988. Symplastic transfer of fluorescent dyes from mesophyll to sieve tube in stripped leaf tissue and partly isolated minor veins of *Commelina benghalensis*. *Plant Physiol.* 88:667–70

253. Van Steveninck, R. F. M. 1976. Cytochemical evidence for ion transport through plasmodesmata. See Ref. 92, pp. 131–147

254. Van Went, J. L., van Aelst, A. C., Tammes, P. M. L. 1975. Anatomy of staminal hairs from *Tradescantia* as a background for translocation studies. *Acta Bot. Neerl.* 24:1–6

255. Vorob'ev, L. N., Tarkhanov, K. A., Vakhmistrov, D. B. 1982. Use of the electrical coupling factor for quantitative estimation of symplastic communications. *Soviet Plant Physiol.* 28:495–502

256. Wade, M. H., Trosko, J. E., Schindler, M. 1986. A fluorescence photobleaching assay of gap junction–mediated communication between human cells. *Science* 232:525–28

257. Wang, X. -D., Canny, M. J. 1985. Loading and translocation of assimilate in the fine veins of sunflower leaves. *Plant Cell Env.* 8:669–85

258. Watanabe, Y., Ooshika, I., Meshi, T., Okada, Y. 1986. Subcellular localization of the 30K protein in TMV-inoculated tobacco protoplasts. *Virology* 152:414–20

259. Weiglin, C., Winter, E. 1988. Studies on the ultrastructure and development of the glandular trichomes of *Cress cretica* L. *Flora (Jena)* 181:19–27

260. Weiner, H., Burnell, J. N., Woodrow, I. E., Heldt, H. W., Hatch, M. D. 1988. Metabolite diffusion into bundle sheath cells from C_4 plants. *Plant Physiol.* 88:815–22

261. Weintraub, M., Ragetli, H. W. J., Leung, E. 1976. Elongated virus particles in plasmodesmata. *J. Ultra. Res.* 56:351–64

262. Wergin, W. P., Elmore, C. D., Hanny, B. W., Inger, B. F. 1975. Ultrastructure of the subglandular cells from the foliar nectaries of cotton in relation to the distribution of plasmodesmata and the symplastic transport of nectar. *Am. J. Bot.* 62:842–49

263. Wille, A. C., Lucas, W. J. 1984. Ultrastructural and histochemical studies on guard cells. *Planta* 160:129–42

264. Willison, J. H. M. 1976. Plasmodesmata: a freeze-fracture view. *Can. J. Bot.* 54:2842–47

265. Willmer, C. M., Sexton, R. 1979. Stomata and plasmodesmata. *Protoplasma* 199:113–24

266. Wolf, S., Deom, C. M., Beachy, R. N., Lucas, W. J. 1989. Movement protein of tobacco mosaic virus modifies plasmodesmatal size exclusion limit. *Science* 246:377–79

267. Wu, J. H., Dimitman, J. E. 1970. Leaf structure and callose formation as determinants of tobacco mosaic virus movement in bean leaves as revealed by UV irradiation studies. *Virology* 40:820–27

268. Xu, A., Meiners, S., Schindler, M. 1990. Investigation of immunological relatedness between plant and animal connexins. See Ref. 194

269. Yada, T., Rose, B., Loewenstein, W. R. 1985. Diacylglycerol downregulates functional membrane permeability. TMB-8 blocks this effect. *J. Membr. Biol.* 88:217–32

270. Yang, S-J., Lou, C-H. 1988. Paramural bodies in callus beside the isolation layer of the graft union. *Acta Bot. Sin.* 30:480–84

271. Zaitlin, M., Hull, R. 1987. Plant virus-host interactions. *Annu. Rev. Plant Physiol.* 38:291–315

272. Zee, S-Y. 1969. The fine structure of differentiating sieve elements of *Vicia faba. Aust. J. Bot.* 17:441–56

273. Zheng, G. -C., Nie, X. -W., Wang, Y. -X., Jian, L. -C., Sun, L. -H., Sun, D. -L. 1985. Cytochemical localisation of ATPase activity during cytomixis in pollen mother cells of David lily—*Lilium davidii* var. Willmottiae and its relation to the intercellular migrating chromatin substance. *Acta Bot. Sin.* 27:26–32

274. Zimmern, D., Hunter, T. 1983. Point mutation in the 30K open reading frame of TMV implicated in temperature-sensitive assembly and local lesion spreading of mutant Ni2519. *EMBO J.* 2:1893–1900

Annu. Rev. Plant Physiol. Plant Mol. Biol. 1990. 41:421–53
Copyright © 1990 by Annual Reviews Inc. All rights reserved

SUNFLECKS AND PHOTOSYNTHESIS IN PLANT CANOPIES*

Robert W. Pearcy

Department of Botany, University of California, Davis, California 95616

KEY WORDS: transient photosynthesis, ribulose-1,5-bisphosphate carboxylase, regulation of
photosynthesis, dynamics, shade adaptation

CONTENTS

INTRODUCTION

Leaves at the very top of a plant canopy or in the most shaded understory sites may experience long periods of light that approximate steady-state conditions.

*****Abbreviations:** c_i, partial pressure of CO_2 in the intercellular air spaces; IRGA, infra-red CO_2 analyzer; k_{cat}, specific activity; L, leaf area index; PFD, photon flux density; PGA, 3-phosphoglyceric acid; PI postillumination; PLI, post-lower-illumination; Rubisco, ribulose-1,5-bisphosphate carboxylase; RuBP, ribulose-1,5-bisphosphate; TP, triose phosphate

However, most leaves are subjected to rapidly alternating periods of sun and shade because of sunflecks. Under these circumstances, a large fraction of CO_2 assimilation may occur under transient light conditions. Early ecophysiological studies (52, 103) recognized the importance of sunflecks to photosynthesis of understory plants, but because of technical limitations (only recently overcome) relatively little attention has been given to how plants respond to them. The environmental and physiological constraints on photosynthetic CO_2 assimilation under transient conditions may be quite different from those under steady-state conditions. Much progress has been made in understanding the controls on photosynthetic performance under steady-state conditions, while those operating on transient photosynthesis have received more limited attention.

In this review I focus on research, mostly undertaken in the last 10 years, into the environmental and physiological controls on the utilization of transient light. The regulatory aspects of photosynthetic carbon metabolism, which play an important role in the dynamic responses of CO_2 assimilation, were recently reviewed in this series (192). The role of sunflecks in the physiological ecology of understory plants was recently reviewed by Chazdon (26).

LIGHT DYNAMICS IN CANOPIES

Theory

Because the theory of photon transport in plant canopies has recently been reviewed in detail (118), only a few points relevant to the occurrence of sunflecks in canopies need be made here. Much more theoretical attention has been given to the statistical prediction of light extinction with depth than to either the spatial or temporal variation of light in canopies. Statistical models of the type first developed by Monsi & Saekai (115) and since refined by many others (e.g. 48, 101, 102, 116, 117, 120, 121, 123) do, however, predict the fraction of a plant's leaf area that at any given canopy depth and time will be in direct sunlight. These models assume that the probability of the solar beam's passing through a gap in a layer of the canopy and hence causing a sunfleck at some deeper location decreases exponentially as the cumulative index of leaf area projected normal to the beam increases. The steepness of the decrease is influenced by the degree of dispersion of the foliage. The probability of sunflecks is somewhat greater for a given leaf area index (L) when the leaves are clumped at the ends of branches, as is usually the case in forests, because of the larger gaps occurring between clumps; in canopies where leaves are more randomly dispersed the sunfleck probability is lower (10, 11). The sizes of sunflecks and their maximum photon flux densities also depend on canopy structure. For sunflecks created by small holes in the canopy, the photon flux density (PFD) is often much lower than that of full sunlight because part of the solar disc is obscured by the gap edges, which cause a penumbral shadow. A gap must have an angular size greater than the approximately $0.5°$ apparent diameter of the solar disk in order to admit the full direct-beam PFD from the sun. Penumbral effects become especially important under tall canopies since the width of the penumbra over which the light grades from the full direct-beam PFD to the shade or umbral PFD is a function of the distance from the gap edge to the surface (38, 110). Penumbral effects make it difficult to define a sunfleck precisely. It is generally necessary to define a sunfleck as occurring any time the PFD exceeds some level that differentiates it from the background shadelight. This level may be as low as 15–50 μmol m^{-2} s^{-1} in forest understories (30, 149) to as high as 200 μmol m^{-2} s^{-1} in crop canopies (140).

Within a canopy, temporal light regimes range from mostly sunlight broken by periods of shade (shadeflecks?) near the top to mostly shade broken by sunflecks near the bottom. The earth's rotation causes sunflecks to move slowly, creating low-frequency fluctuations in the PFD at a given point. Myneni & Impens (116) used a geometrical model of foliage elements of a *Populus* tree to predict the frequency of irradiance changes at single points because of the rotation of the earth. The probability that a given point of observation would be in penumbra or full direct sunlight in this model canopy was 0.52, and on average, irradiance on a leaf element was stable for only 50 s even in the absence of leaf movement. Canopy movement and leaf flutter in the wind cause much higher frequency variations (183). Similar high-frequency variations in PFD occur in aquatic communities because of movement of phytoelements in the water (57, 61, 77, 105) and focusing by waves (39). Cloud movement can cause additional variation in PFD (84, 96), a factor poorly characterized in most plant communities.

There has been essentially no theoretical treatment of canopy movement in the wind and its effect on light fluctuation. Studies of wind-induced flexing of tree trunks show characteristic resonant frequencies (2). As one moves from the trunk to finer branches and twigs, higher frequencies of movement will be added. Leaf-flutter frequencies must depend on the aerodynamic and biomechanical properties of leaves and petioles. Atmospheric turbulence interacts with flexible plant canopies to produce movement compounded of the frequency of gust arrival (e.g. the "wave" in a wheat canopy) and the resonant frequency of the plant (55). Gusts are distinctly periodic and should cause similar periodic fluctuations in PFD in the canopy. The flexing of plant canopies affects the turbulent transport of gasses (56), and it must also play an important role in the dynamics of the light environment experienced by leaves. Canopies of different species differ markedly in their architecture and flexibility, but we do not yet know how these variations influence the characteristics of the resulting sunflecks or whether photosynthetic dynamics have adapted to these characteristics.

Empirical Studies

Most understanding of the dynamics of natural-light regimes under canopies and their importance in the photosynthetic economy of plants comes from direct field measurements. Approaches to the measurement of sunfleck activity have been covered elsewhere (26, 135) and are not considered further here. Direct measurements of the light environment have shown that on clear days 20–80% of the PFD received in the understories of a variety of forests is in the form of sunflecks (15, 27, 30, 60, 132, 133, 137, 150, 200). Field gas-exchange measurements have shown that a similar percentage of the daily CO_2 exchange of leaves in the understory can be attributed to sunfleck utilization (16, 133, 137, 150). Only when photosynthetic capacities are especially low and the background diffuse-light PFD is high (191) does sunfleck utilization make a small contribution to the daily CO_2 exchange in forest understories. Variation in the daily total CO_2 assimilation of the herb *Adenocalulon bicolor* for different days and microsites in a redwood forest understory was closely related to variation in the amount of light received in sunflecks (30). The spatial variation in occurrence of sunflecks can be large even within a small area. Sites less than 0.1 m apart can receive vastly different amounts of sunfleck light on a given day (30, 132), although the amounts may average out to be the same over these distances for longer periods. This spatial heterogeneity is poorly understood but of considerable importance in the understory community. For example, variation in growth rates of tree seedlings in a Hawaiian forest understory correlates with variation in the quantity of light received in sunflecks (minutes per day) at different sites (132).

Characterizations of sunfleck frequencies and durations in forest understories (30, 132, 150) show that most may be less than a few seconds long and separated from others by only brief low-light intervals. These short sunflecks, although often abundant, may contribute a only a small fraction of the daily PFD, in part because they typically reach PFDs of only 50–200 μmol photons $m^{-2} s^{-1}$. The few sunflecks longer than 10 min often reach PFDs of 1500–2000 μmol photons $m^{-2} s^{-1}$ and may contribute more than two thirds of the daily PFD (150, 160). Of course, because of the low light-saturation point for CO_2 assimilation characteristic of shade leaves (17, 18), the extra PFD in these sunflecks can not be well utilized. Because of spatial variations in canopy density, sunflecks are often clustered; several to many may occur in periods of less than an hour or so. Such clusters may be separated from others by low-light intervals also lasting from a few minutes to an hour or more.

Much less attention has been given to the dynamics of light regimes in crop canopies than to those of forest understories. Light regimes in soybean canopies (140) were characterized by sunflecks shorter and brighter than those of forest understories. The number of sunflecks varied from nearly 1800 per day in the upper part of the canopy (L = 0.6) to none for a few locations at the bottom (L = 5.2) At most locations within the canopy, sunflecks contributed 40–90% of the daily PFD, and of this total approximately one third was contributed by sunflecks shorter than 10 s.

Norman & Tanner (122) and Desjardens et al (40) used photosensors connected to high-speed recorders to characterize fluctuations in PFD in crop canopies. Most light fluctuations were of low frequency, but peaks from 5 to 30 Hz were also present that may correspond to the vibrational frequencies of individual leaf elements. Higher frequencies were present in small-leaved canopies [i.e. soybean vs maize (122)] and under windy conditions. The relevance of these high-frequency fluctuations to photosynthesis is questionable because it can be shown that leaves effectively average these variations, yielding a photosynthetic rate identical to that obtained in constant light of the same average irradiance (58).

PHOTOSYNTHESIS IN TRANSIENT LIGHT

Theory

It is useful here to compare the intrinsic properties of dynamic and static systems in relation to the photosynthetic apparatus. A system is static if its output (for the photosynthetic apparatus, the CO_2 exchange rate) depends only on the current input parameters. Most models of photosynthetic gas exchange (e.g. 54, 66, 69, 106) treat it as a static system, which provides an adequate approximation to a steady state in a mathematically tractable form. However, under conditions when the inputs change, a system will exhibit dynamic behavior if it contains dynamic elements. Such elements cause the output of the system to be a function not only of the current conditions but also of its past state. Photosynthetic gas exchange clearly exhibits dynamic behavior manifested as a transient response that requires some time to complete if the environmental inputs are changed. Dynamic systems must be described by differential equations with time as the independent variable. A system can be described as in a dynamic state when the derivative of the rate with respect to time is not zero. When the derivative is zero, steady-state conditions obtain, and the system acts as if it has no dynamic elements. It can again be described by algebraic rather than differential equations.

Whether we view a system or part of a system as dynamic or not depends on the time scale of interest. Energy transfer and charge separation in the photosystems are clearly dynamic processes, but at the time scale relevant to gas exchange (about 0.1 s to hours) they appear instantaneous and hence static. On the other hand, the processes of acclimation and aging are dynamic but occur on a time scale long enough that during an hour or so the photosynthetic rate appears to reach a steady state. Whether or not dynamics need be considered to provide an adequate description of system output depends on how frequently the input transients occur and how much the dynamic responses influence the total output.

What is it about photosynthesis, in a very general sense, that makes it dynamic? The theory underlying dynamic systems is best developed in the context of engineering (see 119, 130, 161) where networks of components can be modeled and their transient behavior understood in terms of the response to pulse, step, or periodic input changes. While the photosynthetic apparatus is much more complex and poorly understood, certain analogies hold reasonably well. Electrical circuits made up solely of resistors and voltage sources are static. A resistance (limitation) and voltage (concentration gradient) analogy is widely applied for the diffusion of gasses and has also been widely extended to enzymatic reactions in photosynthesis (e.g. "carboxylation resistance"). In the latter, the reaction described by its kinetic behavior can also be viewed as static since the change in flux with a step change in substrate concentration will be rapid enough to appear instantaneous. Of course an enzyme behaves as a unique kind of nonlinear "resistor" for which the "resistance" varies as a function of the substrate concentrations. Moreover, regulatory processes can change the effective "resistance" at any substrate concentration. In this sense a kind of transistor (a field-effect transistor) is more analogous since the current (flux) from the source to the drain terminal exhibits saturation as the voltage at the source terminal increases. Regulation can then be viewed as analogous to changing the voltage on the gate terminal of the transistor, which in turn changes the current flow at saturation.

An electrical circuit becomes dynamic if a dynamic element such as a capacitor is added and the voltage is changed. The transient response of such a circuit to a change in voltage is given by the products of the capacitances and resistances. The time constant (time required for 0.632 of the total change to occur) of the transient response increases if the resistance or the capacitance increases. In a photosynthetic system, metabolite pools are functionally equivalent to the charge on a capacitor. Metabolite pools must have some maximum size corresponding to the capacitance, but we do not know exactly how this size is set or how it may vary with conditions. Once the transient is over and a quasi-steady state is reached, the capacitance no longer influences the rate. At this time only the concentration of metabolites and not the charge per se is important in determining the flux.

The other major dynamic elements in photosynthesis are the light-regulated enzymes and the stomata. Compared to the changes in metabolite pool sizes, which have time constants on the order of a few seconds, both stomatal movements and the light regulation of enzymes are slow, with time constants on the order of minutes. Both are often hysteretic, with faster increases than decreases in stomatal conductance or enzyme activity in response to light increases and decreases. As discussed below, the dynamics of both stomatal movement and of light-regulated enzymes are both involved in the slow induction of photosynthetic rate observed after a leaf has been in the dark or low light for an extended period (28, 49, 86, 190). While it is easy to see why the buildup of osmotica in the guard cells and the uptake of water leading to the mechanical opening of stomata may take some time, little is known about why the light regulation of photosynthetic enzymes is such a slow process. It could involve the slow buildup or removal of regulatory metabolites, as determined by still other regulatory systems, or slow conformational changes in the enzymes. In terms of dynamics, systems undergoing conformational changes exhibit resistance and capacitance and thus can be quantified the way we quantify other dynamic systems.

There have been several attempts to model photosynthesis as a dynamic process. Empirical models based on activation of a single enzyme or one or two metabolite pools (64, 78, 185) are clearly oversimplifications that fail to incorporate the range of dynamic responses exhibited by photosynthetic CO_2 exchange. On the other hand, complex dynamic models of the biochemistry of photosynthesis (67, 68, 91, 95, 111, 148) require knowledge of a large number of parameters and are usually applicable only to specific conditions such as saturating CO_2 or saturating light. A semi-mechanistic model for the response of stomatal conductance to sunflecks has been developed (80) that agrees with measurements under most conditions. This model was based on the simplifying assumption of only three dynamic components, whose time constants differed. A dynamic model of photosynthetic CO_2 exchange will need to account for three phenomena: light modulation of photosynthetic enzymes, the variation in intercellular CO_2 pressure (c_i) resulting from the dynamics of stomatal movement, and the variations in those metabolite pools that contribute to the dynamics of CO_2 fixation and photorespiration. Such a model will need to be much simpler than reality, incorporating only the components sufficient to describe adequately the dynamics as well as the steady-state behavior. A model of this type is under development (L. Gross, M. Kirschbaum, and R. Pearcy, in preparation) and should be useful in assessing how key features of the dynamic response control output under different conditions.

Measurement of Transient Photosynthesis

It is possible to follow rapid transients of some components in the photosynthetic apparatus via either light-induced absorption changes or chlorophyll fluorescence (71, 94, 166). However, the output of the system, the assimilation rate, must be followed by measuring the gas exchange. Study of transient gas exchange is limited by the response characteristics of the instruments used. Response times are primarily a function of the volumes of the leaf chamber and the connecting tubing, the flow rate, and the response of the gas analyzers to transients. Currently the biggest limitation is that imposed by the analyzer. The response time of an infrared gas analyzer (IRGA) is determined primarily by the volume of, and the flow through, the sample cell and in some cases by electronic filtering added to reduce the noise in the signal. For transient measurements, electronic filtering must be reduced as much as possible without seriously degrading the signal-to-noise ratio. The flow rates through the leaf chamber must be high, and a sensitive IRGA is therefore required to resolve the small transients in CO_2 concentration. In some IRGAs, a full-scale response in 2 s can be achieved (81, 93). Attempts have been made to extract the true, instantaneous change in CO_2 concentration, and hence the true transient of assimilation, using mathematical models of gas flow through the IRGA cells (65, 143, 146). The gas in the cells is not well mixed, and concentration changes move through as a dispersing front (146). Therefore, approaches based on the fluid dynamics of flow through pipes (e.g. 146) are the most realistic. These models are extremely sensitive to noise in the signal, and artifactual oscillations can result. By and large, these approaches have only marginally improved the resolution of gas exchange measurements. In the future, folded-path IRGAs with small cell volumes (14) or even miniaturized open-path IRGAs (124) mounted within the leaf chamber should allow analysis of faster responses. Fast-responding (0.1-s time constant), ceramic-cell O_2 analyzers have been used for transient measurements (81), but these possess sufficient sensitivity only at low (0.2%) O_2 concentrations.

Delays caused by the remainder of the measuring system are significant but can be dealt with in several ways. Residual delays due to tubing volumes can easily be subtracted because mixing in the tubing is minimal. If the chamber atmosphere is well mixed and the IRGA response is fast, then equations describing the washout (12, 107) can be used to extract the photosynthetic transient from the delays that are a function of the chamber volume (29, 139). The improvements gained by applying washout-kinetics corrections depend primarily on the response time of the analyzer and its signal-to-noise ratio since any noise is accentuated. Alternatively, very thin chambers can be used in which one leaf surface is enclosed and the other is pressed or sealed against a glass water-jacket window (81, 129). Chambers of this design have been useful in dual-channel gas-exchange systems capable of measuring transient responses to either light or gas concentration changes (92–94, 147). If the experimenter pays careful attention to the design of the inlet and outlet, gas moves across the leaf surface as a front yielding a >90% response to a change in assimilation rate in less than 0.5 s (81, 129). Obviously, the use of single-sided chambers is most straightforward with hypostomatous leaves. They can, however, also be used with amphistomatous leaves since the abaxial stomata generally open sufficiently so that c_i is only slightly lower than when gas exchange occurs through both surfaces.

Dynamics of CO_2 Assimilation during Sunflecks

I cover here the dynamic responses to light transients that occur on the time scale of seconds and that seem to involve primarily the dynamics of metabolite pools occurring in response to changes in flux through the light-harvesting system and electron transport. These can be considered, to a first approximation, as distinct from the dynamics of the induction response of photosynthesis that results primarily from the light regulation of enzymes and stomatal movements. However, important interactions between induction and the faster transient responses need to be touched on here as well. The transient responses to lightflecks (pulses of light from artificial sources designed to simulate sunflecks) are illustrated in Figure 1. The response to these lightflecks includes transients involving the rapid acceleration and deceleration of assimilation as light is increased and decreased, and others that involve bursts and gulps of CO_2 uptake and release. These transients overlap sufficiently and are fast enough relative to the response of the measuring instruments to make it difficult to apply a formal analysis of time constants.

Since ribulose-1,5-bisphosphate carboxylase/oxygenase (Rubisco) is the "gatekeeper" between the internal metabolic reactions and the observed assimilation rate (192), transients in assimilation can be related to buildup and declines in pools of ribulose-1,5-bisphosphate (RuBP) as it is utilized by Rubisco and produced from precursor pools by enzymatic reactions in its regeneration path. When a leaf has been in low light for an extended time so that it is not induced, then only a low assimilation rate will be achieved during a lightfleck. However, the decline in assimilation occurring after a lightfleck will be very slow, and most of the CO_2 fixation occurring in response to the lightfleck may actually occur after it has passed. Using the resistance and capacitance analogy, the large time constant for the decline is consistent with slow use of RuBP and its precursors for CO_2 fixation because, owing to low activities of the enzymes, the "resistance" is high. After induction, when activities of Rubisco and other enzymatic steps in the carbon fixation pathway are higher because of light activation, then the decline is much faster, as would be expected if the "resistance" were reduced. Other factors clearly influence the transients and overlap with those associated with CO_2 fixation. After long lightflecks, a distinct post-lightfleck CO_2 burst resulting from metabolism of photorespiratory intermediates (6) can be observed. Under some conditions there may also be small gulps or bursts of CO_2 uptake or

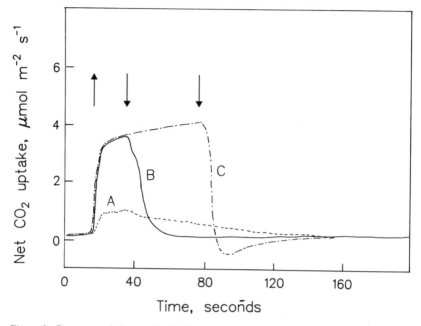

Figure 1   Responses of photosynthetic CO_2 uptake to lightflecks illustrating the differences in the dynamics for lightflecks of different duration and in leaves at different induction states. Response A is to a 20-s duration lightfleck and typical for a leaf that has previously been in low light for greater than 2 hr and is therefore not induced. Response B is also to a 20-s lightfleck but for a leaf that is fully induced. Response C is to a 60-s lightfleck for a fully induced leaf. Note the development of a post-lightfleck CO_2 burst following the 60-s lightfleck but its absence for the 20-s lightflecks. The PFD was increased and decreased at the arrows.

evolution caused by pH changes in the stroma that influence pool sizes of dissolved inorganic carbon (94, 125).

THE EFFICIENCY OF LIGHTFLECK UTILIZATION   Because of the difficulty of separating the transients in assimilation from those of the measuring instruments, an alternative approach to the determination of time constants and the direct characterization of the transients is needed. One method that is independent of system response times is to determine by integration the total carbon gain attributable to a lightfleck and then compare it to an expected amount based on the assumption that the response is instantaneous (29, 139). The ratio of these two has been called the lightfleck utilization efficiency (LUE), which expresses the relative amount that enhance or limit various dynamic responses the carbon gain occurring in response to a lightfleck. The integrations of the assimilation rate must be extended long enough after the lightfleck to include any resulting post-lightfleck losses, such as that due to a

CO_2 burst (187, 188). The amount of CO_2 fixation brought about by the background low light alone should also be subtracted from both so that just the gains and losses due to the lightfleck itself are included.

Lightfleck utilization efficiencies calculated in this way show that there can be considerably higher CO_2 fixation caused by short lightflecks than can be predicted from steady-state measurements. For fully induced leaves of shade plants, LUEs range from 130 to 180% for short (5 s) lightflecks (29, 81, 138, 139). As lightfleck duration was increased, the LUE decreased. This is consistent with there being a certain amount of extra assimilation associated with utilization of lightfleck that is independent of lightfleck duration. However, for long lightflecks (>30 s) LUEs are sometimes less than 100%, probably because other factors that inhibit assimilation increase. LUEs are also much lower in uninduced than in induced leaves and for 1-min lightflecks may only be 30–50%. This occurs because of a limitation on assimilation during the lightfleck itself rather than because of any restriction on post-lightfleck CO_2 assimilation. Although only a few species have been tested, LUEs appear to be higher for shade- than for sun-adapted species and (within a species) for leaves that develop in low as compared to high light (29).

Support for a high-efficiency of use of short, transient light variations also comes from measurements of mean CO_2 uptake or O_2 evolution rates in intermittent light. Mean rates measured in intermittent light for relatively short (0.1 to 10 s) periods typically exceed the mean of the steady-state rates at the same PFDs. Efficiencies of assimilation in intermittent light calculated in a way analogous to the way a LUE is calculated are similar in magnitude to LUEs observed for single lightflecks (62, 90, 97, 139, 151). The highest efficiencies are attained with brief (1 s or less) high-light periods separated by somewhat longer low-light intervals. In some marine macroalgae high efficiencies can be maintained even at low frequencies (47). High efficiencies mean that an equivalent total flux of *high* light is utilized more efficiently when it is given in short intervals than when it is continuous. They do not indicate that the light itself is being utilized more efficiently since the rate under intermittent light can (but rarely does) equal the steady-state assimilation rate at the same *mean* PFD (164). As pointed out by Rabinowich (157), the best utilization of photons for assimilation occurs when they are spread out uniformly over the whole available period—a condition that does not occur naturally in the canopy.

THE CONTRIBUTION OF POST-LIGHTFLECK CO_2 ASSIMILATION   For an LUE to exceed 100%, either the assimilation rate during the lightfleck must be greater than the steady-state rate or there must be a substantial contribution from post-lightfleck CO_2 assimilation. Stitt (177) found higher than steady-state O_2 evolution rates during high-light periods that alternated with low light. The higher rates were attributed to a transient removal of a triose phosphate (TP) utilization limitation, which under steady-state conditions feeds back on the rate of electron transport. The high (50 mbar) CO_2 pressures in these experiments make a TP utilization limitation more likely than at normal ambient (350 μbar) CO_2 pressures (171, 173, 175). Consequently, LUEs at normal CO_2 and O_2 pressures are more likely to result from the continuation of CO_2 fixation for a few seconds after the light decrease.

Post-illumination CO_2 assimilation was first documented by McAlister (109) and since then has been demonstrated to occur in a wide range of higher-plant species and algae (e.g. 93, 112, 113). McAlister attributed it to the buildup of the "CO_2 combining intermediate," and later Laisk et al (92) used the term "assimilatory charge" [originally called by them "assimilatory power" (93), a term also applied to the driving "force" for assimilation (see 72)] to describe the sum of all energetically rich compounds that support continued CO_2 fixation. Laisk and his colleagues have developed novel techniques for quantifying the size of the assimilatory charge and the kinetics of its utilization from transient gas-exchange measurements. They have shown that its magnitude and decay kinetics are generally consistent with the hypothesis that RuBP is the major pool and the kinetics involved are those of Rubisco (93)—especially if the concentrations of other modulators of Rubisco activity are taken into account (92).

Measurements of assimilatory charge in leaves (92, 93, 127) have typically been made for light/dark transitions following long high-light periods, conditions that should maximize its magnitude. To make a significant contribution to the utilization of short lightflecks, the assimilatory charge must rapidly build up to levels large enough to support the net change required for the observed post-lightfleck CO_2 assimilation. Direct evidence for such a buildup was obtained by Sharkey et al (172), who measured photosynthetic metabolite pool sizes just before, at the end of, and 1 min after 5-s lightflecks. In shade-grown leaves of *Alocasia macrorrhiza* and *Phaseolus vulgaris,* pool sizes of RuBP and TP increased sufficiently and then decreased in a manner consistent with use to support post-lightfleck CO_2 assimilation. The net change in metabolite pool sizes was large enough that if all metabolites were used to support post-lightfleck CO_2 fixation the LUEs would have been larger than the measured values. Oxygenation of RuBP and utilization of TP elsewhere would reduce efficiencies so that the agreement with measured LUEs was generally good. For sun leaves, RuBP pools appeared to contribute more than TP pools to the potential for post-lightfleck CO_2 assimilation, whereas in shade leaves larger pools of TP were built up. It may be advantageous to build up TP rather than RuBP since it may prevent the "two kinase" competition (174) resulting from the much higher affinity of phosphoribulokinase than phosphoglycerate kinase for ATP. Such a competition could reduce the buildup of the assimilatory charge. Use of TP requires a supply of ATP. The pool of ATP in chloroplasts is small (178) compared to the amount needed for RuBP regeneration from the TP pool, so post-lightfleck ATP synthesis is required. However, the amounts required, at least in shade leaves, are not in excess of those measured for post-illumination ATP synthesis in isolated chloroplasts (70). Pool sizes of NADPH in chloroplasts are small (182) and therefore contribute relatively little to the assimilatory charge.

Further evidence for the rapid buildup of an assimilatory charge has come from simultaneous measurements of O_2 and CO_2 exchange (Figure 2). Under steady-state conditions, the rates of O_2 and CO_2 exchange are coupled via ATP/ADP and $NADPH_2$/NADP. Under transient conditions, however, these steps can become uncoupled as pools of PGA are reduced to build up the assimilatory charge or, after a lightfleck, as the assimilatory charge is drawn down. Kirschbaum & Pearcy (81), studying *Alocasia* leaves, and Kiirats (79), studying sunflower leaves, observed a burst of O_2 evolution at the beginning of a lightfleck, while CO_2 uptake increased more slowly (Figure 2). After the lightfleck, the O_2 evolution decreased immediately while CO_2 exchange continued at a slowly declining rate for about 10 s. The total CO_2 and O_2 exchange were about the same. For *Alocasia,* the maximum O_2 evolution rate at the peak of the burst was over four times the steady-state light-saturated O_2

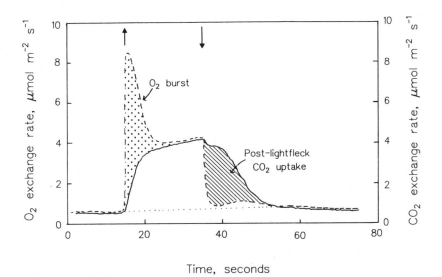

Figure 2   Simultaneous responses of O_2 (dashed line) and CO_2 exchange (solid line) to a 20-s lightfleck showing the transient uncoupling of O_2 evolution and CO_2 uptake. The background PFD was 10 μmol photons $m^{-2} s^{-1}$. and the lightfleck was 475 μmol photons $m^{-2} s^{-1}$. The CO_2 and O_2 pressures were 350 and 1780 μbar, respectively. Redrawn from Ref. 82.

evolution or CO_2 uptake rate. An O_2 burst has also been noted at the beginning of induction both with leaf disc oxygen electrode measurements (156) and photoacoustic techniques (104).

Bursts of O_2 evolution have been attributed to reduction of the plastoquinone pool in the electron transport chain (104). This is unlikely, however, to contribute to a major fraction of the O_2 burst observed during lightflecks or to a significant fraction of the assimilatory charge. The pool of plastoquinone in leaves is on the order of 100 nmol mg^{-1} Chl (3). In broken chloroplasts this is the only pool large enough to account for an O_2 burst (76). Experiments in predarkened, intact sunflower leaves show two O_2 bursts with the smaller first burst (6–8 μmol m^{-2}) associated with plastoquinone reduction and the second (20–34 μmol m^{-2}) with 3-phosphoglyceric acid (PGA) reduction (94). By contrast, transients from low to high light revealed only a single, although kinetically complex, burst. In low light, plastoquinone should remain partially reduced (perhaps mostly in shade leaves) so that the net available pool to accept electrons is smaller than the total. Moreover, PSI activity would be required for its utilization after the lightfleck, which at the limiting light could only occur at the expense of PSII activity, and hence O_2 evolution. No such inhibition of O_2 evolution after lightflecks was apparent (81). An inhibition of O_2 evolution has been observed after high-to-low light transitions following longer exposures to high light in saturating CO_2 (179), but this may result from transient over- and undershoots of NADPH and ATP supply that feed back on electron transport rate. It can be concluded that the O_2 burst at the beginning of a lightfleck is a consequence of accelerated electron transport required for the rapid reduction of PGA and hence buildup of the assimilatory charge.

Factors determining the size and rate of buildup of the assimilatory charge are not well understood. Laisk et al (93) found substantial variation among species and among individual leaves, but no clear ecological or physiological correlates emerge from this data set. The maximum assimilatory charge in sunflower leaves shifted in parallel with mesophyll conductance when leaves were acclimated to high and low PFD (127), presumably reflecting the general increase in pool sizes as photosynthetic capacity increases (170). This parallel shift would not in itself alter lightfleck utilization efficiency. Rapid buildup of the assimilatory charge requires an excess of electron transport capacity over carboxylation capacity, for which there is now good evidence, especially in shade leaves (41, 43, 170, 177). Large PGA pool sizes should favor a large O_2 burst. PGA pool sizes in leaves are relatively high in low light (9, 23) and even in the dark (180). Moreover, for induced leaves, PGA pool sizes are further increased following a high-to-low light transition (154) or a lightfleck (172). Buildup of TP could also feed back and restrict PGA reduction, but mass-action ratios for the PGA-to-TP steps were never far from equilibrium despite large variations in flux (42). The amount of post-lightfleck ATP synthesis may depend at least in part on the buffering capacity of the thylakoids. At present there is no way to assess thylakoid buffering directly in leaves.

THE POST-LIGHTFLECK CO_2 BURST   Leaves of C_3 plants release a burst of CO_2 upon darkening. The burst results from metabolism of a residual pool of photorespiratory metabolites (6). The size of this post-illumination (PI) burst has been studied extensively (22, 33, 44, 143–145) and shown to depend on PFD, CO_2, and O_2 in a manner consistent with the oxygenation of RuBP via Rubisco and the subsequent photorespiratory metabolism of the product, 2-phosphoglycolate. A similar, post-lower-illumination (PLI) burst occurs after high-to-low light transitions (187, 188). The occurrence of a PLI burst following a lightfleck depends on the lightfleck duration and the leaf induction state. No PLI burst is apparent after short (<20 s) lightflecks in either induced or uninduced leaves, even after relatively long (1 min) lightflecks (29). The dynamics of the PLI burst following lightflecks is consistent with the requirement for (*a*) buildup of large pools of photorespiratory metabolites before high rates of CO_2 evolution occur and (*b*) a relatively long time constant for their metabolic processing. Palovsky & Hak (131) have modeled the PI burst using first-order kinetics as an approximation of the generally exponential decay of the apparent rate of CO_2 evolution. Time constants for this decay have been estimated for *Alocasia* leaves at about 15 s (L. Gross, M. Kirschbaum, and R. Pearcy, unpublished observations), and inspection of PI bursts reported in the literature (44, 143, 188) suggests that values for other species may be similar. The decay can be approximated by a first-order reaction, but it is clearly more complex. In the path to CO_2 evolution five separate reactions are involved, along with metabolite traffic among the chloroplast, peroxisome, and mitochondria (6). Support for the hypothesis that pools must build up to large values is provided by the observation that 2–4 min of illumination of *Pelargonium* leaves are required before the PLI burst is maximized (188). At steady-state photosynthesis in high light, large pools of glycolate [1200 nmol C mg^{-1} Chl in sunflower (19)] and glycine [2400 nmol C mg^{-1} Chl in spinach (155)] accumulate. One reason for the long time constant for the decay may be the time required for the substantial metabolite traffic in the photorespiratory pathway.

If the above hypotheses are correct, then the buildup of photorespiratory metabolites in short lightflecks is small and the resulting low rates of CO_2 evolution are spread to undetectable levels. Under steady-state conditions, the rate of photorespiration at a given O_2 and intercellular CO_2 pressure is proportional to the rate of CO_2 assimilation (53). For transient photosynthesis the integral of photorespiration must be approximately proportional to the integral of photosynthesis during the lightfleck. The relationship is only approximate because c_i changes during the lightfleck. The burst will be reduced early in induction because Rubisco is down-regulated. An additional consequence of the lag in CO_2 evolution is that CO_2 uptake rates early in a lightfleck should be higher by an amount equivalent to the photorespiratory CO_2 to be released later on. It has not been possible to detect the higher rates, possibly because other limitations on the acceleration of photosynthesis and instrument response times obscure it. Since this is simply a delay there should be no net effect on the net lightfleck carbon gain except possibly a minor influence due to the dynamics of c_i.

In recent experiments (154, 155), Prinsley and her colleagues have detected an inhibition of CO_2 uptake following a high-to-low light transition under conditions of high CO_2 or low O_2 pressures where photorespiration should be suppressed. The cause of this inhibition is unclear and its occurrence in C_3 plants under normal atmospheric gas concentrations is difficult to ascertain because of interference from the PLI burst. A similar inhibition occurs in the C_4 plant *Zea mays* (45) and develops after lightflecks in the C_4 shade species *Euphorbia forbesii* (139). In these plants, involvement of features of the C_4 cycle may be important (45). For C_3 species, CO_2 inhibition may involve (*a*) a TP utilization limitation on RuBP regeneration due to an overshoot in sucrose synthesis (155) and (*b*) transient inhibition of Rubisco by high PGA levels (154), but neither of these factors wholly explains the observed changes in metabolite pool sizes and fluxes. Whether or not this inhibition represents a significant loss of carbon during lightflecks is unclear.

Limitation of Sunfleck Use by the Photosynthetic Induction Requirement

When a leaf that has been in low light for an extended time is subjected to a sudden increase in PFD, the rate of CO_2 assimilation increases only gradually; 10–30 min or more may be required before a steady-state rate is achieved (28, 86, 134). This response contrasts with the very rapid increase in assimilation, discussed in the previous section, that occurs in leaves previously illuminated at high PFD but then darkened and reilluminated within a few minutes. The slow increase in assimilation occurring upon the initial increase in PFD is the well-known induction requirement of photosynthesis (49, 128, 190), which results from slow light-regulation of photosynthetic enzymes and light-driven stomatal opening. Changes in these factors determine the "readiness" of a leaf to respond to a sudden increase in light and hence are of great importance in determining the utilization of sunflecks. The occurrence of a sunfleck will cause induction to commence and apparently to continue even in the low light between sunflecks. Thus, the occurrence of a sunfleck acts in effect to prime the leaf so that it is better able to utilize subsequent sunflecks.

At the biochemical level, both the light-regulation of Rubisco and the regeneration of RuBP appear to be involved in the induction response but at different times.

LIMITATION BY THE "FAST-INDUCING COMPONENT"   During the first 1–2 min of induction, the biochemical capacity for CO_2 fixation derived from gas exchange is lower than expected on the basis of modeled kinetics of RuBP-saturated Rubisco activity. Thus another limitation may be present at this time (82). In the absence of specifics, this limitation was labeled the "fast-inducing component." This phase was complete in 60–90 s, while deactivation following a return to low light exhibited a half time of about 300 s. The extent of deactivation was dependent on the PFD during the low-light period. The CO_2 dependence of this fast-inducing phase was consistent with a limitation in RuBP regeneration rather than some fast response of Rubisco. Kobza & Edwards (86) found that induction in wheat leaves following a 5-min dark period is limited by RuBP supply, whereas after 30-min or 12-hr dark periods Rubisco activity appeared to be limiting. A limitation in RuBP supply could be caused by inactivation of one of the light-regulated enzymes in the RuBP regeneration path, such as fructose-1,6-bisphosphatase (86, 98, 181) or sedoheptulose-1,7-bisphosphatase (194). However, the dynamics of light-regulation of fructose-1,6-bisphosphatase, for example (86, 98), especially the slower inactivation in the dark than in low light, are not entirely consistent with the dynamics of the fast-inducing component. Other light-regulated enzymes, such as phosphoribulose kinase (5, 99) and the chloroplast coupling factor (89), appear to act more as a switch and are fully active at PFDs equal to those of understory shade. Alternatively, the fast-inducing component may arise from a requirement for buildup of Calvin-cycle metabolite levels (49, 86, 100). This depends on regulation of the branch points (*a*) at the TP translocator for carbon flux out of the chloroplast and (*b*) at hexose-phosphate either for starch synthesis or for RuBP regeneration [see reviews by Woodrow & Berry (192) and Edwards & Walker (49)]. Further work directly comparing the induction loss with enzymatic activities and pool sizes is needed to determine the nature of this limitation.

Since the fast-inducing component has a much shorter time constant for a decrease than Rubisco activity or stomatal conductance, it can become a significant limitation to the use of sunflecks that follow short (2–10 min) low-light periods in leaves that have previously been induced. The fast-inducing component may play an important role when a series of sunflecks results in an increase in stomatal conductance and Rubisco activity but then a period of several minutes intervenes before the next sunfleck. Rubisco activities and stomatal conductances would remain high over this period, and the decrease in the fast-inducing component would attain a greater significance in the overall limitation of the increased assimilation rate in the subsequent lightfleck. These conditions occur frequently in forest understories. A rapid induction loss was observed in a redwood forest understory plant, *Adenocaulon bicolor* (150). This induction loss developed maximally after only 5 min of low light before there was any significant stomatal closure. About 1 min was required to overcome this limitation after a return to high light. This limitation was less apparent after a 10- or 20-min exposure to low light than after just 5 min. The nature of this limitation is unclear, but under some circumstances it will obviously limit sunfleck use by this species. It appears to be less important in *Alocasia* (see 82) than in *A. bicolor*. Why these species vary with respect to the role of this limitation is unclear.

LIMITATION BY LIGHT MODULATION OF RUBISCO ACTIVITY   Evidence that light modulation of Rubisco plays a role during induction was first provided by Perchorowicz et al (142), who reported similar time courses for photosynthesis and Rubisco activity in wheat leaves. Moreover, RuBP pool sizes were large early in induction but then decreased, an observation suggesting that a limitation imposed by Rubisco was being removed. These time courses provide correlative but equivocal evidence for the involvement of light regulation of Rubisco in induction since stomatal opening, which was not followed, should also have occurred. Low intercellular CO_2 pressure due to a stomatal limitation would also cause RuBP pool sizes to be large (9, 170). Comparisons of the time course of the increase in Rubisco activity and the mesophyll conductance in *Alocasia* (134) and *Spinacia oleracea* (193) leaves provide more convincing evidence. In *Alocasia* the time constant for the increase in mesophyll conductance and Rubisco activity were 4.7 and 4.3 min, respectively. In *S. oleracea,* the comparable time constants were 5 min for both. Furthermore, the time courses for the loss of induction and the decline in Rubisco activity were similar and much slower (time constants of 22–30 min) than the decreases in both *S. oleracea* and *Alocasia*.

Short-term light modulation of Rubisco activities in leaves is now known to be under the control of two primary mechanisms. Activation occurs via the ordered binding of CO_2 and Mg^+ to a lysine residue [carbamylation (114)]. In addition, a tight-binding inhibitor, carboxyarabinitol-1-phosphate (CA1P), which binds to the active site of Rubisco, is synthesized in the dark or in low light and degraded in high light (13, 167, 169). Under some conditions, RuBP may also bind to decarbamylated sites that are incapable of catalysis, a binding that subsequently inhibits activation (21, 25). The enzyme Rubisco activase is apparently involved in removal of RuBP and CA1P from the active site via an ATP-dependent reaction (162). There are substantial species (87, 189) and even varietal (20) variations in the degree to which each of these mechanisms contributes to the in vivo light regulation of Rubisco. All species probably have Rubisco activase (163). In some (e.g. *Spinacia oleracea*), CA1P is not detectable; in others (e.g. *Phaseolus vulgaris*), the regulation appears to be controlled almost exclusively by the level of CA1P. In still others (e.g. *Glycine max, Alocasia macrorrhiza*) the regulation is mixed. The net result, however, appears the same regardless of the contributions of each mechanism. This assertion is supported by the close correlation between the rate of photosynthesis per unit of Rubisco protein and the initial specific activity (initial k_{cat}) of Rubisco measured in rapidly prepared extracts from leaves from all three types equilibrated to different PFDs (87). The initial k_{cat} assayed in this fashion should be proportional to the in vivo activity.

It is the rate of modulation of Rubisco activities by these mechanisms that is most important for the dynamic responses of photosynthesis. Kobza & Seemann (88) found no consistent difference in the rates of increase of initial k_{cat} of Rubisco following an increase in light among three species differing in their utilization of CA1P for regulation. Comparison of increases in both mesophyll conductance and initial k_{cat} of Rubisco reported for spinach (193) and *Alocasia* (136) shows a remarkable similarity. Thus it appears there is relatively little variation for the rates of increase of initial k_{cat} even among species adapted to very different light environments. The in vivo regulation of Rubisco is hysteretic, the decrease in initial k_{cat} being slower than the increase. The rate of decline in initial k_{cat} following a decrease in light may depend on both the species and the light transition. The decrease was particularly slow in *Alocasia* when transferred from high to low PFD (168) and in spinach when the transition is to dark (193), but it was faster in spinach following a transition from 1100 to 200 μmol photons m^{-2} s^{-1} (88) rather than to dark. Moreover in *Phaseolus vulgaris*, k_{cat} decreased faster when the transition was to 200 μmol photons m^{-2} s^{-1} than when it was to either higher or lower PFDs (including darkness) (88). The significance of the generally hysteretic response of Rubisco modulation to the utilization of sunflecks is discussed below.

LIMITATION BY STOMATAL CONDUCTANCE   Calculations of c_i from assimilation rates and stomatal conductance during induction have shown that under most circumstances, the changes in c_i were relatively modest, leading to the conclusion that the limitation that is relaxing at least during the first 10 min of induction is biochemical in nature (28, 139, 186). This approach may often underestimate the role of the stomata since it has been shown to be necessary to take into account the small cuticular conductance to water vapor when, as is often the case early in induction, the stomatal conductance is low (83). Moreover, if stomata open in a "patchy" fashion (46, 184) then c_i will also be overestimated and the role of stomata underestimated. The occurrence of patchy stomatal opening during induction has not been systematically evaluated. However, unlike the case in studies of the effects of abscisic acid on photosynthesis where the role of patchiness is best documented (46, 184), there is evidence of a biochemical limitation independent of patchiness during induction. The problem for induction is therefore not whether the limitation on photosynthesis is stomatal or biochemical, but how much each limits assimilation.

The extent to which stomata limit assimilation during induction depends on the value of stomatal conductance at the beginning of induction. Usually, stomatal conductance in the shade is high enough (16, 133, 149) that only a small additional limitation is likely. Nevertheless, the trajectory of stomatal conductance during induction is important in determining the trajectory of assimilation through its effect on c_i. For measurements with *Alocasia* in which the initial stomatal conductances were varied with treatments at different humidities (83), the trajectory of assimilation during induction was hyperbolic when the initial stomatal conductance was high (20–40 mmol m^{-2} s^{-1}), whereas it was distinctly sigmoidal when the initial stomatal conductance was low (5–10 mmol m^{-2} s^{-1}). As a result, the assimilation rate achieved 1 min after the light increase varied by a factor of seven and was positively correlated with stomatal conductance.

Modulation of Induction State in Natural Light Regimes

In forest understories, full induction is likely to occur only during relatively rare sunflecks lasting longer than 10–20 min or where several shorter sunflecks are grouped together. For much of the remaining time, induction state will vary in some complex manner, increasing during periods of sunfleck activity and decreasing during shade periods. The hysteretic nature of both light modulation of Rubisco activities and stomatal conductance is important in determining the time course of induction state and therefore the "readiness" of a leaf to utilize a sunfleck. This hysteresis accounts for the observations (a) that activities of Rubisco are maintained higher in soybean leaves under alternating high/low light than under continuous light of the same mean PFD (59) and (b) that stomatal conductance remains high in shade leaves exposed to periodic sunflecks (137). Stomatal conductances of *Alocasia macrorrhiza* respond to a lightfleck with first a slight lag of 1–2 min and then an increase that may continue 10–20 min after the sunfleck before a slower closing response is initiated (80). When two sunflecks are given in succession the overall effect on opening is additive. These responses are similar to the dynamics of stomatal conductance occurring in response to a blue-light pulse (8, 74, 201). The blue-light response is known to involve a hysteretic activation of electrogenic ion pumping in guard cells (7) that is much faster than the stomatal response. However, it has been possible to model the stomatal responses to lightflecks (80) successfully by coupling a fast but hysteretic component with two slower, nonhysteretic components.

It is doubtful that the stomatal opening occurring in response to a sunfleck is fast enough under most circumstances to have much impact on the carbon gain attributable to that sunfleck itself. However, the carryover and indeed even the increase in stomatal conductance between sunflecks may act to enhance utilization of subsequent sunflecks, especially when combined with the hysteretic behavior of the light regulation of Rubisco activities. Shade-tolerant trees were reported to have faster stomatal opening than shade intolerant trees, but in both cases the closing response was slower than the opening (195). With respect to sunfleck utilization, a faster response would enhance induction during subsequent sunflecks but only when combined with a slow closing response.

These slow stomatal responses to sunflecks extract a penalty in terms of water use efficiency (ratio of the photosynthetic to the transpiration rate) because the stomata remain more open during low-light periods when assimilation is already strongly limited by light. In an understory environment, a lower water-use efficiency may not be too great a cost because (a) the benefit is extra energy and carbon capture, and (b) transpiration rates are restricted anyway by high humidities. There is evidence, however, from species native to higher-light environments that the dynamics of their stomatal behavior may be different from those of species native to understories discussed above. In some species, the dynamic behavior of the stomata may be such that a greater priority appears to be placed on water-use efficiency. Stomatal conductance of herbaceous subalpine species exhibited large and rapid changes during sun-shade transitions that simulated the effect of moving clouds. Under these conditions, water-use efficiencies remained relatively high throughout the cycle (84, 85) despite a cost in terms of carbon gain. Comparative measurements on woody species, however, revealed little decrease in stomatal conductance during the shade period. Consequently, these woody species had lower water-use efficiencies than the herbaceous species but less of a restriction on carbon gain at the beginning of a high-light period. The differences may be related to the greater buffering of water potential in woody plants (85). However, this explanation does not account for the apparent differences between these subalpine herbs and understory herbs. No data are available regarding the time course of the biochemical limitations in these subalpine plants.

At present, only relatively crude estimates can be made of the extent to which induction may limit utilization of sunflecks in the field. Comparison of measured carbon gain in an understory to the predictions of a steady-state model fitted to the light-response curves showed that the model overestimated daily photosynthesis by 10–20% (149). This discrepancy could be explained by a 10–20% (or greater) limitation by induction. The induction response limited the capacity of mid-canopy soybean leaves to respond to a light increase created by parting the canopy (141). The photosynthetic rate achieved 1 min after the canopy was parted was, on average, about 50% of the rate achieved 20 min later. This lag was shown to involve both a stomatal and a Rubisco light-regulation limitation.

ENVIRONMENTAL CONSTRAINTS ON SUNFLECK UTILIZATION

Temperature and Water Stress

Leaves of understory plants in the shade have temperatures that are close to the ambient air temperature. During bright sunflecks, however, leaf temperatures can increase as much as 8–20°C (108, 176, 197, 198), reaching values that in some circumstances cause heat damage and leaf necrosis (158). High radiation loads during prolonged sunflecks, when combined with the large leaf sizes, low stomatal conductance, and low wind speeds characteristic of the understory promote high leaf temperatures during sunflecks. Since shade-grown leaves contain only about 50–100 g H_2O m^{-2} leaf (51) the initial temperature rise can be fast (1–2° C s^{-1}) until energy dissipation increases to balance the increased input. These temperature increases have important consequences for the water relations of understory plants in sunflecks. Measured transpiration rates of understory plants are typically 2–5 times as high in sunflecks as in shade (50, 176). Consequently, water potentials often decline rapidly during a sunfleck but then recover afterwards (50, 196, 199). Wilting of understory plants during sunflecks often occurs even when soil water potentials are high (50). Four of five herbs studied by Smith (176) in a subalpine forest understory exhibited wilting in prolonged sunflecks whereas two shrub species did not. Woodward (196) found rapid decreases in water potentials and shoot extension rates of the understory herb *Circeaea lutetiana* in sunflecks. Decreased shoot-extension rates persisted for at least 30 min after the sunfleck when water potentials had mostly recovered. It is not known if there might be some compensatory increase in extension at some later time (1) following release from short-term water stress.

   Whether the observed wilting is truly stressful in the long run is not known. Clearly, the leaves of these plants persist in spite of multiple stress episodes in a day, perhaps for many days, so it appears there may be little long-term effect. Both stomatal conductance and assimilation can remain relatively high in wilted leaves (31, 169) provided that the wilting periods are not too long (85). Decreases in water potential and hence wilting occur because the rate of water acquisition is insufficient to cope with the high transpiration rates during prolonged sunflecks. In shade-grown plants, biomass is partitioned more towards leaves at the expense of roots, maximizing light capture (17). In evolutionary terms, therefore, selection should have balanced any detrimental effects of wilting, which could be alleviated by increased allocation to roots, against the benefits of increased allocation to leaves. This view is supported by the work of Young & Smith (197) with the congeneric shade species *Arnica cordifolia* and *A. latifolia*, which both exhibit wilting responses in sunflecks. *A. cordifolia* is found in more shaded habitats with shorter, less intense sunflecks, and wilts more rapidly than *A. latifolia* when exposed to prolonged periods of high light. Excavation revealed a more extensive root system in *A. latifolia*. In order to grow in sites with abundant or long sunflecks a plant may need a more extensive root system.

Photoinhibition during Sunflecks

Plants from shaded environments are particularly susceptible to photoinhibition, which reduces the capacity for both light-saturated and light-limited photosynthesis after exposure to bright light (4, 17, 152, 170). The extent and significance of photoinhibition during sunflecks, however, are unclear. No convincing evidence for photoinhibition is apparent in the daily courses of CO_2 exchange under natural sunfleck regimes (133, 137, 149). Sunflecks during these measurements were for the most part brief so that it is possible that any resulting photoinhibition was small and the rate of recovery between sunflecks was sufficient to prevent any cumulative effect. Rate constants for recovery from photoinhibition are greater when photoinhibition is moderate than when it is severe (34). Powles & Björkman (153), using low-temperature fluorescence measurements, found evidence for photoinhibition during relatively long (20 and 50 min) sunflecks in redwood forest understory plants. Leaves of *Oxalis oregona* rapidly fold down during sunflecks to avoid high light. When leaves were restrained to remain horizontal, an average 47% decrease in variable fluorescence (F_V) during sunflecks was observed. By contrast, F_V was only reduced 9% in leaves allowed to fold. Adjacent individuals of *Trillium ovatum*, which did not exhibit leaf folding, had reductions in F_V equivalent to those for the restrained *Oxalis* leaves. In the restrained *Oxalis* leaves, a 30% reduction in the CO_2 assimilation rate occurred in the low light following the photoinhibitory sunfleck. By contrast, no reduction was observed for the unrestrained leaves. Only a few understory species exhibit leaf folding in response to sunflecks. Raven (159) has analyzed the benefits and costs of avoidance of photoinhibition during sunflecks via leaf folding versus the alternative of suffering photoinhibition followed by repair. For short (2-min) sunflecks, repair of damage was energetically favored; for long (30-min) sunflecks, avoidance via leaf folding was favored.

Until a few years ago it was assumed that photoinhibition involved primarily a damage to PS II by excessive light. However, it is now known that regulatory mechanisms (34, 37, 73), possibly involving the zeaxanthin cycle in chloroplast membranes (35), increase nonradiative dissipation of excess energy in leaves exposed to high light and hence may lead to an avoidance of damage. This regulation, however, is still "photoinhibitory" since quantum yields are transiently reduced and the PFD required for saturation should also be increased. However, increased nonradiative dissipation generally relaxes in 30–60 min following a transfer of the leaf back to low light, whereas repair of damage to PSII may require many hours to days (37, 165). Rapid regulation of nonradiative dissipation may also be important in shade plants in maintaining an efficient use of low light while at the same time minimizing photoinhibitory damage in sunflecks. It has been shown recently that nonradiative dissipation increases rapidly in shade-grown leaves following a low-to-high light transition (75). The enzymatic conversion of zeaxanthin to violoxanthin in high light is slow, but if a background level of zeaxanthin is maintained then nonradiative dissipation develops and relaxes much more quickly (36). It remains to be seen how zeaxanthin levels are regulated in shade leaves exposed to natural sunfleck regimes and whether this regulation indeed provides protection against damage to PS II under these conditions.

CONCLUDING REMARKS

Photosynthesis is an unusual biological process because it must respond to large fluctuations in the supply of one of its primary inputs, photons, on short time scales. In most biological systems, homeostatic mechanisms operate to even out the physiological effects of variation in resource supply or environment. For photosynthesis, however, a reasonable expectation, especially in light-limited environments, is that the capture of light energy and conversion to carbohydrate should be maximized. Significant long-term energy storage for use in growth can be achieved only through the end products of photosynthesis. Woodrow & Berry (192) have discussed the conflict between maximizing energy gain and providing stable supplies of carbohydrate for growth, and have shown how regulatory mechanisms in the photosynthetic apparatus achieve a compromise. Optimal allocation of internal resources occurs when the control of the flux is distributed among the various components rather than residing in a single limiting step. Acclimation to different light intensities during growth results in shifts in allocation of resources between light-harvesting and carboxylation capacity consistent with maintaining a balance among limitations (126, 170). However, in transient-light regimes an optimal balance cannot be achieved simply by allocation since it occurs only on a much longer time scale. Regulatory mechanisms appear to function to bring into balance the supply of substrates and capacity for use of substrates in photosynthesis. These processes presumably improve the steady-state efficiency of photosynthesis but how is not entirely clear.

The dynamic responses of photosynthesis are among the most visible manifestations of the complex regulation of carbon metabolism. Steady-state photosynthesis can be described remarkably well by models that entail essentially no regulatory processes (24). Although more information is needed on this point, it may be that the compromises achieved by regulation cost little in terms of the steady-state rate. Because the regulatory steps are slow, however, they impose a time-dependent limitation on photosynthetic rate. If light modulation of Rubisco activity and stomatal conductance were faster, it would appear to be possible to increase the photosynthetic utilization of sunflecks. Ideally, a rapid induction gain and a slow loss could be expected to hold induction state high and hence enhance sunfleck utilization. Rates of stomatal opening and closing vary among species, but more rapid stomatal opening seems to be coupled to a more rapid closing (63). More rapid Rubisco regulation could in theory be achieved by increased Rubisco activase activity or rates of CA1P synthesis and degradation, but the energetic cost of this faster regulation might outweigh any benefits (192). Moreover, there would only be a significant advantage to a faster regulation of Rubisco if the stomata also responded more rapidly. The generally similar increase in Rubisco activity and stomatal conductance may be advantageous in preventing low intercellular CO_2 pressures during high PFD and hence increased photorespiration because of unfavorable internal leaf O_2 to CO_2 ratios.

In forest understories, the effect of the dynamic responses will be strongly dependent on the timing and duration of sunflecks. When sunflecks are frequent but short in duration, post-lightfleck CO_2 fixation may make a substantial contribution to carbon gain. However, when sunflecks are longer and more infrequent, then the induction requirement will probably limit carbon gain. Periods of frequent short sunflecks are typically separated by periods with few or none so that both processes are likely to play an important role.

It is currently unknown what role transient responses play in canopy photosynthesis of crops. Clearly, in a closed canopy most of the leaves are subjected to dynamic light environments, and therefore induction and post-lightfleck CO_2 fixation may play an important role. The great preponderance of studies of crop photosynthesis have concentrated on the upper fully-lit leaves and on steady-state measurements. Moreover, crop models developed to date are inherently steady state. The effects of photosynthetic dynamics on a canopy level are probably within the $\pm$ 10–15% range of uncertainty of these models. While the role of dynamics at a canopy level may therefore seem small, as pointed out by Cowan (32), natural selection has no such prejudice against gains of this magnitude, nor is it necessary that artificial selection be so constrained. Manipulation either of (a) canopy structure to create light environments with different dynamic characteristics or of (b) the dynamic characteristics of the photosynthetic apparatus itself might yield increases in canopy photosynthesis. Before undertaking such an improvement program, however, it will be necessary to understand how the dynamic responses scale up to a whole canopy level so that a reasonable assessment of the potential gains can be made.

ACKNOWLEDGMENTS

This review is based to a large extent on research supported by grants from the National Science Foundation and the US Department of Agriculture, Competitive Grants Program. I thank J. A. Berry, A. J. Bloom, L. J. Gross, J. Kobza, B. D. Moore, and J. R. Seemann for their helpful discussions and comments on the manuscript. Also, the preprints of papers sent by colleagues were much appreciated.

Literature Cited

1. Acevedo, E., Hsiao, T. C., Henderson, D. W. 1971. Immediate and subsequent growth responses of maize leaves to changes in water status. *Plant Physiol.* 48:631–36
2. Amtmann, R. 1985. Data acquisition system for wind induced tree vibration. In *The Forest-Atmosphere Interaction,* ed. B. A. Hutchinson, B. B. Hicks, pp. 149–59. Dordrecht: Reidel
3. Anderson, J. M., Chow, W. S., Goodchild, D. J. 1988. Thylakoid membrane organisation in sun/shade acclimation. *Aust. J. Plant Physiol.* 15:11–26
4. Anderson, J. M., Osmond, C. B. 1987. Shade-sun responses: compromises between acclimation and photoinhibition. In *Photoinhibition,* ed. D. J. Kyle, C. B. Osmond, C. J. Arntzen, pp. 1–38. Amsterdam/London: Elsevier
5. Anderson, L. E. 1986. Light/dark modulation of enzyme activity in plants. *Adv. Bot. Res.* 12:1–45
6. Artus, N. N., Somerville, S. C., Somerville, C. R. 1986. The biochemistry and cell biology of photorespiration. *CRC Crit. Rev. Plant Sci.* 4:121–47
7. Assmann, S. A., Simoncini, L., Schroder, J. 1985. Blue light activates electrogenic ion pumping in guard cell protoplasts of *Vicia faba. Nature* 318:285–87
8. Assmann, S. M. 1988. Enhancement of the stomatal response to blue light by red light, reduced intercellular concentrations of CO_2, and low vapor pressure differences. *Plant Physiol.* 87:226–31
9. Badger, M., Sharkey, T. D., Von Caemmerer, S. 1984. The relationship between steady-state gas exchange of bean leaves and the levels of carbon-reducing-cycle intermediates. *Planta* 160:305–13
10. Baldocchi, D. D., Hutchinson, B. A. 1986. On estimating canopy photosynthesis and stomatal conductance in a deciduous forest with clumped foliage. *Tree Physiol.* 2:155–68
11. Baldocchi, D. D., Matt, D. R., Hutchinson, B. A., Mcmillen, R. T. 1984. Solar radiation within an oak-hickory forest: an evaluation of the extinction coefficients for several radiation components for fully-leafed and leafless periods. *Agric. Forest Meteorol.* 32:307–22
12. Bartholomew, G. A., Vleck, D., Vleck, C. 1981. Instantaneous measurements of oxygen consumption during pre-flight cooling in spingid and saturniid moths. *J. Exp. Biol.* 90:17–32
13. Berry, J. A., Lorimer, G. H., Pierce, J., Seemann, J. R., Meeks, J., Freas, S. 1986. Isolation, identification, and synthesis of carboxyarabinitol-1-phosphate, a diurnal regulator of ribulose bisphosphate carboxylase activity. *Proc. Natl. Acad. Sci. USA* 84:734–38
14. Bingham, G. E., Gillespie, C. H., McQuaid, J. H. 1981. A miniature, battery powered, pyroelectric detector-based differential infra-red absorption sensor for ambient concentrations of carbon dioxide. *Ferroelectrics* 34:15–19
15. Björkman, O., Ludlow, M. 1972. Characterization of the light climate of a Queensland rainforest. *Carnegie Inst. Wash. Yearb.* 71:85–94
16. Björkman, O., Ludlow, M., Morrow, P. 1972. Photosynthetic performance of two rainforest species in their habitat and analysis of their gas exchange. *Carnegie Inst. Wash. Yearb.* 71:94–102
17. Björkman, O. 1981. Responses to different quantum flux densities. In *Ency-*

clopedia of Plant Physiology, New Series, Vol. 12A, ed. O. L. Lange, P. S. Nobel, C. B. Osmond, H. Ziegler, pp. 57–107. Berlin/Heidelberg/New York: Springer-Verlag

18. Boardman, N. K. 1977. Comparative photosynthesis of sun and shade plants. *Annu. Rev. Plant Physiol.* 28:355–77

19. Bourguin, P. W., Fock, H. P. 1983. Effects of irradiance, CO_2, and O_2 on photosynthetic rate and on the levels of ribulose 1,5-bisphosphate and glycollate in sunflower leaves. *Photosynthetica* 17:182–88

20. Bowes, G., Holbrook, G. P. 1988. *Plant Physiol.* 86(4):5 (Abstr.)

21. Brooks, A., Portis, A. R. 1988. Protein-bound ribulose bisphosphate correlates with deactivation of ribulose bisphosphate carboxylase in leaves. *Plant Physiol.* 87:244–49

22. Bulley, N. R., Tregunna, E. B. 1971. Photorespiration and the postillumination burst. *Can. J. Bot.* 49:1277–84

23. Caemmerer, S. V., Edmondson, D. L. 1986. The relationship between steady-state gas exchange, *in vivo* RuP_2 carboxylase activity and some carbon reduction cycle intermediates in *Raphanus sativus. Aust. J. Plant Physiol.* 13:669–88

24. Caemmerer, S. V., Farquhar, G. D. 1981. Some relationships between the biochemistry of photosynthesis and the gas exchange of leaves. *Planta* 153:376–87

25. Cardon, Z. G., Mott, K. A. 1989. Evidence that ribulose 1,5-bisphosphate (RuBP) binds to inactive sites of RuBP carboxylase *in vivo* and an estimate of the rate constant for dissociation. *Plant Physiol.* 89:1253–57

26. Chazdon, R. L. 1988. Sunflecks and their importance to forest understory plants. *Adv. Ecol. Res.* 18:1–63

27. Chazdon, R. L., Fetcher, N. 1984. Photosynthetic light environments in a lowland tropical forest in Costa Rica. *J. Ecol* 72:553–64

28. Chazdon, R. L., Pearcy, R. W. 1986. Photosynthetic responses to light variation in rain forest species. I. Induction under constant and fluctuating light conditions. *Oecologia* 69:517–23

29. Chazdon, R. L., Pearcy, R. W. 1986. Photosynthetic responses to light variation in rain forest species. II. Carbon gain and light utilization during light-flecks. *Oecologia* 69:524–31

30. Chazdon, R. L., Williams, K., Field, C. B. 1988. Interactions between crown structure and light environment in five rainforest *Piper* species. *Am. J. Bot.* 75:1459–71

31. Chiariello, N. R., Field, C. B., Mooney, H. A. 1987. Midday wilting in a tropical pioneer tree. *Funct. Ecol.* 1:3–11

32. Cowan, I. R. 1986. Economics of carbon fixation in higher plants. In *On the Economy of Plant Form and Function*, ed. T. J. Givnish, pp. 133–70. New York/Cambridge: Cambridge Univ. Press

33. Decker, J. P. 1955. A rapid postillumination deceleration of respiration in green leaves. *Plant Physiol.* 30:82–84

34. Demmig, B., Bjorkman, O. 1987. Comparison of the effect of excessive light on chlorophyll fluorescence (77K) and photon yield of O_2 evolution in leaves of higher plants. *Planta* 171:171–84

35. Demmig, B., Winter, K., Krüger, A., Czygan, F.-C. 1987. Photoinhibition and zeaxanthin formation in intact leaves. A possible role of the xanthophyll cycle in the dissipation of excess light energy. *Plant Physiol.* 84:218–24

36. Demmig-Adams, B., Winter, K., Kruger, A., Czygan, F.-C. 1989. Zeaxanthin and the induction and relaxation kinetics of the dissipation of excess excitation energy in leaves in 2% O_2, 0% CO_2. *Plant Physiol.* 90:887–93

37. Demmig, B., Winter, K. 1988. Characterization of three components of non-photochemical fluorescence quenching and their response to photoinhibition. *Aust. J. Plant Physiol.* 15:163–77

38. Denholm, J. V. 1981. The influence of penumbra on canopy photosynthesis. I. theoretical considerations. *Agric. Meteorol.* 25:145–66

39. Dera, J., Gordon, H. R. 1968. Light field fluctuations in the photic zone. *Limnol. Oceanogr.* 13:697–99

40. Desjardens, R. L., Sinclair, T. R., Lemon, E. R. 1973. Light fluctuations in corn. *Agron. J.* 65:904–8

41. Dietz, K.-J., Neimanis, S., Heber, U. 1984. Rate-limiting factors in leaf photosynthesis. II. Electron transport. *Biochim. Biophys. Acta* 767:444–50

42. Dietz, K.-J., Heber, U. 1984. Rate limiting factors in leaf photosynthesis. I. Carbon fluxes in the Calvin Cycle. *Biochim. Biophys. Acta* 767:432–43

43. Dietz, K. J., Schreiber, U., Heber, U. 1985. The relationship between the redox state of Q_a and photosynthesis in leaves at various carbon dioxide, oxygen and light regimes. *Planta* 166:219–26

44. Doehlert, D. C., Ku, M. S. B., Edwards, G. E. 1979. Dependence of the

postillumination burst of CO_2 on temperature, light, CO_2, O_2 concentration in wheat *(Triticum aestivum)*. *Physiol. Plant.* 46:299–306

45. Doncaster, H. D., Acock, M. D., Leegood, R. C. 1989. Regulation of photosynthesis in leaves of C4 plants following a transition from high to low light. *Biochim. Biophys. Acta* 173:176–84

46. Downton, W. J. S., Loveys, B. J., Grant, W. J. R. 1988. Stomatal closure fully accounts for the inhibition of photosynthesis by abcisic acid. *New Phytol.* 108:263–66

47. Dromgoole, F. I. 1987. Light fluctuations and photosynthesis in marine algae. I. Adjustment of rate in constant and fluctuating light regimes. *Funct. Ecol.* 1:377–86

48. Duncan, W. G., Loomis, R. S., Williams, W. A., Hanau, R. 1967. A model for simulating photosynthesis in plant communities. *Hilgardia* 38:181–205

49. Edwards, G., Walker, D. 1983. C_3, C_4: *Mechanisms and Cellular and Environmental Regulation of Photosynthesis.* Berkeley: Univ. Calif. Press. 542 pp.

50. Elias, P. 1983. Water relation pattern of understory species influenced by sunflecks. *Biol. Plant* 25:68–74

51. Elias, P. 1984. Adaptations of understory plants to exist in temperate deciduous forests. In *Being Alive on Land. Tasks for Vegetation Science,* ed. N. S. Margaris, M. Arianoustou-Farragitaki, W. C. Oechel, pp. 157–65. The Hague: Dr. W. Junk

52. Evans, G. C. 1956. An area survey method of investigating the distribution of light intensity in woodlands, with particular reference to sunflecks. *J. Ecol.* 44:391–428

53. Farquhar, G. D., von Caemmerer, S. 1982. Modelling of photosynthetic response to environmental conditions. See Ref. 17, 12B:549–88

54. Farquhar, G., von Caemmerer, S., Berry, J. 1980. A biochemical model of photosynthetic CO_2 assimilation in leaves of C_3 species. *Planta* 149:78–90

55. Finnigan, J. J. 1979. Turbulence in waving wheat. I. Mean statistics and honami. *Boundary-Layer Meteorol.* 16: 213–16

56. Finnigan, J. J. 1985. Turbulent transport in flexible plant canopies. In *The Forest-Atmosphere Interaction,* ed. B. A. Hutchinson, B. B. Hicks, pp. 443–80. Dordrecht: Reidel

57. Gallegos, C. L., Platt, T. 1982. Phytoplankton productivity and water motion in surface mixed layers. *Deep Sea Res.* 29:65–76

58. Gaudillere, J. P. 1977. Effect of periodic oscillations of artificial light emission on photosynthetic activity. *Physiol. Plant.* 41:95–98

59. Gaudillere, J. P., Drevon, J. J., Bernoud, J. P., Jardinet, F., Euvrard, M. 1987. Effects of periodic fluctuations of photon flux density on anatomical and photosynthetic characteristics of soybean leaves. *Photosynth. Res.* 13:81–89

60. Gay, L. W., Knoerr, K. R., Braaten, M. O. 1971. Solar radiation variability on the floor of a pine plantation. *Agric. Meteorol.* 8:39–50

61. Gerard, V. A. 1984. The light environment of a giant kelp forest: influence of *Macrocystis pyrifera* on spatial and temporal variability. *Mar. Biol.* 84:189–95

62. Gloser, J. 1977. Characteristics of CO_2 exchange in *Phragmites communis* Trin. derived from measurements *in situ*. *Photosynthetica* 11:139–47

63. Grantz, D. A., Zeiger, E. 1986. Stomatal responses to light and leaf-air water vapor pressure difference show similar kinetics in sugarcane and soybean. *Plant Physiol.* 81:865–68

64. Gross, L. J. 1982. Photosynthetic dynamics in varying light environments: a model and its application to whole leaf carbon gain. *Ecology* 63:84–93

65. Gross, L. J., Chabot, B. F. 1979. Time course of photosynthetic response to changes in incident light energy. *Plant Physiol.* 63:1033–38

66. Gutschick, V. P. 1984. Photosynthesis model for C_3 leaves incorporating CO_2 transport, propagation of radiation, and biochemistry. I. Kinetics and their parameterization. *Photosynthetica* 18: 549–68

67. Hahn, B. D. 1984. A mathematical model of leaf carbon metabolism. *Ann. Bot.* 54:325–39

68. Hahn, B. D. 1986. A mathematical model of the Calvin cycle: analysis of the steady state. *Ann Bot.* 57:639–53

69. Hall, A. E. 1979. A model of leaf photosynthesis and respiration for predicting carbon dioxide assimilation in different environments. *Planta* 143:299–316

70. Hangarter, R. P., Good, N. E. 1982. Energy thresholds for ATP synthesis in chloroplasts. *Biochim. Biophys. Acta* 430:154–64

71. Harbinson, J., Hedley, C. L. 1989. The kinetics of P-700$^+$ reduction in leaves: a novel *in situ* probe of thylakoid functioning. *Plant. Cell. Env.* 12:357–69

72. Heber, U., Neimanis, S., Dietz, K. J.,

Viil, J. 1987. Assimilatory force in relation to photosynthetic fluxes. In *Progress in Photosynthesis Research*, ed. J. Biggins, 3:293–99. Dordrecht: Martinus Nijhoff

73. Horton, P., Hague, A. 1988. Studies on the induction of chlorophyll fluorescence in isolated barley protoplasts. IV. Resolution of non-photochemical quenching. *Biochim. Biophys. Acta.* 932:107–15

74. Iino, M., Ogawa, T., Zeiger, E. 1985. Kinetic properties of the blue-light response of stomata. *Proc. Natl. Acad. Sci. USA* 82:8019–23

75. Johnson, G., Horton, P., Scholes, J., Grime, P. 1989. Fluorescence response on step changes in irradiance by plants from different light habitats. *Physiol. Plant.* 76:A165

76. Joliot, P., Joliot, A. 1968. A polarographic method for the detection of oxygen production and reduction of Hill reagent by isolated chloroplasts. *Biochim. Biophys. Acta* 153:625–34

77. Joris, C., Bertels, A. 1985. Incubation under fluctuating light conditions provides values much closer to real in situ primary production. *Bull. Mar. Sci.* 37:620–25

78. Kaitala, V., Hari, P., Vapaavuori, E., Salminen, R. 1982. A dynamic model for photosynthesis. *Ann. Bot.* 50:385–96

79. Kiirats, O. 1985. Kinetics of CO_2 and O_2 exchange in sunflower leaves in darklight transitions. In *Kinetics of Photosynthetic Carbon Metabolism in C_3-Plants*, ed. J. Viil, G. Grishina, A. Laisk, pp. 125–31. Tallin: Valgus

80. Kirschbaum, M. U. F., Gross, L. J., Pearcy, R. W. 1988. Observed and modelled stomatal responses to dynamic light environments in the shade plant *Alocasia macrorrhiza*. *Plant Cell Environ.* 11:111–21

81. Kirschbaum, M. U. F., Pearcy, R. W. 1988. Concurrent measurements of O_2 and CO_2 exchange during lightflecks in *Alocasia macrorrhiza* (L.) G. Don. *Planta* 174:527–33

82. Kirschbaum, M. U. F., Pearcy, R. W. 1988. Gas exchange analysis of the fast phase of photosynthetic induction in *Alocasia macrorrhiza*. *Plant Physiol.* 87:818–21

83. Kirschbaum, M. U. F., Pearcy, R. W. 1988. Gas exchange analysis of the relative importance of stomatal and biochemical factors in photosynthetic induction in *Alocasia macrorrhiza*. *Plant Physiol.* 86:782–85

84. Knapp, A. K., Smith, W. K. 1987. Sto-

matal and photosynthetic responses during sun/shade transitions in subalpine plants: influence on water use efficiency. *Oecologia* 74:62–67

85. Knapp, A. K., Smith, W. K., Young, D. R. 1989. Importance of intermittent shade to the ecophysiology of subalpine herbs. *Funct. Ecol.* 3:753–58

86. Kobza, J., Edwards, G. E. 1987. The photosynthetic induction response in wheat leaves: net CO_2 uptake, enzyme activation, and leaf metabolites. *Planta* 171:549–59

87. Kobza, J., Seemann, J. R. 1988. Mechanisms for the light-dependent regulation of ribulose-1,5-bisphosphate carboxylase activity and photosynthesis in intact leaves. *Proc. Natl. Acad. Sci. USA* 85:3815–19

88. Kobza, J., Seemann, J. R. 1989. Light-dependent kinetics of 2-carboxyarabinitol-1-phosphate metabolism and ribulose-1,5-bisphosphate carboxylase activity *in vivo*. *Plant Physiol.* 89:174–79

89. Kramer, D. J., Crofts, A. R. 1989. Activation of the chloroplast ATPase measured by the electrochromic change in leaves of intact plants. *Biochim. Biophys. Acta* 976:28–41

90. Kriedemann, P. E., Torokfalvy, E., Smart, R. E. 1973. Natural occurrence and photosynthetic utilization of sunflecks in grapevine leaves. *Photosynthetica* 7:18–27

91. Laisk, A., Eichelmann, H. 1989. Towards understanding oscillations: a mathematical model of the biochemistry of photosynthesis. *Philos. Trans. R. Soc. London Ser. B* 323:369–84

92. Laisk, A., Kiirats, O., Eichelmann, H., Oja, V. 1987. Gas exchange studies of carboxylation kinetics in intact leaves. See Ref. 72, 4:245–52

93. Laisk, A., Kiirats, O., Oja, V. 1984. Assimilatory power (postillumination CO_2 uptake in leaves). Measurement, environmental dependences and kinetic properties. *Plant Physiol.* 76:723–29

94. Laisk, A., Oja, V., Kiirats, O., Raschke, K., Heber, U. 1989. The state of the photosynthetic apparatus in leaves as analyzed rapid gas exchange and optical methods: the pH of the chloroplast stroma and activation of enzymes in vivo. *Planta* 177:350–58

95. Laisk, A., Walker, D. A. 1986. Control of phosphate turnover as a rate-limiting factor and possible cause of oscillations in photosynthesis: a mathematical model. *Proc. R. Soc. London Ser. B* 227:281–302

96. Lambert, J. L. 1970. Thermal response of a plant canopy to drifting cloud shadows. *Ecology* 51:143–49

97. Lasko, A. H., Barnes, J. E. 1978. Apple leaf photosynthesis in alternating light. *Hort. Sci.* 13:473–74

98. Leegood, R. A., Walker, D. A. 1982. Regulation of fructose-1,6-bisphosphatase activity in leaves. *Planta* 156:449–56

99. Leegood, R. C., Foyer, C. H., Walker, D. A. 1985. Regulation of the Benson-Calvin cycle. In *Photosynthetic Mechanisms and the Environment*, ed. J. Barber, N. R. Baker, pp. 191–258. Amsterdam/New York: Elsevier

100. Leegood, R., Walker, D. 1980. Autocatalysis and light activation of enzymes in relation to photosynthetic induction in wheat chloroplasts. *Arch. Biochem. Biophys.* 200:575–82

101. Lemeur, R. 1973. A method for simulating the direct solar radiation regime in sunflower, Jerusalem artichoke, corn and soybean canopies using actual stand structure data. *Agric. Meteorol.* 12:229–47

102. Lemeur, R., Blad, B. L. 1974. A critical review of light models for estimating the shortwave radiation regime of plant canopies. *Agric. Meteorol.* 14:255–86

103. Lundegarth, L. 1921. Ecological studies in the assimilation of certain forest plants. *Suom. Bot. Tidskr.* 15:46

104. Malkin, S. 1987. Photoacoustic transients from dark-adapted intact leaves: oxygen evolution and uptake pulses during photosynthetic induction—a phenomenological record. *Planta* 171:65–72

105. Marra, J. 1978. Phytoplankton photosynthetic response to vertical movement in a mixed layer. *Mar. Biol.* 46:203–8

106. Marshall, B., Biscoe, P. V. 1980. A model for C_3 leaves describing the dependence of net photosynthesis on irradiance. I. Derivation. *J. Exp. Bot.* 31:29–39

107. Marynick, D. S., Marynick, M. C. 1975. A mathematical treatment of rate data obtained in biological flow systems under nonsteady state conditions. *Plant Physiol.* 56:680–83

108. Masarovicova, E., Elias, P. 1986. Photosynthetic rate and water relations in some forest herbs in spring and summer. *Photosynthetica* 20:187–95

109. McAlister, E. D. 1939. The chlorophyll-carbon dioxide ratio during photosynthesis. *J. Gen. Physiol.* 22:613–36

110. Miller, E. E., Norman, J. M. 1971. A sunfleck theory for plant canopies. I.

Lengths of segments along a transect. *Agron. J.* 63:735–38

111. Milstein, J., Bremmermann, H. J. 1979. Parameter identification of the Calvin photosynthesis cycle. *J. Math. Biol.* 7:99–116

112. Miyachi, S. 1979. Light-enhanced dark CO_2 fixation. In *Encyclopedia of Plant Physiology. Photosynthesis II*, ed. M. Gibbs, E. Latzko, 6:68–76. Berlin/Heidleberg/New York: Springer-Verlag

113. Miyachi, S., Hogetsu, D. 1970. Light-enhanced carbon dioxide fixation in isolated chloroplasts. *Plant Cell Physiol.* 11:927–36

114. Mizioriko, H. M., Lorimer, G. H. 1983. Ribulose-1,5-bisphosphate carboxylase-oxygenase. *Annu. Rev. Biochem.* 52:507–35

115. Monsi, M., Saeki, S. 1953. Über den Lichtfaktor in den Pflanzengesellschaften und seine Bedeutung für die Stoffproducktion. *Jpn. J. Bot.* 14:22–52

116. Myneni, R. B., Impens, I. 1985. A procedural approach for studying the radiation regime of infinite and truncated foliage spaces. II. Experimental results and discussion. *Agric. Forest Meteorol.* 34:3–16

117. Myneni, R. B., Impens, I. 1985. A procedural approach for studying the radiation regime of infinite and truncated foliage spaces. Part III. Effect of leaf and inclination distribution on nonparallel beam radiation penetration and canopy photosynthesis. *Agric. Forest Meteorol.* 34:183–94

118. Myneni, R. B., Ross, J., Asrar, G. 1989. A review on the theory of photon transport in leaf canopies. *Agric. Forest Meteorol.* 45:1–153

119. Nicholson, H., ed. 1980. *Modelling of Dynamical Systems*, Vol. 1. Stevenage, UK: Peter Peregrinus Ltd. 227 pp.

120. Nilson, T. 1971. A theoretical analysis of the frequency of gaps in plant stands. *Agric. Meteorol.* 8:25–38

121. Norman, J. M. 1980. Interfacing leaf and canopy light interception models. In *Predicting Photosynthesis for Ecosystem Models*, ed. J. D. Hesketh, J. W. Jones, 2:49–67. Boca Raton: CRC Press

122. Norman, J. M., Tanner, C. B. 1969. Transient light measurements in plant canopies. *Agron. J.* 61:847–49

123. Norman, J. M., Welles, J. M. 1983. Radiative transfer in an array of canopies. *Agron. J.* 75:481–88

124. Ohtaki, E., Matsui, M. 1982. Infra-red device for simultaneous measurement of atmospheric carbon dioxide and water vapour. *Boundary-Layer Meteorol.* 24:109–19

125. Oja, V., Laisk, A., Heber, U. 1986. Light-induced alkalinization of the chloroplast stroma in vivo as estimated from the CO_2 capacity of intact sunflower leaves. *Biochim. Biophys. Acta* 849:355–65

126. Osmond, C. B., Chow, W. S. 1988. Ecology of photosynthesis in the sun and shade: summary and prognostications. *Aust. J. Plant Physiol.* 15:1–9

127. Osmond, C. B., Oja, V., Laisk, A. 1988. Regulation of carboxylation and photosynthetic oscillations during sunshade acclimation in *Helianthus annus* measured with a rapid-response gas exchange system. *Aust. J. Plant Physiol.* 15:239–51

128. Osterhout, W. J., Hass, J. R. C. 1919. On the dynamics of photosynthesis. *J. Gen. Physiol.* 1:1–16

129. Oya, V. M. 1984. Fast gasiometric device for investigation of leaf photosynthesis kinetics. *Sov. Plant Physiol.* 30:795–802 (trans. from Russian)

130. Palm, W. J. III. 1983. Modeling, Analysis, and Control of Dynamic Systems. New York: Wiley. 740 pp.

131. Palovsky, R., Hak, R. 1988. A model of light-dark transition of CO_2 exchange in the leaf (post-illumination burst of CO_2). Theoretical approach. *Photosynthetica* 22:423–30

132. Pearcy, R. W. 1983. The light environment and growth of C_3 and C_4 tree species in the understory of a Hawaiian forest. *Oecologia* 58:19–25

133. Pearcy, R. W. 1987. Photosynthetic gas exchange responses of Australian tropical forest trees in canopy, gap and understory micro-environments. *Funct. Ecol.* 1:169–78

134. Pearcy, R. W. 1988. Photosynthetic utilization of lightflecks by understory plants. *Aust. J. Plant Physiol.* 15:223–38

135. Pearcy, R. W. 1989. Radiation and light measurements. In *Plant Physiological Ecology: Field Methods and Instrumentation*, ed. R. W. Pearcy, J. R. Ehleringer, H. A. Mooney, P. W. Rundel. New York: Chapman & Hall. 457 pp.

136. Pearcy, R. W. 1989. Regulation of photosynthetic CO_2 assimilation during sunflecks. In *Photosynthesis*, ed. W. R. Briggs, pp. 407–24. New York: Alan R. Liss

137. Pearcy, R. W., Calkin, H. 1983. Carbon dioxide exchange of C_3 and C_4 tree species in the understory of a Hawaiian forest. *Oecologia* 58:26–32

138. Pearcy, R. W., Chazdon, R. L., Kirschbaum, M. U. F. 1987. Photosynthetic

utilization of lightflecks by tropical forest plants. See Ref. 72, 4:257–60

139. Pearcy, R. W., Osteryoung, K., Calkin, H. W. 1985. Photosynthetic responses to dynamic light environments by Hawaiian trees. The time course of CO_2 uptake and carbon gain during sunflecks. *Plant Physiol.* 79:896–902

140. Pearcy, R. W., Roden, J., Gamon, J. A. 1990. Dynamics of sunflecks in relation to canopy structure in a soybean canopy. *Agric. Forest Meteorol.* In press

141. Pearcy, R. W., Seemann, J. R. 1990. The photosynthetic induction state of leaves in a soybean canopy in relation to light regulation of ribulose-1,5-bisphosphate carboxylase and stomatal conductance. *Plant Physiol.* In press

142. Perchorowicz, J. T., Raynes, D. A., Jensen, R. G. 1982. Measurement and preservation of *in vivo* activation of ribulose-1,5,-bisphosphate carboxylase in leaf extracts. *Plant Physiol.* 65:902–5

143. Peterson, R. B. 1983. Estimation of photorespiration based on the initial rate of postillumination CO_2 release. I. A nonsteady-state model for measurement of CO_2 exchange transients. *Plant Physiol.* 73:983–88

144. Peterson, R. B. 1983. Estimation of photorespiration based on the initial rate of postillumination CO_2 release. I. Effects of O_2, CO_2 and temperature. *Plant Physiol.* 73:983–88

145. Peterson, R. B. 1987. Quantitation of the O_2-dependent, CO_2-reversable component of the postillumination CO_2 exchange transient in tobacco and maize leaves. *Plant Physiol.* 84:862–67

146. Peterson, R. B., Ferrandino, F. J. 1984. A numerical approach to measurement of CO_2 exchange transients by infrared gas analysis. *Plant Physiol.* 76:976–78

147. Peterson, R. B., Sivak, M. N., Walker, D. A. 1988. Carbon dioxide-induced oscillations in fluorescence and photosynthesis. *Plant Physiol.* 88:1125–30

148. Pettersson, G., Ryde-Pettersson, U. 1988. A mathematical model of the Calvin photosynthesis cycle. *Eur. J. Biochem.* 175:661–72

149. Pfitsch, W. A., Pearcy, R. W. 1989. Daily carbon gain by *Adenocaulon bicolor*, a redwood forest understory herb, in relation to its light environment. *Oecologia* 80:465–70

150. Pfitsch, W. A., Pearcy, R. W. 1989. Steady-state and dynamic photosynthetic response of *Adenocaulon bicolor* in its redwood forest habitat. *Oecologia* 80:471–76

151. Pollard, D. F. W. 1970. The effect of

rapidly changing light on the rate of photosynthesis in bigtooth aspen *(Populus grandidentata). Can. J. Bot.* 48: 823–29

152. Powles, S. B. 1984. Photoinhibition of photosynthesis induced by visible light. *Annu. Rev. Plant Physiol.* 35:15–44

153. Powles, S. B., Bjorkman, O. 1981. Leaf movement in the shade species *Oxalis oregana.* II. Role in protection against injury by intense light. *Carnegie Inst. Wash. Yearb.* 80:63–66

154. Prinsley, R. T., Dietz, K.-J., Leegood, R. C. 1986. Regulation of photosynthetic carbon metabolism in spinach leaves after a decrease in irradiance. *Biochim. Biophys. Acta* 849:254–63

155. Prinsley, R. T., Hunt, S., Smith, A. M., Leegood, R. C. 1986. The influence of a decrease in irradiance on photosynthetic carbon assimilation in leaves of *Spinacia oleracea* L. *Planta* 167:414–20

156. Prinsley, R., Leegood, R. 1986. Factors affecting photosynthetic induction in spinach leaves. *Biochim. Biophys. Acta* 849:244–53

157. Rabinowitch, E. I. 1956. *Photosynthesis and Related Processes,* Vol. 2, Part 2. New York: Interscience

158. Rackham, O. 1975. Temperature of plant communities as measured by pyrometric and other methods. In *Light As an Ecological Factor II,* ed. G. C. Evans, R. Bainbridge, O. Rackham, pp. 423–50. Oxford: Blackwell

159. Raven, J. A. 1989. Fight or flight: the economics of repair and avoidance of photoinhibition. *Funct. Ecol.* 3:5–19

160. Reifsnyder, W. E., Furnival, G. M. 1970. Power-spectrum analysis of the energy contained in sunflecks. In *Proc. 3rd Forest Microclimate Symp. Kanaskis Exp. Sta., Seebe, Alberta,* pp. 117–18

161. Reswick, J. B., Taft, C. K. 1971. *Introduction to Dynamic Systems.* Englewood Cliffs, NJ: Prentice-Hall. 294 pp.

162. Robinson, S. P., Portis, A. R. 1988. Involvement of stromal ATP in the light activation of ribulose-1,5-bisphosphate carboxylase/oxygenase in intact isolated chloroplasts. *Plant Physiol.* 86:293–98

163. Salvucchi, M. E., Werneke, J. M., Ogren, W. L., Portis, A. R. 1987. Purification and species distribution of Rubisco activase. *Plant Physiol.* 84: 930–36

164. Sangar, J. C., Geiger, W. 1980. Re-evaluation of published data on the relative photosynthetic efficiency of intermittent and continuous light. *Agric. Meteorol.* 22:289–302

165. Schreiber, U., Bilger, W. 1987. Rapid assessment of stress effects on plant leaves by chlorophyll fluorescence measurements. In *Plant Response to Stress. Functional Analysis in Mediterranean Ecosystems,* ed. J. D. Tenhunen, F. M. Catarino, O. L. Lange, W. C. Oechel, pp. 27–53. Berlin/Heidleberg/New York: Springer-Verlag

166. Schreiber, U., Neubauer, C., Klughammer, C. 1989. Devices and methods for room-temperature fluorescence analysis. *Philos. Trans. R. Soc. London Ser. B.* 323:241–51

167. Seemann, J. R., Berry, J. A., Freas, S. M., Krump, M. A. 1986. Regulation of ribulose bisphosphate carboxylase activity *in vivo* by a light-modulated inhibitor of catalysis. *Proc. Natl. Acad. Sci. USA* 82:8024–28

168. Seemann, J. R., Kirschbaum, M. U. F., Sharkey, T. D., Pearcy, R. W. 1988. Regulation of ribulose 1,5-bisphosphate carboxylase activity in *Alocasia macrorrhiza* in response to step changes in irradiance. *Plant Physiol* 88:148–52

169. Seemann, J. R., Kobza, J., Moore, B. D. 1990. Metabolism of 2-carboxyarabinitol 1-phosphate and regulation of ribulose-1,5-bisphosphate carboxylase activity. *Photosynth. Res.* In press

170. Seemann, J. R., Sharkey, T. D., Wang, J. L., Osmond, C. B. 1987. Environmental effects on photosynthetic nitrogen use efficiency, nitrogen partitioning and metabolite pools in leaves of sun and shade plants. *Plant Physiol.* 84:796–802

171. Sharkey, T. D. 1985. O_2 insensitive photosynthesis in C_3 plants. Its occurrence and a possible explanation. *Plant Physiol.* 78:71–75

172. Sharkey, T. D., Seemann, J. R., Pearcy, R. W. 1986. Contribution of metabolites of photosynthesis to post illumination CO_2 assimilation in response to lightflecks. *Plant Physiol.* 82:1063–68

173. Sharkey, T. D., Stitt, M., Heineke, D., Gerhardt, R., Rashke, K., Heldt, H. 1986. Limitation of photosynthesis by carbon metabolism. II. O_2 insensitive CO_2 assimilation results from triose phosphate utilization limitations. *Plant Physiol.* 81:1123–29

174. Sivak, M. M., Walker, D. A. 1986. Summing-up: measuring photosynthesis *in vivo.* In *Biological Control of Photosynthesis,* ed. R. Marcelle, H. Clijesters, M. Van Poucke, pp. 1–31. Dordrecht: Martinus Nijhoff

175. Sivak, M. N., Walker, D. A. 1986. Photosynthesis in vivo can be limited by phosphate supply. *New Phytol.* 102: 499–512

176. Smith, W. K. 1981. Temperature and water relations patterns in subalpine, understory plants. *Oecologia* 48:353–59

177. Stitt, M. 1986. Limitation of photosynthesis by carbon metabolism. I. Evidence for excess electron transport capacity in leaves carrying out photosynthesis in saturating light and CO_2. *Plant Physiol.* 81:1115–22

178. Stitt, M., Lilly, R. M., Heldt, H. W. 1985. Adenine nucleotide levels in the cytosol, chloroplasts and mitochondria of wheat leaf protoplasts. *Plant Physiol.* 70:971–77

179. Stitt, M., Scheibe, R., Feil, R. 1989. Response of photosynthetic electron transport and carbon metabolism to a sudden decrease in irradiance in the saturating or the limiting range. *Biochim. Biophys. Acta* 973:241–49

180. Stitt, M., Wirtz, W., Gerhardt, R., Heldt, H. W., Spencer, C., et al. 1986. A comparative study of metabolite levels in plant leaf material in the dark. *Planta* 166:354–64

181. Stitt, M., Wirtz, W., Heldt, H. 1980. Metabolite levels during induction in the chloroplast and extrachloroplast compartments of spinach protoplasts. *Biochim. Biophys. Acta* 593:85–102

182. Takahama, U., Shimizu-Takahama, M., Heber, U. 1981. The redox state of the NADP system in illuminated chloroplasts. *Biochim. Biophys. Acta* 637:530–39

183. Tang, Y.-T., Washitani, I., Tsuchiya, T. 1988. Fluctuations of photosynthetic photon flux density within a *Miscanthus sinensis* canopy. *Ecol. Res.* 3:253–66

184. Terashima, I., Wong, S.-C., Osmond, C. B., Farquhar, G. D. 1988. Characterization of non-uniform photosynthesis induced by abcisic acid in leaves having different mesophyll anatomies. *Plant Cell Physiol.* 29:385–94

185. Thornley, J. H. M. 1974. Light fluctuations and photosynthesis. *Ann. Bot.* 38:363–73

186. Usuda, H., Edwards, G. E. 1984. Is photosynthesis during the induction period in maize limited by the availability of intercellular carbon dioxide? *Plant Sci. Lett.* 37:41–45

187. Vines, H. M., Armatige, A. M., Chen, S.-S., Tu, Z.-P., Black, C. C. 1982. A transient burst of CO_2 from leaves during illumination at various light intensities as a measure of photorespiration. *Plant Physiol.* 70:629–31

188. Vines, H. M., Tu, Z.-P., Armitage, A. M., Chen, S.-S., Black, C. C. Jr. 1983. Environmental responses of the post-lower illumination CO_2 burst as related to leaf photorespiration. *Plant Physiol.* 73:25–30

189. Vu, C. V., Allen, L. H., Bowes, G. 1984. Dark/light modulation of ribulose bisphosphate carboxylase activity in plants from different photosynthetic categories. *Plant Physiol.* 76:843–45

190. Walker, D. A. 1981. Photosynthetic induction. In *Proceedings of the 5th International Congress on Photosynthesis*, ed. G. Akoyonoglou, 4:189–202. Philadelphia: Balaban Int.

191. Weber, J. A., Jurik, T. W., Tenhunen, J. D., Gates, D. M. 1985. Analysis of gas exchange in seedlings of *Acer saccharum:* integration of field and laboratory studies. *Oecologia* 65:338–47

192. Woodrow, I. E., Berry, J. A. 1988. Enzymatic regulation of photosynthetic CO_2 fixation in C_3 plants. *Annu. Rev. Plant Physiol. Plant Mol. Biol.* 39:533–94

193. Woodrow, I. E., Mott, K. A. 1989. Rate limitation of non-steady state photosynthesis by ribulose 1,5-bisphosphate carboxylase in spinach. *Aust. J. Plant Physiol.* 16:489–500

194. Woodrow, I. E., Walker, D. A. 1980. Light-mediated activation of stromal sedoheptulose bisphosphatase. *Biochem J.* 191:845–49

195. Woods, D. B., Turner, N. C. 1972. Stomatal response to changing light by four species of varying shade tolerance. *New Phytol.* 70:77–84

196. Woodward, F. I. 1980. Shoot extension and water relations of *Circaea lutetiana* in sunflecks. In *Plants and Their Atmospheric Environment*. ed J. Grace, pp. 83–91. Oxford: Blackwell

197. Young, D. R., Smith, W. K. 1979. Influence of sunflecks on the temperature and water relations of two subalpine understory congeners. *Oecologia* 43:195–205

198. Young, D. R., Smith, W. K. 1980. Influence of sunflight on photosynthesis, water relations, and leaf structure in the understory species *Arnica cordifolia*. *Ecology* 61:1380–90

199. Young, D. R., Smith, W. K. 1982. Simulation studies on the influence of understory location on the water and photosynthetic relations of *Arnica cordifolia* Hook. *Ecology* 63:1761–71

200. Zavitkovski, J. 1982. Characterization of the light climate under canopies of intensively-cultivated poplar plantations. *Agric. Meteorol.* 25:245–55

201. Zeiger, E., Iino, M., Ogawa, T. 1985. The blue light responses of stomata: pulse kinetics and some mechanistic implications. *Photochem. Photobiol.* 42:759–63

Annu. Rev. Plant Physiol. Plant Mol. Biol. 1990. 41:455–96

LIGNIN: OCCURRENCE, BIOGENESIS AND BIODEGRADATION

Norman G. Lewis and Etsuo Yamamoto

Departments of Wood Science and Biochemistry, Virginia Polytechnic Institute and State University, Blacksburg, Virginia 24061

KEY WORDS: lignin, biosynthesis, biodegradation, nuclear magnetic resonance spectroscopy

---

CONTENTS

1 INTRODUCTION

Lignin! Although these polymers have been studied extensively for almost a century and a half, the term lignin conjures up different images in different

1040-2519/90/0601-0455$02.00

scientific disciplines. As long ago as 1957, Erdtman noted that "owing to the unusual complexity of the problem, speculation has played a great role in lignin chemistry" (42); speculation continues to play a role today.

Even Schultze's coining of the term lignin (Latin *lignum* = wood) may have marked the beginnings of confusion, since lignins are also found in many nonwoody plants.

Initially, researchers disagreed about whether lignins were true plant constituents or artefacts formed during isolation procedures. Careful studies by the schools of Freudenberg, Kratzl, Neish, and Higuchi, however, established that these substances were plant polymers derived from the hydroxycinnamyl alcohols or monolignols *p*-coumaryl, coniferyl, and sinapyl, shown in Figure 1 (54). Note that the aromatic portions of these phenylpropanoids are described as *p*-hydroxyphenyl (H), guaiacyl (G), and syringyl (S) moieties, respectively, and that lignins are classified according to this distinction.

As regards biosynthetic intermediates and accompanying enzymatic steps, a detailed understanding of monolignol biogenesis has been obtained (72), while the precise cellular location of most of the enzymes has not yet been established. Much also remains unknown about lignin formation and its biodegradation. Several examples will suffice:

First, the mechanism and regulation of transport of monomers (from the cytoplasm into the cell wall) and their subsequent polymerization are not well

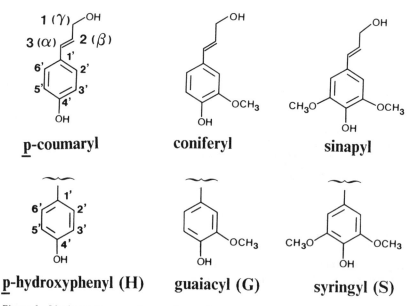

Figure 1 Lignin precursors and aromatic constituents

understood. Further, not only is the precise chemical nature of the monomer undergoing transport uncertain, but so too are the factors controlling the polymerization process and the ultimate structure of lignins in situ.

Second, there remains disagreement about whether certain plant forms (e.g. mosses, algae) contain lignin. In the scientific literature, it is still commonplace to find any insoluble (phenolic) plant polymer described as lignin, even when definitive chemical evidence is lacking. Some authors use the term "lignin-like"—an unfortunate practice that sheds no light on the identity of the substances in question. Moreover, the continued use of inadequate and outdated analytical methodology to characterize both lignin and "lignin-like" polymers further exacerbates this situation.

Third, because we do not know how to isolate lignin in its unaltered form, controversy still rages over whether any lignin-derived preparation adequately represents native lignin structure. For some time methodology has been urgently needed to determine the structure of lignin in vivo and the changes the macromolecule undergoes during delignification. Results obtained from poorly characterized and substantially chemically altered soluble lignin-derived preparations are often extrapolated to explain the chemical and biochemical transformations lignin itself undergoes. In some cases, model substrates (normally phenylpropanoid dimers) are also used, particularly in investigations aimed at elucidating lignin biodegradation mechanisms. This approach has generated considerable information about degradation mechanisms in simple lignin-substructural compounds; but, in view of current confusion over the possible function of extracellular enzymes excreted from the lignin-degrading fungus *Phanerochaete chrysosporium,* use of these compounds to study biodegradation of the lignin polymer may be inappropriate.

1.1 Criteria of Lignin Detection

No definitive criteria have been established for verifying the existence of a lignin polymer in a particular plant species. While numerous physical, chemical, and histochemical techniques are routinely employed to suggest its presence, a systematic approach is rarely, if ever, undertaken. This point is particularly important for nonwoody plants, whose lignins have not been extensively examined. We propose that the following criteria be used:

1. The plant under investigation must possess the biochemical machinery essential for (*a*) synthesis of at least one of the three monolignols (*p*-coumaryl, coniferyl, or sinapyl alcohols) and (*b*) transport of phenylpropanoid monomers from the cytoplasm to the cell wall and their subsequent polymerization.
2. The substance(s) so formed must be matrix components along with the cell

wall carbohydrates, not substances copolymerized with other residues (such as acyl moieties in suberin formation).

3. Histochemical studies must establish the polymer to be a plant cell wall lignin.

4. At least one derivative must be polymeric and must contain substructures consistent with a lignin.

1.2 Lignin Distribution in the Plant Kingdom: A Reevaluation

Before examining evidence for and against the presence of lignin in various plant forms, a discussion of analytical methods is required. It must be emphasized that no single analytical method provides conclusive evidence that lignin is present.

1.2.1 HISTOCHEMICAL STAINING   Staining provides only preliminary evidence of the presence of lignins and of their morphological location in plant tissue. Various reagents (including human urine) have been used in staining (14). The most common are the Cross & Bevan (chlorine water-sodium sulphite), Mäule ($KMnO_4$/HCl/NH_3), Wiesner (phloroglucinol-HCl), and Safranin O reagents (184). The Cross & Bevan test appears to distinguish between lignins rich in guaiacyl (mainly coniferyl alcohol derived) and syringyl (sinapyl alcohol derived) units, giving brown and red colorations, respectively. A positive Wiesner test (red coloration) is generally considered indicative of coniferaldehyde end-groups in guaiacyl lignins, whereas a positive Mäule test (deep rose-red) is thought to result from syringylpropane moieties in syringyl lignin. Safranin O is the least specific reagent, giving a red coloration with phenols.

Erroneous results are often obtained with these reagents, owing to interference from nonlignin substances. For example, some lignans (phenylpropanoid dimers), such as liriodendrin and syringaresinol, produce a strong Mäule reaction (81), and comparable false positives continue to be reported with plants known to be devoid of lignin. Sometimes no staining is observed in plants known to contain lignin [e.g. with syringyl lignins using the Wiesner reagent (184)]. Thus, to minimize ambiguities in testing, at least two different histochemical staining reactions are recommended. However, owing to their limited specificity, histochemical tests should be interpreted with great caution. Indeed, until it has been established using more specific methods [e.g. carbon-13 nuclear magnetic resonance (NMR) spectroscopy] that a tissue does contain lignin, and that a lignin isolate thereof undergoes a similar staining reaction, positive staining should be viewed with skepticism.

1.2.2 DETERMINATION OF TOTAL LIGNIN CONTENTS   Determinations, or more correctly estimates, of lignin content are conducted on organic solvent preextracted tissue using either "wet" chemical techniques [Klason (116), Goering & Van Soest (64), alkali extraction (195), and acetyl bromide (92)] or direct spectroscopic methods [solid state carbon-13 nuclear magnetic resonance (NMR) (80, 122), infrared (IR) (183), and ultraviolet (UV) (11)].

None of the "wet" chemical methods proves that lignin is present. Klason (116) determinations only give a measure of residual plant material following digestion with 72% H_2SO_4. With the Goering & Van Soest (64) method, the tissue is first detergent-washed under acidic conditions, with lignin determinations carried out as before, or the washed fiber is treated with potassium permanganate to solubilize the lignin. In both the hot alkali (195) and acetyl bromide (92) methods, lignin is solubilized from the plant tissue and lignin contents are determined spectrophotometrically by UV analysis. These methods only give useful results providing there are no interferences from lignin-like materials.

Among the spectroscopic methods applicable for direct analysis of plant tissue, solid-state ^{13}C NMR now appears promising. Initial results were not encouraging (80), but refinement of this method by Leary et al (122) demonstrated a fairly good correlation between lignin contents determined by ^{13}C NMR and the Klason method, provided there was no interference from tannins. Estimates of lignin content can be obtained by integrating the ^{13}C NMR resonances due to C_3/C_4 of guaiacyl and C_3/C_5 of syringyl units, and comparing those with the remaining (lignin and carbohydrate) signals. By careful analysis of the ^{13}C NMR spectrum, this method can also give important information about lignin type [e.g. syringyl- or guaiacyl-rich (122)], even to the point of documenting lignin's absence (211). Lignin estimates can also be obtained directly from pulverized plant tissue using UV (183) and IR (11) techniques; these methods will err if interfering materials are present.

1.2.3 ESTIMATE OF INTER-UNIT LINKAGES IN LIGNIN   One of the greatest challenges in lignin chemistry has been to define the nature and relative frequency of inter-unit linkages in native lignin structure. Until recently (see Section 2.6), structural representations of native lignin were extrapolated from findings obtained with soluble lignin-derived products—e.g. those released during harsh chemical treatments such as hydrolysis [H_2O, 100°C, several weeks, (158)], thioacetolysis/alkali extraction (160), and permanganate oxidation (121). From the plethora of products identified, mainly aromatic monomers and dimers in low overall yield, various schemes for typical bonding patterns in lignin were proposed (1, 61, 116, 159). Strictly speaking, such schemes must *not* be taken to represent native lignin structure, since they really only afford limited bonding information about lignin-derived products. Surprisingly, little attention has been given to the possibility of artefact formation during these harsh chemical manipulations. Indeed, owing to concerns such as these, we recently developed a method of examining lignin structure in situ (See Section 2.6).

Today, the most versatile and powerful technique for lignin verification and tentative identification of inter-unit linkages within soluble *lignin-derived* materials, employs solution-state ^{13}C NMR spectroscopy. For a particular lignin isolate, a ^{13}C NMR spectrum gives a series of individual resonances for the polymer. Comparison of these resonances to those of some 30 or so "lignin model compound dimers or substructures" (138), enables various bonding patterns of *p*-hydroxyphenyl, guaiacyl, and syringyl units in lignin-derived isolates to be tentatively identified and substructures assigned (4, 75, 108, 117, 159, 162, 163). Unfortunately, it is not possible to quantify precisely all of the different bonding types and arrive at relative bonding frequencies. However, progress is being made on this front (178); solution-state ^{13}C NMR spectroscopy can now be used to quantify (on a relative basis) the various types of hydroxyl groups (primary, secondary, tertiary, or phenolic) and syringyl:guaiacyl ratios (177, 179).

1.2.4 MONOMER COMPOSITION OF LIGNIN   The monomer composition of lignins from plants can vary depending upon the plant family or morphological region under consideration. Generally, lignins are classified as guaiacyl, guaiacyl-syringyl, or guaiacyl-syringyl-*p*-hydroxyphenyl lignins, according to whether they are from gymnosperms, woody angiosperms, or grasses, respectively. However, great care should be taken with such generalities because (*a*) exceptions to the trends stated above occur (see Sections 1.3.4 and 1.3.5); (*b*) all lignins contain *p*-hydroxyphenyl constituents that are invariably ignored (e.g. in guaiacyl and guaiacyl-syringyl lignins); and (*c*) lignin is frequently heterogeneous, its structure varying with its morphological location.

No degradation technique exists that can establish the exact monomer ratio in the polymer, and only qualitative data can be obtained. Typical analytical methods include alkaline nitrobenzene oxidation (53, 146, 186, 191), sodium hydroxide/cupric oxide oxidation (137), acidolysis (207), thioacidolysis (119), and permanganate oxidation (23).

Alkaline nitrobenzene/CuO-NaOH oxidation   Degradation of isolated lignins (or plant tissues containing them) with these reagents produces simple mixtures of monomeric degradation products, derived *only* from moieties within the polymer that are not involved in inter-unit aromatic carbon-carbon linkages (i.e. only those from so-called uncondensed units). Depending upon the lignin type, alkaline nitrobenzene treatment gives characteristic mixtures of *p*-hydroxybenzaldehyde, vanillin, and syringaldehyde, which are indicative of H, G, and S moieties, respectively. This method is widely used both to verify the presence of lignin and to establish its type.

According to earlier reports (53), alkaline nitrobenzene oxidation of iso-lated gymnosperm lignin yields 20–28% vanillin, whereas angiosperm lignin gives 6–12% vanillin and 30–40% syringaldehyde. However, such high conversions are rarely attained with lignin in plant tissue (186) or even with simple "lignin model compounds" (191). Syringyl:guaiacyl:p-hydroxyphenyl (S:G:H) ratios so obtained should be viewed with caution, since syringyl content as indicated by this method may be as much as three times higher than that evidenced by ^{13}C NMR determinations (146). A similar situation may exist for alkaline cupric oxide treatment which gives, in addition to the aldehydes mentioned above, the corresponding acetophenones (137). The products formed during this treatment are not particularly stable under the conditions employed, as evidenced by model compound experiments which recovered only 29–45% of the products. This may explain the very low product yields (<5% conversion) from a wide variety of plant materials.

Acidolysis/thioacidolysis   These two procedures are used to obtain informa-tion about lignin constituents not involved in inter-unit carbon-carbon link-ages. In the case of acidolysis (dioxane-2N HCl, 9:1, v/v), labile benzylic (and nonbenzylic) aryl-ether linkages are cleaved to afford a mixture of ketonic monomers (so-called Hibbert ketones) (207). These products are used to define the presence (or type) of lignin, even though such conversions occur in low yield. Recently, acidolysis has been largely replaced by thioacidolytic (BF$_3$/C$_2$H$_5$SH) treatments, which result in the conversion of 'uncondensed' monomeric units in lignin into simple diastereomeric mixtures of 1,2,3-trithioethane phenylpropanoid monomers (118). If premethylation of the lignin sample is carried out, this method can be used to estimate the number of alkylaryl ether bonds in lignin, relative to that of free phenolic groups. Once again, overall yields are low.

Permanganate oxidation   This method provides information about inter-unit aromatic carbon-carbon linkages. Chemically, it is a harsh technique (23) involving first treatment with strong alkali to solubilize the lignin (NaOH-CuO, 170°C) and then methylation with dimethyl sulphate. The products so obtained are then oxidized with potassium permanganate/sodium periodate and fully methylated (diazomethane) to produce a series of benzoic acid methyl esters, again in low overall yield. Even simple model compound conversions are only obtained in 40–50% yield, and presumably a more complex situation exists for lignin itself. Identified products from lignin are shown in Figure 2. Artefact formation is undoubtedly important in this treatment.

1.2.5 MOLECULAR WEIGHT DETERMINATIONS   Any investigation designed to confirm the presence of lignin in a particular plant type would be in-

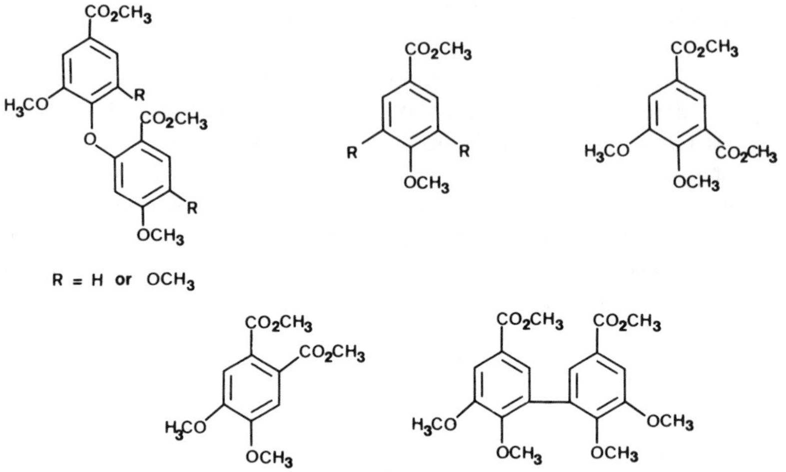

Figure 2   Typical lignin-derived fragments following permanganate oxidation and methylation (23)

complete unless some proof of its polymeric nature was obtained. While this cannot be done for native lignin in situ, many lignin-derivatives, although modified substantially, are polymeric—e.g. as evidenced by gel permeation chromatography (GPC) or ultracentrifugation (67). With respect to GPC, caution should be exercised in making conclusions about the range of molecular sizes, owing to the difficulties encountered with the associative behavior of dissolved lignins (39). Care must also be taken in choice of molecular weight standards, since those most often used (e.g. polystyrene) have hydrodynamic volumes very different from those of lignin fragments. Indeed, only soluble lignin fragments of known molecular weight (as determined by ultracentrifugation) should be used for calibration of GPC columns (67).

1.3 Lignin Occurrence

Application of modern techniques has settled some of the controversial issues (see 74) regarding lignin occurrence.

1.3.1 ALGAE   Russian workers have steadfastly maintained that the brown algae *Cytoseira barbata* (37, 38) and *Fucus vesiculosus* (174) contain lignin in amounts of up to 20% of the total dry plant matter. Evidence first came from isolation of Björkman "lignin" preparations, obtained by extraction of plant material with aqueous dioxane. Further credence was apparently produced by "identification" of lignin-derived products following degradation either by sodium in liquid ammonia or by thioacetolysis. However, identification relied solely upon comparison of chromatographic retention times of products to those of authentic phenylpropanoid standards. These findings were surprising, since various algal lines (172) (except *Dunaliella;* 136) do not contain either phenylalanine ammonia lyase (PAL) or tyrosine ammonia lyase (TAL), the first enzymes committed to the phenylpropanoid pathway. This inconclusive evidence did not go unnoticed, and "*Fucus* lignin" was subjected to greater scrutiny (171, 173). Again a "Björkman" preparation was obtained and subjected to a variety of degradation techniques, including that of sodium in liquid ammonia. Product analysis by gas chromatography mass spectroscopy (GC/MS) provided no evidence of any lignin-derived (phenylpropanoid) structure. Instead, the products obtained, taken together with evidence from the ^{13}C NMR analysis of a "Björkman" preparation from *F. vesiculosus,* revealed the presence of phloroglucinol-derived polymers. Thus, it must be concluded that lignin is *not* a constituent of brown algae.

A similar situation would be envisaged to exist for the green algae. However, a recent report claims that the chlorophycean representative *Coleochaete* contains lignin or "lignin-like" materials (35). The sole evidence for this claim was "autofluorescence" of tissue under UV light and a positive histochemical Mäule test. Since other members of the chlorophyceae family (e.g. *Chlorella pyrenoidosa* and *Ulva lactuca*) are devoid of PAL or TAL (215), much more convincing evidence is needed if green algae are to be included as lignin-synthesizing plants.

1.3.2 MOSSES (MUSCI)   The evidence for lignin in nonvascular plant mosses (e.g. Bryophyta) has also elicited strong differences of opinion and much controversy. For example, it was reported that sphagnum moss contained a complex phenolic substance having a lignin-like UV absorbance spectrum (48). On the other hand, the low methoxyl content of this substance (~1%) and the fact that nitrobenzene oxidation produced only traces of vanillin and syringaldehyde were viewed by others as evidence for lignin's absence (107, 134). Farmer & Morrison (49) subsequently proposed that sphagnum did indeed contain lignin, albeit mainly derived from *p*-hydroxyphenyl units. This view was supported in later studies by Bland et al (9), who also reported that sphagnum contained lignin as well as both *p*-coumaric and ferulic acids.

Mosses belonging to the genera *Dawsonia, Dendroligotrichum,* and *Polytrichum* were also examined. Gametophytes of the first two apparently gave positive lignin-staining reactions with phloroglucinol-HCl, Cross & Bevan, and Mäule tests (192). Wet chemical methods such as Klason lignin and nitrobenzene oxidation determinations supported this view, and lignin contents of ~10% were reported. In the case of *Dendroligotrichum,* it was observed that lignin, or "lignin-like" substances, were located in the cell walls of hydroids (water-conducting cells). This evidence relied upon staining with

Safranin O (a nonspecific test for phenolic groups) and a scanning electron microscopy study of hydroids of stem sections (188).

Miksche and coworkers next examined the "lignin-like" structure of the sphagnum polyphenols. Permanganate oxidation/methylation (see Section 1.2.4) gave a number of compounds that were conceivably lignin degradation products. However, two hydroxybenzofurans (see Figure 3) were also isolated, and their formation could not readily be rationalized as resulting from a typical lignin polymer (44–46). The degradation/methylation experiments were then repeated, now using deuterated dimethyl sulphate [i.e. $(CD_3)_2 SO_4$] as alkylating agent. This experiment indicated that all methyl ether and methyl ester groups in the degradation products were derived from $(CD_3)_2SO_4$. Thus, the polyphenols in sphagnum mosses could, at best, only be p-hydroxyphenylpropyl or caffeoyl derived. Similar polyphenols were present in *Dawsonia* species (151).

Recent work by Wilson et al (211) seems to have resolved these conflicting data. The researchers used both solid-state ^{13}C NMR and pyrolysis GC/MS techniques to examine the mosses *Thamnobyrum pandum, Rhizogonium parramatense, Sphagnum cristatum, Dawsonia superba,* and *Leucobryum candidum*. While pyrolysis GC/MS of uncleaned samples disclosed the presence of lignin-derived degradation products, the amounts were reduced to trace or undetectable levels upon ultrasonic cleaning of the plant tissue. In a similar manner, the ^{13}C NMR spectra showed a substantial decline in aromatic content. The authors speculated that this decline was due to the removal of minute lignin-containing contaminants. Indeed, no evidence whatsoever was obtained for a lignin polymer, based on either p-hydroxyphenyl, coniferyl, or sinapyl alcohol units. Quite to the contrary, aromatic 1,3,5-hydroxybenzene

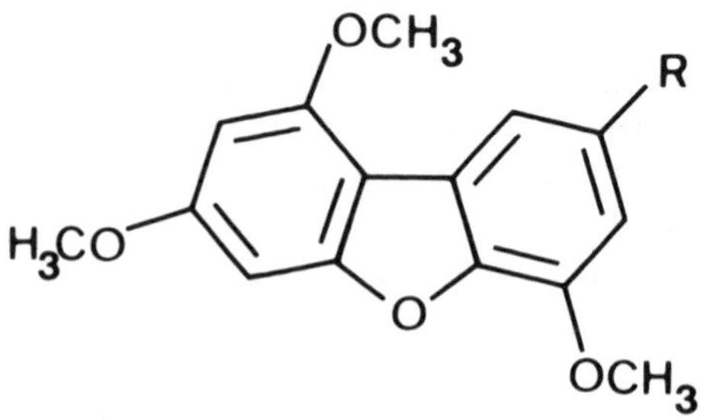

$$R = CH_2CH_2CO_2CH_3 \text{ or } CO_2CH_3$$

Figure 3 Methylated hydroxybenzofurans from permanganate oxidation/methylation of sphagnum

polyphenols were observed, which could be derived from either hydroxyben-zofuran or proanthocyanidin polymers. Although these substances are defi-nitely not lignins, at least a part of their structures may stem from some offshoot of the phenylpropanoid pathway.

1.3.3 PTERIDOPHYTES   It is difficult to draw general conclusions about lignin composition in the Pteridophyta since there are no obvious trends. To date, studies have been rudimentary and have employed either histochemical or chemical degradation methods [e.g. alkaline cupric hydroxide (137, 202) and permanganate (44) oxidation methods]. In the Pteropsida, ferns belonging to the Cyathaceae and Adiantaceae (137) gave products derived mainly from guaiacyl units together with smaller amounts of p-hydroxybenzaldehyde, whereas representatives of the Dennstaedtiaceae (10, 137) mainly contained guaiacyl and syringyl components. In the *Equisetaceae* (Sphenopsida) (91, 137, 209) small quantities of H:G:S lignin-derived products (p-hydroxybenzaldehyde, vanillin, and syringaldehyde) were obtained, whereas in the Lycopsida (91, 137, 202, 209), plants belonging to the *Lycopodiaceae* predominantly gave guaiacyl-like lignin degradation products together with small amounts of p-hydroxybenzaldehyde; syringyl moieties were absent. On the other hand, many examples of the *Selaginellaceae* contained pre-dominantly syringyl units, although with *Psilotum* (Psilopsida) a more "guaiacyl-like" lignin was observed. Interestingly, the aquatic plant *Isoetes* gave no products indicative of lignin (44). Why this aquatic plant does not contain lignin must be determined at the enzymatic level.

1.3.4 GYMNOSPERM LIGNINS   Several representatives of this group (e.g. the spruces and pines have been studied extensively, presumably because of their importance to commercial pulp and paper operations. The ascription to these plants of guaiacyl lignins (i.e. mainly derived from coniferyl alcohol) is both inexact and overgeneralized (164). Exceptions exist where syringyl moieties predominate. For example, in addition to the well-known examples in the Cycadales and Gnetales (45, 150), high syringyl contents in the Coniferales proper [e.g. *Tetraclinis articulata* (Cupressaceae) and *Podocarpus nerrifolius* (Podocarpaceae)] have been documented. Thus great caution should be ex-ercised in defining lignin composition as a function of chemotaxonomy.

In addition to compositional variations between species, structural hetero-geneity of lignins within plants is well documented. Compression wood lignins of *Pinus radiata* and *Tsuga heterophylla* have lower methoxyl and higher lignin contents than those of "normal" wood (7, 8), presumably reflecting a higher p-coumaryl alcohol–derived origin (118). Approximately 90% of p-coumaryl alcohol moieties in compression wood lignin have free phenolic groups, in contrast to the corresponding coniferyl alcohol–derived components, where the bulk (~80%) are involved in ether linkages. The significance of these differences, in terms of lignin structure, has not yet been determined.

Lignin heterogeneity even seems to exist at the subcellular level. For example, in black spruce *(Picea mariana),* the middle lamella and cell-wall corners contain lignin of a higher *p*-hydroxyphenylpropane content than secondary wall layers (210). In general agreement with these findings, Terashima & Fukushima (200) reported that *p*-coumaryl alcohol is first deposited into the middle lamella and cell corners of *Pinus thunbergii* during early stages of cell wall differentiation. Then, guaiacyl units are deposited in two distinct phases, first in the middle lamella and cell corners, which appear to be involved in highly "condensed" inter-unit carbon-carbon linkages. The remaining guaiacyl units are then incorporated into the secondary wall at a late stage of development. Syringyl units, minor constituents of *Pinus thunbergii,* are also deposited in the inner layer of the secondary wall during a late developmental stage. These conclusions were based on results achieved by administration of ^{3}H-labeled monolignol glucosides, *p*-coumaryl alcohol-β-D-glucoside, coniferin, and syringin, and subsequent examination of the cells by autoradiography at different stages of lignification. While these observations generally provide convincing support for the hypothesis that lignin is heterogeneous within the cell wall, caution should be exercised in such interpretation. Substantial interconversion of *p*-coumaryl alcohol-β-D-glucoside and coniferin apparently occurred during these experiments, an unexpected result based on our current knowledge of the enzymes involved in lignification.

Suspension cell cultures of *Pinus taeda* also appear to synthesize lignin (~5% by weight) (55), as evidenced by isolation of a "lignin-like" substance from milled tissue and its subsequent analysis by alkaline nitrobenzene oxidation, UV, IR and ^{1}H NMR analysis. Gel filtration chromatography provided evidence for its polymeric character. Unfortunately, the precise morphological location of the lignin was not described. However, if these cell lines do indeed produce lignin, they may provide an excellent model system in which to study the initiation of the lignification process in gymnosperms: It should be possible to obtain cell lines at varying degrees of differentiation.

1.3.5 ANGIOSPERMS   Like the gymnosperms, woody angiosperms have been examined in considerable detail. Although their lignins are typically described as of the guaiacyl-syringyl type, exceptions to this trend again arise. For example, *Erythrina crista-galli* (Leguminoseae) gave a negative Mäule reaction; syringaldehyde/vanillin S/V ratios were low (0.03–0.11); and methoxyl, IR, and NMR analyses also indicated the presence of a guaiacyl-like lignin (112). Of interest was the observation that *Erythrina crista-galli* shoots had higher S/G ratios (0.15–0.18) than mature tissue—i.e. the syringyl content decreased with age. This finding contrasts with those in most other angiosperms.

Lignin heterogeneity in woody dicots has also received considerable attention in recent years. As for gymnosperms, the most convincing evidence comes from microautoradiography and energy-dispersive X-ray analyses (EDXA), respectively. Terashima and coworkers administered various (^{3}H and ^{14}C) labeled forms of ferulic and sinapic acids to Oxford poplar *(Populus maximowiczii × Populus berolinensis)*, and coniferin and syringin to magnolia *(Magnolius kobus DC)* (199). The distribution of radioactivity from each precursor into the different cellular components (e.g. fiber, vessel, middle lamella) was then estimated by microautoradiography at different developmental stages. With poplar, ferulic acid showed a slight preference for deposition into vessel rather than fiber elements, while the reverse was true for sinapic acid. Surprisingly, some conversion of sinapic acid into ferulic acid apparently occurred; the reasons for this unusual transformation must be established at the enzymatic level. In the case of magnolia, coniferin was preferentially deposited into both vessel and middle lamella lignin during the early stages of lignification, and then into the secondary wall of the fibers at later stages. Syringyl moieties, on the other hand, were deposited mainly into the fiber secondary wall and then the vessels. Eventually, the syringyl content was slightly higher than its guaiacyl counterpart in both vessels and fiber secondary walls, whereas the S/G distribution throughout the middle lamella seemed equitable. No mention was made of the deposition of *p*-coumaryl alcohol into these different morphological regions. These results seem to be in qualitative agreement with chemical degradation studies of differentiating xylem from birch *(Betula maximowiczii)* (41).

Lignin heterogeneity of birch wood (*Betula papyrifera* Marsh) has also been investigated by bromination of ultrathin sections of wood and their subsequent analysis by EDXA (182). In this method, it is proposed that only the lignin is brominated, the syringyl moieties acquiring 1.2 times more bromine than corresponding guaiacyl counterparts. Surprisingly, the contribution of *p*-hydroxyphenylpropane moieties in this lignin was ignored. Estimates of G:S ratios were 12:88 (fiber S$_2$), 88:12 (vessel S$_2$), 49:51 (ray parenchyma S) and >80% G (middle lamella). While supporting the general concept of structural heterogeneity, these findings were only in fair agreement with the same author's previous UV analyses of birchwood, and with the results by Terashima & Fukushima (200).

It should be self-evident that there are significant differences in lignin composition from one subcellular region to another. An explanation of what this means in terms of lignin structure in different morphological regions needs to be determined.

1.3.6 MONOCOTYLEDONS   The most widely studied examples in this group belong to the Poaceae—grasses and cereals. Lignin contents can differ substantially in this class, ranging from 1.2% for herbaceous reed canary grass (18) to 26% for woody bamboo stems *(Phyllostachys herterocycla)* (87, 88). As an added complexity, these lignins not only contain all three monolignols, but they also have ester- and ether-linked hydroxycinnamic acids. For example, in several grasses, esterified *p*-coumarate constitutes 3–10% of the weight of lignin (87); ferulic acid is present in smaller amounts and can be linked through either ether or ester linkages (187).

These hydroxycinnamic acids are found specifically linked not only to lignin but also via ester bonds to polysaccharides in considerable amounts (0.2–1% of dry matter)—e.g. to arabinose in arabinoxylans (79, 212). While ferulic and *p*-coumaric acids are the most abundant of these cell wall–bound hydroxycinnamic acids, 4,4'-dihydroxytruxillic acid (79), 5-hydroxyferulic, diferulic, and sinapic acids may also be linked in a similar manner (166, 212). Note that the relative amounts of these acids can vary considerably depending upon the age and tissue under investigation (212). Indeed, if appropriate precautions (such as saponification) are not taken, these constituents can interfere with lignin determinations. The physiological role of these polysaccharide-bound hydroxycinnamic acids has not been determined.

Several studies have examined lignin heterogeneity in monocotyledons (as previously described for woody gymnosperms and angiosperms). To date, the only convincing evidence has been obtained at the subcellular level. For example, in rice plants *(Oryza sativa)* lignins in the middle lamella of protoxylem vessels and the outer layers of secondary walls of metaxylem were rich in G and H units, whereas those in the compound middle lamella and parenchyma were abundant in G/S moieties (83). The significance of this finding, in terms of lignin structure, is not known.

During maturation of grasses, the syringyl content increases to give a S/V ratio approaching unity at maturity. In a comparable manner, induction of lignin as a response to wounding and fungal infection (in primary leaves) is sometimes accompanied by an increased syringyl content (175), although not always (2).

1.3.7 EVIDENCE FOR LIGNIN IN SUBMERGED AQUATIC ANGIOSPERMS While a number of reports document the presence of lignin in aquatic submerged plants, no rigorous proof has yet been provided.

In the marine angiosperm *Zostera marina* (63), claims for a 7.3% lignin content [by Goering & Van Soest (64) determinations; see Section 1.2.2] are in stark contrast to results obtained from permanganate oxidations, which document its absence (46). In the sea grasses *(Halophila ovalis, Halophila stipulacea,* and *Halodule uninervis)* (6) and in *Posidonia australis* Hook F.

(110), "evidence" for the presence of lignin consisted, in the former group, of UV absorbance of alkali solubles and, in the latter, of histological staining with toluidine blue (a nonspecific test for polyphenols).

Submerged and free-floating monocots and dicots from Chilean waters have also been examined (197). Surprisingly, in both submerged *(Utricularia tenuis* and *Potamogeton lucens)* and free-floating *(Lemna valdiviana)* species, it was concluded that lignin was the major constituent. No evaluation of this claim can be made, since the methodology employed was not described. However, these findings are in direct contradiction to results obtained by permanganate oxidation of *Potamogeton gramineus* L., which indicated the absence of lignin (46).

Evidence for lignin in submerged aquatic plants is not convincing. More definitive proof is necessary if these are to be considered lignin-synthesizing organisms.

1.3.8 LIGNIN IN ANGIOSPERM CULTURES   These have not received very much attention. In the case of *Rosa glauca,* lignin preparations from both suspension and callus cultures have been obtained (179). In both cases, a polymer was obtained that apparently had a much lower methoxyl content than the lignin present in soil-grown plants. Both lignin preparations also gave NMR spectra with several resonances ($\sim$ 20–50 ppm) that are clearly aliphatic perhaps suggesting the additional presence of a suberin-like polymer. In the case of the *R. glauca* suspension culture, this "lignin-like" material was an extracellular secretion.

Soybean *(Glycine max)* suspension cultures also contain a lignin with a higher *p*-hydroxyphenylpropane content than "normal" soybean lignin (161). These results demonstrated that these lignins are not typical guaiacyl-syringyl lignins associated with composite plant material. However, these cultures may be providing important information about lignification during the early stages of cell wall development.

2 LIGNIN BIOGENESIS AND STRUCTURE

The pathway to E-monolignols represents a major branch of the shikimate-chorismate pathway, since large quantities of vascular plant materials are phenylpropanoid derived. Beyond prephenic acid, the enzymes (and many isozymes) required for monolignol formation have been partially, or completely, purified and reasonably well characterized. Products derived from the E-monolignols include mainly lignin and lignans, and presumably to some extent, suberin. Note that phytochemical variations in lignin composition (e.g. guaiacyl, guaiacyl-syringyl) between species have been rationalized as due to differences in the substrate specificities and activities of certain of these enzymes (e.g. ferulate-5-hydroxylase, O-methyltransferases, cinnamyl alcohol dehydrogenase, CoA ligase, and β-glucosidase). A detailed treatment of this topic is beyond the scope of this review; several comprehensive treatments have appeared in recent years (72–74, 84).

More recent developments regarding biochemical processes leading to lignin synthesis, and determination of its structure in situ, are discussed below.

2.1 Arogenic Acid

Until fairly recently, phenylalanine (Phe) and tyrosine (Tyr) were considered to be formed in higher plants exclusively via transamination of phenylpyruvic and p-hydroxyphenylpyruvic acids, respectively (see Figure 4). The finding that arogenic acid, isolated from Agmenellum quadruplicatum and a mutant strain of Neurospora crassa, was an intermediate in Phe/Tyr biosynthesis in Pseudomonas aeruginosa (168) raised the question of whether this pathway was operative in higher plants. Following appropriate studies with various herbaceous species, arogenate was concluded to be the only intermediate in Phe/Tyr biogenesis beyond prephenic acid. This was established for tobacco cultures (Nicotiana silvestris) (57, 97), sorghum (Sorghum bicolor) (28, 193), corn (Zea mays) (20), and spinach (Spinacea oleracea) (95, 97), following detection of enzymatic activity for prephenate aminotransferase (193), arogenate dehydratase (97), and arogenate dehydrogenase (20, 28). Purified arogenate dehydrogenase from sorghum (28) and corn (20) is inhibited by tyrosine, whereas phenylalanine is an effective feedback inhibitor for arogenate dehydratase in tobacco and spinach (97)—i.e. Phe/Tyr levels are regulated by modulation of arogenate dehydratase/arogenate dehydrogenase activities via product inhibition (95). Surprisingly, with mung bean (Vigna radiata) seedlings, both arogenate and phenylpyruvate pathways were observed; the prephenate dehydrogenase from this source had a most unusual cofactor requirement, using NADP rather than NAD (95, 181). Thus a growing body of evidence indicates that arogenate is the exclusive precursor of Phe/Tyr in many plants. The occurrence of this pathway in woody plants would explain an apparent discrepancy in oak, where prephenate dehydrogenase activity was not detected (56).

While all enzymes synthesizing Phe/Tyr have been found in the chloroplast, some evidence suggests that a dual pathway may exist in the cytosolic fraction (96) where the enzymes involved in flavonoid biosynthesis are found. Verification of this dual pathway requires appropriate kinetic data for both chloroplast and cytosolic enzymes, as well as determination of their precise subcellular location [e.g. using immunocytochemical techniques, as applied already to chalcone synthase (90)].

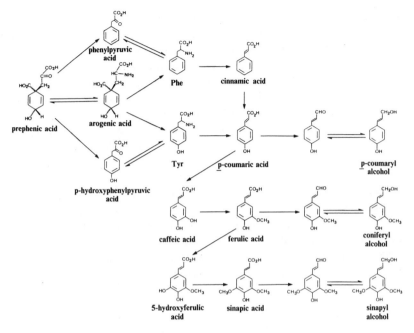

Figure 4   Biosynthesis of E-monolignols

2.2 p-Coumaric Acid to E- and Z-Monolignols

Beyond Phe/Tyr, stereospecific deamination produces cinnamic acid (or, in grasses, *p*-coumaric acid). Of the naturally occurring hydroxycinnamic acids, only *p*-coumaric, ferulic, and sinapic acids are normally viewed as bona fide lignin precursors, as they are converted into the corresponding monolignols without further modification of the aromatic ring (see Figure 4). While pathways to both *p*-coumaric and ferulic acids are well established, that of sinapic acid has only recently been clarified. This was achieved by isolation of ferulate-5-hydroxylase from poplar (68) and 5-hydroxyferulic acid from corn and barley (166). In the absence of evidence to the contrary, sinapic acid can be considered formed directly via methylation of 5-hydroxyferulic acid. Subsequent conversion of *p*-coumaric, ferulic, and sinapic acids (as their CoA derivatives) into the monolignols occurs via a two-step reductive process, producing first the aldehydes.

In most plant material, the quantities of free monolignols present in plant tissue are normally small. An exception to this trend exists in the bark of the Fagaceae where significant quantities of Z (not E) coniferyl and sinapyl alcohols accumulate (126, 155). Labeling experiments with [2-^{14}C] E and Z ferulic acid, E-coniferaldehyde, and E-coniferyl alcohol indicate that this isomerization takes place at the monolignol level (126). Although such interconversions can be achieved photochemically (127), this was not observed in vivo (126), where Z-coniferyl alcohol formation occurred under conditions preventing photochemical isomerization.

All enzymes catalyzing the conversion of *p*-coumaric acid into the three E-monolignols are all readily "cytosol" extractable, and can be fairly readily assayed. At present the only missing enzyme in monolignol biosynthesis is that catalyzing E→Z monolignol formation. However, involvement of these Z isomers in lignin formation in *Fagus grandifolia* has not been established. Nor, for the most part, has the precise intracellular localization of these "cytosolic" enzymes been determined.

Other "cytosolic" enzymes producing the closely related flavonoids are believed to exist as a multi-enzyme membrane-associated complex (196). An obvious question is whether a similar situation exists for monolignols. This would be a means whereby lignification could be regulated at the subcellular level. Indeed, it is important to note that all of the hydroxylation enzymes needed for monolignol formation are membrane associated, with two in the microsomes and *p*-coumarate-3-hydroxylase (phenolase) on the chloroplast lamellae (89).

2.3 Monolignol Transport into the Cell Wall

Following monolignol formation in the cytoplasm, these intermediates are further metabolized into the corresponding glucosides and lignin. Depending upon the plant in question, considerable quantities of monolignols may also be required for lignan and (part of the aromatic component of) suberin. At present it is not established how any of these processes are regulated or controlled.

Lignification itself occurs at a site removed from the cell membrane through which the monomers pass, and is preceded by the deposition of cellulose and other matrix components. The primary transport mechanism of monomers from the cytoplasm can probably be explained in terms of en-domembrane theory—i.e. the monomers are deposited into the cell walls following vesicle association and membrane fusion (212). Some tentative experimental evidence favors this mechanism: When (^{3}H) cinnamic acid was administered to differentiating xylem of wheat *(Triticum aestivum)* roots, radioactivity was located in vesicles derived from the Golgi apparatus and the endoplasmic reticulum, as well as lignifying cell walls (169). Similar observations were made when [ring-^{3}H] Phe was administered to *Cryptomeria japonica* (198). However, these studies afford little precise information about the lignification process, since these intermediates are also involved in other branches of the general phenylpropanoid pathway, leading to flavonoids, suberin, lignans, etc.

Glycosidic derivatives of monolignols are normally viewed as the primary candidates for monolignol transport across cell membranes—e.g. coniferin, formed via intercession of substrate-specific UDPG-glucosyl coniferyl alcohol transferases. At least in the case of *Fagus grandifolia* bark tissue, this transferase exhibits a marked substrate specificity. Z-coniferin is formed from Z- (and not E-) coniferyl alcohol, and this enzyme does not even catalyze the conversion of E-coniferyl alcohol into E-coniferin (213). It needs to be established whether such a strict substrate specificity is found in other plants, particularly those that require the E-monolignols for lignification. Although monolignols can be regenerated in the cell wall by action of a specific β-glucosidase, an exclusive role for E or Z glucosides in monolignol transport has never been proven. On the contrary, they may simply function in a storage capacity; there is some evidence for the latter hypothesis, as the rate of E-coniferin turnover in *Picea abies* does not match that of lignification (147).

Clearly, more definitive experiments need to be devised to establish the chemical nature of the monolignol undergoing transport. Additionally, the mechanism of cell wall "communication" with the cytoplasm for monomer selection needs to be elucidated.

At this juncture, it is also pertinent to discuss briefly the processes believed to accompany both lignan and suberin formation. In lignan biosynthesis, normally two "C_6C_3" phenylpropanoid monomers are coupled to give products of specific stereochemistry and which are often optically active—e.g. (+) pinoresinol in *Forsythia suspensa* (125). These substances are believed to accumulate mainly in vacuolar compartments, and it can be envisaged that the required enzyme complex for their formation is near (or on) the vacuolar membranes. The formation of lignans represents an enigma in phenylpropanoid metabolism, since these compounds are optically active. (Lignin, which contains many lignan-like substructures, is believed not to be.) At present, virtually nothing is known about the enzymology of lignan formation. This represents a significant gap in our knowledge of phenylpropanoid metabolism, since both lignans and lignin are considered to be formed by action of peroxidase/H_2O_2 on monolignols or related compounds.

In the case of suberization, it needs to be established whether monolignols (or other related phenylpropanoids) are transported into the cell wall per se, or are conjugated first as acyl derivatives, prior to passage through the cytoplasmic membrane.

2.4 Lignin Formation

Plants contain a broad spectrum of peroxidase isoforms, and many appear to have different primary structures (26, 29, 102, 114). Several studies have been directed towards identifying specific functions (or roles) for certain individual isozymes or groups thereof—e.g. in lignification, suberization, and plant growth regulator metabolism (58, 59, 206).

For lignification, specific cell wall peroxidases are thought to be required to generate H_2O_2 and monolignol radicals. Several investigators have attempted to correlate the presence of acidic peroxidases with lignification. Most of these studies use either histochemical, electrophoretic, or isoelectric focusing techniques to detect isozymes; occasionally substrate specificity studies are also carried out (22, 26, 34, 59, 94, 101, 170, 180). Isozyme detection is normally carried out by staining with artificial dyes, rather than with natural monolignol substrates. Although suffering from some technical disadvantages (99), isoelectric focusing is preferred to electrophoresis, simply because isozyme separation is based upon a unique property of proteins (i.e. their isoelectric points, pI). For example, with *Nicotiana tabacum*, peroxidase isoforms can be separated into several groups—acidic (pIs 3.5–4.5), weakly acidic to neutral (pIs 4.5–7.5), and basic (pIs 8–11) (114, 142). With electrophoresis, on the other hand, isozymes are separated depending upon whether they migrate to the anode (anionic) or the cathode (cationic). Migration polarity is a rather arbitrary property, being determined not only by the net surface charge of the protein, but also by its Stokes' radius. Hence, varying either gel concentration or electrophoretic buffer pH can lead to differences in mobility. By this method, it is difficult to determine whether a particular isozyme has acidic or basic character.

Lignifying tissue is often histochemically visualized by a staining reaction using artificial substrates, such as syringaldazine, in the presence of H_2O_2. Most of the plants examined for syringaldazine oxidase activity have been angiosperms (22, 34, 78, 93, 94, 101, 170), where Harkin & Obst observed (78) that "usually angiosperms gave more intense broader colorations than gymnosperms . . .". Since gymnosperm lignins are mainly guaiacyl (i.e. primarily coniferyl alcohol–derived), it seems reasonable to infer that peroxidase isozymes from angiosperms may more effectively oxidize sinapyl alcohol. In contrast to efforts spent examining wall-bound peroxidases with artificial substrates such as syringaldazine, little attention has been paid to the use of monolignols as substrates. This unfortunate trend is presumably due to their limited availability and to the lack of suitable methodology to monitor their oxidation. Fortunately, some progress using natural substrates is now being made. Activity staining of gels following electrophoresis has been reported using monolignols (19), and HPLC can be used to determine quantitatively the amounts of monolignols consumed (127).

However, Goldberg et al used syringaldazine to stain the differentiating xylem and phloem vessels of poplar stems (66) and suggested that the isozymes involved migrated rapidly toward the anode on an anionic electrophoresis gel—i.e. they were presumably strongly acidic (94). A direct correlation between syringaldazine activity and lignification per se was not established. Indirect evidence has been obtained that seems to correlate changes in syringaldazine oxidase activity with lignin formation. For example, increases in both quantity of lignin deposited and syringaldazine oxidase activity have been observed upon injury of tomato fruits (51), rubbing young internodes of *Bryonia dioica* (34), and grafting of apricot trees (170). Conversely, in vitrifying carnation tissue, syringaldazine oxidase activity decreased, as did the extent of lignin deposition (101).

The peroxidases of *N. tabacum* have been more extensively studied than those of any other plant, and several attempts have been made to correlate lignification with specific isozymes. For example, Mäder et al demonstrated that peroxidases from *N. tabacum* tissue cultures could be separated into four main groups—two anodic (G_I and G_{II}) and two cathodic migrating groups (G_{III} and G_{IV}) (143, 189). Anodic groups G_I and G_{II} were both located in the cell walls, the G_I group being more strongly acidic than G_{II}. The cathodic isozymes (presumably basic or neutral) were located in the vacuoles and were thus unlikely to be involved in lignification (189). The G_I group more readily oxidizes both *p*-coumaryl and coniferyl alcohols than the G_{II} group, and G_{III} was virtually inactive with either substrate (144). These results suggest the specific involvement of acidic peroxidases in lignin formation in *N. tabacum*.

In *N. tabacum* callus cultures, both the relative efficiencies of G_I, G_{II}, and G_{III} isozyme groups for H_2O_2 production and their K_m values for NADH were also examined. In the absence of stimulators, the G_{III} isozymes had the lowest K_m values for NADH and the fastest reaction rate, thus initially suggesting a specific role for them in H_2O_2 formation (145). However, differences in K_m values were very small among all isozyme groups. Further, when reaction rates were reexamined in the presence of stimulators (Mn^{2+} and phenols), the G_I isozymes ("acidic" and covalently bound to the cell wall) were as effective as the corresponding G_{III} group (141). These results, taken together with the fact that these basic G_{III} (cationic) peroxidases are located in the vacuole, make it difficult to imagine any role for them in lignification.

Lagrimini et al (114) recently obtained and sequenced a cDNA clone of an "anionic" (i.e. acidic) peroxidase from a *N. tabacum* cDNA library that may be identical to one of the G_I isozymes previously discussed. Unfortunately, again only an indirect correlation with lignification was obtained—i.e. the mRNA for this protein was more abundant in heavily lignified stem tissue than in leaf or root tissue.

But an investigation of organogenesis of *N. tabacum* tissue cultures yielded contrary results (99). These investigations suggested that 3–6 isozymes were associated with lignification/tracheary element maturation. Only one isoform was acidic but not cell-wall associated. In a similar vein, when *Zinnia elegans* tracheary elements were induced in culture, only cathodic (i.e. basic isozymes) increased in activity (148). For these reasons, it seems premature to assign any specific role to a peroxidase based simply on its isoelectric point. In order to establish that a specific peroxidase isozyme is involved in lignification, we must determine substrate specificity with monolignols, primary structure, subcellular location, and a temporal correlation with active lignification.

2.5 Regulation of Lignification

Lignification regulation is poorly understood, and much has been explained only phenomenologically. Lignin contents and type vary with species, tissue, developmental stage, and subcellular location. These variations may result from differences in both enzymatic activities of individual monolignol branch point enzymes (e.g. hydroxylation steps to produce caffeic and 5-hydroxy-ferulic acids) and substrate specificity of enzymes from angiosperms and gymnosperms (e.g. O-methyltransferases, β-glucosidases, etc) (72–74, 84). Lignin deposition and composition can also be influenced by wounding, fungal attack, specific enzyme inhibitors, mutations, hormones, metal ions, light (69), and even gravitational loads experienced during growth (30).

In spite of the importance of this major metabolic pathway, no single control (or regulatory) point in lignin biosynthesis has been unambiguously established. Numerous attempts have been made, however, to correlate lignification with the activity of enzymes such as PAL, peroxidase, and cinnamyl alcohol dehydrogenase (CAD). For PAL, it is well established that lignin levels can be reduced via action of specific inhibitors such as AOPP (194) and APEP (113), or increased by fungal elicitors (149, 153, 154, 176). However, the role of PAL as a regulatory enzyme is questionable, since modulation of its activity affects all products from the phenylpropanoid pathway, producing flavonoids and tannins as well as lignin. It has also been difficult to correlate peroxidase activity with lignin formation, since many isozyme forms staining positive are not involved in lignin deposition processes (82). Indeed, there are even instances where a decrease in lignin deposition is accompanied by a general increase in peroxidase activity (152). As far as the third enzyme (CAD) is concerned, it was recently cloned from cell cultures of bean *(Phaseolus vulgaris)* (208). Its activity can be suppressed chemically (70) or activated by fungal elicitor (71). However, it has not yet been established whether CAD serves as an important biochemical marker for lignification or whether regulation is controlled at some preceding stage. Compartmentalization of CAD may be important, because it is required for synthesis not only of monolignols and lignin, but also of lignans and related phenolics—e.g. didehydroconiferyl alcohol, suberin, etc.

Wounding and fungal attack are also accompanied by localized lignification responses at or near the point of insult. This type of lignin deposition is unusual, since it is thought that the complete cell is lignified, both wall and contents (153, 176). It needs to be established how this lignin differs structurally from "normal" lignin. Some investigations indicate that fragments of chitin, chitosan, or some related substance from the fungal cell wall induce this lignin deposition (176). The lignification response to fungal attack cannot be generalized, since wheat apparently produces a lignin of higher syringyl content (175), whereas Japanese radish root (2) responds by forming a lignin of a slightly more guaiacyl character. Precise chemical information about the nature of these wound- or fungus-induced lignins is still incomplete, and the lignin verification criteria described in Section 1.1 have yet to be met. Asada & Matsumoto have indicated that a cell wall–bound material (presumed to be a protein) is responsible for lignin induction by wounding or disease (2). This lignin induction factor is thought to be present in latent form in the cell walls. Its mode of action, vis-à-vis precise enzyme activation, remains to be determined.

Brown-rib corn mutants exhibit a reduced lignin content (120). The 5-hydroxyferulic acid content of these tissues is high, suggesting differences in O-methyltransferase activity. Such differences would be significant since O-methylation may also represent a control point in lignin synthesis.

At present our understanding of the regulation of lignin deposition is poor. Investigations are required to elucidate the mechanisms involved in regulation of lignin during normal development, as well as in defense functions.

2.6 Lignin Structure

We cannot yet isolate lignin in its native (or unaltered) state. Structural representations for lignin have so far relied upon the analysis of soluble lignin-derived preparations, such as milled wood, dioxane, and kraft lignins (1, 61, 116, 159). Because of their similarities to synthetically dehydrogenatively polymerized (DHP) lignins, which are obtained via random polymerization of monolignols in vitro, it has generally been assumed that lignins exist in plant tissue as randomly linked polymers. If this assumption is correct, it sets lignins apart from almost all other naturally occurring products.

The random dehydrogenative polymerization process seems to ignore the fact that lignification in growing plants is a carefully orchestrated process. First, there appears to be a rather precise means of communication between the growing cell wall and the cytoplasm, as regards transport and selection of different monolignols at different developmental stages. Terashima's and Goring's groups observed that different monomers were incorporated into the lignin framework of various subcellular locations at different stages of development—e.g. p-coumaryl alcohol was preferentially deposited in the cell corners during initial stages of lignin deposition. Such temporal and spatial distribution differences may have profound effects on lignin structure in different subcellular locations. Second, lignification occurs within a carbo-hydrate matrix that is initiated at the cell corners and then proceeds along the middle lamella and into the secondary walls. We have suggested that initia-tion of lignification at the cell corners may result from the presence there of "lignification recognition or anchoring sites" such as cell wall–bound hydroxycinnamic acids (212).

A random bonding arrangement for lignin has not been accepted by all investigators. A model involving identical repeating units in spruce lignin was proposed long ago (52) but lacked convincing experimental evidence. Tenta-tive evidence for the ordering of aromatic rings in the secondary wall of black spruce *(Picea mariana)* has been reported, but here, too, additional ex-perimental support is needed (3). Consequently, until the natural lignification process has been fully delineated it would be folly to insist that such a carefully controlled process produces a polymer of totally irregular structure.

For these reasons it was important to develop techniques for studying native lignin structure in situ in plant tissue (130, 132, 133). We needed methods to label (with $> 90\%$ ^{13}C atom %) the atoms most frequently involved in inter-unit linkages in the lignin polymer. Such labeling would enable us to establish precisely the chemical environment of a particular carbon in the polymer in situ using solid-state ^{13}C NMR spectroscopy. This objective was achieved by growing plants hydroponically, under aseptic conditions, in media containing an appropriate ^{13}C-labeled lignin precursor—e.g. [1-^{13}C], [2-^{13}C], [3-^{13}C] phenylalanine, ferulic acid, coniferin, etc. Thus we quickly established that with the fast-growing angiosperm *Leucaena leucocephala* ferulic acid was incorporated intact into root tissue lignin but was not translo-cated into aerial portions. On the other hand, phenylalanine was readily translocated throughout the plant and was more efficiently incorporated into lignin. Similar findings were obtained with wheat.

Thus plants were administered ^{13}C-labeled Phe and then harvested, and the solid-state ^{13}C NMR spectrum of each tissue was determined. From the signals observed, we were able to examine the bonding environment of each individual propane side-chain carbon in lignin in situ—i.e. at sites where most of the lignin inter-unit linkages occur. As shown in Figure 5a–c, the reso-nances noted in *L. leucocephala* were consistent with those obtained for lignin substructures A–E (Figure 6) (130, 132). For example, substructure B domi-nates (see Figure 5b), and significant bonding of the Cα carbon to carbo-hydrate or lignin was also noted (Figure 5c). On the other hand, in wheat, bonding environments were rather dissimilar, and substructure B was hardly evident as an inter-unit linkage (133).

The spectra obtained varied with the plant species under investigation, and none was adequately reflected in a synthetic DHP polymer (128).

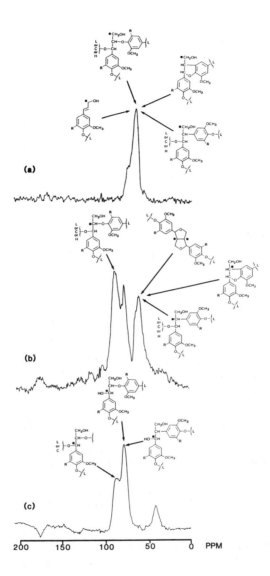

Figure 5 Incorporation of [1-¹³C], [2-¹³C], and [3-¹³C] phenylalanine into *L. leucocephala* stems. L = lignin; C = carbohydrate; R = H or OCH₃

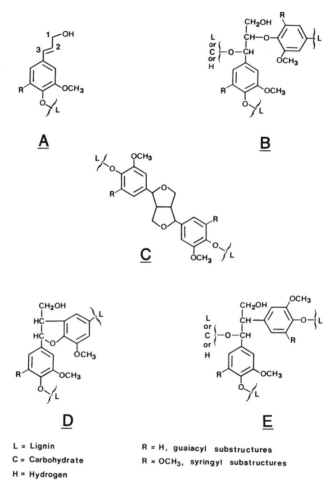

L = Lignin
C = Carbohydrate
H = Hydrogen

R = H, guaiacyl substructures
R = OCH₃, syringyl substructures

Figure 6   Presumed dominant lignin substructural bonding environments

Hence, we have now established that it is possible to study directly precise lignin bonding patterns in intact plant tissue. With further refinement of this method we can determine (*a*) lignin structure in different subcellular regions, (*b*) the contribution of each monomer in the lignin polymer, and (*c*) the range and type of different inter-unit linkages between monomers in the lignin polymer. Until this information is obtained, questions regarding lignin structure in situ cannot be resolved.

3 LIGNIN BIODEGRADATION

During the last 50 years interest in this topic has grown, but progress in elucidating lignin biodegradation mechanisms (particularly as regards identification of enzymes involved) has been painstakingly slow and punctuated by many false starts. Despite clarification of some issues in recent years, much confusion remains.

The range of microorganisms believed capable of degrading lignin (partly or completely) has been described adequately elsewhere (32, 123). Microorganisms able to degrade lignin include the wood-rotting fungi and to a lesser extent some bacteria and the actinomycetes. This discussion is limited to the white-rot fungi, the most widely studied and efficient lignin-degraders.

3.1 Lignin-like Polymers in White-Rot Fungi

Even lignin-degrading white-rot fungi synthesize "lignin-like" materials. For example, in the fruiting bodies of *Polyporus hispidus*, a polymeric substance is deposited that apparently has a function similar to lignin's. Described as fungal "lignin" (15–17, 40), it is thought to be derived by oxidative polymerization of hispidin, presumably formed (at least in part) via the shikimate-phenylpropanoid pathway (Figure 7). Both of these pathways are operational in veratryl alcohol synthesis in the white-rot fungus *Phanerochaete chrysoporium*.

3.2 Criteria of Lignin Degradation

In order to produce meaningful results, studies of lignin degradation must meet the following criteria.

1. An isolated lignin preparation of *known* molecular-weight range, monomer composition, functional group content, and inter-unit bonding patterns must be used to study bioconversion. Unfortunately, this approach is seldom employed. "Model" compound substrates (e.g. phenylpropanoid dimers) should be used to confirm observations with the polymers.

2. The degraded polymer (or fragments thereof) must be characterized fully at different stages of degradation, and appropriate control experiments must

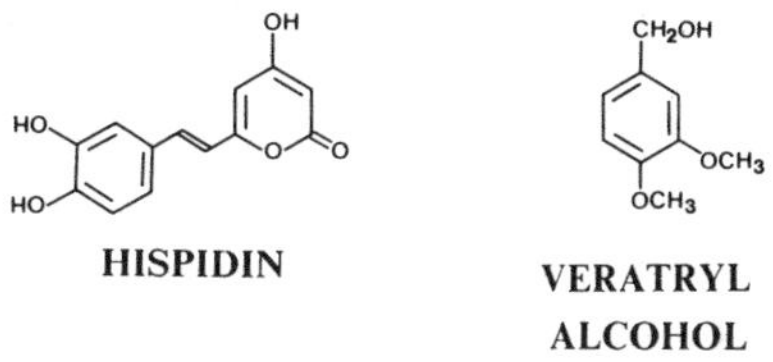

HISPIDIN   **VERATRYL ALCOHOL**

Figure 7   Two metabolites from white-rot fungi

be described in detail. This procedure serves not only to establish the modifications involved but also to determine whether, for example, the polymer is activated in some manner to facilitate depolymerization.

3. For putative enzyme systems, appropriate kinetic data for lignin must be described—e.g. rate of formation of monomeric cleavage products.

3.3 Lignin Biodegradation by White-Rot Fungi

With respect to lignin biodegradation, fungi can be distinguished according to their properties when grown on agar containing either gallic or tannic acid (5, 33). A brown zone surrounding the mycelia is evident with the white-rots, a phenomenon not observed with brown-rot fungi, which do not efficiently degrade lignin. This coloration results from excretion of extracellular phenol oxidase (laccase or peroxidase) activity by the white-rot fungi—e.g. laccase from *Polyporus zonatus* and *Polyporus versicolor* (135), and peroxidases from *Phellinus igniarius* and other white-rots (139). More importantly, phenol-oxidase activity appeared to be temporally correlated with ligninolytic activity, thereby suggesting a direct role in lignin biodegradation (see 31). However, as discussed below, such a role has neither been demonstrated nor shown necessary.

Initial experiments indicated that both laccases and peroxidases played roles in detoxification [e.g. of low-molecular-weight phenolics, such as pinosylvin in gymnosperm heartwood (140)]; the mode of detoxification was envisaged as a "weak polymerization process." Further support for this suggestion was obtained following examination of the effect of both laccase and horseradish peroxidases/H_2O_2 on lignin model substrates (21, 60) and isolated lignin derivatives (98, 214). With a number of simple phenolic lignin models, the primary reaction observed was that of phenolic coupling. Secondary reactions, such as arene side-chain cleavage and quinone formation could occur, but only if benzylic alcohols were completely oxidized to the corresponding aldehydes or ketones (21). It is of interest that neither enzyme had any significant effect on nonphenolic model compounds (60), nor did they depolymerize various lignin derivatives (98, 214). In fact, the opposite was observed—i.e. apparent polymerization of a milled-wood lignin sample (214). Gierer & Opara (60) concluded that the function of the extracellular peroxidase and laccase "consists in detoxifying low-molecular-weight phenolic compounds which may be released from lignin during its fungal decomposition, and in maintaining the metabolic balance around the fungi. No experimental evidence for a direct involvement of these enzymes in the microbial degradation of lignin was obtained." These authors did not rule out the possibility that "the above oxidative coupling reactions may constitute the first step of a reaction sequence, in which peroxidase and laccase in concert with other (unknown) enzymes participate in the microbial lignin break-down." (See below.)

Thus the suspected involvement of both laccase and peroxidase/H_2O_2 in lignin biodegradation lost favor, even though a much later report indicated that lignins could be degraded by horseradish peroxidase/H_2O_2 in organic media (36). This claim was not substantiated in experiments conducted in our laboratory (131).

Attention was next directed towards "non-enzymatic" degradation of lignin involving activated oxygen species such as OH·, ·OOH, H_2O_2, and $O_2 \dot{-}$ (77). In these studies, the most convincing findings were obtained with hydroxyl radicals (OH·) from Fenton's reagent, which produced a substantial de-polymerization of lignin (111).

This notion of "non-enzymatic" degradation held prominence until an apparent (and widely publicized) breakthrough occurred in 1983, when it was reported that the partial depolymerization of lignin had been achieved with a concentrated extracellular fluid from *Phanerochaete chrysosporium* cultures (62, 201). The investigators thought that this effect was due to a H_2O_2-requiring oxygenase, which they called ligninase (201) and diarylpropane oxygenase (62). This preparation was also able to engender $C\alpha C\beta$ cleavage of nonphenolic lignin model compounds, a reaction viewed to be involved in lignin biodegradation. The term ligninase is in conformity with that first used in 1937 by Bose & Sarkar (12) to describe extracellular lignin-degrading enzyme(s).

In the light of subsequent findings, it would be remiss not to discuss both of these oxidase papers in greater detail. First, the lignin substrates used by Tien & Kirk (201) corresponded to a ^{14}C-methyl iodide/K_2CO_3 methylated aqueous acetone extract of spruce wood (*Picea engelmanii* Parry) and a similarly treated milled-wood lignin preparation from birch (*Betula verrucosa* L). It is not obvious, at least to us, how well an aqueous acetone extract of spruce wood represents lignin, or how it differs from (oligomeric) lignans, stilbenes, or other phenolics. Nevertheless, as suggested by comparison of gel-permeation chromatograms of the methylated spruce extract before and after treatment with the *concentrated, crude extracellular fluid* from *P. chrysosporium,* approximately 22% depolymerization had occurred (data reported), whereas that with birch milled wood lignin only afforded 6% de-polymerization (no data reported). This was a surprising observation since it is well documented that angiosperm ("guaiacyl-syringyl") lignins are more rapidly biodegraded by white-rot fungi than their gymnosperm ("guaiacyl") counterparts (47). Other points of interest were that (*a*) following incubation of the methylated spruce extract, the sample required *filtration* prior to gel permeation chromatography; (*b*) no indication of recovered radioactivity prior to column loading was disclosed; and (*c*) the enzymatic activity considered responsible for this effect was assigned to a single gel-electrophoretic band, because veratraldehyde was liberated, albeit at *severely* attenuated rates, from the [^{14}C]methylated spruce extract and "lignin model compounds" (201). In the report by Glenn et al (62), a similar treatment resulted in the claim that ~10% degradation of a [ring-^{14}C]-labeled lignin, presumably *phenolic,* had occurred, as evidenced by GPC (no data reported).

These preliminary reports spawned a flurry of activity worldwide, which within three years profoundly altered this view of lignin biodegradation. First, ligninase (diarylpropane oxygenase) was shown *not* to be an oxidase, but instead a series of heme-containing peroxidases, now called lignin peroxidases. With very low pIs and molecular weights of ~41–42 kDa (see 105), these peroxidases seem to resemble closely the peroxidases isolated by Kruger & Pfeil (109) from the white-rot *P. igniarius.* Other peroxidases have also been isolated from *P. chrysosporium* that exhibit a curious manganese-dependence (65, 167). These manganese-dependent peroxidases are also being examined for their ability to degrade lignin. Since they readily convert phenols into phenoxy radicals, they are likely to undergo (re)polymerization reactions, although they may have a role in the generation of extracellular H_2O_2 (see 104). Second, *all* subsequent experiments with (partially) purified lignin peroxidases/H_2O_2 only engendered *further polymerization* of several distinct lignin derivatives (76, 100). No depolymerization of lignin has been reported with lignin peroxidase preparations since the original claims with crude extracellular fluid from *P. chrysosporium.* Third, lignin peroxidases and H_2O_2 convert specifically labeled (1-^{13}C, 2-^{12}C and 3-^{13}C) coniferyl alcohols into polymeric substances (185), arguably identical to synthetic dehydrogenatively polymerized (DHP) lignins obtained via the action of HRP/H_2O_2 (128). Finally, ligninolytic activity was observed with *P. chrysosporium* under conditions where lignin peroxidase was not detected (124).

Thus, once again attention has turned to the polymerizing abilities of fungal peroxidases. [The terms ligninase, lignin peroxidase(s) (LiP), and manganese-dependent peroxidases (Mn-P) should no longer be used because there is no evidence that they play a specific role in lignin degradation. Indeed, these nonspecific enzymes may readily act on other phenolics, such as tannins, flavonoids, and lignans. The term fungal peroxidases, therefore, seems more appropriate.] The fungal peroxidases would have been of only passing interest, if it were not for the results obtained with nonphenolic dimeric "lignin model substrates." In the presence of H_2O_2, "lignin peroxidases" engender a number of "model compound" conversions; these include $C\alpha C\beta$ oxidation, β-O-aryl and aromatic ring cleavage, 1,2-diol addition across side-chain double bonds, benzylic hydroxylation, and $C\alpha$ oxidations [see reviews by Higuchi (84–86), Kirk & Farrell (105), Umezawa (203), Umezawa & Higuchi (204), and references therein]. An example is shown in Figure 8, using a lignin model compound "dimer" (106), which in our opinion better represents a potential lignin substructure than most others employed.

Umezawa & Higuchi (205) next reported that a synthetic dehydrogenative polymer obtained from coniferyl alcohol and a β-O-aryl trimer underwent reactions similar to those of model substrates—e.g. β-O-aryl and aromatic ring cleavage and Cα oxidation (Figure 9). However, these transformations occurred only with the high-molecular-weight portion of the DHP preparation (Sephadex LH-20 excluded; nonphenolic?) and not the lower-molecular-weight portion (presumably phenolic). No data were presented showing either the effect of lignin peroxidase/H$_2$O$_2$ on the molecular weight profiles of the samples or the yields of the degradation products. It must be emphasized that in all other studies where phenolic moieties have been used, only oxidative coupling reactions have been documented (e.g. 103).

On the basis of these findings with artificial substrates, debate continues that fungal peroxidases are responsible for lignin depolymerization and breakdown, in spite of their efficacy as regards polymerization of phenol-containing lignin fragments. To account for this obvious disparity, suggestions have been made that low-molecular-weight cleavage products are rapidly assimilated by the fungus before polymerization can occur (124, 190). In this way, the net balance of reactions would be towards degradation. This suggestion has not been corroborated. It was also proposed that cellobiose:quinone oxidoreductase might prevent free-radical formation, thereby preventing net polymerization, but this proposal was shown to be incorrect. This enzyme had no effect upon the apparent product size distribution arising from (*a*) the lignin-peroxidase-catalyzed polymerization of either guaiacol or a DHP from coniferyl alcohol, or (*b*) phenoxy radical formation from acetosyringone (165). Thus to this point there is no convincing evidence that purified lignin peroxidases (or manganese peroxidases), in the presence of H$_2$O$_2$, can engender the depolymerization of lignin samples.

Let us now attempt to correlate the original conclusions of Gierer & Opara (60) and with recent findings in order to develop a working hypothesis for lignin biodegradation. Fenn & Kirk (50) observed recently that "lignin model

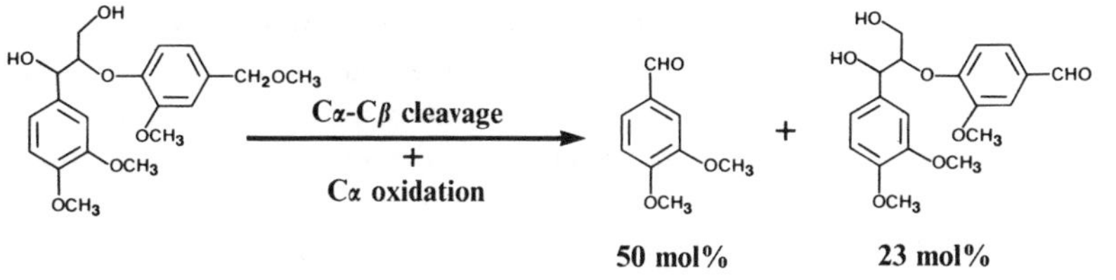

Figure 8   Bioconversion of a β-O-aryl "lignin model dimer" by lignin peroxidase/H$_2$O$_2$ (106)

DEHYDROGENATION COPOLYMER OF

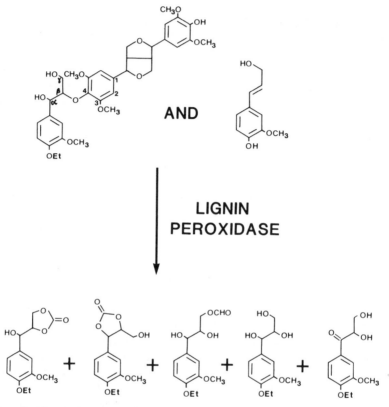

Figure 9 Monomeric products identified following treatment of a DHP copolymer, obtained from a synthetic trimer and coniferyl alcohol, with lignin peroxidase/H_2O_2 (205).

compounds" containing α-carbonyl functionalities were not readily degraded by *P. chrysosporium*. Interestingly, many monomeric α-carbonyl-containing lignin biodegradation products are formed by this organism that retain an intact propanoid side chain (25). In contrast, lignin samples that have been selectively oxidized at Cα to the corresponding ketones are more rapidly assimilated by *P. chrysosporium* than their nonoxidized counterparts (50). This observation, while a surprise to those investigators, provides support to the hypothesis of S. Sarkanen (personal communication) that one role of lignin peroxidase is to introduce α-carbonyl functionalities into the polymer, thereby facilitating depolymerization by an enzymatic depolymerization process yet to be determined. This suggestion is in accordance with Gierer & Opara's (60) view that fungal peroxidase(s) only play a cooperative role in lignin biodegradation. It would also explain the small increase in ligninolytic activity observed when lignin peroxidases were added to ligninolytic cultures of *P. chrysosporium* lacking detectable lignin peroxidase activity (124).

More clues to the process of lignin degradation may be forthcoming from detailed examination of wood-decayed lignins. To date, such studies have been very preliminary (see 24). However, one observation of the effect of *P. chrysosporium* on spruce wood may warrant further investigation (27, 178). Following a sixteen-week treatment with fungus, the lignin rendered soluble from the decayed spruce with methanol afforded two fractions, one acidic and one phenolic. The acidic fraction had two very different molecular-weight distributions, each of which had clearly undergone significant fungus-catalyzed oxidative modifications (i.e. partial side-chain cleavage giving α-carboxylic acids, Cα-oxidation, and apparently considerable oxidative C$_5$–C$_5$ coupling). Such modifications were not noted with the phenolic fraction, which more closely resembled a milled-wood lignin-like preparation. This suggests that the phenolic fraction was rendered soluble by a process not requiring fungal peroxidase(s). However, no firm conclusions can be drawn at this point since no control experiments were described (27).

It has also been proposed that laccase may be intimately involved in lignin biodegradation—e.g. with *Coriolus versicolor* (13, 86, 156, 157). With phenolic constituents, laccase can readily generate free-radical species (and thus appropriate coupling products), as well as Cα oxidation, demethylation, and CαCβ/Cα arene cleavage reactions. Nonphenolic model compounds can also be oxidized by laccase, but only in the presence of kraft lignin or ABTS (2,2' azinobis-3-ethylbenzthiazoline-6-sulphonate) (13). The mechanism for the oxidation has not yet been delineated. No convincing evidence has been obtained with isolated laccase that this enzyme depolymerizes either isolated lignins or lignins in situ.

CONCLUDING REMARKS

Progress toward understanding lignin's occurrence, synthesis, structure, and biodegradation has been slow but (with new approaches and methodology) steady. The application of both solution and solid-state carbon-13 NMR spectroscopy in the analysis of lignins is clarifying a number of controversial issues. The criteria listed in Section 2.1.1 must be met before the presence of lignins in plants (particularly nonwoody species) can be verified un-ambiguously.

The heterogeneity of the polymers at the subcellular level and the poor correlation of their bonding environments in situ with those of synthetic DHP preparations provide a strong case against an "uncontrolled" polymerization process. Indeed, lignification in plants is probably very carefully orchestrated. Stable-isotope carbon-13 labeling studies continue to elucidate in situ bonding environments, helping to identify differences in lignin structure at the subcellular and tissue level (e.g. fibers and vessels) and to determine the effect of either mechanical stress or fungal attack.

Most of the enzymes involved in both E and Z monolignol biosynthesis have been isolated. In future it should be possible to identify key control or regulatory points and to determine how E or Z monomers are selected and transported into the cell wall. Whether specific peroxidases are involved in lignification is at present unknown. Such uncertainties now seem ripe for clarification at the molecular level.

The mechanisms involved in lignin biodegradation still need extensive clarification. Recurrent claims that ligninases (now lignin peroxidases) and laccases are involved in this process have yet to be supported by studies using the polymer itself (either in situ or with lignin-derived isolates). As mentioned, the only reported case of lignin depolymerization resulted from use of a crude extracellular fluid from *P. chrysosporium,* not an isolated (or purified) enzyme. Sarkanen's hypothesis that one role of lignin peroxidase is to introduce Cα carbonyl functionalities into the polymer should not be dismissed, particularly since Fenn & Kirk's (50) study indicated that such functionalities increased the rate at which lignin was assimilated by the fungus. More attention needs to be devoted to identifying other enzymes involved in lignin degradation—e.g. those that specifically catalyze β-O-aryl bond cleavages.

ACKNOWLEDGMENTS

Financial support from the US Department of Energy (DE-FG05-88-ER13883) is gratefully acknowledged, as are the valuable discussions with S. Sarkanen.

Literature Cited

1. Adler, E. 1977. Lignin chemistry—past, present and future. *Wood Sci. Technol.* 11:169–218
2. Asada, Y., Matsumoto, I. 1987. Induction of disease resistance in plants by a lignification-inducing factor. In *Molecular Determinants of Plant Diseases,* ed. S. Nishimura, et al, pp. 223–33. Tokyo: Jpn. Sci. Soc. Press; Berlin: Springer-Verlag
3. Atalla, R. H., Agarwal, U. P. 1985. Raman microprobe evidence for lignin orientation in the cell walls of native woody tissue. *Science* 227:636–38
4. Bardet, M., Gagnaire, D., Nardin, R., Robert, D., Vincendon, M. 1986. Use of ^{13}C enriched wood for structural NMR investigation of wood and wood

components, cellulose and lignin, in solid and in solution. *Holzforschung* 40 (Suppl.):17–24
5. Bavendamm, W. 1928. Über das Vorkommen und den Nachweis von Oxydasen bei holzzerstörenden Pilzen. *Z. Pflanzenkrank. Pflanzenschutz* 38:257–76
6. Baydoun, E. A.-H., Brett, C. T. 1985. Comparison of cell wall compositions of the rhizomes of three seagrasses. *Aquat. Bot.* 23:191–96
7. Bland, D. E. 1958. Chemistry of reaction wood. I. Lignins of *Eucalyptus goniocalyx* and *Pinus radiata. Holzforschung* 12:36–43
8. Bland, D. E. 1961. The chemistry of reaction wood. III. The milled wood lig-

nins of *Eucalyptus goniocalyx* and *Pinus radiata*. *Holzforschung* 15:103–6

9. Bland, D. E., Logan, A., Menshun, M., Sternhell, S. 1968. The lignin of *Sphagnum*. *Phytochemistry* 7:1373–77

10. Bohm, B. A., Tryon, R. M. 1967. Phenolic compounds in ferns. I. A survey of some ferns for cinnamic and benzoic acid derivatives. *Can. J. Bot.* 45:585–93

11. Bolker, H. I., Somerville, N. G. 1962. Ultraviolet spectroscopic studies of lignins in the solid state. I. Isolated lignin preparations. *Tappi* 45:826–29

12. Bose, S. R., Sarkar, S. N. 1937. Enzymes of some wood-rotting polypores. *Proc. R. Soc. London Ser. B* 123:193–213

13. Bourbonnais, R., Paice, M. G. Oxidation of non-phenolic substrates: an expanded role for laccase in lignin biodegradation. Submitted

14. Brauns, E. F., Alexander, D. 1960. *The Chemistry of Lignin*, Suppl. Vol. Ch. 4, pp. 33–61. New York: Academic

15. Bu'Lock, J. D. 1967. Fungal metabolites with structural function. In *Essays in Biosynthesis and Microbial Development*, pp. 1–18. New York: Wiley & Sons

16. Bu'Lock, J. D., Leeming, P. R., Smith, H. G. 1962. Pyrones. Part II. Hispidin, a new pigment and precursor of fungus "lignin". *J. Chem. Soc.* 400:2085–89

17. Bu'Lock, J. D., Smith, H. G. 1961. A fungus pigment of novel type, and the nature of fungus "lignin." *Experientia* 17:553–56

18. Burritt, E. A., Bittner, A. S., Street, J. C., Anderson, M. J. 1984. Correlations of phenolic acids and xylose content of cell wall with *in vitro* dry matter digestibility of three maturing grasses. *J. Dairy Sci.* 67:1209–13

19. Butler, M. J., Lachance, M. A. 1987. The use of N,N,N',N'-tetramethylphenylenediamine to detect peroxidase activity on polyacrylamide electrophoresis gels. *Anal. Biochem.* 162:443–45

20. Byng, G. S., Whitaker, R., Flick, C., Jensen, R. A. 1981. Enzymology of L-tyrosine biosynthesis in corn *(Zea mays)*. *Phytochemistry* 20:1289–92

21. Caldwell, E. S., Steelink, C. 1969. Phenoxy radical intermediates in the enzymatic degradation of lignin model compounds. *Biochim. Biophys. Acta* 184:420–31

22. Castillo, F. J., Greppin, H. 1986. Balance between anionic and cationic extracellular peroxidase activities in *Sedum album* leaves after ozone exposure. Analysis by high-performance liquid chromatography. *Physiol. Plant.* 68:201–8

23. Chen, C. L. 1988. Characterization of lignin by oxidative degradation: use of gas chromatography—mass spectrometry technique. *Methods Enzymol.* 161:110–36

24. Chen, C. L., Chang, H.-M. 1985. Chemistry of lignin biodegradation. See Ref. 84, Ch. 19, pp. 535–56

25. Chen, C. L., Chang, H.-M., Kirk, T. K. 1983. Carboxylic acids produced through oxidative cleavage of aromatic rings during degradation of lignin in spruce wood by *Phanerochaete chrysosporium*. *J. Wood Chem. Technol.* 3:35–37

26. Chibbar, R. N., van Huystee, R. B. 1986. Immunochemical localization of peroxidase in cultured peanut cells. *J. Plant Physiol.* 123:477–86

27. Chua, M. G. S., Chen, C.-L., Chang, H.-M., Kirk, T. K. 1982. ^{13}C NMR spectroscopic study of spruce lignin degraded by *Phanerochaete chrysosporium*. *Holzforschung* 36:165–72

28. Connelly, J. A., Conn, E. E. 1986. Tyrosine biosynthesis in *Sorghum bicolor:* isolation and regulatory properties of arogenate dehydrogenase. *Z. Naturforsch. Teil C* 41:69–78

29. Conroy, J. M., Borzelleca, D. C., McDonnell, L. A. 1982. Homology of plant peroxidases: an immunochemical approach. *Plant Physiol.* 69:28–31

30. Cowles, J. R., Le May, R., Jahns, G., Scheld, W. H., Peterson, C. 1989. Lignification in young plant seedlings grown on earth and aboard the space shuttle. See Ref. 129, Ch. 15, pp. 203–13

31. Cowling, E. B. 1961. Comparative biochemistry of the decay of sweetgum sapwood by white-rot and brown-rot fungi. *USDA Tech. Bull. No.* 1258. 79 pp.

32. Crawford, R. L. 1981. *Lignin Biodegradation and Transformation*. New York: Wiley & Sons

33. Davidson, R. W., Campbell, W. A., Blaisdell, D. J. 1938. Differentiation of wood-decaying fungi by their reactions on gallic or tannic acid medium. *J. Agric. Res.* 57(9):683–95

34. De Jaegher, G., Boyer, N., Gaspar, T. 1985. Thigmomorphogenesis in *Bryonia dioica:* changes in soluble and wall peroxidases, phenylalanine ammonia-lyase activity, cellulose, lignin content and monomeric constituents. *Plant Growth Regul.* 3:133–48

35. Delwiche, C. F., Graham, L. E., Thomson, N. 1989. Lignin-like compounds

and sporopollenin in *Coleochaete*, an algal model for land plant ancestry. *Science* 245:399–401

36. Dordick, J. S., Marletta, M. A., Klibanov, A. M. 1986. Peroxidases depolymerize lignin in organic media but not water. *Proc. Natl. Acad. Sci. USA* 83:6255–57

37. Dovgan, I. V., Medvedeva, E. I. 1983. Change in the structural elements of the lignin of the brown alga *Cytoseira barbata* at different ages. *Chem. Nat. Compd.* 19(1):81–84

38. Dovgan, I. V., Medvedeva, E. I., Yanishevskaya, E. N. 1983. Cleavage of the lignins of the alga *Cytoseira barbata* by thioacetic acid. *Chem. Nat. Compd.* 19(1):84–87

39. Dutta, S., Garver, T. M., Sarkanen, S. 1989. Models of association between kraft lignin components. See Ref. 67, Ch. 12, pp. 155–76

40. Edwards, R. L., Lewis, D. G., Wilson, D. V. 1961. Constituents of the higher fungi. Part I. Hispidin, a new 4-hydroxy-6-styryl-2-pyrone from *Polyporus hispidus*. (*Bull.*) *Fr. J. Chem. Soc.* pp. 4995–5002

41. Eom, T. J., Meshitsuka, G., Nakano, J. 1987. Chemical characteristics of lignin in the differentiating xylem of a hardwood. II. *Mokuzai Gakkaishi* 33(7):576–81

42. Erdtman, H. 1957. Outstanding problems in lignin chemistry. *Ind. Eng. Chem.* 49(9):1385–86

43. Erickson, M., Miksche, G. E. 1974. On the occurrence of lignin or polyphenols in some mosses and liverworts. *Phytochemistry* 13:2295–99

44. Erickson, M., Miksche, G. E. 1974. Charakterisierung der Lignine von Pteridophyten durch oxidativen Abbau. *Holzforschung* 28:157–59

45. Erickson, M., Miksche, G. E. 1974. Charakterisierung der Lignine von Gymnospermen durch Oxidativen Abbau. *Holzforschung* 28:135–38

46. Erickson, M., Miksche, G. E., Somfai, I. 1973. Charakterisierung der Lignine von Angiospermen durch oxydativen Abbau. II. Monokotylen. *Holzforschung* 27(5):147–50

47. Faix, O., Mozuch, M. D., Kirk, T. K. 1985. Degradation of gymnosperm (guaiacyl) vs. angiosperm (syringyl/guaiacyl) lignins by *Phanerochaete chrysosporium*. *Holzforschung* 39:203–8

48. Farmer, V. C. 1953. The lignin of *Sphagnum*. *Research* 6(8):A1-A3

49. Farmer, V. C., Morrison, R. I. 1964. Lignin in sphagnum and phragmites and in peats derived from these plants. *Geochim. Cosmochim. Acta* 28:1537–46

50. Fenn, P., Kirk, T. K. 1984. Effects of Cα-oxidation in the fungal metabolism of lignin. *J. Wood Chem. Technol.* 4(2):131–48

51. Fleuriet, A., Deloire, A. 1982. Aspects histochimiques et biochimiques de la cicatrisation des fruits de Tomate blessés. (Histochemical and biochemical aspects of cicatrization of tomato fruit lesions.) *Z. Pflanzenphysiol.* 107:259–68

52. Forss, K., Fremer, K.-E. 1983. Comments on the nature of coniferous lignin. *J. Appl. Polym. Sci.* (Appl. Polym. Symp.) 37:531–46

53. Freudenberg, K., Lautsch, W., Engler, K. 1940. Die Bildung von Vanillin ans Fichtenlignin. *Chem. Ber.* 73(B):167–71

54. Freudenberg, K., Neish, A. C. 1968. *Constitution and Biosynthesis of Lignin, Molecular-Biology Biochemistry and Biophysics*, Vol. 2. Berlin: Springer-Verlag. 129 pp.

55. Fukuda, T., Mott, R. L., Harada, C. 1988. Studies on tissue culture of tree cambium. XI. Characterization of lignin in suspension-cultured cells of loblolly pine. *Mokuzai Gakkaishi* 34(2):149–54

56. Gadal, P., Bouyssou, H. 1973. Allosteric properties of chorismate mutase from *Quercus pedunculata*. *Physiol. Plant* 28:7–13

57. Gaines, C. G., Byng, G. S., Whitaker, R. J., Jensen, R. A. 1982. L-tyrosine regulation and biosynthesis via arogenate dehydrogenase in suspension-cultured cells of *Nicotiana silvestris* Speg. et. Comes. *Planta* 156:233–40

58. Gaspar, Th., Penel, C., Castillo, F. J., Greppin, H. 1985. A two-step control of basic and acidic peroxidases and its significance for growth and development. *Physiol. Plant.* 64:418–23

59. Gaspar, Th., Penel, C., Thorpe, T., Greppin, H. 1982. *Peroxidases 1970–1980. A Survey of Their Biochemical and Physiological Roles in Higher Plants.* Geneve: Univ. Geneve

60. Gierer, J., Opara, A. E. 1973. Studies on the enzymatic degradation of lignin. The action of peroxidase and laccase on monomeric and dimeric model compounds. *Acta Chem. Scand.* 27:2909–22

61. Glasser, W. G., Glasser, H. R. 1981. The evaluation of lignin's chemical structure by experimental and computer simulation techniques. *Paperi Ja Puu* 63(2):71–83

62. Glenn, J. K., Morgan, M. A., Mayfield, M. B., Kuwahara, M., Gold, M. H.

1983. An extracellular H_2O_2-requiring enzyme preparation involved in lignin biodegradation by the white-rot basidiomycete *Phanerochaete chrysosporium.* *Biochem. Biophys. Res. Commun.* 114:1077–83

63. Godshalk, G. L., Wetzel, R. G. 1978. Decomposition of aquatic angiosperms. III. *Zostera marina* L. and a conceptual model of decomposition. *Aquat. Bot.* 5:329–54

64. Goering, H. K., Van Soest, P. J. 1970. Forage fiber analyses (apparatus, reagents, procedures and some applications). *USDA Agric. Res. Serv., Agric. Handb.* 379. 20 pp.

65. Gold, M. H., Wariishi, H., Akileswaran, L., Mino, Y., Loehr, T. M. 1987. Spectral characterization of Mn-peroxidase, an extracellular heme enzyme from *Phanerochaete chrysosporium.* In *Lignin Enzymic and Microbial Degradation,* ed. E. Odier, pp. 113–18. Paris: INRA Publ.

66. Goldberg, R., Catesson, A.-M., Czaninski, Y. 1983. Some properties of syringaldazine oxidase, a peroxidase specifically involved in the lignification process. *Z. Pflanzenphysiol.* 110:267–79

67. Goring, D. A. I. 1989. The lignin paradigm. *ACS Symp. Ser.* 397:1–10

68. Grand, C. 1984. Ferulic acid-5-hydroxylase: a new cytochrome P-450 dependent enzyme from higher plant microsomes involved in lignin synthesis. *FEBS Lett.* 169:7–11

69. Grand, C., Boudet, A. M., Ranjeva, R. 1982. Natural variations and controlled changes in lignification process. *Holzforschung* 36:217–23

70. Grand, C., Sarni, F., Boudet, A. M. 1985. Inhibition of cinnamyl alcohol dehydrogenase activity and lignin synthesis in poplar (*Populus* × *euramericana* Dode) tissues by two organic compounds. *Planta* 163:232–37

71. Grand, C., Sarni, F., Lamb, C. J. 1987. Rapid induction by fungal elicitor of the synthesis of cinnamyl alcohol dehydrogenase, a specific enzyme of lignin synthesis. *Eur. J. Biochem.* 169:73–77

72. Grisebach, H. 1981. Lignins. In *The Biochemistry of Plants. A Comprehensive Treatise, Secondary Plant Products,* ed. P. K. Stumpf, E. E. Conn, 7:457–78. New York: Academic

73. Gross, G. G. 1977. Biosynthesis of lignin and related monomers. *Rec. Adv. Phytochem.* 11:141–84

74. Gross, G. G. 1979. Recent advances in the chemistry and biochemistry of lignin. *Rec. Adv. Phytochem.* 12:177–220

75. Guittet, E., Lallemand, J. Y., Lapierre, C., Monties, B. 1985. Applicability of the ^{13}C NMR "inadequate" experiment to lignin, a natural polymer. *Tetrahedron Lett.* 26(22):2671–74

76. Haemmerli, S. D., Leisola, M. S. A., Fiechter, A. 1986. Polymerization of lignins by ligninases from *Phanerochaete chrysosporium.* *FEBS Microbiol. Lett.* 35:33–36

77. Hall, P. 1980. Enzymatic transformations of lignin: 2. *Enzyme Microb. Technol.* 2:170–76

78. Harkin, J. M., Obst, J. R. 1973. Lignification in trees: indication of exclusive peroxidase participation. *Science* 180:296–98

79. Hartley, R. D., Ford, C. W. 1989. Phenolic constituents of plant cell walls and wall biodegradability. See Ref. 129, Ch. 9, pp. 137–45

80. Hatfield, G. R., Maciel, G. E., Erbatur, O., Erbatur, G. 1986. Qualitative and quantitative analysis of solid lignin samples by carbon-13 nuclear magnetic resonance spectrometry. *Anal. Chem.* 59 (1):172–79

81. Hathway, D. E. 1962. The lignans. In *Wood Extractives and Their Significance to the Pulp and Paper Industries,* ed. W. E. Hillis, Ch. 4, pp. 159–90. New York: Academic

82. Hazell, P., Murray, D. R. E. 1982. Peroxidase isozymes and leaf senescence in sunflower, *Helianthus annuus* L. *Z. Pflanzenphysiol.* 108:87–92

83. He, L., Terashima, N. 1989. Formation and structure of lignin in monocotyledons. II. Deposition and distribution of phenolic acids and their association with cell wall polymers in rice plants (*Oryza sativa*). *Mokuzai Gakkaishi* 35:123–29

84. Higuchi, T. 1985. Biosynthesis of lignin. In *Biosynthesis and Biodegradation of Wood Components,* ed. T. Higuchi, pp. 141–62. Orlando, FL: Academic

85. Higuchi, T. 1985. Degradative pathways of lignin model compounds. See Ref. 84, pp. 557–78

86. Higuchi, T. 1989. Mechanisms of lignin degradation by lignin peroxidase and laccase of white-rot fungi. See Ref. 129, pp. 482–502

87. Higuchi, T., Ito, Y., Kawamura, I. 1967. *p*-Hydroxyphenylpropane component of grass lignin and role of tyrosine ammonia lyase in its formation. *Phytochemistry* 6:875–81

88. Higuchi, T., Ito, Y., Shimada, M., Kawamura, J. 1967. Chemical properties of milled wood lignin of grasses. *Phytochemistry* 6:1551–56

89. Hrazdina, G., Wagner, G. 1985. Meta-

bolic pathways as enzyme complexes: evidence for the synthesis of phenylpropanoids and flavonoids on membrane associated enzyme complexes. *Arch. Biochem. Biophys.* 237:88–100

90. Hrazdina, G., Zobel, A. M., Hoch, H. C. 1987. Biochemical, immunological, and immunocytochemical evidence for the association of chalcone synthase with endoplasmic reticulum membranes. *Proc. Natl. Acad. Sci. USA* 84:8966–70

91. Ibrahim, R. K., Towers, G. H. N., Gibbs, R. D. 1962. Syringic and sinapic acids as indicators of differences between major groups of vascular plants. *Bot. J. Linn. Soc.* 58:223–30

92. Iiyama, K., Wallis, A. F. A. 1988. An improved acetyl bromide procedure for determining lignin in woods and wood pulps. *Wood Sci. Technol.* 22:271–80

93. Imberty, A., Goldberg, R., Catesson, A. M. 1984. Tetramethylbenzidine and *p*-phenylenediamine-pyrocatechol for peroxidase histochemistry and biochemistry: two new, non-carcinogenic chromogens for investigating lignification process. *Plant Sci. Lett.* 35:103–8

94. Imberty, A., Goldberg, R., Catesson, A. M. 1985. Isolation and characterization of *Populus* isoperoxidases involved in the last step of lignin formation. *Planta* 164:221–26

95. Jensen, R. A. 1986. Tyrosine and phenylalanine biosynthesis: relationship between alternative pathways, regulation and subcellular location. *Rec. Adv. Phytochem.* 20:57–82

96. Jensen, R. A., Morris, P., Bonner, C. A., Zamir, L. O. 1989. Biochemical interface between aromatic amino acid biosynthesis and secondary metabolism. See Ref. 129, Ch. 6, pp. 89–107

97. Jung, E., Zamir, L. O., Jensen, R. A. 1986. Chloroplasts of higher plants synthesize L-phenylalanine via L-arogenate. *Proc. Natl. Acad. Sci. USA* 83:7231–35

98. Kaplan, D. C. 1979. Reactivity of different oxidases with lignins and lignin model compounds. *Phytochemistry* 18:1917–19

99. Kay, L. E., Basile, D. V. 1987. Specific peroxidase isoenzymes are correlated with organogenesis. *Plant Physiol.* 84:99–105

100. Kern, H. W., Kirk, T. K. 1987. Influence of molecular size and ligninase pretreatment on degradation of lignins by *Xanthomonas* sp. Strain 99. *Appl. Environ. Microbiol.* 53:2242–46

101. Kevers, C., Gaspar, Th. 1985. Soluble, membrane and wall peroxidases, phenylalanine ammonia-lyase, and lignin changes in relation to vitrification of carnation tissues cultured *in vitro. J. Plant. Physiol.* 118:41–48

102. Kim, S. S., Wender, S. H., Smith, E. C. 1980. Comparison of tryptic peptide maps of eight isoperoxidases from tobacco tissue cultures. *Phytochemistry* 19:169–71

103. Kirk, T. K. 1987. Lignin-degrading enzymes. *Philos. Trans. R. Soc. London Ser. A* 321:461–74

104. Kirk, T. K. 1988. Lignin degradation by *Phanerochaete chrysosporium. ISI Atlas Sci: Biochem.* 1(1):71–76

105. Kirk, T. K., Farrell, R. L. 1987. Enzymatic 'combustion': the microbial degradation of lignin. *Annu. Rev. Microbiol.* 41:465–505

106. Kirk, T. K., Tien, M., Kersten, P. J., Mozuch, M. D., Kalyanaraman, B. 1986. Ligninase of *Phanerochaete chrysosporium. Biochem. J.* 232:279–87

107. Kratzl, K., Eibl, J. 1951. Chemical and botanical evidence for lignification. *Mitt. Oesterr. Ges. Holzforsch.* 3:76–77

108. Kringstad, K. P., Mörck, R. 1983. ^{13}C NMR spectra of kraft lignins. *Holzforschung* 37:237–44

109. Krüger, G., Pfeil, E. 1976. Darstellung und Eigenschaften einer Peroxidase aus *Phellinus igniarius. Arch. Microbiol.* 109:175–79

110. Kuo, J., Cambridge, M. L. 1978. Morphology, anatomy and histochemistry of the Australian seagrasses of the genus *Posidonia König* (Posidonicea). II. Rhizome and root of *Posidonia australis* Hook F. *Aquat. Bot.* 5:191–206

111. Kutsuki, H., Gold, M. H. 1982. Generation of hydroxyl radical and its involvement in lignin degradation by *Phanerochaete chrysosporium. Biochem. Biophys. Res. Commun.* 109:320–27

112. Kutsuki, H., Higuchi, T. 1978. The formation of lignin of *Erythrina cristagalli. Mokuzai Gakkaishi* 24(9):625–31

113. Laber, B., Kiltz, H.-H., Amrhein, N. 1986. Inhibition of phenylalanine ammonia lyase *in vitro* and *in vivo* by (1-amino-2-phenylethyl) phosphonic acid, the phosphonic analogue of phenylalanine. *Z. Naturforsch. Teil C* 41:49–55

114. Lagrimini, L. M., Burkhart, W., Moyer, M., Rothstein, S. 1987. Molecular cloning of complementary DNA encoding the lignin-forming peroxidase from tobacco: molecular analysis and tissue-specific expression. *Proc. Natl. Acad. Sci. USA* 84:7542–46

115. Lagrimini, L. M., Rothstein, S. 1987. Tissue specificity of tobacco peroxidase isozymes and their induction by wounding and tobacco mosaic virus infection. *Plant Physiol.* 84:438–42

116. Lai, Y. Z., Sarkanen, K. V. 1971. Isolation and structural studies. See Ref. 184, Ch. 5, pp. 165–240

117. Lapierre, C., Gaudillere, J. P., Monties, B., Guittet, E., Rolando, C., Lallemand, J. Y. 1983. Enrichissement photosynthétique en carbone 13 de lignines de peuplier: caractérisation préliminaire par acidolyse et RMN 13C. *Holzforschung* 37:217–24

118. Lapierre, C., Monties, B., Rolando, C. 1988. Thioacidolyses of diazomethane-methylated pine compression wood and wheat straw *in situ* lignins. *Holzforschung* 42(6):409–11

119. Lapierre, C., Monties, B., Rolando, C. 1986. Thioacidolysis of poplar lignins. Identification of monomeric syringyl products and characterization of guaiacyl-syringyl lignin fractions. *Holzforschung* 40:113–18

120. Lapierre, C., Tollier, M.-T., Monties, B. 1988. Mise en évidence d'un nouveau type d'unité constitutive dans les lignines d'un mutant de Maïs bm3. *C. R. Acad. Sci. III* 307:723–28

121. Larsson, S., Miksche, G. E. 1971. Gaschromatographische Analyse von Ligninoxydationsprodukten. V. Zwei trimere Abbauprodukte aus Fichtenlignin. *Acta Chem. Scand.* 25:673–79

122. Leary, G. J., Newman, R. H., Morgan, K. R. 1986. A carbon-13 nuclear magnetic resonance study of chemical processes involved in the isolation of Klason lignin. *Holzforschung* 40(5):267–72

123. Leisola, M. S. A., Fiechter, A. 1985. New trends in lignin biodegradation. *Adv. Biotechnol. Process.* 5:59–89

124. Leisola, M. S. A., Haemmerli, S. D., Waldner, R., Schoemaker, H. E., Schmidt, H. W. H., Fiechter, A. 1988. Metabolism of a lignin model compound, 3,4-dimethoxybenzyl alcohol by *Phanerochaete chrysosporium. Cell. Chem. Technol.* 22:267–77

125. Lewis, N. G., Davin, L., Umezawa, T., Yamamoto, E. 1990. On the biosynthesis of (+) pinoresinol in *Forsythia suspensa.* Submitted to *Phytochemistry*

126. Lewis, N. G., Dubelsten, P., Eberhardt, T. L., Yamamoto, E., Towers, G. H. N. 1987. The E/Z isomerization step in the biosynthesis of Z-coniferyl alcohol in *Fagus grandifolia. Phytochemistry* 26:2729–34

127. Lewis, N. G., Inciong, Ma. E., Dhara, K. P., Yamamoto, E. 1989. High-performance liquid chromatographic separation of E- and Z-monolignols and their glucosides. *J. Chromatogr.* 479:345–52

128. Lewis, N. G., Newman, J., Just, G., Ripmeister, J. 1987. Determination of bonding patterns of ^{13}C specifically enriched dehydrogenatively polymerized lignin in solution and solid state. *Macromolecules* 20:1752–56

129. Lewis, N. G., Paice, M. G., eds. 1989. In *Plant Cell Wall Polymers: Biogenesis and Biodegradation. ACS Symp. Ser.* 399. 676 pp.

130. Lewis, N. G., Razal, R. A., Dhara, K. P., Yamamoto, E., Bokelman, G. H., Wooten, J. B. 1988. Incorporation of $[2-^{13}C]$ ferulic acid, a lignin precursor, into *Leucaena leucocephala* and its analysis by solid state ^{13}C NMR spectroscopy. *J. Chem. Soc. Chem. Commun.* pp. 1626–28

131. Lewis, N. G., Razal, R. A., Yamamoto, E. 1987. Lignin degradation by peroxidase in organic media: a reassessment. *Proc. Natl. Acad. Sci. USA* 84:7925–27

132. Lewis, N. G., Razal, R. A., Yamamoto, E., Bokelman, G. H., Wooten, J. B. 1989. ^{13}C specific labelling of lignin in intact plants. See Ref. 129, Ch. 12, pp. 169–81

133. Lewis, N. G., Yamamoto, E., Wooten, J. B., Just, G., Ohashi, H., Towers, G. H. N. 1987. Monitoring biosynthesis of wheat cell-wall phenylpropanoids *in situ. Science* 237:1344–46

134. Lindberg, B., Theander, O. 1952. Studies on sphagnum peat. II. Lignin in sphagnum. *Acta Chem. Scand.* 6:311–12

135. Lindeberg, G., Fåhraeus. 1952. Nature and formation of phenol oxidases in *Polyporus zonatus* and *Polyporus versicolor. Physiol. Plant.* 5:277–83

136. Löffelhardt, W., Ludwig, B., Kindl, H. 1973. Thylakoid-gebundene L-Phenylalanin-Ammoniak-Lyase. *Hoppe-Seyler's Z. Physiol. Chem.* 354:1006–12

137. Logan, K. J., Thomas, B. A. 1985. Distribution of lignin derivatives in plants. *New Phytol.* 99:571–85

138. Lüdemann, H.-D., Nimz, H. 1973. Carbon-13 nuclear magnetic resonance spectra of lignins. *Biochem. Biophys. Res. Commun.* 52(4):1162–69

139. Lyr, H. 1955. Vorkommen von Peroxydase bei holzzerstörenden Basidiomyceten. *Planta* 46:408–13

140. Lyr, H. 1962. Detoxification of heart-

wood toxins and chlorophenols by higher fungi. *Nature* 195:289–90

141. Mäder, M., Amberg-Fisher, V. 1982. Role of peroxidase in lignification of tobacco cells. Oxidation of nicotinamide adenine dinucleotide and formation of hydrogen peroxide by cell wall peroxidases. *Plant Physiol.* 70:1128–31

142. Mäder, M., Bopp, M. 1976. Neue Vorstellungen zum Problem der Isoperoxidasen anhand der Trennung durch disk-elektrophorese und isoelektrische Fokussierung. *Planta* 128:247–53

143. Mäder, M., Meyer, Y., Bopp, M. 1975. Lokalisation der Peroxidase-Isoenzyme in Protoplasten und Zellwänden von *Nicotiana tabacum* L. *Planta* 122:259–68

144. Mäder, M., Nessel, A., Bopp, M. 1977. Über die physiologische Bedeutung der Peroxidase-Isoenzymgruppen des Tabaks anhand einiger biochemischer Eigenschaften. II. pH-Optima, Michaelis-Konstanten, Maximale Oxidationsraten. *Z. Pflanzenphysiol.* 82:247–60

145. Mäder, M., Ungemach, J., Schloss, P. 1980. The role of peroxidase isoenzyme groups of *Nicotiana tabacum* in hydrogen peroxide formation. *Planta* 147:467–70

146. Manders, W. F. 1987. Solid state ^{13}C NMR determination of the syringyl/guaiacyl ratio in hardwoods. *Holzforschung* 41(1):13–18

147. Marcinowski, S., Grisebach, H. 1978. Enzymology of lignification: Cell-wallbound β-glucosidase for coniferin from spruce *(Picea abies)* seedlings. *Eur. J. Biochem.* 87:37–44

148. Masuda, H., Fukuda, H., Komamine, A. 1983. Changes in peroxidase isoenzyme patterns during tracheary element differentiation in a culture of single cells isolated from the mesophyll of *Zinnia elegans. Z. Pflanzenphysiol.* 112:417–26

149. Maule, A. J., Ride, J. P. 1976. Ammonia-lyase and O-methyl transferase activities related to lignification in wheat leaves infected with *Botrytis. Phytochemistry* 15:1661–64

150. Miksche, G. E., Yasuda, S. 1977. Über die Lignine der Blätter und Nadeln einiger Angiospermen und Gymnospermen. *Holzforschung* 31(2):57–59

151. Miksche, G. E., Yasuda, S. 1978. Lignin of "giant" mosses and some related species. *Phytochemistry* 17:503–4

152. Miller, A. R., Roberts, L. W. 1986. Is there a relationship between tracheary element formation and lignification in soybean *(Glycine max* var. Wayne) callus cultures? *Can. J. Bot.* 64:2716–18

153. Moerschbacher, B. M. 1989. Lignin biosynthesis in stem rust infected wheat. See Ref. 129, Ch. 27, pp. 370–82

154. Moerschbacher, B. M., Noll, U. M., Flott, B. F., Reisener, H. J. 1988. Lignin biosynthetic enzymes in stem rust infected, resistant and susceptible near isogenic wheat lines. *Physiol. Mol. Plant Pathol.* 33:33–46

155. Morelli, E., Rej, R. N., Lewis, N. G., Just, G., Towers, G. H. N. 1986. Cis-monolignols in *Fagus grandifolia* and their possible involvement in lignification. *Phytochemistry* 25(7):1701–5

156. Morohoshi, N., Shibuya, Y., Murayama, A., Katayama, Y., Haraguchi, T. 1989. Properties of the phenoloxidases secreted by *Coriolus versicolor* in wood meal medium. I. *Mokuzai Gakkaishi* 35:342–47

157. Morohoshi, N., Wariishi, H., Muraiso, C., Nagai, T., Haraguchi, T. 1987. Degradation of lignin by the extracellular enzymes of *Coriolus versicolor.* IV. Properties of three laccase fractions fractionated from the extracellular enzymes. *Moruzai Gakkaishi* 33:218–25

158. Nimz, H. 1966. Der Abbau des Lignins durch schonende Hydrolyse. *Holzforschung* 20:105–9

159. Nimz, H. 1974. Beech lignin—proposal of a constitutional scheme. *Angew. Chem. Int. Ed.* 13(5):313–21

160. Nimz, H., Das, K. 1971. Durch Abbau mit thioessigsäure-erhaltene dimere Abbauphenole des Buchenlignins. *Chem. Ber.* 104:2359–80

161. Nimz, H., Ebel, J., Grisebach, H. 1975. On the structure of lignin from soybean cell suspension cultures. *Z. Naturforsch. Teil C* 30:442–44

162. Nimz, H. H., Robert, D., Faix, O., Nemr, M. 1981. Carbon-13 NMR spectra of lignins. 8. Structural differences between lignins of hardwoods, softwoods, grasses and compression wood. *Holzforschung* 35:16–26

163. Nimz, H. H., Tschirner, U., Stähle, M., Lehmann, R., Schlosser, M. 1984. Carbon-13 NMR spectra of lignins. 10. Comparison of structural units in spruce and beech lignin. *J. Wood Chem. Technol.* 4:265–84

164. Obst, J. R., Landucci, L. L. 1986. The syringyl content of softwood lignin. *J. Wood. Chem. Technol.* 6(3):311–27

165. Odier, E., Mozuch, M. D., Kalyanaraman, B., Kirk, T. K. 1988. Ligninase-mediated phenoxy radial formation and polymerization unaffected by cellobiose: quinone oxidoreductase. *Biochimie* 70:847–52

166. Ohashi, H., Yamamoto, E., Lewis, N. G., Towers, G. H. N. 1987. 5-Hydroxyferulic acid in *Zea mays* and *Hordeum vulgare* cell walls. *Phytochemistry* 26:1915–16

167. Paszczynski, A., Huynh V.-B., Crawford, R. 1986. Comparison of ligninase-I and peroxidase M-2 from the white-rot fungus *Phanerochaete chrysosporium*. *Arch. Biochem. Biophys.* 244:750–65

168. Patel, N. L., Pierson, D. L., Jensen, R. A. 1977. Dual enzymatic routes to L-tyrosine and L-phenylalanine via pretyrosine in *Pseudomonas aeruginosa*. *J. Biol. Chem.* 252:5839–46

169. Pickett-Heaps, J. D. 1968. Further ultrastructural observations on polysaccharide localization in plant cells. *J. Cell. Sci.* 3:55–64

170. Quessada, M.-P., Macheix, J.-J. 1984. Caractérisation d'une peroxydase impliquée spécifiquement dans la lignification, en relation avec l'incompatibilité au greffage chez l'abricotier. *Physiol. Veg.* 22:533–40

171. Ragan, M. A. 1984. *Fucus* "lignin": a reassessment. *Phytochemistry* 23(9): 2029–32

172. Ragan, M. A., Chapman, D. J. 1978. *A Biochemical Phylogeny of the Protists*. New York: Academic

173. Ragan, M. A., Glombitza, K. W. 1986. Phlorotannins, brown algal polyphenols. *Prog. Phycol. Res.* 4:129–241

174. Reznikov, V. M., Mikhaseva, M. F., Zilbergleit, M. A. 1978. The lignin of the alga *Fucus vesiculosus*. *Chem. Nat. Comp.* 14(5):554–56

175. Ride, J. P. 1975. Lignification in wounded wheat leaves in response to fungi and its possible role in resistance. *Physiol. Plant Pathol.* 5:125–34

176. Ride, J. P., Barber, M. S., Bertram, R. E. 1989. Infection-induced lignification in wheat. See Ref. 129, Ch. 26, pp. 361–69

177. Robert, D., Brunow, G. 1984. Quantitative estimation of hydroxyl groups in milled wood lignin from spruce and in a dehydrogenation polymer from coniferyl alcohol using ^{13}C NMR spectroscopy. *Holzforschung* 38:85–90

178. Robert, D., Chen, C.-L. 1987. Structural analysis of biodegraded lignins from spruce wood decayed by *Phanerochaete chrysosporium* by quantitative ^{13}C NMR spectroscopy. In *Proc. 4th Int. Symp.* Wood and Pulping Chem. (EUCEPA Paris), 1:139–45

179. Robert, D., Mollard, A., Barnoud, F. 1989. ^{13}C NMR qualitative and quantitative study of lignin structure synthesised in *Rosa glauca* calluses. *Plant. Physiol. Biochem.* 27(2): 297–304

180. Ros Barcelo, A., Muñoz, R., Sabater, F. 1989. Subcellular location of basic and acidic soluble isoperoxidases in *Lupinus. Plant Sci.* 63:31–38

181. Rubin, J. L., Jensen, R. A. 1979. Enzymology of L-tyrosine biosynthesis in mung bean (*Vigna radiata* [L] Wilczek). *Plant Physiol.* 64:727–34

182. Saka, S., Goring, D. A. I. 1988. The distribution of lignin in white birch wood as determined by bromination with TEM-EDXA. *Holzforschung* 42(3):149–53

183. Sarkanen, K. V., Chang, H.-M., Allan, G. G. 1967. Species variation in lignins. II. Conifer lignins. *Tappi* 50:583–87

184. Sarkanen, K. V., Ludwig, C. H. 1971. Definition and nomenclature. In *Lignins:Occurrence, Formation, Structure and Reactions*, ed. K. V. Sarkanen, C. H. Ludwig, Ch. 1, pp. 1–18. New York: Wiley-Interscience. 916 pp.

185. Sarkanen, S., Razal, R. A., Piccariello, T., Yamamoto, E., Leisola, M. S. A., Lewis, N. G. Lignin peroxidase: towards a clarification of its role in vivo. Submitted

186. Scalbert, A., Monties, B., Guittet, E., Lallemand, J. Y. 1986. Comparison of wheat straw lignin preparations. I. Chemical and spectroscopic characterizations. *Holzforschung* 40:119–27

187. Scalbert, A., Monties, B., Lallemand, J.-Y., Guittet, E., Rolando, C. 1985. Ether linkage between phenolic acids and lignin fractions from wheat straw. *Phytochemistry* 24:1359–62

188. Scheirer, D. C. 1973. Hydrolysed walls in the water-conducting cells of *Dendroligotrichum (Bryophyta):* histochemistry and ultrastructure. *Planta* 115: 37–46

189. Schloss, P., Walter, C., Mäder, M. 1987. Basic peroxidases in isolated vacuoles of *Nicotiana tabacum* L. *Planta* 170:225–29

190. Schoemaker, H. E., Meijer, E. M., Leisola, M. S. A., Haemmerli, S. D., Waldner, R., Sanglard, D., Schmidt, H. W. H. 1989. Oxidation and reduction in lignin biodegradation. See Ref. 129, Ch. 33, pp. 454–71

191. Schultz, T. P., Templeton, M. C. 1986. Proposed mechanism for the nitrobenzene oxidation of lignin. *Holzforschung* 40(2):93–97

192. Siegel, S. M. 1969. Evidence for the presence of lignin in moss gametophytes. *Am. J. Bot.* 56(2):175–79

193. Siehl, D. L., Connelly, J. A., Conn, E.

E. 1986. Tyrosine biosynthesis in *Sorghum bicolor:* characteristics of prephenate aminotransferase. *Z. Naturforsch. Teil C* 41:79–86

194. Smart, C. C., Amrhein, N. 1985. The influence of lignification on the development of vascular tissue in *Vigna radiata* L. *Protoplasma* 124:87–95

195. Stafford, H. A. 1960. Differences between lignin-like polymers formed by peroxidation of eugenol and ferulic acid in leaf sections of *Phleum. Plant Physiol.* 35:108–14

196. Stafford, H. A. 1974. The metabolism of aromatic compounds. *Annu. Rev. Plant Physiol.* 25:459–86

197. Steubing, L. C., Ramirez, C., Alberdi, M. 1980. Energy content of water and bog-plant associations in the region of Valdivia, Chile. *Vegetatio* 43:153–61

198. Takabe, K., Fukazawa, K., Harada, H. 1989. Deposition of cell wall components in conifer tracheids. See Ref. 129, Ch. 4, pp. 47–66

199. Terashima, N. 1989. An improved radiotracer method for studying formation and structure of lignin. See Ref. 129, Ch. 10, pp. 148–59

200. Terashima, N., Fukushima, K. 1989. Biogenesis and structure of macromolecular lignin in the cell wall of tree xylem as studied by microautoradiography. See Ref. 129, Ch. 11, pp. 160–68

201. Tien, M., Kirk, T. K. 1983. Lignin-degrading enzyme from the hymenomycete *Phanerochaete chrysosporium* Burds. *Science* 221:661–62

202. Towers, G. H. N., Maass, W. G. 1965. Phenolic acids and lignins in the Lycopodiales. *Phytochemistry* 4:57–66

203. Umezawa, T. 1988. Mechanisms for chemical reactions involved in lignin biodegradation by *Phanerochaete chrysosporium. Wood Res.* 75:21–79

204. Umezawa, T., Higuchi, T. 1989. Aromatic ring cleavage by lignin peroxidase. See Ref. 129, Ch. 36, pp. 503–18

205. Umezawa, T., Higuchi, T. 1989. Cleavages of aromatic ring and β-O-4 bond of synthetic lignin (DHP) by lignin peroxidase. *FEBS Lett.* 242:325–29

206. van Huystee, R. B. 1987. Some molecular aspects of plant peroxidase biosynthetic studies. *Annu. Rev. Plant Physiol.* 38:205–19

207. Wallis, A. F. A. 1971. Solvolysis by acids and bases. See Ref. 184, Ch. 9, pp. 345–72

208. Walter, M. H., Grima-Pettenati, J., Grand, C., Boudet, A. M., Lamb, C. J. 1988. Cinnamyl alcohol dehydrogenase, an enzyme specific for lignin synthesis: cDNA cloning and mRNA induction by fungal elicitor. *Proc. Natl. Acad. Sci. USA* 85:5546–50

209. White, E., Towers, G. H. N. 1967. Comparative biochemistry of the lycopods. *Phytochemistry* 6:663–67

210. Whiting, P., Goring, D. A. I. 1982. Chemical characterization of tissue fractions from the middle lamella and secondary wall of black spruce tracheids. *Wood Sci. Technol.* 16:261–67

211. Wilson, M. A., Sawyer, J., Hatcher, P. G., Lerch, H. E. III. 1989. 1,3,5-hydroxybenzene structures in mosses. *Phytochemistry* 28(5):1395–1400

212. Yamamoto, E., Bokelman, G. H., Lewis, N. G. 1989. Phenylpropanoid metabolism in cell walls: an overview. See Ref. 129, Ch. 5, pp. 68–88

213. Yamamoto, E., Inciong, Ma. E., Davin, L., Lewis, N. G. 1990. Formation of *cis*-coniferin in cell free extracts of *Fagus grandifolia* Ehrh Bark. *Plant Physiol.* In press

214. Young, M., Steelink, C. 1973. Peroxidase-catalyzed oxidation of naturally-occurring phenols and hardwood lignins. *Phytochemistry* 12:2851–61

215. Young, M. R., Towers, G. H. N., Neish, A. C. 1966. Taxonomic distribution of ammonia-lyases for L-phenylalanine and L-tyrosine in relation to lignification. *Can. J. Bot.* 44:341–49

Annu. Rev. Plant Physiol. Plant Mol. Biol.. 1990. 41:497–526

PHENOLIC SIGNALS IN COHABITATION: IMPLICATIONS FOR PLANT DEVELOPMENT

David G. Lynn and Mayland Chang

Searle Chemistry Laboratory, The University of Chicago, Chicago, IL 60637

KEY WORDS: *Striga asiatica*, phenolic compounds, quinones, *Agrobacterium*, *Rhizobium*

CONTENTS

INTRODUCTION

The cells of eukaryotic organisms must not only respond to environmental signals, they must also coordinate and organize the responses of partner cells

497

1040-2519/90/0601-0497$02.00

and, on some level, control their development. Determining the mechanisms that control the information exchange between these cells is still a major challenge for biology. Early experiments of Spemann and Mangold (151) and later a simple model of Turing (163) suggested that a concentration gradient emanating from a source cell might be sufficient to organize the development of most embryonic fields (42, 64). As appealing as the model was, only recently have attempts to characterize these morphogenetic signals or the inducer molecules that form such gradients been successful.

Such inducer molecules might be more readily discovered in a biological interaction where one organism has coupled its developmental program with another. We suggest that the parasitic angiosperms, where both host and parasite are higher plants, offer such an opportunity. The parasitic plants are found in at least 20 different families (114). Most are hemiparasites, photosynthetic and capable of maturing to seed set without a host; but a few, including *Striga asiatica* (Scrophulariaceae), absolutely require early host attachment for continued development. The initiation of germination, host attachment, and host penetration by *Striga* are dependent on very precise host-derived signals (30, 101).

The experimental evidence gathered thus far suggests that these signals are simple phenolic compounds, compounds that have been repeatedly suggested, but never experimentally proven, to be involved in plant development (66, 82; 110). The more recent discoveries that these same phenolic compounds were involved in plant-microbe recognition has suggested a general role for this class of compounds in signal transduction (149).

In this review, we address the "information content" of this class of signal compound through a comparison of the role of the signals in the interaction of *Striga*, *Agrobacterium*, and *Rhizobium* with their hosts. In each case we review the xenognostic character, the potential for spatial definition, the controls on the biosynthetic production, and the mechanisms of detection of these compounds. We show that the information provided to the plant cell by these compounds is remarkably precise, and that it is sufficient to play a critical role in plant development.

STRIGA

The *Striga* species are among the most extensively studied of the parasitic plants, at least in part because these parasites severely reduce grain production in Asia and Africa (119). The almost exclusive parasitism of grasses by many of the *Striga* species has suggested that host recognition occurs primarily at the level of *Striga* seed germination, and specific host-exuded signals have been characterized for *Striga asiatica* (32). Once germinated, however, this species is parasitically competent for only about 5 days; after that period it is no longer capable of developing the host attachment organ, the haustorium. The signal molecules that induce haustorial development are also host derived (31). Both of these signalling events control host recognition, but these signal molecules also provide important temporal and spatial information necessary for *Striga's* development. While the further events of parasite/host integration also appear to be controlled by elaborate signal transduction mechanisms that are less well understood, at this point, some general ideas have emerged about the initiation of germination and haustorial formation (see Figure 1); and those are reviewed here.

Germination Signals

Worsham & Eagley have recently published a thorough review of *Striga* seed germination requirements and the 50 years of research on host-exuded germination stimulants (179, 180). Early studies showed that the parasite seeds germinated only in the immediate vicinity of the host roots (23, 142). These findings suggested that potential germination stimulants provided not only information about host recognition but also geographical information about the relationship of host to parasite. Over the intervening years a range of compounds were identified that were capable of breaking *Striga* seed dormancy. For the most part, these compounds can be classified within plant hormone groupings including ethylene, the gibberellins, and the cytokinins (180 and references therein).

   One of the first germination-inducing compounds identified in root exudates was the unusual sesquiterpene, strigol (39). This compound was isolated from cotton, an established trap crop and nonhost for *Striga* and one of the few exudates that contained a stable stimulant (38). Extensive synthetic efforts based on the original work of Heather et al (73) identified a range of active strigol analogues (120, 165), implicating two separate molecular domains as responsible for the activity. The analogues also suggested the possible involvement of autoxidation events in the biological activity, an observation consistent with the previous demonstration that mild treatments with NaOCl and H_2SO_4 (55, 75) could induce germination. It was not altogether surprising, then, that the first host-derived germination stimulant proved to be the hydroquinone shown in structure 1, the activity of which was strictly controlled by the oxidation state of the compound (32).

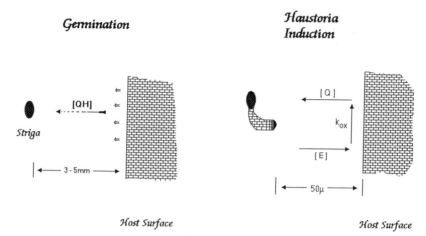

Germination   **Haustoria Induction**

Figure 1 Diagrammatic representation of the mechanisms of the early responses involved in *Striga*/host recognition. Germination: [QH], concentration of the hydroquinone, Structure 1; dashed arrow, direction of decreasing concentration. Haustorium Induction: [Q], quinone concentration, Structure 3; [E], enzyme concentration; k_{ox}, oxidation rate constant; double-headed arrow specifies distance.

The structural features of the hydroquinone were consistent with those previously defined for the natural stimulants (180). Structure 1 contains a carbon-carbon double bond, and since the quinone, structure 2, was exuded together with the hydroquinone, there was always a detectable carbon-oxygen double bond. The most significant feature, however, was that the hydroquinone was found to be labile under autoxidative conditions. This lability was accentuated in basic solutions, an instability previously shown for the *Zea mays* stimulant (159, 182). Methods have now been developed to "image" the hydroquinone based on its ability to reduce the indicator methylene blue (58, 59). In agar, the hydroquinone was found to be exuded constantly along the entire root surface of young sorghum seedlings for at least one month in culture. The competing autoxidation rate resulted in a steady-state zone of *Striga* germination potential in close proximity to the host root surface (Figure 1). The extent to which this zone develops in soil is not yet known, but models with the pure hydroquinone in agar define a *Striga* germination zone identical to that seen around the host (58).

The *Striga* germination response was found to require a precise exposure time of ≥5hr to this extremely labile hydroquinone. Shorter exposures produced no emergence. When seedlings were reexposed 2 hr after an initial 2-hr exposure, they required the normal 5 hr of incubation with the stimulant. These early periods of signal exposure had previously been shown to be phytochrome responsive. Red-light treatment inhibited germination only during the first few hours of stimulant exposure (181). While the mechanistic details for the induction of germination are not understood, the time dependence for the exposure to this labile compound seems to provide an important selective advantange for *Striga*. Germination is an irreversible commitment initiating the 5-day window in which *Striga* must attach to its host (30). The requirement for a persistent exposure to a labile host-derived compound in effect provides a mechanism for increasing the "resolution" of host detection.

Taken together, these results suggest that the *Striga* seeds half-maximally germinate when exposed to 10–100nM hydroquinone. Because of the long exposure time that is required and the oxidative lability of the stimulant, the level of the activity is highly dependent on the amount of autoxidation stabilizer present in the assay (58). The finding that the quinone is exuded along the entire length of the sorghum seedling and not associated with any particular region of the root (58), together with the observation that haustorial attachment and penetration appear indiscriminate (5, 135), suggests that the primary role of the hydroquinone is in the initial host selection. While there is some controversy over how widely these germination factors may occur in other hosts (118), related quinones have now been detected, isolated, and characterized from the exudates of a number of *Striga* host plants (G. Fate, D. G. Lynn, unpublished). The contradictions may be due to the fact that some *Striga* hosts, such as *Zea mays,* seem to exude stimulant only 3–6 mm behind the meristematic tip (159, 181) and produce significantly less material. Whether these related but structurally different stimulants can account for the species-specific host recognition (90, 127) seen in some *Striga* "strains" has yet to be determined.

Haustorium Induction Signals

The structure and development of the haustorium have been dealt with in several reviews (90, 115, 169) and are not covered in detail here. The first molecules capable of haustorium induction (105, 155) were discovered based on the assumption that such compounds would be exuded from the host plants. By screening fractions from a commercially available plant exudate, gum tragacanth, against axenic cultures of the hemiparasite *Agalinis purpurea* (Scrophulariaceae), the compounds shown in Structures 4 and 5 were identified.

3

2,6-DMBQ

These host-recognition compounds (dubbed xenognosins, from the Greek *xenos* meaning foreign or guest and *gignoskein* meaning to recognize) were apparently both phenylpropanoid derived. Several synthetically prepared analogues of the xenognosins (81) established the m-methoxy phenol and the propene double bond of xenognosin A (Structure 4) as critical features for activity (155). Work in other laboratories (8, 28, 29, 49) has shown the xenognosins to be phytoalexins and plant stress metabolites. The identification of the xenognosins then provided the first experimental support that simple phenolic compounds could serve as specific signals inducing the developmental commitment of the parasitic angiosperms.

The exudation of such phenylpropanoid derivatives was, however, not seen in the *Agalinis'* host plants (101, 156, 157). When 2,6-dimethoxy-*p*-benzoquinone (2,6-DMBQ, Structure 3) was discovered as an effective inducer of haustorial development in *Striga asiatica,* an obligate parasite within the same family (31), as well as in *Agalinis,* a general picture began to emerge.

Such benzoquinones are biosynthesized in plants (37) and by pathogenic fungi (164) via the oxidation of phenylpropanoid precursors. Histochemical staining of root surfaces (70) identified H_2O_2-requiring oxidase activity with both host and parasite but also detected oxidative capability in both *Striga asiatica* and *Agalinis purpurea* that required no exogenous oxidant (31). The quinone was not detected in axenic cultures of either the host or the parasite. However, if the host (sorghum or maize) was mildly abraded with glass wool (5 sec, 4°C), the removed material served as a potent haustorium inducer (30). These findings suggested that parasite-derived enzymes might be capable of removing the haustorium-inducing signal from the host root surface (Figure 1).

The hypothesis was tested by adding the host-surface material to the *Striga* cultures and monitoring with HPLC for the conversion to the quinone. Figure 2 shows the HPLC analysis (30) of the growth media of two-day old *Striga* seedlings as a function of the time of the exposure. Both the host surface material and a model compound, syringic acid [previously shown to induce haustoria in *Striga hermonthica* (106)], appeared to be converted efficiently into the quinone (Figure 3). The quinone did not accumulate but did appear in

Table 1 Correlation of haustorium-inducing activity with enzyme removal

| | % Haustoria[a] | | 2,6-DMBQ produced (μM) |
| | 2,6-DMBQ | Syringic acid | Striga |
Number of washes	(10 μM)	(100 μM)	Seedlings
0	98.1 ± 1.9	84.4 ± 4.0	1.31 ± 0.12
1	98.1 ± 1.9	19.6 ± 3.6	0.68 ± 0.03
2	93.0 ± 3.5	15.5 ± 1.2	0.62 ± 0.03
3	100.0 ± 0	2.8 ± 2.8	0.42 ± 0.12
3 + P.o. laccase[b]		97.2 ± 2.8	36.2 ± 1.30
3 = horseradish peroxidase[c]		11.0 ± 1.0	0.65 ± 0.05

[a] Activity is expressed as the percentage of seedlings that develop haustoria. Experiments represent the average of two replicates on one-day-old post-germinated seedlings.
[b] Crude *Pyricularia oryzae* laccase (EC 1.10.3.2).
[c] Horseradish peroxidase (EC 1.11.17).

the media just preceding the initiation of haustorium development (dashed lines). If the oxidative activity was removed by washing the seedlings, no quinone was produced from the syringic acid or from the host-derived surface material, and haustorium development was not induced (30). The activity of the quinone is unaffected by these treatments (Table 1).

These washing experiments allowed for the physical removal of the oxidative enzyme. While the nature of the enzyme involved is not yet known, its detection has suggested an explanation of the structure/activity data. The haustorium inducers fall into three general structural categories, the flavanoids (e.g. Structures 4 and 5), the *p*-hydroxy acids (e.g. Structures 6 and 7), and the quinones (e.g. 2,6-DMBQ, Structure 3). The oxidative decarboxylation of the *p*-hydroxy acids into the quinones can be catalyzed by peroxidases (89, 134), hydroxylases (25), and laccases (95, 123). These enzymes are found in many plant species and are commonly exuded by pathogenic microorganisms for the degradation of plant cell walls (84, 107, 133). The oxidative carbon-carbon bond cleavage might be catalyzed by either of two mechanisms: (*a*) the addition of O_2 *para* to a phenoxy radical in a manner

5

6
"Ferulic acid" type

7
"Syringic acid" type
R = H, OH, alkyl

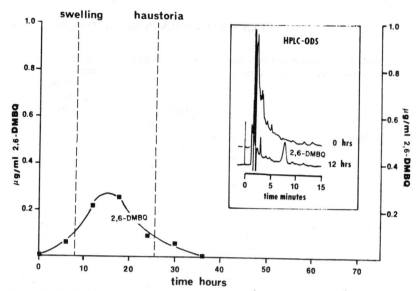

Figure 2 Metabolism of sorghum root surface material (500 μg/ml) by *Striga asiatica* seedlings (30 each). The HPLC traces (ODS, HOAc/MeOH/H₂O, 0.8/22.0/72.2) of the sorghum root surface material in the absence of the seedlings (0 hr) and after a 12-hr incubation are shown in the insert. Four replicates were averaged for each point.

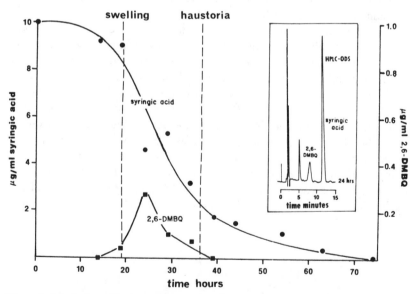

Figure 3 Metabolism of syringic acid (10 μg/ml) by *Striga asiatica* seedlings (30 each). The HPLC trace of syringic acid at 24 hr incubation with the seedlings is shown in the insert. Each point represents the average of two replicates and the concentrations are estimated with the use of synthetic samples.

similar to autoxidation or, (*b*) the addition of H_2O to the further oxidized carbocation. Either process could generate intermediates that would lead to aryl/C_α bond cleavage via a retro-Claisen condensation and ultimately result in the 4e⁻ oxidation to the quinone (see Scheme 1). Such processes could degrade both the flavanoids and the aryl acids to the quinones and resolve the structure/activity data into the substrate specificity of the enzyme and the structural features of the quinones necessary for haustorial induction.

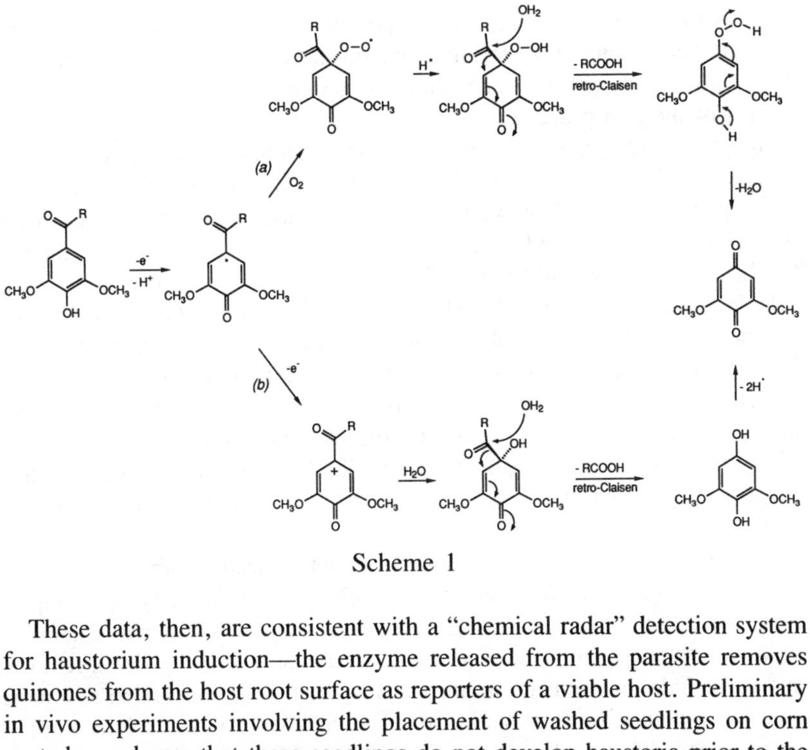

Scheme 1

These data, then, are consistent with a "chemical radar" detection system for haustorium induction—the enzyme released from the parasite removes quinones from the host root surface as reporters of a viable host. Preliminary in vivo experiments involving the placement of washed seedlings on corn roots have shown that these seedlings do not develop haustoria prior to the addition of the fungal laccase preparation (30; the lack of activity of the peroxidase shown in Table 1 and in vivo suggests that there is no obvious source of H_2O_2 in the washed seedlings). Such an hypothesis finds some support in fungal pathogenesis. Shaykh et al (145) showed that *Fusarium solini* exuded cutinase activity following induction with host wall-derived hydroxy acids, the products of cutinase digestion. The strategy provides an effective method for parasitic commitment only to viable host surfaces. The characterizations of the parasite enzyme and its substrates are required to further support the hypothesis. In that regard, laccases have been found in a number of plant exudates (14, 84), and it may prove possible to use these methods in the characterization of the parasite enzyme and the further characterization of the substrate specificity.

Mechanism of Induction

There are clear similarities in the structures of the compounds controlling germination and haustorium induction suggesting possible mechanistic similarities between the two processes. Germination is induced with much lower concentrations of the stimulant than is haustorium development, but both processes require similar extended exposures to the signal (148). Synthetic analogues exploiting the differing oxidation states and substitution patterns of these compounds provide an additional strategy for comparing the mechanisms of the two processes.

Simple benzoquinones occur widely in plant tissues and have a broad spectrum of biological effects (24, 33, 71, 79, 86, 93, 144, 183). Specific benzoquinones are well known for their role in electron transport processes in both oxidative phosphorylation and photosynthesis. One obvious mechanism for these quinones would be through alterations in the energetic charge of the cell. In fact, the 2,6- and 2,3-DMBQ isomers, both active haustorium inducers, have been shown to inhibit mitochondrial electron transport whereas the 2,5-DMBQ isomer does not induce haustoria or inhibit respiration (71, 131). Generally, millimolar concentrations of these quinones are required for the inhibition of respiration whereas haustorium induction requires micromolar concentrations. The active quinones do inhibit haustorium induction at higher concentrations, and this inhibition may be more directly caused by the inhibition of respiration.

It is also possible that the quinones are metabolized to other active components, a suggestion consistent with the observation that they do disappear from the culture medium during haustorium induction (Figures 2 and 3). However, the exposure-time requirements were not altered by signal removal and immediate replacement, arguing against metabolic conversion to a more active compound, at least within the bathing media (148). More work is clearly needed before the mechanistic details of these activations are understood.

RHIZOBIUM

Technological advances in molecular genetics have opened new approaches to the study of Rhizobium/Leguminosae (Fabaceae) symbiosis. The last few years have seen a dramatic increase in our understanding of this interaction. These recent discoveries have been extensively reviewed (see, for example, 50, 100) and are only summarized here.

Many highly coordinated metabolic events leading to the establishment of symbiosis have been documented (6). The fast-growing *Rhizobium* species typically have large plasmids, one or more of which carry genes required for symbiosis. These genes have been lumped into several functional categories, the *nod* genes being those required for the early events of host nodulation. Mutations in four contiguous and highly homologous genes, designated *nod-DABC*, generally result in strains unable to induce the early host response of root hair curling (Hac⁻ phenotype) (52) and cortical cell division in the host (53).

Gene fusions with the *E. coli lacZ* reporter have defined *nodD* as a constitutively expressed gene and a series of other *nod* genes, including *nodABC*, as genes that require both *nodD* and a plant factor for expression (76, 113, 138). These constructs provided a facile assay streamlining the isolation of the responsible plant factors—now characterized as hydroxylated flavones and flavanones of the general type shown in Figure 4 (60, 88, 121, 132, 184).

Not only was the discovery of these factors as phenolics somewhat unexpected, but these are not compounds thought to be uniquely associated with host plants and able to define the narrow host range seen in some *Rhizobium* species. Several experiments have now suggested how this host specificity may be defined.

Both Djordjevic et al (51) and Peters & Long (122) have shown that the inducer compounds are exuded specifically between the meristem and the root hair zone, the area of the root that had been shown to be most responsive to nodulation in soybean (10, 27). This restricted location of the exudation, the release of specific inhibitors [common phenolic plant products that appear to be exuded (60, 122)], and the expression of specific repressor proteins (87) may all work in concert to define both host range and the optimal spatial domain for the bacterial response (50).

The detection of these phenolics by multiple *nodD* alleles also appears to contribute to differential flavanoid recognition (68, 150). Whether the *nodD* products function as transcriptional activating proteins (74, 87, 100) or

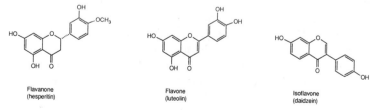

Figure 4 Examples of the structural classes found to be active in the induction of *nod* gene expression.

through some interaction within the cytoplasmic membrane, a site where the flavanoids accumulate (130), is still being debated.

Rhizobia that have been activated with the flavanoids release an as yet unknown low-molecular-weight factor(s) that induces the root hair curling and cortical cell divisions of the plant (7, 9). The morphological similarities between the plant responses during nodulation and in the formation of lateral haustoria (90) are striking. In each case the organogenesis is initiated from meristematic tissue along the root and ultimately gives rise to a highly vascularized structure. It is interesting that an oxidative carbon-carbon bond cleavage of the flavanoids, analogous to that catalyzed by the *Striga* enzyme, would generate simple benzoquinones. Hesperitin has recently been shown to be converted into a compound able to induce haustorium formation in *Striga* (G. Fate, K. M. Hess, and D. G. Lynn, unpublished), and some of the benzoquinones that induce haustorium formation in *Striga* can induce root hair curling in white clover (B. Solheim and T. V. Bhuvaneswari, personal communication). Thus the early responses of the *Rhizobium* host plant may result from rhizobial catabolism of the plant-derived flavanoids into simple benzoquinones.

We have no evidence yet of rapid flavanoid catabolism by rhizobia (130), and the biosynthesis of additional factors may be more important. Oxidative enzymes capable of the metabolism of phenyl propanoids are produced and even exuded by bacteria, fungi, and higher plant cells (84, 107). While some of these enzymes can metabolize flavanoids (14), their expression in the white rot fungus *Polyporus versicolor* was induced by flavanoids that were not substrates for the induced enzymes. Another reasonable possibility would involve flavanoid induction of the bacterial expression of enzymes capable of oxidatively producing simple quinones. Through a mechanism analogous to haustorium induction in *Striga,* the release of these enzymes would be capable of degrading the plant root surface to the root-hair-curling factor and even the cell-division factor. The recent experiments suggesting that the nodA and nodB proteins are involved with the production of a low-molecular-weight factor that activates protoplast divisions would be consistent with this hypothesis (143).

Successful symbiosis certainly depends on the attachment of the bacteria to the plant cells. While this attachment is thought to be mediated by plant-secreted lectins (45–48), it is not yet clear how the release of the root hair curling factor is synchronized with bacterial attachment. The removal of host surface components may restrict the root-hair-curling factor only to the point of contact between the bacteria and the plant, as has been claimed for *Striga* (31, 148).

Following attachment, the infection thread develops via host-derived processes (43, 44, 136), allowing movement of the bacteria from the root hair cell into the newly dividing cortical cells (43). The host does recognize the *Rhizobial* penetration and at this point has released a factor that restricts further nodulation in the immediate area of the infection (26, 27, 124). A simple derivative of nicotinic acid isolated several years ago induced specific G2 cellular arrest in pea and other legumes (57, 104). This specific arrest may predispose the cell for a rapid developmental change (98). The arrest was recently shown to be antagonized by a simple amino acid analog (103). Compounds like these, as well as the simple quinones, are produced from substrates readily available to both the plant and the bacteria, providing readily available and easily controlled signals for the symbiosis.

The successful bacteria then interface with the cortical cells and initiate the formation of bacteroids. In analogy to the closely related *Agrobacterium tumefaciens*, this event may mark the true transition between a parasitic infection and a symbiotic relationship (50) but may also be viewed as a final conquest by the plant of the invading parasite.

AGROBACTERIUM

The Bacterial Response

The interrelationship between this pathogenic bacterium and its host plant cells has been extensively reviewed (13, 109, 185), almost annually for the past 10 years, and therefore is only briefly summarized here. Powerful molecular biological approaches have greatly clarified the understanding of this plant-pathogen interaction (80), and it was really this understanding that allowed for the realization of plant genetic engineering (34, 35).

Agrobacterium tumefaciens initiates parasitic expression in the presence of wounded host plant tissue. A small segment of its tumor-inducing plasmid (Ti) is excised, transferred (T-DNA), and ultimately incorporated into the plant genome. The construction of a Tn3-*lacZ* transposon system and its integration with the Ti plasmid allowed for visualization of the expression of the loci responsible for virulence (*vir*) (152, 154). These reporter gene constructs provided a bioassay for the identification of the signal molecules initiating the pathogenic response and led to the isolation of acetosyringone and α-hydroxy acetosyringone as the major inducing products exuded from wounded tobacco cells (153). A range of other phenolics have now been characterized that induce *vir* gene expression (Figure 5). They bear the common structural motifs of simple aromatic methoxyphenols containing a carbon-carbon or carbon-oxygen double bond *para* to the hydroxyl group. These structures are remarkably similar to the phenols that are oxidatively converted to the quinones in *Striga* (31).

The *vir* region is composed of six complementation groups, two of which, *virA* and *virG*, are constitutively expressed (154). Sequence analysis of these

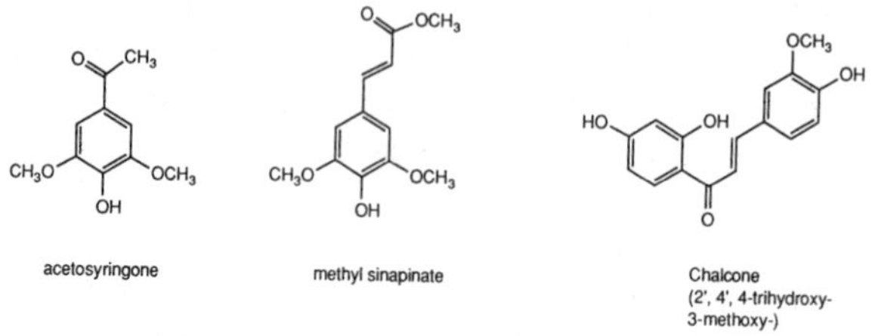

Figure 5 Examples of the structural classes found to be active in the induction of *vir* gene expression.

genes has found them to be homologous to other prokaryotic two-component sensory/regulatory systems (137). The *virA* product is an inner membrane protein (96, 108) and therefore is most likely involved in sensing the phenolic compounds and relaying that information to the regulatory *virG* protein.

The activation and transfer events in this interaction may involve more sophisticated controls than predicted by this model. For example, *virA*, *virG*, *virB*, and *virD* are essential while *virC* and *virE* enhance the efficiency of plant transformation (154). In addition to these plasmid genes, at least four different chromosomal regions are directly involved in the attachment of *Agrobacterium* to plant cells (see 185) and at least five additional proteins are induced by the plant signals (56). A direct interrelationship of *vir* induction with stress responses has been suggested (168), and more recently it has been shown that other low-molecular-weight signals, including products of the transformed plant cell, can greatly enhance *vir* induction (166).

The *Agrobacterium* host range is, for the most part, widely dispersed throughout the dicotyledonous plants. Nevertheless, host selection is still a critical event (11). The commitment to a parasitic mode involves the mobilization of a diverse array of molecular responses within the bacterium. While the commitment appears (e.g. as in *Striga*) not to be irreversible, it nevertheless involves a significant commitment of resources. A signal molecule must be able to reveal a cell that is receptive to transformation—i.e. one with a cell wall that can be penetrated for effective transfer and integration of the DNA, and one that is responsive to the transferred genes.

As Atsatt had previously suggested for the parasitic plants (3, 4), Stachel et al (153) suggested that *Agrobacterium* had evolved to recognize these phenolic compounds as characteristic of the wounded cells and the production of these compounds by the plant was related to defence and/or cell wall repair. However, these same compounds had previously been recognized as signal molecules critical to *Striga's* development (101, 106). Even more striking is the direct overlap between compounds that activate *vir* expression and those that can replace cytokinin in the induction of cell division in tobacco (12, 102; see below). Such results may be better rationalized as indicative of a direct connection between the wound-induced production of these compounds and wound-induced cellular replication.

The Transformed Cell

As Smith and Jensen proposed and Braun and others later demonstrated, the crown gall system serves as "an experimental model for studies designed to uncover the fundamental concepts that underlie the tumorous state generally"—that of cell division and differentiation (16). *Agrobacterium* is now known to transfer genes encoding the enzymes necessary for synthesis of both auxins and cytokinins. Early reports of Wood (177, 178) identified factors from *Agrobacterium*-transformed *Vinca* cells, distinctly different from the known growth factors, that were capable of stimulating the growth of non-transformed tissue. Both the auxins and the cytokinins are pleiotropic. The possibility of identifying secondary signals activated by the cytokinins was the basis of the experimental approach that identified the dehydrodiconiferyl glucosides (DCGs, Structure 8) as inducers of plant cell division (12, 102).

8
(DCG)

The finding of a phenylpropanoid as the active constituent of these tumor lines was not expected. In fact, the original data were consistent with a nitrogen heterocyclic base as a major part of the structure (179). A total synthesis of Structure 8 has now been completed (M. D. Dudley, D. G. Lynn, unpublished), and bioassays on the synthetic sample have proven the original claims of growth-promoting activity (R. Teutonico and A. N. Binns, unpublished).

The roles of phenylpropanoids in plants are diverse. They serve as an integral part of the structural matrix of lignin (63, 116), function in plant protection as phytoalexins (85, 174) and allelopathic agents (62, 101), and dramatically inhibit plant cell growth (82). Both dehydrodiconiferyl alcohol (174) and the monomeric dihydroconiferyl alcohol have been isolated from plant tissue. The monomer has been isolated from lettuce seedlings as a gibberellin synergist in the induction of hypocotyl elongation (146); it stimulates the auxin-induced elongation of cucumber hypocotyl sections (140). The same compound has been found in sycamore sap (125), and it stimulates cell division in the soybean callus and tobacco callus (12, 102) assays and the radish leaf senescence assay (92). It is therefore not surprising, in retrospect, that the DCGs are the compounds originally detected by Wood & Braun.

Consistent with earlier suggestions that cell wall fragments might play some role in development (174), the diastereomeric relationship of the agly-cone of the identified DCGs suggested that they may be cell wall derived (12, 102). Biosynthetic studies with transformed *Vinca* cells have shown that these compounds are made by the direct dimerization of coniferyl alcohol (J. D. Orr and D. G. Lynn, unpublished) and are therefore derived directly from the shikimic acid pathway (69). The mRNAs for cinnamyl alcohol dehydro-genase, an enzyme involved in the production of compounds similar to coniferyl alochol, are among the first to accumulate in bean cell cultures induced with fungal elicitors (67, 172). Whether these genes are rapidly expressed on wounding, controlling DCG production at the transcriptional level, has yet to be determined.

The biosynthetic production of the DCGs coincides with the initial round of cell divisions in fresh tobacco pith explants grown on auxins and cytokinins (R. Teutonico and A. N. Binns, unpublished). These studies suggest that the compounds are at least markers and possibly constitutive inducers of cell division. The more recent finding that the aglycone of the DCGs is a potent inducer of *vir* expression (R. Morris, P. Zambryski, and A. N. Binns, personal communication), even though this structure does not strictly fit with the structures of the other inducers, would be consistent with a role for the aglycone as an indicator of wound-induced cell division.

INFORMATION CONTENT OF THE MOLECULAR SIGNALS

The Xenognosins

In each of the cases mentioned above, a simple compound initiates an organism's transition to a parasitic state. All such compounds could be termed "xenognosins" in that they carry functional information about the compatibil-ity between two organisms, a function in many respects not unlike those involved in self/non-self compatibility reactions (78). The remarkable feature of the interactions discussed above is not that these seemingly divergent "parasites" should utilize the same class of substances as recognition mole-cules (in some cases they use identical molecules) but that these molecules should be plant phenolics. Frequently suggested but never proven to have a hormonal or signal transduction role in the plant, phenolics now appear to play a critical role in the associations of three different organisms with their host plants.

Host recognition in *Striga* appears to be controlled initially by the irreversible germination event. The sorghum hydroquinone, a polyketide-derived phenolic (G. Fate and D. G. Lynn, unpublished), has features common to exuded materials from a wide range of *Striga* host plants. Related compounds have also been found in other monocots (177) and dicots (176), but there is no evidence that they are exuded from the roots of these plants as they are from sorghum. Nevertheless, the structural differences in the germination stimulant from host to host, combined with the specific exudation in an area where the *Striga* seeds would be responsive, may be sufficient to explain not only host selection but also the very selective response of certain *Striga* strains (127).

The haustorium inducers (originally termed xenognosins) control the second stage of host recognition. The evidence suggests that they are released from the host root surface by parasite enzymes and therefore can carry information about the nature of the host root surface. In the case of *Striga,* the details of this information must await the further characterization of the host substrate and the parasite oxidase.

Studies of the molecular events involved in host recognition by *Rhizobium* offer great potential. The flavanoids stimulate *nod* gene expression at the same nanomolar concentrations required for the *Striga* germination stimulant (58). These compounds are widely dispersed across plant families, and the taxonomic value they possess is not consistent with the observed *Rhizobium* host associations, but there appear to be other important factors controlling the recognition event. The signal compounds are specifically exuded, they define a region of susceptible host cells, and they are specifically and very sensitively detected by the bacteria. *Nod* gene expression appears not to be an irreversible commitment for *Rhizobium,* as is the case with *Striga,* but it is the first stage in parasitic/symbiotic expression. Studies on the control of host production of these compounds as well as studies exploiting bacterial genetics to show how they are detected should be very profitable.

Although the host range for *Agrobacterium* is in most cases much broader than those of *Striga* and *Rhizobium,* as reflected by the wider range of suitable molecular signals, the information content of the xenognosins appears very similar. Host selectivity can be controlled at least in part by the interaction of the *vir*A product and the phenolic (108), and the Agrobacterium/host interaction may prove the simplest for the investigation of protein/phenolic binding and signal transduction.

Geographic Distribution

Of central importance to each of these interactions is attachment to the host. The data presented with the *Striga*/sorghum interaction suggest that the physical instability of the germination stimulant restricts parasitic initiation to the immediate vicinity of the host. This response appears even more restrictive in some other *Striga* hosts. For example, in *Zea mays,* production of the germination stimulant, like that of the *Rhizobium* stimulant, is only found 3–6 mm behind the root tip (159, 181).

The lifetime of the *Striga* seedling, as determined by the loss of haustorial competence, is ∼5 days (30, 135). This limit appears to result from reduction in oxidase production and, later, the depletion of energy stores (30). The zone of *Striga* seed germination around agar-grown sorghum seedlings corresponds approximately to the distance that the seedling could traverse in the 5-day period. A steady-state zone of the hydroquinone is maintained at a concentration sufficient to induce germination within this zone (58). It is not clear to what extent this model will hold up in other systems, but it does provide a simple mechanism for spatial control.

The mechanism proposed for the second level of recognition, haustorium induction, is consistent with a process that occurs only at the surface of the host (30, 31). This mechanism greatly restricts the distance between host and parasite before the parasitic transition. Suboptimal concentrations of 2,6-DMBQ ($<10^{-6}$ M) induce chemotropic growth of the parasite radicle (G. Fate and D. G. Lynn, unpublished). This chemotropism implies that if the oxidation product did accumulate, it could function to recruit the seedlings. In this regard, exuded substrates, either the flavanoids or the DCGs, both of which serve as substrates for the *Striga* enzyme, could direct the growth of the seedling towards the viable host. Further experiments are necessary to establish whether these elaborate mechanisms of initiation and chemotropism could combine to result in the observed attachment efficiency of *Striga*.

These two molecular signals for *Striga* and the mechanisms of their release appear to very effectively provide information about the spatial relationship of the host to the parasite. This information may be more important to *Striga*, since germination is an irreversible commitment (58). Nevertheless, host cell attachment is of primary importance to each system, and the spatial domain of the response should not be minimized.

With *Agrobacterium*, host cell attachment is both highly specific and critical to the transformation event (for review see 11). The bacterium must attach to a cell competent for cell division. This requirement was pointed out by Armin Braun (17, 18) many years ago when he discovered a "window of competence" during which the cells were susceptible to *Agrobacterium* transformation, a period coincident with a round of wound-induced cell divisions (19, 99). These results have been supported more recently by suspension culture (1) and protoplast studies (61, 183), suggesting that bacterial attachment may be more directly related to wound-induced cell division.

In controlled experiments where such phenolics are added to wounded tobacco tissue, the compounds are destroyed rapidly (J. D. Orr and D. G. Lynn, unpublished). Whether this degradation is enzymatic, due to autoxidation, or a combination of both is not known, but the fact that the half-life is

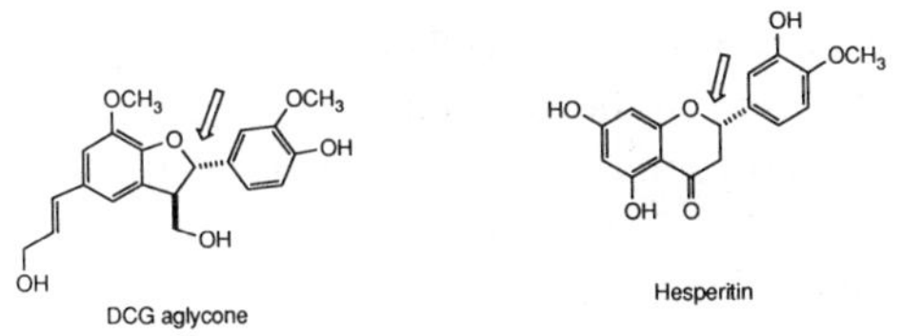

Figure 6   Arrow indicates the ketal carbon in both structures.

short suggests the potential for the same spatial information that was found with *Striga*. In that regard, the phenolics induce chemotaxis (2), directing bacteria up the concentration gradient from the receptive cell. The demonstration that the DCGs were found only in rapidly dividing tobacco cells (12, 102), were themselves labile when exposed to wounded plant tissue, and were active in the induction of *vir* expression (A. Binns, P. Zambryski, R. Morris, personal communication) suggests that these compounds may signal viable, replicative cells in a spatially dependent manner.

The attachment of *Rhizobium* to the root hair cell is an essential prerequisite of infection thread formation and ultimate bacterial integration. Bacterial mutants unable to construct the exopolysaccharide properly do not invade the plant normally (94, 112). The flavanoid signal appears to be produced only in the responsive region of the root, and the flavanones contain the same phenolic benzyl phenyl ether functionality of the DCGs (Figure 6). This vinylogous hemiketal or ketal structure is very labile and would be expected to have a limited lifetime in the soil. In addition, phenols are readily oxidized by tissue peroxidases and by the enzymes exuded by *Striga,* which suggests that these compounds may have a limited lifetime in the cellular matrix as well. Further work may establish the importance of a spatial response to the overall efficiency of this association.

Site of Production

Insight into the kinds of information provided by these signals may increase with understanding of their biosynthetic production. In sorghum, the polyketide *Striga* germination stimulant is produced along the entire root surface of the young seedling. Whether the synthesis occurs in the epidermal layer or deeper in the root is not known, nor is the reason for its production apparent.

The most obvious origin of acetosyringone is through the decarboxylation of a shikimate-derived phenylpropanoid. Many of these phenolics have been known for years to be incorporated into cell walls (69), and the possibility had been suggested that this outermost protective boundary for the plant cell, like the plasma membrane of the animal cells, provided a source of signal molecules (139). The structures of the DCGs suggested that they may also be wall derived (12, 102); however, there is now evidence that the DCGs are derived directly via a phenylpropanoid biosynthetic pathway (J. D. Orr and D. G. Lynn, unpublished). Both of these products then may originate from branch points off the phenylpropanoid pathway that normally leads to wall-bound phenolics.

In many ways, these branches are not unlike the pathways leading to the flavanoids. The compounds are of mixed biosynthetic origin involving both phenylpropanoid and polyketide pathways. These pathways have been studied primarily in relation to plant defense, and multiple isozymes of the committed steps of the phenylpropanoid pathway have been found. In some cases, these pathways are transcriptionally regulated and have been shown to be both developmentally and environmentally controlled (69, 91). Whether this same control is exerted on the production of these phenolic signals is not yet known.

Rhizobium, not unlike *Agrobacterium,* must induce *nod* gene expression in the presence of plant cells that are competent for symbiosis. In that regard, the structural similarity between the inducing flavanones and the DCGs that are produced in proliferating tissues as possible secondary messengers for cytokinins is of interest. At least in some tissues the flavanoids are biosynthesized within the endoplasmic reticulum, the site of phenylpropanoid biosynthesis (170). The DCGs are produced coincident with or just preceding a round of cell divisions (R. Teutonico and A. N. Binns, unpublished data) and may be exuded into the surrounding media during that process. The release of such compounds may be important to the developmental status of certain regions of the plant root. Whether the DCGs might be sufficient to induce *nod* gene expression or whether the flavanoids might be indicative of the developmental status of the tissue is not yet known.

Signal Detection

Another way of viewing the information provided in these signals is to investigate how the molecules might be detected. A very interesting system has recently been described (128, 129) where the levels of endogenous salicylic acid control the thermogenic organs of *Sauromatum guttatum* (voodoo lily). Earlier studies with salicylic acid had shown that exogenous application stimulated flowering in certain plants (36, 83, 117), induced disease resistance (175), inhibited ethylene biosynthesis (97), and inhibited the maintenance of H_2O and salt relations (65, 126). In the case of the voodoo lily, the salicylic acid was termed the calorigen because it activated the cyanide-resistant redox pathway sufficiently to raise the temperature of the flower by more than 10°C above the ambient conditions. The mechanism of the activation of this redox pathway by this phenol is not known.

The findings that quinones of differing redox states induce both germination and haustorium formation in an exposure time–dependent manner have not been easily explained. It was suggested many years ago that plasma membrane–localized redox chemistry may play an important role in development (22). More recent experiments have suggested that such a process occurs widely in both plants, animals, and fungi (40, 41). Preliminary experiments (C. E. Smith, T. Ruttledge, D. G. Lynn, unpublished) have been consistent with the involvement of such redox chemistry in the formation of haustoria. The redox processing of these quinones may explain the requirement for the extended exposure-time dependence of the response (148).

Both the germination and haustorium induction events can be viewed as the activation of a tissue that is predisposed to receiving the appropriate signal. It is not yet known how general these quinones may be in their ability to induce plant cell division in susceptible tissue. These same quinones, however, have recently been shown to induce rapid rounds of cell divisions (J. Verbeke, personal communication) when applied to the very responsive epidermal cells of the carpel primordia in the developing flower of *Catharanthus rosea* (167, 171). Whether the response of these epidermal cells is a specific effect or an artifact of cells that are obviously predisposed for a rapid developmental change (147) has yet to be determined. The activation of the plant cells by *Rhizobium* may be a similar situation of a cell population poised for rapid growth (98), a growth that may be initiated by similar quinones.

Another interesting perspective is provided by the finding that flavanoids serve as endogenous regulators of auxin transport (77) and thereby maintain hormone homeostasis. Cytokinin exposure has been shown to stimulate the production of the DCG growth factors (12, 102). Thus the phenolics may provide endogenous information about the hormonal status and the growth potential of the tissue. An easily testable model now exists for how these phenols are detected by *Agrobacterium*. Whether similar detection mechanisms exist in plants and how such detection may translate into the observed physiological changes can now be studied.

CONCLUSIONS

In this review we have summarized the information exchange occurring between the parasitic plant *Striga asiatica* and its host plants. The early developmental program of *Striga* is synchronized with the metabolic/developmental events of its host. Specific phenolic compounds induce germination in the immediate vicinity of the host root. Haustorium development is initiated at the host-root surface. This ability of one plant to couple its development so closely to another makes this parasitic strategy possible. The more recent finding that similar compounds induced the expression of parasitic genes in pathogenic microbes immediately raised the question about these compounds as general signals. Here we have attempted to show that there are common requirements of the signal molecules in several systems and have more generally questioned the role of such compounds in plant development.

Any signal molecules controlling a developmental commitment must provide very specific information. Phenolic compounds have now been shown (a) to provide information about specific hosts in highly selective host/ parasite interactions; (b) to possess inherent physical lability, limiting their lifetime within certain environments—i.e. providing spatial and temporal information; (c) to be biosynthetically produced from branches of preexisting pathways; and (d) to be detected with high resolution so as to reduce the chance of an improper commitment. These attributes taken together suggest that such a class of compounds might provide sufficient information to control the development of eukaryotic tissues. In that regard, it is interesting that the two low-molecular-weight molecules claimed to play morphogenetic roles in animal tissues and slime molds have some of the same general features.

In avian development, grafting the posterior portion of one limb bud to the anterior portion of another causes a duplication of the ultimate developmental pattern. Retinoic acid (Structure 9), when applied to the anterior portion of embryonic chick limb buds, produced the same duplication pattern (158, 161). This compound has now been shown to exist naturally in a concentration gradient across the developing bud (160) and to have similar effects in the development of other tissues, such as the central nervous system of *Xenopus* (54). In each case, reduction of the acid greatly reduces the activity.

In the slime mold, *Dictyostelium discoideum*, differentiation of slime mold cells into prespore and prestalk usually occurs only when the amoebae aggregate to form a mound of 10^5 cells (72, 141). Isolated amoebae can also differentiate into stalk cells when incubated in a salt media containing cAMP and a dialysate collected from high-density cultures (15, 162). The phenolic compound DIF-1/[1-(2,5-dichloro-2,6-dihydroxy-4-methoxyphenyl)-1-hexanone,] (Structure 10) (111) causes the activity of this dialysate. This slime mold morphogen has been quantified both intra- and extracellularly with respect to development times (21), and a concentration gradient has been reported (20). Again this compound has been shown to be highly sensitive to autoxidation.

The above two compounds, both of which are morphogenetically active, are low-molecular-weight highly unsaturated compounds whose activity is very sensitive to oxidation state. Clearly, one would expect the molecules for compounds that must diffuse through avian tissue during embryogenesis,

9
(retinoic acid)

10
(DIF - 1)

through an unorganized array of *Dictyostelium* cells, and within plant tissue to be quite different. Within this context, the structural similarities between these compounds seem remarkable. These findings can be viewed as supportive of the original theories of a reaction-diffusion model for development where the diffusion gradients of simple compounds are capable of organizing tissue differentiation. To the extent that such a signal transduction mechanism is substantiated in plant tissues, it may be possible to define precisely the role of such gradients in developing the tissues and possibly even use these signals to define morphologically the destinies of the cells.

Acknowledgments

DGL dedicates this review to the memory of his sister, Melanie Elizabeth Lynn, whose love for science and the world around her will always be an inspiration. We are grateful to our students and colleagues without whom our work would not have been possible, to Professors Andy Binns and Fred Ruddat, and to Gwendolyn Fate, and Tom Ruttledge for carefully reading the manuscript, and to the NIH (GM-33585), The Dow Chemical Company, Ciba Geigy Corporation, The Frasch Foundation, and USAID for financial support.

Literature Cited

1. An, G. 1985. High efficiency of transformation of cultured tobacco cells. *Plant Physiol.* 79:568–70

2. Ashby, A. M., Watson, M. D., Shaw, C. H. 1987. A Ti-plasmid determined function is responsible for chemotaxis of *Agrobacterium tumefaciens* toward the plant wound product acetosyringone. *FEMS Microbiol. Lett.* 41:189–92

3. Atsatt, P. R. 1977. The insect herbivore as a predictive model in parasitic seed plant biology. *Am. Nat.* 111:579–86

4. Atsatt, P. R., Hearn, T. F., Nelson, R. T. 1978. Chemical induction and repression of haustoria in *Orthocarpus purpurascens* (Scrophulariaceae). *Ann. Bot.* 42:1147–84

5. Baird, W. V., Riopel, J. L. 1983. Experimental studies of the attachment of the parasitic angiosperm *Agalinis purpurea* to a host. *Protoplasma* 118:206–18

6. Bauer, W. D. 1981. The infection of legumes by rhizobia. *Annu. Rev. Plant Physiol.* 32:407–49

7. Bauer, W. D., Bhuvaneswari, T. V., Calvert, H. E., Law, I. J., Malik, N. A. S., Vesper, S. J. 1985. Recognition and infection in slow-growing rhizobia. In *Nitrogen Fixation Research Progress,* ed. H. J. Evans, P. J. Bottomley, W. E. Newton, pp. 247–53. Amsterdam: Nijhoff

8. Bell, A. A. 1981. Biochemical mechanisms of disease resistance. *Annu. Rev. Plant Physiol.* 32:21–81

9. Bhuvaneswari, T. V., Solheim, B. 1985. Root hair deformations in the white clover / *Rhizobium trifolii* symbiosis. *Physiol. Plant.* 63:25–34

10. Bhuvaneswari, T. V., Turgeon, G. B., Bauer, W. D. 1980. Early events in the infection of soybeans (*Glycine max* L. Merr). *Plant Physiol.* 66:1027–31

11. Binns, A. N. 1989. *Agrobacterium*-mediated gene delivery and the biology of host range limitations. *Physiol. Plant.* In press

12. Binns, A. N., Chen, R. H., Wood, H. N., Lynn, D. G. 1987. Cell division promoting activity of naturally occurring dehydrodiconiferyl glucosides: Do cell wall components control cell division? *Proc. Natl. Acad. Sci. USA* 84:615–19

13. Binns, A. N., Thomashow, M. F. 1980. Cell biology of *Agrobacterium* infection and transformation of plants. *Annu. Rev. Microbiol.* 42:575–606

14. Bligny, R., Douce, R. 1983. Excretion of laccase by syccamore *(Acer pseudoplatanus)* cells. *Biochem. J.* 209:489–96
15. Bonner, J. T. 1970. Induction of stalk cell differentiation by cAMP in the cellular slime mold *Dictyostelium discoideum. Proc. Natl. Acad. Sci. USA* 65:110–13
16. Braun, A. C. 1982. A history of the crown gall problem. In *Molecular Biology of Plant Tumors,* ed. G. Kahl, J. Schell, pp. 155–209. New York: Academic. 615 pp.
17. Braun, A. C. 1952. Conditioning of the host cell as a factor in the transformation process in crown gall. *Growth* 16:65–74
18. Braun, A. C., Mandle, R. J. 1948. Studies on the inactivation of the tumor inducing principle in crown gall. *Growth* 12:255–69
19. Braun, A. C., Stonier, T. 1958. Morphology and physiology of plant tumors. *Protoplasmatalogia* 10(5a):1–93
20. Brookman, J. J., Jermyn, K. A., Kay, R. R. 1987. Nature and distribution of the morphogen DIF in the *Dictyostelium* slug. *Development* 100:119–24
21. Brookman, J. J., Town, C. D., Jermyn, K. A., Kay R. R. 1982. Developmental regulation of a stalk cell differentiation-inducing factor in *Dictyostelium discoideum. Dev. Biol.* 91:191–96
22. Brooks, M. M. 1947. Activation of eggs by oxidation reduction indicators. *Science* 106:320
23. Brown, R., Edwards, M. 1944. The germination of the seed of *Striga lutea*. Host influence and the progress of germination. *Ann. Bot.* 8:131–48
24. Bungenberg de Jong, H. L., Klaar, W. J., Vliegenthart, J. A. 1955. Glycosides and their importance in wheat germ. In *Third International Brotkongress,* pp. 29–33. Hamburg
25. Buswell, J. A., Eriksson, K.-E. 1988. Vanillate hydroxylase from *Sporotrichum pulverulentum Methods Enzymol.* 161:274–81
26. Caetano-Anolles, G., Bauer, W. D. 1988. Feedback regulation of nodule formation in *alfalfa. Planta* 175:546–57
27. Calvert, H. E., Pence, M. K., Pierce, M., Malik, N. S. A., Bauer, W. D. 1984. Anatomical analysis of the development and distribution of *Rhizobium* infections in soybean roots. *Can. J. Bot.* 62:2375–84
28. Carlson, R. E., Dolphin, D. H. 1981. Chromatographic analysis of isoflavonoid accumulation in stressed *Pisum sativum. Phytochemistry* 20:2281–84
29. Carlson, R. E., Dolphin, D. H. 1982. *Pisum sativum* strees metabolites: Two cinnamylphenols and a 2'methoxychalcone. *Phytochemistry* 21:1733–36
30. Chang, M. 1986. *Semiochemicals involved in host recognition in* Striga asiatica. Phd thesis. Univ. Chicago
31. Chang, M., Lynn, D. G. 1986. The haustorium and the chemistry of host recognition in parasitic angiosperms. *J. Chem. Ecol.* 12:561–79
32. Chang, M., Netzly, D. H., Butler, L. G., Lynn, D. G. 1986. Chemical regulation of distance: characterization of the first natural host germination stimulant for *Striga asiatica. J. Am. Chem. Soc.* 108:7858–60
33. Chiji, H., Najao, A., Veda, S., Izawa, M., Ochi, T., Anada, S., Hara, K. 1980. Sparingly water-soluble germination inhibitors in the balls of sugar beets. *Tensai Kaiho* 22:61–68
34. Chilton, M. D. 1983. A vector for introducing new genes into plants. *Sci. Am.* 248:50–59
35. Chilton, M. D., Saiki, R. K., Yadav, N., Gordon, M. P., Quetier, F. 1980. T-DNA from *Agrobacterium* Ti plasmid is in the nuclear DNA fraction of crown gall tumor cells. *Proc. Natl. Acad. Sci. USA* 77:4060–64
36. Cleland, C. F. 1974. Isolation of flower-inducing and flower-inhibitory factors from aphid honeydew. *Plant Physiol.* 54:899–903
37. Conn, E. E., ed. 1986. The shikimate acid pathway. *Rec. Adv. Phytochem.* 20:1–374
38. Cook, C. E., Whichard, L. P., Wall, M. E., Egely, G. H. 1966. Germination of witchweed *(Striga lutea* Lour.): Isolation and properties of a potent stimulant. *Science* 196:1189–90
39. Cook, C. E., Whichard, L. P., Wall, M. E., Egely, G. H., Coggon, P., et al. 1972. Germination stimulants. II. The structure of strigol—a potent seed germination stimulant for witchweed *(Striga lutea* Lour.). *J. Am. Chem. Soc.* 94:6198–99
40. Crane, F. L., Morre, D. J., Low, H., eds. 1988. *Plasma Membrane Oxidoreductases in Control of Animal and Plant Growth. NATO ASI Ser.,* Vol. 157. New York: Plenum. 443 pp.
41. Crane, F. L., Sun, I. L., Clark, M. G., Grebing, C., Low, H. 1985. Transplasma-membrane redox systems in growth and development. *Biochim. Biophys. Acta* 811:233–64
42. Crick, F. 1970. Diffusion in embryogenesis. *Nature* 225:420–22

43. Dart, P. J. 1974. The infection process. In *The Biology of Nitrogen Fixation*, ed. A. Quispal, pp. 381–489. Amsterdam: New Holland
44. Darvill, A. G., Albersheim, P. 1984. Phytoalexins and their elicitors–a defense against microbial infection in plants. *Annu. Rev. Physiol.* 35:243–75
45. Dazzo, F. B., Hollingsworth, R. I., Sherwood, J. E., Abe, M., Hrabak, E. M., et al. 1985. Recognition and infection of clover root hairs by *Rhizobium trifolii*. In *Nitrogen Fixation Research Progress*, ed. H. J. Evans, P. J. Bottomley, W. E. Newton, pp. 239–45. Amsterdam: Nijhoff
46. Dazzo, F. B., Truchet, G. L. 1983. Interaction of lectins and their saccharide receptors in the *Rhizobium*-legume symbiosis. *J. Membr. Biol.* 73:1–16
47. Dazzo, F. B., Truchet, G. L., Sherwood, J. E., Hrabak, W. M., Gardiol, A. E. 1982. Alteration of the trifolin A binding capsule of *Rhizobium trifolii* 0403 by enzymes released from clover roots. *Appl. Environ. Microbiol.* 44:478–90
48. Dazzo, F. B., Urbano, M. R., Brill, W. J., 1979. Transient appearance of lectin receptors on *Rhizobium trifolii. Curr. Microbiol.* 2:15–20
49. Dewick, P. M. 1975. Pterocarpan biosynthesis: 2'-Hydroxy-isoflavone and isoflavanone precursors of demethylhomopterocarpin in red clover. *J. Chem. Soc. Chem. Commun.*, 656–8
50. Djordjevic, M. A., Gabriel, D. W., Rolfe, B. G. 1987. Rhizobium—the refined parasite of legumes. *Annu. Rev. Phytopathol.* 25:145–68
51. Djordjevic, M. A., Redmond, J. W., Batley, M., Rolfe, B. G. 1987. Clovers secrete specific phenolic compounds which either stimulate or repress *nod* gene expression in *Rhizobium trifolii. EMBO J.* 6:1173–79
52. Downie, J. A., Knight, C. D., Johnston, A. W. B., Rossen, L. 1985. Identification of genes and gene products involved in the nodulation of peas by *Rhizobium leguminosarum. Mol. Gen. Genet.* 198:255–62
53. Dudley, M. E., Jacobs, T. W., Long, S. R. 1987. Microscopic studies of cell divisions induced in alfalfa roots by *Rhizobium meliloti. Planta* 171:289–301
54. Durston, A. J., Timmermans, J. P. M., Hage, W. J., Hendriks, H. F. J., de Vries, N. J., Heideveld, M., Nieuwkoop, P D. 1989. Retinoic acid causes and anteroposterior transformation in the developing central nervous system. *Nature* 340:140–44
55. Egley, G. H. 1972. Influence of the seed envelope and growth regulators upon seed dormancy in witchweed (*Striga lutea* Lour.). *Ann. Bot.* 36:755–70
56. Engstrom, P., Zambryski, P., Van Montagu, M., Stachel, S. 1987. Characterization of *Agrobacterium tumefaciens* virulence proteins produced by the plant factor acetosyringone. *J. Mol. Biol.* 197:635–46
57. Evans, L. S., Almeida, M. S., Lynn, D. G., Nakanishi, K. 1979. Chemical characterization of a hormone that promotes cell arrest in G2 in complex tissues. *Science* 203:1122–23
58. Fate, G., Chang, M., Lynn, D. G. 1990. Control of germination in *Striga asiatica*: The chemistry of spatial definition. *Plant Physiol.* In press
59. Fate, G., Lynn, D. G. 1990. Molecular diffusion coefficients: Experimental determination and demonstration. *J. Chem. Ed.* In press
60. Firmin, J. L., Wilson, K. E., Rossen, L., Johnston, A. W. B. 1986. Flavonoid activation of nodulation genes in *Rhizobium* reversed by other compounds present in plants. *Nature* 324:90–92
61. Firoozabody, E., Galbraith D. W. 1984. Presence of a plant cell wall is not required for transformation of *Nicotiana* by *Agrobacterium tumefaciens. Plant Cell Tissue Organ Cult.* 3:175–84
62. Friend, J. 1979. Phenolic substances and plant disease. *Rec. Adv. Phytochem.* 12:557–88
63. Freudenberg, K., Neish, A. C. 1968. *Constitution and Biosynthesis of Lignin.* New York: Springer. 129 pp.
64. Gierer, A., Meinhardt, H. 1972. A theory of biological pattern formation. *Kybernetik* 12:30–39
65. Glass, A., Dunlop, J. 1974. Influence of phenolic acids on ion uptake. *Plant Physiol.* 54:855–58
66. Goodwin, T. W., Mercer, E. I. 1983. *Introduction to Plant Biochemistry*, pp. 528–66. Oxford: Pergamon. 677 pp. 2nd ed.
67. Grand, C., Sarni, F., Lamb, C. J. 1987. Rapid induction by fungal elicitor of the synthesis of cinnamyl-alcohol dehydrogenase, a specific enzyme of lignin synthesis. *Eur. J. Biochem.* 169:73–77
68. Gyorgypal, Z., Iyer, N., Kondorosi, A. 1988. Three regulatory *nod*D alleles of diverged flavanoid specificity are involved in host-dependent nodulation by *Rhizobium meliloti. Mol. Gen. Genet.* 212:85–92

69. Hahlbrock, K., Scheel, D. 1989. Physiology and molecular biology of phenylpropanoid metabolism. *Annu. Rev. Plant Physiol. Plant Mol. Biol.* 40:347–69

70. Harkin, J. M., Obst, J. R. 1973. Syringaldazine, an effective reagent for detecting laccases and peroxidases in fungi. *Experientia* 29:381–87

71. Hausen, B. M., Simattupang, M. H., Kingreen, J. C. 1972. Untersuchungen zur Überempfindlichkeit gegen Sucupira und Palisanderholz. *Berufs-Dermatosen.* 20:1–7

72. Hayashi, M., Takeuchi, I. 1976. Quantitative studies on cell differentiation during morphogenesis of the cellular slime mold *Dictyostelium discoideum. Dev. Biol.* 50:302–9

73. Heather, J. B., Mithal, R. S. D., Sih, C. J. 1976. Synthesis of the witchweed seed germination stimulant (±)strigol. *J. Am. Chem. Soc.* 98:3661–69

74. Henikoff, S., Haughn, G. W., Calvo, J. M., Wallace, J. C. 1988. A large family of bacterial activator proteins. *Proc. Natl. Acad. Sci. USA* 85:6602–6

75. Hsiao, A. I., Worsham, A. D., Moreland, D. E. 198. Effects of sodium hypochlorite and certain plant growth regulators on germination of witchweed *(Striga asiatica)* seeds. *Weed Sci.* 29:98

76. Innes, R. W., Kuempel, P. L., Plazinski, J., Canter-Cremers, H., Rolfe, B., Djordjevic, M. A. 1985. Plant factors induce the expression of nodulation and host-range genes in *Rhizobium trifolii. Mol. Gen. Genet.* 201:426–32

77. Jacobs, M., Rubery, P. H. 1988. Naturally occurring auxin transport regulators. *Science* 241:346–49

78. Jahnen, W., Batterham, M. P., Clark, A. E., Moritz, R. L., Simpson, R. J. 1989. Identification, isolation, and N-terminal sequencing of style glycoproteins associated with self-incompatibility in *Nicotiana alata. Plant Cell* 1:493–99

79. Jones, E., Ekundayo, O., Kingston, D. G. I. 1981. Plant anticancer agents. XI. 2,6-Dimethoxybenzoquinone as a cytotoxic constituent of *Tibouchina pulchra. J. Nat. Prod.* 44:493–95

80. Kahl, G., Schell, J., eds. 1982. *Molecular Biology of Plant Tumors.* New York: Academic. 615 pp.

81. Kamat, V. S., Graden, D. W., Lynn, D. G., Steffens, J. C., Riopel, J. L. 1982. A versatile total synthesis of xenognosin. *Tetrahedron Lett.* 1541–44

82. Kefeli, V. I., Dashek, W. V. 1984. Non-hormonal stimulants and inhibitors of plant growth and development. *Biol. Rev.* 59:273–88

83. Khurana, J. P., Maheshwari, S. C. 1980. Some effects of salicylic acid on growth and flowering in *Spirodela polyrrhiza* SP_{20}. *Plant Cell Physiol.* 21:923–27

84. Kirk, T. K., Farrell, R. L. 1987. Enzymatic "Combustion": the microbial degradation of lignin. *Annu. Rev. Microbiol.* 41:465–505

85. Kistler, H. C., VanEtten, E. D. 1984. Regulation of pisatin demethylation in *Nectria haematococca* and its influence on pisatin tolerance and virulence. *J. Gen. Microbiol.* 130:2605–13

86. Kodaira, H., Ishikawa, M., Komoda, Y., Nakajima, T. 1983. Isolation and identification of anti-platelet aggregation principles from the bark of *Fraxinus japonica* Blume. *Chem. Pharmacol. Bull.* 31:2262–68

87. Kondorosi, E., Gyuris, J., Schmidt, J., John, M., Duda, E., et al. 1989. Positive and negative control of *nod* gene expression in *Rhizobium meliloti* is required for optimal nodulation. *EMBO J.* 8:1331–40

88. Kosslak, R. M., Bookland, R., Barkei, J., Paaren, H. E., Appelbaum, E. R. 1987. Induction of *Bradyrhizobium japonicum* common nod genes by isoflavones isolated from *Glycine max.* Proc. Natl. Acad. Sci. USA 84:7428–32

89. Krisnangkura, K., Gold, M. H. 1979. Peroxidase catalysed oxidative decarboxylation of vanillic acid to methoxy-*p*-hydroquinone. *Phytochemistry* 18:2019–21

90. Kuijt, J. 1969. *The Biology of Parasitic Flowering Plants.* Berkeley, CA: Univ. Calif. Press

91. Lamb, C. J., Lawton, M. A., Dron, M., Dixon, R. A. 1989. Signals and transduction mechanisms for activation of plant defences against microbial attack. *Cell.* 56:215–24

92. Lee, T. S., Purse, J. G., Pryce, R. J., Horgan, R., Wareing, P. F. 1961. Dihydroconiferyl alcohol—a cell division factor from *Acer* species. *Planta* 152:571–77

93. Lehle, F. R., Putnam, A. R. 1983. Allelopathic potential of sorghum *(Sorghum bicolor).* Isolation of seed germination inhibitors. *J. Chem. Ecol.* 9:1223–34

94. Leigh, J. A., Reed, J. W., Hanks, J. F., Hirsh, A. M., Walker, G. C. 1987. *Rhizobium meliloti* mutants that fail to succinylate their calcoflour-binding exo-

polysaccharide are defective in nodule invasion. *Cell* 51:579–87

95. Leonowicz, A., Edgehill, R. U., Bollag, J.-M. 1984. The effect of pH on the transformation of syringic and vanillic acids by the laccases of *Rhizoctonia praticola* and *Trametes versicolor*. *Arch. Microbiol.* 137:89–96

96. Leroux, B., Yanofsky, M. F., Winans, S. C., Ward, J. E., Zeigler, S. F., Nester, E. W. 1987. Characterization of the *vir*A locus of *Agrobacterium tumefaciens:* a transcriptional regulator and host range determinant. *EMBO J.* 6:849–56

97. Leslie, C. A., Romani, R. J. 1986. Salicylic acid: a new inhibitor of ethylene biosynthesis. *Plant Cell Rep.* 5:144–46

98. Libbenga, K. R., Torrey, J. G. 1973. Hormone-induced endoreduplication prior to mitosis in cultured pea root cortex cells. *Am. J. Bot.* 60:293–99

99. Lipetz, J. 1966. Crown gall tumorigenesis. II. Relations between wound healing and the tumorigenic response. *Cancer Res.* 26:1597–605

100. Long, S. R. 1989. *Rhizobium*-legume nodulation: Life together in the underground. *Cell* 56:203–14

101. Lynn, D. G. 1985. The involvement of allelochemicals in the host selection of parasitic angiosperms. In *The Chemistry of Allelopathy: Biochemical Interactions Among Plants*, ed. A. C. Thompson, pp. 55–81. ACS Symp. Ser. No. 2268. Washington, DC: Am. Cem. Soc. 470 pp.

102. Lynn, D. G., Chen, R. H., Manning, K. S., Wood, H. N. 1987. The structural characterization of endogenous factors from *Vinca rosea* crown gall tumors that promote cell division of tobacco cells. *Proc. Natl. Acad. Sci. USA.* 84:615–19

103. Lynn, D. G., Jaffe, K., Cornwall, M., Tramontano, W. 1987. Characterization of an endogenous factor controlling the cell cycle of complex tissues. *J. Am. Chem. Soc.* 109:5858–59

104. Lynn, D. G., Nakanishi, K., Patt, S. L., Occolowitz, J. L., Almeida, S., Evans, L. S. 1978. Isolation and characterization of the first mitotic cycle hormone that regulates cell proliferation. *J. Am. Chem. Soc.* 100:7759–60

105. Lynn, D. G., Steffens, J. C., Kamat, V. S., Graden, D. W., Shabanowitz, J., Riopel, J. L. 1981. Isolation and characterization of the first host recognition substance for parasitic angiosperms. *J. Am. Chem. Soc.* 103:1868–70

106. MacQueen, M. 1984. Haustorial induc-

ing activity of several simple phenolic compounds. In Proc. Third Int. Symp. Parastic Weeds, pp. 118–22, Aleppo, Syria, ICARDA

107. Mayer, A. M. 1987. Polyphenol oxidases in plants—recent progress. *Phytochemistry* 26:11–20

108. Melchers, L. S., Regensburg-Tuïnk, T. J. G., Bourret, R. B., Sedee, N. J. A., Schilperoort, R. A., Hooykaas, P. J. J. 1989. Membrane topology and functional analysis of the sensory protein *vir*A of *Agrobacterium tumefaciens. EMBO J.* 8:1919–25

109. Memelink, J., Sylvia de Pater, B., Hoge, J. H. C., Schilperoort, R. A. 1987. T-DNA Hormone biosynthesis genes: phytohormones and gene expression in plants. *Dev. Genet.* 8:321–37

110. Milborrow, B. V. 1984. Inhibitors. In *Advanced Plant Physiology*, ed. M. B. Wilkins, pp. 76–110. London: Pitman Publishing Ltd. 514 pp.

111. Morris, H. R., Taylor, G. W., Maasento, M. S., Jermyn, K. A., Kay, R. R. 1987. Chemical structure of the morphogen differentiation-inducing factor from *Dictyostelium discoideum. Nature* 328:811–14

112. Muller, P., Hynes, M., Kapp, D., Niehaus, K., Puhler, A. 1988. Two classes of *Rhizobium meliloti* infective mutants differ in exopolysaccharide production and in coinoculation properties with nodulation mutants. *Mol. Gen. Genet.* 211:17–26

113. Mulligan, J. T., Long, S. R. 1985. Induction of *Rhizobium meliloti nod*C expression by plant exudate requires *nod*D. *Proc. Natl. Acad. Sci. USA* 82:6609–13

114. Musselman. L. J. 1980. The biology of *Striga, Orobanche,* and other rootparasitic weeds. *Annu. Rev. Phytopathol.* 18:463–89

115. Musselman. L. J., Dickison, W. C. 1975. The structure and development of the haustorium in parasitic Scrophulariaceae. *Bot. J. Linn. Soc.* 70:183

116. Nakatsubo, F., Kirk, T. K., Shimada, M., Higuchi, T. 1981. Metabolism of a phenylcoumarin substructure lignin model compound in lignolytic cultures of *Phanerochaete chrysosporium. Arch. Microbiol.* 128:416–20

117. Nanda, K. K., Kumar, S., Sood, V. 1976. Effect of gibberellic acid and some phenols on flowering of *Impatiens balsamina,* a quantitative short-day plant. *Physiol. Plant.* 38:53–56

118. Netzly, D. H., Riopel, J. L., Ejeta, G., Butler, L. G. 1988. Germination stimulants of witchweed from hydrophobic

root exudates of sorghum *(Sorghum bicolor)*. *Weed Sci.* 36:441–46

119. Nour, J., Press, M., Stewart, G., Tuohy, J. 1986. Africa in the grip of witchweed. *New Sci.*, January 9th, pp. 44–48

120. Pepperman, A. B., Connick, W. J., Vail, S. L., Worsham, A. D., Pavlista, A. D., Moreland, D. E. 1982. Evaluation of precursors and analogues of strigol as witchweed *(Striga asiatica)* seed germination stimulants. *Weed Sci.* 30:561–66

121. Peters, N. K., Frost, J. W., Long, S. R. 1986. A plant flavone, luteolin, induces expression of *Rhizobium meliloti* nodulation genes. *Science* 223:977–79

122. Peters, N. K., Long, S. R. 1988. Alfalfa root exudates and compounds which promote or inhibit induction of *Rhizobium meliloti* nodulation genes. *Plant Physiol.* 88:396–400

123. Pickard, M. A., Westlake, D. W. S. 1970. Fungal metabolism of flavanoids. Purification, properties and substrate specificity of an inducible laccase from *Polyporus versicolor* PRL 572. *Can. J. Biochem.* 48:1351–58

124. Pierce, M., Bauer, W. D. 1983. A rapid regulatory response governing nodulation in soybean. *Plant. Physiol.* 73:286–90

125. Purse, J. G., Horgan, R., Horgan, J. M., Wareing, P. F. 1976. Cytokinins of sycamore spring sap. *Planta* 132:1–8

126. Rai, V. K., Sharma, S. S., Sharma, S. 1986. Reversal of ABA-induced stomatal closure by phenolic compounds. *J. Exp. Bot.* 37:129–34

127. Rao, M. J. V., Musselman, L. J. 1987. Host specificity in *Striga spp.* and physiological strains. In *Parasitic Weeds in Agriculture*, ed. L. J. Musselman, 1:13–25. Boca Raton: CRC Press

128. Raskin, I., Ehmann, A., Melander, W. R., Meeuse, B. J. D. 1987. Salicylic acid: a natural inducer of heat production in *Arum* lilies. *Science* 237:1601–2

129. Raskin, I., Tumer, I. M., Melander, W. R. 1989. Regulation of heat production in the inflorescences of an *Arum* lily by endogenous salicylic acid. *Proc. Natl. Acad. Sci. USA* 86:2214–18

130. Recourt, K., van Brussel, A. A. N., Driessen, A. J. M., Lugtenberg, B. J. J. 1989. Accumulation of a *nod* gene inducer, the flavanoid naringenin, in the cytoplasmic membrane of *Rhizobium leguminosarum* biovar *viciae* is caused by the pH-dependent hydrophobicity of naringenin. *J. Bacteriol.* 171:4370–77

131. Redfearn, E. R., Whittaker, P. A. 1962. The inhibitory effects of quinones on the succinic oxidase system of the respiratory chain. *Biochim. Biophys. Acta* 56:440–44

132. Redmond, J. R., Batley, M., Djordjevic, M. A., Innes, R. W., Kuempel, P., Rolfe, B. G. 1986. Flavones induce the expression of the nodulation genes in *Rhizobium. Nature* 323:632–35

133. Reinhammer, B. 1984. Laccase. In *Copper Proteins and Copper Enzymes*, ed. E. Lontie, 3:1–35. Boca Raton: CRC Press 240 pp.

134. Renganathan, V., Miki, K., Gold, M. H. 1986. Role of molecular oxygen in lignin peroxidase reactions. *Arch. Biochem. Biophys.* 246:155–61

135. Riopel, J. L., Baird, Wm. V. 1987. Morphogenesis of the early development of primary haustoria in *Striga asiatica*. In *Parasitic Weeds in Agriculture*, ed. L. J. Musselman, 1:107–125. Boca Raton: CRC Press.

136. Robertson, J. G., Littleton, P. 1982. Coated and smooth vesicles in the biogenesis of cell walls, plasma membranes, infection threads and peribacteroid membranes in root hairs and nodules in white clovers. *J. Cell. Sci.* 58:63–78

137. Ronson, C. W., Nixon, B. T., Ausabel, F. M. 1987. Conserved domains in bacterial regulatory proteins that respond to environmental stimuli. *Cell.* 49:579–81

138. Rossen, L., Shearman, C. A., Johnston, A. W. B., Downie, J. A. 1985. The *nodD* gene of *Rhizobium leguminosarum* is autoregulatory and in the presence of plant exudate induces the expression of *nod*ABC genes. *EMBO J.* 4:3369–73

139. Ryan, C. A. 1987. Oligosaccharide signalling in plants. *Annu., Rev. Cell Biol.* 3:295–317

140. Sakurai, N., Shibata, K., Kamisaka, S. 1974. Stimulation of auxin-induced elongation of cucumber hypocotyl sections by dihydroconiferyl alcohol. Dihydroconiferyl alcohol inhibits indole-3-acetic acid degradation in vivo and in vitro. *Plant Cell Physiol.* 15:709–16

141. Sampson, J. 1976. Cell patterning in migrating slugs of *Dictyostelium discoideum*. *J. Embryol. Exp. Morphol.* 36:663–68

142. Saunders, A. R. 1933. Studies in phanerogamic parasitism with particular reference to *Striga lutea* Lour. *Dept. Agric. S. Afr. Bull.* 128:1–57

143. Schmidt, J., Wingender, R., John, M., Wieneke, U., Schell, J. 1988. *Rhizobium meliloti nodA* and *nodB* genes are

involved in generating compounds that stimulate mitosis of plant cells. *Proc. Natl. Acad. Sci. USA.* 85:8578–82

144. Schultz, K. H., Garbe, I., Hausen, B. M., Simptupang, M. H. 1979. The sensitizing capacity of naturally occurring quinones. Experimental studies on guinea pigs. *Arch. Dermatol. Res.* 264: 275–86

145. Shaykh, C. M., Soliday, C. L., Kolattukudy, P. E. 1977. Proof of the production of cutinase by *Fusarium solini* f. *pisi* during penetration into its host *Pisum sativum. Plant Physiol.* 60:170–72

146. Shibata, K., Kubota, T., Kamisaka, S. 1974. Dihydroconiferyl alcohol as a gibberellin synergist in inducing lettuce hypocotyl elongation. An assessment of structure-activity relationships. *Plant Cell Physiol.* 15:191–94

147. Siegel, B. A., Verbeke, J. A. 1989. Diffusible factors essential for epidermal cell redifferentiation in *Catharanthus roseus. Science* 244:580–81

148. Smith, C. E., Dudley, M. D., Lynn, D. G. 1990. Control of the vegetative parasitic transition in *Striga asiatica. Plant Physiol.* In press

149. Smith, C. E., Orr, J. D., Lynn, D. G. 1989. Chemical communication and the control of development. In *Natural Product Isolation: separation methods for anti-microbials, antivirals, and enzyme inhibitors,* ed. G. H. Wagman, R. Cooper, pp. 561–97. Amsterdam: Elsevier. 619 pp.

150. Spaink, H. P., Wijffelman, C. A., Pees, E., Okker, R. J. H., Lugtenberg, B. J. J. 1987. *Rhizobium* nodulation *nod*D as a determinant of host specificity. *Nature* 328:337–40

151. Spemann, H., Mangold, H. 1924. Über Induktion von Embryonalanlagan durch Implantation artfremder Organisatoren. *Arch. Mikr. Anat. Entw. Mech.* 100: 599–638

152. Stachel, S. E., An, G., Flores, C., Nester, E. W. 1985. A Tn3 *lacZ* transposon for the random generation of β-galactosidase gene fusions: application to the analysis of gene expression in *Agrobacterium. EMBO J.* 4:891–98

153. Stachel, S. E., Messens, E., Van Montagu, M., Zambryski, P. 1985. Identification of the signal molecules produced by wounded plant cells that activate T-DNA transfer in *Agrobacterium tumefaciens. Nature* 318:624–629

154. Stachel, S. E., Nester, E. W. 1986. The genetic and transcriptional organization of the *vir* region of the A6 Ti plasmid of *Agrobacterium tumefaciens. EMBO J.* 5:1445–54

155. Steffens, J. C., Lynn, D. G., Kamat, V. S., Riopel, J. L. 1982. Molecular specificity of haustorial induction in *Agalinis purpurea* (L.) Raf. (Scrophulariaceae). *Ann. Bot.* 50:1–7

156. Steffens, J. C., Lynn, D. G., Riopel, J. L. 1986. An haustoria inducer for the root parasite *Agalinis purpurea. Phytochemistry* 25:2291–98

157. Steffens, J. C., Roark, J. L., Lynn, D. G., Riopel, J. L. 1983. Host recognition in parasitic angiosperms: use of correlation spectroscopy to identify long-range coupling in an haustoria inducer. *J. Am. Chem. Soc.* 105:1669–71

158. Summerbell, D. 1983. The effect of local application of retinoic acid to the anterior margin of the developing chick limb. *J. Embryol. Exp. Morphol.* 78: 269–89

159. Sunderland, N. C. 1960. The production of *Striga* and *Orobanche* germination stimulants by maize roots. I. The number and variety of stimulants. *J. Exp. Bot.* 11:236–45

160. Thaller, C., Eichele, G. 1987. Identification and spatial distribution of retinoids in the developing chick limb bud. *Nature* 327:625–28

161. Tickle, C., Alberts, B., Wolpert, L., Lee, J. 1982. Local application of retinoic acid to the limb bud mimics the action of the polarizing region. *Nature* 296:564–66

162. Town, C. D., Gross, J. D., Kay, R. R. 1976. Cell differentiation without morphogenesis in *Dictyostelium discoideum. Nature* 262:717–19

163. Turing, A. 1952. The chemical basis of morphogenesis. *Philos. Trans. R. Soc. Ser. B.* 237:32

164. Umezawa, T., Nakatsubo, F., Higuchi, T. 1982. Lignan degradation by *Phanerochaete chrysosporium:* Metabolism of a phenolic phenylcoumarin substrate model compound. *Arch. Microbiol.* 131:124–28

165. Vail, S. L., Dailey, O. D., Connick, W. J., Pepperman, A. B. 1985. Strigol syntheses and related structure-bioactivity studies. In *The Chemistry of Allelopathy: Biochemical Interactions Among Plants,* ed. A. C. Thompson, pp. 445–56. ACS Symp. Ser. No. 2268. Washington, DC: Am. Chem. Soc. 470 pp.

166. Veluthambi, K., Krishnan, M., Gould, J. H., Smith, R. H., Gelvin, S. B. 1989. Opines stimulate induction of the *Agrobacterium tumefaciens* Ti plasmid. *J. Bacteriol.* 171:3696–3703

167. Verbeke, J. A. 1989. Stereological analysis of ultrastructural changes during induced epidermal cell redifferentiation in developing flowers of *Catharanthus roseus* (Apocynaceae). *Am. J. Bot.* 76:952–57

168. Vernade, D., Herrera-Estrella, A., Wang, K., Van Montagu, M. 1988. Glycine betaine allows enhanced induction of the *Agrobacterium tumefaciens vir* genes by acetosyringone at low pH. *J. Bacteriol.* 170:5822–29

169. Visser, J., Dorr, I. 1987. The haustorium. In *Parasitic Weeds in Agriculture*, ed. L. J. Musselman, 1:91–106. Boca Raton: CRC Press

170. Wagner, G. J., Hrazdina, G. 1984. Endoplasmic reticulum as a site of phenylpropanoid and flavanoid metabolism in *Hippeastrum*. *Plant Physiol.* 74:901–6

171. Walker, D. B. 1975. Postgential fusion in *Catharanthus roseus* (Apocynaceae). I. Light and scanning electron microscope study of gynoecial ontogeny. *Am. J. Bot.* 62:457–67

172. Walter, M. H., Grima-Pettenati, J., Grand, C., Boudet, A. M., Lamb, C. J. 1988. Cinnamyl-alcohol dehydrogenase, a molecular marker specific for lignin synthesis: cDNA cloning and mRNA induction by fungal elicitor. *Proc. Natl. Acad. Sci. USA* 85:5546–50

173. Weiler, E. W., Spanier, K. 1981. Phytohormones in the formation of crown gall tumors. *Planta* 153:326–37

174. Weinges, K., Muller, R., Kloss, P., Jaggy, H. 1970. Isolierung und Konstitutionsaufklärung eines optisch aktiven Dehydro-diconiferylalkohols aus den Samen der Mariendistel *Silybum marianum* (Gaertn.). *Liebigs Ann. Chem.* 736:170–72

175. White, R. F. 1979. Acetylsalicylic acid induces resistance to tobacco mosaic virus in tobacco. *Virology* 99:410–12

175. Windholz, M., Budavari, S., eds. 1983. Entry 9697, p. 1414. In *The Merck Index*. Rahway: Merck. 1463 pp. 10th ed.

176. Wong, S. M., Pezzuto, J., Fong, H. S., Farnsworth, N. R. 1985. Isolation, structural elucidation and chemical synthesis of 2-hydroxy-3-octadecyl-5-methoxy-1,4-benzoquinone (Irisoquin), a cytotoxic constituent of *Iris miisouriensis*. *J. Pharm. Sci.* 74:1114–16

177. Wood, H. N. 1964. The characterization of naturally occurring kinins from crown gall tumor cells of *Vinca rosea* L. In *Regulateurs Naturels de la Croissance Vegetale*, 123:97–102. Paris: Colloq. Int. Cent. Nat. Rech. Sci.

178. Wood, H. N. 1970. Revised identification of the chromophore of a cell division factor from crown gall tumor cells of *Vinca rosea* L. *Proc. Natl. Acad. Sci. USA* 67:1283–87

179. Worsham, A. D. 1987. Germination of witchweed seeds. In *Parasitic Weeds in Agriculture*, ed. L. J. Musselman, 1: 45–61. Boca Raton: CRC Press

180. Worsham, A. D., Egley, G. H. 1989. Physiology of witchweed seed dormancy and germination in *Striga asiatica* (L.) O. Kuntze (witchweed). In *Cooperative Research and Control Programs in the United States*, ed. P. F. Sand. Champaign, IL: Weed Sci. Soc. Am. Monogr. Ser. In press

181. Worsham, A. D., Moreland, D. E., Klingman, G. C. 1964. Characterization of the *Striga asiatica* germination stimulant from *Zea mays* L. *J. Exp. Bot.* 15:556–67

182. Wullems, G. J., Molendijk, L., Ooms, G., Schilperoort, R. A. 1981. Differential expression of crown gall tumor markers in transformants obtained after *in vivo Agrobacterium tumefaciens*–induced transformation of cell wall regenerating protoplasts derived from *Nicotiana tobacum*. *Proc. Natl. Acad. Sci. USA.* 78:4344–48

183. Yokota, M., Zenda, H., Kosuge, T., Yamamoto, T., Torigoe, Y. 1978. Studies on isolation of naturally occurring biologically active principles. V. Antifungal constituents in *Betulae* cortex. *Yakugaku Zasshi* 98:1607–12

184. Zaat, S. I. J., Wijffelman, C. A., Spaink, H. P., Van Brussel, A. A. N., Okker, R. J. H., Lugtenberg, G. J. J. 1987. Induction of the *nod*A promoter of *Rhizobium leguminosarum* Sym plasmid PRL1JI by plant flavanones and flavones. *J. Bacteriol.* 169:198–204

185. Zambryski, P., Tempe, J., Schell, J. 1989. Transfer and function of t-DNA genes from *Agrobacterium* Ti and Ri plasmids in plants. *Cell.* 56:193–201

Annu. Rev. Plant Physiol. Plant Mal. Biol. 1990. 41:527–52

THE EFFECTS OF PLANT TRANSPOSABLE ELEMENT INSERTION ON TRANSCRIPTION INITIATION AND RNA PROCESSING

C. F. Weil and S. R. Wessler

Botany Department, University of Georgia, Athens, Georgia 30602

KEY WORDS: Ac/Ds, Spm, Tam, splicing, methylation

CONTENTS

INTRODUCTION

Our understanding of the molecular biology of plant transposable elements has increased immensely over the past 8 years. It is no longer possible to

1040-2519/90/0601-0527$02.00

summarize all the molecular aspects of plant elements in one review, as was done previously in this series by Freeling (36). In the present review we focus on the influence of elements on the transcription and processing of transcripts from the genes in which they reside. Extensive studies have shown that the effects of insertion and excision are not restricted to the inhibition and restoration, respectively, of gene expression. Rather, interactions between gene and element sequences during the course of plant development are frequently far more complex. The sequences introduced by insertion or altered by excision have the ability to (a) impose new temporal and spatial patterns of gene expression (21, 68, 95), (b) drastically alter RNA processing (115), and (c) influence epigenetic control of genes during plant development (32).

Genetic analysis of plant transposable elements provided insights into the nature of these interactions prior to any understanding of introns and exons, regulatory proteins, or DNA modifications. The conclusions from some key studies can be summarized as follows: (a) different alleles containing the same element can have a diversity of phenotypes (11, 73); (b) element insertion does not always abolish gene expression, and element excision does not always restore expression; (c) stable derivatives arising from element excision can display novel spatial and temporal patterns of gene expression (35); and (d) reversible changes in an element's activity phase can be correlated with reversible effects on expression of the resident gene (66, 68).

Since 1982, most of the transposable element alleles and their derivatives in maize and *Antirrhinum majus* have been either cloned or analyzed at the level of Southern blots. Characterizations of the element, the position of insertion within the resident gene, and the mutant gene products have provided molecular explanations for most of the phenomena listed above. To cover the subject adequately, it is first necessary to summarize aspects of element structure and coding capacity relevant to this review. For more in-depth treatments of these topics, the reader is referred to several comprehensive reviews of element genetics (31, 67, 79) and molecular biology (33, 70, 114).

OVERVIEW OF PLANT ELEMENT STRUCTURE

Spm/dSpm

The *Suppressor/mutator (Spm)* family is comprised of the 8.3-kb *Spm* element and smaller elements, called *dSpm* (defective *Spm*), which arise by internal deletion of *Spm* sequences (63, 77, 78, 85, 92, 94) (Figure 1a). The *Enhancer (En)* element was described independently (110) and has been shown to be nearly identical to *Spm* (63, 77). Insertion of *dSpm* elements into genes does not always abolish gene expression. Genetic analysis of the interaction between these leaky *dSpm* alleles and *trans*-acting *Spm* elements led to the

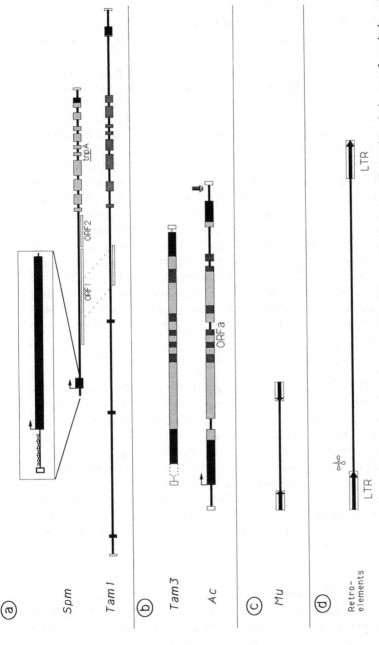

Figure 1 Transposable element structures. Transcription start sites are indicated by the thin black arrows. Inverted terminal repeats for each element are designated by white boxes, exon sequences by black boxes, and the open reading frames within exons as shaded areas. (*a*) *Spm* and *Tam1*. The inset shows the region upstream of the transcription start site; *tnpA*-binding sites are indicated with white arrows. Intron sequences in *Tam1* similar to ORF1 of *Spm* are shown. (*b*) *Ac* and *Tam3*. Regions of ORFa and the open reading frame of *Tam3* (ORFTam3) that are more similar are more darkly shaded; these regions are colinear within the two open reading frames. The three splice donor sites of *Ac* (also shared with *Ds*) are indicated by the three vertical arrows. (*c*) *Mu*. (*d*) A typical retro-element. The long terminal direct repeats (LTRs) are shown. The tRNA-like primer site is indicated by the cloverleaf structure.

identification of two genetically distinct *Spm*-encoded functions (67). The *suppressor (sp)* function suppresses the residual gene expression from leaky *dSpm* alleles. The *mutator (m)* function is required for element excision, and is probably the transposase.

The termini of all *Spm* and *dSpm* elements contain 13-bp inverted repeats that are flanked by a 3-bp direct repeat of host sequence generated upon insertion (94). Excision of *Spm* and most other plant transposable elements usually does not restore the wild-type target sequence; most often, all or part of the target duplication is left behind (80, 93, 106). The major transcript of *Spm* is 2.5 kb long and is processed from a pre-mRNA that spans most of the 8.3-kb element (Figure 1a) (77). This transcript encodes the TNPA protein, which is probably the *trans*-acting suppressor (24, 41). The product of *tnpA* may also act as a positive regulatory factor of *Spm* transcription (63). *Spm* encodes minor transcripts that probably arise from alternative processing of the pre-mRNA (62, 77). These large minor transcripts retain two open reading frames, "ORF1" and "ORF2", found in the first intron of the major transcript. Elements with frameshift mutations engineered into ORF1 and ORF2 lose their ability to transpose from constructs transformed into tobacco, suggesting these open reading frames may be part of the transposase (62). Consistent with this, elements with deletions that remove parts of ORF1 and ORF2 [*Spm-weak (Spm-w)* elements] retain *sp* function but are deficient in *m* function (41, 63).

Tam1

The *Tam1* (Transposon Antirrhinum *majus*) element of *A. majus* shares several structural features with *Spm* elements: (*a*) 12/13 bp of the inverted repeat termini are identical (9); (*b*) both generate 3-bp direct repeats of host sequence upon insertion (9); and (*c*) the major *Tam1* transcript is also spliced from a large pre-mRNA, and the size and organization of *Tam1* exons are similar to those of *Spm* (Figure 1a) (101). Although protein products of *Tam1* have not yet been characterized, an open reading frame with regions homologous to ORF1 of *Spm* at the DNA (56%) and amino acid (50–70%) levels occurs in a large intron of the major *Tam1* mRNA (Figure 1a) (101). Conservation of these ORF1 sequences may indicate that they encode important structural domains of the transposase (101).

Ac/Ds

The *Ac/Ds* family is comprised of the 4.6-kb *Activator (Ac)* element and a heterogeneous group of *Dissociation (Ds)* elements (30, 106). *Ac* elements encode everything necessary for their own transposition and for the transposition of the non-autonomous *Ds* elements (65). *Ac* and *Ds* elements have 11-bp imperfect inverted repeats at their termini, flanked by an 8-bp direct repeat of host sequence generated upon insertion (30, 106). Part or all of this direct repeat remains after *Ac* or *Ds* excision (76, 106, 118). *Ds* elements fall into three distinct structural categories: (*a*) deletion derivatives of *Ac* (30); (*b*) *Ds2* elements, which are approximately 1.5 kb long, contain less than 1 kb of the *Ac* terminal regions, and have internal sequences unrelated to *Ac* (69, 110); and (*c*) *Ds1* elements, which are approximately 400 bp long (87, 106, 118). *Ds1* homology to *Ac* includes only the terminal inverted repeats and ≈20 bp of sequence adjacent to the 3' terminus of *Ac* (76). The 20-bp sequence contains splice donor sites (27) (S. Wessler, in preparation) and one copy of a transposase binding sequence (53). The remainder of *Ds1* sequences contain approximately 75% A and T residues and are unrelated to *Ac* sequences.

Ac encodes a 3.5-kb transcript that is spliced from a larger pre-mRNA (54). This transcript contains one long open reading frame (ORFa) that encodes the *Ac* transposase (103). The ORFa protein, overproduced in a baculovirus vector (46), binds in vitro to a subterminally repeated *Ac* sequence, 5'-CCGTTT-3' (53).

Tam3

Tam3 elements share several structural features with *Ac* (Figure 1b): (*a*) the 7 bp at the ends of both elements are identical (100); (*b*) both generate 8-bp direct repeats upon insertion (100); and (*c*) *Tam3* contains a single intronless transcription unit that, when compared to *Ac,* has colinear regions with 50–65% amino acid identity (101).

Mutator

Robertson first characterized lines of maize that had a dramatically increased forward mutation rate ($\approx 10^{-4}$) (83). Most of the mutations isolated from these *Mutator* lines are genetically unstable and are caused by the insertion of a family of elements that have highly conserved terminal inverted repeats of approximately 200 bp (3, 4, 104, 107). At least eight *Mu* elements *(Mu1–8),* which differ in their internal sequences, have been described (3, 17, 88, 89, 107, 108). These *Mu* elements generate 9-bp target duplications upon insertion (4).

The autonomous member of the *Mu* element family has not yet been cloned. For this reason, there is no molecular description of a transposase or other element-encoded factors that might bind to the *Mu* termini.

Retro-elements

Some stable mutations in several plant species have been shown to result from the insertion of large, retrovirus-like elements (25, 45, 49, 50) (M. Varagona, M. Purugganan, and S. Wessler, in preparation). These elements have been defined on the basis of their resemblance to retro-elements found in mammals, *Drosophila*, and yeast (Figure 1d) (5, 7, 111). Most retro-elements have structural features presumably required for their transposition via an RNA intermediate. Two of these features are long terminal direct repeats (LTRs) and a tRNA-like site necessary for priming reverse transcription.

INSERTION INTO EXONS AND INTRONS

The excision of transposable elements from scorable genes is a dynamic process easily visualized as patterns of somatic mutability in plant tissues. A far less visible but equally dynamic situation involves the processing of transposable element and DNA insertion sequences from pre-mRNA. Many recent studies have focused on the alterations in RNA processing caused by the insertion of *Ac/Ds*, *Spm/dSpm*, and *Mu* elements into exons or introns (27, 52, 81, 84, 98, 112, 115, 116). The results of these studies are summarized in Figure 2 and are considered below.

Spm/dSpm

dSpm alleles of several loci, including *bz1, a1,* and *a2* (Table 1) have intermediate or wild-type phenotypes if an autonomous *Spm* element is not also in the genome. An *Spm* element can, in *trans,* suppress this residual expression and promote excision of the *dSpm* element. Studies on the *bz-m13* allele demonstrated that despite the presence of a 2.2-kb *dSpm* element inserted antiparallel in exon 2 of the *bz* gene, kernels had 5–10% of nonmutant UFGT activity and contained *bz* transcripts produced by alternative splicing of the normal donor site of intron one (D) to either of two acceptor sites within the *dSpm* element (A1 or A2, Figure 2) (81). Splicing to A1, located 2 bp within the *dSpm* terminus, produces an in-frame transcript that is probably responsible for the UFGT activity (52, 81). The transcript resulting from D-to-A2 splicing would not be expected to encode an active UFGT enzyme. The *bz-m13CS9* allele is a derivative of *bz-m13* that has sustained a 1340-bp deletion in the *dSpm* element (85), which removes the unproductive A2 acceptor site (81). More frequent use of A1 is probably responsible for the higher level of UFGT expression (67% of wild-type) encoded by this allele. This *dSpm* element also excises infrequently (85), probably because the deletion that removes A2 also removes part of the subterminal repeats, which have been implicated as a *cis*-requirement for transposition (92). Raboy et al (81) have suggested that *bz-m13* and its derivative *bz-m13CS9* may illustrate how a transposable element evolves into an intron. The 1340-bp deletion makes the *bz-m13CS9* element more intron-like than its 2.2-kb progenitor in two ways: it permits a higher level of UFGT activity and it transposes at a lower frequency.

Whether insertion of *dSpm* is parallel or antiparallel has a dramatic effect

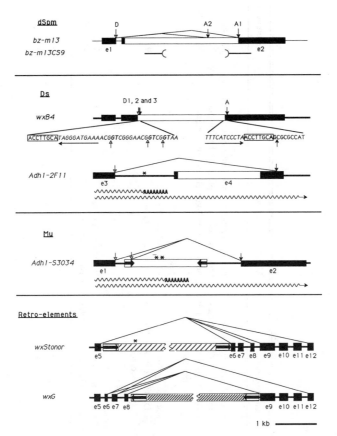

Figure 2 Splicing of transposable elements from transcripts. Exons are shown as black boxes and designated e1, e2, etc. Inserted elements are shown as white or shaded boxes. Splice donor sites are indicated by vertical open arrows, splice acceptor sites by vertical closed arrows, and polyadenylation sites by asterisks.

on posttranscriptional processing. The *bz-m13* alleles illustrate that *dSpm* elements of 2.2 kb or smaller, inserted antiparallel, permit transcriptional readthrough and may subsequently be spliced (52, 81). In contrast, the *wx-m8* allele contains the identical 2.2-kb *dSpm* element in the parallel orientation in exon 10 of the *waxy (wx)* gene (93, 94). In the absence of *Spm*, transcription initiated at the *wx* promoter terminates within the element, producing a truncated polyadenylated transcript (43). An interesting observation is that in the presence of *Spm* in *trans,* both the truncated *wx-m8* transcripts (43) and the leaky expression of the *bz-m13* alleles are suppressed. These results provide a molecular explanation for the suppressor function of *Spm*. It has

534   WEIL & WESSLER

Table 1 Genes discussed in this review

Name	Gene product	Mutant phenotype[a]	References
Maize			
alcohol dehydrogenase 1 (Adh1)	alcohol dehydrogenase	cannot germinate under anaerobic conditions	37, 40
anthocyaninless (A1)	dihydroflavonol-4-reductase	colorless aleurone; green or brown plant	71
bronze (Bz1)	UDP-Glucose : flavonoid glucosyltransferase (UFGT)	bronze-colored kernel and plant parts	34
high chlorophyll fluorescence 106 (hcf*106)	—	pale green plant	57
shrunken (Sh1)	sucrose synthase	collapsed endosperm	12
waxy (Wx)	UDP-Glucose starch glucosyltransferase	waxy endosperm	97
Antirrhinum			
nivea (niv)	chalcone synthase	colorless flowers	120
pallida (pal)	dihydroflavonol-4-reductase	ivory flowers	59

[a] Phenotypes are from Ref. 19

been proposed that *sp* function results from the binding of TNPA protein to the ends of *dSpm* elements, thus preventing transcription from continuing through element sequences (41). This model for *sp* function predicts that insertions into untranscribed regions need not be suppressible. This is indeed the case for *a1-m2* that contains a *dSpm* inserted into 5' flanking sequences of the *a1* gene (63, 96).

Ac/Ds

Although there are few leaky *Ds* mutations, there are five examples of *Ds* alleles that encode wild-type-sized proteins and mRNAs despite insertions in exons (28, 76, 91, 116, 117). Four of these alleles, *wxB4, adh1Fm-335, wx-m9,* and *wx-m1* (Table 2), utilize a similar mechanism to remove most of the *Ds* sequences from pre-mRNA (Figure 2). Unlike *dSpm* splicing from *bz-m13*, where the element is removed with the *bz* intron, the splicing of *Ds* sequences involves the creation of a new intron. In each case donor sites located adjacent to the inverted repeat (Figure 2, D1, 2, 3) are spliced to acceptor sites that are either part of the 8-bp target duplication or are adjacent to this sequence. The splicing of *Ds* from pre-mRNA does not produce normal *Wx* or *Adh1* transcripts; in all cases part of the *Ds* terminus persists in the mature transcript (27, 116) (S. Wessler, in preparation). However, it has been noted that alternative splicing of D1, D2, and D3 to the same acceptor generates three different reading frames (117). This feature may increase the likelihood of producing a functional protein.

What is perhaps most remarkable about *Ds* splicing is that a similar mechanism underlies the splicing of highly diverse *Ds* elements. Whereas *wx-m9* and *wxB4* contain *Ac*-like and *Ds2* elements (30, 110), respectively, both *wx-m1* and *adh1-Fm335* contain *Ds1* elements (106, 116). As mentioned above, *Ds1* elements only share about 35 bp of homology with *Ac,* including the inverted repeat termini and most of the 20-bp sequence that contains the 3 splice donor sites (76, 106). For the *wx-m1* allele, all three donor sites are utilized in vivo (S. Wessler, in preparation). Conservation of only the donor splice sites suggests that the ability to be spliced may confer a selective advantage on *Ds* elements. Alternatively, the 20-bp sequence containing the donor sites may contain *cis*-elements required for transposition. In this regard, it may be significant that an *Ac* transposase binding sequence is present within this 20-bp region (53).

Splicing of the *Ds2* element in the *adh1-2F11* allele does not utilize donor sites within the element as in the *Ds* examples mentioned above (Figure 2) (98). The presence of a 1.3-kb *Ds2* element in exon 4 leads to the splicing of the normal *Adh1* donor from intron 3 to a cryptic acceptor downstream of the insertion site (Figure 2). Less abundant poly A+ transcripts that terminate within intron 3 are also detected.

The effect of element orientation on RNA processing events is the same for

Table 2 Alleles caused by plant transposable element insertions within genes

Allele	Element	Orientation	Position	Comments	References
bz1-m13	dSpm	antiparallel[a]	exon 2	5–10% wild-type activity levels; Spm-suppressible; element spliced from pre-mRNA; acceptor site within element	81, 85
bz1-m13CS9	dSpm	antiparallel	exon 2	deletion derivative of bz-m13; 69% wild-type activity; smaller, fewer sectors	52, 81, 85
wx-m8	dSpm	parallel	exon 11	transcripts terminate within element	43, 94
wxB4	Ds2	antiparallel	exon 13	element spliced from pre-mRNA; no Wx activity	110–119
adh1-Fm335	Ds1		exon 4	element spliced from pre-mRNA; 10% wild-type levels of Adh activity	27, 76, 106
wx-m9	Ds	antiparallel	exon 10	deletion derivative of Ac; 10% of wild-type Wx activity; element spliced from pre-mRNA	30, 80, 116
wx-m1	Ds1		exon 9	element spliced from pre-mRNA; no Wx activity	118[b]
adh1-2F11	Ds2	antiparallel	exon 4	element spliced from pre-mRNA with upstream and down-stream sequences, no splicing signals within element	28, 69, 98

wx-m6	Ds	parallel	exon 8	transcripts terminate within element	30[b]
wxB3	Ac	parallel	exon 10	transcripts terminate within element	[c]
adh1-S3034	Mu1		intron 1	20–40% wild-type activity; 40% wild-type mRNA; element removed with intron, alternative splicing detected	84, 104, 112[d]
adh1-S4477	Mu1		intron 1	50–70% wild-type activity; insertion 341 bp downstream of adh1-3034, element in same orientation	84[d]
wxStonor	retro	parallel	intron 6	element spliced from pre-mRNA, alternative splicing detected; leaky activity results from splicing	119[e]
wxG	retro	antiparallel	intron 8	element spliced from pre-mRNA, alternative splicing detected; leaky activity results from splicing	119[e]

[a] Elements inserted in the parallel orientation are transcribed in the same direction as the host gene; elements inserted in the antiparallel orientation are transcribed in the opposite direction.
[b] S. Wessler, unpublished
[c] G. Baran et al, unpublished
[d] D. Ortiz and J. Strommer, in preparation
[e] M. Varagona, M. Purugganan, and S. Wessler, in preparation

dSpm and *Ac/Ds* elements. Like *dSpm*, *Ds* elements inserted in the antiparallel orientation are spliced whereas parallel insertion results in truncated transcripts. In two instances parallel insertion of a 2.2-kb *Ds* element *(wx-m6)* and a 4.6-kb *Ac* element *(wxB3)* into exons of the *wx* gene result in transcripts that terminate at polyadenylation sites within the respective elements (G. Baran and S. Wessler, unpublished). The effects of *Ds* insertions on transcription and RNA processing are not affected by the presence or absence of *Ac* elements in the genome. This is in marked contrast to the effect in *trans* of *Spm* on *dSpm* insertions but is consistent with genetic evidence that *Ac* elements lack suppressor function.

Mutator

Mu1 insertions into the first intron of the *Adh1* gene have been characterized in several labs. The *Mu* allele *Adh1-S3034* has 20–40% of Adh activity and 40% of wild-type levels of *Adh1* mRNA despite the presence of a *Mu1* element 71 bp from the donor splice site of intron 1 (Figure 2) (38, 104). Two studies of *Adh1-S3034* transcription produced disparate results concerning the effect of the insertion on transcription initiation and processing. Vayda & Freeling (112) concluded that the *Mu* insertion had no effect on chromatin structure or transcription initiation but, in some manner, impeded transcript elongation. In contrast, Rowland & Strommer (84) measured relative levels of run-off transcripts and concluded that the insert affected the level of transcription initiation and did not affect RNA processing.

Recently Ortiz and Strommer (personal communication) sequenced cDNAs from *Adh1-S3034* and characterized several transcripts that arise by alternative processing. In most cases the *Mu* element is spliced with the surrounding intron, producing a wild-type transcript. However, transcripts truncated by polyadenylation within *Mu* were also detected (Figure 2). Preliminary characterization of another *Mu* allele, *Adh1-S4477*, indicates that this allele may have a higher level of ADH activity than *Adh1-S3034* (50–70% for *Adh1-S4477* vs 20–40% for *Adh1-S3034*) because there is more transcriptional readthrough and less polyadenylation within the element (84) (D. Ortiz and J. Strommer, personal communication). Although both alleles contain insertions in intron 1 that are in the same orientation, the *Mu1* element in *Adh1-S4477* resides 341 bp downstream from the *Adh1-S3034* insertion site. Ortiz and Strommer speculate that the shorter distance between the *Adh1-S4477* element and the splice acceptor site favors splicing over polyadenylation within *Mu* sequences and results in more wild-type transcripts. Competition between splicing and polyadenylation may also explain why the *adh-2F11 (Ds)* allele encodes transcripts truncated within the intron preceding the insertion site (Figure 2) (98). Failure to recognize the normal splice acceptor may delay the processing of intron 3 and provide the time necessary to recognize cryptic polyadenylation sites in this intron.

Retro-elements

Four leaky mutations of the maize *wx* gene are caused by large insertions of retrovirus-like elements into the transcription unit (119) (M. Varagona, M. Purugganan, and S. Wessler, in preparation). Transcripts from two of these alleles, *wxStonor* and *wxG,* have been characterized in detail, and the results are summarized in Figure 2. These alleles contain different elements that are inserted in opposite orientations relative to *wx* transcription. In addition, the *wxStonor* element is inserted in the splice acceptor site of intron 5, whereas the *wxG* element is inserted in the middle of intron 8. Despite these differences, the processing of these elements from pre-mRNA shares two features. First, wild-type transcripts are produced by the splicing of the element with the surrounding intron. Second, the presence of the elements leads to alternative splicing of *wx* pre-mRNA (Figure 2). Alternative splicing generally involves the use of natural *wx* donor and acceptor splice sites and results in the loss of some exons from mature transcripts.

Although there are a few examples of *Drosophila* and mammalian elements that are spliced with the surrounding introns, the remarkable diversity of splicing events in maize is unprecedented. The splicing of maize transposable and retro-elements from pre-mRNA occurs whether insertion is into exons or introns and involves the use of natural and cryptic donors and acceptors in the element and the gene. Differences between plant and animal systems may indicate that the *cis*-requirements for splicing of plant introns are more relaxed. In support of this notion, Goodall & Filipowicz (44) have demonstrated that plant introns, unlike animal introns, do not require a stretch of polypyrimidine residues prior to the acceptor site.

INSERTIONS INTO 5' FLANKING SEQUENCES

Interactions between transposable elements and the regulatory sequences of genes can lead to alterations in the level of transcription and the temporal and spatial distribution of transcripts. A gene's perception of *trans*-acting regulatory proteins can be altered when critical *cis*-elements in 5' flanking sequences are added, deleted, or rearranged by the insertion or excision of transposable elements.

Antirrhinum majus

The *pallida (pal)* and *nivea (niv)* loci encode enzymes in the anthocyanin biosynthetic pathway (Table 1). The ability to detect subtle changes in the intensity and/or distribution of anthocyanin pigments in the flowers of *A. majus* has facilitated the identification of many interesting mutations. Of four well-characterized *Tam*-induced mutant alleles, three are caused by insertions into the promoters of the *niv* and *pal* genes (Table 3) (8, 59, 100). It has been

Table 3 Alleles caused by plant transposable element insertions 5' of the target gene

Allele	Element	Orientation	Position	Comments	References
niv-rec53	Tam1	Antiparallel	-47	insertion is upstream of TATA box; null with full-red revertant spots	8
niv-5311				pale color; Tam1 excised leaving 66-bp deletion that includes TATA box; transcription reduced 50-fold and starts 20 bp farther downstream	99
niv-5312				pale color; Tam1 excised leaving 28-bp deletion extending up to but not including TATA box; transcription reduced 4-fold; uses normal initiation site	99
niv-5313				pale color; Tam1 excised leaving 5-bp deletion and 2-bp insertion; transcript reduced 65%; uses normal initiation site	99
niv-5314				pale color; Tam1 excised leaving 15-bp deletion; transcription reduced 74%; uses normal initiation site	99
niv-46	Tam1	Antiparallel	-47	5 bp deleted from distal terminus of element; stable; very pale pigmentation	47
pal-rec2	Tam3	Antiparallel	-70	insertion 41 bp upstream of TATA box; 5-bp direct duplication of target site; null with full-red revertant spots	21, 59
pal-32 pal-33 pal-35G				Tam3 excised leaving 19-, 10-, and 9-bp deletions, respectively; less pigment expressed in floral lobes than in floral tubes	21, 22
pal-518				same as pal-32, etc, but leaves 13-bp deletion to left of Tam3 excision site; pigment wild-type in floral tube, reduced in floral lobes	21, 22
pal-15				98-bp deletion and 7 bp added from 125 upstream of pal; pigment restricted to base of floral tube	21, 22

Allele	Element	Orientation	Position	Description	Reference
pal-41				Tam3 excised and all DNA 5' of insertion site replaced with DNA from 5 map units upstream; pigment of floral tube has unusual pattern	21, 82
pal-42	Tam3	Antiparallel	-70	same rearrangement as in pal-41 but Tam3 element still present and 17 bp of proximal element terminus and 8 bp of adjacent DNA deleted; some pal expression	22, 82
pal-510	Tam3	Antiparallel	-70	3 bp deleted from proximal terminus of Tam3; some pal expression restored	21, 22
Bz-wm	Ds1		-63	mutable, leaky bz; also has 3-bp insertion in exon 2 from previous Ac excision	85, 86
Bz' (wm)-1				Bz revertant of Bz(wm); Ds1 excised leaving 6-bp insertion	105
Bz' (wm)-2				Bz revertant of Bz(wm); Ds1 excised leaving 8-bp insertion	105
Adh1-3F1124	Mu3		-31	tissue-specific decrease in Adh1 expression	16, 17
sh1-9026	Mu1		-2	mutable, leaky; transcription initiates within distal Mu1 end	72
hcf*106	Mu1		-1	pale green, photosynthesis deficient; suppressed when Mu1 methylated, and hcf*106 transcription initiates in proximal Mu1 end; orientation opposite Mu1 of sh-9026	57, 58[a]
a-m2 (7991A1)	Spm	Antiparallel	-99	expression dependent on Spm activity; insertion 15 bp upstream of presumed CAAT box	62
a-m2 (8004)	dSpm	Antiparallel	-99	deletion derivative of Spm at a-m2: reduced a expression, still dependent on Spm activity	62
a-m2 (8167B1)	Spm	Antiparallel	-99	reduced a expression; no deletion of Spm sequences; effects on a expression probably due to methylation of Spm	1, 62

[a] A. Barkan, personal communication

suggested that *Tam3* might show target site preference for a short sequence found in the promoter region of these coordinately induced genes (21).

In most cases excision of the *Tam*1 and 3 elements from the *niv* and *pal* promoters restores wild-type full-red flowers (9, 60, 99, 100). However, selection for rare mutant phenotypes has led to the isolation of excision events that produce more drastic alterations in promoter sequences. The *niv-rec53* : :*Tam1* allele has a *Tam1* element inserted at position −47 (8, 9). Four pale-red revertants of *niv-rec53* : :*Tam 1* (*niv-5311* through *niv-5314*, Table 3) have sustained deletions at the *Tam1* excision site and have reduced levels of *niv* mRNA (47, 99). The palest allele, *niv-5311,* has a 66-bp deletion that removes the TATA box and reduces the level of steady-state transcripts by 50-fold. Although the wild-type transcription start is still present, transcripts initiate 20 bp downstream of the normal transcription start. Derivatives containing deletions of 5, 15, and 28 bp upstream of the TATA box exhibit the normal transcription start and have 25–35% of wild-type levels of *niv* transcripts. These data led Sommer et al (99) to suggest that the TATA box may not only specify the transcription initiation site but may also influence the frequency of initiation.

A similar allelic series has been derived from the *pal-rec2* : :*Tam3* allele by excision of *Tam3* from its position approximately 70 bp upstream from the *pal* transcription start (21). The new alleles, *pal-32, -33, -35G,* and *-518* (Table 3), have reduced levels of *pal* transcript and contain small deletions at the site of *Tam3* excision (21). These deletions remove a sequence, CACGTG, which is specifically recognized by a nuclear protein, CG-1, that is conserved in several plant species, including *A. majus* (102). Unlike the *niv* deletions mentioned above, these *pal* derivatives display alterations in the spatial distribution of pigmentation and *pal* mRNA; there is a more severe reduction in floral lobes than in floral tubes (22). The disproportionate reduction in *pal* transcripts suggests that there are independent *cis*-elements in *pal* upstream sequences regulating transcription initiation in these floral parts (22). The 100-bp deletion of the *pal-15* derivative (Table 3) apparently removes both *cis*-elements because neither floral part is pigmented (22). Differential regulation of lobe and tube expression is also inferred from the *delila* mutation, which acts in *trans* to reduce *pal* and *niv* transcripts in tubes but has no effect on lobe expression (22).

More complex rearrangements resulting from *Tam3* excision can also produce new *pal* alleles with novel pigmentation patterns. In the *pal-41* allele, *Tam3* and all sequences 5' of the insertion site have been replaced by sequences from 5 map units upstream (21, 82). Despite these gross alterations, *pal* transcription is 25% of wild-type levels and initiates at the normal start (82). The molecular basis for the novel pattern of anthocyanin distribution remains unclear, because the foreign sequences near *pal* do not contain a floral promoter.

The presence of *Tam1* and *Tam3* elements in the promoters of the *pal* and *niv* genes apparently abolishes transcription initiation, because both alleles display a null background phenotype. Three derivatives of these alleles, *pal-42*, *pal-510* and *niv-46* (Table 3), have each sustained a very small deletion which includes one terminus of the element (20, 47, 82). In each case the element's ability to transpose has been severely reduced and the block in downstream gene expression has been somewhat relieved. Because these derivatives are expressed, it has been suggested that the element does not passively block gene expression. Rather, downstream gene transcription may be blocked by the formation of a transposition complex that contain both termini brought together by the transposase or some other DNA binding protein(s) (22). The *niv-46* derivative of *niv53 : : Tam1* represents the strongest argument in favor of this model. This allele has some *niv* expression despite the fact that it differs from its null progenitor by a 5-bp deletion in the upstream terminus; the deletion is about 14 kb from the *niv* transcription start. Further support for the role of transposition complexes in gene suppression may also come from studies of reversibly inactive elements (see below).

Maize

There are several examples of tissue specific alterations in gene expression following insertion of *Ds* or *Mu* elements into promoter sequences. The presence of a 406-bp *Ds1* element at position -63 in the *Bz (wm)* allele reduces *bz* transcripts to 6% and 46% of wild-type levels in husks and seedlings, respectively (86, 105). Excision of *Ds1* produced two new alleles that continue to display tissue-specific effects. *Bz'*(wm)-1 and *Bz'*(wm)-2 have 6- and 8-bp transposon footprints at the site of *Ds1* excision. Both show reduced steady-state *bz* mRNA in husks compared to seedlings; however, the *Bz'*(wm)-2 has 8 times more *bz* transcripts in husk tissues. Sullivan et al (105) have suggested that this region may contain a *cis*-element that regulates husk expression. Furthermore, they speculate that the addition of 8 bp results in higher expression than 6 bp because 8 bp is closer in size to a full helical turn. With the recent cloning of two putative transcriptional activators that regulate *bz* expression (the products of the *c1* and *R* genes) (23, 26, 56, 74, 75) it should be possible to test whether these proteins bind to this region.

Of over 100 *Adh1* mutants characterized to date, only the *Adh1-3F1124* allele displays altered organ-specific expression (16). The unstable mutant *Adh1-3F1124* conditions normal ADH1 activity and *Adh1* mRNA in pollen but only 6% of wild-type levels in seed and anaerobically induced seedlings (16, 17). *Adh1-3F1124* results from insertion of a 1.85-kb *Mu3* element at -31 of the *Adh1* gene (17). The host sequences duplicated upon insertion contain the TATA box, which now flanks the *Mu3* element. Although it is known that *cis*-elements upstream from the site of *Mu3* insertion are required for correct *Adh1* expression (29, 113), the effect of the *Mu* sequences on transcription initiation remains unknown.

Insertion of *Mu1* elements into 5' flanking sequences can lead to transcription initiation at sites within the conserved *Mu* termini. The *sh-9026* allele contains a *Mu1* element at −2 of the *sh* gene and encodes *sh* transcripts of a 4.4 and 3.0 kb (72). The 4.4-kb transcript initiates 8 bp inside the *Mu* element and probably contains all of *Mu* (1.4 kb) fused to a normal *sh* mRNA (3.0 kb). Thus, *sh* regulatory sequences upstream of the insertion site are probably promoting transcription initiation within the distal *Mu* terminus. Consistent with this interpretation is the observation that the chimeric transcript, like the normal *sh* transcript, is anaerobically induced.

The *hcf*106* allele contains a *Mu1* insertion at position −1 of a gene that affects chloroplast biogenesis (57) (A. Barkan, personal communication). Unlike *sh-9026*, sequences within *Mu* presumably promote transcription initiation within the proximal *Mu* terminus, resulting in a transcript of near-wild-type size (A. Barkan, personal communication). Although both *hcf*106* and *sh-9026* contain *Mu1* insertions, their opposite orientation relative to the adjacent transcription unit may be responsible for the dramatically different transcription starts within *Mu* sequences.

The promoter insertions best characterized genetically are the *a-m2* allele and its derivatives (Table 3). In contrast to the situation in most mutations containing *dSpm* insertions, where residual gene expression is suppressed in *trans* by *Spm*, *a-m2* expression is apparently induced (66). The antiparallel insertion of a *dSpm* element at −99 of the *A1* gene displaces the normal *A1* upstream sequences with sequences upstream of the major *Spm* transcription unit (63, 96). The *dSpm* derivatives have a colorless phenotype; however, the presence of *Spm* in *trans* results in anthocyanin accumulation. Masson et al (63) have suggested that *A1* gene expression conditioned by *a-m2* alleles results in part from the binding of an *Spm*-encoded product to sequences upstream of the *Spm* transcription start site. However, sequences within exon 1 of *Spm* also regulate *A1* expression, because deletion of these sequences, but not the *Spm* upstream region, results in the reduced *A1* expression displayed by the *a-m2(8004)* allele (63).

EPIGENETIC CHANGES

Changes in the phenotype of an allele with no alteration of its genotype are referred to as epigenetic (48). The reversible inactivation of plant transposable elements represents an example of epigenetic change and has been correlated with methylation of element sequences (1, 14, 18, 23, 61, 90). Similar correlations between the methylation state of a gene and its activity have been documented in nonplant species. For example, inactivation of mammalian X chromosome genes is followed by increased DNA methylation on the inactive X (55, 109, 121). Furthermore, some mammalian genes expressed in a tissue-specific manner are methylated in gametes and in early development and are demethylated in tissues where they are expressed (6, 39). In plants, the correlation between DNA methylation of the maize elements *Spm, Mu,* and *Ac,* and the inactivation of these elements, represent the only extensively documented cases relating DNA modification and gene expression. Methylation of transposable element sequences not only influences element expression but can also affect expression of adjacent genes.

As described in the previous section, the *Spm*-dependent expression of the *a-m2* allele allows the isolation of colorless derivatives resulting from deletions in *Spm* sequences (63). In contrast with these *dSpm* alleles, another class of colorless derivatives arise when the *Spm* element at *a* switches to an inactive phase (63, 66). *Spm* inactivation is associated with methylation of element DNA (1) and is reversible. Alleles of *a-m2* that contain an inactive *Spm* element resemble *dSpm*-containing alleles in two ways: both *a* gene expression and element excision occur only in the presence of a *trans*-acting *Spm* or *Spm-w* element. However, unlike *dSpm* alleles, *a-m2* alleles caused by *Spm* inactivation can regain *a* gene expression when the element cycles back to an active phase.

The observation that either deletion or methylation of *Spm* sequences can lead to *a-m2* derivatives with reduced *A1* gene expression suggests that the affected sequences may influence transcription initiation. It has been proposed that these sequences bind an *Spm*-encoded positive regulatory factor that activates both *Spm* and *a* transcription (63). For two reasons, the best candidate for such a factor is the TNPA protein: (*a*) *Spm-w* elements, which encode only TNPA, can *trans*-activate *a* expression from *a-m2* alleles (63, 66); and (*b*) sequences that bind TNPA in vitro are specifically methylated when *Spm* is inactive (1)—binding that is inhibited by DNA methylation (42). Thus far, *a* gene expression has been estimated indirectly from the intensity of kernel and plant pigmentation. Proof that *Spm* sequences act as enhancers of *a* gene expression will require direct measurements of transcripts from the *a-m2* alleles.

In contrast with the *a-m2* alleles, inactivation of a transposable element can suppress the mutant phenotype caused by its insertion. The *hcf*106* mutation of maize conditions a pale green plant deficient in photosynthesis (2). As mentioned earlier, this allele contains a *Mu1* element inserted at position −1 of a gene that normally encodes a 1.2-kb transcript (57) (A. Barkan, personal communication). No transcript is detected in pale green, mutant tissue. Dark green sectors have been observed on the leaves of mutant plants, and analysis of these sectors shows that the *Mu* element is still present but is now methylated (58). In addition, transcription of *hcf*106* is restored in the dark green sectors; however, the position of the transcription start is displaced upstream to a site within the proximal end of the *Mu1* element (A. Barkan, personal communication). These data suggest that *Mu1* sequences may now serve as a new *hcf*106* promoter when the element is methylated. Precisely how methylation of the *Mu1* element increases *hcf*106* expression is not known, but parallels with the *A. majus* alleles *pal-42, pal-510,* and *niv-46* may be relevant. In those examples, deletion of sequences at the element termini permit expression of *pal* or *niv*, possibly by preventing the ends of the inserted *Tam* element from coming together (22). Methylation of the element at *hcf*106* may also prevent interactions of the *Mu* ends, freeing *Mu* sequences to promote *hcf*106* transcription.

The *Mu* element may also allow *hcf*106* to respond to a pattern of developmental cues normally occurring during plant development but not normally recognized by the gene. For example, Martienssen et al (58) have observed that in some *hcf*106* plants the proportion of dark green tissue in a leaf increases with the developmental age of that leaf, eventually resulting in a leaf that appears to be wild-type. These data indicate that methylation of the *Mu* element at *hcf*106* increases during plant development and that the expression pattern of *hcf*106* in all leaves is subverted by this pattern of increasing methylation.

Similar epigenetic changes appear to underlie the unstable phenotypes associated with other alleles in maize. Examples include transposable element alleles of the *vp1* (64) and *R* (15) loci (*vp1-mum2* and *R-m*, respectively), which display concentric sectors of colorless and pigmented aleurone tissue. This phenotype is more consistent with epigenetic changes than with a series of excision and reinsertion events. Reversible epigenetic changes in other plant species may also be the products of the interaction between transposable elements and adjacent genes. Genetic studies in soybean suggest that variegated and solid seed-coat phenotypes interconvert, resembling the cycling phase changes of maize transposable elements (13).

CONCLUDING REMARKS

The insertion of transposable elements into plant genes sets the stage for a variety of interactions that can lead to alterations in normal transcriptional processes. The complex nature of these interactions, as summarized in this review, reflects the insertion of one intricate set of regulatory and processing signals into another. Although the interactions appear on the surface to be diverse, common underlying themes are beginning to emerge. Knowledge of the position of insertion (5' flanking sequences, intron or exon), the relative orientation of element and gene, and the methylation state of the element can lead to educated guesses regarding changes in the expression pattern of the adjacent gene.

For example, *Ds* elements inserted into exons in antiparallel orientation are usually spliced in a similar manner and frequently lead to near-wild-type transcripts. *Ds* is not unique in this regard, as recent studies indicate that *Mu*, *dSpm*, and retro-elements can also be processed from transcripts. This phenomenon has thus far been demonstrated only in maize. This may be a statistical accident or may reflect fundamental differences in splicing requirements in this species or in the selection pressure on maize transposable elements.

Studies on eukaryotic-gene regulation indicate that the overall pattern of transcription initiation results from the additive effects of *cis*-regulatory sequences. This building-block approach to gene regulation explains why transposable elements can reduce or alter transcription initiation and produce new spatial patterns of expression. Excision of transposable elements from promoters can delete or rearrange *cis*-regulatory sequences. In addition, transposon insertion into 5' flanking sequences can place the terminally located element promoters near the TATA box of normal genes. This juxtaposition can dramatically alter gene expression because the protein binding sites involved in the transcription and transposition of the element can now become new *cis*-acting modules that influence the expression of the adjacent gene.

The finding that changes in the methylation of element sequences during development affects adjacent gene expression suggests that the transcription complex can sense changes in methylation and demonstrates that this may be a common mechanism regulating plant gene expression. In this regard it is not surprising that epigenetic changes in the expression of certain normal genes during plant development [such as imprinting (51) and paramutation (10) displayed by certain alleles of the maize *R* locus] resemble the epigenetic changes plant elements impose on adjacent genes.

Thus, most transcriptional interactions between elements and genes are probably governed by processes common to normal plant genes. Molecular and genetic characterization of these underlying mechanisms has and will continue to assist us in understanding transcription regulation in plants.

Acknowledgments

We thank Alice Barkan, Daniel Ortiz, and Judy Strommer for communicating results prior to publication. We thank Michael Purugganan for critical reading of the manuscript. This work was supported in part by NIH grant GM32528 to SRW.

Literature Cited

1. Banks, J. A., Masson, P., Fedoroff, N. 1988. Molecular mechanisms in the developmental regulation of the maize *Suppressor-mutator* transposable element. *Genes Dev.* 2:1364–80
2. Barkan, A., Miles, D., Taylor, W. C. 1986. Chloroplast gene expression in nuclear photosynthetic mutants of maize. *EMBO J.* 5:1421–27
3. Barker, R. F., Thompson, D. V., Talbot, D. R., Swanson, J., Bennetzen, J. L. 1984. Nucleotide sequence of the maize transposable element *Mu1*. *Nucleic Acids Res.* 12:5955–67
4. Bennetzen, J. L., Swanson, J., Taylor, W. C., Freeling, M. 1984. DNA insertion in the first intron of maize *Adh1* affects message levels: cloning of progenitor and mutant *Adh1* alleles. *Proc. Natl. Acad. Sci. USA* 81:4125–28
4a. Berg, D. E., Howe, M. M., eds. 1989. *Mobile DNA*. Washington, DC: Am. Soc. Microbiol.
5. Bingham, P. M., Zachar, Z. 1989. Retrotransposons and the FB transposon from *Drosophila melanogaster*. See Ref. 4a, pp. 485–502
6. Bird, A. P. 1987. CpG islands as gene markers in the vertebrate nucleus. *Trends Genet.* 3:342–47
7. Boeke, J. D. 1989. Transposable elements in *Saccharomyces cerevisiae*. See Ref. 4a, pp. 335–74
8. Bonas, U., Sommer, H., Harrison, B. J., Saedler, H. 1984. The transposable element *Tam1* of *Antirrhinum majus* is 17 kb long. *Mol. Gen. Genet.* 194:138–43
9. Bonas, U., Sommer, H., Saedler, H. 1984. The 17-kb *Tam1* element of *Antirrhinum majus* induces a 3-bp duplication upon integration into the chalcone synthase gene. *EMBO J.* 5:1015–19
10. Brink, R. A. 1973. Paramutation. *Annu. Rev. Genet.* 7:129–52
11. Brink, R. A., Williams, E. 1973. Mutable *R-Navajo* alleles of cyclic origin in maize. *Genetics* 73:273–96
12. Burr, B., Burr, F. A. 1982. *Ds* controlling elements of maize are large and dissimilar insertions. *Cell* 29:977–86
13. Chandlee, J. M., Vodkin, L. O. 1989. Unstable expression of a soybean gene during seed coat development. *Theor. Appl. Genet.* 77:587–94
14. Chandler, V. L., Walbot, V. 1986. DNA modification of a maize transposable element correlates with loss of activity. *Proc. Natl. Acad. Sci. USA* 83:1767–71

15. Chang, M. T., Neuffer, M. G. 1987. *Mr R-m* controlling element system in maize. *J. Hered.* 78:163–70
16. Chen, C.-H., Freeling, M., Merckelbach, A. 1986. Enzymatic and morphological consequences of *Ds* excision from maize *Adh1*. *Maydica* 31:93–108
17. Chen, C.-H., Oishi, K. K., Kloeckner-Gruissem, B., Freeling, M. 1987. Organ-specific expression of maize *Adh1* is altered after a *Mu* transposon insertion. *Genetics* 116:469–77
18. Chomet, P. S., Wessler, S., Dellaporta, S. L. 1987. Inactivation of the maize transposable element *Activator (Ac)* is associated with its DNA modification. *EMBO J.* 6:295–302
19. Coe, E. J. Jr., Neuffer, M. G., Hoisington, D. A. 1988. The genetics of corn. In *Corn and Corn Improvement*, ed. G. F. Sprague, pp. 81–258. Madison, WI: Am. Soc. Agronomy
20. Deleted in proof
21. Coen, E. S., Carpenter, R., Martin, C. 1986. Transposable elements generate novel spatial patterns of gene expression in *Antirrhinum majus*. *Cell* 47:285–96
22. Coen, E. S., Robbins, T. P., Almeida, J., Hudson, A., Carpenter, R. 1989. Consequences and mechanisms of transposition in *Antirrhinum majus*. See Ref. 4a, pp. 413–36
23. Cone, K. C., Burr, F. A., Burr, B. 1986. Molecular analysis of the maize anthocyanin regulatory locus *C1*. *Proc. Natl. Acad. Sci. USA* 83:9631–35
24. Cuypers, H., Dash, S., Peterson, P. A., Saedler, H., Gierl, A. 1988. The defective *En-I102* element encodes a product reducing the mutability of the *En/Spm* transposable element system in *Zea mays*. *EMBO J.* 7:2953–60
25. Day, A., Schirmer-Rahire, M., Kuchka, M. R., Mayfield, S. P., Rochaix, J.-D. 1988. A transposon with an unusual arrangement of long terminal repeats in the green alga *Chlamydomonas reinhardtii*. *EMBO J.* 7:1917–27
26. Dellaporta, S. L., Greenblatt, I., Kermicle, J. L., Hicks, J. B., Wessler, S. R. 1987. Molecular cloning of the maize *R-nj* allele by transposon tagging with *Ac*. *Stadler Symp.* 18:263–82
27. Dennis, E. S., Sachs, M. M., Gerlach, W., Beach, L., Peacock, W. J. 1988. The *Ds1* transposable element acts as an intron in the mutant allele *Adh1-Fm335* and is spliced from the message. *Nucleic Acids Res.* 16:3315–28
28. Doring, H. P., Freeling, M., Hake, S.,

Johns, M. A., Kunze, R., et al. 1984. A *Ds* mutation of the *Adh1* gene in *Zea mays*. *Mol. Gen. Genet.* 193:199–204

29. Ellis, J. G., Llewellyn, D. J., Dennis, E. S., Peacock, W. J. 1987. Maize *Adh1* promoter sequences control anaerobic regulation: addition of upstream promoter elements from constitutive genes is necessary for expression in tobacco. *EMBO J.* 6:11–16

30. Fedoroff, N., Wessler, S., Shure, M. 1983. Isolation of the transposable maize controlling elements *Ac* and *Ds*. *Cell* 35:235–42

31. Fedoroff, N. V. 1983. Controlling elements in maize. In *Mobile Genetic Elements*, ed. J. A. Shapiro, pp. 1–63. New York: Academic

32. Fedoroff, N. V. 1989. About maize transposable elements and development. *Cell* 56:181–91

33. Fedoroff, N. V. 1989. Maize transposable elements. See Ref. 4a, pp. 375–411

34. Fedoroff, N. V., Furtek, D. B., Nelson, O. E. 1984. Cloning of the *bronze* locus in maize by a simple and generalizable procedure using the transposable element *Activator*. *Proc. Natl. Acad. Sci. USA* 81:3825–29

35. Fincham, J. R. S., Harrison, B. 1967. Instability at the *pal* locus in *Antirrhinum majus*. II. Multiple alleles produced by mutation of one original unstable allele. *Heredity* 22:211–27

36. Freeling, M. 1984. Plant transposable elements and DNA insertions. *Annu. Rev. Plant Physiol.* 35:277–98

37. Freeling, M., Bennett, D. C. 1985. Maize *Adh1*. *Annu. Rev. Genet.* 19:297–323

38. Freeling, M., Cheng, D. S.-K., Alleman, M. 1982. Mutant alleles that are altered in quantitative organ-specific behavior. *Dev. Genet.* 3:179–96

39. Gardiner-Garden, M., Frommer, M. 1987. CpG islands in vertebrate genomes. *J. Mol. Biol.* 196:261–82

40. Gerlach, W. L., Pryor, A. J., Dennis, E. S., Ferl, R. J., Sachs, M. M., et al. 1982. cDNA cloning and induction of the alcohol dehydrogenase gene *(Adh1)* of maize. *Proc. Natl. Acad. Sci. USA* 79:2981–85

41. Gierl, A., Cuypers, H., Lütticke, S., Pereira, A., Schwarz-Sommer, Zs., et al. 1988. Structure and function of the *En/Spm* transposable element of *Zea mays*. See Ref. 69b, pp. 115–20

42. Gierl, A., Lütticke, S., Saedler, H. 1988. TnpA product encoded by the transposable element *En-1* of *Zea mays* is a DNA binding protein. *EMBO J.* 7:4045–53

43. Gierl, A., Schwarz-Sommer, Z., Saedler, H. 1985. Molecular interactions between the components of the *En-I* transposable system of *Zea mays*. *EMBO J.* 4:579–83

44. Goodall, G. J., Filipowicz, W. 1989. The AU-rich sequences present in the introns of plant nuclear pre-mRNAs are required for splicing. *Cell* 58:473–83

45. Grandbastien, M.-A., Spielmann, A., Caboche, M. 1989. *Tnt1*, a mobile retroviral-like transposable element of tobacco isolated by plant cell genetics. *Nature* 337:376–80

46. Hauser, C., Fusswinkel, H., Li, J., Oellig, C., Kunze, R., et al. 1988. Overproduction of the protein encoded by the maize transposable element *Ac* in insect cells by a baculovirus vector. *Mol. Gen. Genet.* 214:373–78

47. Hehl, R., Sommer, H., Saedler, H. 1987. Interaction between the *Tam1* and *Tam2* transposable elements of *Antirrhinum majus*. *Mol. Gen. Genet.* 207:47–53

48. Holliday, R. 1987. The inheritance of epigenetic defects. *Science* 238:163–70

49. Jin, Y.-K., Bennetzen, J. L. 1989. The structure and coding properties of *Bs1*, a maize retrovirus-like transposon. *Proc. Natl. Acad. Sci. USA* 86:6235–39

50. Johns, M. A., Mottinger, J., Freeling, M. 1985. A low copy number, copia-like transposon in maize. *EMBO J.* 4:1093–1102

51. Kermicle, J. L. 1978. Imprinting of gene action in maize endosperm. In *Maize Breeding and Genetics*, ed. D. B. Walden, pp. 357–71. New York: Wiley & Sons

52. Kim, H. Y., Schiefelbein, J. W., Raboy, V., Furtek, D. B., Nelson, O. E. 1987. RNA splicing permits expression of a maize gene with a *defective Suppressor-mutator* transposable element insertion in an exon. *Proc. Natl. Acad. Sci. USA* 84:5863–67

53. Kunze, R., Starlinger, P. 1989. The putative transposase of transposable element *Ac* from *Zea mays L*. interacts with subterminal sequences of *Ac*. *EMBO J.* 8:3177–86

54. Kunze, R., Stochaj, U., Laufs, J., Starlinger, P. 1987. Transcription of transposable element *Activator (Ac)* of *Zea mays L*. *EMBO J.* 6:1555–63

55. Lock, L. E., Takagi, N., Martin, G. R. 1987. Methylation of the *Hprt* gene on the inactive X occurs after chromosome inactivation. *Cell* 48:39–46

56. Ludwig, S. R., Habera, L. F., Dellaporta, S. L., Wessler, S. R. 1989. *Lc*, a member of the maize *R* gene family

responsible for tissue-specific anthocyanin production, encodes a protein similar to transcriptional activators and contains the myc-homology region. *Proc. Natl. Acad. Sci. USA* 86:7092–96

57. Martienssen, R. A., Barkan, A., Freeling, M., Taylor, W. C. 1989. Molecular cloning of a maize gene involved in photosynthetic membrane organization that is regulated by Robertson's Mutator. *EMBO J.* 8:1633–39

58. Martienssen, R. A., Barkan, A., Taylor, W. C., Freeling, M. 1989. Somatically heritable switches in the DNA modification of Mu transposable elements monitored with a supressible mutant in maize. *Genes Dev.* In press

59. Martin, C., Carpenter, R., Sommer, H., Saedler, H., Coen, E. S. 1985. Molecular analysis of instability in flower pigmentation of *Antirrhinum majus*, following isolation of the *pallida* locus by transposon tagging. *EMBO J.* 4:1625–30

60. Martin, C., MacKay, S., Carpenter, R. 1988. Large-scale chromosomal restructuring is induced by the transposable element *Tam 3* at the *nivea* locus of *Antirrhinum majus*. *Genetics* 119:171–84

61. Martin, C., Prescott, A., Lister, C., MacKay, S. 1989. Activity of the transposon *Tam3* in *Antirrhinum* and tobacco: Possible role of DNA methylation. *EMBO J.* 8:997–1004

62. Masson, P., Rutherford, G., Banks, J. A., Fedoroff, N. 1989. Essential large transcripts of the maize *Spm* transposable element are generated by alternative splicing. *Cell* 58:755–65

63. Masson, P., Surosky, R., Kingsbury, J. A., Fedoroff, N. V. 1987. Genetic and molecular analysis of the *Spm-dependent a-m2* alleles of the maize *a* locus. *Genetics* 177:117–37

64. McCarty, D. R., Carson, C. B., Lazar, M., Simonds, S. C. 1989. Transposable element induced mutations of the *viviparous-1* gene in maize. *Dev. Genet.* 10:473–81

65. McClintock, B. 1951. Chromosome organization and genic expression. *Cold Spring Harbor Symp. Quant. Biol.* 16:13–47

66. McClintock, B. 1962. Topographical relations between elements of control systems in maize. *Carnegie Inst. Wash. Yearb.* 61:448–61

67. McClintock, B. 1965. The control of gene action in maize. *Brookhaven Symp. Biol.* 18:162–84

68. McClintock, B. 1967. Development of the maize endosperm as revealed by

clones. In *The Clonal Basis of Development,* ed. S. Subtelny, I. M. Sussex, pp. 217–37. New York: Academic

69. Merckelbach, A., Doring, H. P., Starlinger, P. 1986. The aberrant *Ds* element in the *adh1-2F11 : :Ds* allele. *Maydica* 31:109–22

69a. Müller-Neumann, M., Yoder, J. I., Starlinger, P. 1984. The DNA sequence of the transposable element *Ac* of *Zea mays* L. *Mol. Gen. Genet.* 198:19–24

69b. Nelson, O. E. Jr., ed. 1988. *Plant Transposable Elements*. New York: Plenum

70. Nevers, P., Shepherd, N. S., Saedler, H. 1985. Plant transposable elements. *Adv. Bot. Res.* 12:103–203

71. O'Reilly, C., Shepherd, N. S., Pereira, A., Schwarz-Sommer, Z., Bertram, I., et al. 1985. Molecular cloning of the *a1* locus of *Zea mays* using the transposable elements *En* and *Mu1*. *EMBO J.* 4:877–82

72. Ortiz, D. F., Rowland, L. J., Gregerson, R. G., Strommer, J. N. 1988. Insertion of *Mu* into the *Shrunken 1* gene of maize affects transcriptional and post-transcriptional regulation of *Sh1* RNA. *Mol. Gen. Genet.* 214:135–41

73. Orton, E. R., Brink, R. A. 1966. Reconstitution of the variegated pericarp allele in maize by return of *Modulator* to the *P* locus. *Genetics* 53:7–16

74. Paz-Ares, J., Ghosal, D., Wienand, U., Peterson, P. A., Saedler, H. 1987. The regulatory *c1* locus of *Zea mays* encodes a protein with homology to myb proto-oncogene products and with structural similarities to transcriptional activators. *EMBO J.* 6:3553–58

75. Paz-Ares, J., Wienand, U., Peterson, P. A., Saedler, H. 1986. Molecular cloning of the *c* locus of *Zea mays:* a locus regulating the anthocyanin pathway. *EMBO J.* 5:829–33

76. Peacock, W. J., Dennis, E. S., Gerlach, W. L., Sachs, M. M., Schwartz, D. 1984. Insertion and excision of *Ds* controlling elements in maize. *Cold Spring Harbor Symp. Quant. Biol.* 49:347–54

77. Pereira, A., Cuypers, H., Gierl, A., Schwarz-Sommer, Z., Saedler, H. 1986. Molecular analysis of the *En/Spm* transposable element system of *Zea mays*. *EMBO J.* 5:835–41

78. Pereira, A., Schwarz-Sommer, Z., Gierl, A., Bertram, I., Peterson, P. A., Saedler, H. 1985. Genetic and molecular analysis of the *Enhancer (En)* transposable element system of *Zea mays*. *EMBO J.* 4:17–23

79. Peterson, P. A. 1987. Mobile elements

in plants. *CRC Crit. Rev. Plant Sci.* 6:105–208

80. Pohlman, R., Fedoroff, N. V., Messing, J. 1984. The nucleotide sequence of the maize controlling element *Activator*. *Cell* 37:635–43

80a. Pohlman, R. F., Fedoroff, N. V., Messing, J. 1984. Correction nucleotide sequence of *Ac*. *Cell* 39:417

81. Raboy, V., Kim, H.-Y., Schiefelbein, J. W., Nelson, O. E. Jr. 1989. Deletions in a *dSpm* insert in a maize *bronze-1* allele alter RNA processing and gene expression. *Genetics* 122:695–703

82. Robbins, T. P., Carpenter, R., Coen, E. S. 1989. A chromosome rearrangement suggests that donor and recipient sites are associated during *Tam3* transposition in *Antirrhinum majus*. *EMBO J.* 8:5–13

83. Robertson, D. S. 1978. Characterization of a mutator system in maize. *Mutat. Res.* 51:21–28

84. Rowland, L. J., Strommer, J. N. 1985. Insertion of an unstable element in an intervening sequence of maize *Adh1* affects transcription but not processing. *Proc. Natl. Acad. Sci. USA* 82:2875–79

85. Schiefelbein, J. W., Raboy, V., Fedoroff, N. V., Nelson, O. E. 1985. Deletions within a *defective Suppressor-mutator* element in maize affect the frequency and developmental timing of its excision from the *bronze* locus. *Proc. Natl. Acad. Sci. USA* 82:4783–87

86. Schiefelbein, J. W. 1988. Molecular characterization of *Suppressor-mutator (Spm)*-induced mutations at the *bronze-1* locus in maize: the *bz-m13* alleles. See Ref. 69a, pp. 261–78

87. Schiefelbein, J. W., Furtek, D. B., Dooner, H. K., Nelson, O. E. 1988. Two mutations in a maize *bronze-1* allele caused by transposable elements of the *Ac-Ds* family alter the quantity and quality of the gene product. *Genetics* 120:767–77

88. Schnable, P. S., Peterson, P. A. 1989. Genetic evidence of a relationship between two maize transposable element systems: *Cy* and *Mutator*. *Mol. Gen. Genet.* 215:317–21

89. Schnable, P. S., Peterson, P. A., Saedler, H. 1989. The *bz-rcy* allele of the *Cy* transposable element system of *Zea mays* contains a *Mu*-like element insertion. *Mol. Gen. Genet.* 217:459–63

90. Schwartz, D., Dennis, E. 1986. Transposase activity of the *Ac* controlling element in maize is regulated by its degree of methylation. *Mol. Gen. Genet.* 205:476–82

91. Schwartz, D., Echt, C. 1982. The effect of *Ac* dosage on the production of multi-ple forms of the Wx protein. *Mol. Gen. Genet.* 187:410–13

92. Schwarz-Sommer, Z., Gierl, A., Berndtgen, R., Saedler, H. 1985. Sequence comparison of states of *a1-m1* suggests a model of *Spm* action. *EMBO J.* 4:2439–43

93. Schwarz-Sommer, Z., Gierl, A., Cuypers, H., Peterson, P. A., Saedler, H. 1985. Plant transposable elements generate the DNA sequence diversity needed in evolution. *EMBO J.* 4:591–97

94. Schwarz-Sommer, Z., Gierl, A., Klosgen, R. B., Wienand, U., Peterson, P. A., Saedler, H. 1984. The *Spm (En)* transposable element controls the excision of a 2 kb DNA insert at the *wx-m8* locus of *Zea mays*. *EMBO J.* 3:1021–1028

95. Schwarz-Sommer, Z., Saedler, H. 1987. Can plant transposable elements generate novel regulatory systems? *Mol. Gen. Genet.* 209:207–9

96. Schwarz-Sommer, Z., Shepherd, N., Tacke, E., Gierl, A., Rohde, W., et al. 1987. Influence of transposable elements on the structure and function of the *A1* gene of *Zea mays*. *EMBO J.* 6:287–94

97. Shure, M., Wessler, S., Fedoroff, N. 1983. Molecular identification and isolation of the *waxy* locus in maize. *Cell* 35:225–33

98. Simon, R., Starlinger, P. 1987. Transposable element *Ds2* of *Zea mays* influences polyadenylation and splice site selection. *Mol. Gen. Genet.* 209:198–99

99. Sommer, H., Bonas, U., Saedler, H. 1988. Transposon-induced alterations in the promoter region affect transcription of the chalcone synthase gene of *Antirrhinum majus*. *Mol. Gen. Genet.* 211:49–55

100. Sommer, H., Carpenter, R., Harrison, B. J., Saedler, H. 1985. The transposable element *Tam3* of *Antirrhinum majus* generates a novel type of sequence alteration upon excision. *Mol. Gen. Genet.* 199:225–31

101. Sommer, H., Hehl, R., Krebbers, E., Piotrowiak, R., Lonning, W.-E., et al. 1988. Transposable elements of *Antirrhinum majus*. See Ref. 69b, pp. 227–36

102. Staiger, D., Kaulen, H., Schell, J. 1989. A CACGTG motif of the *Antirrhinum majus* chalcone synthase promoter is recognized by an evolutionarily conserved nuclear protein. *Proc. Natl. Acad. Sci. USA* 86:6930–34

103. Starlinger, P., Baker, B., Coupland, G., Kunze, R., Laufs, J., et al. 1988. Stud-

ies on transposable element *Ac* of *Zea mays*. See Ref. 69b, pp. 91–100

104. Strommer, J. N., Hake, S., Bennetzen, J., Taylor, W. C., Freeling, M. 1982. Regulatory mutants of the maize *Adh1* gene caused by DNA insertions. *Nature* 300:542–44

105. Sullivan, T. D., Schiefelbein, J. W., Nelson, O. E. Jr. 1989. Tissue-specific effects of maize *bronze* gene promoter mutations induced by *Ds1* insertion and excision. *Dev. Genet.* 10:412–24

106. Sutton, W. D., Gerlach, W. L., Schwartz, D., Peacock, W. J. 1983. Molecular analysis of *Ds* controlling element mutations at the *Adh1* locus of maize. *Science* 223:1265–68

107. Talbert, L. E., Patterson, G. I., Chandler, V. L. 1989. *Mu* transposable elements are structurally diverse and distributed throughout the genus *Zea*. *J. Mol. Evol.* 29:28–39

108. Taylor, L. P., Walbot, V. 1987. Isolation and characterization of a 1.7 kb transposable element from a *Mutator* line of maize. *Genetics* 117:297–307

109. Toniolo, D., Martini, G., Migeon, B. R., Dono, R. 1988. Expression of the G6PD locus on the human X chromosome is associated with demethylation of three CpG islands within 100 kb of DNA. *EMBO J.* 7:401–6

110. Varagona, M., Wessler, S. 1989. Implications for the *cis*-requirements for *Ds* transposition based on the sequence of the *wxB4 Ds* element. *Mol. Gen. Genet.* In press

111. Varmus, H., Brown, P. 1989. Retroviruses. See Ref. 4a, pp. 53–109

112. Vayda, M. E., Freeling, M. 1986. Insertion of the *Mu1* transposable element into the first intron of maize *Adh1* interferes with transcript elongation but does not disrupt chromatin structure. *Plant Mol. Biol.* 6:441–54

113. Walker, J. C., Howard, E. A., Dennis, E. S., Peacock, W. J. 1987. DNA sequences required for anaerobic expression of the maize alcohol dehydrogenase gene. *Proc. Natl. Acad. Sci. USA* 84:6624–28

114. Wessler, S. R. 1988. Phenotypic diversity mediated by the maize transposable elements *Ac* and *Spm*. *Science* 242:399–405

115. Wessler, S. R. 1989. The splicing of maize transposable elements from pre-mRNA—a minireview. *Gene* 82:127–33

116. Wessler, S. R., Baran, G., Varagona, M. 1987. The maize transposable element *Ds* is spliced from RNA. *Science* 237:916–18

117. Wessler, S. R., Baran, G., Varagona, M. 1988. Alterations in gene expression mediated by DNA insertions in the *waxy* gene of maize. See Ref. 69b, pp. 293–303

118. Wessler, S. R., Baran, G., Varagona, M., Dellaporta, S. L. 1986. Excision of *Ds* produces *waxy* proteins with a range of enzymatic activities. *EMBO J.* 5:2427–32

119. Wessler, S. R., Varagona, M. 1985. Molecular basis of mutations at the *waxy* locus of maize: Correlation with the fine structure genetic map. *Proc. Natl. Acad. Sci. USA* 82:4177–81

120. Wienand, U., Sommer, H., Schwarz, Z., Shepherd, N., Saedler, H., et al. 1982. A general method to identify plant structural genes among genomic DNA clones using transposable element–induced mutations. *Mol. Gen. Genet.* 187:195–201

121. Wolf, S. F., Migeon, B. R. 1985. Clusters of CpG dinucleotides implicated by nuclease hypersensitivity as control elements of housekeeping genes. *Nature* 314:467–69

Annu. Rev. Plant Physiol. Plant Mol. Biol. 1990. 41:553–75

THE HEAVY METAL–BINDING PEPTIDES OF PLANTS

J. C. Steffens

Department of Plant Breeding and Biometry, Cornell University, Ithaca, New York 14853

KEY WORDS: phytochelatin, metal toxicity, metallothionein, γ-glutamyl peptide, glutathione

CONTENTS

INTRODUCTION

This review is concerned with the response of plants to heavy-metal exposure. Heavy metals such as zinc and copper are required by biological systems as

1040-2519/90/0601-0553$02.00

structural and catalytic components of proteins and enzymes, and as cofactors essential to normal growth and development. In excess, these micronutrients and related heavy metals such as cadmium, mercury, nickel, and lead become extremely toxic to cells. Toxic or lethal levels of heavy metals are experienced by plants growing near mining or smelting operations, industrial and municipal waste disposal sites, on some natural soil types, and on some agricultural soils. While plant growth may be severely restricted by heavy metals, some plant species possess a unique ability to adapt rapidly and evolve tolerance to toxic or lethal levels of heavy metals. The ecological adaptation of plant communities to chronically high heavy-metal exposure stands as a classical example of rapid evolution under extreme selection pressures (1, 67, and references therein). Despite the excellent work detailing the population response of plants to heavy-metal toxicity, the biochemical and molecular bases of adaptation and tolerance of plant communities to high metal levels are poorly understood.

Here I review one mechanism used by plants to alleviate stresses imposed by exposure to heavy-metal excess—the synthesis of metal-binding polypeptides whose apparent function is to sequester and detoxify excess metal ions. Failure to synthesize these peptides results in growth inhibition or cell death. These heavy metal-binding polypeptides are known as phytochelatins. The role of phytochelatins in plant metal tolerance has been the subject of several recent reviews (39a, 48b, 48c, 59a).

Nomenclature

The heavy metal–binding polypeptides of plants are those whose synthesis is induced by heavy metals, and which possess the generalized structure (γ-Glu-Cys)$_n$-Gly, where $n = 2$–7. The heavy metal–binding polypeptides of plants have been variously referred to by the trivial names *cadystins* (26), *phytochelatins* (11), *γ-glutamyl peptide* (46), *poly-(γ-glutamylcysteinyl) glycines* (19), and *Cd-peptide* (45). At a recent meeting,[1] a nomenclature section was convened to determine whether the varying terms could be unified. The term *cadystins* holds priority, but it is not informative in terms of function and suggests restriction to yeasts. *Phytochelatins,* on the other hand, does suggest the chelating function of the molecules and approximates the phylogenetic distribution of the peptides in nature. However, differences of opinion also exist about whether the prefix *phyto-* adequately includes the two yeasts in which these peptides also occur. At any rate, a consensus had not been reached at the time of this writing, and the terminology used in this review

[1]UCLA Colloquium on Metal Ion Homeostasis: Molecular Biology and Chemistry, Frisco, Colorado, April 10–16, 1989

reflects my own preferences. I extend my apologies to those who find this terminology objectionable.

BACKGROUND

Toxic metal ions are thought to enter the cell by means of the same uptake systems that move such physiologically important ions as Cu and Zn (54). In animals, excess intracellular heavy metals induce biosynthesis of proteins called metallothioneins. Fungal, invertebrate, insect, and mammalian metallothioneins are low-molecular-weight, cysteine-rich proteins (approximately 30% Cys) that usually lack aromatic amino acid residues. Cysteine residues are typically present in the protein as a Cys-X-Cys motif (where X is an amino acid other than Cys), or in Cys-Cys clusters.

Metallothioneins are distinguished by a highly conserved primary structure and are grouped structurally into two types. Type I resembles equine renal metallothionein; it includes most vertebrate forms and some fungal metallothioneins [e.g. *Neurospora* and *Agaricus* (29, 32)]. Type II metallothioneins, such as are found in *Saccharomyces cerevisae,* do not share extensive sequence homology with Type I metallothioneins yet characteristically possess a high amount of Cys and, like Type I, bind metals via cysteine thiolate ligands. Synthesis of metallothioneins is induced at the transcriptional level, and cells selected for heavy-metal tolerance may exhibit amplification of metallothionein-genes (16). Metallothionein synthesis in vertebrates is under developmental control as well as control by exogenous metals and is induced by a number of other agents, such as corticosteroids. The broad range of metallothionein inducers has led some to question whether metal homeostasis and detoxification are the primary roles of these proteins (21). For further details on these interesting and highly conserved proteins the reader is referred to Hamer (16) and Kagi & Kojima (20a).

Early efforts at understanding the detoxification of excess heavy metals by plants depended in large part on models based on metallothioneins. The degree of structural and regulatory conservation exhibited by metallothioneins from a broad array of organisms suggested that proteins similar in structure and function to metallothioneins would occur in plants. Identification of such sequestration proteins would immediately open up avenues for molecular biological investigations of metal tolerance and detoxification in plants. Initial studies of plants exposed to heavy metals revealed the apparent presence of metallothionein-like proteins (2, 4, 42, 43, 61, 28, 65). The metal inducibility of these proteins was demonstrated in cabbage and tobacco leaves (63, 64) and in maize (41). These studies suggested that metallothionein-like proteins are involved in the metabolism of heavy metals by plants.

PHYTOCHELATINS: PRIMARY SEQUENCE

More detailed studies of the inducible metal-binding polypeptides in plants showed that in contrast to animal systems, plants possess nonprotein metal-binding polypeptides that, although differing in structure and biosynthesis, may be functionally analogous to metallothioneins (11, 19, 55). The metal-binding polypeptides of plants possess the unusual structure (γ-Glu-Cys)$_n$-Gly, where $n = 2$–11. In comparison, glutathione (GSH) possesses the structure γ-Glu-Cys-Gly in which the peptide bond is formed between the γ-, or side-chain carboxylate of glutamic acid rather than the α-carboxylate utilized in peptide bonds of polypeptides whose synthesis is ribosome-dependent (Figure 1).

These polypeptides were originally described by Murasugi et al (33, 34) in the fission yeast *Schizosaccharomyces pombe,* where they were found to be

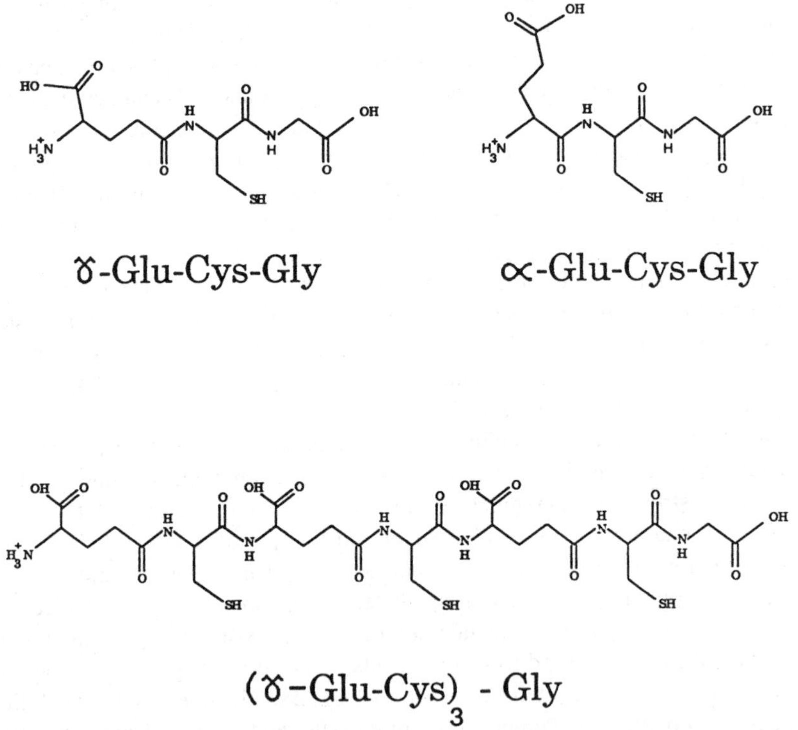

δ-Glu-Cys-Gly α-Glu-Cys-Gly

$(\delta$-Glu-Cys$)_3$ - Gly

Figure 1   Structures of glutathione and the metal-binding polypeptide (γ-glutamylcysteinyl) $_3$-glycine. Peptide bonds between glutamate and cysteine utilize the side chain (or γ) carboxylate of glutamate (dotted box), rather than the α-carboxylate characteristically found in proteins whose synthesis is ribosome dependent.

inducible by Cd and were given the name *cadystin*. Several structures were originally proposed, culminating in the identification and total synthesis of (γ-Glu-Cys)$_2$-Gly and (γ-Glu-Cys)$_3$-Gly (25, 26). Subsequently Grill and coworkers (11), working with plant cell cultures, reported a similar series of metal-binding polypeptides, (γ-Glu-Cys)$_{3-7}$-Gly. The term *phytochelatins* was proposed for these peptides based on their widespread occurrence in plants and their apparent function in sequestration of metal ions (11).

OCCURRENCE

The data on the occurrence of phytochelatins is derived primarily from the work of Grill et al (11, 12, 15). This phylogenetic survey has established that phytochelatins are invariably expressed as constituents of plants exposed to heavy metals. No other thiol-rich, heavy metal–binding constituents other than phytochelatins were detectable in the many plants assayed (estimated 0.01% of all known plants). The ability to synthesize phytochelatins in response to heavy metals is conserved from the Orchidales, the most advanced group of higher plants, to the red, green, and brown algae (10). Thus, there is a fundamental biochemical dichotomy between animals and plants in that animals sequester metals by means of metallothioneins, and plants employ the polypeptides (γ-Glu-Cys)$_n$-Gly for the same purpose. Most fungi appear to utilize metallothioneins for this function. However, *S. pombe* and *Torulopsis glabrata* (also known as *Candida glabrata*), which possess phytochelatins, are known exceptions (26, 66). The regulation of metal resistance in *T. glabrata* is exceptionally interesting (31). In response to excess Cu, this yeast synthesizes two metallothionein-like proteins possessing 30 mol% Cys and two repeats of the Cys-X-Cys motif typical of metallothioneins. However, in response to Cd, *T. glabrata* synthesizes (γ-Glu-Cys)$_2$-Gly and the des-Gly peptide (γ-Glu-Cys)$_2$. The differential regulation by Cd is unique in *T. glabrata,* as phytochelatins in other organisms are typically inducible by both Cd and Cu.

IDENTIFICATION AND ASSAY OF PHYTOCHELATINS

Due to the acidic nature of phytochelatin complexes formed with metals, phytochelatins are highly soluble and easily extracted from cells. Substantial purification is achieved by passage over strong anion exchange columns. Complex purity is usually achieved by chromatography on Sephadex G50 or similar column supports. Individual peptide components can be purified by denaturation in dilute acid (0.1 N HCl or 0.1% trifluoroacetic acid) and chromatography on reversed-phase HPLC columns using acetonitrile/water gradients in 0.1% trifluoroacetic acid (11, 44, 55). Covalent chromatography on thiol Sepharose supports has also been described (19).

Because of the unusual repeating γ-glutamyl peptide linkages of phytochelatins a number of approaches have been taken to establish their primary sequence. Phytochelatins do not sequence by conventional Edman protein sequencing methodology, although Grill et al (11) employed a sequencing strategy involving γ-glutamyl transferase and manual Edman degradation. Mass spectrometry provides an efficient means to sequence and establish the location of γ-glutamyl residues directly (55). γ-Glutamyl linkages have also been identified by ^{13}C-NMR of peptides obtained from cell cultures labeled with $(1,5-^{13}C_2)$-Glu (19).

Minimal criteria for identification of purified phytochelatins are amino acid compositions consisting of Glu, Cys, and Gly, with Glu and Cys being equimolar and present at 2–7 times the level of Gly. While amino acid compositions differing from the above have been reported for a variety of putative plant metal-binding "proteins," until sequences or rigorous criteria for purity are established, reports of plant metal-binding peptides other than phytochelatins must be considered uncertain.

Changes in phytochelatin levels are most conveniently and sensitively assayed by HPLC using sulfhydryl derivatization to generate a sensitive chromophore. Post-column derivatization with Ellman's reagent (*bis*-dithionitrobenzoic acid) permits detection at 410 nm (11). Pre-column sulfhydryl derivatization with *p*-chloromercuribenzoic acid (50) provides a 254-nm chromophore. Pre-column derivatization with the sulfhydryl reagent mono-bromobimane provides pmol sensitivity for thiols by fluorescence detection (37, 55).

METAL ION SPECIFICITY AND PHYTOCHELATIN INDUCTION

In mammalian systems, metallothionein synthesis is induced during normal development and differentiation, by corticosteroids, and by an array of other agents and oxidative stresses, in addition to heavy metals (8). This has led some to question whether the primary cellular role of metallothioneins is trace metal homeostasis (21). Phytochelatins are distinctive in that heavy metals are the primary inducers. Cadmium, Pb, Zn, Sb, Ag, Ni, Hg, Cu, Sn, Au, Bi, Te, and W all induce phytochelatins (13). Among the common metals, Cd is the strongest inducer, while Zn appears to be weak, requiring very high levels for induction (Table 1). In general this difference parallels that between the relative toxicities of Zn and Cd (17).

In addition, the arsenate (AsO_4^{3-}) and selenate (SeO_3^{2-}) anions have been shown to be inducers of phytochelatin synthesis (13). Glutathione forms

Table 1 Induction of phytochelatins by heavy metal ions.

Salt		Phytochelatin[a] (μmol/g[b])			Total γ-glutamyl-cysteine in phytochelatin (μmol/g[b])
Formula	Concentration (μM)	$n = 2$	$n = 3$	$n = 4$	
Cd(NO$_3$)$_2$	100	1.27	2.91	2.30	20.5
Pb(NO$_3$)$_2$	1000	1.78	2.28	0.25	11.4
ZnSO$_4$	1000	1.51	1.68	0.12	8.5
SbCl$_3$	200	0.94	1.72	0.37	8.5
AgNO$_3$	50	1.07	1.90	0.08	8.2
Ni(NO$_3$)$_2$	100	1.53	0.82	0.07	5.8
Hg(NO$_3$)$_2$	10	1.28	0.55	0.02	4.3
Na$_2$HAsO$_4$	20	1.40	0.34	0	3.8
CuSO$_4$	50	0.88	0.41	0.04	3.1
SnSO$_4$	100	0.86	0.33	0.03	2.8
NaSeO$_3$	100	0.75	0.22	0.07	2.4
AuCl	50	0.71	0.17	0.03	2.0
Bi(NO$_3$)$_3$	100	0.69	0.18	0	1.9
TeCl$_4$	10	0.58	0.13	0.05	1.8
WCl$_6$	100	0.42	0.09	0	1.1
None	—	0	0	0	0

R. serpentina cell suspension cultures were exposed to the indicated salts for three days. A zinc- and copper-free medium (14) was used in all experiments. No phytochelatin was detected in cells exposed to 2 mM Ca(NO$_3$)$_2$, Al(NO$_3$)$_3$, FeSO4, MgSO$_4$, MnCl$_2$, or NaCl; to 0.1 mM NaMoO$_4$, CsCl, or CR(NO$_3$)$_3$; to 0.05 mM UO(NO$_3$)$_2$; or to 0.02 mM VOSO$_4$. Reprinted from Ref. 17 with permission.
[a] Individual species with n γ-glutamylcysteine units per molecule
[b] Dry weight of cells

complexes with these ions and reduces them to the tridentate (GS)$_3$As and bidentate (GS)$_2$Se compounds (15, 20). Whether these elements actually form complexes with phytochelatins, or simply induce their synthesis, is unknown. The fact that they are known to form complexes with GSH may suggest that depletion of the GSH pool may be part of the signal involved in inducing phytochelatin synthesis.

An interesting observation is that while Cd (a potent inducer of phytochelatins) also induces heat shock proteins (5, 53), heat shock does not induce phytochelatin biosynthesis. Also, whereas CuSO$_4$ is an effective elicitor of both phytoalexins and phytochelatins, elicitor treatments such as ultraviolet radiation (UV) and fungal cell wall preparations do not induce phytochelatins. Cold shock, alteration of hormonal levels, and oxidative stresses such as H$_2$O$_2$ and return from N$_2$-induced anoxia also do not appear to induce phytochelatins in cell cultures (J. C. Steffens, unpublished; 13). Thus, although many plant stress responses are elicited by a number of inducers, phytochelatin biosynthesis is tightly regulated by the availability of metal ions.

BIOSYNTHESIS

The similarity of phytochelatins to GSH (Figure 1) indicates that they cannot be primary gene products and that their biosynthesis shares a common enzymology with GSH. The available evidence suggests that GSH is synthesized in plants as it is in animals (22, 23, 48, 48a). The first committed and rate-limiting step of GSH biosynthesis, Glu + Cys → γ-Glu-Cys, is catalyzed by γ-glutamylcysteine synthetase (EC 6.3.2.2), using ATP hydrolysis to drive the reaction. γ-Glutamylcysteine synthetase is inhibited by the transition-state analog inhibitor buthionine sulfoximine (BSO). The second step of GSH biosynthesis, γ-Glu-Cys + Gly→γ-Glu-Cys-Gly, is catalyzed by glutathione synthetase (EC 6.3.2.3) and is also driven by ATP.

The biosynthesis of phytochelatins is inhibited by BSO (55, 56, 11, 45, 50). Buthionine sulfoximine inhibition provided the first evidence that phytochelatins are necessary for heavy-metal detoxification. Cells treated with BSO are unable to synthesize phytochelatins and become susceptible to growth inhibition by heavy metals at concentrations below those normally inhibiting growth in the absence of BSO (55, 56). Thus this is one of the few plant stress responses in which the association between the stress *response* and stress *adaptation* is easily demonstrated. However, it has not been ruled out that BSO potentiation of metal toxicity is due in part to the absence of GSH itself in BSO-treated cells. In hepatoma cells, excess Cu is initially sequestered in a labile complex with GSH, and the metal ions are subsequently transferred from this complex to metallothionein by an unknown mechanism (9). Hence, the sensitivity to metals conferred by BSO may be in part due to elimination of an initial GSH complexation step leading to decreased free metal concentrations in the time before significant phytochelatin accumulation occurs.

The time-course of phytochelatin induction also implicates GSH in phytochelatin biosynthesis. Challenge of *S. pombe* or tomato cell cultures with Cd ion provokes the synthesis of phytochelatins, concomitant with a rapid decrease in the level of GSH (11, 39, 50; Figure 2). In the first hours after metal exposure the rate and extent of GSH disappearance is nearly equal to the rate of γ-glutamylcysteine incorporation into phytochelatin, and is not affected by buthionine sulfoximine (11). In intact maize seedlings treated with Cd a similar effect is seen, with GSH in roots being depleted more rapidly than GSH in shoots after exposure to Cd. Root GSH declines by nearly one half within two hours of exposure to Cd, recovering to initial levels in 24 hr. In contrast the shoot GSH only decreases by about 10% 2 hr after exposure and fully recovers by 24 hr (39).

Plant cells provided with GSH as the sole sulfur source appear to take up the molecule intact (47, 48). In vivo experiments show that exogenous GSH

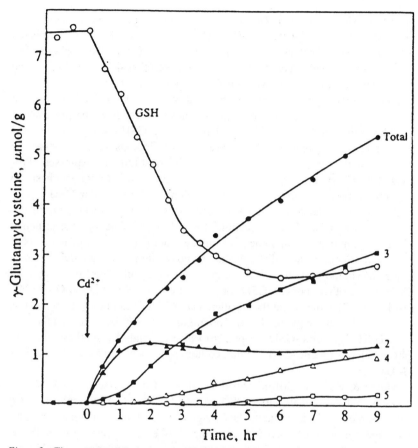

Figure 2 Time course of phytochelatin induction and glutathione consumption after administration of 200 μM Cd(NO$_3$)$_2$ to *Rauvolfia serpentina* cell suspension culture. Quantities of glutathione (GSH), total phytochelatin (●), and individual phytochelatins with n (number of γ-glutamylcysteine units per molecule) = 2 (▲), 3 (■), 4 (△), or 5 (□) are expressed as μmol of γ-glutamylcysteine per g (dry weight) of cells. Data from reference 13 with permission of the author.

supports phytochelatin synthesis, presumably as a direct precursor (50). In addition, buthionine sulfoximine-inhibited phytochelatin synthesis can be overcome by exogenous GSH, but not by Cys (50). Pulse-chase studies with *Datura innoxia* cells show that ^{35}S-Cys is rapidly assimilated into GSH, and upon metal exposure the radioactivity of ^{35}S-GSH is incorporated within an hour into (γ-Glu-Cys)$_2$-Gly (3). Surprisingly, although the kinetics of accumulation suggest a precursor/product relationship between (γ-Glu-Cys)$_2$-Gly and the $n+1$ extended forms, the radioactivity of the (γ-Glu-Cys)$_2$-Gly pool was not observed to chase into the expected longer products over the course of 32 hr. Thus, in *Datura,* substantially large pools of newly synthesized $(\gamma\text{-Glu-Cys})_n$-Gly are unavailable in vivo to undergo the expected elongation reactions that would lead to the $(\gamma\text{-Glu-Cys})_{n+1}$-Gly peptides. The lack of processivity seen in vivo may represent rapid formation of the metal-binding complex, preventing further polymerization.

Phytochelatin synthesis can be observed within 5–15 min of exposure to excess metals (49, 50). Phytochelatin synthesis proceeds in the presence of cycloheximide until the GSH pool is depleted (50). GSH levels do not recover in cycloheximide treated cells. These in vivo findings suggested that the enzyme catalyzing phytochelatin synthesis is constitutively expressed. Prolonged phytochelatin synthesis may require de novo protein synthesis to support formation of the precursors γ-glutamylcysteine and/or GSH.

Additional evidence for the participation of GSH in phytochelatin biosynthesis derives from the identification of $(\gamma\text{-Glu-Cys})_n$-$\beta$-Ala as the principal metal-binding peptides of some plants in the order Fabales. Some members of this group of legumes are distinguished by possession of γ-glutamylcysteinyl-β-alanine (homo-glutathione) as the predominant cellular thiol rather than by GSH, while others possess GSH, or a mixture of GSH and homo-GSH (24). Grill et al (12) showed that these patterns are maintained in the phytochelatins synthesized by these plants. For example, *Lathryus ochrus,* which synthesizes both GSH and homo-GSH, responds to heavy metals by synthesizing both phytochelatins $(\gamma\text{-Glu-Cys})_{2-4}$-Glu and homo-phytochelatins $(\gamma\text{-Glu-Cys})_{2-4}$-$\beta$-Ala.

Mutoh & Hayashi (36) described a series of Cd-hypersensitive *S. pombe* mutants that were unable to synthesize $(\gamma\text{-Glu-Cys})_n$-Gly in response to Cd (36). Some of these mutants lack γ-glutamylcysteine synthetase or glutathione synthetase activity, providing further evidence that both γ-glutamylcysteine and GSH are precursors to phytochelatins. In addition, Cd-hypersensitive mutants were characterized that possessed wild-type levels of both γ-glutamylcysteine synthetase and glutathione synthetase and are presumed to be mutant in the enzyme responsible for synthesis of phytochelatins (phytochelatin synthase). Thus, in addition to providing the first genetic evidence that phytochelatins are essential for Cd tolerance, Mutoh & Hayashi's work (36) supported in vivo studies that associated glutathione metabolism with phytochelatin synthesis.

Although in vivo studies were not able to distinguish the biosynthetic mechanism employed by phytochelatin synthase, this problem was recently illuminated by the demonstration that phytochelatin synthase is a specific γ-glutamylcysteine dipeptidyl transpeptidase that catalyzes the transfer of γ-glutamylcysteine from GSH to another molecule of GSH, forming $(\gamma\text{-Glu-Cys})_2$-Gly (12b). In addition to transferring γ-glutamylcysteine from GSH to an acceptor GSH molecule, phytochelatin synthase also catalyzes the stepwise addition of γ-glutamylcysteine from GSH to phytochelatin oligomers ranging from (γ-Glu-Cys)$_2$-Gly to (γ-Glu-Cys)$_5$-Gly. Time-course analysis of Cd-stimulated phytochelatin synthase activity in the presence of GSH showed immediate appearance of (γ-Glu-Cys)$_2$-Gly, and after successive 20-min lags, the appearance of (γ-Glu-Cys)$_3$-Gly and (γ-Glu-Cys)$_4$-Gly. Similarly, incubation of the enzyme with purified (γ-Glu-Cys)$_2$-Gly, GSH and Cd led to the immediate synthesis of (γ-Glu-Cys)$_3$-Gly, followed by the appearance of (γ-Glu-Cys)$_4$-Gly. Synthesis stops when the Cd in the reaction becomes complexed by the newly synthesized phytochelatin; addition of Cd stimulates further synthesis. In the absence of GSH, phytochelatin synthase was also shown to catalyze the transfer of γ-glutamylcysteine from (γ-Glu-Cys)$_2$-Gly to another molecule of (γ-Glu-Cys)$_2$-Gly, forming (γ-Glu-Cys)$_3$-Gly and subsequently, (γ-Glu-Cys)$_4$-Gly.

Phytochelatin synthase purified to homogeneity possesses a molecular weight of 25,000 under denaturing conditions. Under non-denaturing conditions, phytochelatin synthase activity elutes from gel permeation columns primarily at M_r 95,000. Activity also elutes with a M_r of 50,000, suggesting that the enzyme is a tetrameric protein that dissociates during purification to a dimeric form that retains activity. The enzyme purified from *Silene cucubalus* has a K_m for glutathione of 6.7 mM, and for the gluthathione-S-monobromobimane adduct, 1.5 mM.

Phytochelatin synthase is activated by the heavy metal cations Cd, Ag, Bi, Pb, Zn, Cu, Hg, and Au and is completely inactive in their absence. Relative ability of the different metal ions to activate phytochelatin synthase in vitro was found to closely mirror their ability in vivo to induce phytochelatin synthesis (10, 11, 13, 17). Ability of the in vivo anionic inducers selenate and arsenate to activate phytochelatin synthase was not reported (12b). Phytochelatin synthesis ceases immediately after addition of EDTA or metal-free phytochelatins, suggesting an elegant feedback loop in which metal ions activate the enzyme, and when sufficient phytochelatins have been synthesized to complex free metal ions, activity ceases (12b).

Phytochelatin synthase was demonstrated in cell-free extracts of *Silene cucubalus, Podophyllum peltatum, Escholtzia californica, Beta vulgaris,* and *Equisetum giganteum.* In agreement with in vivo studies (49, 50) the synthase was shown to be a constitutive enzyme in these plants.

TURNOVER

The primary evidence for turnover of phytochelatins is provided by Grill et al (14). Cell cultures inoculated onto normal tissue culture media undergo a transient induction and accumulation of phytochelatins that ceases when the media become depleted of free Cu and/or Zn. Cessation of phytochelatin accumulation is followed by a decrease in the cellular phytochelatin titer in the late exponential- to stationary-growth phase of the cells, a decrease taken to represent phytochelatin degradation (14). Phytochelatin titers increase dramatically after exposure of cells to large amounts of Cd, and this accumulation phase is followed by maintenance of steady-state levels and then a rapid decrease in phytochelatin levels (Figure 2, *upper panel*). Either the metal-binding capacity of the phytochelatins must increase during the time that total phytochelatin decreases in the cell, or the metal becomes associated with other ligands as phytochelatins are degraded. Based on the reaction mechanism of phytochelatin synthase (12b), the des-Gly phytochelatins often found as components of metal-binding complexes are likely to represent phytochelatin degradation products (66). However, no other changes in length distribution of phytochelatins have been identified during the time that phytochelatins are thought to be degraded.

COMPARTMENTALIZATION

Because GSH is primarily extravacuolar, and GSH biosynthetic enzymes are localized in the plastid and cytoplasm (22, 23), phytochelatin synthesis is also presumed to be cytoplasmic. It has been generally assumed that phytochelatins are maintained as soluble cytoplasmic complexes. However, recent work has demonstrated the vacuolar compartmentalization of phytochelatins (R. Vogeli-Lange and G. Wagner, in preparation). Tobacco leaf protoplasts from plants exposed to low Cd concentrations contained all the intracellular Cd and phytochelatin in vacuoles when protoplasts and vacuoles were compared on the basis of equal α-mannosidase activities. Rather than existing as a relatively static means to sequester metal ions, phytochelatins may be a component of a shuttle system for the transfer of metals from the cytoplasm to the vacuole. Despite the co-occurrence of phytochelatin and Cd in the vacuole of tobacco, there is no indication at this time whether the two remain associated in this compartment. Vacuolar pH will exert a profound effect upon the state of the complex. Depending on peptide lengths and degree of sulfide incorporation, low pH can cause accretion of phytochelatin complexes. Inclusions of S-rich Cd granules in *Agrostis* roots may be taken as evidence for phytochelatin accretion, although their distribution is not strictly vacuolar (40). Also, depending on amount of labile sulfur, Cd can be displaced from phytochelatins at pH values of 5–3.5 (6). Vacuolar pH ranges from 3.5 to 6.0 (27). Displacement of Cd by protons may allow association of the metal with other ligands. The concentration of organic acids (citrate, malate) or phytic acid (inositol hexaphosphate) in vacuoles can be high, and these may act as effective metal ligands at vacuolar pH values (62, 67). Recycling of the phytochelatin could be accomplished by transport of the apopeptide out of the vacuole, or by degradation and resynthesis of the component amino acids.

THE STRUCTURE OF PHYTOCHELATIN COMPLEXES WITH METALS

Metal Binding

Although it is clear that many different metals induce phytochelatin synthesis (13), binding of metals by phytochelatins has only been demonstrated for Cu, Zn, Pb, and Cd; Zn binding is difficult to demonstrate because of low affinity of the ligand for the Zn ion (15, 45). Unfortunately, comparative binding affinities for various combinations of metals and peptide lengths are not available. Complexes composed of (γ-Glu-Cys)$_{2-7}$-Gly are extremely heterogeneous because of multiple peptide components and the many possible combinations of complexes that can arise from these mixtures. Thus, metal stoichiometry is frequently reported on the basis of mols sulfhydryl rather than mols peptide. Ratio of Cys-SH to metal in phytochelatin complexes is 2:1 for Cd, Zn, and Pb. Copper, which is present in (γ-Glu-Cys)$_n$-Gly complexes as Cu(I), is bound to sulfhydryls at a ratio of 1:1 (15).

Ligands

The primary metal ligand in phytochelatin complexes is the cysteine thiol. There is no evidence to suggest that carboxylate or amino groups participate in this function. The prominent charge-transfer transition exhibited at 254 nm in native phytochelatin complexes indicates that Cd-thiolate ligands are present in phytochelatins. Cotton effect extrema at 254 nm in circular dichroism spectra are reminiscent of Cd metallothionein spectra and indicative of similar Cd-thiolate ligands (11). Peptides reconstituted with Zn also exhibit Zn-thiolate UV transitions (46). Ultraviolet excitation of the S. pombe Cu complex results in luminescence at 619 nm. Luminescence spectra indicate Cu(I) coordination characteristic of a site shielded from interactions with solvent (66). Binding of metals by phytochelatins is strongly pH dependent, with low pH favoring protonation of thiolate ligands and displacement of metal ions. The Cd ion is 50% displaced from phytochelatin complexes at pH 5. In contrast, the same fraction of Cu is displaced from the S. pombe complex at pH 1.3 (44).

Significance of Peptide Heterogeneity

Metal complexation occurs equally well with full-length and des-Gly (lacking the C-terminal glycine) peptides, suggesting that the C-terminal Gly is not essential for complex formation (30). In reconstitution experiments that generated complexes composed of peptides of unique n values, peptides of n = 2–4 were all capable of forming complexes that displayed the 260-nm transition and luminescence characteristics of Cu(I)-thiolate coordination. Copper binding capacity is less, and Cu more reactive in complexes consisting of unique n peptides than in the heterogeneous native Cu complexes. This indicates that one function of peptide heterogeneity in Cu complexes is conferral of added stability to the complex (30).

Molecular Mass of Binding Complexes

Phytochelatin complexes elute as broad peaks from gel permeation columns. The M_r of Cu and Cd phytochelatin complexes varies from about 3,000 to 10,000 depending upon ionic strength (13, 44, 66), a property that probably contributed to the confusing variety of molecular masses initially reported for plant metal-binding "proteins" (13). The lower M_r observed at high ionic strength suggests that complexes possess a trimeric or tetrameric peptide stoichiometry (13, 44, 66). The high M_r observed at low ionic strength has been suggested to result both from electrostatic repulsion of the negatively charged free Glu carboxylates of the polypeptides and from complex aggregation (66). Polyglutamate displays similar ionic strength effects on its M_r (44). The fact that the luminescence of the Cu complex is not affected by ionic strength suggests that changes in conformation around the metal thiolate center are not induced by ionic strength changes, and therefore that ionic strength changes are unlikely to effect assembly or reassembly of the metal-binding complex. Examination of leading and trailing edges of Cu complexes eluting from gel permeation columns reveals that peptides with both high and low values of n elute in both regions, indicating that metal complexes are ordinarily composed of peptides mixtures varying in n, and that aggregation phenomena or variation in complex composition is responsible for the broad range of complex molecular masses (66). Incorporation of varying amounts of sulfide or sulfite ion (see below) may also contribute to the size heterogeneity of phytochelatin complexes.

Labile Sulfide

Acidification of purified phytochelatin complexes results in the evolution of H_2S, indicating the presence of labile sulfide ion (35, 55). Complexes containing sulfide ion exhibit higher M_r. One effect of sulfide in the phytochelatin complex is the stabilization of the pH at which dissociation of metal occurs. Cd complexes lacking sulfide are half dissociated at pH 5, whereas the sulfide-containing complexes only become half dissociated below pH 4 (44, 46). Presence of sulfide in Cd complexes leads to a characteristic UV transition in the region 305–318 nm whose maximum is dependent on quantity of sulfide in the complex. Incorporation of sulfide ion in the complex and the resulting higher stability and metal-binding capacity may increase the effectiveness of these peptide complexes as a mechanism for sequestration of toxic metals.

High-sulfide complexes of the yeasts *C. glabrata* and *S. pombe* possess properties of quantum semiconductor crystallites (6). Electron micrographs and X-ray diffraction of the complexes reveal a crystallite particle diameter of ~ 16–18 Å, which appears to be constituted of about 30 (γ-Glu-Cys)$_2$-Gly or (γ-Glu-Cys)$_2$ peptides stabilizing a core of about 85 CdS molecules. Diffraction suggests that the CdS crystallite core approximates a six-coordinate structure different from that possessed by bulk CdS. These properties lead to the prominent 305–318-nm UV transitions of sulfide-containing complexes. These UV transitions occur at higher energies than would be expected from bulk CdS, a property that is characteristic of quantum semiconductors. The position of these transitions reflects the size of the CdS core of the particle. Similar to other semiconductor crystallite particles, these structures exhibit photoinduced electron transfer to methyl viologen when irradiated at the CdS transition. Reduction of methyl viologen is accompanied by loss of sulfide ion. Whether this sulfide is oxidized to the level of sulfite or sulfate during this process is unknown.

Labile Sulfite

Whether the properties of yeast metal-binding complexes closely resemble those of higher plants is uncertain. An important and overlooked aspect of labile sulfur in phytochelatin complexes is the presence of sulfur in oxidation states other than sulfide. In addition to labile sulfide ion, labile sulfite is also a constituent of phytochelatin complexes from tomato cells (Figure 2; N. T. Eannetta, J. C. Steffens, in preparation). Like sulfide, sulfite ion is labile to both EDTA and acidification. In phytochelatin complexes isolated from tomato cell suspensions, sulfite is present at levels several times higher than is sulfide ion. Although sulfite ion has not been reported in the yeasts studied by Winge et al, it may be significant that the ratios of sulfide/Cys-SH in complexes from higher plants are much lower than those found in yeasts.

Levels of sulfite and sulfide ion in phytochelatin complexes vary with the time after exposure to metal (Figure 3). The highest densities of labile sulfur ions occur after phytochelatin levels decrease. Phytochelatin complexes containing labile sulfur ions are more stable and resistant to proteolytic degradation than are those lacking these constituents. The increasing ratio of labile sulfur ions to cysteine thiol seen in Figure 3 may represent protection from degradation afforded by inclusion of these ions in the complex.

The unusual properties of phytochelatin complexes—their possession of sulfite and sulfide ions—may be related to functions other than metal

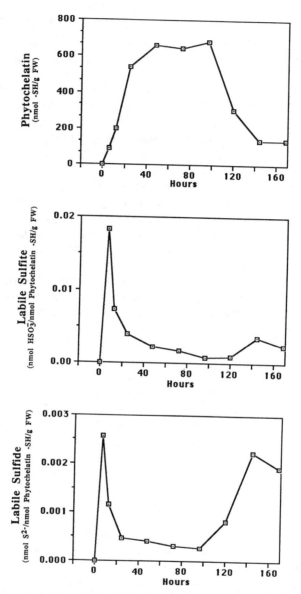

Figure 3   *Upper panel:* Accumulation of phytochelatins as a function of time after exposure to 500 μM $CdCl_2$. *Middle panel:* Ratio of labile sulfite ion ($HSO_3{}^-$) to phytochelatin sulfhydryl in metal complexes isolated from cells assayed at same time points as upper panel. *Lower panel:* Ratio of labile sulfide ion (S^{2-}) to phytochelatin sulfhydryl in metal complexes isolated as above.

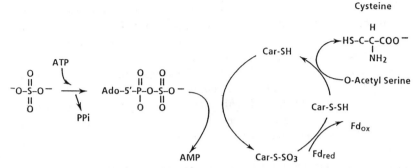

Figure 4 Assimilatory pathway of sulfur metabolism. Sulfate is activated by ATP sulfurylase, then transferred to a sulfhydryl of the sulfur carrier (Car-SH) in a carrier-bound thiosulfonate linkage. Electrons from ferredoxin then reduce the bound sulfonate to sulfide. Carrier-bound sulfur can be released as sulfite or sulfide by acidification. Bound sulfide is ultimately used to form cysteine from O-acetylserine.

sequestration. Plants reduce sulfate to the level of Cys using bound intermediates, presumably because this reduces the susceptibility of reduced intermediates to oxidation (51, 52, 18) (Figure 4). Reduction of sulfite to sulfide occurs on an unidentified polypeptide "sulfur carrier" whose properties are strikingly similar to those of the phytochelatins. The S-carrier molecular weight is about 1200; possesses one or more thiols to which APS sulfotransferase donates the sulfo group; and contains only Glu, Cys, and Gly (60). The S-carrier is further identified by its possession of the intermediates of sulfate reduction, sulfite and sulfide. Because these inorganic forms of sulfur are bound to thiol groups of the carrier (Figure 4), they can be exchanged and volatilized by acidification—hence the name "acid labile sulfur."

Prolonged phytochelatin synthesis requires large amounts of Cys. Phytochelatin synthesis, ATP sulfurylase, and APS sulfotransferase activities are coordinately regulated by Cd exposure (38). In cells growing on Cd, labile sulfite and sulfide are associated exclusively with phytochelatin complexes (N. T. Eannetta and J. C. Steffens, in preparation). The association of labile sulfide and sulfite with phytochelatins, along with structural similarities between the phytochelatins and the proposed S-carrier, raise the possibility that phytochelatins possess a dual function of sulfate reduction and sequestration of excess metals (55). However, most of the labile sulfite in phytochelatin complexes can be released from the complex with EDTA, indicating that sulfite ions are more likely bridged by metal atoms rather than bound directly to cysteine as the organic thiosulfonate product expected from action of APS sulfotransferase. APS sulfotransferase has little specificity for thiol acceptors. If the labile sulfur in phytochelatins does originate from APS sulfotransferase, it cannot be ruled out that the abundance of phytochelatins in metal-stressed cells simply presents the sulfate assimilation apparatus with an alternative acceptor molecule. Both APS sulfotransferase and ATP sulfurylase activities are localized primarily in plastids (3a); there is as yet no evidence that phytochelatins occur in plastids. The biochemical pathway that results in the assimilation of sulfate to the level of sulfite and sulfide in these complexes may be different from that known in plastids. While it is clear that sulfite and sulfide levels vary in phytochelatins depending on time after exposure to metals, pulse-chase studies are needed to determine whether the labile sulfur of phytochelatin complexes is used to support cysteine biosynthesis.

ROLE OF PHYTOCHELATINS IN METAL-TOLERANT PLANTS AND CELLS

The "phytochelatin response," or synthesis of heavy metal–binding polypeptides, is one of the few examples in plant stress biology in which it can be readily demonstrated that the stress response (phytochelatin synthesis) is truly an adaptive stress response. The extent to which this response accounts for the differential tolerance of metal-tolerant plants from metal-contaminated sites or cell cultures selected for heavy-metal tolerance is not clear. In one line of tomato cells selected for Cd tolerance, phytochelatins are accumulated at higher levels than in normal cells (55, 56). In the absence of Cd these cells are insensitive to growth inhibition by buthionine sulfoximine and possess four times more activity of γ-glutamylcysteine synthetase than normal cells (57). Since buthionine sulfoximine binds strongly at the active site of γ-glutamylcysteine synthetase, these results suggest that overproduction of this enzyme may be a mechanism to support increased utilization of γ-glutamylcysteine into phytochelatin. However, a second Cd-tolerant tomato cell line does not overproduce phytochelatins, although buthionine sulfoximine still renders the cells susceptible to growth inhibition by Cd (17). Despite the higher affinity of phytochelatins for Cu than Cd (44), Huang et al (17) have shown that this cell line is only slightly tolerant of Cu and is not cross-tolerant to growth inhibition by Pb, Zn, Hg, and Ag. Rates of metal-complex formation may be an important determinant of tolerance. Delhaize et al (7) have shown that while Cd-tolerant and Cd-sensitive *Datura innoxia* cell suspensions accumulate phytochelatins at similar rates, the rate of metal-complex formation was significantly higher in the tolerant cell line.

Phytochelatins are present in plants with evolved tolerances to heavy metals. Metal-binding peptides were detectable in roots of both Cu-tolerant and nontolerant ecotypes of *Agrostis gigantea*, and peptide accumulation occurred at higher rates in Cu-tolerant ecotypes than in nontolerant forms in response to Cu exposure (38a). However, nontolerant ecotypes accumulated higher levels of metal-binding polypeptides. Plants from metal-tolerant and nontolerant populations of *Silene cucubalus* both possessed phytochelatins (15, 61a). In plants growing on Zn-rich mine tailings, the amount of phytochelatin present was not sufficient to sequester all the Zn present in the plant (15). In *S. cucubalus* exposed to Cu, only 30–35% of the total Cu was complexed to phytochelatins (61a). Tolerant ecotypes did not possess a higher proportion of Cu bound to phytochelatin, nor were levels of phytochelatin higher in the tolerant plants. Cu-tolerant *Mimulus guttatus* possessing cross-tolerance to Cd produces phytochelatins as the primary Cu- and Cd-binding components of roots (49b). In addition, buthionine sulfoximine in the presence of Cd is lethal, whereas buthionine sulfoximine in the presence of Cu greatly decreases root growth. Therefore, like the situation described by Huang et al (17), tolerance to several metals in metal-tolerant plants can be shown to involve phytochelatins, but tolerance to all metals does not result from phytochelatins alone (49b, 59b). Overproduction of phytochelatins in plants chronically exposed to heavy metals seems an unlikely mechanism for metal tolerance, owing to the energy required for sulfate reduction to support phytochelatin synthesis.

CONCLUSIONS

Phytochelatins appear to be the primary metal-binding polypeptides of plants, and the enzyme catalyzing their biosynthesis, phytochelatin synthase, is constitutively expressed in plants. Although the sporadic occurrence of toxic levels of metal ions in the biosphere seems unlikely to have exerted sufficient selection for evolution of a heavy-metal detoxification system in plants (48a), it is evident that phytochelatins play a central role in the detoxification of excess metals. Phytochelatins are also involved in trace-metal homeostasis, and their participation in detoxification of excess metals may be a consequence of this homeostatic function. Although phytochelatins are present in tissues of metal-tolerant ecotypes, the bases for evolved metal tolerances do not appear to involve mechanisms as simple as phytochelatin overproduction. Phytochelatin complexes are heterogeneous in their peptide composition and contain labile sulfur in addition to heavy metals. Understanding the function and biosynthesis of labile sulfur in phytochelatin complexes may illuminate the evolution of this pathway in plants.

Although the reasons for the apparent dichotomy in metal sequestration by animals (metallothioneins) and plants (phytochelatins) remain elusive, two recent reports of authentic metallothionein-like proteins in plants suggest that phytochelatins may not enjoy an exclusive role as the heavy metal–binding polypeptides of plants. The first report described an apparent Zn metallothionein, the E_c (early cysteine-labeled) protein, which was first identified as a Cys-rich protein of wheat embryos (27a). While homology of E_c to known metallothioneins is not high, it does possess Cys-X-Cys repeats and binds five mol Zn per mol E_c. E_c appears to act as a Zn storage protein in wheat seeds, being degraded upon germination (27a). In a second case, Tomsett et al have reported isolation of a cDNA possessing homology to both type I and II metallothioneins (59b). The cDNA encodes a protein of 72 amino acids possessing two domains, each of which contains 3 Cys-X-Cys sequences. This plant metallothionein was found by differential screening of Cu-induced roots of Cu-tolerant *Mimulus guttatus*. Determination of its role in the evolution of metal tolerance in *Mimulus*, and of its occurrence and regulation in other species, will be met with great interest. Although phytochelatins may be involved in many aspects of plant heavy-metal homeostasis, there are many challenges to be met in reaching a full understanding of the ways plants metabolize these elements.

Literature Cited

1. Antonovics, J., Bradshaw, A. D., Turner, R. G. 1971. Heavy metal tolerance in plants. *Adv. Ecol. Res.* 7:1–85
2. Bartolf, M., Brennan, E., Price, C. A. 1980. Partial characterization of a cadmium-binding protein from the roots of cadmium treated tomato. *Plant Physiol.* 66:438–41
3. Berger, J. M., Jackson, P. J., Robinson, N. J., Lujan, L. D., Delhaize, E. 1989. Precursor-product relationships of poly (γ-glutamylcysteinyl) glycine biosynthesis in *Datura innoxia*. *Plant Cell Rep.* 7:632–35
3a. Brunold, C., Suter, M. 1989. Localization of enzymes of assimilatory sulfate reduction in pea roots. *Planta* 179:228–34
4. Casterline, J. L. Jr., Barnett, N. M. 1982. Cadmium binding components in soybean plants. *Plant Physiol.* 69:1004–7
5. Czarnecka, E., Edelman, L., Schoffl, F., Key, J. L. 1984. Comparative analysis of physical stress responses in soybean seedlings using cloned heat shock cDNAs. *Plant Mol. Biol.* 3:45–58
6. Dameron, C. T., Reese, R. N., Mehra, R. K., Kortan, A. R., Carroll, P. J., et al. 1989. Biosynthesis of cadmium sulphide quantum semiconductor crystallites. *Nature* 338:596–98
7. Delhaize, E., Jackson, P. J., Lujan, L. D., Robinson, N. J. 1989. Poly(γ-glutamylcysteinyl) glycine synthesis in *Datura innoxia* and binding with cadmium. *Plant Physiol.* 89:700–6
8. Durnam, D. M., Palmiter, R. D. 1984. Induction of metallothionein-I mRNA in cultured cells by heavy metals and iodoacetate: evidence for gratuitous inducer. *Mol. Cell Biol.* 4:484–91
9. Freedman, J. H., Ciriolo, M. R., Peisach, J. 1989. The role of glutathione in copper metabolism and toxicity. *J. Biol. Chem.* 264:5598–5605
10. Gekeler, W., Grill, E., Winnacker, E.-L., Zenk, M. H. 1988. Algae sequester heavy metals via synthesis of phytochelatin complexes. *Arch. Microbiol.* 150:197–202
10a. Gekeler, W., Grill, E., Winnacker, E.-L., Zenk, M. H. 1989. Survey of the plant kingdom for the ability to bind heavy metals through phytochelatins. *Z. Naturforsch.* 44c:361–69
11. Grill, E., Winnacker, E.-L., Zenk, M. H. 1985. Phytochelatins: the principal heavy-metal complexing peptides of higher plants. *Science* 230:674–76
12. Grill, E., Gekeler, W., Winnacker, E.-L., Zenk, M. H. 1986. Homophytochelatins are heavy metal-binding peptides of homo-glutathione containing Fabales. *FEBS Lett.* 205:47–50
12a. Grill, E., Winnacker, E.-L., Zenk, M. H. 1986. Synthesis of seven different homologous phytochelatins in metal-exposed *Schizosaccharomyces pombe* cells. *FEBS Lett.* 197:115–20
12b. Grill, E., Löffler, S., Winnacker, E.-L., Zenk, M. H. 1989. Phytochelatins, the heavy-metal binding peptides of plants, are synthesized from glutathione by a specific γ-glutamylcysteine dipeptidyl transpeptidase (phytochelatin synthase). *Proc. Natl. Acad. Sci. USA* 86:6838–42

13. Grill, E., Winnacker, E.-L., Zenk, M. H. 1987. Phytochelatins, a class of heavy-metal-binding peptides from plants, are functionally analogous to metallothioneins. *Proc. Natl. Acad. Sci. USA* 84:439–43

14. Grill, E., Thumann, J., Winnacker, E.-L., Zenk, M. H. 1988. Induction of heavy-metal binding phytochelatins by inoculation of cell cultures in standard media. *Plant Cell Rep.* 7:375–78

15. Grill, E. 1989. Phytochelatins in plants. In *Metal Ion Homeostasis: Molecular Biology and Chemistry*, ed. D. H. Hamer, D. R. Winge, pp. 283–300. New York: Alan R. Liss, Inc.

16. Hamer, D. H. 1986. Metallothionein. *Annu. Rev. Biochem.* 55:913–51

17. Huang, B., Hatch, E., Goldsbrough, P. B. 1987. Selection and characterization of cadmium tolerant cells in tomato. *Plant Sci.* 52:211–21

18. Huxtable, R. J. 1986. *Biochemistry of Sulfur*. NY: Plenum. 445 pp.

19. Jackson, P. J., Unkefer, C. J., Doolen, J. A., Watt, K., Robinson, N. J. 1987. Poly (γ-glutamylcysteinyl) glycine: its role in cadmium resistance in plant cells. *Proc. Natl. Acad. Sci. USA* 84:6619–23

20. Jocelyn, P. C. 1972. *Biochemistry of the SH Group: The Occurrence, Chemical Properties, Metabolism and Biological Function of Thiols and Disulphides*. London/New York: Academic 404 pp.

20a. Kagi, J. H. R., Kojima, Y., ed. 1987. *Metallothionein II*. Basel: Birkhäuser. 755 pp.

21. Karin, M. 1985. Metallothioneins: proteins in search of function. *Cell* 41:9–10

22. Klapheck, S., Latus, C., Bergmann, L. 1987. Localization of glutathione synthetase and distribution of glutathione in leaf cells of *Pisum sativum* L. *J. Plant Physiol.* 131:123–31

23. Klapheck, S., Zopes, H., Levels, H. G., Bergmann, L. 1988. Properties and localization of the homoglutathione synthetase from *Phaseolus coccineus* leaves. *Physiol. Plant* 74:733–39

24. Klapheck, S. 1988. Homoglutathione: isolation quantification and occurrence in legumes. *Physiol. Plant.* 74:727–32

25. Kondo, N., Isobe, M., Imai, K., Goto, T. 1983. Structure of cadystin, the unit-peptide of cadmium-binding peptides induced in a fission yeast, *Schizosaccharomyces pombe*. *Tetrahedron Lett.* 24:925–28

26. Kondo, N., Wada-Nakagawa, C., Hayashi, Y. 1984. Cadystin A and B, major unit peptides comprising cadmium-binding peptides induced in a fission yeast—separation, revision of structures and synthesis. *Tetrahedron Lett.* 25:3869–72

27. Kurkdjian, K., Guern, J. 1989. Intracellular pH: measurement and importance in cell activity. *Annu. Rev. Plant Physiol. Plant Mol. Biol.* 40:271–303

27a. Lane, B., Kajioka, R., Kennedy, T. 1987. The wheat-germ E_c protein is a zinc-containing metallothionein. *Biochem. Cell Biol.* 65:1001–5

28. Lerch, K., Beltramini, M. 1983. *Neurospora* copper metallothionein: molecular structure and biological significance. *Chemica Scripta* 21:109–15

29. Lolkema, P. C., Donker, M. H., Schouteen, A. J., Ernst, W. H. O. 1984. The possible role of metallothioneins in copper tolerance of *Silene cucubalus*. *Planta* 162:174–79

30. Mehra, R. K., Winge, D. R. 1988. Cu(1) binding to the *Schizosaccharomyces pombe* γ-glutamyl peptides varying in chain lengths. *Arch. Biochem. Biophys.* 265:381–89

31. Mehra, R. K., Tarbet, E. B., Gray, W. R., Winge, D. R. 1988. Metal-specific synthesis of two metallothioneins and γ-glutamyl peptides in *Candida glabrata*. *Proc. Natl. Acad. Sci. USA* 85:8815–19

32. Munger, K., Lerch, K. 1985. Copper metallothionein from the fungus *Agaricus bisporus*: chemical and spectroscopic properties. *Biochemistry* 24:6751–56

33. Murasugi, A., Wada, C., Hayashi, Y. 1981. Cadmium-binding peptide induced in fission yeast, *Schizosaccharomyces pombe*. *J. Biochem.* 90:1561–64

34. Murasugi, A., Wada, C., Hayashi, Y. 1981. Purification and unique properties in UV and CD spectra of Cd-binding peptide 1 from *Schizosaccharomyces pombe*. *Biochem. Biophys. Res. Commun.* 103:1021–28

35. Murasugi, A., Wada, C., Hayashi, Y. 1983. Occurrence of acid-labile sulfide in cadmium-binding peptide 1 from fission yeast. *J. Biochem.* 93:661–64

36. Mutoh, N., Hayashi, Y. 1988. Isolation of mutants of *Schizosaccharomyces pombe* unable to synthesize cadystin, small cadmium-binding peptides. *Biochem. Biophys. Res. Commun.* 151:32–39

37. Newton, G. L., Dorian, R., Fahey, R. C. 1981. Analysis of biological thiols: derivatization with monobromobimane and separation by reverse-phase high performance liquid chromatography. *Anal. Biochem.* 114:383–87

38. Nussbaum, S., Schmutz, D., Brunold, C. 1988. Regulation of assimilatory sulfate reduction by cadmium in *Zea mays* L. *Plant Physiol.* 88:1407–10

38a. Rauser, W. E. 1984. Copper-binding protein and copper tolerance in *Agrostis gigantea. Plant Sci. Lett.* 33:239–47

39. Rauser, W. E. 1987. Changes in glutathione content of maize seedlings exposed to cadmium. *Plant Sci.* 51:171–75

39a. Rauser, W. E. 1990. Phytochelatins. *Annu. Rev. Biochem.* 59:61–86

40. Rauser, W. E., Ackerley, C. A. 1987. Localization of cadmium in granules within differentiating and mature root cells. *Can. J. Bot.* 65:643–46

41. Rauser, W. E., Glover, G. 1984. Cadmium-binding protein in roots of maize. *Can. J. Bot.* 62:1645–50

42. Rauser, W. E., Hartmann, H., Weser, U. 1983. Cadmium-thiolate protein from the grass *Agrostis gigantea. FEBS Lett.* 164:102–4

43. Rauser, W. E., Curvetto, N. E. 1980. Metallothionein occurs in roots of *Agrostis* tolerant to excess copper. *Nature* 287:563–64

44. Reese, R. N., Mehra, R. K., Tarbet, E. B., Winge, D. R. 1988. Studies on the γ-glutamyl Cu-binding peptide from *Schizosaccharomyces pombe. J. Biol. Chem.* 263:4186–92

45. Reese, R. N., Wagner, G. J. 1987. Properties of tobacco *(Nicotiana tabacum)* cadmium-binding peptides(s). *Biochem. J.* 241:641–47

46. Reese, R. N., Winge, D. R. 1988. Sulfide stabilization of the cadmium-γ-glutamyl peptide complex of *Schizosaccharomyces pombe. J. Biol. Chem.* 263:12832–35

47. Rennenberg, H., Uthemann, R. 1980. Effects of L-methionine-S-sulfoximine on growth and glutathione synthesis in tobacco suspension cultures. *Z. Naturforsch.* 35:945–51

48. Rennenberg, H. 1982. Glutathione metabolism and possible biological roles in higher plants. *Phytochemistry* 21:2771–81

48a. Rennenberg, H. 1987. Aspects of glutathione function and metabolism in plants. In *Plant Molecular Biology*, ed. D. Von Wettstein, N. H. Chua, pp. 279–92. New York: Plenum

48b. Robinson, N. J., Jackson, P. J. 1986. "Metallothionein-like" metal complexes in angiosperms; their structure and function. *Physiol. Plant.* 67:499–506

48c. Robinson, N. J. 1990. Metal binding polypeptides in plants. In *Heavy Metal Tolerance in Plants*, ed. A. J. Shaw, pp. 195–214. Boca Raton: CRC Press

49. Robinson, N. J., Ratliff, R. L., Anderson, P. J., Delhaize, E., Berger, J. M., Jackson, P. J. Biosynthesis of poly(γ-glutamylcysteinyl)glycines in cadmium-tolerant *Datura innoxia* (Mill.) cells. *Plant Sci.* 56:197–204

49b. Salt, D. E., Thurman, D. A., Tomsett, A. B., Sewell, A. K. 1989. Copper phytochelatins of *Mimulus guttatus. Proc. R. Soc. Lond. Ser. B* 236:79–89

50. Scheller, H. V., Huang, B., Hatch, E., Goldsbrough, P. B. 1987. Phytochelatin synthesis and glutathione levels in response to heavy metals in tomato cells. *Plant Physiol.* 85:1031–35

51. Schiff, J., Hodson, R. C. 1973. The metabolism of sulfate. *Annu. Rev. Plant Physiol.* 24:381–414

52. Schmidt, A. 1982. Assimilation of sulfur. In *On the Origins of Chloroplasts*, ed. J. A. Schiff, pp. 179–97. New York: Elsevier North Holland Inc.

53. Schoffl, F., Key, J. L. 1982. An analysis of mRNAs for a group of heat shock proteins of soybean using cloned cDNAs. *J. Mol. Appl. Genet.* 1:301–14

54. Silver, S. 1983. Bacterial interactions with mineral cations and anions: good ions and bad. In *Biomineralization and Biological Metal Ion Accumulation*, ed. P. Westbroek, E. W. deJong, pp. 439–57. Amsterdam: D. Reidel

55. Steffens, J. C., Hunt, D. F., Williams, B. G. 1986. Accumulation of nonprotein metal-binding polypeptides (γ-glutamyl-cysteinyl)$_n$-glycine in selected cadmium-resistant tomato cells. *J. Biol. Chem.* 261:13879–82

56. Steffens, J. C., Williams, B. G. 1987. Molecular biology of heavy metal tolerance in tomato. In *Plant Biology*, ed. D. J. Nevins, R. A. Jones, 4:109–18. New York: Alan R. Liss, Inc.

57. Steffens, J. C., Williams, B. G. 1989. Increased activity of γ-glutamylcysteine synthetase in DMSO permeabilized Cd^r cells. See Ref. 15, pp. 359–66

58. Steinkamp, R., Rennenberg, H. 1985. Degradation of glutathione in plant cells: evidence against the participation of a γ-glutamyltranspeptidase. *Z. Naturforsch.* 40:29–33

59. Steinkamp, R., Schweihofen, B., Rennenberg, H. 1987. γ-Glutamyl cyclotransferase in tobacco cell suspension cultures: catalytic properties and subcellular localization. *Physiol. Plant.* 69:499–503

59a. Tomsett, A. B., Thurman, D. A. 1988.

Molecular biology of metal tolerances of plants. *Plant Cell Environ.* 11:383–94

59b. Tomsett, A. B., Salt, D. E., De-Miranda, J., Thurman, D. A. 1989. Metallothioneins and metal tolerance. *Aspects Appl. Biol.* 22:365–72

60. Tsang, M. L., Schiff, J. A. 1978. Studies of sulfate utilization by algae 18. Identification of glutathione as a physiological carrier in assimilatory sulfate reduction by *Chlorella. Plant Sci. Lett.* 11:177–83

61. Tukendorf, A., Baszynski, T. 1985. Partial purification and characterization of copper-binding protein from roots of *Avena sativa* grown on excess copper. *J. Plant Physiol.* 120:57–63

61a. Verkleij, J. A. C., Koevoets, P., van't Riet, J., van Rossenberg, M. C., Bank, R., Ernst, W. H. O. 1989. The role of metal-binding compounds in the copper tolerance mechanism of *Silene cucubalus*. See Ref. 15, pp. 347–57

62. Wagner, G. J., Krotz, R. M. 1989. Perspectives on Cd and Zn accumulation, accommodation, and tolerance in plant cells: the role of Cd-binding peptide versus other mechanisms. See Ref. 15, pp. 325–336

63. Wagner, G. J. 1984. Characterization of a cadmium-binding complex of cabbage leaves. *Plant Physiol.* 76:797–805

64. Wagner, G. J., Trotter, M. A. 1982. Inducible cadmium binding complexes of cabbage and tobacco. *Plant Physiol.* 69:804–9

65. Weigel, H. J., Jager, H. J. 1980. Subcellular distribution and chemical form of cadmium in bean plants. *Plant Physiol.* 65:480–82

66. Winge, D. R., Reese, R. N., Mehra, R. K., Tarbet, E. B., Hughes, A. K., Dameron, C. T. 1989. Structural aspects of metal-γ-glutamyl peptides. See Ref. 15, pp. 300–13

67. Woolhouse, H. W. 1983. Toxicity and tolerance in the responses of plants to metals. In *Encyclopedia of Plant Physiology*, NS, Vol. 12C, ed. O. L. Lange, P. S. Nobel, C. B. Osmond, H. Ziegler, pp. 245–300. New York: Springer-Verlag

AUTHOR INDEX

SUBJECT INDEX

succinctum
  membrane potential, 34
  turgor pressure regulation,
    30, 39
Lannoye, Bob, 13-14
Lathraea, 135, 137
  clandestina
    glutamine synthetase, 138
    Rubisco activity, 140
Laties, George, 15
Lespedeza sericea, 131
Light
  nitrate utilization regulation,
    246-47
Lignan, 458, 473
Lignin, 353, 455-97, 511
  analysis
    acidolysis/thioacidolysis,
      461
    alkaline nitrobenzene/CuO-
      NaOH oxidation, 460-
      61
    histochemical staining, 458
    inter-unit linkages, 459-60
    molecular weight de-
      termination, 461-62
    monomer composition, 460-
      61
    permanganate oxidation,
      461-62
    total lignin content de-
      termination, 458-59
  biodegradation, 481
    criteria, 481-82
    lignin-like polymers in
      white-rot fungi, 481
    white-rot fungi, 482-87
  biosynthesis and structure,
    469-70
    arogenic acid, 470
    p-coumaric acid to E-and
      Z-monolignols, 471-72
    formation, 473-76
    lignification regulation,
      476-77
    monolignol transport in cell
      walls, 472-73
    phenylalanine incorporation,
      479
    structure, 477-80
    substructural bonding en-
      vironments, 480
  conclusions, 487-88
  criteria for detection, 457-58
  distribution, 458-62
  introduction, 455-57
    precursors and aromatic
      constituents, 456
    occurrence and composition,
      462
      algae, 462-63
      angiosperm cultures, 469

angiosperms, 466-67
gymnosperms, 465-66
monocotyldeons, 468
mosses, 463-65
pteridophytes, 465
submerged aquatic an-
  giosperms, 468-69
Lignin peroxidase, 484-87
Ligninase, 483
Loranthus europaeus
  haustorial resistances, 135
  nitrogen requirement, 137
Lucas, Bill, 16

M

MacRobbie, Enid, 5-6, 20
Manganese, 258-59, 263, 266-
  68
  cofactor in Photosystem II,
    260-62
  inhibitory amines as
    probes, 264-65
Mannitol, 35, 37, 40, 132
  accumulation
    parasitic angiosperms, 135
    regulation, 38
Mauro, Alex, 11
Meares, Patrick, 11
Medicarpin, 131
Membrane
  mitotic apparatus, 281-82
Membrane potential, 33-34
Membrane transport systems,
  kinetic modeling, 77-107
  conclusions, 102
  general approaches
    carriers and channels, 78-
      79
    practical aspects, 79-80
    introduction, 77-78
  models of carriers from their
    electrical properties
    experimental applications,
      85-91
    experimental approach, 80-
      81
    four-state reaction kinetic
      model, 82
    ligands effect in higher-
      state models, 84-85
    pseudo-two-state model,
      81-84
  models of carriers from unidi-
    rectional fluxes
    analytical methods, 91-92
    experimental applications,
      95-98
    experimental approach, 91
    gradient driven transport,
      92-93

random binding and dual
  isotherms, 93-95
reaction kinetic models for
  H$^+$-coupled transport,
    93
models of ionic channels
  gating, 100-1
  unitary currents, 98-100
Mestome sheath, 386
Metallothioneins, 555, 558
  induction, 558
  occurrence, 557
  plants, 571-72
Metaphase
  chromosome movements at
    prometaphase, 293-94
  microtubule dynamics, 294-95
  spindle formation, 290-93
    kinetochore-microtubule
      association, 291-93
    microtubule rearrangements,
      291
    nuclear envelope break-
      down, 290-91
6-Methoxymellein, 359
m-Methoxyphenol, 130
Microplasmodesmata, 381
Microtubules, 279-80, 295-96
  anaphase A
    microtubule inhibitors
      effects, 299
    models for force produc-
      tion, 299-301
    organization changes, 297-
      99
  anaphase B, 302
    dynamics at metaphase, 294-
      95
    kinetochore association, 291-
      93
    phragmoplast formation, 303-
      5
    rearrangements in prometa-
      phase, 291
Mistletoe
  see Parasitic angiosperms,
    physiology and biochem-
    istry
Mitchell, Peter, 12
Mitotic spindle structure and
  function, 277-315
  anaphase, 295
    anaphase A, 297-301
    anaphase B, 301-303
    chromosome to pole move-
      ment kinetics, 297
    disjunction, 295
    energetics, 296
    general anaphase events,
      295-96
    kinetochore based models,
      301

CUMULATIVE INDEXES

CONTRIBUTING AUTHORS, VOLUMES 33–41

CHAPTER TITLES, VOLUMES 33–41

ORGANELLES AND CELLS

ANNUAL REVIEWS INC.

A NONPROFIT SCIENTIFIC PUBLISHER

4139 El Camino Way
P.O. Box 10139
Palo Alto, CA 94303-0897 • USA

ORDER FORM

ORDER TOLL FREE
1-800-523-8635
(except California)

Annual Reviews Inc. publications may be ordered directly from our office; through booksellers and subscription agents, worldwide; and through participating professional societies. Prices subject to change without notice.

ARI Federal I.D. #94-1156476

- **Individuals:** Prepayment required on new accounts by check or money order (in U.S. dollars, check drawn on U.S. bank) or charge to credit card—American Express, VISA, MasterCard.
- **Institutional buyers:** Please include purchase order.
- **Students:** $10.00 discount from retail price, per volume. Prepayment required. Proof of student status must be provided (photocopy of student I.D. or signature of department secretary is acceptable). Students must send orders direct to Annual Reviews. Orders received through bookstores and institutions requesting student rates will be returned. You may order at the Student Rate for a maximum of 3 years.
- **Professional Society Members:** Members of professional societies that have a contractual arrangement with Annual Reviews may order books through their society at a reduced rate. Check with your society for information.
- **Toll Free Telephone orders:** Call 1-800-523-8635 (except from California) for orders paid by credit card or purchase order and customer service calls only. California customers and all other business calls use 415-493-4400 (not toll free). Hours: 8:00 AM to 4:00 PM, Monday-Friday, Pacific Time. **Written confirmation** is required on purchase orders from universities before shipment.
- **FAX: 415-855-9815 Telex: 910-290-0275**

Regular orders: Please list below the volumes you wish to order by volume number.
Standing orders: New volume in the series will be sent to you automatically each year upon publication. Cancellation may be made at any time. Please indicate volume number to begin standing order.
Prepublication orders: Volumes not yet published will be shipped in month and year indicated.
California orders: Add applicable sales tax.
Postage paid (4th class bookrate/surface mail) **by Annual Reviews Inc.** Airmail postage or UPS, extra.

ANNUAL REVIEWS SERIES		Prices Postpaid per volume USA & Canada/elsewhere	Regular Order Please send:	Standing Order Begin with:
			Vol. number	Vol. number
Annual Review of ANTHROPOLOGY				
Vols. 1-16	(1972-1987)	$31.00/$35.00		
Vols. 17-18	(1988-1989)	$35.00/$39.00		
Vol. 19	(avail. Oct. 1990)	$39.00/$43.00	Vol(s). _____	Vol. _____
Annual Review of ASTRONOMY AND ASTROPHYSICS				
Vols. 1, 4-14, 16-20	(1963, 1966-1976, 1978-1982)	$31.00/$35.00		
Vols. 21-27	(1983-1989)	$47.00/$51.00		
Vol. 28	(avail. Sept. 1990)	$51.00/$55.00	Vol(s). _____	Vol. _____
Annual Review of BIOCHEMISTRY				
Vols. 30-34, 36-56	(1961-1965, 1967-1987)	$33.00/$37.00		
Vols. 57-58	(1988-1989)	$35.00/$39.00		
Vol. 59	(avail. July 1990)	$39.00/$44.00	Vol(s). _____	Vol. _____
Annual Review of BIOPHYSICS AND BIOPHYSICAL CHEMISTRY				
Vols. 1-11	(1972-1982)	$31.00/$35.00		
Vols. 12-18	(1983-1989)	$49.00/$53.00		
Vol. 19	(avail. June 1990)	$53.00/$57.00	Vol(s). _____	Vol. _____
Annual Review of CELL BIOLOGY				
Vols. 1-3	(1985-1987)	$31.00/$35.00		
Vols. 4-5	(1988-1989)	$35.00/$39.00		
Vol. 6	(avail. Nov. 1990)	$39.00/$43.00	Vol(s). _____	Vol. _____

Annual Review of **COMPUTER SCIENCE**
Vols. 1-2 (1986-1987)............... $39.00/$43.00
Vols. 3-4 (1988, 1989-1990)........... $45.00/$49.00   Vol(s). _____   Vol. _____

Annual Review of **EARTH AND PLANETARY SCIENCES**
Vols. 1-10 (1973-1982)............... $31.00/$35.00
Vols. 11-17 (1983-1989)............... $49.00/$53.00
Vol. 18 (avail. May 1990)........... $53.00/$57.00   Vol(s). _____   Vol. _____

Annual Review of **ECOLOGY AND SYSTEMATICS**
Vols. 2-18 (1971-1987)............... $31.00/$35.00
Vols. 19-20 (1988-1989)............... $34.00/$38.00
Vol. 21 (avail. Nov. 1990)........... $38.00/$42.00   Vol(s). _____   Vol. _____

Annual Review of **ENERGY**
Vols. 1-7 (1976-1982)............... $31.00/$35.00
Vols. 8-14 (1983-1989)............... $58.00/$62.00
Vol. 15 (avail. Oct. 1990)........... $62.00/$66.00   Vol(s). _____   Vol. _____

Annual Review of **ENTOMOLOGY**
Vols. 10-16, 18 (1965-1971, 1973)
20-32 (1975-1987)............... $31.00/$35.00
Vols. 33-34 (1988-1989)............... $34.00/$38.00
Vol. 35 (avail. Jan. 1990)........... $38.00/$42.00   Vol(s). _____   Vol. _____

Annual Review of **FLUID MECHANICS**
Vols. 2-4, 7-19 (1970-1972, 1975-1987)....... $32.00/$36.00
Vols. 20-21 (1988-1989)............... $34.00/$38.00
Vol. 22 (avail. Jan. 1990)........... $38.00/$42.00   Vol(s). _____   Vol. _____

Annual Review of **GENETICS**
Vols. 1-21 (1967-1987)............... $31.00/$35.00
Vols. 22-23 (1988-1989)............... $34.00/$38.00
Vol. 24 (avail. Dec. 1990)........... $38.00/$42.00   Vol(s). _____   Vol. _____

Annual Review of **IMMUNOLOGY**
Vols. 1-5 (1983-1987)............... $31.00/$35.00
Vols. 6-7 (1988-1989)............... $34.00/$38.00
Vol. 8 (avail. April 1990)........... $38.00/$42.00   Vol(s). _____   Vol. _____

Annual Review of **MATERIALS SCIENCE**
Vols. 1, 3-12 (1971, 1973-1982)........... $31.00/$35.00
Vols. 13-19 (1983-1989)............... $66.00/$70.00
Vol. 20 (avail. Aug. 1990)........... $70.00/$74.00   Vol(s). _____   Vol. _____

Annual Review of **MEDICINE**
Vols. 9, 11-15 (1958, 1960-1964)
17-38 (1966-1987)............... $31.00/$35.00
Vols. 39-40 (1988-1989)............... $34.00/$38.00
Vol. 41 (avail. April 1990)........... $38.00/$42.00   Vol(s). _____   Vol. _____